ASTROCHEMISTRY VII: THROUGH THE COSMOS
FROM GALAXIES TO PLANETS

IAU SYMPOSIUM 332

COVER ILLUSTRATION:

Top: ALMA antennas and centre of the Galaxy, credit: ESO/José Francisco Salgado (josefrancisco.org) Middle: The Central Molecular Zone (CMZ) in methanol (at 48.37 GHz) with the Mopra radio telescope. Bottom, spectrum of Sgr B2 LMH with the ATCA, around the methanol line showing the complex chemistry.

IAU SYMPOSIUM PROCEEDINGS SERIES

INTERNATIONAL ASTRONOMICAL UNION

UNION ASTRONOMIQUE INTERNATIONALE

International Astronomical Union

IAU

ASTROCHEMISTRY VII: THROUGH THE COSMOS FROM GALAXIES TO PLANETS

PROCEEDINGS OF THE 332nd SYMPOSIUM OF THE INTERNATIONAL ASTRONOMICAL UNION HELD IN PUERTO VARAS, CHILE MARCH 20–24, 2017

Edited by

MARIA CUNNINGHAM

Schools of Physics, University of New South Wales

TOM MILLAR

School of Maths and Physics, Queen's University Belfast

and

YURI AIKAWA

Department of Astronomy, University of Tokyo

CAMBRIDGE
UNIVERSITY PRESS

CAMBRIDGE UNIVERSITY PRESS
University Printing House, Cambridge CB2 8BS, United Kingdom
1 Liberty Plaza, Floor 20, New York, NY 10006, USA
10 Stamford Road, Oakleigh, Melbourne 3166, Australia

First published 2018

Printed in the UK by Bell & Bain, Glasgow, UK

Typeset in System LaTeX 2ε

A catalogue record for this book is available from the British Library Library of Congress Cataloguing in Publication data

This journal issue has been printed on FSC™-certified paper and cover board. FSC is an independent, non-governmental, not-for-profit organization established to promote the responsible management of the world's forests. Please see www.fsc.org for information.

ISBN 9781107192577 hardback
ISSN 1743-9213

Table of Contents

Interstellar chemistry and the life cycles of stars

Laboratory astrochemistry and modelling

Complex organic molecules

Preface

The 7th IAU Symposium on Astrochemistry: Through the Cosmos from Galaxies to Planets was held on 19-24th March 2017 in Puerto Varas, southern Chile, a beautiful spot on the edge of Lake Lanquihue and overlooked by the snow-capped peak of the Osorno volcano. The location was chosen especially because of the influence that the Atacama Large Millimeter/submillimeter Array (ALMA) has had and is having on our discipline. ALMA, which is situated in the Atacama Desert at an altitude of 5000m, some 2500km north of Puerto Varas, has opened up almost the entire Molecular Universe with exquisite sensitivity and spatial resolution and is allowing us to probe the fundamental physical properties of stars, galaxies and planets and the material between them.

The astrochemistry community has come a long way in the years since the first IAU Symposium on astrochemistry in Goa in 1985 in understanding how molecules respond to local conditions and thereby how to use them to provide information on temperatures, densities, dynamics and ionization. Molecules are crucial to our understanding of the 'hidden' phases of both low-mass and high-mass star formation, the latter of which determine the properties of the massive stars that shape the evolution of galaxies, at all epochs, due to the energetic events that shape their births and deaths. Molecules are also key tools in exploring planet formation in low-mass protoplanetary disks and provide hints as to the type of molecular complexity that may have encouraged the beginnings of biology on Earth. Neglecting isotopologues, we have now detected some 200 different molecules beyond our Solar System, ranging in complexity from species such as methyl ethyl ether, $C_2H_5OCH_3$, and benzonitrile, C_6H_5CN, to the famous C_{60} and C_{70} 'buckyballs'. The nature of the detected species, including cations, anions, free radicals, unsaturated and saturated hydrocarbons, and their seeming propensity to exist in favoured types of regions, many at temperatures as low as 10-20K and in which collision time scales can be weeks or months, provides important information. They tell us immediately that the dominant chemistry must involve exothermic ion-neutral and radical-radical reactions together with a rich solid-state chemistry driven through the interaction of ultraviolet photons and cosmic-ray particles with ice-covered interstellar grains.

At the same time as new and re-instrumented observational facilities such as ALMA, NOEMA, APEX, SOFIA, Rosetta and the Herschel Space Observatory are revealing new views of our Molecular Universe, there has been a concerted effort among physical chemists to provide the large amount of fundamental data required to interpret these observations. This critical collaboration was reinforced at the XXIXth IAU General Assembly in 2015 by the creation of two new IAU Commissions - B5 on Laboratory Astrophysics and H2 on Astrochemistry - and this Symposium, IAU S332, is the first fruit of what we plan to be a close and fruitful engagement between us. Nationally, too, the importance of close collaboration between the chemical and astronomical communities has been recognised through, for example, the creation of the Astrochemistry Sub-Division of the American Chemical Society, the UK's Astrophysical Chemistry Working Group, a joint activity of the Royal Society of Chemistry and the Royal Astronomical Society, and the Dutch Astrochemistry Network in the Netherlands.

This synthesis of astronomical observation, laboratory experiment and theoretical modelling has driven a close and effective cooperation between the astronomical and chemical communities for over 30 years and the proceedings of this Symposium make it plain that it is vital to continue these efforts. Not only does such collaboration lead to a better understanding of astrophysical phenomena but also to advances in fundamental atomic and molecular physics and chemistry. The importance of tunnelling in both gas phase and ice mantle reactions at 10K and of spin chemistry in determining D/H ratios in molecules are just two of the ways in which microscopic quantum effects influence the macroscopic properties of our Universe.

It was thus with a degree of excitement and anticipation that some 169 scientists from 25 countries gathered at the Hotel Cumbres, Puerto Varas, where they were welcomed by an opening address from Ewine van Dishoeck in which she formulated some 14 challenges for astrochemistry in the coming decades. Following this, and inspired by some 21 keynote and invited speakers, participants undertook their own astrochemical journey through the cosmos.

At the Toledo Symposium, there was a panel discussion entitled 'On to ALMA'. In Puerto Varas, it was immediately clear that we are firmly in the ALMA Age with talks on how ALMA has provided new insights into high-redshift and ultra-luminous galaxies, dwarf galaxies such as the LMC and SMC, and photon-dominated regions in our own Milky Way. Here we were impressed by the ALMA detections of lines at redshifts up to $z=7$ and its detections of thousands of molecular lines from the ULIRG, Arp 220.

There were several talks and posters on low-mass star formation including a talk on an ALMA study of the rotating, infalling envelope of IRAS 16293-2422 Source A that demonstrated the power of chemistry to trace different regions of the infall, in particular the position of the centrifugal barrier at which the gas composition appears to be dominated by the evaporation products of ice mantles. This work highlights the importance of ALMA to our field as the chemistry associated with protostellar collapse may be a useful tool to isolate important phases, such as the young disk. This work is still in its infancy and we can look forward to significant advances in the coming years.

The subject of ice mantles is a very active area of theoretical and experimental research worldwide and theoretical models of surface chemistry were summarised, in particular in relation to the formation of complex organic molecules (COMs). Laboratory astrophysics is playing an important role in understanding gas-phase chemistry as well as surface and bulk processes in ices and the Symposium featured many talks and posters on these subjects. The interaction of ices with soft X-rays was reviewed and it was shown that soft X-ray irradiation of cosmic ice mixtures induced a complex chemistry that resulted in the formation of the nucleobase adenine and peptide bonds, amongst others. One of the major challenges of ice chemistry is to identify the primary products of photolysis and to determine photodesorption yields. A new experiment was described in which the photolysis products of methanol ice were detected using gas-phase millimetre and sub-millimetre spectroscopy. This opens up an exciting new way in which yields can be measured directly and in which products that cannot be differentiated using conventional mass spectroscopic techniques can be identified. This is certain evidence that laboratory science is moving in directions that continue to be central to astrochemistry.

Molecules are also important in the formation of dust grains in supernovae, circumstellar envelopes and the interstellar medium (ISM) and we had talks on ALMA observations of late-type AGB stars. Here detections of species such as TiO, TiO_2, SiO, AlO and AlOH are aimed at understanding how dust precursors help nucleation and growth. Observations were complemented by laboratory evidence that SiO molecules could grow to nanoscale amorphous clusters in liquid He drops at 0.37K, giving experimental support to the hypothesis that most silicate dust must form in the ISM given its fast destruction time scale in supernovae explosions.

A significant part of the program was devoted to low-mass star formation and protoplanetary disks (PPDs) including the importance of molecules in tracing the dynamics in such regions. There were also several PPD talks on subjects such as the means to determine the water snowline, observations of COMs, gaps, and rings, grain growth and dynamics, disk masses and carbon depletion, dust in transition disks, and isotopic chemistry. One of the highlights of this session was the amazing images from ALMA where it is clear that the chemistry associated with planet formation is slowly being revealed.

Comets are an important part of our astronomical heritage and our journey through the cosmos took us to the fabulous and detailed results emerging from the Rosetta Mission, in particular, to the molecular 'zoo' detected at comet 67P/CG. A range of results were presented, from the observation that the comet does not have a solar Xe isotopic distribution, that its O_2 is likely inherited from interstellar ice, to the conclusion that comets did not supply the Earth's water. The audience was impressed at the vast array of molecules detected by Rosetta and we suspect that many will be keen to take advantage of the new research opportunities in this exciting field.

Exoplanet atmospheres are becoming of increasing importance in astronomy and there are now many groups using astrochemical techniques to interpret their observations. A major task identified for those interested in exoplanet atmospheres is to develop a community similar to that current in astrochemistry. The chemical composition of exoplanetary atmospheres is an area in which astrochemists will continue to make important contributions and we expect that this topic will play a larger role in future Symposia.

The programme explicitly included a number of invited talks on future facilities, including the JWST and ALMA Band 1, and concluded with an extensive and passionate debate on whether the C_{60}^+ ion was responsible for five of the diffuse interstellar bands, with arguments against suggesting that atmospheric contamination had not been effectively removed from the astronomical spectra and that non-LTE stellar lines were also blended with C_{60}^+ features.

In addition to the oral presentations, some 116 papers were presented as posters in two sessions and poster prizes awarded to six individuals, three from each session: Ko-Ju Chuang, Anna Miotello, Victor Rivilla, Richard Teague, Marie van de Sande and Merel van't Hoff.

Ewine van Dishoeck also gave a special presentation on 'Women in Astronomy', summarising the results of the recent IAU survey on this topic. She encouraged participants, as we do readers, to contact her with any ideas on how the IAU could grow the number of its female members.

The organisers would like to thank a number of institutions for providing organisational and financial support: in particular, the IAU for the award of travel grants to 34 young scientists, some 19 nationalities based in 16 countries; NRAO; ALMA; ESO; Queen's University Belfast; the University of New South Wales; the Universidad de Chile, and the journal Molecular Astrophysics.

The Local Organising Committee, which was ably chaired by Gautier Mathys, who brought his exceptional organisational skills and experience to this key role, comprised of Ann Edmunds, Maria Gomez, Natalia Inostroza, Paulina Jirón, Diego Mardones, Sergio Martin, Lars Nyman, and Ursala Throm. They performed their duties both before and during the Symposium in a highly effective manner and the fact that the Symposium ran so smoothly is a testament to their professional approach that ensured attendees had a stress-free experience.

Ted Bergin, Maria Cunningham and Tom Millar, as Co-Chairs, were greatly aided in their duties by the other members of the SOC and would like to express their thanks to them: Susanna Aalto, Yuri Aikawa, Jacob Bean, Dominique Bockelée-Morvan, Paola Caselli, Pierre Cox, Jes Jørgensen, Sun Kwok, Farid Salama, Stephan Schlemmer and Satoshi Yamamoto.

In closing, we are already looking forward to Astrochemistry VIII. At the time of the next Symposium, the Mid-InfraRed Instrument (MIRI) on the James Webb Space Telescope (JWST) will be making ground-breaking observations crucial for understanding the interstellar medium, including molecules, such as H_2 and CO_2, without permanent dipoles, and solid-state ices on dust grain mantles. These observations can only be taken from space. And by 2020, ALMA will be operating with full capabilities, including the ability to measure polarisation and hence magnetic fields in molecular gas. ALMA will also have a working 7-mm band, opening up a new, high-resolution window on the many complex molecules that have their fundamental frequencies within this band, helping solve the puzzle of how molecular complexity develops in star forming regions.

Maria Cunningham (University of New South Wales, Australia)
Tom Millar (Queen's University Belfast, UK)
Yuri Aikawa (University of Tokyo, Japan)
Editors
14 February 2018

Editors

Maria Cunningham
School of Physics
University of New South Wales

Tom Millar
School of Maths and Physics,
Queen's University Belfast

Yuri Aikawa
Department of Astronomy
University of Tokyo

Local Organising Committee

Gautier Mathys (Chair)	ALMA/ESO, Chile
Maria Cunningham	University of New South Wales, Australia
Ann Edmunds	ALMA, Chile
María Eugenia Gómez	ESO, Chile
Natalia Inostroza	Universidad Autónoma de Chile, Chile
Paulina Jirón	ESO, Chile
Diego Mardones	Universidad de Chile, Chile
Sergio Martin	ALMA, Chile
Tom Millar	Queen's University Belfast, UK
Lars Nyman	ALMA, Chile
Ursula Throm	ALMA, Chile

Scientific Organising Committee

Tom Millar (Co-Chair)	Queen's University Belfast, UK
Ted Bergin (Co-Chair)	University of Michigan, USA
Susanne Aalto	Chalmers University of Technology, Sweden
Yuri Aikawa	University of Tokyo, Japan
Jacob Bean	University of Chicago, USA
Dominique Bockelee-Morvan	Observatoire Paris, Meudon, France
Paola Caselli	MPI for Extraterrestrial Physics, Germany
Pierre Cox	ALMA, Chile
Maria Cunningham (Secretary)	University of New South Wales, Australia
Jes Jorgensen	Niels Bohr Institute, University of Copenhagen, Denmark
Sun Kwok	University of Hong Kong, China
Farid Salama	NASA-Ames Research Center, USA
Stephan Schlemmer	University of Cologne, Germany
Satoshi Yamamoto	University of Tokyo, Japan

CONFERENCE PHOTOGRAPH

Photo 1. Conference photo, show attendees at the conference venue, Hotel Cumbres, Puerto Varas, Chile.

Photo 2. Student prize winners for best talks and posters.

Photo 3. IAU Commission H2 (Astrochemistry) President, Tom Millar, giving the after dinner speech at the conference dinner.

Participants

First Name	Last Name	Affiliation/ Organisation	Country	email
Claudia	Agliozzo	Universidad Andres Bello	Chile	c.agliozzo@gmail.com
Mohamad	Ali-Dib	University of Toronto	Canada	m.alidib@utoronto.ca
Kathrin	Altwegg	Physikalisches Institut, University of Bern	Switzerland	altwegg@space.unibe.ch
Dana	Anderson	California Institute of Technology	USA	deanders@caltech.edu
Manuel	Aravena	Universidad Diego Portales	Chile	manuel.aravenaa@mail.udp.cl
David	Arias-Olivares	Universidad Andres Bello	Chile	ndariaso@gmail.com
Elizabeth	Artur de la Villarmois	Niels Bohr Institute	Denmark	eliartur87@gmail.com
Chris	Arumainayagam	Wellesley College	USA	carumain@wellesley.edu
Sean	Ayling	University of Sussex	Great Britain	s.ayling@sussex.ac.uk
Arnaud	Belloche	Max-Planck-Institut für Radioastronomie	Germany	belloche@mpifr-bonn.mpg.de
Edwin	Bergin	University of Michigan	USA	ebergin@umich.edu
Jennifer	Bergner	Harvard University	USA	jennifer.bergner@cfa.harvard.edu
John	Black	Chalmers University of Technology	Sweden	john.black@chalmers.se
Dominique	Bockelée	Observatoire de Paris	France	Dominique.Bockelee@obspm.fr
Heloisa	Boechat	Universidade Federal do Rio de Janeiro	Brazil	heloisa@astro.ufrj.br
Arthur	Bosman	Leiden University	Netherlands	bosman@strw.leidenuniv.nl
Leonardo	Bronfman	DAS, U Chile	Chile	leo@das.uchile.cl
Hannah	Calcutt	University of Copenhagen, Niels Bohr Institute	Denmark	hmcalcutt@gmail.com
Jan	Cami	CPSX, The University of Western Ontario; SETI Institute, California	Canada	jcami@uwo.ca
Ewen	Campbell	University of Basel	Switzerland	ewen.campbell@unibas.ch
Roberto	Capuzzo Dolcetta	Dep. of Physics, Sapienza, UniversitÃ di Romar	Italy	roberto.capuzzodolcetta@uniroma1.it
John	Carpenter	Joint ALMA Observatory	Chile	john.carpenter@alma.cl
Brandon	Carroll	Caltech	USA	pcarroll@caltech.edu
Simon	Casassus	U Chile	Chile	scasassus@u.uchile.cl
Paola	Caselli	Max-Planck-Institute for Extraterrestrial Physics	Germany	caselli@mpe.mpg.de
Emmanuel	Caux	IRAP/CNRS-UPS, Toulouse	France	emmanuel.caux@irap.omp.eu
Paolo	Cazzoletti	Max Planck Institute for Extraterrestrial Physics	Germany	pcazzoletti@mpe.mpg.de
Qiang	Chang	Xinjiang Astronomical Observatory, Chinese Academy of Sciences	China	changqiang@xao.ac.cn
Steven	Charnley	NASA Goddard Space Flight Center	USA	steven.b.charnley@nasa.gov
Tao	Chen	Leiden Observatory	Netherlands	chen@strw.leidenuniv.nl
Allan	Cheung	University of Hong Kong	Hong Kong	hrsccsc@hku.hk
Ko-Ju	Chuang	Leiden Observatory	Netherlands	chuang@strw.leidenuniv.nl
Ilse	Cleeves	Harvard-Smithsonian CfA	USA	ilse.cleeves@cfa.harvard.edu
Ilsa	Cooke	University of Virginia	USA	irc5zb@virginia.edu
Martin	Cordiner	NASA Goddard	USA	martin.cordiner@nasa.gov
Pierre	Cox	Joint ALMA Observatory	Chile	pierre.cox@alma.cl
Maria	Cunningham	University of New South Wales	Australia	maria.cunningham@unsw.edu.au
Herma	Cuppen	Institute for Molecules and Materials, Radboud University	Netherlands	h.cuppen@science.ru.nl
Victor	de Souza Bonfim	Universidade do Vale do Paraíba – UNIVAP	Brazil	victordsb@gmail.com
Leen	Decin	KU Leuven	Belgium	leen.decin@kuleuven.be
Otoniel	Denis Alpizar	Universidad Autónoma de Chile	Chile	otonieldenisalpizar@gmail.com
Natalia	Drake	Observatorio Nacional-MCTIC, Rio de Janeiro	Brazil	drake@on.br
Maria	Drozdovskaya	Center for Space and Habitability, Universität Bern	Switzerland	maria.drozdovskaya@csh.unibe.ch
Gwenaelle	Dufour	NASA Goddard Space Flight CEnter	USA	gwenaelle.c.dufour@nasa.gov
Anne	DUTREY	University of Bordeaux, France	France	anne.dutrey@u-bordeaux.fr
Yuji	Ebisawa	University of Tokyo	Japan	ebisawa@taurus.phys.s.u-tokyo.ac.jp
Christian	Eistrup	Leiden Observatory	Netherlands	eistrup@strw.leidenuniv.nl
Stefano	Facchini	MPE	Germany	facchini@mpe.mpg.de

First Name	Last Name	Affiliation/ Organisation	Country	email
Felipe	Fantuzzi	Universidade Federal do Rio de Janeiro	Brazil	felipe.fantuzzi@iq.ufrj.br
Edith	Fayolle	Harvard-Smithsonian Center for Astrophysics	USA	efayolle@cfa.harvard.edu
Gary	Fuller	University of Manchester	Great Britain	G.fuller@manchester.ac.uk
Kenji	Furuya	University of Tsukuba	Japan	furuya@ccs.tsukuba.ac.jp
Pedro	García-Lario	ESAC/ESA, Madrid, Spain	Spain	Pedro.Garcia.Lario@esa.int
Rob	Garrod	University of Virginia	USA	rgarrod@virginia.edu
Lisseth	Gavilan	LATMOS - Université de Versailles	France	lisseth.gavilan@latmos.ipsl.fr
Maryvonne	Gerin	LERMA	France	Maryvonne.gerin@ens.fr
Javier	Goicoechea	Instituto de Ciencia de Materiales de Madrid	Spain	jr.goicoechea@icmm.csic.es
Marina	Gomes Rachid	Universidade do Vale do Paraíba	Brazil	marina.g.rachid@gmail.com
DAVID ENRIQUE	GREEN TRIPP	INSTITUTO DE ASTRONOMÍA. UNAM	Mexico	dgreen@astro.unam.mx
Viviana	Guzman	Joint ALMA Observatory (JAO)	Chile	viviana.guzman@alma.cl
Andrés	Guzmán	Universidad de Chile	Chile	aguzman@das.uchile.cl
Alvaro	Hacar	Leiden Observatory	Netherlands	hacar@strw.leidenuniv.nl
Satoshi	Hamano	Kyoto Sangyo University	Japan	hamano@cc.kyoto-su.ac.jp
Nanase	Harada	ASIAA	Taiwan	harada@asiaa.sinica.edu.tw
Jinhua	He	CASSACA; Yunnan Observatories, CAS (YNAO)	Chile	jinhuahe@ynao.ac.cn
Thomas	Henning	Max Planck Institute for Astronomy	Germany	henning@mpia.de
Antonio	Hernández-Gómez	IRyA, Universidad Nacional Autónoma de México; IRAP, U Toulouse III Paul Sabatier	Mexico	a.hernandez@crya.unam.mx
Aya	Higuchi	RIKEN	Japan	aya.higuchi@riken.jp
Mark	Hollands	University of Warwick	Great Britain	M.Hollands@warwick.ac.uk
Tien-Hao	Hsieh	Academia Sinica Institute of Astronomy and Astrophysics (ASIAA)	Taiwan	shawinchone@gmail.com
Jane	Huang	Harvard-Smithsonian Center for Astrophysics	USA	jane.huang@cfa.harvard.edu
Jean	Huang	Wellesley College	USA	jhuang6@wellesley.edu
Muneaki	Imai	University of Tokyo	Japan	imai@taurus.phys.s.u-tokyo.ac.jp
Violette	Impellizzeri	NRAO/ALMA	Chile	vimpelli@alma.cl
Nick	Indriolo	Space Telescope Science Institute	USA	nindriolo@stsci.edu
Natalia	Inostroza	Universidad Autonoma de Chile	Chile	natalia.inostrozapino@gmail.com
Steffen	Jacobsen	Centre for Star and Planet Formation & Niels Bohr Institute, University of Copenhagen	Denmark	steffen.jacobsen@nbi.ku.dk
Edward	Jenkins	Princeton University Observatory	USA	ebj@astro.princeton.edu
Paul	Jones	University of New South Wales	Australia	paulcojones@gmail.com
Jes	Jorgensen	Niels Bohr Institute, University of Copenhagen	Denmark	jeskj@nbi.ku.dk
Juris	Kalvans	Engineering Research Institute "Ventspils International Radio Astronomy Center", Ventspils UC	Latvia	juris.kalvans@venta.lv
Agata	Karska	Center for Astronomy, Nicolaus Copernicus University, Torun, Poland	Poland	agata.karska@umk.pl
Habib	Khosroshahi	School of Astronomy, IPM	Iran	habib@ipm.ir
Jacek	Krelowski	Organisation: Nicolaus Copernicus University, Torun	Poland	jacek@umk.pl
Lars E.	Kristensen	Center for Star and Planet Formation, Niels Bohr Institute, University of Copenhagen	Denmark	lars.kristensen@nbi.ku.dk
Yi-Jehng	Kuan	National Taiwan Normal University	Taiwan	kuan@ntnu.edu.tw
Xiaohu	Li	ASIAA & NAOC	Taiwan	xiaohu@asiaa.sinica.edu.tw
Niels	Ligterink	Sackler Laboratory for Astrophysics / Leiden Observatory	Netherlands	ligterink@strw.leidenuniv.nl

First Name	Last Name	Affiliation/ Organisation	Country	email
Harvey	Liszt	National Radio Astronomy Observatory	USA	hliszt@nrao.edu
Ryan	Loomis	Harvard Smithsonian Center for Astrophysics	USA	rloomis@cfa.harvard.edu
John	Lopez	UNSW Australia	Australia	j.lopez@unsw.edu.au
Suzanne	Madden	CEA Saclay	France	madden@discovery.saclay.cea.fr
Nikku	Madhusudhan	University of Cambridge	Great Britain	nmadhu@ast.cam.ac.uk
Diego	Mardones	Universidad de Chile	Chile	diego@das.uchile.cl
Sergio	Martin	JAO ESO	Chile	smartin@eso.org
Jesus	Martin Pintado	CAB (INTA CSIC)	Spain	jmartin@cab.inta-csci.es
Sarah	Massalkhi	Instituto de Ciencia de Materiales de Madrid, CSIC	Spain	sarah.massalkhi@csic.es
Gautier	Mathys	ESO/JAO	Chile	gmathys@eso.org
Melissa	McClure	European Southern Observatory	Germany	mmcclure@eso.org
Brett	McGuire	NRAO/Harvard CfA	USA	bmcguir3@gmail.com
Christopher	Merchantz	University of Michigan	USA	cjmerch@umich.edu
Hugo	Messias	ALMA JAO	Chile	hugo.messias@alma.cl
Stefanie	Milam	NASA Goddard Space Flight Center	USA	stefanie.n.milam@nasa.gov
Tom	Millar	Queen's University Belfast	UK	Tom.Millar@qub.ac.uk
Anna	Miotello	Leiden Observatory	Netherlands	miotello@strw.leidenuniv.nl
Oscar	Morata	Academia Sinica Institute of Astronomy and Astrophysics (ASIAA)	Taiwan	omorata@asiaa.sinica.edu.tw
Nadia	Murillo	Leiden Observatory, Leiden University	Netherlands	nmurillo@strw.leidenuniv.nl
Sudeep	Neupane	Department of Astronomy, University of Chile	Chile	sneupane@das.uchile.cl
Yuri	Nishimura	The University of Tokyo	Japan	yuri@taurus.phys.s.u-tokyo.ac.jp
Hideko	Nomura	Tokyo Institute of Technology	Japan	nomura@geo.titech.ac.jp
Shota	Notsu	Department of Astronomy, Graduate School of Science, Kyoto University	Japan	snotsu@kusastro.kyoto-u.ac.jp
Lars-Ake	Nyman	Joint ALMA Observatory/ESO	Chile	lars.nyman@alma.cl
Karin	Öberg	Harvard University	USA	koberg@cfa.harvard.edu
Hans	Olofsson	Dept. of Earth and Space Sciences, Chalmers	Sweden	hans.olofsson@chalmers.se
Jan	Orkisz	IRAM	France	orkisz@iram.fr
Yoko	Oya	The University of Tokyo	Japan	oya@taurus.phys.s.u-tokyo.ac.jp
Laurent	Pagani	CNRS; Observatoire de Paris	France	laurent.pagani@obspm.fr
Tyler	Pauly	Cornell University	USA	tap74@cornell.edu
Els	Peeters	Western University, Canada & SETI Institute, USA	Canada	epeeters@uwo.ca
Jamila	Pegues	Harvard-Smithsonian Center for Astrophysics	USA	jamila.pegues@cfa.harvard.edu
Neil	Phillips	ESO / Joint ALMA Observatory	Chile	Neil.Phillips@alma.cl
Sergio	Pilling	UNIVAP - Univ. do Vale do Paraiba	Brazil	sergiopilling@yahoo.com.br
Rafael	Pinotti	Observatório do Valongo UFRJ	Brazil	rpinotti@astro.ufrj.br
Ellen	Price	Harvard CFA	USA	ellen.price@cfa.harvard.edu
Anna	Punanova	Max-Planck-Institut für extraterrestrische Physik	Germany	punanova@mpe.mpg.de
David	Quénard	Queen Mary University of London - Astronomy Unit	United Kingdom	d.quenard@qmul.ac.uk
Anthony	Remijan	NRAO/ALMA	USA	aremijan@nrao.edu
Thomas	Rice	University of Michigan	USA	tsrice@umich.edu
Victor M.	Rivilla	Osservatorio Astrofisico di Arcetri, OAA-INAF	Italy	ryvendel@gmail.com
Monica	Rubio	Departamento de Astronomia, Universidad de Chile	Chile	mrubio@das.uchile.cl
Farid	Salama	ASA Ames Research Center	USA	farid.salama@nasa.gov
Vachail	Salinas	Leiden Observatory	Netherlands	vachail@gmail.com
Deborah	Schmidt	University of Arizona	USA	drschmidt@email.arizona.edu
Kamber	Schwarz	University of Michigan	USA	kamberrs@umich.edu
Takashi	Shimonishi	Tohoku University	Japan	shimonishi@astr.tohoku.ac.jp
Ian	Sims	University of Rennes 1	France	ian.sims@univ-rennes1.fr
Peter	Strittmatter	U Arizona	USA	pstrittm@email.arizona.edu
Taiki	Suzuki	SOKENDAI	Japan	taiki.suzuki@nao.ac.jp
Jonathan	Tan	University of Florida	USA	jctan.astro@gmail.com
Luis Fernando	Tapia Schiavon	Nuclear Sciences Institute at the National Autonomous University of Mexico	Mexico	l.tapia@crya.unam.mx
Vianney	Taquet	Leiden Observatory	Netherlands	taquet@strw.leidenuniv.nl
Richard	Teague	Max Planck Institute for Astronomy	Germany	teague@mpia.de

First Name	Last Name	Affiliation/ Organisation	Country	email
Chenoa	Tremblay	ICRAR-Curtin University	Australia	chenoa.tremblay@postgrad.curtin.edu.au
Lukasz	Tychoniec	Leiden Observatory	Netherlands	tychoniec@strw.leidenuniv.nl
Valeska	Valdivia	Organisation: Service d'Astrophysique - CEA/Saclay	France	valeska.valdivia@cea.fr
Merel	van 't Hoff	Leiden Observatory	Netherlands	vthoff@strw.leidenuniv.nl
Marie	Van de Sande	Instituut voor Sterrenkunde, KU Leuven	Belgium	marie.vandesande@kuleuven.be
Nienke	van der Marel	Institute for Astronomy, University of Hawaii	USA	marel@hawaii.edu
Matthijs	van der Wiel	ASTRON	Netherlands	wiel@astron.nl
Ewine	van Dishoeck	Leiden Observatory / MPE	Netherlands	ewine@strw.leidenuniv.nl
Thomas	Vidal	Organisation: Laboratoire d'Astrophysique de Bordeaux	France	thomas.vidal@u-bordeaux.fr
Ruud	Visser	European Southern Observatory	Germany	rvisser@eso.org
Catherine	Walsh	University of Leeds	Great Britain	c.walsh1@leeds.ac.uk
Yoshimasa	Watanabe	The University of Tokyo	Japan	nabe@taurus.phys.s.u-tokyo.ac.jp
Nadine	Wehres	I. Physical Institute, Universit of Cologne	Germany	wehres@ph1.uni-koeln.de
Susanna	Widicus Weaver	Emory University	USA	swidicu@emory.edu
Eric	Willis	University of Virginia	USA	ew2zb@virginia.edu
Eva	Wirström	Chalmers University of Technology	Sweden	eva.wirstrom@chalmers.se
Paul	Woods	Nature Astronomy	Great Britain	paul.woods@nature.com
Alwyn	Wootten	NRAO	USA	awootten@nrao.edu
Satoshi	Yamamoto	U Tokyo	Japan	yamamoto@taurus.phys.s.u-tokyo.ac.jp
Jiaerken	Yeshengbieke	Xinjiang Astronomical Observatory, Chinese academy of sciences	China	jarken@xao.ac.cn
Kento	Yoshida	The Univ. of Tokyo/RIKEN	Japan	yoshida@taurus.phys.s.u-tokyo.ac.jp
Sarolta	Zahorecz	Konkoly Observatory / Osaka Prefecture University	Hungary/ Japan	zahorecz.sarolta@csfk.mta.hu
Ke	Zhang	University of Michigan	USA	kezhang@umich.edu
Jianjun	Zhou	Xinjiang Astronomical Observatory, Chinese Academy of Sciences	China	zhoujj@xao.ac.cn
Lei	Zhu	CASSACA	Chile	lzhupku@gmail.com
Igor	Zinchenko	Institute of Applied Physics of the Russian Academy of Sciences	Russia	zin@appl.sci-nnov.ru
Hans	Zinnecker	Deutsches SOFIA Institut	Germany	zinnecker@dsi.uni-stuttgart.de

Overview of astrochemistry in 2017

Astrochemistry VII – Through the Cosmos from Galaxies to Planets
Proceedings IAU Symposium No. 332, 2017
M. Cunningham, T. Millar & Y. Aikawa, eds.

doi:10.1017/S1743921317011528

Astrochemistry: overview and challenges

Ewine F. van Dishoeck[1,2]

[1]Leiden Observatory, Leiden University
P.O. Box 9513, NL-2300 RA, Leiden, the Netherlands
email: ewine@strw.leidenuniv.nl

[2]Max Planck Institute for Extraterrestrial Physics, Garching, Germany

Abstract. This paper provides a brief overview of the journey of molecules through the Cosmos, from local diffuse interstellar clouds and PDRs to distant galaxies, and from cold dark clouds to hot star-forming cores, protoplanetary disks, planetesimals and exoplanets. Recent developments in each area are sketched and the importance of connecting astronomy with chemistry and other disciplines is emphasized. Fourteen challenges for the field of Astrochemistry in the coming decades are formulated.

Keywords. astrochemistry, molecular data, molecular processes, ISM: molecules, stars: formation, planets: formation, comets, techniques: spectrocopic, galaxies: ISM

1. Introduction

Astrochemistry, also known as molecular astrophysics, is 'the study of the formation, destruction and excitation of molecules in astronomical environments and their influence on the structure, dynamics and evolution of astronomical objects' as stated by the pioneer of this field, Alexander Dalgarno in 2008. This definition covers not only the chemical aspects of the field, but also recognizes that molecules are excellent diagnostics of the physical conditions and processes in the regions where they reside. Moreover, they actively contribute to the physical state of the gas by being important coolants. From an astronomical perspective, dense molecular clouds are important as the nurseries of new generations of stars and planets, some of which may even harbor life. From the pure chemistry perspective, interstellar space provides a unique environment in which molecular behavior can be studied under extreme conditions. It is this combination that makes Astrochemistry such a fascinating research field, for both astronomers and chemists alike.

More than 200 different molecules have been detected in interstellar space (Fig. 1). The main questions in the field of Astrochemistry therefore include: how, when and where are these molecules produced and excited? What do they tell us about temperatures, densities, gas masses, ionization rates, radiation fields, and dynamics of the clouds? How are they cycled through the various phases of stellar evolution, from birth to death? How far does chemical complexity go? And, most far-reaching, can interstellar molecules become part of new planetary systems and form the building blocks for life elsewhere in the Universe?

The interdisciplinary aspects of this field will be emphasized throughout this overview as a two-way street: Astrochemistry needs basic data on molecular spectroscopy and chemical processes, but it also inspires new chemical physics through studies of different classes of molecules and reactions that are not normally considered on Earth. Astrochemistry is a 'blending of astronomy and chemistry in which each area enriches the other in a mutually stimulating interaction' (Dalgarno 2008). The field is becoming increasingly multidisciplinary: besides chemistry, physics and mathematics, which have traditionally

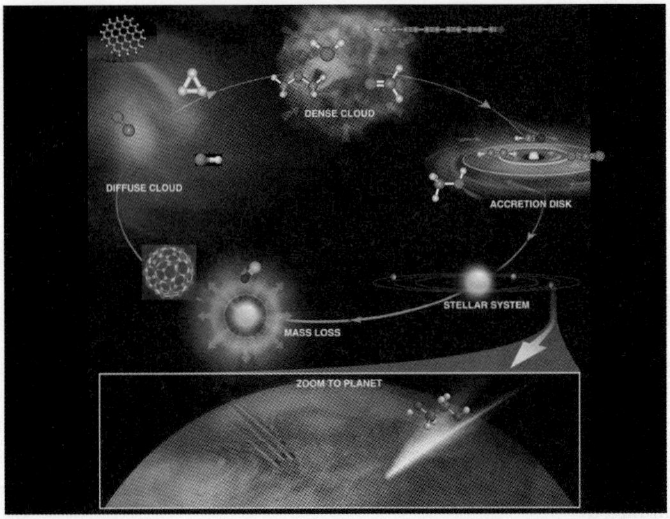

Figure 1. Lifecycle of gas and dust in interstellar space. Some characteristic molecules at each of the star- and planet formation and stellar death stages are indicated. Image by Bill Saxton (NRAO/AUI/NSF) and molecule pictures from *the Astrochymist* (`www.astrochemyst.org`; this website also contains a list of detected molecules in space).

been an integral part of the field, knowledge about biology, geochemistry and informatics becomes important as well.

In the following, a brief overview of recent developments in the large range of topics addressed in this Symposium will be given, together with a challenge for each of these topics. References will be limited to just a few per topic. Recent overviews of astrochemistry with extensive reference lists can be found in Caselli & Ceccarelli (2012); Tielens (2013); van Dishoeck (2014) and the 2013 Chemical Reviews special issue, as well as in the individual papers in this volume.

2. Observational facilities

Progress in Astrochemistry is driven by new observational data. Molecules can be observed through their electronic, vibrational or rotational transitions at optical/UV, infrared and millimeter wavelengths, respectively. Astrochemistry has been fortunate that several of the most powerful new telescopes deployed over the past decade have been particularly well suited to observe interstellar molecules: the *Spitzer Space Telescope* and the *Herschel Space Observatory* at mid- and far-infrared wavelengths, a number of ground-based 8m optical/infrared telescopes equipped with high resolution spectrometers, and various single-dish and interferometers at millimeter wavelengths with increasingly sensitive broad-band detectors, culminating in the Atacama Large Millimeter/submillimeter Array (ALMA). The *Stratospheric Observatory for Infrared Astronomy* (SOFIA) is contributing unique data, and the *James Webb Space Telescope* (JWST), to be launched late 2018, will be the next big jump in mid-infrared capabilities.

On the longer term horizon the situation looks less favorable for Astrochemistry, however. ALMA is expected to be still going strong for decades, and at least one Extremely Large Telescope (ELT) will be equipped with a high resolution infrared spectrometer. However, several new flagship space missions in the 2020s–2035s are dedicated to large area cosmological and galaxy evolution surveys, high energy astrophysics and to the emerging field of gravitational wave astrophysics. Exoplanet atmosphere studies will likely be a strong driver for future satellites, but their modest spectral resolution instruments have limited applications in other areas of Astrochemistry.

Challenge 1: To secure new facilities for Astrochemistry in the 2030–2040 timeframe.

3. Laboratory astrophysics

The availability of accurate atomic and molecular data is another prerequisite for Astrochemistry. Note that the term 'laboratory astrophysics' implies experiments as well as quantum chemical and molecular dynamics calculations. Over the past decade, several new techniques and new groups have entered the field, ensuring new efforts to address the many chemical-physics questions and the transfer of (often unique) expertise to the younger generation. At the same time, traditional techniques and data gathering remain important as well, even if some of those experiments are not as 'flashy' as using, for example, He-droplets to study ultracold reactions. The key for a fruitful interaction is for astronomers to ask the right questions and have enough chemical physics knowledge of what can, and what cannot, be done in the laboratory. Chemists, in turn, have to sometimes let go of perfectionism, since astronomers are often content at the factor of 2 uncertainty level. Also, astronomers just want a number rather than a deep understanding of the underlying molecular processes. This 'tension' needs to be recognized by both sides, and, importantly, also by the funding agencies. Equally important, astronomers need to properly cite the relevant data used in Astrochemistry to support laboratory scientists.

Spectroscopy: The most basic information that astronomers need is spectroscopy from UV to millimeter wavelengths. Techniques range from classical absorption set-ups to cavity ringdown spectroscopy, (chirped-pulse) Fourier transform microwave spectroscopy and THz time domain spectroscopy. Transition frequencies and strengths of (sub)mm transitions are summarized in the Jet Propulsion Laboratory catalog (JPL) (Pickett *et al.* 1998) and the Cologne Database for Molecular Spectroscopy (CDMS) (Endres *et al.* 2016) †. Databases for vibrational transitions at infrared wavelengths include the HITRAN (Rothman *et al.* 2009) and the EXOMOL (Tennyson *et al.* 2016) line lists‡.

Spectroscopy of large samples of PAHs has been mostly carried out with matrix-isolation experiments, although selected PAHs have also been measured in the gas phase (Boersma *et al.* 2014). Spectroscopic databases of solids include those for ices in Leiden, NASA-Ames, Goddard and Madrid ¶; and the Heidelberg-Jena-St. Petersburg database of optical constants for silicates (Henning 2010). Information on carbonaceous material is more scattered in the literature.

Rate coefficients: Chemical models require rates for thousands of two-body reactions under space conditions. There are only a handful types of reactions, however. In the gas-phase, these are radiative association and associative detachment (formation of bonds); ion-molecule, neutral-neutral and charge-transfer reactions (rearrangement of bonds); and dissociative recombination and photodissociation processes (destruction of bonds).

Gas-phase processes have been summarized by Smith (2011) and Sims, this volume, and many recent developments are described in the 2014 release of the KIDA database (Wakelam *et al.* 2015). There has been good progress on improving many rate coefficients for ion-molecule and neutral-neutral reactions over a range of temperatures. Radiative association, however, is an example of a process whose rate coefficients continue to have large uncertainties due to their low values and the difficulties of measuring or computing

† `spec.jpl.nasa.gov` and `www.astro.uni-koeln.de/cdms/catalog`
‡ `www.cfa.harvard.edu/hitran` and `www.exomol.com`
¶ `icedb.strw.leidenuniv.nl`

them. Photodissociation and photoionization rates of atoms and molecules exposed to different radiation fields have recently been updated by Heays *et al.* (2017).

Rate coefficients for collisional excitation of molecules continue to be essential for the quantitative non-LTE excitation analysis of the observed molecular lines. The vast majority of these rates come from theoretical calculations and the field is fortunate to have a steady stream of new results from quantum chemists (Roueff & Lique 2013; Dubernet *et al.* 2013; Wiesenfeld *et al.* 2016). With new instrumental developments on the horizon, it may be time for experiments to catch up in this area.

Ice chemistry: The last two decades have seen significant progress in ice astrochemistry studies, with modern surface science techniques at ultra-high vacuum conditions now used to quantitatively study various chemical processes in ices. Summaries of activities across the world are given by Allodi *et al.* (2013); Linnartz *et al.* (2015). Molecules such as CH_3OH, H_2O, CO_2 have been demonstrated to form at low temperatures on surfaces through atom addition reactions, validating reaction schemes postulated 30 years earlier (Tielens & Hagen 1982). Binding energies are key to many astrochemical models and more work on astrochemically relevant species and mixtures is needed (Collings *et al.* 2004; Cuppen *et al.* 2017).

Many laboratory experiments across the world have been performed using some form of 'processing' that ices are expected to undergo in space: heating, UV irradiation, and exposure to energetic particles and/or electrons (Öberg 2016). All of them produce complex molecules, in some cases even biologically interesting species like amino acids, sugars, and amides, and often in similar amounts. The main differences in outcome may be related to whether or not the strong CO or N_2 bonds can be broken, especially if there is no other source of atomic C or N in the ice mixture. For example, UV radiation longward of 1200 Å does not photodissociate CO or N_2. Otherwise the results seem to mainly depend on the amount of energy deposited in ice, not on which form. In that respect it is useful to recall that UV radiation deposits 10× more energy per molecule in the ice than cosmic rays for typical cosmic ray fluxes in dark clouds (Shen *et al.* 2004). While products are readily identified, the step-by-step kinetics and mechanisms are still poorly constrained in most of these experiments, so the results cannot be readily incorporated into astrochemical models.

Molecular dynamics techniques can also be extended to solid-state processes, as demonstrated by the detailed study of the photodesorption of water ice and its isotopologs, as well as CO_2 ice formation (Arasa *et al.* 2015, 2013). Such models provide much insight into the processes occuring in and on the ice.

Challenge 2: To continue to bring chemists, physicists and astronomers together to characterize and quantify molecular processes that are at the heart of Astrochemistry, to have open lines of communication to prioritize needs, and to convince funding agencies to continue supporting this interdisciplinary research.

4. Astrochemical models

Compilations of reaction rate coefficients together with codes that solve the coupled differential equations to compute abundances include the UMIST 2013 database (McElroy *et al.* 2013), and the KIDA database (Wakelam *et al.* 2012) †. The latter website includes the Nahoon (formally known as Ohio State) gas-grain chemistry code by Herbst and co-workers. The UCLCHEM code is now also publicly released (Holdship *et al.* 2017). These databases and codes contain primarily two-body reactions because of the

† `www.udfa.net` and `kida.obs.u-bordeaux1.fr`

low densities in interstellar space. Three-body reactions become significant at densities above $\sim 10^{13}$ cm^{-3} such as encountered in the inner midplanes of protoplanetary disks and the atmospheres of stars and exoplanets. Here the chemistry may approach that in thermodynamic equilibrium (LTE).

It is important to recognize that astrochemical databases are put together and updated on a 'best effort' basis. Following the good example of atmospheric chemistry, some databases (UMIST, KIDA) have an (estimated) uncertainty associated with each rate coefficient which can then be propagated in the network. Examples of pure gas-phase networks show that even the abundances of simple molecules like H_2O, SO or CH, have an uncertainty of a factor of ~3 just from the uncertainties in individual rate coefficients under dark cloud conditions (Wakelam *et al.* 2010); for larger molecules, the cumulative effect of many reactions will be even larger. This illustrates the level of agreement one can expect between different models, and between models and observations.

Gas-grain chemistry networks introduce another level of complexity. The overall efficiency depends on the probability that the atoms or molecules stick to the grains upon collision, their mobility on the surface, the probability that molecule formation occurs, and finally the probability that the molecule is released back into the gas phase. The standard treatment of the interplay between gas phase and grain surface chemistry is through rate equations (Hasegawa & Herbst 1993; Garrod *et al.* 2008). This approach is known to be inadequate under some conditions, however, especially for models with very small grains and only a few species per grain. Many alternative approaches are being considered such as the modified rate equations, Monte Carlo, Master equation, and hybrid methods, each with their advantages and drawbacks.

Chemistry occurs not just at the surface but also deep inside the ice, since UV photons typically penetrate at least 50 monolayers. The atoms and radicals following photodissociation (or some other form of energetic processing) are created with excess energy and are thus initially highly mobile, reacting with neighboring species and overcoming energy barriers. However, they quickly lose this energy on pico-second timescales and become trapped. Alternatively, these radicals become mobile and find each other once the ice temperature increases from ~10 K to 20–40 K. A general concern is how to translate laboratory data to parameters that can be used in grain surface chemistry. Laboratory experiments provide rates as function of H or UV fluence or some other parameter over timescales of hours; astronomical applications involve timescales $> 10^5$ yr (Cuppen *et al.* 2017).

Which of these gas-phase or gas-grain reactions dominates the formation of a molecule depends on the astrophysical situation. For example, water is primarily formed by ion-molecule reactions in diffuse clouds; by high temperature neutral-neutral reactions in shocks; and by grain-surface chemistry in dense cores (van Dishoeck *et al.* 2013).

Challenge 3: To build realistic gas-grain models from microscopic to macroscopic scales, including translation of laboratory ice chemistry experiments into parameters that can be adopted in models.

5. Diffuse and translucent clouds

Astrochemistry started nearly a century ago with optical absorption spectroscopy of atoms and molecules in diffuse clouds along the lines of sight to bright stars. These clouds, and their somewhat denser translucent counterparts, have visual extinctions of a few mag and are in the interesting regime where the transition of most elements from atomic to molecular form takes place (van Dishoeck & Black 1989; Snow & McCall

2006). Because UV photons and cosmic rays can penetrate the clouds to ionize atoms and dissociate molecules, timescales to reach equilibrium are short (few thousand yr) and the chemistry is dominated by ion-molecule reactions. Physical parameters such as density, temperature and UV field are well constrained and even the H_2 column density can be measured directly through UV absorption lines. Thus, these clouds form the best testbed for *precision Astrochemistry*, at the factor of 2 or better level. At the same time, because so many parameters are well determined, they also serve as diagnostics of the physical structure and processes in the interstellar medium (ISM).

Herschel-HIFI data have given a new boost to this field (Gerin *et al.* 2016). Pure rotational lines of molecules have been seen in absorption throughout the Galaxy along the lines of sight toward distant far-infrared sources. Since most of the population is in the lowest level(s), the conversion from absorption to column density is accurate and straight-forward. H_2 cannot be observed directly at far-infrared wavelengths, but HF has proven to be a reliable tracer of its column density because of its particularly simple chemistry. Small hydrides like H_2O are found to have abundences of 5×10^{-8} with respect to H_2, values that are indeed well explained by low-temperature ion-molecule chemistry.

One of the main *Herschel* surprises has been the discovery of strong OH^+ and H_2O^+ absorption along most galactic lines of sight and even in high-redshift galaxies. Also the first noble gas molecule, $^{36}ArH^+$, has been detected. These species point to an H_2-poor phase of the interstellar medium containing molecules (typically $H/H_2 > 10$) that had not been recognized before. As such, the abundances of these molecules form important tests for hydrodynamical simulations of the large scale ISM. They are also excellent diagnostics of the cosmic ray ionization rate in the low density ISM, giving a median value of $\zeta_H = 1.8 \times 10^{-16}$ s^{-1} that agrees well with previous results from H_3^+ data (Indriolo *et al.* 2015). The origin of this H_2-poor phase is interesting by itself: in contrast with other molecules, timescales for the formation of H_2 itself are very long, up to 10^7 yr. Both analytical and hydrodynamical models for the H/H_2 structure under non steady-state conditions are now being studied (Bialy *et al.* 2017) †.

On the other hand, there are lingering discrepancies and mysteries even for these simple clouds: the high abundance of CH^+ (and SH^+) requires introduction of additional parameters to capture the effects of turbulent heating at low (column) densities (Godard *et al.* 2014); the discovery of increasingly complex molecules with abundances similar to those in dark clouds is difficult to explain (Liszt *et al.* 2005); the need to form grains in the diffuse ISM to solve the dust lifetime problem implies active grain chemistry (Krasnokutski *et al.* 2014); and the Diffuse Interstellar Bands remain unidentified (except plausibly C_{60}^+, Campbell *et al.* 2016), pointing to the presence of a reservoir of large carbonaceous molecules. It may well take another century to solve these questions!

Challenge 4: To obtain a full inventory of the chemical constituents of diffuse and translucent clouds, and explain – at the same time – their chemical simplicity and complexity.

6. Photon-Dominated Regions (PDRs)

Photon-Dominated or Photo-Dissociation Regions (PDRs) are clouds exposed to intense FUV radiation (10^3-10^5 times the standard interstellar radiation field, ISRF) controlling their heating and chemistry (Hollenbach & Tielens 1997). The diffuse and translucent clouds discussed above are low density, low FUV PDR examples. Since most of the

† Non steady-state (or time-dependent) chemistry is often (incorrectly) denoted as non-equilibrium chemistry by hydrodynamical modelers.

molecular gas in galaxies is in the form of PDRs, studies of their structure are highly relevant for understanding spatially unresolved large scale observations. Locally, the layered structure of PDRs, with PAH emission and carbonaceous radicals peaking closer to the UV source than saturated molecules, has been beautifully confirmed by recent observations of the Orion Bar and Horsehead Nebula (Guzmán *et al.* 2014; Cuadrado *et al.* 2017). Such studies are also highly relevant for analyzing emission from the surfaces of protoplanetary disks, which are even higher density, higher FUV PDR cases.

X-rays can also affect the chemistry, although the chemical structure of XDRs shows similarities with that of PDRs since X-rays produce UV radiation through the interaction of the secondary electrons with H_2. X-rays, like cosmic rays, can however penetrate much deeper into clouds, heating them over a larger volume (Meijerink *et al.* 2007). While the individual heating and cooling processes in PDRs and XDRs have been identified decades ago, it is important to realize that the resulting temperature structures differ considerably from model to model (Röllig *et al.* 2007). This in turn affects the predicted excitation and emission of molecules in those zones such as the high-J H_2 and CO rotational lines (Visser *et al.* 2012). These uncertainties limit their diagnostic potential and are often not acknowledged by the community.

Large aromatic molecules are readily observed in PDRs, consisting most likely of a mix of neutral and ionized PAHs that emit strongly in mid-infrared bands, although there is still an active discussion on the fraction of aliphatic moieties in the larger carbonaceous species. PAHs constitute about 5–10% of the available carbon, with perhaps trace amounts of nitrogen (Tielens 2008). Fullerenes, C_{60} and C_{70}, have also been identified (Cami *et al.* 2010), and there is evidence for the conversion of large PAHs to fullerenes when exposed to intense FUV radiation (Berné & Tielens 2012). Large PAHs are stable against FUV radiation (except some stripping of H atoms), but smaller PAHs can be destroyed piecemeal, creating C_2 or C_2H_2 fragments. This, in turn, triggers a top-down chemistry in which small carbonaceous molecules in PDRs may be produced more efficiently from destruction of larger species rather than bottom-up starting from carbon atoms (Zhen *et al.* 2014). JWST will be particularly powerful in imaging these chemical transition zones in PDRs. Destruction of small carbonaceous grains in the PDR layers of disks by UV and atomic O has also been invoked to explain why Earth is so poor in carbon (Anderson *et al.* 2017).

Challenge 5: To quantify the importance of top-down versus bottom-up chemistry in the production of carbon-bearing molecules.

7. Evolved stars

The chemistry in circumstellar envelopes combines many important elements of Astrochemistry. From inside to ouside: LTE chemistry in the stellar atmospheres, non-LTE high temperature shock-induced chemistry in the inner envelope, the formation of dust grains as the gas cools, grain surface chemistry, and photodissociation and ion-molecule chemistry at the outer edge (Decin *et al.* 2010). Because the wind-accelerated material is moving outward, the time-dependent chemistry and dynamics need to be coupled. Circumstellar envelopes are good laboratories to test different types of chemistry, because they tend to be either carbon- (C/O>1) or oxygen-rich (C/O<1). The C/O>1 situation, which results in very bright lines of carbon-chain molecules, occurs under only a few other astrophysical situations, most notably protoplanetary disks (see § 12).

The C/O dichotomy is not completely black-and-white, however. *Herschel* confirmed that H_2O is present in a number of carbon-rich envelopes, as found in earlier SWAS data, but ruled out the interpretation that the water emission orginates in a cold zone

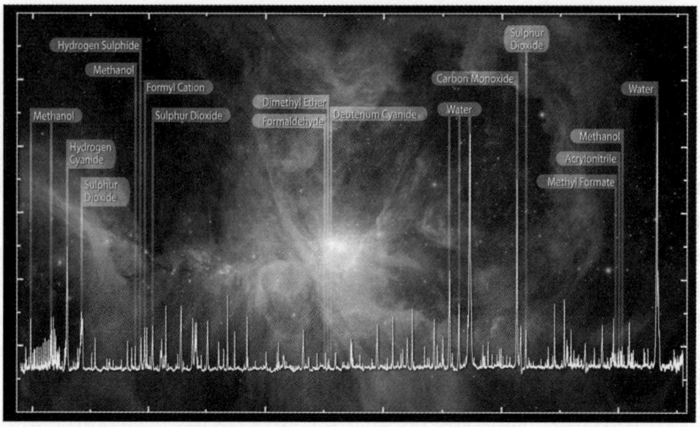

Figure 2. *Herschel*-HIFI spectrum of the Orion KL region around 550 GHz, showing strong emission from water and organic molecules, superposed on a *Spitzer* image (NASA/JPL-Caltech/S.T. Megeath) (Bergin *et al.* 2010).

of sublimating Kuiper-Belt-like icy planetesimals. More likely, the presence of H_2O under carbon-rich conditions points to a clumpy envelope with photodissociation breaking up CO and liberating oxygen. ALMA now allows these various chemical zones in many different molecules to be imaged, zooming in on the dust formation zone. Together with future high-resolution infrared spectra, these data will form critical tests of the so-far poorly constrained models of dust formation, not only in our own Milky Way but also for sources in nearby galaxies with lower metallicities such as the LMC and SMC. Dust destruction in the (reverse) shocks may also be seen. This will be of fundamental importance for understanding the dust life cycle in the Universe, including the surprising presence of large amounts of dust at high redshifts: does most of our dust come from AGB stars or from supernova remnants?

Almost exactly 30 years prior to this symposium, on Feb. 24 1987, SN 1987A exploded in the Large Magellanic Cloud. I remember watching the supernova shine brightly from a hotel in Puerto Varas where we were vacationing in March 1987 after an observing run at La Silla. It is exciting to see ALMA now detecting the cold dust formation directly in SN 1987A, as well as ^{29}SiO in the ejecta (Matsuura *et al.* 2017). SiO was actually first observed by its strong infrared emission a year after the explosion and chemically modeled at that time by Liu & Dalgarno (1994).

Challenge 6: To nail down the dust formation and destruction processes and their efficiencies in the envelopes of evolved low- and high-mass stars, for different metallicities.

8. Extragalactic astrochemistry

Astrochemistry and cosmology go hand-in-hand: the first stars in the Universe could not have formed without at least some H_2 and HD being present as coolants to allow cloud collapse (Galli & Palla 2013). This 'dawn of chemistry' (term coined by Dalgarno & Lepp 1987) at redshifts z more than 1000 involves primarily H, D and He and has been well studied. It also highlights the importance of accurate rate coefficients for critical reactions (at the better than 50% level) (Kreckel *et al.* 2010). Early Universe chemistry remains outside the realm of direct observational tests, however.

Chemistry in high-redshift galaxies out to $z \approx 10$ can be observed by different techniques. Quasar absorption line observations at optical and millimeter wavelengths probe mostly low-density gas in intervening galaxies, and reveal H_2 and a suite of molecules including OH^+, H_2O^+ and ArH^+ (Muller *et al.* 2016) that have been seen in diffuse galactic clouds (§5). Thus, *Herschel* flew just in time: without those data, astrochemists would not have been able to make sense of these extragalactic detections.

Far-infrared and millimeter emission line data probing the denser ISM show a wide variety of molecules including more complex species (González-Alfonso *et al.* 2012; Costagliola *et al.* 2015). Indeed, some of the ALMA spectra of nearby star-forming galaxies are as rich as those of Orion were 30 years ago with the poorer sensitivity and spatial resolution available at that time.

Most studies of extragalactic molecular clouds have been carried out at kpc scales. ALMA now allows surveys down to scales of $\lesssim$ 20 pc, and exceptionally probing down to 2 pc in nearby galaxies (Schruba *et al.* 2017). Lensed systems provide an opportunity to image high-z galaxies on sub-kpc scales (ALMA Partnership *et al.* 2015). There is a large range in 'metallicities' in these galaxies, with carbon and oxygen underabundant by factors of 5 to $>$10 compared with solar abundances. In contrast, most galactic astrochemistry studies have been done at $<$0.1–1 pc scales in clouds with close to solar metallicities. Thus, there is a huge gap between galactic and extragalactic molecular cloud studies that needs to be bridged (Kennicutt & Evans 2012); so far, this has not yet been done even for CO, let alone for other molecules. It is somewhat embarrassing that after nearly 50 years of millimeter observations, the observed distributions and abundance ratios in galaxies are such a puzzle. Cloud-scale surveys in the Milky Way and nearby galaxies on 1–20 pc scales in key tracers of different physical structures (e.g., cold gas: N_2H^+; warmer gas: SO, CH_3OH; dense gas: HCN, HCO^+) will therefore become increasingly important to place the extragalactic data in context (Pety *et al.* 2017; Watanabe *et al.* 2017).

Challenge 7: To bridge the gap between subpc galactic and kpc extragalactic astrochemical studies as functions of metallicity out to the highest redshifts, and to use molecular observations of calibrated tracers to unveil a new understanding of star formation in the early Universe.

9. Dense prestellar cores and cold protostellar envelopes

Dense cold clouds prior to star formation, with typical temperatures of 10 K and densities of $10^4 - 10^5$ cm^{-3}, exhibit a variety of chemical characteristics, including long carbon chain molecules, ice formation and heavy deuterium fractionation (Bergin & Tafalla 2007). Most of these chemical signatures are transferred to the protostellar stage where they are observed in the outer parts of the collapsing envelope (Fig. 1).

Ice formation. Infrared observations of ice features toward reddened background stars show that formation of water ice starts in the translucent cloud phase at A_V of a few mag, when densities are of order 10^3 cm^{-3} and dust temperatures $\lesssim$ 15 K (Boogert *et al.* 2015). The formation of water ice is now well characterized in the laboratory and models, and tested against observations of cold water gas and intermediate products such as HO_2 and H_2O_2 (see review by van Dishoeck *et al.* 2013). CH_4, NH_3 and some CO_2 ice are also made in this early phase, with the amount of ice rapidly increasing as the cores become more concentrated. All of these molecules are made on the grain surfaces, from reactions of atomic O, C and N with atomic hydrogen; they are not accreted from the gas.

At densities around 10^5 cm^{-3}, the timescales for freeze-out become shorter than the lifetime of the core, and CO – the dominant form of volatile carbon at these high densities – rapidly depletes from the gas onto the grains. This 'catastrophic' freeze-out of CO results in a separate 'water-poor' or 'apolar' ice phase, which has been observed directly in CO ice profiles. The CO-rich ice can subsequently react with atomic H to form H_2CO and CH_3OH, a process that has been demonstrated to be effective at low temperatures in the laboratory (Hidaka *et al.* 2009; Fuchs *et al.* 2009). Various other routes can transform CO to CO_2 and more complex organic ices.

One of the main observational surprises has been the recent detection of various complex organic molecules in cold cores (Bacmann *et al.* 2012; Öberg *et al.* 2010). The issue is not their production, since several laboratory experiments have now demonstrated that they can be formed at low temperatures without the need for heating. The main puzzle is their return to the gas phase at dust temperatures well below that for thermal desorption. Non-thermal processes such as photodesorption, cosmic-ray induced spot heating and reactive desorption have been proposed, and are being studied theoretically and in the laboratory, but they remain difficult to quantify and may lead to desorption of only fragments (Ivlev *et al.* 2015; Minissale *et al.* 2016; Bertin *et al.* 2016). Chemical models often still use a 'fudge factor' to prevent molecules from being fully frozen out: in older models this used to be a sticking probability of 90–99% rather than 100%. In modern models including detailed grain surface chemistry, a reactive desorption probability of ~1% is often introduced to explain observations (Garrod *et al.* 2007).

Challenge 8: To identify and quantify the mechanisms by which molecules, including the more complex ones, are desorbed (intact) from the grain surface in cold clouds.

Carbon-chain molecules. A number of cold cores (but not all) show relatively strong lines of carbon-chain molecules such as HC_3N, HC_5N, C_4H and CCS, with TMC-1CP still the iconic example (Kaifu *et al.* 2004). Other cores are more abundant in species like NH_3. One interpretation is that cores rich in carbon chains have only recently contracted from the diffuse cloud phase and are thus still rich in atomic carbon, which can be inserted into small molecules to make longer chains (Suzuki *et al.* 1992). Another explanation is their environment: if these cores are embedded in filaments or fibers from which they are still accreting fresh atomic carbon-rich material, they could build carbon chains through the same process. In the evolutionary explanation, the statistics on cores with or without carbon-chain molecules could be used to infer timescales for this phase.

Cold carbon-chain chemistry should not be confused with the warm carbon-chain molecule chemistry (WCCC) associated with some low-mass protostars at a later stage of evolution (Sakai & Yamamoto 2013). Here one explanation is sublimation of CH_4 ice at moderate dust temperatures, triggering a gas-phase chemistry that also leads to carbon-chain molecules. Alternatively, UV-irradiated outflow cavity walls can be rich in carbon-bearing molecules such as c-C_3H_2 (Murillo *et al.*, subm.).

Deuterium fractionation. Cold clouds have high abundances of deuterated molecules such as DCO^+, DCN and HDCO, with ratios to their undeuterated counterparts at least three orders of magnitude higher than the overall [D]/[H] ratio of $\sim 2 \times 10^{-5}$. Even doubly- and triply-deuterated molecules such as D_2CO and ND_3 have been detected (Ceccarelli *et al.* 2014). This huge fractionation can be explained by two effects. First, the lower zero-point vibrational energy of deuterated molecules makes their production reactions exothermic. In cold cores, most of the fractionation is initiated by the $H_3^+ + HD \rightarrow H_2D^+ + H_2$ reaction, which liberates 230 K. Correct calculation of the H_2D^+ abundance requires explicit treatment of the nuclear spin states (ortho and para) of all the species involved in the reactions (Sipilä *et al.* 2010). At slightly elevated temperatures, the CH_2D^+ ion, formed by reaction of CH_3^+ + HD, becomes more effective in controlling the DCN/HCN ratio.

The second boost in deuterium fractionation occurs in the densest coldest phase when CO, the main destroyer of both H_3^+ and H_2D^+, freezes out (Roberts *et al.* 2003). The increased atomic D/H ratio in the gas also leads to enhanced deuteration of grain surface species like H_2CO, NH_3 and CH_3OH (Tielens 1983; Furuya *et al.* 2016). Thus, the amount of deuteration in principle provides information on the formation history. Similarly, analysis of the size of the freeze-out zones of pre- and protostellar cores can be used to infer a lifetime of this heavy freeze-out phase of about 10^5 yr (Jørgensen *et al.* 2005).

Challenge 9: To use chemical signatures to constrain physical structure, evolutionary stage and the amount of time spent in certain cloud phases.

10. Shocks, jets and outflows

Shocks with velocities of tens to hundreds of km s^{-1} are common in the ISM, triggered by a variety of physical phenomena such as supernova explosions, cloud-cloud collisions as well as jets and winds from young stars which sweep up surrounding material in outflows. In the thin shocked layer, gas temperatures increase to a few thousand K, thus triggering chemical reactions with energy barriers that cannot proceed in cold gas. Well-known examples are the O + H_2 and OH + H_2 reactions leading to the formation of water, each of which has an energy barrier of about 2000 K. The ubiquitous strong and broad water line profiles revealed by *Herschel*-HIFI in star-forming regions are a testimony to this high-temperature chemistry (Kristensen *et al.* 2012; San José-García *et al.* 2016).

Shocks can also sputter ice mantles, further enhancing H_2O and other ice mantle molecules like CH_3OH into the gas, and at high velocity even the grain cores themselves (Bachiller & Tafalla 1999). Thus, they can reveal ice mantle material in dark clouds far away from the driving source (Arce *et al.* 2008; Lefloch *et al.* 2017). The release of Si and S leads to strong emission from molecules like SiO, SO and SO_2 associated with jets. Kinematical signatures in the H_2O line profiles demonstrate the presence of multiple types of shocks that have not yet been spatially resolved. The observed low H_2O abundance, high OH/H_2O ratio and the presence of hydrides like CH^+ point to the importance of UV irradiation of the (pre-)shocked gas and outflow cavity walls in the chemistry (Karska *et al.* 2014). All of these processes are now being imaged at high angular resolution by ALMA, especially close to the base of the outflows where they originate and first impact the surrounding envelope.

Challenge 10: To characterize the chemical and physical structure of outflows, especially near the launching point of the jets and (disk) winds that drive them.

11. Warm protostellar envelopes, hot cores

The gas and dust close to a protostar are heated by its luminosity, with temperatures increasing from 10 K in the outer envelope to a few hundred K in the innermost region. This can result in numerous chemical changes (Herbst & van Dishoeck 2009): radicals become mobile in and on icy surfaces and recombine to form even more complex molecules (so-called first generation complex molecules) and increased UV and X-rays trigger further chemistry in the ice and gas. When dust temperatures become high enough for ices to sublimate, molecules do so presumably in a sequence according to their binding energies. Once dust temperatures of $\sim$100 K are reached, even the strongly-bound water and methanol ice sublimate, together with any minor molecules trapped in them, resulting in particularly rich gas-phase millimeter spectra (Fig. 2). This inner 100 K zone is called the 'hot core'. Here high-temperature gas-phase reactions between sublimated molecules can result in 'second generation' complex organic molecules (Charnley *et al.* 1992).

Observationally, most chemical surveys of star-forming regions have still been performed with single-dish millimeter telescopes equipped with broadband receivers. Over the past decade, pioneering high-frequency surveys have been carried out with *Herschel* of well-known sources such as Orion (Crockett *et al.* 2015). With ALMA, a new era has started, since ALMA can image each line *and* spatially resolve the hot core region. Also, ALMA has increased sensitivity to complex molecules by 1–2 orders of magnitude, and it can study solar-mass protostars on solar-system scales, not just the most luminous high-mass sources. A 'sweet' example is the ALMA-PILS survey of the low-mass protobinary IRAS16293-2422, where an increasing number of complex molecules is found that had previously only been seen in SgrB2 and Orion (Jørgensen *et al.* 2016). This includes prebiotic molecules like glycolaldehyde and ethelyne glycol, and most recently methyl isocyanate – a precursor to peptides (Martín-Doménech *et al.* 2017; Ligterink *et al.* 2017). In SgrB2, the quest for the 'molecular bricks of life' has now succeeded in the detection of the first chiral molecule, propylene oxide (McGuire *et al.* 2016), and branched cyanides such as iso-propylcyanide and N-methyl formamide (Belloche *et al.* 2017).

Detection of even more complex molecules is becoming hard, however, even with ALMA, because of the confusion limit. Whether lower frequency searches with the future SKA or ngVLA provide indeed the advantages that are claimed still needs to be demonstrated. Also, the current focus is still on the hunt for individual species in a few sources; the field should soon move to larger samples with ALMA and address the deeper underlying questions as to when, where and how certain classes of complex molecules are produced.

Both single-dish and interferometric surveys show that not all protostars have line-rich spectra, even though there is evidence for warm dense dust in their immediate surroundings (e.g., from SED fitting) (Fayolle *et al.* 2015). Thus, strictly speaking, not all 'hot cores' have a 'hot core chemistry'. Both evolutionary effects and geometry, such as the presence of a large cold disk on similar scales as the hot core, as well as beam dilution (for single-dish data) have been suggested as explanations. Dynamics may also play a role, especially if infall is fast and molecules spend only a short time in the hot core region and/or are exposed to excess UV *en route* from cloud to disk (Drozdovskaya *et al.* 2016). Since the time scales of many of the relevant chemical and dynamical processes in star formation are similar, there may be subtle effects at play that can lead to large differences. Coupling chemistry with full (magneto) hydrodynamical models of star formation is still computationally challenging and allows investigation of only a limited number of cases (Hincelin *et al.* 2016).

While complex molecules are traditionally associated only with protostellar hot cores, they are actually now found at all stages of star formation: in cold cores (§ 9), shocks (§ 10), near PDRs (§ 6) and in protoplanetary disks (§ 12). This points to the increasing importance of forming these molecules under cold conditions as zero-generation ices without the need for 'energetic' or UV processing, rather than as first- or second generation species. Cold gas-phase chemistry involving fragments coming off the ice may play a role in cold clouds and high temperature gas-phase chemistry could still be important for selected complex species in hot cores (Taquet *et al.* 2016). The chemical specificity, i.e., the fact that some complex molecules like $HCOOCH_3$ and CH_3OCH_3 are more abundant than others, must hold clues on the relative importance of the various reaction pathways (Tielens 2013).

Challenge 11: To identify the main formation routes of complex molecules in dense clouds, to push detections to even higher levels of complexity including prebiotic species like amino acids, and to assess how dynamics and geometry during star formation can affect their abundances.

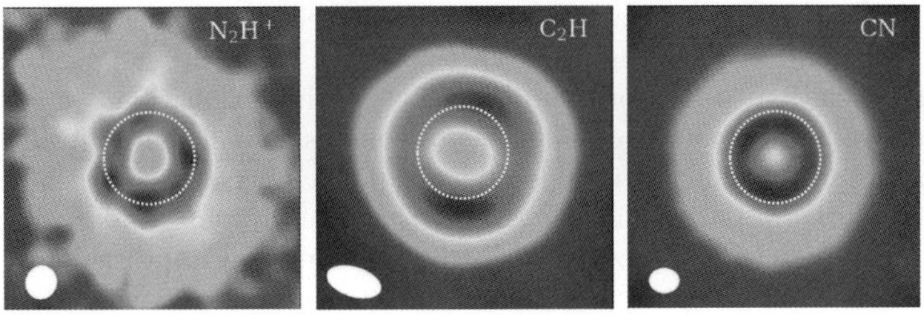

Figure 3. Molecular rings observed with ALMA in the TW Hya protoplanetary disk, showing different locations for each species. The white dotted circle indicates the location of the N_2H^+ ring; C_2H peaks outside and CN inside this ring (Figure by P. Cazzoletti, using data from Qi *et al.* 2013; Bergin *et al.* 2016; Teague *et al.* 2017; each image is $6'' \times 6''$, $1''$=54 AU).

12. Protoplanetary disks

Protoplanetary disks have become one of the centerpieces of Astrochemistry, providing the initial chemical conditions for planet formation. They remain difficult to study both observationally and theoretically, however. Observationally, signals are very weak since they are small (typically less than $1''$ on the sky) and their mass is only 1% of that of the collapsing cloud. Theoretically, they are a challenge since they cover a huge range of densities and temperatures in at least two dimensions, from >1000 K in the inner disk and upper layers, to $\leqslant$10 K in the outer midplane, and from densities of $> 10^{13}$ cm^{-3} in the inner midplane down to 10^5 cm^{-3} in the upper outer layers. Impinging UV radiation fields from the central star can be as high as 10^5 times the ISRF in the surface layers at 10 AU, setting up a PDR-like structure in the vertical direction. Thus, different types of chemistry are important in different parts of the disks. Moreover, gas and dust are largely decoupled (except for the smallest grains): dust grains grow to pebble size (few cm), settle to the midplane and drift in radially (Fig. 4). If they encounter a pressure bump, dust traps can form where particles can grow to even larger, planetesimal sizes. Gas/dust ratios can therefore differ significantly from 100.

The decreasing temperature in the radial direction sets up a range of snowlines, i.e., radii where molecules freeze-out onto the grains, defined as the half-gas, half-ice point. Because of the vertical temperature gradient, the 2D snow surfaces are actually curved, but usually only midplane snowline radii are cited. Snowlines are thought to play a significant role in planet formation, since ice coating of grains enhances the solid mass and promotes coagulation of grains to larger particles, effects which are particularly prominent just outside the snowline. They also control the composition of the icy planetesimals and gas from which exoplanetary atmospheres are built (§13, 14).

Pre-ALMA observations of disks at millimeter and far-infrared wavelengths detected mostly simple molecules originating from the intermediate height warm layers (see Bergin *et al.* 2007; Henning & Semenov 2013, for reviews), and confirmed the gas-dust temperature decoupling predicted by disk models (Bruderer *et al.* 2012). ALMA is now starting to probe deeper into the disk and spatially resolve the emission, revealing rings of molecular emission that could be associated with snowlines. The most obvious example is N_2H^+, tracing the CO snowline because its abundance is enhanced when CO is frozen out (Fig. 3). However, many other molecular rings are observed at different locations, with each of them having different explanations: CO snowlines (N_2H^+, DCO^+), CO photodesorption (DCO^+), chemistry (DCN, DCO^+ inner ring), UV (CN), C/O ratios (C_2H, C_3H_3), gas gaps or cavities (CO isotopologues, CS). Only two complex molecules, CH_3CN and CH_3OH, have been detected so far, and only barely. Even the brightest disks do not show rich line forests such as found for hot cores. Deep observations of these and other species in a much larger sample of disks are needed to understand what the different molecular images are telling us about the gaseous physical and chemical structure of disks and its relation to that of the dust. Whether time allocation committees can be convinced to give large amounts of time to this topic is a challenge in itself.

The important water snowline is generally out of reach because it typically occurs at a few AU for T Tauri stars, too small for ALMA to image. Systems undergoing luminosity outbursts such as V883 Ori (Cieza *et al.* 2016) or young disks with enhanced accretion luminosity (Harsono *et al.* 2015), for which the water snowlines have moved out to tens of AU, may offer the best opportunity for direct imaging. More generally, determining the chemical composition and structure of young embedded disks may be highly relevant for comparison with comets, if planetesimal formation indeed starts early (§ 13).

The warm gas in the upper layers of the inner disk ($<$10 AU) emits strongly at mid-infrared wavelengths. Indeed, a dense forest of lines due to simple molecules has been detected by *Spitzer* and ground-based infrared telescopes, providing a glimpse of the chemistry in that region (Pontoppidan *et al.* 2014). Here the high temperatures can completely 'reset' the chemistry. A big advantage of mid-IR spectroscopy is that key abundant molecules without a dipole moment such as CO_2 and C_2H_2 can be observed, together with H_2O, OH and HCN, allowing tests of high temperature chemistry. The degree to which these data can probe the disk midplane chemistry and ice sublimation, relevant for planet formation models, is still unclear however, since the mid-infrared dust and line emission is highly optically thick. JWST will be poised to provide much deeper searches for CH_4, NH_3 and minor species, detect isotopologs to constrain line optical depth, and follow the inner disk chemistry (at least in the surface layers) from the youngest embedded disks to the debris disk stage, and across the stellar type range.

Challenge 12: To make a chemical inventory of disks (from inner to outer, surface to midplane, and young to old) and relate observed molecular structures to underlying gas and dust structures.

13. Planet building blocks: comets, planetesimals

ALMA and high contrast infrared imaging of disks are providing striking pictures of the first steps of planet formation: rings, gaps, cavities, asymmetric structures, and / or spiral arms. Just as for the molecular rings there are many possible interpretations of these structures (§ 12), but they do indicate that growth of dust grains to pebbles and planetesimals is taking place, starting even in the embedded phase.

How does this affect the chemistry? First hints came from the very deep *Herschel*-HIFI integrations of H_2O lines in disks, which revealed surprisingly weak features (Hogerheijde *et al.* 2011; Du *et al.* 2017). The favorite interpretation is that most water is locked up in large icy bodies in the midplane and no longer participates in the chemistry at intermediate disk layers where the gaseous lines can be observed. The second clue comes from weak CO isotopologue lines in most disks (Favre *et al.* 2013; Ansdell *et al.* 2016). These lines are reduced in strength not just due to the well understood processes of freeze-out and isotope selective photodissociation (Miotello *et al.* 2017), but also by chemistry transforming CO into CH_3OH, CO_2 or hydrocarbons (Yu *et al.* 2017). Alternatively, the rapid planetesimal growth invoked for H_2O also locks up significant amounts of CO and volatile carbon in various forms.

Thus, the overall picture of the outer disk is that much of the volatile oxygen and carbon is locked up in ices in large bodies. Since the dominant H_2O and CO_2 ices have more oxygen than carbon, the ice is oxygen rich, with an overall C/O ratio that is lower

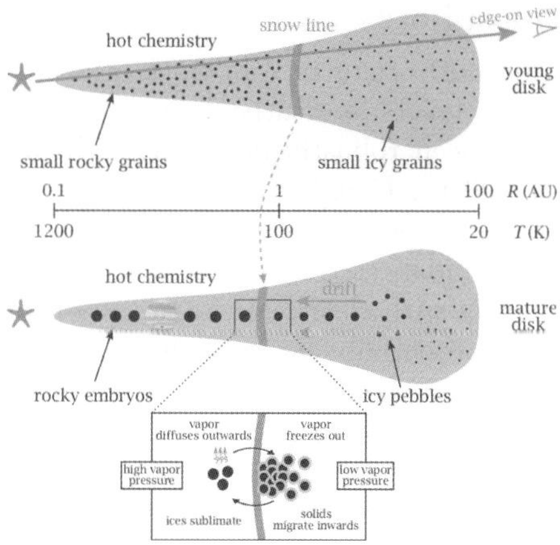

Figure 4. Sketch of the importance of grain growth, radial drift of icy pebbles and snowlines in disk chemistry and in building planets and their atmospheres (Figure by R. Visser, after Sato *et al.* 2016).

than the interstellar or solar abundance (Öberg *et al.* 2011). In contrast, the gas is carbon rich even though it is overall depleted in carbon and oxygen. Once C/O>1, the same situation occurs as for carbon-rich AGB stars: the gas has high abundances of small hydrocarbon molecules. This can indeed explain the strong observed C_2H and c-C_3H_2 emission in some disks (Bergin *et al.* 2016) (Fig. 3), although midplane chemistry tends to evolve toward gaseous C/O significantly less than 1 (Eistrup *et al.* 2017). JWST will be able to observationally constrain the ice composition in planet-forming zones, but only for a handful of near edge-on disks and then only in the intermediate layers.

So how do we determine the disk chemical composition just prior to planet formation? Much of the chemistry of planet formation is unfortunately hidden from our view, including the bulk C- and O-containing species. One option is to observe warmer, younger disks where less of the material is frozen out and fewer planetesimals have formed. The forming disks around Class 0 sources, like those for IRAS16293-2422, may well provide the most detailed information of chemistry on solar-system scales, even if the material has not yet settled into a Keplerian structure. That chemistry, in turn, may have been set already to a large degree in the dense cold core just prior to and during collapse, arguing for an 'inheritance' rather than a 'reset' scenario, at least in the outer disk.

Alternatively, comets and other icy bodies in our own solar system provide clues to the chemistry in our natal solar nebula disk, if they are largely unchanged since their formation 4.5 billion years ago. Comparisons between cometary and interstellar abundances have shown some tantalizing similarities (Mumma & Charnley 2011), but there is a large scatter on both axes. The *Rosetta* mission provides an unprecedented view of the chemical composition of one comet, 67P/C-G (Altwegg, this volume), which can be compared with that inferred for protostars such as IRAS16293-2422. However, this is just one comet and one protostar; ultimately, one would like at least 10 Rosetta-type missions to comets originating in different parts of our solar system! More bright comets like Hale-Bopp that can be probed with ALMA and infrared instruments, including determining their D/H ratio in water, will provide further insight into their origin.

<u>Challenge 13:</u> To determine the (bulk) chemical composition and origin of planet-forming material (inheritance or reset) and relate that to what is found for icy bodies in our own Solar System.

14. Exoplanetary atmospheres

The chemical composition and origin of exoplanet atmospheres, from Super-Earths to mini-Neptunes and Jovian planets, is clearly the next frontier for Astrochemistry. So far, only CO and H_2O have been robustly detected, with hints of CH_4, NH_3 and/or HCN (Madhusudhan *et al.* 2016). Clouds and hazes often obscure optical and near-infrared features, especially for the lower mass planets. Even when features are detected, inferring accurate absolute abundances is a challenge, although relative abundances are more reliable. For those planets with clear features, a C/O ratio < 1 has generally been inferred, with so far only one potentially carbon-rich planet.

Exoplanet chemistry models are being developed by various groups, and vary in sophistication from pure LTE chemistry to the inclusion of kinetic chemistry, especially photochemistry in the upper parts of the atmosphere (Venot *et al.* 2015). Titan is a good example of a body in our own solar system to test nitrogen-rich photochemistry †, at the same time also highlighting its complexity.

One of the ultimate goals is to link the planetary atmosphere composition with its formation history in the natal protoplanetary disk. On the one hand, the changing C/O ratio with disk radius could provide such a probe of the formation location. However, the route from disk gas+dust to a mature planet is long and involves many steps, each of them with significant uncertainties (Mordasini *et al.* 2016). For example, are the heavy elements (i.e., other than H and He) in a giant planet atmosphere accreted mostly from the gas or delivered by icy pebbles? How does the migration history affect the outcome? Moreover, the disk structure and chemistry is not static but evolving, modifying the C/O 'step function', as does the growth and drift of particles (Piso *et al.* 2015; Eistrup *et al.* 2017). Thomas Henning, at the Kavli Exo-atmospheres meeting July 2017, lists at least 11 challenges for the field to overcome, most notably our poor understanding of mass and angular momentum transport in disks (see also Hartmann *et al.* 2017). It is also clear that individual molecule abundances will be fully reset in giant planet atmosphere, preserving only the overall C/O, C/N, O/H etc. abundance ratios. And even those could be affected if part of the atmosphere material is cycled to the planetary core. Thus, as exciting as this topic is, there are some sobering notes.

The atmospheres of rocky terrestrial planets may have an even more complicated history. Well outside the water snowline, the planets are built up largely from planetesimals that are roughly half rock and half ice. When these planets move inward, water becomes liquid, resulting in ocean planets or water worlds. Inside the snowline, the planets are usually thought to be very dry. Computing the atmospheric composition of terrestrial exoplanets is significantly more complex than that of giant exoplanets and requires consideration of many additional processes, including even plate tectonics (Kaltenegger 2017). Because they are very hot initially, they lose most of their original volatiles through outgassing and atmospheric erosion. A secondary atmosphere can however be formed through impacts of (icy) planetesimals in the *late* stages of their formation, including comets such as 67P/C-G which can deliver water and organics to the young planet. The survival of these organics depends on planetesimal size and impact speed, and whether the volatile material can perhaps be shielded by a protective layer on the parent body. If they do, then there would indeed be a direct link between interstellar molecules and the building blocks for life on new planets.

† Photochemistry in the chemical sense, i.e., the fragments resulting from photodissociation are produced in excited states that react immediately to products before relaxing to the ground state. Within Astrochemistry, the term photochemistry should not be used for gas phase photodissociation.

Figure 5. Alexander Dalgarno in 2003.

Finding signs of life on an alien habitable planet is an ultimate goal of exoplanet science, and of astronomy as a whole, and is the driver for the biggest future telescopes and space missions. Biomarker molecules have been identified whose detection could indicate the presence of life on other planets. Key molecules are O_2, O_3, CH_4, N_2O and CH_3Cl. However it is also clear that a single molecule such as O_3 is not sufficient but that a combination of features, including at least one oxidizing and reducing species, is needed (Kaltenegger 2017; Catling *et al.* 2017).

In this context, it is interesting that several biomarker molecules, most notably O_2 (Bieler *et al.* 2015) and CH_3Cl (a molecule produced on Earth primarily by biological and industrial processes) (Fayolle *et al.* 2017), have recently been detected in comet 67P/C-G. For O_2, the abundance of 4% with respect to water makes it the fourth most abundant ice species. If comets are representative of the icy planetesimals from which terrestrial planet atmospheres are built, one should be careful of such 'false positives' that can have an abiotic origin.

Challenge 14: To determine exoplanetary atmosphere compositions and to characterize the chemical changes along the many steps to planet formation, necessary to relate exoplanetary atmosphere compositions to their birth sites in disks.

15. Conclusions

Looking out over the beautiful Llanquihue lake at the snowline on the Osorno volcano, the quote attributed to Rabindranath Tagore becomes very appropriate: *'You can't cross the sea merely by standing and staring at the water'*. The field of Astrochemistry is vibrant and poised to address major topics in astronomy. It forms the scientific basis for one of mankind's biggest questions: 'are we alone?'. To deliver on this promise, however, Astrochemistry has to tackle the questions and challenges outlined above, do the hard (not always glorious) work and provide the hard numbers. The field also has to build the right ships (i.e., new telescopes, new laboratory experiments, new modeling tools), get the right crew (i.e., excellent young people with diverse backgrounds), and make the right connections (i.e., with neighboring fields) to be able to steer the ship in the right direction and cross the water. This will take time but having this long term vision ensures that there will be exciting developments in the coming decades, leading up to Astrochemistry XV!

Acknowledgments. This overview is dedicated to Alexander Dalgarno (1928–2015), who initiated the IAU working group on Astrochemistry and organized the first two symposia in this series. His pioneering work to bring molecular physics into astronomy inspired many young scientists (including this author) to enter our field. The broad scope of these meetings reflect the spirit of his work, having contributed to all aspects of Astrochemistry, and bringing people from across the world together.

The responses from IAU S332 participants to my question on the future of Astrochemistry, as well as comments on this draft by E. Bergin, G. Blake, E. Herbst, T. Millar, C. Walsh and members of my group, are much appreciated. The writing of this review was supported by EU A-ERC grant 291141 CHEMPLAN.

References

Allodi, M. A., Baragiola, R. A., Baratta, G. A., *et al.* 2013, *Space Sci. Rev.*, 180, 101
ALMA Partnership, Vlahakis, C., Hunter, T. R., *et al.* 2015, *ApJL*, 808, L4
Anderson, D. E., Bergin, E. A., Blake, G. A., *et al.* 2017, *ApJ*, 845, 13
Ansdell, M., Williams, J. P., van der Marel, N., *et al.* 2016, *ApJ*, 828, 46
Arasa, C., Koning, J., Kroes, G.-J., Walsh, C., & van Dishoeck, E. F. 2015, *A&A*, 575, A121
Arasa, C., van Hemert, M. C., van Dishoeck, E. F., & Kroes, G. J. 2013, J. Phys. Chem. A, 117, 7064
Arce, H. G., Santiago-García, J., Jørgensen, J. K., Tafalla, M., & Bachiller, R. 2008, *ApJL*, 681, L21
Bachiller, R. & Tafalla, M. 1999, in NATO ASIC Proc. 540: The Origin of Stars and Planetary Systems, ed. C. J. Lada & N. D. Kylafis, 227
Bacmann, A., Taquet, V., Faure, A., Kahane, C., & Ceccarelli, C. 2012, *A&A*, 541, L12
Belloche, A., Meshcheryakov, A. A., Garrod, R. T., *et al.* 2017, *A&A*, 601, A49
Bergin, E. A., Aikawa, Y., Blake, G. A., & van Dishoeck, E. F. 2007, in Protostars and Planets V, ed. B. Reipurth, D. Jewitt, K. Keil (Tucson: Univ. of Arizona), 751–766
Bergin, E. A., Du, F., Cleeves, L. I., *et al.* 2016, *ApJ*, 831, 101
Bergin, E. A., Phillips, T. G., Comito, C., *et al.* 2010, *A&A*, 521, L20
Bergin, E. A. & Tafalla, M. 2007, *ARA&A*, 45, 339
Berné, O. & Tielens, A. G. G. M. 2012, *Proc. Natl. Acad. Sci.*, 109, 401
Bertin, M., Romanzin, C., Doronin, M., *et al.* 2016, *ApJL*, 817, L12
Bialy, S., Burkhart, B., & Sternberg, A. 2017, *ApJ*, 843, 92
Bieler, A., Altwegg, K., Balsiger, H., *et al.* 2015, *Nature*, 526, 678
Boersma, C., Bauschlicher, Jr., C. W., Ricca, A., *et al.* 2014, *ApJS*, 211, 8
Boogert, A. C. A., Gerakines, P. A., & Whittet, D. C. B. 2015, *ARA&A*, 53, 541
Bruderer, S., van Dishoeck, E. F., Doty, S. D., & Herczeg, G. J. 2012, *A&A*, 541, A91
Cami, J., Bernard-Salas, J., Peeters, E., & Malek, S. E. 2010, *Science*, 329, 1180
Campbell, E. K., Holz, M., & Maier, J. P. 2016, *ApJL*, 826, L4
Caselli, P. & Ceccarelli, C. 2012, *A&A Review*, 20, 56
Catling, D. C., Krissansen-Totton, J., Kiang, N. Y., *et al.* 2017, ArXiv 1705.06381
Ceccarelli, C., Caselli, P., Bockelée-Morvan, D., *et al.* 2014, Protostars & Planets VI, ed. Beuther, H., Klessen, R., Dullemond, K., Henning, Th. (Tucson: Univ. Arizona Press), 859–882
Charnley, S. B., Tielens, A. G. G. M., & Millar, T. J. 1992, *ApJL*, 399, L71
Cieza, L. A., Casassus, S., Tobin, J., *et al.* 2016, *Nature*, 535, 258
Collings, M. P., Anderson, M. A., Chen, R., *et al.* 2004, *MNRAS*, 354, 1133
Costagliola, F., Sakamoto, K., Muller, S., *et al.* 2015, *A&A*, 582, A91
Crockett, N. R., Bergin, E. A., Neill, J. L., *et al.* 2015, *ApJ*, 806, 239
Cuadrado, S., Goicoechea, J. R., Cernicharo, J., *et al.* 2017, *A&A*, 603, A124
Cuppen, H. M., Walsh, C., Lamberts, T., *et al.* 2017, *Space Sci. Rev.*in press
Dalgarno, A. 2008, *ARA&A*, 46, 1

Dalgarno, A. & Lepp, S. 1987, in IAU Symposium, Vol. 120, Astrochemistry, ed. M. S. Vardya & S. P. Tarafdar, 109–118
Decin, L., Agúndez, M., Barlow, M. J., *et al.* 2010, *Nature*, 467, 64
Drozdovskaya, M. N., Walsh, C., van Dishoeck, E. F., *et al.* 2016, *MNRAS*, 462, 977
Du, F., Bergin, E. A., Hogerheijde, M., *et al.* 2017, *ApJ*, 842, 98
Dubernet, M.-L., Alexander, M. H., Ba, Y. A., *et al.* 2013, *A&A*, 553, A50
Eistrup, C., Walsh, C., & van Dishoeck, E. F. 2017, A&A in press
Endres, C. P., Schlemmer, S., Schilke, P., Stutzki, J., & Müller, H. S. P. 2016, J. Mol. Spectrosc., 327, 95
Favre, C., Cleeves, L. I., Bergin, E. A., Qi, C., & Blake, G. A. 2013, *ApJL*, 776, L38
Fayolle, E. *et al.* 2017, Nature Astronomy, in press
Fayolle, E. C., Öberg, K. I., Garrod, R. T., van Dishoeck, E. F., & Bisschop, S. E. 2015, *A&A*, 576, A45
Fuchs, G. W., Cuppen, H. M., Ioppolo, S., *et al.* 2009, *A&A*, 505, 629
Furuya, K., van Dishoeck, E. F., & Aikawa, Y. 2016, *A&A*, 586, A127
Galli, D. & Palla, F. 2013, *ARA&A*, 51, 163
Garrod, R. T., Wakelam, V., & Herbst, E. 2007, *A&A*, 467, 1103
Garrod, R. T., Widicus Weaver, S. L., & Herbst, E. 2008, *ApJ*, 682, 283
Gerin, M., Neufeld, D., & Goicochea, J. 2016, *ARA&A*, 54, in press
Godard, B., Falgarone, E., & Pineau des Forêts, G. 2014, *A&A*, 570, A27
González-Alfonso, E., Fischer, J., Graciá-Carpio, J., *et al.* 2012, *A&A*, 541, A4
Guzmán, V. V., Pety, J., Gratier, P., *et al.* 2014, Faraday Discussions, 168, 103
Harsono, D., Bruderer, S., & van Dishoeck, E. F. 2015, *A&A*, 582, A41
Hartmann, L., Ciesla, F., Gressel, O., & Alexander, R. 2017, ArXiv 1708.03360
Hasegawa, T. I. & Herbst, E. 1993, *MNRAS*, 263, 589
Heays, A. N., Bosman, A. D., & van Dishoeck, E. F. 2017, *A&A*, 602, A105
Henning, T. 2010, *ARA&A*, 48, 21
Henning, T. & Semenov, D. 2013, Chem. Rev., 113, 9016
Herbst, E. & van Dishoeck, E. F. 2009, *ARA&A*, 47, 427
Hidaka, H., Watanabe, M., Kouchi, A., & Watanabe, N. 2009, *ApJ*, 702, 291
Hincelin, U., Commerçon, B., Wakelam, V., *et al.* 2016, *ApJ*, 822, 12
Hogerheijde, M. R., Bergin, E. A., Brinch, C., *et al.* 2011, *Science*, 334, 338
Holdship, J., Viti, S., Jiménez-Serra, I., Makrymallis, A., & Priestley, F. 2017, *AJ*, 154, 38
Hollenbach, D. J. & Tielens, A. G. G. M. 1997, *ARA&A*, 35, 179
Indriolo, N., Neufeld, D. A., Gerin, M., *et al.* 2015, *ApJ*, 800, 40
Ivlev, A. V., Röcker, T. B., Vasyunin, A., & Caselli, P. 2015, *ApJ*, 805, 59
Jørgensen, J. K., Schöier, F. L., & van Dishoeck, E. F. 2005, *A&A*, 435, 177
Jørgensen, J. K., van der Wiel, M. H. D., Coutens, A., *et al.* 2016, *A&A*, 595, A117
Kaifu, N., Ohishi, M., Kawaguchi, K., *et al.* 2004, *Pub. Astron. Soc. Japan*, 56, 69
Kaltenegger, L. 2017, *ARA&A*, 55, 433
Karska, A., Kristensen, L. E., van Dishoeck, E. F., *et al.* 2014, *A&A*, 572, A9
Kennicutt, R. C. & Evans, N. J. 2012, *ARA&A*, 50, 531
Krasnokutski, S. A., Rouillé, G., Jäger, C., *et al.* 2014, *ApJ*, 782, 15
Kreckel, H., Bruhns, H., Čížek, M., *et al.* 2010, *Science*, 329, 69
Kristensen, L. E., van Dishoeck, E. F., Bergin, E. A., *et al.* 2012, *A&A*, 542, A8
Lefloch, B., Ceccarelli, C., Codella, C., *et al.* 2017, *MNRAS*, 469, L73
Ligterink, N. F. W., Coutens, A., Kofman, V., *et al.* 2017, *MNRAS*, 469, 2219
Linnartz, H., Ioppolo, S., & Fedoseev, G. 2015, Int. Rev. Phys. Chem., 34, 205–237
Liszt, H., Lucas, R., & Pety, J. 2005, in IAU Symposium, Vol. 231, Astrochemistry: Recent Successes and Current Challenges, ed. D. C. Lis, G. A. Blake, & E. Herbst, 187–196
Liu, W. & Dalgarno, A. 1994, *ApJ*, 428, 769
Madhusudhan, N., Agúndez, M., Moses, J. I., & Hu, Y. 2016, *Space Sci. Rev.*, 205, 285
Martín-Doménech, R., Rivilla, V. M., Jiménez-Serra, I., *et al.* 2017, *MNRAS*, 469, 2230
Matsuura, M., Indebetouw, R., Woosley, S., *et al.* 2017, *MNRAS*, 469, 3347

McElroy, D., Walsh, C., Markwick, A. J., *et al.* 2013, *A&A*, 550, A36
McGuire, B. A., Carroll, P. B., Loomis, R. A., *et al.* 2016, *Science*, 352, 1449
Meijerink, R., Spaans, M., & Israel, F. P. 2007, *A&A*, 461, 793
Minissale, M., Dulieu, F., Cazaux, S., & Hocuk, S. 2016, *A&A*, 585, A24
Miotello, A., van Dishoeck, E. F., Williams, J. P., *et al.* 2017, *A&A*, 599, A113
Mordasini, C., van Boekel, R., Mollière, P., Henning, T., & Benneke, B. 2016, *ApJ*, 832, 41
Muller, S., Müller, H. S. P., Black, J. H., *et al.* 2016, *A&A*, 595, A128
Mumma, M. J. & Charnley, S. B. 2011, *ARA&A*, 49, 471
Öberg, K. I. 2016, Chem. Rev., 116, 9631
Öberg, K. I., Boogert, A. C. A., Pontoppidan, K. M., *et al.* 2011, *ApJ*, 740, 109
Öberg, K. I., Bottinelli, S., Jørgensen, J. K., & van Dishoeck, E. F. 2010, *ApJ*, 716, 825
Pety, J., Guzmán, V. V., Orkisz, J. H., *et al.* 2017, *A&A*, 599, A98
Pickett, H. M., Poynter, I. R. L., Cohen, E. A., *et al.* 1998, *J. Quant. Spectrosc. Rad. Transfer*, 60, 883
Piso, A.-M. A., Öberg, K. I., Birnstiel, T., & Murray-Clay, R. A. 2015, *ApJ*, 815, 109
Pontoppidan, K. M., Salyk, C., Bergin, E. A., *et al.* 2014, Protostars & Planets VI, ed. Beuther, H., Klessen, R., Dullemond, K., Henning, Th. (Tucson: Univ. Arizona Press), 363–385
Qi, C., Öberg, K. I., Wilner, D. J., *et al.* 2013, *Science*, 341, 630
Roberts, H., Herbst, E., & Millar, T. J. 2003, *ApJL*, 591, L41
Röllig, M., Abel, N. P., Bell, T., *et al.* 2007, *A&A*, 467, 187
Rothman, L. S., Gordon, I. E., Barbe, A., *et al.* 2009, *J. Quant. Spectrosc. Rad. Transfer*, 110, 533
Roueff, E. & Lique, F. 2013, Chem. Rev., 113, 8906
Sakai, N. & Yamamoto, S. 2013, Chemical Reviews, 113, 8981
San José-García, I., Mottram, J. C., van Dishoeck, E. F., *et al.* 2016, *A&A*, 585, A103
Sato, T., Okuzumi, S., & Ida, S. 2016, *A&A*, 589, A15
Schruba, A., Leroy, A. K., Kruijssen, J. M. D., *et al.* 2017, *ApJ*, 835, 278
Shen, C. J., Greenberg, J. M., Schutte, W. A., & van Dishoeck, E. F. 2004, *A&A*, 415, 203
Sipilä, O., Hugo, E., Harju, J., *et al.* 2010, *A&A*, 509, A98
Smith, I. W. M. 2011, *ARA&A*, 49, 29
Snow, T. P. & McCall, B. J. 2006, *ARA&A*, 44, 367
Suzuki, H., Yamamoto, S., Ohishi, M., *et al.* 1992, *ApJ*, 392, 551
Taquet, V., Wirström, E. S., & Charnley, S. B. 2016, *ApJ*, 821, 46
Teague, R., Semenov, D., Gorti, U., *et al.* 2017, *ApJ*, 835, 228
Tennyson, J., Yurchenko, S. N., Al-Refaie, A. F., *et al.* 2016, J. Mol. Spectrosc., 327, 73
Tielens, A. G. G. M. 1983, *A&A*, 119, 177
Tielens, A. G. G. M. 2008, *ARA&A*, 46, 289
Tielens, A. G. G. M. 2013, *Rev. Mod. Phys.*, 85, 1021
Tielens, A. G. G. M. & Hagen, W. 1982, *A&A*, 114, 245
van Dishoeck, E. F. 2014, Faraday Discussions, 168, 9
van Dishoeck, E. F. & Black, J. H. 1989, *ApJ*, 340, 273
van Dishoeck, E. F., Herbst, E., & Neufeld, D. A. 2013, Chemical Reviews, 113, 9043
Venot, O., Hébrard, E., Agúndez, M., Decin, L., & Bounaceur, R. 2015, *A&A*, 577, A33
Visser, R., Kristensen, L. E., Bruderer, S., *et al.* 2012, *A&A*, 537, A55
Wakelam, V., Herbst, E., Le Bourlot, J., *et al.* 2010, *A&A*, 517, A21
Wakelam, V., Herbst, E., Loison, J.-C., *et al.* 2012, *ApJS*, 199, 21
Wakelam, V., Loison, J.-C., Herbst, E., *et al.* 2015, *ApJS*, 217, 20
Watanabe, Y., Nishimura, Y., Harada, N., *et al.* 2017, *ApJ*, 845, 116
Wiesenfeld, L., Thi, W.-F., Caselli, P., *et al.* 2016, ArXiv 1610.00438
Yu, M., Evans, II, N. J., Dodson-Robinson, S. E., Willacy, K., & Turner, N. J. 2017, *ApJ*, 841, 39
Zhen, J., Castellanos, P., Paardekooper, D. M., Linnartz, H., & Tielens, A. G. G. M. 2014, *ApJL*, 797, L30

Astrochemistry in External Galaxies

Astrochemistry VII – Through the Cosmos from Galaxies to Planets
Proceedings IAU Symposium No. 332, 2017
M. Cunningham, T. Millar & Y. Aikawa, eds.

doi:10.1017/S1743921317006755

High-Temperature Chemistry in External Galaxies

Nanase Harada

Academia Sinica Institute of Astronomy and Astrophysics
Taipei, Taiwan
email: harada@asiaa.sinica.edu.tw

Abstract. In external galaxies, some galaxies have higher activities of star formation and central supermassive black holes. The interstellar medium in those galaxies can be heated by different mechanisms such as UV-heating, X-ray heating, cosmic-ray heating, and shock/mechanical heating. Chemical compositions can also be affected by those heating mechanisms. Observations of many molecular species in those nearby galaxies are now possible with the high sensitivity of Atacama Large Millimeter/sub-millimeter Array (ALMA). Here I cover different chemical models for those heating mechanisms. In addition, I present recent ALMA results of extragalactic astrochemistry including our results of a face-on galaxy M83 and an infrared-luminous merger NGC 3256.

Keywords. galaxies: ISM galaxies: active ISM: molecules ISM: abundances astrochemistry

1. Astrochemistry in External Galaxies

Some external galaxies are going through violent activities that Milky Way currently does not experience. Star formation rates can vary orders of magnitude among galaxies. Active galactic nuclei (AGNs) emit high intensity of high-energy photons such as X-rays. Merger events are thought to be important processes in the course of galaxy evolution. Such phenomena can affect the properties of molecular clouds in those galaxies. Elevated gas temperatures are one of those properties. Central regions of starburst galaxies and active galaxies tend to have higher gas kinetic temperature than Milky Way sources (Table 1). There are various heating mechanisms of the molecular medium. Starburst galaxies may have higher UV radiation field, while AGNs emit high luminosity of X-rays. After the starburst activity, supernova explosions can also create high flux of cosmic-rays. Both starburst and AGNs can cause high degree of turbulence or shocks. Those heating sources affect the chemical composition in different ways although some effects are similar.

Observing molecular tracers of those heating mechanisms has recently become easier. After the commission of Atacama Large Millimeter/sub-millimeter Array (ALMA) in 2011, its early science has already shown its capability to widely expand the potential of spatially-resolved astrochemical studies in external galaxies. ALMA has increased the number of detectable species, and its arcsecond/sub-arcsecond resolution allows comparisons of various locations in galaxies in a scale of giant molecular clouds (GMCs). A larger number of galaxies can be studied in astrochemical point of view with ALMA, and comparisons between various galaxy types become possible. Since some galactic nuclei are highly obscured, it would be useful for a wider astronomical community if some molecular species are found to be tracers of star formation or AGN activities.

Table 1. Gas kinetic temperatures in selected galaxies or Galactic sources traced by excitations of ammonia (NH_3) or formaldehyde (H_2CO). IRDC stands for an infrared dark cloud. Two temperatures are listed if multi-components are derived from their observations. Beam sizes are listed because hot components tend to be compact, and higher angular resolution observations are more likely to pick up hot components if there are any. References are a: Ragan *et al.* (2011), b: Ginsburg *et al.* (2016), c: Mangum *et al.* (2013), d: Ao *et al.* (2011)

Source	Temperature (K)	Galaxy type	Molecule	Beam size (pc)	Reference
Milky Way IRDC	8-13	—	NH_3	0.08 - 0.16	a
Milky May CMZ	60 / > 100	—	H_2CO	1.2	b
NGC 253	78 ± 22 / > 100	Starburst	NH_3	500	c
Arp 220	234 ± 52	Starburst (+ AGN?)	NH_3	12000	c
NGC 1068	$80 \pm 20/140 \pm 30$	AGN	NH_3	1200	d

In this proceedings, I will summarize previous theoretical models for each heating mechanisms and observational results especially ALMA results in galaxies affected by those heating sources.

2. Heating Mechanisms and Theoretical Models

2.1. *Photon-dominated regions*

Starburst galaxies may have orders of magnitude higher star formation rate than the Milky Way. In those galaxies, the interstellar radiation field could also increase by orders of magnitude. In the vicinity of massive stars, high-energy UV-photons create ionized HII regions. Further out, after photons with energies above the ionization energy of hydrogen (13.6 eV) are absorbed, there are photon-dominated regions (PDRs) where atoms with lower ionization energy than that of hydrogen (e.g., carbon) can be ionized. The interstellar medium (ISM) in PDRs is heated by photo-electric effect, de-excitation of vibrationally excited species after the photopumping of H_2, and IR pumping of the OI line and its collisional de-excitation (Tielens 2005). Those UV-photons not only ionize, but dissociate molecules. Therefore, more ions or radicals are found in PDRs. The rate coefficients of photo-dissociation or photo-ionization scales as $k \propto exp(-bA_V)$ with typical values of $b = 1 - 4$, the effects of UV-photons are not significant when the molecular clouds are shielded by columns of $N_H \gtrsim 2 \times 10^{22}$ cm^{-2}. Many PDR models using a one-dimensional slab have been developed over decades (Röllig *et al.* 2007, and references therein) to include complex physics to radiative transfer and thermal balance. In extragalactic observations of molecular lines where the spatial scale is at least GMC-scale, how much emission comes from PDRs needs to be investigated with the consideration of the density probably distribution of the ISM (see also Section 3.1). PDR models including 3-dimensional structure of the molecular clouds (e.g., Levrier *et al.* 2012) are useful, and such models in the context of starburst galaxies are needed.

2.2. *X-ray dominated regions*

X-rays from AGNs are another possible heating source if the galactic nuclei are active. They can heat the medium by Coulomb heating, or ionization or excitation that eventually lead to heating the gas (Maloney *et al.* 1996). X-ray heating is more efficient than UV-photon heating given the same amount of energy (Meijerink & Spaans 2005). Beside heating and ionization, it can dissociate molecules via internally generated UV-photons. X-rays can also produce doubly-ionized ions such as C^{++} and S^{++}, and may increase the fractional abundances of species produced from those species (Abel *et al.* 2008). Since

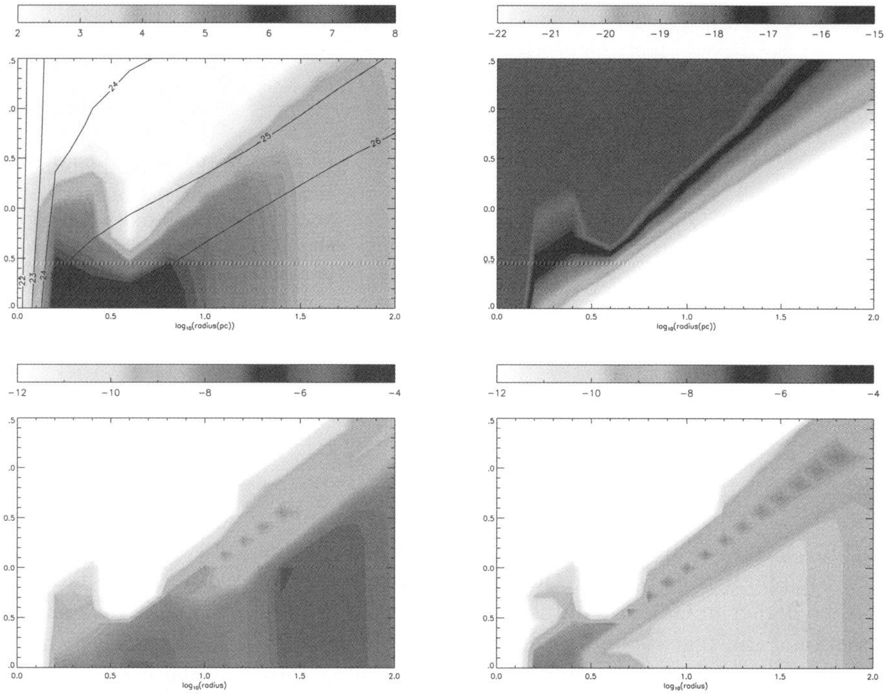

Figure 1. (*Top left*) The density structure of the AGN disk model used in Harada *et al.* (2013). Column densities intervening X-rays from the central AGN are also shown as black contours. (*top right*) Values of ζ_X/n where ζ_X is the X-ray ionization rate and n is the density. The influence of X-rays should roughly scale as ζ_X/n (Lepp & Dalgarno 1996). Fractional abundances of HCN in an axisymmetric model around an AGN in Harada *et al.* (2013) at (*bottom left*) a dynamical time and (*bottom right*) steady-state.

the photo-absorption cross sections are 1.0×10^{-24} cm^{-2} for 10 keV photon with the energy dependence of $E^{-\gamma}$ where $\gamma = 2.6 - 3$, X-rays penetrate up to $\sim 10^{24} - 10^{25}$ cm^{-2} (Maloney *et al.* 1996). This means that in highly-obscured galactic nuclei, the radius of influence of X-rays will be dependent on the morphology of the surrounding ISM. Meijerink & Spaans (2005) and Meijerink *et al.* (2007) has compared the effects of PDR and XDR using a steady-state model in a one-dimensional slab model to derive fractional abundances and intensities of molecular lines.

Although one-dimensional models are useful in grasping the dependence of the chemistry on the X-ray strength, how far the effects of X-rays can extend can be modeled should be examined though a multi-dimensional model. Time dependance is also important for some species. Harada *et al.* (2013) conducted a chemical model with a time-dependent gas-phase chemical model in axisymmetric disk models implied and modified from the density structure in Thompson *et al.* (2005). In these models, the importance of time-dependence was shown for species such as HCN and CN (Figure 1). Although the steady-state is achieved in a relatively short time scale for high X-ray flux regions, the accurate comparisons between the regions with and without strong influence of X-rays should be made with the consideration of time dependence. These models also showed the variation of the distribution of XDRs in different density structure of models.

2.3. *Cosmic-ray-dominated regions*

Cosmic rays from supernova explosions cause ionization further into molecular clouds than the UV-photons or X-rays, up to a column of $N_{\rm H} \sim 10^{26}$ cm^{-2} (Umebayashi &

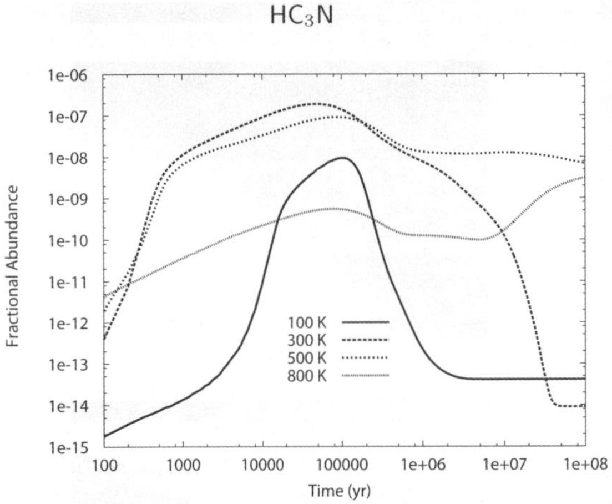

Figure 2. Fractional abundances of HC_3N at various gas kinetic temperatures calculated with a gas-phase chemical model including high-temperature reactions in Harada *et al.* (2010).

Nakano 1981). Similar to X-rays, cosmic rays can also internally generate UV-photons which can dissociated molecules. A typical value of a cosmic-ray ionization rate in dense cores in the spiral arm part of the Galaxy ranges $(1 - 5) \times 10^{-17}\,s^{-1}$. This value is found to be higher in Galactic diffuse clouds because low-energy cosmic rays have higher cross sections, causing about one order of magnitude higher ionization rate in diffuse clouds while they become attenuated before reaching into the dense clouds. (Indriolo *et al.* 2009; Padovani *et al.* 2009). In starburst galaxies, star formation rates per unit volume are higher than the Galaxy, and cosmic-ray ionization rates are also expected to increase due to the higher supernova rates. According to Papadopoulos (2010), for the case of extreme starburst galaxies such as ultra-luminous infrared galaxies (ULIRGs; $L_{\rm IR} > 10^{12} L_{\odot}$), cosmic-ray energy density can be 3-4 orders of magnitude higher than the Galaxy, and the cosmic-ray heating alone can raise the gas kinetic temperature to 80 - 160 K. Meijerink *et al.* (2006) compared the scenarios between an XDR model and a CRDR+PDR model, to find the elevated CI emission in an XDR model. Bayet *et al.* (2011) also modeled the chemistry in the CRDR+PDR model, but for larger number of molecular species. Their results showed that OH, H_2O, H_3^+ , H_3O^+ and OH^+ have high fractional abundances in high cosmic-ray ionization rate ($\zeta > 10^{-13}$ s^{-1}). Meijerink *et al.* (2011) modeled the CRDR with the consideration of mechanical-heating-dominated regions (MDR; see Section 2.4), and also proposed the use of similar species : OH^+, OH, H_2O^+, H_2O, and H_3O^+.

2.4. *Mechanical Heating*

Increase of temperature can also be caused by "mechanical heating", shocks or turbulence due to supernova explosions or stellar winds. When the kinetic temperature of gas increases, rates of some reactions with high barriers increase significantly ($k \propto exp(-\gamma/T_{kin})$, where γ is the barrier of the reaction). A well-known example is efficient production of water through accelerated reactions of $O + H_2 \longrightarrow OH + H$ and $OH + H_2 \longrightarrow H_2O + H$. Harada *et al.* (2010) found that this efficient production of water can cause an effective C-rich condition because the dominant form of oxygen become water, and the fractional abundances of O, O_2 decrease. Reactions between carbon-chains with atomic oxygen are one of the main destruction routes for the destruction of carbon-chain molecules. Thus, this lack of atomic oxygen creates an effective C-rich condition, which leads to an enhancement of fractional abundances of carbon-containing species such as HCN, HC_3N, and C_2H_2 (Figure 2). Loenen *et al.* (2008) examined the effects of mechanical heating on the line intensities of HCN, HNC, and HCO^+ in a steady-state model. HNC is also known to be sensitive to the elevation of the gas temperature because of a reaction HNC + H $\longrightarrow$ HCN + H although the exact value of the barrier is still in debate (Graninger *et al.* 2014). Kazandjian *et al.* (2012, 2016) extended their work by connecting the mechanical heating rate to the star formation rate, and by considering the probability density distribution function. The work mentioned above assumed constant increase of temperature. However, if the shock heating contribution is significant, the temperature fluctuates as a function of time. The case of constant temperature increase and the case of shock heating may affect reactions with barriers differently, therefore, the chemistry may be different between these two cases. Such study of temperature fluctuation in the context of GMC needs to be investigated.

Besides the elevation of the temperature, shock waves affect the chemistry by sputtering ice species from mantles, or by sputtering from grain cores. SiO is a well-known example of grain core sputtering caused by strong shocks (e.g., Martin-Pintado *et al.* 1992). For the ice sputtering, CH_3OH is an example of possible weak shock tracer from ice sputtering although non-thermal desorption such as cosmic-ray or UV-induced desorption can also increase methanol in the gas-phase. At the same time, there is a possibility that methanol gets destroyed by fast shocks ($v_s > 15$ km/s) (Suutarinen *et al.* 2014).

3. Astrochemical Observation in Nearby Galaxies

Statistical studies of molecular lines in various galaxy types can highlight the difference caused by different heating mechanisms. In fact, some molecular intensities / abundances are suggested to be diagnostics of galaxy properties. For example, HCN/HCO^+ ratios are proposed to be enhanced in AGN-containing galaxies as opposed to starburst galaxies (Kohno *et al.* 2001; Izumi *et al.* 2013). In addition, Aladro *et al.* (2011, 2015) claimed that CH_3CCH is enhanced in starburst galaxies than in AGN-containing galaxies, especially in an old starburst galaxy such as M82 in their survey of nearby galaxies using IRAM 30-meter telescope. Another possible diagnostic molecule is HC_3N. Costagliola *et al.* (2011) found the enhancement of HC_3N/HCN in compact obscured galactic nuclei. Above mentioned studies done with single-dish observations are ideal to follow-up with high-capability interferometers such as ALMA or NOEMA. In fact, those high-sensitivity interferometers can discover new molecular tracers. Current status of observational efforts is summarized below.

3.1. *GMC-scale Chemistry*

Although extreme regions of extragalactic sources are interesting, an important benchmark should first be made in the Galaxy for the better interpretation. Even with ALMA, the beam size to observe nearby galaxies in multiple species are at least a few tens of parsec, about the size of GMCs. However, previous Galactic astrochemical observations have focused on sub-pc scale star-forming regions. Thus, before the chemistry in extreme regions are discussed, the chemistry in relatively quiescent regions in a GMC-scale must be understood. To obtain this "normal" chemical abundances in a GMC-scale, observational effort has been recently made for Galactic spiral-arm GMCs such as W49 (Nagy *et al.* 2015) in the 1-mm band, Orion B (Pety *et al.* 2017), W51 (Watanabe *et al.* 2017), Orion A (Kauffmann *et al.* 2017), and W3 (OH) (Nishimura *et al.* 2017) in the 3-mm band. One of the important results many of these studies found is that even molecules with high critical densities such as HCN has a rather extended emission. Although HCN has been used as a "dense gas" tracer in many of extragalactic studies (e.g., Gao & Solomon 2004), some of the densities of the gas emitting HCN may even be $n \sim 10^3$ cm^{-3} (Kauffmann *et al.* 2017). For a template of a non-starburst galactic nucleus, mapping observations of Galactic Center CMZ by Jones *et al.* (2012, 2013) are also important as benchmarks. Similar large-scale observations in higher frequency bands are needed to obtain accurate column densities and to compare the chemistry in ALMA observations of nearby galaxies.

3.2. *Starburst Galaxies*

Starburst activities in galactic nuclei can be induced by merging of two galaxies, interaction with another galaxy, and certain galactic structures such as a bar to induce gas inflow into the center of galaxies. Connection between the galactic structures and the chemistry is first proposed by Meier & Turner (2005) in their multi-species observations of a starburst galaxy IC 342 in the 3-mm band. They proposed the picture where methanol emission tracers shocked regions due to the orbit intersection or due to the density waves, PDR tracers such as CCH is enhanced in nuclear starburst regions. They also proposed a similar picture in their ALMA observations of a nearby starburst galaxy NGC 253 (Meier *et al.* 2015).

More recently, Harada *et al.* (in preparation) observed a nuclear region of a face-on starburst galaxy M83 in multiple molecular tracers. The observed area includes the regions where the bar orbit intersects with an inner orbit, the circumnuclear ring. Those galactic bars are said to promote inflow of the gas into the central regions, and promote star formation (Ho *et al.* 1997). In these observations, methanol emission is enhanced at the entering point of gas into the circumnuclear ring (Cloud A in Fig 3d) compared with ^{13}CO tracing overall molecular gas (Cloud A in Fig 3c). Although methanol emission can be caused by cosmic-ray induced photo-desorption, it is unlikely that cosmic-ray ionization rate is very different only in two GMCs in the orbit. Methanol emission can also come from hot cores, but such contribution should be compact, and methanol emission in the GMC scale is known to be extended (e.g., Watanabe *et al.* 2017). Slightly downstream in the central orbit, N_2H^+ emission is enhanced (Cloud A & B in Fig 3e). N_2H^+ is normally seen in dense cores in the Galaxy. Further down in the orbit, there is a peak of CN emission (Cloud B & C in Fig 3f), which could be enhanced by PDRs. From this picture, we propose a scenario where shocks at the orbit intersection compress the gas creating dense gas clumps eventually to induce star formation.

3.3. *AGN-containing Galaxies*

Influence of an AGN is initially thought to appear most prominently as X-ray dominated regions of molecular clouds. Yet, observational trends such as enhanced HCN/HCO^+ ratios cannot be explained by XDR models. Observational results also seem to indicate that the elevated intensity HCN/HCO^+ ratios in AGNs are not from XDRs. NGC 1068 is a well-studied AGN-containing galaxy. High-resolution observations of NGC 1068 with ALMA by García-Burillo *et al.* (2014) and Viti *et al.* (2014) showed that the intensity ratio of HCN(1-0)/HCO^+(1-0) decreases within the central ~ 30 pc. Also in NGC 1068, observations by Takano *et al.* (2014) and Nakajima *et al.* (2015) detected HC_3N, HNCO, CN, CS, SO, CH_3CN, and CH_3OH in their ALMA cycle 0 observations. In their observations, HC_3N had a similar fractional abundance to that of a Galactic cold dense core TMC-1. A relatively large molecule such as HC_3N is prone to dissociation by radiation. This relatively high HC_3N fractional abundance means that a few 100-pc scale circumnuclear disk (CND) in NGC 1068 is not a giant XDR, but contains a significant amount of dense shielded gas from radiation. A relatively high fractional abundances of CH_3OH in their observations may result from shocks. Results by García-Burillo *et al.* (2010) and Kelly *et al.* (2017) are also consistentin claiming a strong influence of shock. The elevated

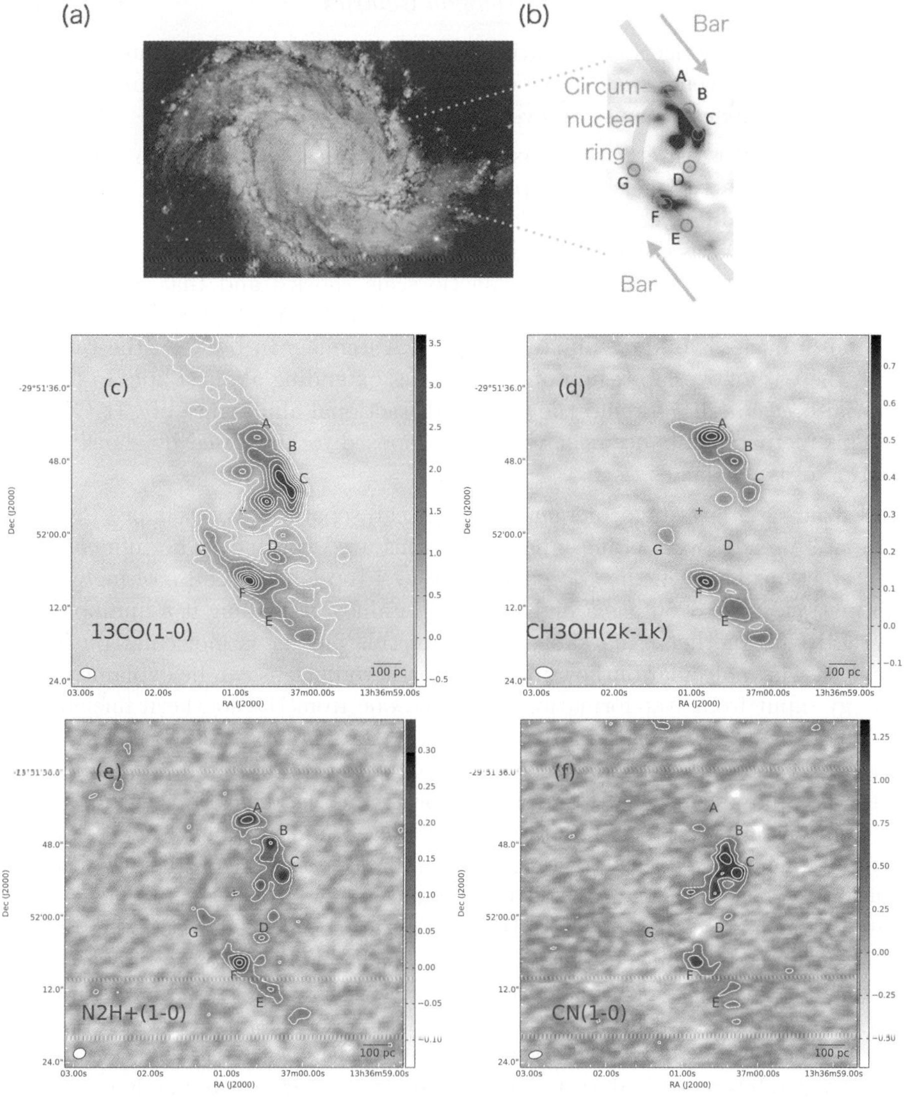

Figure 3. (a) *Hubble* image of M83 (credit: NASA, ESA, and the Hubble Heritage Team (STScI/AURA)) with the area shown in the panel (b) indicated as a rectangle. (b) A schematic image of the bar orbit and the orbit of circumnuclear ring overlaid on the ^{13}CO (1-0) map. Velocity-integrated intensity maps of (c) ^{13}CO (1-0), (d) CH_3OH (2_k-1_k), (e) N_2H^+(1-0), and (f) CN($1_{3/2}$-$0_{1/2}$) in the circumnuclear ring of M83 obtained from ALMA data 2015.1.00175.S (PI: Harada) except for CN, which is taken from the ALMA archival data 2011.0.00772.S. The units of those figures are Jy/beam km/s, and contours are plotted for every 3σ.

intensity HCN/HCO^+ ratio in a few 100-pc scale CND is likely to be caused by shocks. A low-luminosity AGN NGC 1097 also has a high intensity ratio of HCN/HCO^+. In NGC 1097, Martín *et al.* (2015) compared the chemical compositions of the nuclear region and a kpc-scale circumnuclear ring. They found elevated abundances of SiO, HNCO, and HC_3N in the nuclear region, indicating the possibility of shocks.

3.4. *Merging Galaxies*

Merging events can induce star formation by promoting the gas inflow into the central regions. Such merger events can cause a change in chemical compositions in the interaction regions of two galaxies or cause an evolution during the merger stages.

NGC 4038/4039 ("Antennae" galaxies) are the closest mergers with the distance of 22 Mpc, whose galactic nuclei are separated by $\sim$ 7.3 kpc. Ueda *et al.* (2016) observed Antennae galaxies in several molecular species. In their observations, enhancement of CH_3OH and HNCO was found in a location close to the northern nucleus. They concluded that this enhancement is caused by galactic scale shocks, and that other production mechanisms of gas-phase methanol are unlikely. VV 114 is a galaxy merger system whose separation of nuclei is 6 kpc, similar to that of Antennae. In between the two nuclei of VV 114, there is an overlap region of two galaxies extending about 4 kpc. In this overlap region, Saito *et al.* (2017) found the increased fractional abundance of CH_3OH in a kpc scale. This methanol enhancement was also proposed to be caused by shock due to the interaction of two galaxies.

NGC 3256 is a late-stage merger with nuclear separation of 850 pc. Parts of two galaxies are already overlapping, and spiral arms are expected to be influenced by the tidal interactions between the two galaxies (Fig. 4 top left). Because of merger-induced star formation, NGC 3256 has a high infrared luminosity, making it a luminous-infrared galaxy (LIRG). Another remarkable feature of this merger is molecular outflows from both nuclei. The molecular outflow from the northern nucleus can be explained only with the energy input from star formation while the one from the southern nucleus is likely to be caused by an AGN (Sakamoto *et al.* 2014). Harada *et al.* (2018) conducted multi-species observations of NGC 3256 with ALMA in the 3-mm and 1.3-mm band. As we compare the column density ratios over ^{13}CO or CS in several locations within NGC 3256, most variations are seen for SiO and CH_3OH (Fig. 4 middle panels). Those species are likely to be tracing the shocked regions. SiO abundance ratios are particularly enhanced in the outflow regions (OS) †, and CH_3OH abundances show similar enhancement. Although the desorption mechanisms of CH_3OH can also be cosmic-ray or UV-photon induced photo-desorption or chemical desorption, those mechanisms cannot account for increased abundances in particular locations. Other molecules such as CCH, CN, HC_3N, and N_2H^+ do not have significant variations within NGC 3256. When compared with other galactic nuclei, the chemistry of NGC 3256 is very similar to that of NGC 253, despite the fact that NGC 3256 has much higher star formation rates ($15M_\odot$/yr in northern nucleus and $8M_\odot$/yr in southern nucleus) than NGC 253 ($2M_\odot$/yr). On the other hand, the chemistry of Arp 220 is very different (Fig. 4 bottom panels).

Arp 220 is another galaxy merger with even less separation of galactic nuclei ($\sim$ 400 pc). Its chemistry is very distinct. It has high fractional abundances of HC_3N and complex organic molecules compared with local starburst galaxies (Martín *et al.* 2011, see also Martín *et al.* in this volume). High excitation condition in this galaxy also allows even detections of vibrationally-excited lines with a lower energy level $E_L > 1000$ K (Martín *et al.* 2016).

† In this analysis of the outflow position (OS), we integrated the velocity in the entire velocity range around this galaxy, not only the high-velocity components. Although it is likely that most of the emission is associated with the outflow, there may be some contamination from components not associated with the outflow.

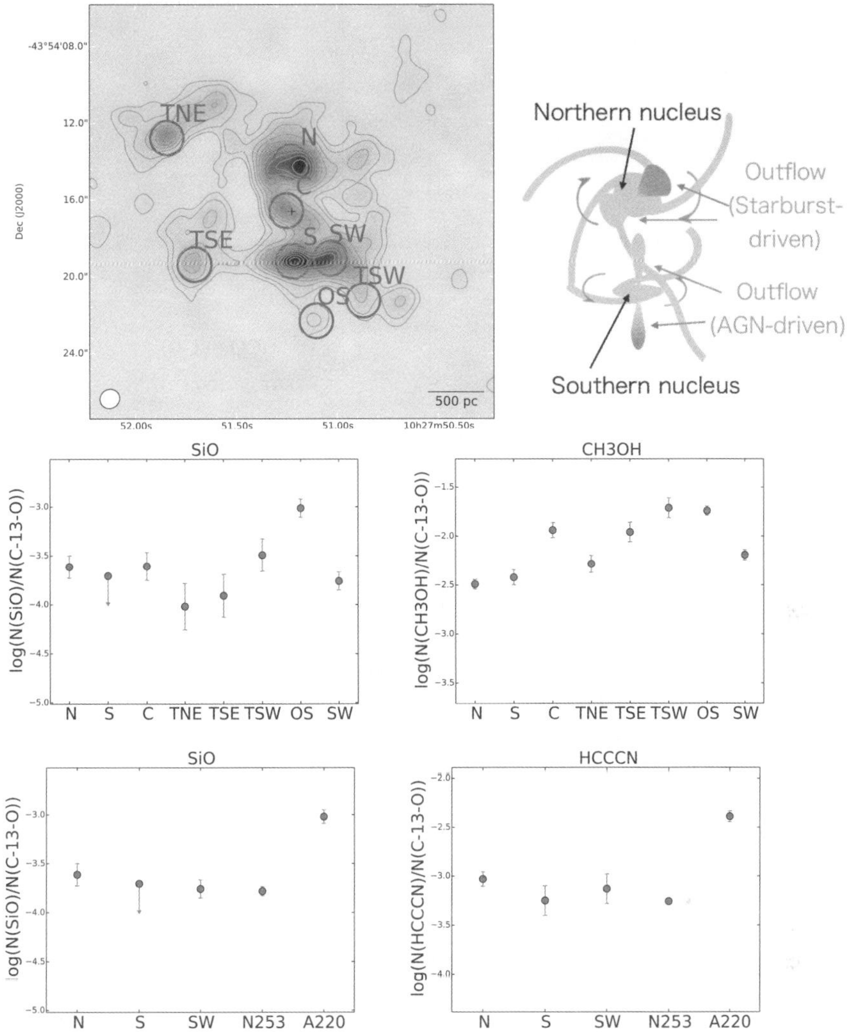

Figure 4. (*Top left*) A velocity-integrated intensity map of $^{13}CO(2\text{-}1)$ in NGC 3256 Harada *et al.* (2018). Each position is labeled as N: Northern nucleus, S: Southern nucleus, C: central position, TNE: tidal arm northeast, TSE: tidal arm southeast, TSW: tidal arm southwest, OS: Outflow south, SW: southern nucleus west. (*top right*) A cartoon picture of morphology of NGC 3256. (*middle left*) Column density ratios of SiO over ^{13}CO in locations indicated in the top left panel. (*middle right*) Same as the middle left panel, but of CH_3OH. (*bottom left*) Column density ratios of SiO over ^{13}CO in northern nucleus, southern nucleus, southern galaxy west of NGC 3256, a local starburst NGC 253, and Arp 220. The data of NGC 253 and Arp 220 are taken from Aladro *et al.* (2015). (*bottom right*) Same as the bottom left panel, but of HC_3N.

3.5. *Compact Obscured Nuclei*

There are galaxies with very compact, highly obscured, and high-luminosity energy sources in their galactic nuclei. Because of the obscuration, confirmation of their dominant energy sources, whether they are AGNs or starbursts, is difficult. Arp 220 mentioned above is an example of such galactic nuclei, but those compact obscured nuclei are not limited to galaxy mergers. For example, NGC 4418 does not have an obvious evidence of being a merger, yet its chemistry has a great deal of similarity with that of Arp 220.

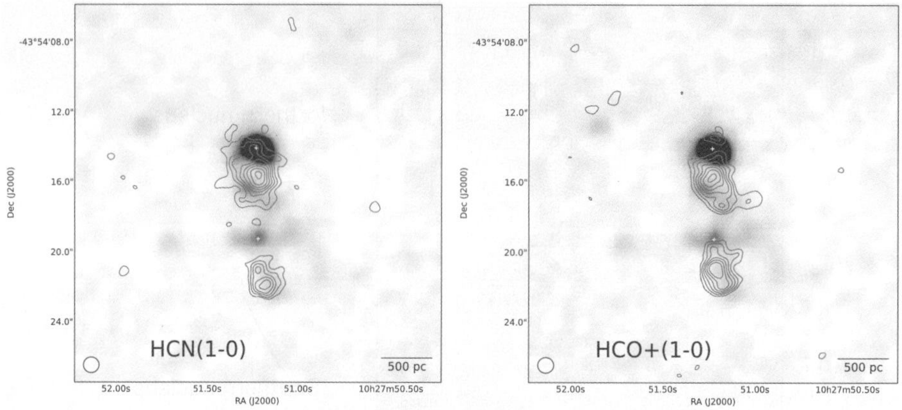

Figure 5. (*Left*) An intensity map of HCN(1-0) integrated for a red-shifted velocity range (+225 to +375 km/s from the systemic velocity) is shown in red contours, while a blue-shifted velocity range (-435 to -195 km/s from the systemic velocity) is shown in blue contours (Harada *et al.* 2018). Those velocity ranges are not associated with the nuclear disks of northern and southern galaxies. (*Right*) Same as the left panel, but for HCO^+ (1-0).

The fractional abundance of HC_3N in NGC 4418 is equivalent to that of Arp 220, and is more than one order of magnitude compared with local starburst galaxies (Costagliola *et al.* 2015). Strong vibrationally-excited lines are also detected (Sakamoto *et al.* 2010). Although the chemistry in the compact obscured nuclei is unique and interesting, understanding the exact mechanism of triggering such chemistry is difficult due to the high degree of obscuration, and needs further investigation.

3.6. *Outflow Chemistry*

Molecular outflow from galactic nuclei can be caused both by starburst and AGN activities, and can cause quenching of further star formation by expelling the relatively dense and cold gas. Those outflows are traced by high-velocity components shifted from the systemic velocities of galaxies. CO outflows are detected in many galaxies so far, both in starburst galaxies and in AGN-containing galaxies. Although outflow features are usually weak, detections of outflow components are not limited to CO. In an AGN-containing galaxy NGC 1068, high-velocity components of CN, SiO, HCN, and HCO^+ are detected (García-Burillo *et al.* 2010, 2014). Outflow components of HCN and HCO^+ lines were detected in Mrk 231, which is thought to be caused by a quasar-driven wind (Aalto *et al.* 2012). A follow-up work of this Mrk 231 outflow by Lindberg *et al.* (2016) found a high abundance ratio of $HCN/HCO^+ \gtrsim 1000$. This enhancement is likely caused by a shock in the outflow. In the outflows of NGC 3256 mentioned above, CN (Sakamoto *et al.* 2014), HCN, and HCO^+ (Harada *et al.* 2018) are also detected (Fig. 5). In a starburst galaxy NGC 253, HCN, CN, HCO^+, and CS are also detected in the outflow (Walter *et al.* 2017).

Those species have higher critical densities than CO, and were unexpected because outflows are thought to be generally hot and tenuous. Detailed study using these molecules would be useful in determining physical and chemical state of outflow.

4. Summary

As shown here, ALMA has already provided rich information in astrochemical observations in external galaxies, and has shown its potential for even further studies. However, quantitative interpretation of those observations in connection with starburst or AGN activities are still challenging. Detailed case studies in a well-studied source or statistical studies of selected molecular lines in various galaxy types would observationally help our understanding of extragalactic astrochemistry. Although there are many sophisticated chemical models of PDR, XDR, MDR, and CRDR, the complexity of GMC-scale observations poses challenge, and needs to be included in future models.

References

Aalto, S., Garcia-Burillo, S., Muller, S., *et al.* 2012, *A&A*, 537, A44

Aladro, R., Martín-Pintado, J., Martín, S., Mauersberger, R., & Bayet, E. 2011, *A&A*, 525, A89

Aladro, R., Martín, S., Riquelme, D., *et al.* 2015, *A&A*, 579, A101

Abel, N. P., Federman, S. R., & Stancil, P. C. 2008, *ApJL*, 675, L81

Ao, Y., Henkel, C., Braatz, J. A., *et al.* 2011, *A&A*, 529, A154

Bayet, E., Williams, D. A., Hartquist, T. W., & Viti, S. 2011, *MNRAS*, 414, 1583

Costagliola, F., Aalto, S., Rodriguez, M. I., *et al.* 2011, *A&A*, 528, A30

Costagliola, F., Sakamoto, K., Muller, S., *et al.* 2015, *A&A*, 582, A91

Gao, Y. & Solomon, P. M. 2004, *ApJ*, 606, 271

García-Burillo, S., Usero, A., Fuente, A., *et al.* 2010, *A&A*, 519, A2

García-Burillo, S., Combes, F., Usero, A., *et al.* 2014, *A&A*, 567, A125

Ginsburg, A., Henkel, C., Ao, Y., *et al.* 2016, *A&A*, 586, A50

Graninger, D. M., Herbst, E., Öberg, K. I., & Vasyunin, A. I. 2014, *ApJ*, 787, 74

Harada, N., Herbst, E., & Wakelam, V. 2010, *ApJ*, 721, 1570-1578

Harada, N., Sakamoto, K., Martín, S., *et al.* 2018, *ApJ*, 855, 49

Harada, N., Thompson, T. A., & Herbst, E. 2013, *ApJ*, 765, 108

Ho, L. C., Filippenko, A. V., & Sargent, W. L. W. 1997, *ApJ*, 487, 591

Indriolo, N., Fields, B. D., & McCall, B. J. 2009, *ApJ*, 694, 257

Izumi, T., Kohno, K., Martín, S., *et al.* 2013, *PASJ*, 65, 100

Jones, P. A., Burton, M. G., Cunningham, M. R., *et al.* 2012, *MNRAS*, 419, 2961

Jones, P. A., Burton, M. G., Cunningham, M. R., Tothill, N. F. H., & Walsh, A. J. 2013, *MNRAS*, 433, 221

Kauffmann, J., Goldsmith, P. F., Melnick, G., *et al.* 2017, *A&AL*, in press, arXiv:1707.05352

Kazandjian, M. V., Meijerink, R., Pelupessy, I., Israel, F. P., & Spaans, M. 2012, *A&A*, 542, A65

Kazandjian, M. V., Pelupessy, I., Meijerink, R., *et al.* 2016, *A&A*, 595, A124

Kelly, G., Viti, S., García-Burillo, S., *et al.* 2017, *A&A*, 597, A11

Kohno, K., Matsushita, S., Vila-Vilaró, B., *et al.* 2001, The Central Kiloparsec of Starbursts and AGN: The La Palma Connection, 249, 672

Lepp, S. & Dalgarno, A. 1996, *A&A*, 306, L21

Levrier, F., Le Petit, F., Hennebelle, P., *et al.* 2012, *A&A*, 544, A22

Lindberg, J. E., Aalto, S., Muller, S., *et al.* 2016, *A&A*, 587, A15

Loenen, A. F., Spaans, M., Baan, W. A., & Meijerink, R. 2008, *A&A*, 488, L5

Martin-Pintado, J., Bachiller, R., & Fuente, A. 1992, *A&A*, 254, 315

Maloney, P. R., Hollenbach, D. J., & Tielens, A. G. G. M. 1996, *ApJ*, 466, 561

Martín, S., Krips, M., Martín-Pintado, J., *et al.* 2011, *A&A*, 527, A36

Martín, S., Kohno, K., Izumi, T., *et al.* 2015, *A&A*, 573, A116

Martín, S., Aalto, S., Sakamoto, K., *et al.* 2016, *A&A*, 590, A25

Mangum, J. G., Darling, J., Henkel, C., *et al.* 2013, *ApJ*, 779, 33

Meier, D. S. & Turner, J. L. 2005, *ApJ*, 618, 259

Meier, D. S., Walter, F., Bolatto, A. D., *et al.* 2015, *ApJ*, 801, 63

Meijerink, R. & Spaans, M. 2005, *A&A*, 436, 397

Meijerink, R., Spaans, M., & Israel, F. P. 2006, *ApJL*, 650, L103

Meijerink, R., Spaans, M., & Israel, F. P. 2007, *A&A*, 461, 793

Meijerink, R., Spaans, M., Loenen, A. F., & van der Werf, P. P. 2011, *A&A*, 525, A119
Nagy, Z., van der Tak, F. F. S., Fuller, G. A., & Plume, R. 2015, *A&A*, 577, A127
Nakajima, T., Takano, S., Kohno, K., *et al.* 2015, *PASJ*, 67, 8
Nishimura, Y., Watanabe, Y., Harada, N., *et al.* 2017, *ApJ*, 848, 17
Padovani, M., Galli, D., & Glassgold, A. E. 2009, *A&A*, 501, 619
Papadopoulos, P. P. 2010, *ApJ*, 720, 226
Pety, J., Guzmán, V. V., Orkisz, J. H., *et al.* 2017, *A&A*, 599, A98
Ragan, S. E., Bergin, E. A., & Wilner, D. 2011, *ApJ*, 736, 163
Röllig, M., Abel, N. P., Bell, T., *et al.* 2007, *A&A*, 467, 187
Saito, T., Iono, D., Espada, D., *et al.* 2017, *ApJ*, 834, 6
Sakamoto, K., Aalto, S., Evans, A. S., Wiedner, M. C., & Wilner, D. J. 2010, *ApJL*, 725, L228
Sakamoto, K., Aalto, S., Combes, F., Evans, A., & Peck, A. 2014, *ApJ*, 797, 90
Suutarinen, A. N., Kristensen, L. E., Mottram, J. C., Fraser, H. J., & van Dishoeck, E. F. 2014, *MNRAS*, 440, 1844
Takano, S., Nakajima, T., Kohno, K., *et al.* 2014, *PASJ*, 66, 75
Tielens, A. G. G. M. & Hollenbach, D. 1985, *ApJ*, 291, 722
Tielens, A. G. G. M. 2005, The Physics and Chemistry of the Interstellar Medium, by A. G. G. M. Tielens, pp. . ISBN 0521826349. Cambridge, UK: Cambridge University Press, 2005.,
Thompson, T. A., Quataert, E., & Murray, N. 2005, *ApJ*, 630, 167
Ueda, J., Watanabe, Y., Iono, D., *et al.* 2017, *PASJ*, 69, 6
Umebayashi, T. & Nakano, T. 1981, *PASJ*, 33, 617
Walter, F., Bolatto, A. D., Leroy, A. K., *et al.* 2017, *ApJ*, 835, 265
Viti, S., García-Burillo, S., Fuente, A., *et al.* 2014, *A&A*, 570, A28
Watanabe, Y., Nishimura, Y., Harada, N., *et al.* 2017, *ApJ*, 845, 116

Astrochemistry VII – Through the Cosmos from Galaxies to Planets
Proceedings IAU Symposium No. 332, 2017
M. Cunningham, T. Millar & Y. Aikawa, eds.

doi:10.1017/S1743921317007700

Ultraluminous Extragalactic Chemistry

Sergio Martín Ruiz[1,2]

[1]European Southern Observatory,
Alonso de Córdova 3107, Vitacura, Casilla 19001, Santiago 19, Chile

[2]Joint ALMA Observatory,
Alonso de Córdova 3107, Vitacura, Casilla 19001, Santiago 19, Chile
email: smartin@eso.org

Abstract. At a distance of 77 Mpc, the Ultralumious galaxy Arp 220 is the closest extragalactic equivalent to Galactic hot cores. The low resolution SMA survey showed a highly excited confusion limited spectrum. The new ALMA snapshot spectral scan opens the possibility of chemically resolve the two nuclei at unprecedented sensitivity. When completed, it will be the widest survey ever done towards an extragalactic object. The model of Band 6 and 7 data already shows the chemical similarities between the interacting nuclei which may provide clues on the similar heating sources. Vibrationally excited transitions may be tracing the deeply embedded dust obscured active nuclei and/or hot compact star burst. This vibrational emission is the brightest ever measured in an extragalactic object, and even so compared with Galactic hot cores. In fact, the eastern one is the brightest in such vibrational emission. Water mega-maser emission also points towards a very compact sources likely related to star forming clumps within both Arp 220 nuclei.

Keywords. ISM: molecules, galaxies: abundances, galaxies: active, galaxies: starburst, galaxies: individual (Arp 220)

1. Introduction

During the last decade the increased receiver bandwidth in state of the art millimeter and submillimeter telescopes has allowed a rapid increase in the observations of unbiased line surveys in the extragalactic interstellar medium. Table 2 in Martín (2011) summarized some of the ongoing observational efforts, some of which have been recently published (Aladro *et al.* 2011, 2015; Davis *et al.* 2013; Watanabe *et al.* 2014, 2016). The increasing number of such unbiased surveys allows comparative studies between the nuclear region of galaxies dominated by different types of activity (Aladro *et al.* 2015).

Obviously, the advent of high sensitivity instruments like ALMA is having a transformational impact in the field of extragalactic astrochemistry both through unbiased line surveys (Costagliola *et al.* 2015; Muller *et al.* 2014b,a, 2016) and deep multi molecular observations towards nearby galaxies (Meier *et al.* 2015; Martín *et al.* 2015; Nakajima *et al.* 2015; Tosaki *et al.* 2017).

These high sensitivity observations are demonstrating the potential to investigate deeply buried galactic nuclear regions through unbiased molecular observations. In particular it may an unique tool to the study the obscured nuclei of luminous and ultraluminous galaxies. These galaxies are radiating most of their energy as thermal dust emission in the infrared and they are the dominant contribution to the star formation rate at $z > 2$ (Madau & Dickinson 2014). Some of these objects are observed to be extremely compact where most of their luminosity is observed to emerge within a few tens of parsecs nuclear region (Aalto *et al.* 2015b; Costagliola *et al.* 2015; Martín *et al.* 2016). Their nuclear power source, embedded inside obscurations of $A_v > 1000$ mag and thus obscured at almost everywavelength, prevent us from determining their nature. In most cases, the presence of a dominating AGN or a hot starburst in an intense and compact mode cannot be differentiated (Aalto *et al.* 2015b).

2. ULIRG chemistry of embedded nucleus: Arp 220

Arp 220, at a distance of ~ 77 pc, is the prototype of ULIRG, in a later merging state showing two deeply buried nuclei separated by ~ 1" (Sakamoto *et al.* 1999). The nuclei are optically thick outside the range of $5-350$ GHz (Barcos-Muñoz *et al.* 2015). Arp 220 has been observed with almost any instrument available but the nature of its nuclear powering source is still debated. This galaxy has been the subject of many recent studies with ALMA during its first observing cycles (Rangwala *et al.* 2015; Scoville *et al.* 2015, 2017; Zschaechner *et al.* 2016; Martín *et al.* 2016; König *et al.* 2017).

The presence of an AGN in its nuclei has been claimed based on the detected hard X-ray point source towards the western nucleus and softer emission towards the eastern one Clements *et al.* (2002), as well as the large equivalent width of the Fe 6.7 keV line (Teng *et al.* 2009). However its starburst contribution is evidenced by the dozens of SN remnants towards both nuclei, with a SN rate consistent with the SFR derived from FIR observations (Lonsdale *et al.* 2006; Parra *et al.* 2007). Similarly the extended faint soft X-ray emission may be starburst driven (McDowell *et al.* 2003).

In order to establish whether the effect of the inner powering source had an imprint on the molecular composition in the ISM Martín *et al.* (2011) carried out an unbiased line survey at 1.3 mm with the SMA which resulted in no obvious evidence from an AGN on the observed molecular abundances. In fact, the vibrationally excited transitions of HC_3N, together with the large abundance of water based on the isotopologues ratio $H_2^{18}O/C^{18}O$, were claimed to be evidences of a compact concentration of $\sim 10^6$ hot cores within the inner 700 pc of Arp 220.

It is worth mentioning that the unbiased line survey carried out with ALMA towards the also compact obscured nuclei in the LIRG NGC 4418 showed molecular abundances resembling closely those of Arp 220 (Costagliola *et al.* 2015).

3. Dissecting the two nuclei of Arp 220 with ALMA

Since the resolution achieved with the SMA observations was around 700 pc, it was not possible to disentangle the emission from the individual nuclei. Thus we made use of the unprecedented sensitivity and resolution of ALMA to study the individual molecular emission from the two nuclei in Arp 220. At a resolution of $0.5''$, we scanned the atmospheric bands 6 and 7 in Cycle 1, and the bands 3 and 4 in Cycle 3. The observations covered the frequency range from 86 to 363 GHz, which, when completed, will result in the widest unbiased scan towards an extragalactic source. It is worth noting that the scan consisted of snapshot observations of 2-4 min at the higher bands and 8-20 min in the lower, in order to achieve an homogeneous noise level of 2mJy and 1mJy, respectively.

Figure 1 shows the observed spectrum towards the western nucleus and the modeled synthetic spectrum fitted to the data. The large line width of ~ 300km s^{-1} results in a heavily blended spectrum where the model includes 1331 transitions with flux densities above 1 σ. The eastern nucleus, though generally fainter, still shows a heavily blended spectrum.

As shown in the Figure 1, modeling the data is particularly tricky due to the fact that almost all spectral features show a heavy absorption in the center. This was clearly shown in the study of HCN and HCO^+ emission by Martín *et al.* (2016) which showed

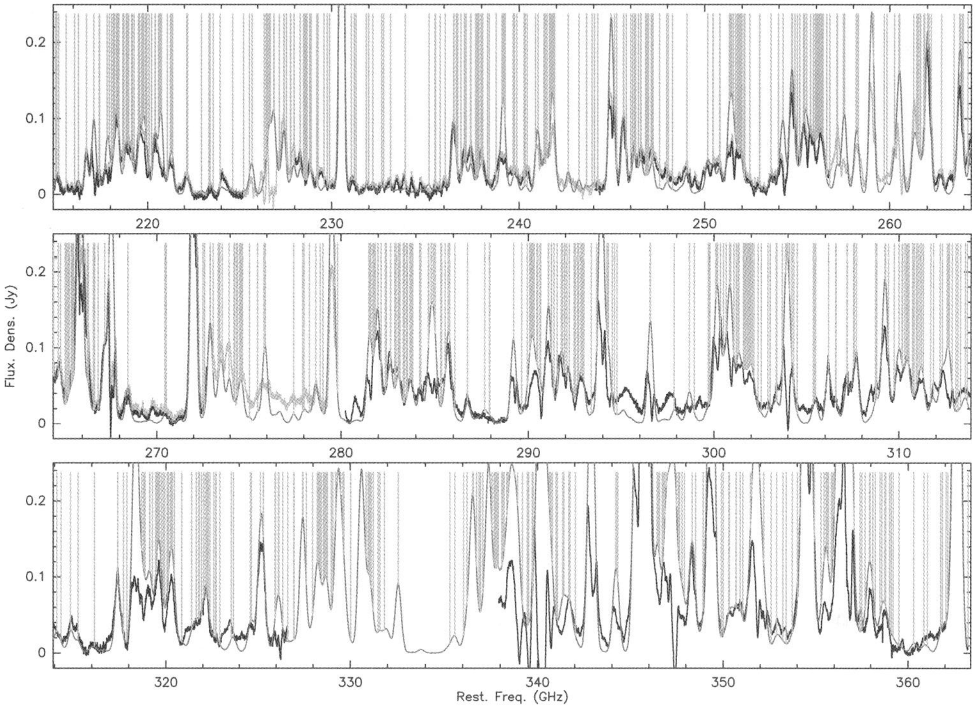

Figure 1. Full band 6 and 7 spectral coverage with ALMA towards its western nucleus. Black line shows the observed spectrum at 0.65″ while the green section shows observation at coarser resolution. The model is shown in red and the grey vertical lines show the location of all the spectral features used in the model (Martín *et al.* In preparation)

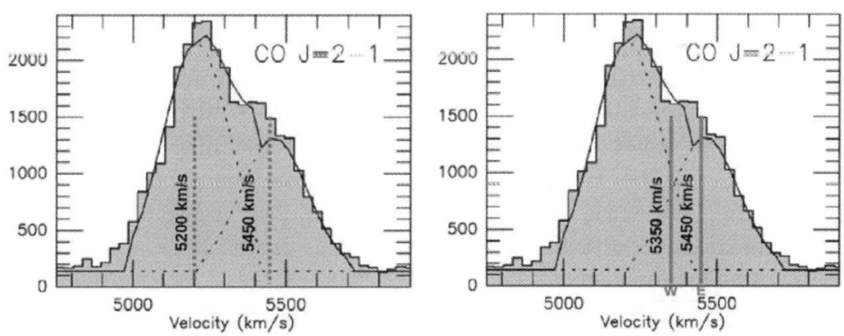

Figure 2. (LEFT) Double Gaussian fit used by Martín *et al.* (2011) to disentangle the two Arp 220 nuclei in velocity. This velocity identification may not be adequate as shown in the right panel. (RIGHT) Systemic velocity of the two nuclei as resulting from an absorbed Gaussian fit to the individual spectra of the two nucleus.

multiple absorption systems in the line of sight of both nuclei where approximately 70% of the line emission is absorbed. Thus the model fit was performed making use of the profile outside the absorbed region which implies a significant level of incertitude and explains why most of model in Figure 1 falls above the observed spectrum.

The fit of absorbed Gaussian profiles to these lines was in agreement with previous estimates of the individual systemic velocities of both nuclei. It clearly shows that, as sometimes done in the literature, the use of a double Gaussian fit to unresolved spectral

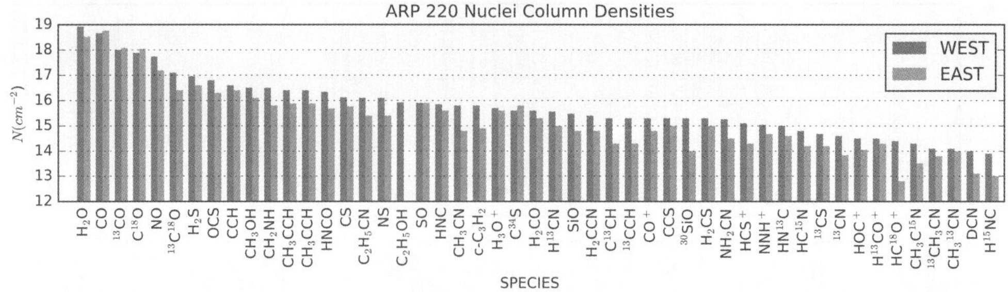

Figure 3. Comparison between the absolute column densities of all detected species in the two nuclei of Arp 220. Figure extracted from Martín *et al.* In prep.

observations as a way to separate the two nuclei in velocities (Greve *et al.* 2009; Martín *et al.* 2011) may not be adequate. This difference is shown in Figure 2.

Based on the modeled spectra, column densities of 49 molecular species, including some isotopologues, were calculated for the individual nucleus. Figure 3 shows a preliminary comparison based on these derived column densities. Within a factor of 3, no significant or obvious molecular trend is found to differentiate the nuclei. This result indicates that either there is no molecular fingerprint due to the putative AGN in the western nucleus or that such is already diluted at 240 pc resolution.

4. The vibe of Arp 220

One of the most interesting spectral features in the Arp 220 scan are the brightness of its vibrationally excited transitions.

After the first detection of vibrationally excited absorption of HCN by Salter *et al.* (2008), a number of vibrational transitions of HCN and HC_3N have been reported towards LIRGs (Sakamoto *et al.* 2010; Costagliola & Aalto 2010; Costagliola *et al.* 2015) and ULIRGs (Martín *et al.* 2011; Imanishi & Nakanishi 2013; Aalto *et al.* 2015a,b). These transitions require of mid-IR photons stemming from heated dust with $T_d > 100$ K to be efficiently pumped (Sakamoto *et al.* 2010; Aalto *et al.* 2015b). All the sources where this emission is been observed share a large luminosity within a very compact nuclear region (Aalto *et al.* 2015b). However, though a very compact concentration of gas and a high temperature gradient is required to explain the observed emission, it is uncertain whether this vibrational transitions are excited in the surroundings of an AGN or due to a compact hot star forming event.

Though the sample of detected sources is still limited, a lot of observational efforts are ongoing to expand the range of sources and luminosities (Imanishi & Nakanishi 2014; Imanishi *et al.* 2016).

Martín *et al.* (2016) presented for the first time high resolution observations of the vibrational emission where the two nuclei were resolved from each other. The emission is observed to be at the systemic velocity of each nucleus, and although not resolved, it appears to be coming from a region smaller than the resolved extended continuum emission.

It is worth noting that, as shown in Figure 4, the nuclei in Arp 220 show the highest ratio between the luminosity of the vibrational emission of HCN and the infrared luminosity. This ratio, measured with both the $J = 3 - 2$ and $4 - 3$ HCN vibrational transitions, is higher than any other extragalactic source and one order of magnitude above what is observed in Galactic hot cores. Also interesting is the fact that among the two nuclei, the eastern one is showing the highest ratio. So if this emission is AGN

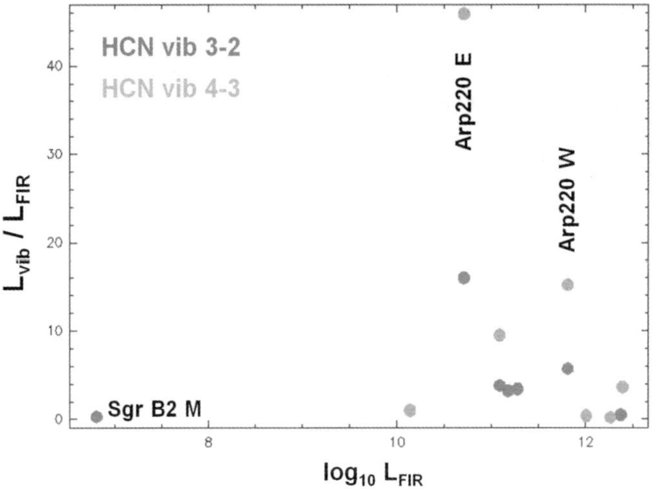

Figure 4.

powered, this effect is more prominent towards the eastern nucleus and to a lesser degree in the western, where the putative AGN is placed.

5. Resoved water "mega-maser" in Arp 220

The ALMA spectral scan of Arp 220 also covered the 325 GHz transition of water. The water emission has been claimed to be of maser origin despite the lack of line variability and showing a line shape similar to any other species (Cernicharo *et al.* 2006). König *et al.* (2017) compiled the 325 GHz in the ALMA survey together with the 22 GHz VLA detection by Zschaechner *et al.* (2016) and the Band 5 science verification data. All these observation had a similar resolution of $\sim 0.7''$ which was enough to separate the emission from both nuclei.

Line blending in Arp 220 is particularly an issue. Indeed all 3 transitions are significantly affected by blending (see figures in Zschaechner *et al.* 2016; König *et al.* 2017). The 22 GHz was seen to emit right within the NH_3 profile. The ammonia contribution was subtracted by using the line profile of nearby ammonia transitions. In the case of the 180 GHz and 325 GHz, the de-blending was carried out by subtracting the synthetic profile fitted to the whole spectral scan. In the case of the 325 GHz transition, the line is falling within the range of the scan. However, the modeled spectra had to be extrapolated down to the 180 GHz transition, but still Figure 1 in König *et al.* (2017) shows that the agreement between the "preliminary" model and the observed profile is good, in particular towards the western nucleus. Additionally the de-blended line profiles show a good agreement in shape between the 180 and 325 GHz transition.

Despite the non variability of the water emission is confirmed over a one decade period, line intensities do not agree with a thermal origin. Also, the modeling suggest a very compact origin of the emission, which could stem from a large number of star-forming clumps. These clumps might be the origin of an isotropically distributed maser emission, required to explain the non variability of the line profile, in both nuclei of Arp 220.

References

Aalto, S., Garcia-Burillo, S., Muller, S., *et al.* 2015a, *A&A*, 574, A85

Aalto, S., Martín, S., Costagliola, F., *et al.* 2015b, *A&A*, 584, A42
Aladro, R., Martín, S., Martín-Pintado, J., *et al.* 2011, *A&A*, 535, A84
Aladro, R., Martín, S., Riquelme, D., *et al.* 2015, *A&A*, 579, A101
Barcos-Muñoz, L., Leroy, A. K., Evans, A. S., *et al.* 2015, *ApJ*, 799, 10
Cernicharo, J., Pardo, J. R., & Weiss, A. 2006, *ApJ*, 646, L49
Clements, D. L., McDowell, J. C., Shaked, S., *et al.* 2002, *ApJ*, 581, 974
Costagliola, F. & Aalto, S. 2010, *A&A*, 515, A71
Costagliola, F., Sakamoto, K., Muller, S., *et al.* 2015, *A&A*, 582, A91
Davis, T. A., Heiderman, A., Evans, N. J., & Iono, D. 2013, *MNRAS*
Greve, T. R., Papadopoulos, P. P., Gao, Y., & Radford, S. J. E. 2009, *ApJ*, 692, 1432
Imanishi, M. & Nakanishi, K. 2013, *AJ*, 146, 91
Imanishi, M. & Nakanishi, K. 2014, *AJ*, 148, 9
Imanishi, M., Nakanishi, K., & Izumi, T. 2016, *ApJ*, 825, 44
König, S., Martín, S., Muller, S., *et al.* 2017, *A&A*, 602, A42
Lonsdale, C. J., Diamond, P. J., Thrall, H., Smith, H. E., & Lonsdale, C. J. 2006, *ApJ*, 647, 185
Madau, P. & Dickinson, M. 2014, *ARAA*, 52, 415
Martín, S. 2011, in: J. Cernicharo & R. Bachiller (eds.), *The Molecular Universe*, Proc. IAU Symposium No 280 p. 351
Martín, S., Aalto, S., Sakamoto, K., *et al.* 2016, *A&A*, 590, A25
Martín, S., Kohno, K., Izumi, T., *et al.* 2015, *A&A*, 573, A116
Martín, S., Krips, M., Martín-Pintado, J., *et al.* 2011, *A&A*, 527, A36
McDowell, J. C., Clements, D. L., Lamb, S. A., *et al.* 2003, *ApJ*, 591, 154
Meier, D. S., Walter, F., Bolatto, A. D., *et al.* 2015, *ApJ*, 801, 63
Muller, S., Black, J. H., Guélin, M., *et al.* 2014a, *A&A*, 566, L6
Muller, S., Combes, F., Guélin, M., *et al.* 2014b, *A&A*, 566, A112
Muller, S., Müller, H. S. P., Black, J. H., *et al.* 2016, *A&A*, 595, A128
Nakajima, T., Takano, S., Kohno, K., *et al.* 2015, *PASJ*, 67, 8
Parra, R., Conway, J. E., Diamond, P. J., *et al.* 2007, *ApJ*, 659, 314
Rangwala, N., Maloney, P. R., Wilson, C. D., *et al.* 2015, *ApJ*, 806, 17
Sakamoto, K., Aalto, S., Evans, A. S., Wiedner, M. C., & Wilner, D. J. 2010, *ApJ*, 725, L228
Sakamoto, K., Scoville, N. Z., Yun, M. S., *et al.* 1999, *ApJ*, 514, 68
Salter, C. J., Ghosh, T., Catinella, B., *et al.* 2008, *AJ*, 136, 389
Scoville, N., Murchikova, L., Walter, F., *et al.* 2017, *ApJ*, 836, 66
Scoville, N., Sheth, K., Walter, F., *et al.* 2015, *ApJ*, 800, 70
Teng, S. H., Veilleux, S., Anabuki, N., *et al.* 2009, *ApJ*, 691, 261
Tosaki, T., Kohno, K., Harada, N., *et al.* 2017, *PASJ*, 69, 18
Watanabe, Y., Sakai, N., Sorai, K., Ueda, J., & Yamamoto, S. 2016, *ApJ*, 819, 144
Watanabe, Y., Sakai, N., Sorai, K., & Yamamoto, S. 2014, *ApJ*, 788, 4
Zschaechner, L. K., Ott, J., Walter, F., *et al.* 2016, *ApJ*, 833, 41

Astrochemistry VII – Through the Cosmos from Galaxies to Planets
Proceedings IAU Symposium No. 332, 2017
M. Cunningham, T. Millar & Y. Aikawa, eds.

doi:10.1017/S1743921317007724

Determining the effect of a non-uniform AGB outflow on its chemistry

M. Van de Sande[1], J. O. Sundqvist[1], T. J. Millar[2] and L. Decin[1]

[1]Department of Physics and Astronomy, Institute of Astronomy, KU Leuven, Celestijnenlaan 200D, 3001 Leuven, Belgium
email: marie.vandesande@kuleuven.be

[2]School of Mathematics and Physics, Queens University Belfast, University Road, Belfast BT7 1NN, UK

Abstract. The molecular composition of the stellar outflows of AGB stars is determined by the stellar elemental carbon-to-oxygen abundance ratio, together with the physical circumstances in the innermost region of the outflow. Near the stellar surface, thermal equilibrium (TE) can be assumed. This leads to a certain molecular composition with a O- or C-rich signature. However, several molecular species have been detected that are not expected to be present in the inner region under the assumption of TE chemistry. As a solution to explain the presence of these unexpected species, non-equilibrium chemistry in the inner region of the outflow has been proposed. The outflows of AGB stars are generally not spherically symmetric or homogeneous, which influences the penetration of interstellar UV photons throughout the outflow. We investigate the effect of a clumpy, non-homogeneous outflow on the composition of the inner region by introducing a simple porosity formalism in our chemical model.

Keywords. astrochemistry, molecular processes, stars: AGB and post-AGB, circumstellar matter, stars: mass loss

1. Introduction

The final life stage of low- to intermediate-mass stars, the asymptotic giant branch (AGB) phase, is characterised by strong mass loss via a stellar outflow. This outflow removes the outer layers of the star and creates a large circumstellar envelope. The elemental carbon-to-oxygen abundance ratio of the central AGB star determines the molecular composition of the outflow, together with the physical conditions in its innermost region. Thermal equilibrium (TE) can be assumed at the stellar surface, leading to a certain molecular composition with a O- or C-rich nature. However, several molecular species have been detected that are not expected to be present in the inner region under the assumption of TE chemistry, such as the C-rich CO_2, HCN, CS, and CN in the outflows of O-rich stars (Omont *et al.* (1993), Bujarrabal *et al.* (1994), Justtanont *et al.* (1996)) and warm H_2O for a C-rich outflow (Decin *et al.* (2010)). NH_3 has been observed with abundances higher than predicted by TE chemistry in both O- and C-rich stars (Keady *et al.* (1993), Menten *et al.* (1995)). As a solution to the presence of these unexpected species, pulsation-induced non-equilibrium chemistry in the inner region of the outflow has been proposed (Duari *et al.* (1999), Cherchneff (2006), Gobrecht *et al.* (2016)). These models are able to account for the abundances of most unexpected species, but are unable to account for e.g. NH_3 in O-rich stars (Gobrecht *et al.* (2016)).

The outflows of AGB stars are generally not spherically symmetric. Both large-scale structures, such as spirals (e.g. Maercker *et al.* (2012)) and disks (e.g. Kervella *et al.* (2014)), and small-scale density inhomogeneities (e.g. Khouri *et al.* (2016), Agúndez *et al.* (2017)) have been observed. We investigate the potential effects of small-scale density

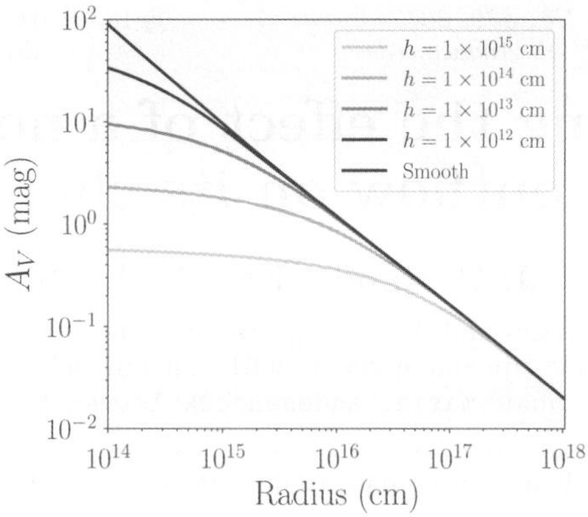

Figure 1. Visual extinction throughout a clumpy outflow with mass-loss rate 10^{-5} $M_\odot$ yr^{-1} where all material is located inside of clumps (void inter-clump component). Result for a smooth outflow is shown in black. Results for different values of the porosity length h are shown in blue.

inhomogeneities or clumps within the outflow on the composition of the inner region. This is done by introducing a porosity formalism in our chemical model.

2. Methods

2.1. *Chemical model*

The chemical reaction network used in our chemical model is the most recent release of the UMIST Database for Astrochemistry (UDfA), RATE12 (McElroy *et al.* (2013)), a gas-phase only network. The chemical model itself is based on the UDfA CSE model. This is a one-dimensional model describing a uniformly expanding, spherically symmetric outflow with a constant mass-loss rate. The physical parameters and parent species of the model are taken from Agúndez *et al.* (2010). We have modified the temperature profile of the code to a power-law, and have implemented the porosity formalism (see Sect. 2.2).

2.2. *Porosity formalism*

The porosity formalism used in this work (Owocki *et al.* (2004), Owocki *et al.* (2006), Sundqvist *et al.* (2012), Sundqvist *et al.* (2014)) provides us with a mathematical framework that describes the resultant effect of a stochastic distribution of clumps in an outflow on the intensity of the external radiation field throughout the outflow. This description is included in the code by changing the optical depth τ of a smooth outflow to the effective optical depth $\tau_{\rm eff}$ of a clumpy one, which then accounts for potential light-leakage through porous channels within the clumpy outflow. Our porosity formalism also divides the outflow into two separate components: an overdense clump component and a rarefied inter-clump component. The division between the two components is determined via the volume filling factor $f_{\rm vol}$, which is the fraction of the total volume of the outflow occupied by clumps, and the density contrast $f_{\rm ic}$, which establishes the density of the inter-clump component relative to the spatially averaged density of the outflow. A void inter-clump component corresponds to $f_{\rm ic} = 0$.

Porosity affects the visual extinction throughout the outflow. This is shown in Fig. 1 for an outflow where all material is located inside of clumps (void inter-clump medium). The porosity length h represents the local mean free path between two clumps. Using the formalism, we can hence take both the larger density of the clumps and the decrease in visual extinction throughout the outflow into account.

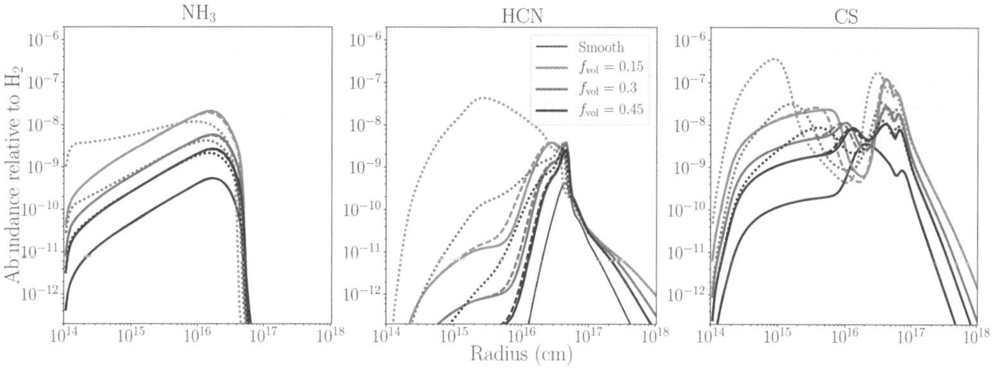

Figure 2. Abundance of NH_3, HCN, and CS relative to H_2 throughout a one-component O-rich outflow with a mass-loss rate of 10^{-5} $M_\odot$ yr^{-1} for different clump volume filling factors f_{vol}. Solid black line: calculated abundance for a smooth, uniform outflow. Solid coloured line: $l_* = 5 \times 10^{12}$ cm. Dashed coloured line: $l_* = 1 \times 10^{13}$ cm. Dotted coloured line: $l_* = 5 \times 10^{13}$ cm.

Table 1. Column density of NH_3, HCN, and CS in a smooth O-rich outflow with a mass-loss rate of 10^{-5} $M_\odot$ yr^{-1}, together with column density ratios relative to the smooth outflow for specific one-component outflows. The corresponding abundance profiles are shown in Fig. 2.

Species	NH_3			HCN			CS		
Smooth outflow	1.8e+12			4.2e+10			4.2e+12		
f_{vol}	0.15	0.30	0.45	0.15	0.30	0.45	0.15	0.30	0.45
$l_* = 5 \times 10^{12}$ cm	3.3e+02	4.2e+01	1.2e+01	2.1e+02	3.6e+01	1.3e+01	5.3e+02	6.2e+01	1.7e+01
$l_* = 1 \times 10^{13}$ cm	3.2e+02	4.2e+01	1.2e+01	2.4e+02	3.9e+01	1.3e+01	5.3e+02	6.2e+01	1.7e+01
$l_* = 5 \times 10^{13}$ cm	1.2e+03	4.9e+01	1.2e+01	4.5e+04	1.4e+02	2.3e+01	8.4e+03	2.7e+02	3.5e+01

3. Effect on chemistry

In Fig. 2, we show the effect of clumpiness on the abundance profiles of NH_3, HCN, and CS in an O-rich outflow with a mass-loss rate of 10^{-5} $M_\odot$ yr^{-1}. All material of the outflow is located inside of clumps, the inter-clump medium is void ($f_{ic} = 0$). The central star has a radius of 5×10^{13} cm. Our model is calculated from 2 R_* onwards.

It is clear that both the overdensity of the clumps and the porosity-associated decrease in visual extinction have an effect on the chemistry. For smaller values of f_{vol}, the overdensity of the clumps is larger, resulting in an overall larger abundance of the species. For larger values of h, more interstellar UV photons are able to reach the inner wind and release the deficient element through photodissociation. This results in the production of HCN and NH_3 closer to the central star. Using the porosity formalism, we are therefore able to produce both HCN and NH_3 in the inner region of the outflow.

A clumpy outflow not only influences the abundances in the inner region, but also affects the chemistry throughout the outflow. This can be seen in the abundance profile of CS: clumpiness of the outflow clearly alters its formation and destruction pathways in the intermediate wind, resulting in a different shape of the abundance profile.

The change in column density relative to a smooth outflow is given in Table 1. We find that the column density increases with decreasing f_{vol} for all three species. The relative increase in column density is larger for smaller values of f_{vol}: for $f_{vol} = 0.15$, the column density can be up to two orders of magnitude larger relative to a smooth outflow. The size of the clumps l_*, and therefore the porosity length h, affects the column density only when the UV field throughout the outflow is large enough to influence the chemistry and increase the abundance in the inner region. In that case, the column density increases by an order of magnitude relative to a smooth outflow for the same value of f_{vol}.

In a forthcoming paper, we will additionally describe the effect of a clumpy outflow on its chemistry for a two-component medium and for several mass-loss rates, and this for both an O- and C-rich chemistry.

References

Agúndez, M., Cernicharo, J., & Guélin, M. 2010, *ApJ*, 724, L133
Agúndez, M., Cernicharo, J., Quintana-Lacaci, G. *et al.* 2017, *A&A*, 601, A4
Bujarrabal, V., Fuente, A., & Omont, A. 1994, *A&A*, 285, 247
Cherchneff, I. 2006, *A&A*, 456, 1001
Decin, L., Agúndez, M., Barlow, M. J. *et al.* 2010, *A&A*, 467, 64
Duari, D., Cherchneff, I., & Willacy, K. 1999, *A&A*, 341, L47
Gobrecht, D., Cherchneff, I., Sarangi, A. *et al.* 2016, *A&A*, 585, A6
Justtanont, K., de Jong, T., Helmich, F. P. *et al.* 1996, *A&A*, 315, L127
Keady, J. J. & Ridgway, S. T. 1993, *A&A*, 564, A88
Kervella, P., Montargès, M., Ridgway, S. T. *et al.* 2014, *ApJ*, 406, 199
Khouri, T., Maercker, M., Waters, L. B. F. M. *et al.* 2016, *A&A*, 591, A70
Maercker, M., Mohamed, S., Vlemmings, W. H. T. *et al.* 1995, *Nature*, 490, 232
McElroy, D., Walsh, C., Markwick, A. J. *et al.* 2013, *A&A*, 550, A36
Menten, K. M. & Alcolea, J. 1995, *ApJ*, 448, 416
Omont, A., Lucas, R., Morris, M., & Guilloteau, S. 1993, *A&A*, 267, 490
Owocki, S. P., Gayley, K. G., & Shaviv, N. J. 2004, *ApJ*, 616, 525
Owocki, S. P. & Cohen, D. H. 2006, *ApJ*, 648, 565
Sundqvist, J. O., Owocki, S. P., Cohen, D. H. *et al.* 2012, *MNRAS*, 420, 1553
Sundqvist, J. O., Puls, J., & Owocki, S. P. 2014, *A&A*, 568, A59

Astrochemistry VII – Through the Cosmos from Galaxies to Planets
Proceedings IAU Symposium No. 332, 2017
M. Cunningham, T. Millar & Y. Aikawa, eds.

doi:10.1017/S1743921317006858

Modeling CO, CO_2 and H_2O Ice Abundances in the Envelopes of Young Stellar Objects in the Magellanic Clouds

Tyler Pauly[1] **and Robin T. Garrod**[2]

[1]Cornell Center for Astrophysics and Planetary Science and Department of Astronomy, Cornell University, Ithaca, NY 14853, USA
email: tap74@cornell.edu

[2]Departments of Chemistry and Astronomy , University of Virginia, Charlottesville, VA 22904-4319, USA

Abstract. Massive young stellar objects (MYSOs) in the Magellanic Clouds (MCs) show infrared absorption features corresponding to significant abundances of CO, CO_2 and H_2O ice along the line of sight, with the relative abundances of these ices varying between sources in the Magellanic Clouds and the Milky Way. We use our gas-grain chemical code MAGICKAL, with multiple grain sizes and grain temperatures, and further expand it with a treatment for increased interstellar radiation field intensity to model the elevated dust temperatures observed in the MCs. We also adjust the elemental abundances used in the chemical models, guided by observations of HII regions in these metal-poor satellite galaxies. With a grid of models, we are able to reproduce the relative ice fractions observed in MC MYSOs, indicating that metal depletion and elevated grain temperature are important drivers of the MYSO envelope ice composition. The observed shortfall in CO in the Small Magellanic Cloud can be explained by a combination of reduced carbon abundance and increased grain temperatures. The models indicate that a large variation in radiation field strength is required to match the range of observed LMC abundances.

Keywords. astrochemistry — ISM: abundances — ISM: molecules — Magellanic Clouds

1. Introduction

Mid-infrared spectral observations of embedded young stellar objects (YSOs) in the Milky Way (MW) and the nearby Magellanic Clouds (MCs) have found a wealth of solid-state features, showing high column densities of ices such as H_2O, CO, CO_2, and CH_3OH. H_2O is the most abundant ice, with a typical column density of $\sim 10^{-4}$ with respect to total hydrogen abundance. In MW YSOs, CO_2 is next most abundant, at an average value of CO_2:H_2O $\simeq 0.2$ Boogert & Ehrenfreund (2004). CO and CH_3OH ices follow at lower abundance, though with nearly an order of magnitude of variation between lines of sight (Gerakines *et al.* 1999; Gibb *et al.* 2004). In contrast, observations of MYSOs in the nearby Magellanic Clouds show a marked difference in ice abundances. Shimonishi *et al.* (2010) and Oliveira *et al.* (2011) detected H_2O, CO and CO_2 ice in MYSOs in the Large Magellanic Cloud (LMC); they found that LMC sources had elevated CO_2 ice or depleted H_2O ice when compared to galactic counterparts, with an average value for CO_2:H_2O of 0.32. Oliveira *et al.* (2011, 2013) found only an upper limit for CO ice in all Small Magellanic Cloud (SMC) sources studied, with abundances (with respect to their H_2O columns) a factor of three to ten lower than their galactic counterparts. Oliveira *et al.* (2011) and Shimonishi *et al.* (2016) provided additional near-infrared spectra of

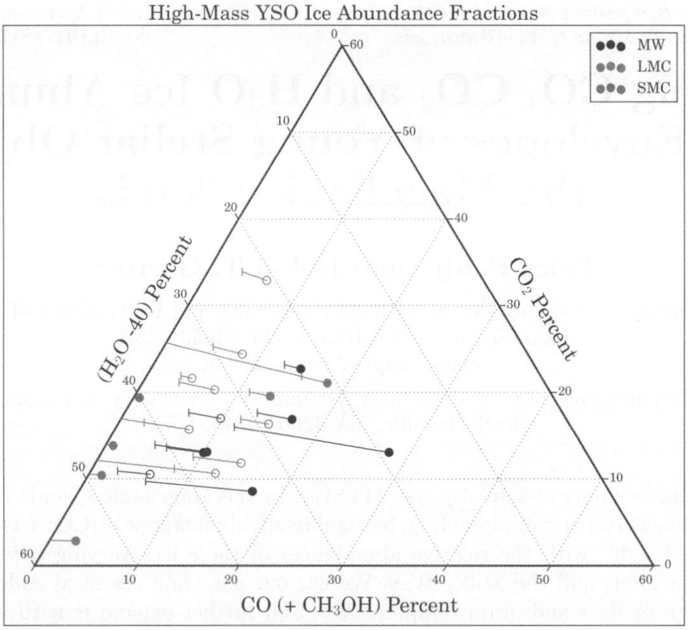

Figure 1. The relative abundances of the four major ices in massive young stellar objects in the Milky Way (black), LMC (red), and SMC (blue). SMC sources have only upper limits on CO and no information on CH_3OH; SMC points show the composition at the upper limit value with a line drawn to zero CO abundance. For LMC and Milky Way sources, the vertical tick shows the H_2O:CO:CO_2 composition, while the circles show the composition including CH_3OH ice as H_2O:(CO+CH_3OH):CO_2. For LMC sources with only an upper limit on CO, lines have been drawn to zero CO abundance, while sources with an upper limit on CH_3OH ice have an open circle. (Ternary figure style from Harper *et al.* 2015)

a sample of LMC MYSOs, with detections or upper limits for CH_3OH ice towards all sources studied.

These ices are found in the dense, cold envelopes surrounding the highly luminous central source, and they hold information on their past collapse history via e.g. the polar to apolar ratio of the CO and CO_2 ice features (Gibb *et al.* 2000). They are processed to some extent by the internal radiation source, though the source-to-source variation of this processing is expected to be large. Local environment also plays a role, with changes in the nearby interstellar radiation field (ISRF) or the cosmic ray ionization rate affecting gas and grain surface chemistry. Additionally, variations in the underlying elemental abundances of the collapsing cloud will influence the general chemistry and ice column densities.

We collate observations of MYSOs for which H_2O, CO, CO_2 and CH_3OH detections or upper limits are available, excluding CH_3OH for SMC sources (toward which no measurements of CH_3OH have yet been achieved). Figure 1 shows the total observation sample we will use for model comparison, plotted in a ternary H_2O:CO_2:CO ice diagram. The ternary plot describes the relative abundances of this three-component ice system; importantly, we also consider methanol (CH_3OH) ice, a key component for galactic YSOs and now detected toward some LMC MYSOs. To include this fourth component on a ternary diagram, we include a second point for those sources with methanol detections or upper limits; these points show the fractional abundance of H_2O:CO_2:(CO+CH_3OH). This pairing choice of (CO+CH_3OH) is chemically motivated, as CH_3OH is primarily formed from the successive hydrogenation of CO on grain surfaces (Watanabe & Kouchi

2002; Cuppen *et al.* 2009). The figure shows a transition in composition between environments, with some blending between some LMC and galactic sources.

Using the single-point free-fall collapse model detailed by Garrod & Pauly (2011) and Pauly & Garrod (2016), we investigate parameters responsible for the chemical variation amongst MYSOs in the galaxy, LMC and SMC. We take the elemental abundances and dust temperatures of the forming YSO to be the parameters of interest for this model study.

2. Model Methods

We use the gas-grain chemical code MAGICKAL and its associated chemical network, first presented by Garrod (2013) and updated by Pauly & Garrod (2016) to include a grain-size distribution consisting of five grains. The model features 475 gas-phase species and 200 grain surface species with a network of roughly 9000 reactions and processes. Grain surface species are tracked in two separate phases, surface and mantle; the surface species participate in desorption, reaction and diffusion across grain sites, while the ice mantle is treated as a separate phase that is coupled to the surface. Bulk diffusion in the mantle ice is treated explicitly, allowing reactions within the mantle, as well as exchange between surface and mantle components; however, for the low temperatures involved in this work we treat the mantle phase as inert except for the transfer of surface material into the bulk, as the mantle grows. The model uses the modified-rate approach detailed in Garrod (2008) (method "C") to account for possible stochastic effects in the surface chemistry. The chemical network also includes photodissociation and photoionization processes, with photons sourced either from the ambient field or the cosmic ray-induced UV field.

We model the evolution of dust temperature as a function of the visual extinction and dust radius, following methods outlined in Garrod & Pauly (2011). We add an additional variable in a model-dependent interstellar radiation field (ISRF). We approximate the ISRF in various environments by modifying the multi-component fit from Zucconi, Walmsley, & Galli (2001) for the Milky Way. The fit includes contributions from three discrete stellar black-body populations, both hot and cool diffuse dust components, and the cosmic microwave background. To simulate variation in ISRF intensity in the Magellanic Clouds, we scale the stellar components uniformly, from the base factor of 1.0 to 3.0 in increments of 0.5. The resulting dust temperatures shift by 3-5 K at $A_V \sim 1$ and 1.5 K at $A_V \sim 10$, with the smallest grains having the highest temperature. The increase in A_V during the collapse process results in a general cooling and a flattening of the temperature distribution with respect to grain size; the temperature difference between the smallest and largest grains shrinks from 5 K to <1 K.

To model the ISM of the metal-poor galaxies, we deplete the heavy elemental abundances in the initial setup of our models. The Magellanic Clouds have bulk metallicity of $Z_{LMC} \sim 0.4Z_{\odot}$ and $Z_{SMC} \sim 0.2Z_{\odot}$ (Russell & Dopita 1992). Kurt & Dufour (1998) collated observations of six SMC HII regions to find the mean abundances of carbon, nitrogen and oxygen. Peimbert (2003) collected a UV-visible spectrum of 30 Doradus in the LMC; with 269 identified emission lines, they calculate the total abundance of carbon, nitrogen and oxygen. These two studies provide us with ISM compositions to model the metal-depleted environments associated with star formation in the Magellanic Clouds. The abundance values are shown in Table 1; these abundances will be referred to as MW, LMC, and SMC.

Element	MW	LMC	SMC
H	5.000(-5)	5.000(-5)	5.000(-5)
H_2	0.499975	0.499975	0.499975
O	3.200(-4)	2.140(-4)	1.047(-4)
C^+	1.400(-4)	6.310(-5)	1.585(-5)
N	7.500(-5)	1.12(-5)	2.820(-6)
C/O Ratio	0.438	0.295	0.151

Table 1. Elemental abundances, listed with respect to total hydrogen number density, n_H. The first column represents galactic abundances, taken from Garrod & Pauly (2011). The second column is used as representative values for the LMC, taken from Peimbert (2003). The final column is the most depleted abundances considered and are taken from the sample of SMC HII regions in Kurt & Dufour (1998).

3. Results & Discussion

We computed a grid of fifteen models by multiplying the stellar component of the ISRF with values of [1.0, 1.5, 2.0, 2.5, 3.0] and varying the elemental abundances between initial abundance setups MW, LMC, and SMC. H_2O is the dominant ice component in nearly all models as expected, following observations. The collapse is complete by $\sim 8 \times 10^5$ ($10^{5.9}$) years; for models with 1.0, 1.5 or 2.0 stellar intensity, this causes dust temperatures to drop below an efficiency threshold for producing CO_2 from CO + OH, identified by Garrod & Pauly (2011). CO mobility on the grain surface is sufficiently slowed at temperatures below $\sim$ 12 K; by this point, the fractional abundance of CO grows above that of CO_2. Models with 2.5 or 3.0 stellar intensity never drop below this temperature threshold, and as a result high CO_2 ice abundances are found throughout those models.

CH_3OH ice is formed via the hydrogenation of surface CO, which is only present after temperatures drop below the 12 K threshold. For MW models with abundant CO ice, the efficiency of CH_3OH formation appears low, with the abundance ratio of CH_3OH:CO ranging from 1:2 to 1:5. However, in SMC models with low CO surface abundance, surface CH_3OH can be equal in abundance to CO, and these molecules are similarly abundant throughout those model runs.

The hydrides CH_4 and NH_3 appear to track closely the elemental abundance of their atomic parent, with some dependence on temperature shown for models with galactic elemental abundances.

Figure 3 shows the model results overlaid on the observations. For each model we plot two points per panel; for the left panel, the smaller leftward point shows the ice composition at $8{\times}10^5$ years, while the large rightward point shows the composition at 10^6 years. The right panel shows the same models at times of 10^6 years and $5{\times}10^6$ years. The choice of time spent at post-collapse density is arbitrary, and because physical conditions do not change significantly during this time, the composition follows a roughly straight line between these points. The MW and SMC abundance models appear to match observations more closely at earlier times, while the LMC observations have variation such that model matches are found at both early and late times.

Models with MW abundances show compositions enriched in CO_2 and [CO+CH_3OH], with the relative enrichment between the two set by the stellar flux parameter. As no observations lie near the high stellar flux galactic abundance models, these models do not appear to represent any observed MYSO and can be ignored.

LMC MYSOs demonstrate large variation in composition, from extreme [CO+CH_3OH] depletion (as in SMC sources) to highly enriched in CO_2. Models with LMC-like elemental abundances fall across the ensemble of LMC MYSOs, though there is considerable overlap

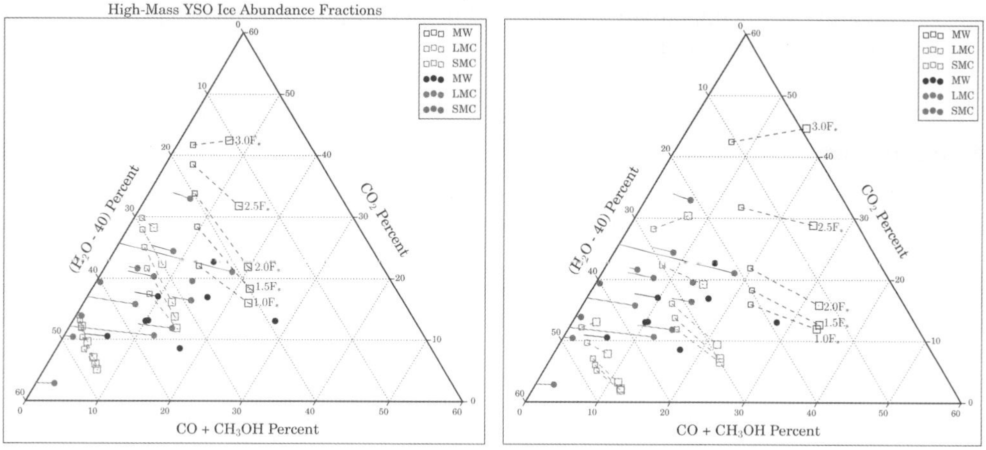

Figure 2. These panels overlay the model results on the observations from Figure 1. With matching colors for model elemental abundances to source environment (MW, LMC, SMC), models with increasing stellar flux parameter move upwards on the plot, with galactic chemical abundance models labeled. For each model, two points are plotted; for the top panel, the smaller leftward square shows the ice composition at the time of collapse completion ($\sim 8\times10^5$ years), while the larger rightward square shows the composition at 10^6 years. For the lower panel, the leftward square shows the composition at 10^6 years, while the rightward square shows the values for 5×10^6 years. (Ternary figure style from Harper *et al.* 2015)

between LMC and galactic sources. The models separate cleanly on this plot, implying additional effects not addressed in the model setup. Variation in local metallicity may cause blending, as metal-poor MW YSOs may appear chemically similar to LMC MYSOs. Parameters beyond our model, such as variation in collapse speed or ice processing may also play a role. Models are able to fit LMC observations at the full range of stellar flux parameter tested.

Models with the most depleted elemental abundances fall near the observed SMC MYSOs, matching the low (undetected) CO abundance and presence of CO_2. The models lying closest to observed YSO abundances have high stellar flux values, though the models cannot fully reproduce the spread in CO_2 abundances and typically overproduce [CO+CH_3OH]. Of note, the composition of SMC models shows a roughly equal abundance of surface CO and CH_3OH, while observational upper limits exist only for CO. Tightening the abundance constraints on these two species would provide strong evidence for the validity of our [CO+CH_3OH] surface chemistry.

The increased CH_3OH abundance relative to CO in SMC models is an unexpected result. CH_3OH formation requires CO surface residence times to be longer than the mantle deposition time scale, to allow sufficient time for hydrogenation. In this way, the balance between CO and CH_3OH is determined primarily by the accretion rate of elements heavier than hydrogen. With long CO surface residence times, CO_2 production will also be increased if dust temperatures are above the necessary threshold for CO surface mobility.

The models produce a reasonable fit to observations, though a general trend exists in overproduction of [CO+CH_3OH]. This may not be a simple model issue but instead a comparison of model results to observations in different physical regimes. These MYSOs are highly luminous objects, and thermal processing of the envelope is likely to have occurred in many sources. In this case, the most volatile ices may be under-abundant due to evaporation when compared to the final model output, which ends prior to a grain heating and ice evaporation phase.

4. Summary

Our results suggest that gas-grain models of cold cloud collapse can produce ice mantle abundances that match reasonably well to observations in a variety of environments. We conclude that:

- The values of ISRF intensity and elemental abundances chosen provide an adequate distribution of ice abundances that cover the observed ice abundances in YSOs. Models with strongly enhanced ISRF intensity at MW elemental abundances are excluded, while SMC models with enhanced ISRF are preferred.
- LMC models lie near observed YSOs for every value of the ISRF intensity modeled, characterizing the large spread in LMC YSO ice abundances. This may be indicative of large local fluctuations in the LMC ISRF.
- The ISRF intensity strongly affects the relative abundance of CO_2 to CO/CH_3OH, with higher ISRF values leading to CO_2 enhancement. This is caused by a temperature threshold for CO mobility on grain surfaces, leading to efficient production of CO_2 at dust temperatures $\gtrsim 12$ K.
- Increasing model elemental abundances (and corresponding C/O ratio) decreases the H_2O abundance against the other ices; this is evidenced by model values moving parallel to the H_2O ternary axis with changes in elemental abundance.
- Our models indicate that the lack of CO in SMC sources is most likely caused by a combination of low elemental abundances and high ISRF intensity.
- CH_3OH abundance is found to be enhanced in low-metallicity environments. This will be important for complex organic molecule production in LMC and SMC hot core models.

We leave some issues to be addressed in future work. Thermal processing of the ice is important for matching observed ice abundances, and it is not included in these models. We find significant growth in the [dust+mantle] radius, which affects both the dust temperature and surface chemistry; however, we assume a Q_{abs} of carbonaceous dust for temperature calculations, though the Q_{abs} of ice will differ. We also use a grain size distribution found for silicate grains; this could be resolved by using values for silicate or carbonaceous grains throughout, or by attempting to model both populations.

Future models could investigate the dependence on cosmic ray ionization rate, a parameter with large variation across the LMC. The rate of collapse may also be important, as it sets the heavy atom accretion rate. Follow-up models will address behavior in collapse to higher densities ($\sim 10^7$ cm^{-3}), including a warm-up phase for comparison to a newly detected hot core in the LMC (Shimonishi *et al.* 2016).

References

Boogert, A. C. A. & Ehrenfreund, P. 2004, *ASPC*, 309, 547
Cuppen, H. M., van Dishoeck, E. F., Herbst, E., & Tielens, A. G. G. M. 2009, *A&A*, 508, 275
Garrod, R. T. 2008, *A&A*, 491, 239
Garrod, R. T. & Pauly, T. 2011, *ApJ*, 735, 15
Garrod, R. T. 2013, *ApJ*, 765, 60
Gerakines, P. A., Whittet, D. C. B., Ehrenfreund, P., Boogert, A. C. A., Tielens, A. G. G. M., Schutte, W. A., Chiar, J. E., van Dishoeck, E. F., Prusti, T., Helmich, F. P., & de Graauw, T. 1999, *ApJ*, 522, 357

Gibb, E. L., Whittet, D. C. B., Schutte, W. A., Boogert, A. C. A., Chiar, J. E., Ehrenfreund, P., Gerakines, P. A., Keane, J. V., Tielens, A. G. G. M., van Dishoeck, E. F., & Kerkhof, O. 2000, *ApJ*, 536, 347

Gibb, E. L., Whittet, D. C. B., Boogert, A. C. A., & Tielens, A. G. G. M. 2004, *ApJS*, 151, 35

Harper, M., Weinstein, B., Simon, C., chebee7 i, Swanson-Hysell, N., The Gitter Badger, Greco, M., & Zuidhof, G. 2015, *doi* = 10.5281/zenodo.34938

Kurt, C. M. & Dufour, R. J. 1998, *RMxAC*, 7, 202

Oliveira, J. M., van Loon, J. T., Sloan, G. C., Indebetouw, R., Kemper, F., Tielens, A. G. G. M., Simon, J. D., Woods, P. M., & Meixner, M. 2011, *MNRAS*, 411, 36

Oliveira, J. M., van Loon, J. T., Sloan, G. C., Sewiło, M., Kraemer, K. E., Wood, P. R., Indebetouw, R., Filipović, M. D., Crawford, E. J., Wong, G. F., Hora, J. L., Meixner, M., Robitaille, T. P., Shiao, B., & Simon, J. D. 2013, *MNRAS*, 428, 3001

Pauly, T. & Garrod, R. T. 2016, *ApJ* 817, 146

Peimbert, A. 2003, *ApJ* 584, 735

Russell, S. C. & Dopita, M. A. 1992, *ApJ* 384, 508

Shimonishi, T., Onaka, T., Kato, D., Sakon, I., Ita, Y., Kawamura, A., & Kaneda, H. 2010, *A&A*, 514, 12

Shimonishi, T., Dartois, E., Onaka, T., & Boulanger, F. 2016, *A&A* 585 107

Shimonishi, T., Onaka, T., Kawamura, A., & Aikawa, Y. 2016, *ApJ* 827, 72

Watanabe, N. & Kouchi, A. 2002, *ApJL* 571, 173

Zucconi, A., Walmsley, C. M., & Galli, D. 2001, *A&A* 376, 650

Chemistry in disks and planets

Astrochemistry VII – Through the Cosmos from Galaxies to Planets
Proceedings IAU Symposium No. 332, 2017
M. Cunningham, T. Millar & Y. Aikawa, eds.

doi:10.1017/S1743921317009541

Zooming in on the Chemistry of Protoplanetary Disks with ALMA

L. Ilsedore Cleeves

Harvard-Smithsonian Center for Astrophysics
60 Garden St MS-51
Cambridge, MA 02138, USA
email: ilse.cleeves@cfa.harvard.edu

Abstract. During the first few ~Myr of a young stars life, it is encircled by a disk made up of molecular gas, dust, and ice – the building blocks for future planetary systems. How/when these disks form planets and what sets the planets initial compositions remain key outstanding questions in disk science. In recent years, major leaps in sensitivity and spatial resolution afforded by the Atacama Large Millimeter/Submillimeter Array (ALMA) have revolutionized our understanding of protoplanetary disks chemical composition and physical properties, revealing in some cases complex radial, vertical, and azimuthal structure in the dust and gas. In this contribution, I review recent observational results and new theoretical puzzles, and how these fit into a newly emerging picture of the disk environment.

Keywords. planetary systems: protoplanetary disks, astrochemistry

1. Introduction

Protoplanetary disks are complex systems of gas, dust, and ice, that over the course of millions of years assemble new planetary systems. Understanding their compositions can thus inform our understanding of the initial composition of "typical" planetesimals. In the six years between the previous astrochemistry symposium in 2011, IAUS 280: "The Molecular Universe" in Toledo, Spain, and the present symposium, the 2017 IAUS 332: "Through the Cosmos from Galaxies to Planets" in Puerto Varas, Chile, we have gone from the pre-ALMA to the post-ALMA era. ALMA has brought order(s) of magnitude improvement in spatial resolution and in sensitivity. And though ALMA has not (yet) substantially increased the observed molecular inventory in disks (as of the meeting, only adding $\sim 25\%$ to our known list of radio-detected molecules); instead, ALMA's sensitivity has enabled us i) to make high spatial and spectral resolution maps with unprecedented signal to noise and ii) to measure the rarer isotopologues that peer deeper into gas where the main isotopologues are thick.

Of the most striking results out of ALMA is the the prevalence of ring-like structure. Some of the most famous rings are those seen in the continuum as either fine-ringed structure (e.g., ALMA Partnership *et al.* 2015; Andrews *et al.* 2016) or large inner gaps/asymmetries (e.g., van der Marel *et al.* 2013). However, the chemistry has offered perhaps even more complex puzzles regarding the varied distribution of emission between molecules for the same disk, and between different disks for the same molecule.

Understanding the nature of this highly structured molecular emission is a highly active area of research. Generally interpretations fall into three broad camps, which include i) changes in the "true" abundances of molecules with distance from the star and/or height, ii) changes in the excitation conditions (temperatures/densities), and iii) deficits in bulk H_2 gas within cleared out "rings" (e.g., Isella *et al.* 2016; Teague *et al.* 2017). Moreover many of these interpretations can operate even within the same disk. Thus observing

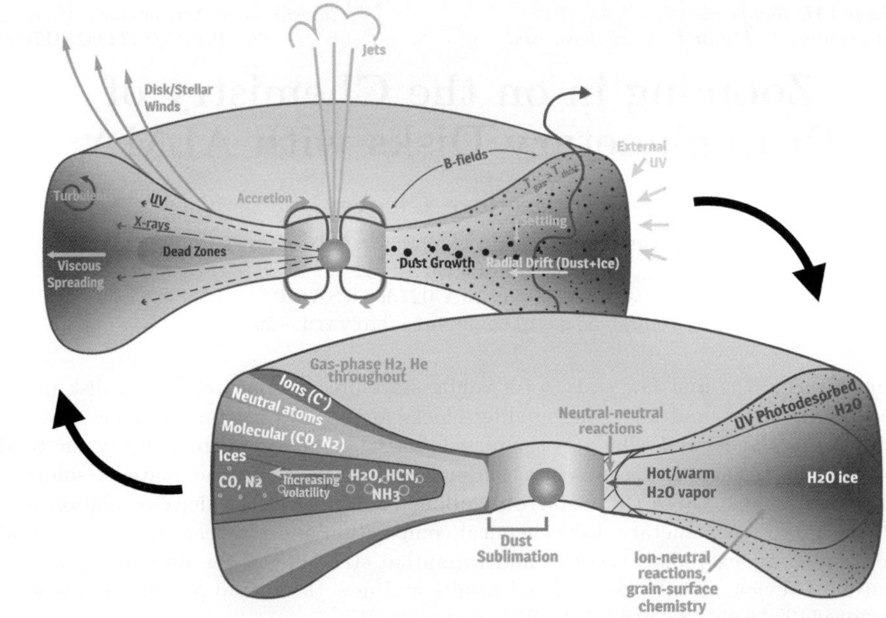

Figure 1. Cartoon illustrating the complex, co-existing chemical and physical processes that are thought to occur in protoplanetary disks. From Cleeves (2015, PhD thesis).

many molecules at relatively high ($\gtrsim 0.3''$) resolution is necessary to disentangle, for example, scenario i) vs. iii).

To begin interpreting these ring-like structures, we can utilize our knowledge of the sensitive links between the local chemical composition and the physical conditions of the gas and dust (e.g., Bergin *et al.* 2007). At the most basic level, the temperature of the gas and dust determine where molecules predominantly reside in the gas or ice phase. In the midplane, the radial position at which this transition occurs is called the snow line (see Sections 2.1 and 2.2). The gas density of the disk regulates the rate at which reactions can occur, but also provides a source of shielding that can shadow the deeper layers of the disk from radiation. The dust itself provides a substrate to facilitate the formation of water and complex organics in the ice-phase chemistry; however, it too grows and evolves via coagulation and fragmentation over the lifetime of the disk (Testi *et al.* 2014), impacting the chemistry in recently discovered ways (see Sections 2.2 and 3). Simultaneously, much of the grain-surface chemistry and broader gas-phase chemistry occurring in cold gas is sensitive to the ionization state of the disk, which can come from many sources (e.g., Glassgold *et al.* 1997; Cleeves *et al.* 2013, 2014), and thus understanding the interaction of the disk gas with the varied ionization sources is critical for understanding the active chemical state of the gas and its history (see Section 4).

The present contribution was designed to give an overview and an update of the new results coming out of ALMA since the last astrochemistry IAU symposium. It is not designed to be a detailed review, but rather an attempt to link together various clues that have been coming out over the last six years into just one possible broader narrative. This narrative will undoubtedly change, and has only continued to evolve over the short months between the IAUS 332 meeting in March and these proceedings. Here I attempt to focus on what my contribution covered at time of meeting, but have incorporated new information where necessary.

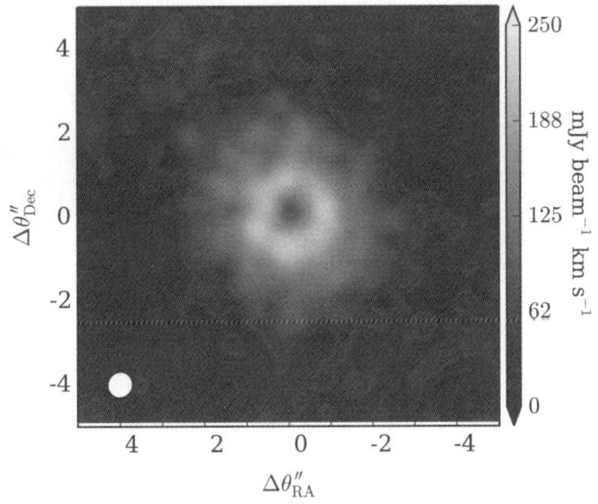

Figure 2. Observed distribution of N_2H^+ $J = 4-3$ in TW Hya. Data from Qi *et al.* (2013).

2. A Prevalence of Molecular Patterns

2.1. *N_2H^+ and its relationship with CO*

Perhaps one of the most notable chemical rings is that of N_2H^+ seen in the $J = 4-3$ rotational transition in the TW Hya disk reported in Qi *et al.* (2013). The N_2H^+ shows a clear ring-like distribution (see Figure 2) with an inner radius of $\sim$ 30 AU. The chemistry of N_2H^+ is such that it is rapidly destroyed in the presence of CO. As a consequence, N_2H^+ has been historically used as a marker of CO freeze-out in the dense interstellar medium (Charnley 1997; Bergin *et al.* 2002). Thus the Qi *et al.* (2013) paper interpreted the ring of N_2H^+ as a marker of the TW Hya disk's CO snow line, which was also in reasonable agreement with model temperature estimates of the disk at these radii (Qi *et al.* 2013, 2015).

More recently, measurements with ALMA of trace species in Schwarz *et al.* (2016) found a steep drop off of CO intensity at $\sim 17-23$ AU, interior to the N_2H^+ ring. These results were more recently supported by $^{13}C^{18}O$ observations reported in Zhang *et al.* (2017), which place the break at a radius of 20.5 ± 1.3 AU. Both papers attribute the steep drop off CO isotopologue emission at this location also to the CO snow line. This spatial discrepancy between the N_2H^+ transition and the CO transition is in part due to the inescapable nature of the disk temperature gradients, which are not purely radial. Instead, the vertical increase in temperature with height in the disk (see Figure 3) causes the region of CO freeze-out to occupy a wedge, with N_2H^+ in the simplest case bounding its borders until N_2 freeze-out commences. More detailed modeling of the chemistry (e.g., Aikawa *et al.* 2015; van't Hoff *et al.* 2017) shows that the distribution of N_2H^+ is complicated by additional factors, including the desorption rates of CO vs. N_2, CO abundance, and disk ionization.

The face-on nature of the disk of the TW Hya complicates the analysis further. Future work on different, more inclined disks, coupled with models, or additional constraints on the radial temperature profile and gas surface density will hopefully shed additional light on the nature of the observed radial transitions and their relationship with CO freeze-out.

2.2. *Double ring CO emission*

New sensitive observations of CO isotopologues have also revealed non-monotonic brightness distributions with distance from the star. More specifically, a secondary ring of CO isotopologue has now been seen in multiple disks (e.g., Huang *et al.* 2016; Nomura *et al.*

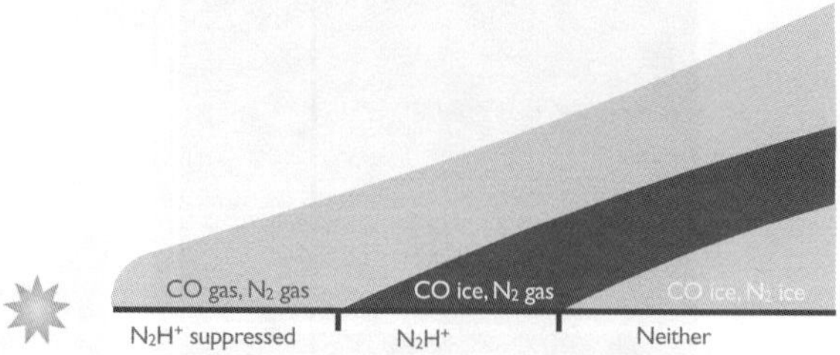

Figure 3. Schematic of the "classical" picture of the relationship between N_2H^+ and the CO snow line. N_2 is expected to stay in the gas down to lower temperatures than CO, leading to a layer between the region of CO freeze out and N_2 freeze out where N_2H^+ is expected to occupy. Additional factors beyond these that impact the N_2H^+ distribution are described in Aikawa *et al.* (2015) and van't Hoff *et al.* (2017).

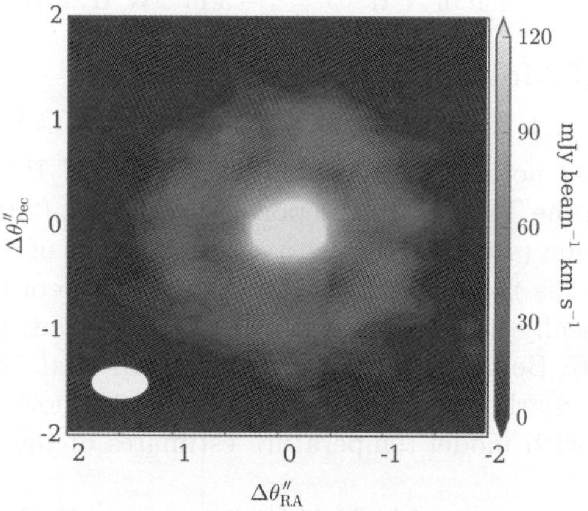

Figure 4. Observed distribution of $C^{18}O$ $J = 3-2$ in TW Hya, color stretched to highlight the outer ring. Data from Schwarz *et al.* (2016).

2016; Schwarz *et al.* 2016), see Figure 4 for the case of TW Hya from Schwarz *et al.* (2016). The $C^{18}O$ $J = 3-2$ emission is centrally peaked, steeply drops off beyond ~ 20 AU, but then rises again at around $\sim 50-60$ AU. This brightening of CO corresponds to the edge of the millimeter emission, which sharply drops off at this point (i.e., the edge is not a "sensitivity" edge but a true deficit of grains, see, e.g., Andrews *et al.* 2012; Hogerheijde *et al.* 2016). Essentially the millimeter emission traces the location of "large" grains (those that have undergone growth). For comparison, the small grains as traced by scattered light extend out to ~ 200 AU (Trilling *et al.* 2001), which is similar to the radius of the gas as traced by ^{12}CO (Andrews *et al.* 2012).

The re-emergence of CO at these distant radii provides insight into the conditions in the gas locally, namely either there is a temperature transition enabling thermal desorption (essentially a secondary CO snow line) or an opacity transition that allows external (or scattered internal) UV photons to directly photodesorb CO ices.

Cleeves (2016) modeled the disk temperature and UV opacity structure for models that include a truncated distribution of large grains with a continuous distribution of entrained small grains following the gas (see Figure 5). It was found that when millimeter grains are compact, the dust outside of this region can be directly heated by the reprocessed radiation originating from the upper warm layers of the disk. This same reduction in dust opacity also permits UV radiation to reach the outer disk midplane. Which of these two

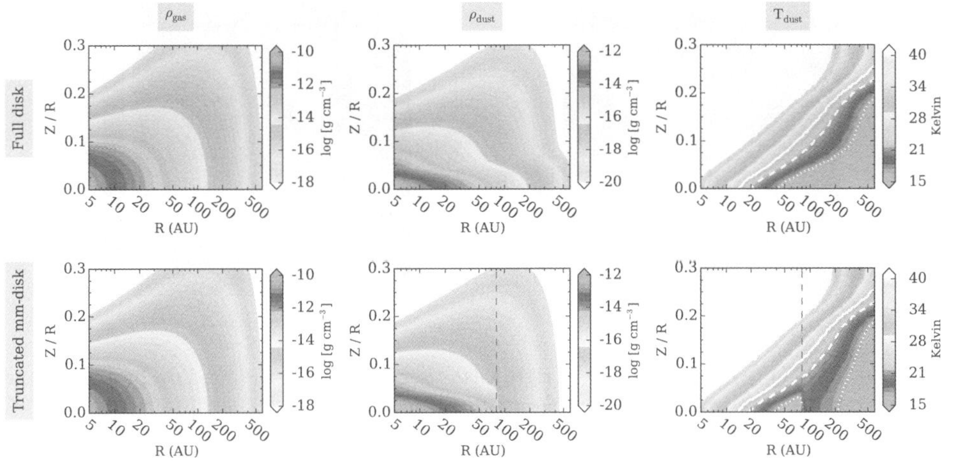

Figure 5. Gas density (left), dust density (center) and temperature (right) for models without and with a sharp radial transition in the millimeter dust disk (top and bottom respectively). Reproduced with permission from Cleeves (2016).

processes dominate depends on the size of the millimeter grain disk, e.g., more compact disks tend to favor efficient thermal desorption, while extended disks do not become sufficiently warm and non-thermal mechanisms are expected to dominate.

2.3. *Deuterated species*

In the pre-ALMA era, observational studies of isotope fractionation were only possible on disk-averaged scales, and often using optically thick (bright) species. It is now feasible to image disks on scales corresponding to fractions of an arcsecond (or < 50 AU for disks at the distance of Taurus). ALMA is beginning to map nitrogen isotopic fractionation (Guzmán *et al.* 2015, 2017; Hily-Blant *et al.* 2017) and deuterium fractionation (Huang *et al.* 2017; Salinas *et al.* 2017).

Rings are again commonly seen, and perhaps even more striking is that often the distribution of the less abundant species does not follow that of the more abundant species. As one example, Figure 6 shows velocity integrated emission from $H^{13}CO^+$ and DCO^+ toward the IM Lup protoplanetary disk (observations from Öberg *et al.* 2015a). The $H^{13}CO^+$ shows a single broad ring, while the DCO^+ exhibits two narrow rings, almost in diametric opposition to the $H^{13}CO^+$. In the pre-ALMA era, disk integrated fluxes were often used for isotope ratios, but as can be seen here, how those values translate to real isotopic ratios is not always straightforward.

Generally, deuterium fractionation is expected to be a temperature driven process; lower temperatures tend to favor reactions that incorporate the heavy species over the light ones (e.g., review by Millar 2003). If one visually traces out D/H with distance from the star from Figure 6, it is clear that the profile transitions from a high D/H, to a low D/H in between the DCO^+ ring, finally returning back to a high value. This could be a non-monotonic radial temperature effect, or instead it could point to multiple fractionation pathways becoming dominant under different conditions (see Huang *et al.* 2017).

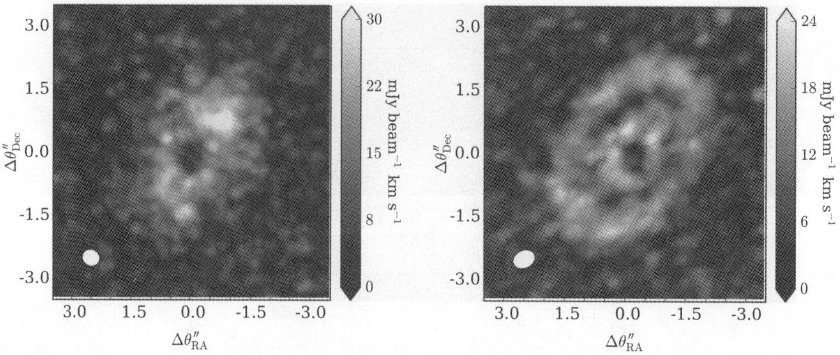

Figure 6. $H^{13}CO^+$ (left) and DCO^+ (right) as observed toward the IM Lup protoplanetary disk. Data from Öberg *et al.* (2015a).

3. Volatile Depletion

3.1. *Missing oxygen and carbon?*

The *Herschel* Space Observatory gave early indications that the volatile reservoir of oxygen in disks may be reduced relative to interstellar gas. More specifically water was found to be under abundant in disks by one to two orders of magnitude (Bergin *et al.* 2010; Hogerheijde *et al.* 2011; Du *et al.* 2017) though with some uncertainty in the specific values due to uncertainty in the chemistry and radiative transfer (Kamp *et al.* 2013). Gas mass measurements from atomic oxygen yielded masses far lower than that expected from the dust for a gas to dust mass ratio of $\sim$ 100 (Thi *et al.* 2010). This pointed to two possible scenarios, a true gas depletion (H_2) or a depletion of oxygen in the observable layers. Part of the solution came from *Herschel* observations of HD, which pointed to a H_2 gas-rich disks more consistent with measurements from the dust (Bergin *et al.* 2013). It was later found with ALMA that CO was likewise depleted (Schwarz *et al.* 2016), supporting earlier measurements from disk averaged SMA observations (Favre *et al.* 2013; Cleeves *et al.* 2015). Furthermore, Schwarz *et al.* (2016) found that the CO abundance remained low even within a 20 AU radius of the star, the region where thermal desorption should return CO to the gas if it is present. Observations of atomic carbon with APEX also supported that the carbon itself was missing, rather than CO being preferentially dissociated (Kama *et al.* 2016).

McClure *et al.* (2016) reported *Herschel* HD measurements for six more disks, enabling additional volatile abundance determinations. They compared CO depletion factors for those disks whose trace CO isotopologues have been measured, and found that the younger disk, DM Tau, had relatively mild CO depletion, up to a factor of five relative to interstellar. Comparing these values, the depletion in DM Tau's CO is far more mild than that in the older disk TW Hya; however, both disks have similarly high degrees of water depletion (Bergin *et al.* 2010; Hogerheijde *et al.* 2011). This may hint at the timescales for sequestration of oxygen versus carbon, but the sample size is still too small to make definitive statements.

3.2. *Hydrocarbons*

One surprising discovery regarding the disk molecular inventory was the overall brightness of hydrocarbons, specifically C_2H and $c{-}C_3H_2$. C_2H was first detected and imaged with the SMA in TW Hya, and was found to occupy a bright single ring of emission (Kastner *et al.* 2015). ALMA imaged and confirmed the ring-like geometry of C_2H in TW Hya, and found the ring nests outside of the millimeter thermal dust edge (Bergin *et al.* 2016).

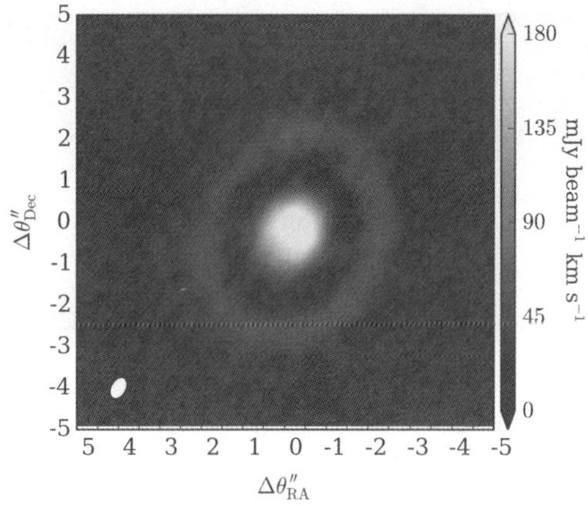

Figure 7. C_2H $N = 3 - 2$ (integrated over all hyperfine components) as observed in the DM Tau protoplanetary disk. The center is saturated by the color scheme in order to enhance the visibility of the faint outer ring. Data from Bergin *et al.* (2016).

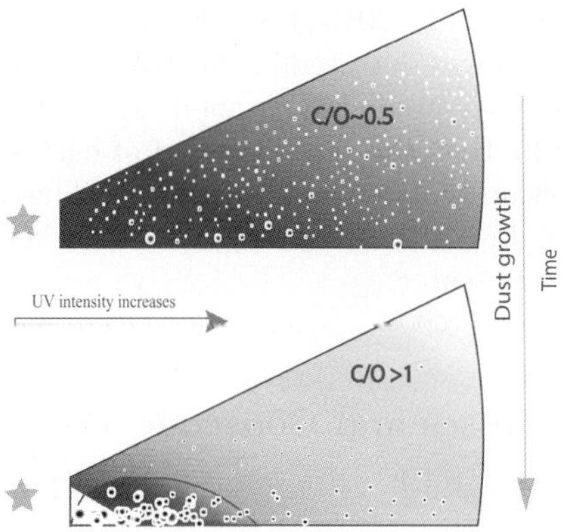

Figure 8. Evolution of the C/O ratio with time by dust growth, settling, and dust radial drift. Oxygen is preferentially sequestered as water ice on cold grains. Figure credit Ke Zhang.

The same work also imaged the DM Tau C_2H distribution, which was found to occupy not one but two separate rings (see Figure 7). The inner ring appears just outside of the brightest peak of dust emission, suggesting possible dust extinction effects impacting the observable line emission, i.e., a "false" hole. The second ring, however, nests outside of DM Tau's disk of millimeter emission. Thus there appears to be some relationship between the hydrocarbon emission and a lack of large grains.

Additional chemical modeling of C_2H found that the brightness of line required large column densities ($\gtrsim 10^{14}$ cm^{-2}). To reach the necessary abundances required a severe depletion of oxygen (i.e., C/O $\gg 1$; Bergin *et al.* 2016). Correspondingly, the brightness of C_2H naturally fits with the picture of a lack of oxygen (water) relative to carbon as described in Section 3. Thus the relationship with the dust grain edge may suggest that the water/oxygen mass has been sequestered into the larger grains closer to the star, leaving a region of high C/O in the outer disk and upper layers (see Figure 8).

3.3. *Organics*

The sensitivity improvements of ALMA also enabled the first detection of organics in a protoplanetary disk. Perhaps more surprising, however, was that the first organics found

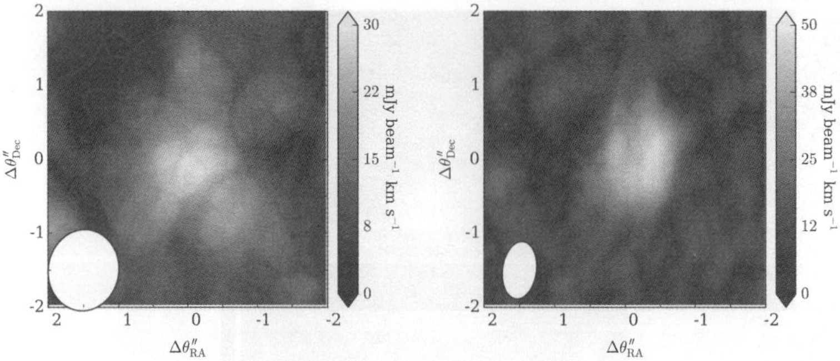

Figure 9. Methyl cyanide (left) and cyanoacetylene (right) observed toward the MWC 480 disk with ALMA. Data are from Öberg *et al.* (2015b).

were not the "usual suspects" as were expected from observations of protostars, e.g., methanol (e.g., Herbst & van Dishoeck 2009, and references therein). Instead, the first complex species observed in disks were methyl cyanide (CH_3CN) and cyanoacetylene (HC_3N) as reported in Öberg *et al.* (2015b), see Figure 9.

Shortly thereafter, methanol was detected in the TW Hya disk (Walsh *et al.* 2016), but it's abundance was strongly over-predicted by models. As methanol forms by hydrogen addition to CO ice, it may be that the sequestration of carbon and/or oxygen from CO starves off the local production of observable methanol. This does not imply that disks are globally methanol poor, but rather the upper layers where methanol comes off the grain show a deficit, similar to water.

4. Disk Energetic Processes and Consequences

4.1. *Disk ionization*

Ionization plays an important role in both the physical properties and the volatile composition of protoplanetary disks. High energy photons/particles with energies exceeding $E > 13.6$ eV are able to act on the bulk H_2 and helium gas, impacting e.g., disk heating and coupling efficiency to the magnetic fields. Chemically, ionization impacts both gas and grain abundances in disks (Cleeves *et al.* 2014; Eistrup *et al.* 2016). At low temperatures ($\lesssim 100$ K) gas-phase chemistry proceeds through ion-neutral reactions. Simultaneously, much of the rich interstellar grain-surface chemistry requires a source of atomic hydrogen, where ionization of H_2 is a key production term.

The key sources of high energy ionization include X-rays, cosmic rays, and short-lived radionuclides. X-rays from the central star are obstructed by the inner disk except for those at very high energies (> 5 keV). Cosmic rays are thought to be the most important ionization source for the X-ray-shielded outer disk; however, in the presence of stellar/disk winds or magnetic fields, CRs are easily deflected (Turner & Drake 2009; Cleeves *et al.* 2013). Radionuclides have uncertain abundances depending on the local history of star formation, especially massive stars, and how these materials are delivered to the disk (e.g., Adams 2010).

In Cleeves *et al.* (2014), we identified that observations of HCO^+, $H^{13}CO^+$ and N_2H^+ can be used in concert to measure the ionization rate and its sources, in particular X-rays versus cosmic rays. We applied these techniques to observations of the well-constrained TW Hya disk with the SMA and ALMA to measure its disk-averaged ionization rate (Cleeves *et al.* 2015). We found low CR models (a factor of $\gtrsim 100$ reduction) systemati-

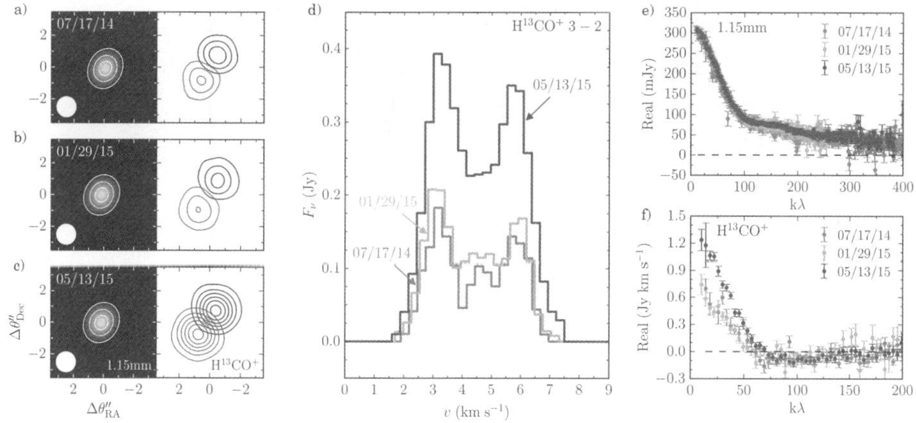

Figure 10. $H^{13}CO^+$ $J = 3 - 2$ toward the IM Lup protoplanetary disk as measured at three different epochs with ALMA. The velocity integrated line flux doubles relative to a constant continuum flux, which is visible in the image (left), spectral (middle), and visibility (right) domains. Figure taken with permission from Cleeves *et al.* (2017).

cally fit the data better. Thus the outer disk was far more ionization poor than expected by conventional physical models.

4.2. *Time dependent ionization chemistry?*

To extend upon our study of TW Hya, we obtained spatially resolved data to study the ionization properties of the IM Lup protoplanetary disk. However, upon inspection we discovered that the flux of one of the key molecular ions, $H^{13}CO^+$, was not constant in time over the year the ALMA observations were carried out (see Figure 10). Initial models presented in Cleeves *et al.* (2017) show that variability in the incident X-ray flux (as expected from X-ray flares from young stars), can cause short-term chemical changes in the atmospheric abundances of the disk. The impact on $H^{13}CO^+$ is lessened in the shielded parts of the disk where densities are high (quenching the enhancement in the ionization fraction) and the X-ray fluxes are low.

While these short-term events may not impact the bulk chemistry in the densest parts of the disk, time-domain observations of molecular ions in disk may provide new insights into the ionization fraction in the upper layers of the disk where the disappearance time of the effect is directly related to the local electron density.

4.3. *Gas kinematics and turbulence*

The low global ionization rates, and particularly the lack of evidence for cosmic ray ionization in the outer disk from Cleeves *et al.* (2015), is expected to reduce the efficiency of coupling between the gas and disk magnetic fields. Such coupling is necessary for the magnetorotational instability (MRI; Balbus & Hawley 1991) to operate as a source of viscosity for disk accretion.

MRI is expected to generate additional non-thermal motions that should be detectable with high spectral resolution observations (Simon *et al.* 2015). Flaherty *et al.* (2015) carried out parameterized modeling to isolate non-thermal motions in the HD163296 disk. They find an upper limit for line broadening due to turbulence of $< 3\%$ of the local sound speed. These limits are an order of magnitude lower than theory would predict, suggesting that MRI is less efficient beyond 30 AU than previously considered. Recent work on the 3D structure of turbulence by Flaherty *et al.* (2017) support this picture, and add that turbulence is low throughout the vertical extent of the disk. Note

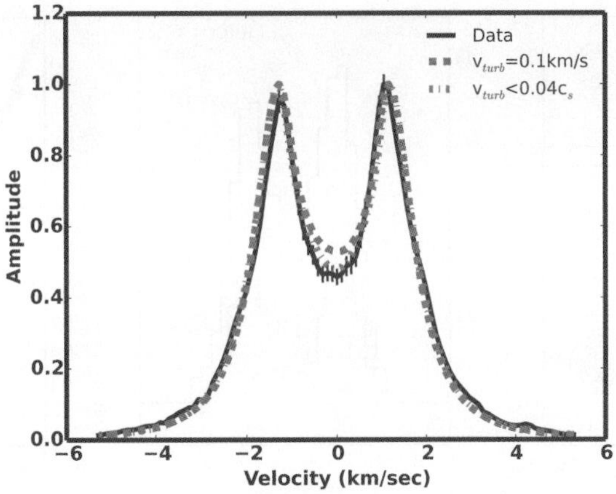

Figure 11. Normalized disk integrated spectra of CO $J = 3-2$ toward HD163296. The data are fit better by models with little to no turbulence (red dashed line) than with turbulence typical of MRI (blue dashed line). Reproduced with permission from Flaherty *et al.* (2015).

additional measurements by Teague *et al.* (2016) who use optically thick lines to constrain temperature find a larger value for turbulent broadening; however, they state that even these should be considered upper limits given the systematics in the data.

5. An emerging understanding and implications

As each result comes out, we can begin to assemble a broader picture of the protoplanetary disk physical and chemical environment over the first few Myr. Early observations of depleted oxygen and carbon from the SMA, ALMA, and *Herschel* pointed to rapid sequestration of volatiles potentially in the form of ices on larger grains. These findings imply that vertical mixing cannot be very efficient at dredging up material in the outer disk (beyond 10 AU), otherwise the upper layers would be replenished in their observable volatile content. This picture is consistent with measured low ionization rates in TW Hya, which can reduce the efficiency of MRI driven turbulence, likewise explaining the strong upper limits to non-thermal motions observed.

As such, a consistent picture is emerging between the physical nature of the disk and its chemical content. A less vigorously mixed disk tends to be less chemically altered, which may act to preserve more of the initial composition of the disk at formation. However it should be emphasized that the results described in this contribution come from disks spanning a wide variety of ages, physical environments, and stellar masses. This "emerging picture" is only just the beginning and will undoubtedly be continuously revised with new data. Furthermore, we are still at a phase where we do not yet fully understand the "massive" disks toward which we are observationally biased. Future chemical surveys in the coming years, especially those targeting lower mass disks, will hopefully tell us whether the intuition we are building towards the massive end holds in general for planet formation across the disk/stellar mass spectrum.

References

Adams, F. C. 2010, *ARA&A*, 48, 47
Aikawa, Y., Furuya, K., Nomura, H., & Qi, C. 2015, *ApJ*, 807, 120
ALMA Partnership, Brogan, C. L., Pérez, L. M., Hunter, T. R., *et al.* 2015, *ApJL*, 808, L3
Andrews, S. M., Wilner, D. J., Hughes, A. M., *et al.* 2012, *ApJ*, 744, 162

Andrews, S. M., Wilner, D. J., Zhu, Z., Birnstiel, T., Carpenter, J. M., Pérez, L. M., Bai, X.-N., Öberg, K. I., Hughes, A. M., Isella, A., & Ricci, L. 2016, *ApJL*, 820, L40

Balbus, S. A. & Hawley, J. F. 1991, *ApJ*, 376, 214

Bergin, E. A., Aikawa, Y., Blake, G. A., & van Dishoeck, E. F. 2007, *Protostars and Planets V*, 751

Bergin, E. A., Alves, J., Huard, T., & Lada, C. J. 2002, *ApJL*, 570, L101

Bergin, E. A., Cleeves, L. I., Gorti, U., Zhang, K., Blake, G. A., Green, J. D., Andrews, S. M., Evans, II, N. J., Henning, T., Öberg, K., Pontoppidan, K., Qi, C., Salyk, C., & van Dishoeck, E. F. 2013, *Nature*, 493, 644

Bergin, E. A., Du, F., Cleeves, L. I., Blake, G. A., Schwarz, K., Visser, R., & Zhang, K. 2016, *ApJ*, 831, 101

Bergin, E. A., Hogerheijde, M. R., Brinch, C., Fogel, J., Yıldız, U. A., Kristensen, L. E., van Dishoeck, E. F., Bell, T. A. *et al.* 2010, *A&A*, 521, L33

Charnley, S. B. 1997, *MNRAS*, 291, 455

Cleeves, L. I. 2015, *PhD thesis*, University of Michigan

—. 2016, *ApJL*, 816, L21

Cleeves, L. I., Adams, F. C., & Bergin, E. A. 2013, *ApJ*, 772, 5

Cleeves, L. I., Bergin, E. A., & Adams, F. C. 2014, *ApJ*, 794, 123

Cleeves, L. I., Bergin, E. A., Öberg, K. I., Andrews, S., Wilner, D., & Loomis, R. 2017, *ApJL*, 843, L3

Cleeves, L. I., Bergin, E. A., Qi, C., Adams, F. C., & Öberg, K. I. 2015, *ApJ*, 799, 204

Du, F., Bergin, E. A., Hogerheijde, M., van Dishoeck, E. F., Blake, G., Bruderer, S., Cleeves, I., Dominik, C., Fedele, D., Lis, D. C., Melnick, G., Neufeld, D., Pearson, J., & Yıldız, U. 2017, *ApJ*, 842, 98

Eistrup, C., Walsh, C., & van Dishoeck, E. F. 2016, *A&A*, 595, A83

Favre, C., Cleeves, L. I., Bergin, E. A., Qi, C., & Blake, G. A. 2013, *ApJL*, 776, L38

Flaherty, K. M., Hughes, A. M., Rose, S. C., Simon, J. B., Qi, C., Andrews, S. M., Kóspál, Á., Wilner, D. J., Chiang, E., Armitage, P. J., & Bai, X.-n. 2017, *ApJ*, 843, 150

Flaherty, K. M., Hughes, A. M., Rosenfeld, K. A., Andrews, S. M., Chiang, E., Simon, J. B., Kerzner, S., & Wilner, D. J. 2015, *ApJ*, 813, 99

Glassgold, A. E., Najita, J., & Igea, J. 1997, *ApJ*, 480, 344

Guzmán, V. V., Öberg, K. I., Huang, J., Loomis, R., & Qi, C. 2017, *ApJ*, 836, 30

Guzmán, V. V., Öberg, K. I., Loomis, R., & Qi, C. 2015, *ApJ*, 814, 53

Herbst, E. & van Dishoeck, E. F. 2009, *ARA&A*, 47, 427

Hily-Blant, P., Magalhaes, V., Kastner, J., Faure, A., Forveille, T., & Qi, C. 2017, *A&A*, 603, L6

Hogerheijde, M. R., Bekkers, D., Pinilla, P., Salinas, V. N., Kama, M., Andrews, S. M., Qi, C., & Wilner, D. J. 2016, *A&A*, 586, A99

Hogerheijde, M. R., Bergin, E. A., Brinch, C., *et al.* 2011, *Science*, 334, 338

Huang, J., Öberg, K. I., & Andrews, S. M. 2016, *ApJL*, 823, L18

Huang, J., Öberg, K. I., Qi, C., Aikawa, Y., Andrews, S. M., Furuya, K., Guzmán, V. V., Loomis, R. A., van Dishoeck, E. F., & Wilner, D. J. 2017, *ApJ*, 835, 231

Isella, A., Guidi, G., Testi, L., Liu, S., Li, H., Li, S., Weaver, E., Boehler, Y., Carperter, J. M., De Gregorio-Monsalvo, I., Manara, C. F., Natta, A., Pérez, L. M., Ricci, L., Sargent, A., Tazzari, M., & Turner, N. 2016, *Physical Review Letters*, 117, 251101

Kama, M., Bruderer, S., Carney, M., *et al.* 2016, *A&A*, 588, A108

Kamp, I., Thi, W.-F., Meeus, G., Woitke, P., Pinte, C., Meijerink, R., Spaans, M., Pascucci, I., Aresu, G., & Dent, W. R. F. 2013, *A&A*, 559, A24

Kastner, J. H., Qi, C., Gorti, U., Hily-Blant, P., Oberg, K., Forveille, T., Andrews, S., & Wilner, D. 2015, *ApJ*, 806, 75

McClure, M. K., Bergin, E. A., Cleeves, L. I., van Dishoeck, E. F., Blake, G. A., Evans, II, N. J., Green, J. D., Henning, T., Öberg, K. I., Pontoppidan, K. M., & Salyk, C. 2016, *ApJ*, 831, 167

Millar, T. J. 2003, *Space Sci. Rev.*, 106, 73

Nomura, H., Tsukagoshi, T., Kawabe, R., Ishimoto, D., Okuzumi, S., Muto, T., Kanagawa, K. D., Ida, S., Walsh, C., Millar, T. J., & Bai, X.-N. 2016, *ApJL*, 819, L7

Öberg, K. I., Furuya, K., Loomis, R., Aikawa, Y., Andrews, S. M., Qi, C., van Dishoeck, E. F., & Wilner, D. J. 2015a, *ApJ*, 810, 112

Öberg, K. I., Guzmán, V. V., Furuya, K., Qi, C., Aikawa, Y., Andrews, S. M., Loomis, R., & Wilner, D. J. 2015b, *Nature*, 520, 198

Qi, C., Öberg, K. I., Andrews, S. M., Wilner, D. J., Bergin, E. A., Hughes, A. M., Hogherheijde, M., & D'Alessio, P. 2015, *ApJ*, 813, 128

Qi, C., Öberg, K. I., Wilner, D. J., D'Alessio, P., Bergin, E., Andrews, S. M., Blake, G. A., Hogerheijde, M. R., & van Dishoeck, E. F. 2013, *Science*, 341, 630

Salinas, V. N., Hogerheijde, M. R., Mathews, G. S., Öberg, K. I., Qi, C., Williams, J. P., & Wilner, D. J. 2017, ArXiv e-prints

Schwarz, K. R., Bergin, E. A., Cleeves, L. I., Blake, G. A., Zhang, K., Öberg, K. I., van Dishoeck, E. F., & Qi, C. 2016, *ApJ*, 823, 91

Simon, J. B., Hughes, A. M., Flaherty, K. M., Bai, X.-N., & Armitage, P. J. 2015, *ApJ*, 808, 180

Teague, R., Guilloteau, S., Semenov, D., Henning, T., Dutrey, A., Piétu, V., Birnstiel, T., Chapillon, E., Hollenbach, D., & Gorti, U. 2016, *A&A*, 592, A49

Teague, R., Semenov, D., Gorti, U., Guilloteau, S., Henning, T., Birnstiel, T., Dutrey, A., van Boekel, R., & Chapillon, E. 2017, *ApJ*, 835, 228

Testi, L., Birnstiel, T., Ricci, L., Andrews, S., Blum, J., Carpenter, J., Dominik, C., Isella, A., Natta, A., Williams, J. P., & Wilner, D. J. 2014, *Protostars and Planets VI*, 339

Thi, W.-F., Mathews, G., Ménard, *et al.* 2010, *A&A*, 518, L125

Trilling, D. E., Koerner, D. W., Barnes, J. W., Ftaclas, C., & Brown, R. H. 2001, *ApJL*, 552, L151

Turner, N. J. & Drake, J. F. 2009, *ApJ*, 703, 2152

van der Marel, N., van Dishoeck, E. F., Bruderer, S., *et al.* 2013, *Science*, 340, 1199

van't Hoff, M. L. R., Walsh, C., Kama, M., Facchini, S., & van Dishoeck, E. F. 2017, *A&A*, 599, A101

Walsh, C., Loomis, R. A., Öberg, K. I., Kama, M., van 't Hoff, M. L. R., Millar, T. J., Aikawa, Y., Herbst, E., Widicus Weaver, S. L., & Nomura, H. 2016, *ApJL*, 823, L10

Zhang, K., Bergin, E. A., Blake, G. A., Cleeves, L. I., & Schwarz, K. R. 2017, *Nature Astronomy*, 1, 0130

Astrochemistry VII – Through the Cosmos from Galaxies to Planets
Proceedings IAU Symposium No. 332, 2017
M. Cunningham, T. Millar & Y. Aikawa, eds.

doi:10.1017/S1743921317009802

Chemical evolution in planet-forming regions. Impact on volatile abundances and C/O ratios of planet-building material

Christian Eistrup[1], Catherine Walsh[1,2] and Ewine F. van Dishoeck[1,3]

[1]Leiden Observatory, Niels Bohrweg 2, 2333RA Leiden, Netherlands
email: eistrup@strw.leidenuniv.nl

[2]School of Physics and Astronomy, E C Stoner Building, University of Leeds, Leeds, UK
[3]Max-Planck-Institut für extraterrestrische Physik, P.O. Box 1312, D-85741, Garching, Germany

Abstract. Connecting the observed composition of exoplanets to their formation sites often involves comparing the atmospheric C/O ratio to a disk midplane model with a fixed chemical composition. In this scenario chemistry during the planet formation era is not considered. However, kinetic chemical evolution during the lifetime of the gaseous disk can change the relative abundances of volatile species, thus altering the C/O ratios of planetary building blocks. In our chemical evolition models we utilize a large network of gas-phase, grain-surface and gas-grain interaction reactions, thus providing a comprehensive treatment of chemistry. The results show that, if sufficient ionisation is present, then chemistry does alter the C/O ratios of gas and ice during the epoch of planet(esimal) formation. This modifies the picture of C/O ratios in disk midplanes defined simply by volatile ice lines in a midplane of fixed chemical composition. Chemical evolution thus needs to be addressed when predicting the makeup of planets and their atmospheres.

Keywords. Protoplanetary disks, Planet formation, Kinetic chemistry, Chemical composition

1. Introduction

Since the first detection of an exoplanet 22 years ago, the field of exoplanetary science has moved from detection to also characterisation. Several studies show detections of molecular absorption in the transmission spectra of exoplanet atmostpheres (Seager & Deming 2010;Snellen *et al.* 2010;Birkby *et al.* 2013;Fraine *et al.* 2014;Crossfield 2015;Sing *et al.* 2016). Atmospheric abundances of molecules such as CO and H_2O are used to infer the elemental C/O ratio of the atmospheres, which in turn points to the elemental composition of the material that went into forming the planets, in the protoplanetary disk midplanes. Thereby, a connection between observed planets and the disk midplanes where they formed can be established, shining light on planet formation.

Such a connection is, however, not straightforward to make. Several different processes take place during the evolution of the planet-forming midplanes of disks, which could all affect how, when, and from what material planets form. Such processes include dynamical effects such as grain settling from upper layers to midplane, grain growth, grain radial drift and accumulation of grains into larger bodies, and eventually into planets. These effects have been studied extensively (see e.g. Lambrechts & Johansen 2012;Alibert *et al.* 2013;Ali-Dib *et al.* 2014;Madhusudhan *et al.* 2014;Bitsch & Johansen,and references therein) and the chemical composition of the material have been considered in order to track which chemical abundances the planets are made up of.

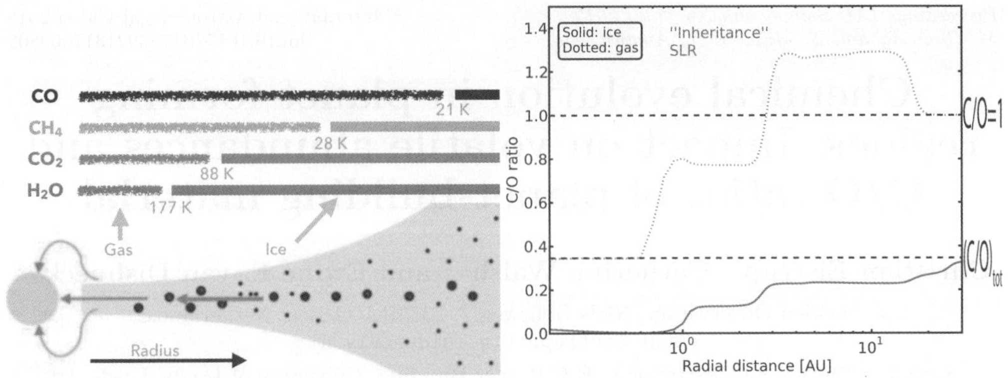

Figure 1. Left panel: modified disk illustration from Williams & Cieza (2011), showing the regions of the disk midplane (indicated by the radius) where four selected volatile species are gas or ice, with freeze-out temperatures depending on the species' binding energies. Right panel: the C/O elemental ratios of gas and ice as function of radius from Eistrup *et al.* (2016), as only defined by icelines. Dashed lines indicate C/O=1 as well as the total assumed C/O ratio of 0.34. Comparable to Fig. 1 in Öberg *et al.* (2011).

The volatile composition of the planet-forming material is usually taken to be fixed. That is, certain abundances of volatiles species such as H_2O, CO, CO_2 and CH_4 have been assumed, and depending on their binding energies (freeze-out temperatures), these species are assumed to be either in the gas-phase, or ices on grain surfaces, depending on the radial temperature profile in the disk midplane. The iceline of a species is the radius marking the transition from gas to ice for that species (see Fig. 1, left panel), and based on icelines an impression of the C/O elemental ratios in gas and ice of the midplane material can be obtained, as shown in Fig. 1, right panel.

However, most models of planet formation do not allow chemical evolution of the material. That is, chemical reactions that can alter relative abundances of the chemical species, or even produce new species, are not included in the models. This work focuses on chemical evolution in the disk midplane, and how this evolution might modify the view of a chemically static midplane with C/O ratios in gas and ice defined by icelines only.

2. Allowing chemical evolution in the disk midplane

We have used the physical structure for a disk midplane (temperature, density and ionisation level, as functions of radius) that is evolving in time from Alibert *et al.* 2013, i.e. the disk is cooling and losing mass over time. Temperature and ionisation levels at different timesteps are featured in Fig. 2. The initial disk mass is 0.1 Minimum Mass Solar Nebula (see Eistrup *et al.* 2016), and we considered radii out to 30 AU. We utilised an extensive gas-phase, gas-grain and grain-surface chemical network. The gas-phase chemistry is from the latest release of the UMIST Database for Astrochemistry (McElroy *et al.* 2013) termed RATE12. The rates for gas-grain interactions and grain-surface chemistry are calculated as described in Walsh *et al.* (2015), and references therein.

We assumed a constant gas-to-dust ratio of 100, as well as spherical grains of constant size 0.1 μm. The initial chemical abundances are assumed those from the "inheritance" scenario in Table 1 in Eistrup *et al.* (2016). A high level of ionisation (including the contribution from short-lived radionuclides and cosmic rays) is adopted. We evolve the midplane chemistry for up to 7 Myr.

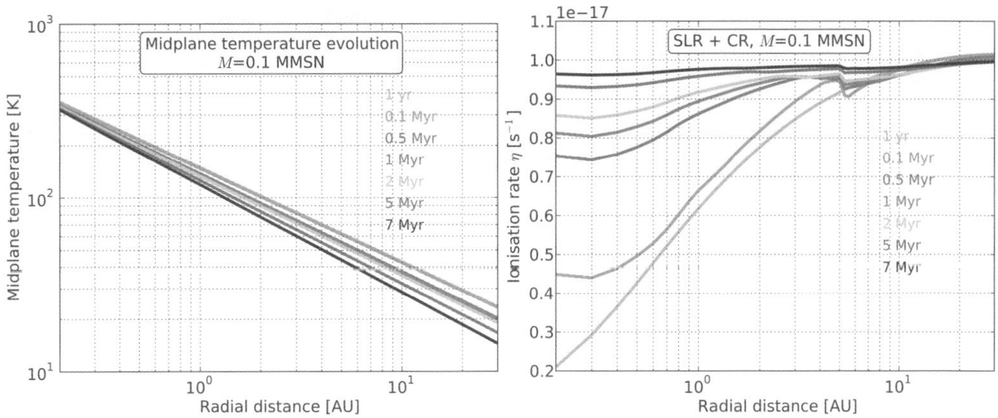

Figure 2. Midplane temperature (left) and ionisation rate (right) as functions of disk radius for different evolutionary times.

3. Effects of chemical evolution

For the conditions assumed here, in particular the high level of ionisation, large changes to the initial chemical composition are seen at all radii. Chemical evolution becomes important after a few times 10^5 yrs (which is evident at 5 AU radius as seen in Fig. 3, left panel), and continues until at least 7 Myr. By 7 Myr, chemical steady state is reached in the outer icy parts of the disk midplane (outside the O_2 iceline). This means that the fingerprint of the initial chemical abundances has been erased.

Chemical evolution changes the relative abundances of the initial species, but it also causes production of new species. Most prominently, O_2 gas is abundantly produced between the H_2O and the O_2 icelines, and HCN and C_2H_6 ices becomes important carriers of elemental carbon in the cold outer disk midplane. The changes to the chemical composition also affect the elemental carbon and oxygen ratios in gas and ice. This evolution is featured in Fig. 3, left panel. Here, the orange profile by 0.1 Myr is identical to the plot in Fig. 1 right panel, as the C/O ratio without chemical evolution, defined by icelines alone.

Fig. 3 right panel shows that chemical evolution acts to significantly lower the gas C/O ratio, so it drops from ~1.2 initially to 0.3 by 7 Myr at 5 AU. Alongside that, the C/O ratio in the ice at 5 AU increases to 0.5 by 7 Myr, making the ice between 3 and 10 AU more carbon rich than the gas (opposite of the initially condition, where the gas was more carbon rich than the ice). Overall, by 7 Myr the C/O ratios in both gas and ice are below 0.5 (given that the total assumed C/O ratio is 0.34), showing that with chemical evolution included, C/O ratios may change significantly over the disk lifetime.

The results presented here suggest that coupling the chemical makeup of a disk midplane to the chemical makeup of the planets that form out of it may not be straightforward. Indeed, if planet formation lasts for longer than a few times 10^5 yrs in the disk midplane and a high level of ionisation is present, then the chemical abundances and elemental ratios in gas and ice will change alongside with planet formation. Henceforth, chemical evolution needs to be addressed when predicting the chemical makeup of planets and their atmospheres, as well as when inferring formation locations of exoplanets based on their observed atmospheric elemental ratios.

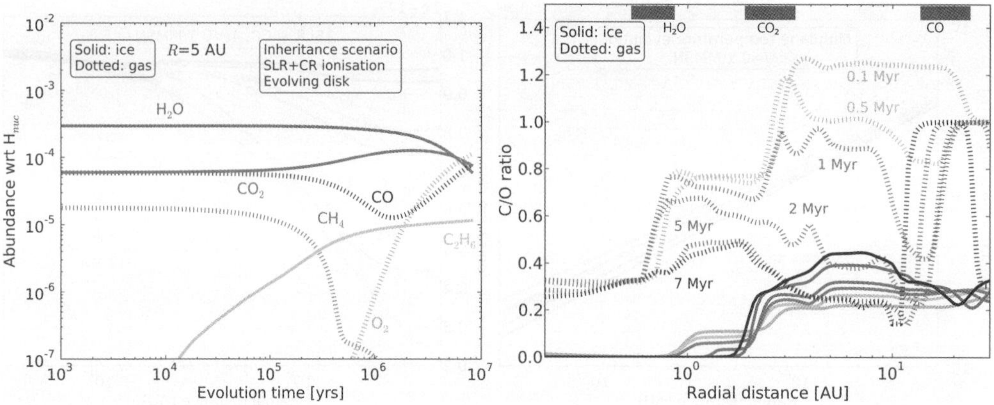

Figure 3. Left panel: time evolution of key volatile species at 5 AU in the disk midplane. Right panel: evolution of the elemental C/O ratios in gas and ice caused by the evolution in the chemistry, at different timesteps. The grey boxes in the right panel indicate the radial regions that the iceline of a volatile move over due to the evolution of the physical conditions.

References

Alibert, Y., Carron, F., Fortier, A., *et al.* 2013, *A&A*, 558, A109
Ali-Dib, M., Mousis, O., Petit, J.-M., *et al.* 2014, *ApJ*, 785, 125
Birkby, J. L., de Kok, R. J., Brogi, M., *et al.* 2013, *MNRAS*, 436, L35
Bitsch, B. & Johansen, A. 2016, *A&A*, 590, A101
Crossfield, I. J. M. 2015, *PASP*, 127, 941
Eistrup, C., Walsh, C., & van Dishoeck, E. F. 2016, *A&A*, 595, A83
Fraine, J., Deming, D., Benneke, B., *et al.* 2014, *Nature*, 513, 526
Walsh, C., Nomura, H., & van Dishoeck, E. 2015, *A&A*, 582, A88
Williams, J. P. & Cieza, L. A. 2011, *ARA&A*, 49, 67
Öberg, K. I., Murray-Clay, R., & Bergin, E. A. 2011, *ApJL*, 743, L16
Lambrechts, M. & Johansen, A. 2012, *A&A*, 544, A32
Madhusudhan, N., Amin, M. A., Kennedy, G. M. 2014, *ApJL*, 794, L12
McElroy, D., Walsh, C., Markwick, A. J., *et al.* 2013, *A&A*, 550, A36
Seager, S. & Deming, D. 2010, *AR&A*, 48, 631
Sing, D. K., Fortney, J. J., Nikolov, N., *et al.* 2016, *Nature*, 529, 59
Snellen, I. A. G., de Kok, R. J., de Mooij, , *et al.* 2010, *Nature*, 465, 1049

Astrochemistry VII – Through the Cosmos from Galaxies to Planets
Proceedings IAU Symposium No. 332, 2017
M. Cunningham, T. Millar & Y. Aikawa, eds.

doi:10.1017/S1743921317007591

Chemical Change in the Disk Forming Region of IRAS 16293–2422 Studied with ALMA

Yoko Oya

Department of Physics, The University of Tokyo,
Bunkyo-ku, Tokyo, 113-0033, Japan
email: oya@taurus.phys.s.u-tokyo.ac.jp

Abstract. We have analyzed rotational spectral line emission of OCS, CH_3OH, $HCOOCH_3$, and H_2CS observed toward the low-mass Class 0 protostar IRAS 16293–2422 (Source A and B) at a sub-arcsecond resolution with ALMA. Significant chemical differentiation is found at a 50 au scale. OCS is found to trace the infalling-rotating envelope, while COM distributions are concentrated around the inner part of the envelope. The kinematic structure in Source A is explained with a ballistic model, and the protostellar mass and the radius of the centrifugal barrier are evaluated to be $\sim 0.75\ M_\odot$ and ~ 50 au, respectively. This study has revealed that the centrifugal barrier plays a central role not only in the disk formation but also in the associated chemical evolution.

Keywords. ALMA, IRAS 16293–2422, Molecules, Disk formation

1. Introduction

1.1. *Chemical Diversity of Protostellar Envelopes*

In star-formation studies, disk formation is an important subject. However, it is still controversial *when and how disks are formed from envelopes*, and also, *how molecules in envelopes are delivered into disks.* To answer these questions, we are conducting a chemical analysis of disk forming regions. It is well known that the chemical composition of the protostellar envelopes shows a significant diversity (Sakai & Yamamoto 2013). One distinct case is warm carbon-chain chemistry, which is rich in unsaturated carbon-chain molecules. Another distinct case is hot corino chemistry, which is rich in saturated organic molecules. How such a chemical diversity evolves into the disk is an interesting issue for astrochemistry. At the same time, chemical diagnostics can give us important information on physical processes in disk formation. Here, we report about such a chemical analysis toward a very famous source, IRAS 16293–2422.

1.2. *IRAS 16293–2422*

This source is a well-studied low-mass Class 0 protostar in Ophiuchus (d = 120 pc; Knude & Hog 1998). It is rich in saturated complex organic molecules (COMs), such as $HCOOCH_3$ and $(CH_3)_2O$, in the vicinity of the protostars (e.g., Schöier *et al.* 2002; Cazaux *et al.* 2003; Bottinelli *et al.* 2004a; Kuan *et al.* 2004; Pineda *et al.* 2012; Jørgensen *et al.* 2012, 2016). With this characteristic chemical composition, this source is very famous as the prototypical hot corino source (Ceccarelli 2004; Bottinelli *et al.* 2004a). Toward Source A, a rotating motion is reported (e.g. Pineda *et al.* 2012; Favre *et al.* 2014). We analyzed an archival data toward this source in ALMA Cycle 1 (ADS/JAO.ALMA #2012.1.00712.S) to investigate this rotating motion in detail. We analyzed the spectral lines of OCS (J=19–18), CH_3OH ($11_{0,11}$–$10_{1,10}$; A^+), $HCOOCH_3$ ($19_{9,10}$–$19_{8,11}$; E),

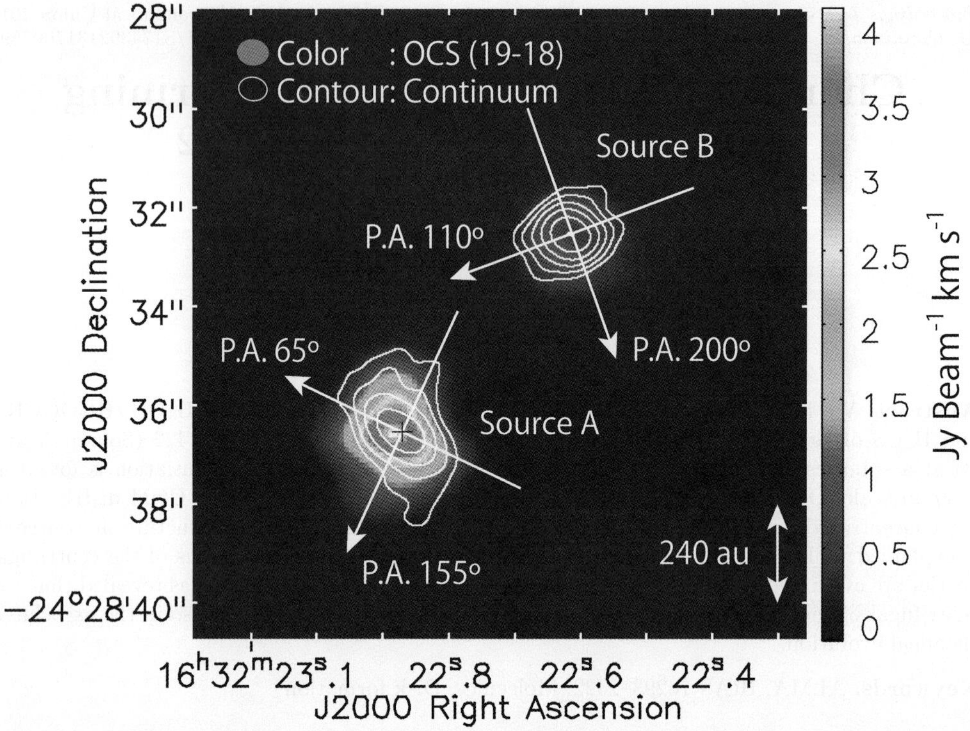

Figure 1. Integrated intensity map of the OCS (color; J=19–18) and the 1.2 mm continuum map (contours). The contour levels for the continuum are 10, 20, 40, 80, 160, and 320σ, where the rms noise level is 2 mJy beam^{-1}. White arrows on Source A represent the direction along its disk/envelope system (P.A. 65°) and one perpendicular to it (P.A. 155°). A black cross represents the protostellar position of Source A. White arrows on Source B represent the direction along its disk/envelope system (P.A. 110°) and one perpendicular to it (P.A. 200°).

and H_2CS ($7_{0,7}$–$6_{0,6}$; $7_{2,5}$–$6_{2,4}$; $7_{4,4}$–$6_{4,3}$, $7_{4,3}$–$6_{4,2}$) in Band 6 ($\sim$ 250 GHz). A typical resolution is $0\farcs5$ ($\sim$ 60 au).

2. Envelope and Disk Structures in Source A

2.1. *Envelope Traced by OCS*

Figure 1 shows the integrated intensity map of the OCS (J=19–18) line. The white arrows represent the direction along which the mid-plane of the disk/envelope system is extended (P.A. 65°) and the line perpendicular to it (P.A. 155°). We prepared the position-velocity (PV) diagrams of OCS (Figures 2a, b) to analyze the kinematic structure of the gas in Source A. In the diagram prepared along the disk/envelope system (Figure 2a), there is a velocity gradient indicating the rotating motion. In addition, a velocity gradient perpendicular to the envelope can also be seen (Figure 2b), which indicates the infall motion. Thus, the kinematic structure of OCS cannot be explained by the Keplerian motion, which does not have the infall motion. In order to interpret these kinematic structures, we consider the infall and rotation motion in the envelope.

Figure 3 shows a schematic illustration of a ballistic model of an infalling-rotating envelope (Oya *et al.* 2014). In this model, the gas is falling and rotating under the gravity of the central protostar. Because of energy and angular momentum conservation, the gas cannot fall inward of a certain radius, or the 'perihelion'. Thus, the distribution of the gas has a hole in this model. This position is called as a 'centrifugal barrier'. This model considers only the central part (a few 100 au) of a whole envelope. It does not involve the Keplerian disk component, which probably exists inside the centrifugal barrier. We calculated the kinematic structure of the envelope with this model.

The results of the infalling-rotating envelope model are shown in Figures 2(a, b). This model seem to reproduce the kinematic structure of OCS along the two directions simultaneously. Comparing the observation with this model, we evaluated the protostellar mass (M) and the radius of the centrifugal barrier ($r_{\rm CB}$) to be from 0.5 to 1.0 $M_\odot$ and from 40 to 60 au, respectively, assuming the inclination angle (i) of 60° (0° for a face-on configuration). This protostellar mass is slightly larger than a previous report by Favre *et al.* (2014) (0.49 $M_\odot$ derived from the $C^{17}O$ observation).

2.2. *Chemical Differentiation*

We also investigated some other molecular lines toward IRAS 16293–2422 Source A: the CH_3OH ($11_{0,11}$–$10_{1,10}$; A^+), $HCOOCH_3$ ($19_{9,10}$–$19_{8,11}$; E), and H_2CS ($7_{0,7}$–$6_{0,6}$) lines (Figures 2c–h). The kinematic structures revealed by these lines are apparently different from that of the OCS line. First, CH_3OH and $HCOOCH_3$ have more compact distributions concentrated around the centrifugal barrier, as seen in the SO line in L1527 (Sakai *et al.* 2014a,b). On the other hand, H_2CS traces the envelope component as OCS does. In addition, H_2CS shows high velocity-shift components concentrated to the protostellar position.

The results of the infalling-rotating envelope model are shown by blue contours in Figure 2. In these models, the physical parameters (e.g., M of 0.75 $M_\odot$, $r_{\rm CB}$ of 50, and i of 60°) are common for all the molecular lines, except for the outer radius of the envelope. Smaller outer radii are appropriate for the CH_3OH (80 au) and $HCOOCH_3$ (55 au) cases. This means that the emitting regions of these lines have ring-like structures around the centrifugal barrier. In addition, the high velocity-shift components of H_2CS can be explained by the Keplerian motion with the protostellar mass derived from the OCS analysis (Figures 2g, h).

The above chemical differentiation can be schematically illustrated like Figure 4(a); OCS traces the envelope, while CH_3OH and $HCOOCH_3$ highlight the centrifugal barrier. These molecules are possibly processed or trapped onto the dust surface while they are being delivered into the disk. On the other hand, H_2CS traces the Keplerian disk component inside the centrifugal barrier in addition to the envelope component. This chemical change is very important in understanding the chemical evolution from the envelope to the disk.

2.3. *Physical/Chemical Changes across the Centrifugal Barrier*

Why is the emitting region different from molecule to molecule? To address this question, we derived the gas kinetic temperature in the envelope component, the centrifugal barrier, and the disk component. The result is shown in Table 1. The gas kinetic temperature increases from the envelope to the centrifugal barrier, and it goes down again in the disk. If the mid-plane temperature is as low as the gas kinetic temperature in the disk (~ 80 K), COMs would be adsorbed onto dust grains. Therefore, this change in the temperature could be the reason why COMs are bright only at the centrifugal barrier.

We also evaluated the abundance ratio of CH_3OH and $HCOOCH_3$. Assuming the rotation temperature to be around the derived gas kinetic temperatures, the ratios are derived as shown in Table 1. The high abundance ratio of $N(HCOOCH_3)$ / $N(CH_3OH) \sim (0.8 - 8.7)$ is consistent with a previous report (Bottinelli *et al.* 2007). More interestingly,

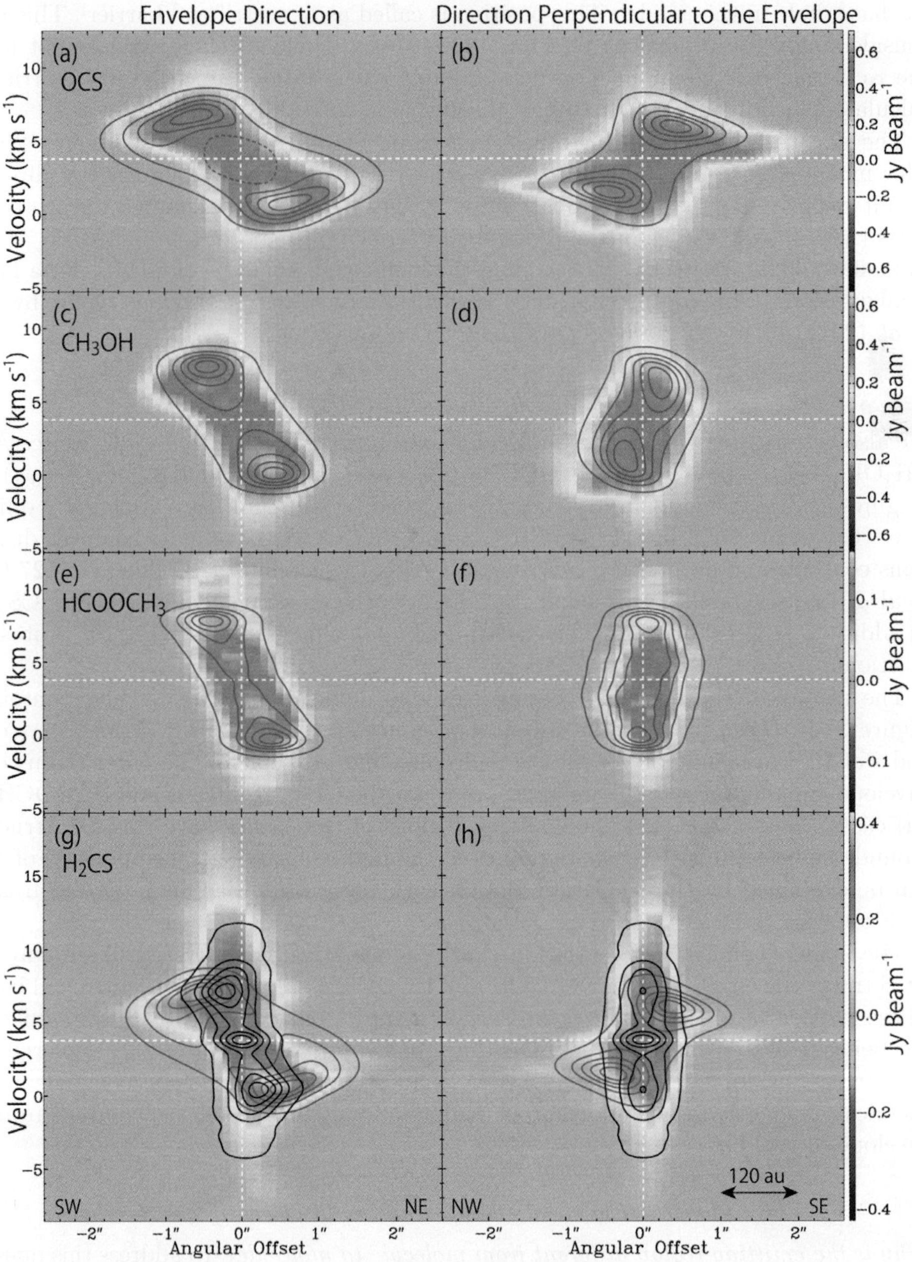

Figure 2. PV diagrams of the OCS (a, b; J=19–18; 231.061 GHz), CH_3OH (c, d; $11_{0,11}$–$10_{1,10}$; A^+; 250.507 GHz), $HCOOCH_3$ (e, f; $19_{9,10}$–$19_{8,11}$; E; 232.597 GHz), and H_2CS (g, h; $7_{0,7}$–$6_{0,6}$; 240.267 GHz) lines. The position axes are along the disk/envelope system (a, c, e, g; P.A. 65°) and the direction perpendicular to it (b, d, f, h; P.A. 155°). The vertical white dotted lines represent the protostellar position, and the horizontal ones represent the systemic velocity of Source A (∼ 3.8 km s^{-1}; e.g., Favre *et al.* 2014). Blue contours represent the results of the infalling-rotating envelope model with the following physical parameters: M of 0.75 $M_\odot$, r_{CB} of 50, and i of 60°. The scale height of the envelope is assumed to increase as the distance from the protostar with the extending angle of 45°. The outer radius of the envelope model is set to be 180, 80, 55, and 150 au for the OCS, CH_3OH, $HCOOCH_3$, and H_2CS lines, respectively, so as to demonstrate the molecular distributions. Black contours in panels (g, h) represent the Keplerian model with the protostellar mass of 0.75 $M_\odot$ and the inclination angel of 60°.

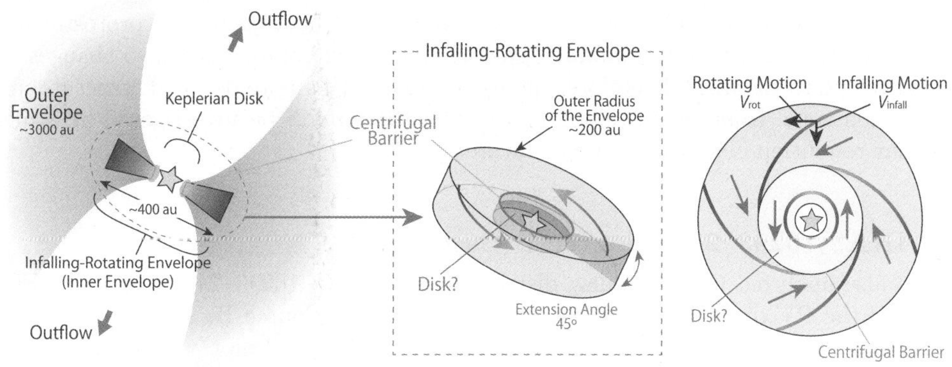

Figure 3. Schematic illustration of the gas components around Source A. The outer envelope part represented by light blue regions in the left figure is resolved out in these observations. The gas components are roughly divided into the infalling-rotating envelope and the disk component by the centrifugal barrier.

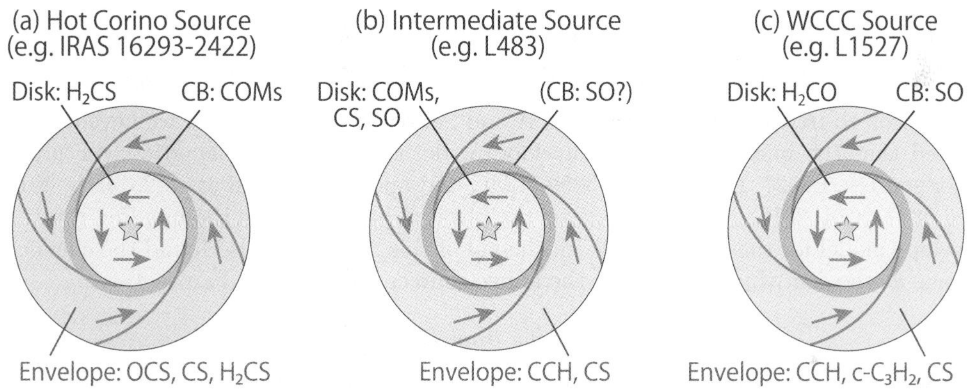

Figure 4. Schematic illustration of the chemical differentiation across the centrifugal barrier.

Table 1. Physical Parameters

	Envelope	Centrifugal Barrier	Disk
Gas Kinetic Temperature[a]	70 – 110	110 – 140	70 – 90
Abundance Ratio[b]	0.8 ± 0.5	4.2 ± 0.3	8.7 ± 1.3

[a] In K. The gas kinetic temperatures are derived from the intensity ratio of the H_2CS ($7_{0,7}$–$6_{0,6}$; $7_{2,5}$–$6_{2,4}$) lines by using RADEX code (van der Tak *et al.* 2007). The assumed ranges for the H_2 density and the column density of H_2CS are from 10^7 to 10^9 cm^{-3} and from 10^{13} to 10^{15} cm^{-2}, respectively. The error is estimated only from the statistical error and does not contain the calibration error, because it will be almost canceled in the intensity ratios.
[b] The abundance ratio of $HCOOCH_3$ to CH_3OH derived under the LTE approximation. The quoted errors represent 3σ, which are estimated from the rms noise level.

the abundance ratio systematically increases from the envelope to the disk. Its origin is puzzling, and is subject to chemical modeling of the disk forming region.

2.4. *Summary for Source A*

In this analysis, the basic structure of the envelope is found to be reproduced by a simple ballistic model. Moreover, we find that significant physical and chemical changes are occurring around the centrifugal barrier. First, the gas kinetic temperature is highest at the centrifugal barrier. This can be caused by thermal heating by the protostar and a weak accretion shock by the infalling gas. The chemical composition also changes from the envelope to the disk. Therefore, the next target is the more detailed structure in the 'transition zone' from the envelope to the disk, which will be investigated at a higher angular resolution observation in the future.

3. IRAS 16293–2422 Source B Case

We also analyzed the molecular distributions in IRAS 16293–2422 Source B. Figure 5 shows the PV diagrams centered at the protostar in Source B. All of the molecular lines show an inverse P-Cygni profile, which implies the infall motion. The distributions of the OCS and H_2CS lines are extended in a 400 au scale. On the other hand, that of the CH_3OH line is more concentrated in a 200 au scale. This chemical differentiation is similar to the Source A case. Detailed analyses of their kinematic structure by the infalling-rotating envelope model are in progress.

4. Discussion

4.1. *Infalling-Rotating Envelope in Low-Mass Protostellar Sources*

In addition to IRAS 16293–2422 Source A and Source B studied above, we have demonstrated that the infalling-rotating envelope model explains the kinematic structure of other sources, L1527, IRAS 15398–3359, TMC–1A, and L483 (Sakai *et al.* 2014a,b, 2016; Oya *et al.* 2014, 2015, 2017). We derived the physical parameters in the infalling-rotating envelope model for these sources. With the aid of this model, we also found the chemical change at the centrifugal barrier in the above sources, as shown in Figure 4.

4.2. *Transition Zone from Envelopes to Disks*

Generally, the disk radius is thought to be close to the centrifugal radius, where the Keplerian velocity equals to the rotation velocity. This radius is the twice the radius of the centrifugal barrier. However, the simple ballistic model well reproduces the kinematic structure in IRAS 16293–2422 Source A and the other sources mentioned above. This means that the gas keeps falling toward the centrifugal barrier to some extent. However, our model is too simplified, and we need to consider various effects, such as the gas pressure, the magnetic field effect, the self-gravity, and the time variation of the angular momentum of the gas. They will affect where the infall motion of the gas practically vanishes. Considering these complex situations, it is still unknown *what is occurring between the centrifugal radius and the centrifugal barrier.* This is a really interesting problem left for the future studies.

4.3. *Chemical Diversity in Disk Forming Regions*

In the binary system, IRAS 16293–2422, we found a drastic change in the chemical composition of the gas around the centrifugal barrier. Such a chemical differentiation is also seen in other low-mass protostellar sources with different chemical characteristics (Figure 4). These chemical changes at the centrifugal barrier will affect the initial condition for chemical evolution in the disk formation, and eventually, the material environment of the possible planets formed in the future. The chemical differentiation in the disk forming region provides us with an essential clue to understanding the total chemical processes during the disk formation.

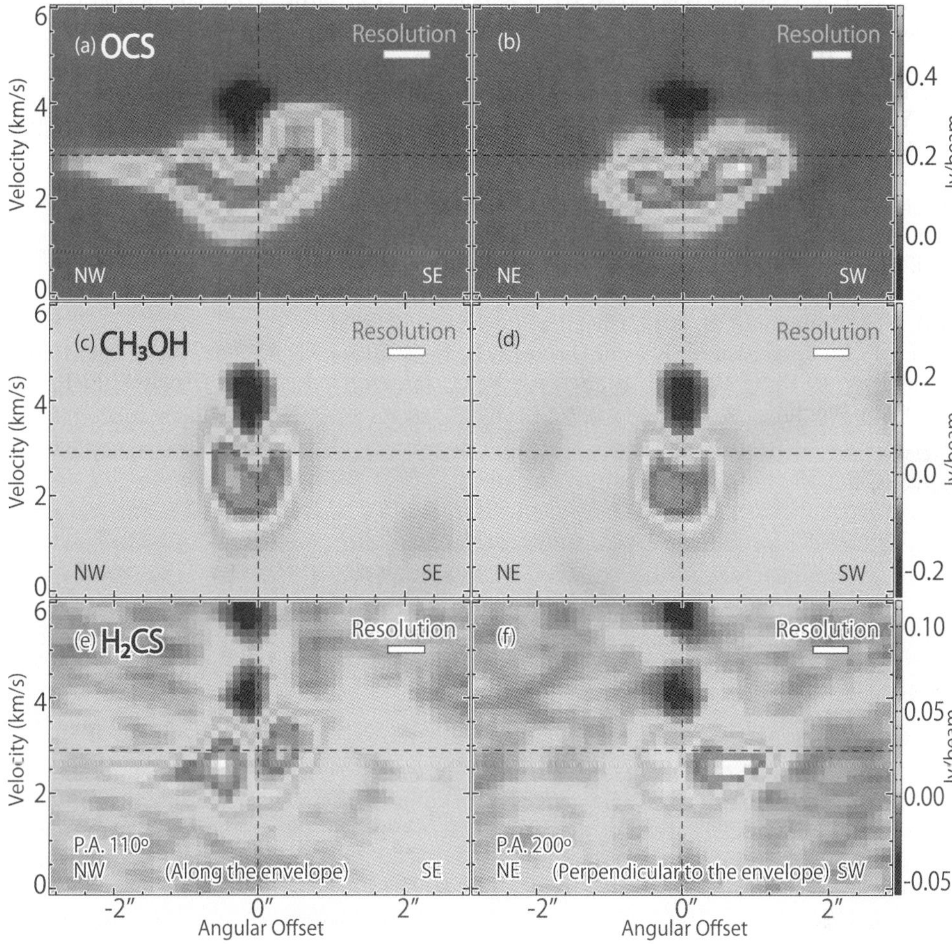

Figure 5. PV diagrams of the OCS (a, b; J=19–18), CH_3OH (c, d; $5_{1,5}$–$4_{1,4}$; A^+), and H_2CS (e, f; $7_{0,7}$–$6_{0,6}$) lines. The position axes are along the disk/envelope system (P.A. 110°) and the direction perpendicular to it (P.A. 200°) in Source B. The vertical black dotted lines represent the protostellar position, while the horizontal ones represent the systemic velocity of Source B (~ 2.9 km s^{-1}).

Acknowledgement

The author thanks Nami Sakai, Ana López-Sepulcre, Yoshimasa Watanabe, Cecilia Ceccarelli, Bertrand Lefloch, Cécile Favre, and Satoshi Yamamoto for invaluable discussions.

References

Bottinelli, S., Ceccarelli, C., Neri, R., *et al.* 2004a, ApJ, 617, L69

Bottinelli, S., Ceccarelli, C., Williams, J. P., & Lefloch, B. 2007, A&A, 463, 601

Cazaux, S., Tielens, A. G. G. M., Ceccarelli, C., *et al.* 2003, ApJ, 593, L51

Ceccarelli, C. 2004, Star Formation in the Interstellar Medium: In Honor of David Hollenbach, 323, 195

Favre, C., Jørgensen, J. K., Field, D., *et al.* 2014, ApJ, 790, 55

Jørgensen, J. K., Favre, C., Bisschop, S. E., *et al.* 2012, ApJ, 757, L4

Jørgensen, J. K., van der Wiel, M. H. D., Coutens, A., *et al.* 2016, A&A, 595, A117
Knude, J. & Hog, E. 1998, A&A, 338, 897
Kuan, Y.-J., Huang, H.-C., Charnley, S. B., *et al.* 2004, ApJ, 616, L27
Oya, Y., Sakai, N., Sakai, T., *et al.* 2014, ApJ, 795, 152
Oya, Y., Sakai, N., Lefloch, B., *et al.* 2015, ApJ, 812, 59
Oya, Y., Sakai, N., López-Sepulcre, A., *et al.* 2016, ApJ, 824, 88
Oya, Y., Sakai, N., López-Sepulcre, A., *et al.* 2017, ApJ, 837, 174
Pineda, J. E., Maury, A. J., Fuller, G. A., *et al.* 2012, A&A, 544, L7
Sakai, N., Sakai, T., Hirota, T., *et al.* 2014a, Nature, 507, 78
Sakai, N., Oya, Y., Sakai, T., *et al.* 2014b, ApJ, 791, L38
Sakai, N., Oya, Y., López-Sepulcre, A., *et al.* 2016, ApJ, 820, L34
Sakai, N. & Yamamoto, S. 2013, Chemical Reviews, 113, 8981
Schöier, F. L., Jørgensen, J. K., van Dishoeck, E. F., & Blake, G. A. 2002, A&A, 390, 1001
van der Tak, F. F. S., Black, J. H., Schöier, F. L., Jansen, D. J., & van Dishoeck, E. F. 2007, A&A, 468, 627

Astrochemistry VII – Through the Cosmos from Galaxies to Planets
Proceedings IAU Symposium No. 332, 2017
M. Cunningham, T. Millar & Y. Aikawa, eds.

doi:10.1017/S1743921317007402

Detection of submillimeter-wave [C I]emission in gaseous debris disks of 49 Ceti and β Pictoris

Aya E. Higuchi[1], **Aki Sato**[2], **Takashi Tsukagoshi**[2], **Nami Sakai**[1], **Kazunari Iwasaki**[3], **Munetake Momose**[2], **Hiroshi Kobayashi**[4], **Daisuke Ishihara**[4], **Hidehiro Kaneda**[4] **and Satoshi Yamamoto**[5]

[1]The Institute of Physical and Chemical Research (RIKEN), 2-1, Hirosawa, Wako-shi, Saitama 351-0198, Japan, email: aya.higuchi@riken.jp

[2]College of Science, Ibaraki University, Bunkyo 2-1-1, Mito 310-8512, Japan

[3]Department of Earth and Space Science, Graduate School of Science Osaka University, 1-1 Machikaneyama-cho, Toyonaka-shi, Osaka 560-0043, Japan

[4]Department of Physics, Nagoya University, Furo-cho, Chikusa-ku, Nagoya, Aichi, 464-8602, Japan

[5]Department of Physics, The University of Tokyo, Hongo, Bunkyo-ku, Tokyo 113-0033, Japan

Abstract. We have detected [C I] $^3P_1$–$^3P_0$ emissions in the gaseous debris disks of 49 Ceti and β Pictoris with the 10 m telescope of the Atacama Submillimeter Telescope Experiment, which is the first detection of such emissions. The line profiles of [C I] are found to resemble those of CO(J=3–2) observed with the same telescope and the Atacama Large Millimeter/submillimeter Array. This result suggests that atomic carbon (C) coexists with CO in the debris disks, and is likely formed by the photodissociation of CO. Assuming an optically thin [C I] emission with the excitation temperature ranging from 30 to 100 K, the column density of C is evaluated to be (2.2 ± 0.2) × 10^{17} and (2.5 ± 0.7) × 10^{16} cm^{-2} for 49 Ceti and β Pictoris, respectively. The C/CO column density ratio is thus derived to be 54 ± 19 and 69 ± 42 for 49 Ceti and β Pictoris, respectively. These ratios are higher than those of molecular clouds and diffuse clouds by an order of magnitude. The unusually high ratios of C to CO are likely attributed to a lack of H_2 molecules needed to reproduce CO molecules efficiently from C. This result implies a small number of H_2 molecules in the gas disk; i.e., there is an appreciable contribution of secondary gas from dust grains.

Keywords. (stars:) planetary systems: protoplanetary disks, ISM: atoms, radio lines: general

1. Introduction

Debris disks are optically thin circumstellar dust components around main-sequence stars, and are crucial to our understanding of the formation of planetary bodies, such as planets, comets, and asteroids. Debris disks stand between gas-rich protoplanetary disks and mature planetary systems, and their investigation sheds light on the late stages of planet formation. Almost all the gas in debris disks is thought to have dissipated, and dust grains smaller than 10 μm are therefore expected to be quickly blown out by stellar radiation pressure from the central star (especially at $\geqslant 10L_\odot$) (e.g., Wyatt 2008). However, recent observational studies at infrared wavelengths have shown that such small grains survive in debris disks (e.g., Ishihara *et al.* 2016). One possible mechanism by which small grains remain for a long time is that the gas component remains in debris disks to a moderate extent.

Several debris disks harboring a gas component have been discovered in survey observations at optical, infrared, and radio wavelengths. The gas components around β Pictoris and HD 32297 were first detected via Ca II and Na I absorption lines (Slettebak 1975; Hobbs *et al.* 1985; Redfield 2007). For HD 172555, η Tel, and AU Mic, the gas component was identified from [O I], [C II], and fluorescent H_2 emissions, respectively (France *et al.* 2007; Riviere-Marichalar *et al.* 2012, 2014). [O I] and [C II] emissions were also detected for β Pictoris (Brandeker *et al.* 2016). Some debris disks are known to have submillimeter-wave CO emission; e.g., 49 Ceti (Dent *et al.* 2005; Hughes *et al.* 2008), β Pictoris (Fernández *et al.* 2006; Dent *et al.* 2014), HD 21997 (Moór *et al.* 2011. 2013; Kóspál *et al.* 2013), HD 131835 (Moór *et al.* 2015), and HD 141569 (White *et al.* 2016).

Why debris disks can possess a gas component is not well understood. One possibility is that such gas has a primordial origin; i.e., the gas is the remnants of a protoplanetary gas disk. The other possibility is that the gas has a secondary origin, for which several different mechanisms can be considered; e.g., the sublimation of dust grains (Kobayashi *et al.* 2011) or planetesimals (Lagrange *et al.* 1998), photo-sputtering of dust grains (Grigorieva *et al.* 2007), collisional vaporization of dust grains (Czechowski & Mann 2007), and collision of comets or icy planetesimals (Zuckerman & Song 2012). Such secondary gas is thought to be mainly composed of CO and H_2O (Mumma & Charnley 2011), while only a small amount of H_2 is expected.

The present study proposes a new approach to tackle the origin of the gas in debris disks using the submillimeter-wave fine-structure line of atomic carbon, [C I]. Debris disks can generally be regarded as a photodissociation region (PDR; e.g., Tielens & Hollenbach 1985) as shown by the detection of [O I] and [C II] (e.g., Brandeker *et al.* 2016). In the PDR, the abundance of atomic carbon (C) is determined by the balance among the photodissociation of CO, photoionization of C, electron recombination of C^+, and reactions between C^+ and H_2, and it is thus sensitive to both the ultraviolet (UV) radiation field and H_2 density as well as metallicity. The [C I] line has therefore been widely used to diagnose molecular clouds (e.g., Ikeda *et al.* 1999; Shimajiri *et al.* 2013; Krips *et al.* 2016). This method could be applied to debris disks. Although [C II] and CO emissions have been observed in a few debris disks, no detection of [C I] emission has been reported. With this in mind, we conducted sensitive observations of [C I] emission toward two representative gaseous debris disks; i.e., those of 49 Ceti and β Pictoris.

2. Result

2.1. *CO(J=3–2) and [C I] $^3P_1$–$^3P_0$ spectra*

2.1.1. *49 Ceti*

Figure 1(a) shows the CO(J=3–2) spectrum observed toward 49 Ceti. The root-mean-square (rms) noise level of the CO line is 5.8 mK (at a resolution of 1.1 km s^{-1}) in T_B. We detected the CO line emission with a peak intensity of 26.3 mK (4.5 σ) in T_B toward 49 Ceti. The integrated intensity of CO is derived to be 0.15 K km s^{-1} (5.9 σ). The spectrum shows a double-peak profile with a central LSR velocity of 1 km s^{-1}. To verify our detection of the CO emission with the ASTE telescope, we compare its spectrum with that prepared from the ALMA archival data and with that observed with SMA/JCMT (Hughes *et al.* 2008; Dent *et al.* 2005). Figure 1(b) shows the ALMA spectrum of CO(J=3–2) (dotted lines) toward 49 Ceti, where we integrated the CO emission over the area within the 3 σ contour of the ALMA map. To compare the ASTE spectrum with the ALMA spectrum, the intensity of the ASTE spectrum is scaled by the inverse of a beam filling factor, f_d (see Section 3.2), of 54. The spectral features and

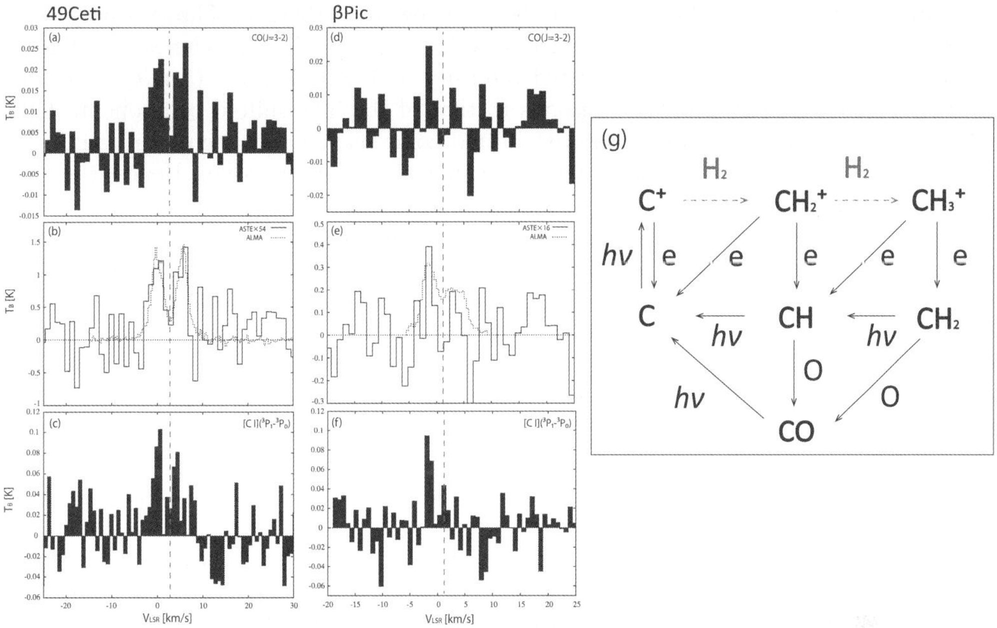

Figure 1. (a) and (d) : CO spectrum observed with the ASTE toward 49 Ceti (a) and β Pictoris (d). (b) and (e) : Comparison with the ALMA spectrum for 49 Ceti (b) and β Pictoris (e). (c) and (f) : [C I] spectrum observed with the ASTE toward 49 Ceti (c) and β Pictoris (f). The vertical red line indicates the systemic velocity of the gas disk (Dent *et al.* 2005). (g) Basic carbon chemistry producing CO in debris disks.

the systemic velocities of the two spectra are consistent. Our spectrum is also consistent with that observed with SMA/JCMT (Dent *et al.* 2005; Hughes *et al.* 2008).

Figure 1(c) shows the [C I] $^3P_1$–$^3P_0$ spectrum toward 49 Ceti. The rms noise level of the [C I] line is 26.4 mK (at a resolution of 0.76 km s^{-1}) in $T_{\rm B}$. The peak intensity is 103 mK (3.9 σ), while the integrated intensity is 0.45 K km s^{-1} (10.2 σ). The [C I] line also has a double-peak profile with a central LSR velocity of 3 km s^{-1}. The double-peak profile is similar to that of CO(J=3–2) observed with the ASTE telescope. The similarity of the spectral shapes of the [C I] and CO emissions suggests that the two emissions come from the same region. The intensity of [C I] is higher than that of CO(J=3–2). This is the first [C I] line detection at 492 GHz toward debris disks.

2.1.2. *β Pictoris*

Figure 1(d) shows the CO(J=3–2) spectrum observed toward β Pictoris. The rms noise level of the CO line is 7.7 mK (at a resolution of 1.1 km s^{-1}) in $T_{\rm B}$. The peak intensity of the CO emission is 24.5 mK (3.2 σ) in $T_{\rm B}$, while the integrated intensity is 0.047 K km s^{-1} (2.2 σ). Because the statistical significance is poor, we compare the spectrum with the ALMA spectrum (Dent *et al.* 2014) for confirmation. Figure 1(e) shows the spectrum of CO(J=3–2) obtained with the ALMA toward β Pictoris (dotted lines), where we integrated the CO emission over the area within the 3-σ contour of the ALMA map. Here, the intensity of the ASTE spectrum is scaled by an inverse of the beam filling factor of 16 for comparison. In the ALMA spectrum, the blue-shifted emission (-5 to 1 km s^{-1}) is brighter than the redshifted emission (1 to 5 km s^{-1}). This feature can barely be seen in the ASTE spectrum. Although the signal-to-noise ratio is poor, the CO line seems to be marginally detected with the ASTE telescope.

Figure 1(f) shows the [C I] spectrum observed toward β Pictoris. The rms noise level of the [C I] line is 24 mK (at a resolution of 0.76 km s^{-1}) in T_B. The peak intensity is 94.4 mK (3.9 σ), while the integrated intensity is 0.18 K km s^{-1} (5.1 σ). The LSR velocity of the intensity peak corresponds to that of the blue-shifted component of the CO(J=3–2) spectrum. Although the red-shifted component is not seen, the overall line profile of the [C I] emission is consistent with that of the CO emission. Hence, the [C I] emission is marginally detected for β Pictoris. The similarity of the spectral line profiles between CO(J=3–2) and [C I] emissions suggests that the two emissions come from the same region. As in the case of 49 Ceti, the [C I] intensity from β Pictoris is higher than that of CO(J=3–2). Note that Kral *et al.* (2016) recently reported the non-detection of [C I] emission toward β Pictoris according to APEX (Atacama Pathfinder Experiment) observations. However, this non-detection is due to the insufficient sensitivity of their observation, with an rms noise level of 139 mK (at a resolution of 0.046 km s^{-1}).

2.2. *Optical Depth of the [C I] Emission and Column Density of C*

The optical depth of the [C I] emission and the column density of C were derived using the equations reported by Oka *et al.* (2001) and Tsukagoshi *et al.* (2015). Here, the beam filling factor f_d was derived from the ratio of the solid angle of the disk to that of the telescope beam (Ω_A): $f_d = \Omega_d/\Omega_A$. The solid angle Ω_d was estimated by assuming that the extent of the [C I] emission is identical to that of the CO disk observed by SMA and ALMA (e.g., Hughes *et al.* 2008; Dent *et al.* 2014).

Because the dust temperature derived from the model fitting of the spectral energy distribution is 62 K for 49 Ceti and 60 K for β Pictoris (e.g., Roberge *et al.* 2013; Cataldi *et al.* 2014), we assumed the range of T_{ex} to be from 30 to 100 K. The optical depths of the [C I] and CO lines are lower than 0.18 for 49 Ceti and lower than 0.05 for β Pictoris, even for the lowest assumed excitation temperature (30 K), and we thus adopted the optically thin condition in the following analysis. Assuming the local thermodynamic equilibrium, we derived the column density of atomic carbon, N(C), as (Oka *et al.* 2001)

$$N(\mathrm{C}) = 1.98 \times 10^{15}\ Q(T_{\mathrm{ex}}) \exp\left(\frac{23.6}{T_{\mathrm{ex}}}\right) f_{\mathrm{d}}^{-1} \int T_{\mathrm{B}}\, dv\ [\mathrm{cm}^{-2}], \tag{2.1}$$

where $Q(T_{ex})$ is the ground-state partition function for atomic carbon. The column densities of C are evaluated to be from $(2.1 \pm 0.2)\times10^{17}$ to $(2.2 \pm 0.2)\times10^{17}$cm^{-2} and from $(2.4 \pm 0.5)\times10^{16}$ to $(2.7 \pm 0.5)\times10^{16}$cm^{-2} for 49 Ceti and β Pictoris, respectively, assuming the above temperature range. The total mass of atomic carbon was derived by integrating the column density over the solid angle of the source. Rough estimations were (4.0–4.4)$\times10^{-3} M_\oplus$ and (1.5–1.8)$\times10^{-4} M_\oplus$ for 49 Ceti and β Pictoris, respectively, assuming the above temperature range. Our result is an order of magnitude lower than previous estimates made from [C II] observations for β Pictoris: $7.1\times10^{-3} M_\oplus$ (Cataldi *et al.* 2014) and $2\times10^{-3} M_\oplus$ (Kral *et al.* 2016).

Under the local thermodynamic equilibrium and optically thin condition, the CO column density was calculated as (e.g., Oka *et al.* 2001)

$$N(\mathrm{CO}) = 4.64 \times 10^{12}\ T_{\mathrm{ex}} \exp\left(\frac{33.2}{T_{\mathrm{ex}}}\right) f_{\mathrm{d}}^{-1} \int T_{\mathrm{B}}\, dv\ [\mathrm{cm}^{-2}]. \tag{2.2}$$

The CO column densities were derived to range from $(3.3 \pm 0.6)\times10^{15}$ to $(5.1 \pm 0.9)\times 10^{15}$cm^{-2} and from $(3.2 \pm 1.5)\times10^{14}$ to $(4.9 \pm 2.3)\times10^{14}$cm^{-2} for 49 Ceti and β Pictoris, respectively, assuming the above temperature range. Our results are consistent with the column densities derived from observations of absorption lines of [C I] and CO at

UV wavelengths for β Pictoris by Roberge *et al.* (2000): $N(\mathrm{C})=(2\text{–}4)\times10^{16}$ cm^{-2} and $N(\mathrm{CO})=(6.3\pm0.3)\times10^{14}$ cm^{-2}.

3. Discussion

3.1. *[C I] emission from debris disks*

In this study, a submillimeter-wave [C I] emission was detected for the first time towards debris disks. The [C I] emission is most likely associated with the rotating gas disk around the star, judging from the line profiles (Section 3.1). The [C I]/CO(J=3–2) integrated intensity ratios corrected for beam filling factors are 1.8 ± 0.4 and 2.3 ± 1.2 for 49 Ceti and β Pictoris, respectively.

The [C I] emission toward a protoplanetary disk has recently been reported. Tsukagoshi *et al.* (2015) detected the [C I] line towards the vicinity of DM Tau. Kama *et al.* (2016) detected [C I] in six of 37 T-Tauri and Herbig Ae/Be stars, which included the first unambiguous detections of the [C I] line in TW Hya and HD 100546. The spectral profiles of the [C I] emission from two protoplanetary disks (DM Tau and TW Hya) have a single peak, suggesting that atomic carbon is not associated with the disk component. For HD 100546, the [C I] emission has a double-peak profile, suggesting that atomic carbon extends mainly to the outermost part of the disk. Note that the [C I] emission is fainter than the CO emission in both cases, although the observed transition of CO is different from ours. The [C I]/CO(J=4–3) integrated intensity ratio toward DM Tau has been reported to be 0.9 (Tsukagoshi *et al.* 2015), while the [C I]/CO(J=6–5) ratio is between 0.05 and 0.3 (Kama *et al.* 2016).

In contrast to the case for protoplanetary disks, extensive [C I] observations have been made toward molecular clouds. In the Orion A molecular cloud, the [C I]/CO(J=3–2) integrated intensity ratios range from 0.1 in the Orion KL region to 1.2 at the southern end (Ikeda *et al.* 1999). For W3(OH) and AFGL 333, the [C I]/CO(J=3–2) integrated intensity ratio has been reported to be from 0.4 to 0.9 (Sakai *et al.* 2006). The ratios for 49 Ceti and β Pictoris are thus higher than those for molecular clouds.

3.2. *C/CO ratio in debris disks*

The C/CO column density ratio is derived to be 54 ± 19 and 69 ± 42 for 49 Ceti and β Pictoris, respectively, where the errors quoted include the rms noise of the spectra and the uncertainty due to the assumed range of the excitation temperature. Although the C/CO column density ratio has not been derived for the protoplanetary disks, it would be lower than the above ratios in the debris disks, owing to the lower [C I]/CO intensity ratio.

We next compare the ratios with those reported for molecular clouds (e.g., Orion and W3) and diffuse clouds. The C/CO column density ratio ranges from 0.2 to 2.5. The ratio is relatively constant between values of 0.1 and 0.2 toward the interior of Orion molecular clouds (e.g., Ikeda *et al.* 1999). The C/CO column density ratio in diffuse clouds mostly ranges from 0.4 to 2.5. The average ratio is 1.2 in molecular clouds at high Galactic latitude (Ingalls *et al.* 1997). The ratio has also been reported to be from 0.2 to 1.1 in MCLD 123.5+24.9 (Bensch *et al.* 2003). The ratio is higher in diffuse clouds than in the interior of molecular clouds owing to photodissociation by UV radiation. In comparison with the C/CO ratios observed for molecular clouds and diffuse clouds, the C/CO ratio is unusually high for 49 Ceti and β Pictoris.

3.3. *Origin of the high C/CO ratio in the debris disk*

We now discuss why the C/CO ratio is an order of magnitude higher than the ratios for diffuse clouds and molecular clouds. Because the visual extinction, A_v, in the debris disks is estimated to be 0.01 or lower, the debris disks can be regarded as PDRs. As for the UV radiation source, we have to consider radiation from the central star in addition to interstellar UV radiation. Because we focus on the photodissociation of CO and the photoionization of C, we only consider far-ultraviolet (FUV) photons ($11 < h\nu < 13.6$ eV) in our discussion. Hence, the ratio of the FUV photon density relative to that for the standard interstellar condition, which is denoted χ, is used to define the UV strength (Drain & Bertoldi 1996). The χ value at 50 au is evaluated to be 10 for 49 Ceti and 1 for β Pictoris.

In PDRs, the main destruction process of CO is photodissociation, having a time scale of ~100 yr. We note that, if CO is continuously destroyed in the debris disk and never reproduced, CO has to be supplied by planetesimal collisions or other mechanisms with a continuous rate of $\sim 10^{18}$ kg yr^{-1} (Dent *et al.* 2014). It seems controversial whether this production rate is realistic during the debris disk phase. Meanwhile, CO can be reproduced by the chemical reactions shown in Figure 1(g). Although this figure presents a simplified network of the production and destruction of CO, it is apparent that H_2 is essential for the reproduction of CO. The H_2 number density, $[H_2]$, is thus an important factor controlling the C/CO ratio. We roughly estimated the number density of atomic carbon from N(C) to be $\sim 10^2$ cm^{-3} by assuming a disk size of ~50 au. Because CO is a minor gas component in debris disks, most carbon in the gas phase is C^+ and C owing to photodissociation and photoionization. Hence, the H_2 density could be as high as 10^6 cm^{-3}, if the abundance of CO relative to H_2 in interstellar clouds ($\sim 10^{-4}$) is applied to the abundance of atomic carbon.

PDR models have extensively been applied for molecular clouds and protoplanetary disks (e.g., Tielens & Hollenbach 1985; Warin *et al.* 1996; Rolling *et al.* 2007). However, PDR models for debris disks have not been reported to the best of our knowledge. We thus conducted a chemical model calculation using the simplified toy model (Figure 1(g)). For H_2 density of 10^6 cm^{-3} and χ of 1, the C/CO ratio is close to unity. However, CO production becomes less efficient if the carbon abundance relative to H_2 is higher (i.e., there is less H_2). We found that the C/CO ratio can be 70 and 900 for $[H_2]/[C_{tot}]$ ratios of 70 and 7, respectively, where $[C_{tot}]$ is the total number density of C^+, C, and all carbon-bearing molecules. In these calculations, we used the rate coefficients tabulated in the UMIST database for astrochemistry (McElroy *et al.* 2013). Although the chemical model is preliminary, the observed high C/CO ratio suggests a low $[H_2]/[C_{tot}]$ ratio. The H_2 mass for β Pictoris was estimated to be 0.02 $M_\oplus$ using the $N(H_2)/N(C)$ ratio of 140, assuming equal C^+ and C abundances. This H_2 mass is well below the upper limit ($<$ 17 $M_\oplus$) reported by Chen *et al.* (2007).

The number of H_2 molecules in the debris disks could be small, because the CO gas in the debris disk is not always of interstellar origin. A large amount of C in the debris disks suggests that the secondary CO gas that evaporated from the dust grains makes an appreciable contribution, as pointed out by Cataldi *et al.* (2014) and Dent *et al.* (2014).

References

Bensch, F., Leuenhagen, U., Stutzki, J., & Schieder, R. 2003, *ApJ*, 591, 1013
Brandeker, A., Cataldi, G., Olofsson, G., *et al.* 2016, *A&A*, 591, A27
Cataldi, G., Brandeker, A., Olofsson, G., *et al.* 2014, *A&A*, 563, A66
Chen, C. H., Li, A., Bohac, C., *et al.* 2007, *ApJ*, 666, 466

Czechowski, A. & Mann, I. 2007, *ApJ*, 660, 1541
Dent, W. R. F., Greaves, J. S., & Coulson, I. M. 2005, *MNRAS*, 359, 663
Dent, W. R. F., Wyatt, M. C., Roberge, A., *et al.* 2014, *Science*, 343, 1490
Draine, B. T. & Bertoldi, F. 1996, *ApJ*, 468, 269
Fernández, R., Brandeker, A., & Wu, Y. 2006, *ApJ*, 643, 509
France, K., Roberge, A., Lupu, R. E., Redfield, S., & Feldman, P. D. 2007, *ApJ*, 668, 1174
Grigorieva, A., Artymowicz, P., & Thébault, P. 2007, *A&A*, 461, 537
Hobbs, L. M., Vidal-Madjar, A., Ferlet, R., Albert, C. E., & Gry, C. 1985, *ApJl*, 293, L29
Hughes, A. M., Wilner, D. J., Kamp, I., & Hogerheijde, M. R. 2008, *ApJ*, 681, 626-635
Ikeda, M., Maezawa, H., Ito, T., *et al.* 1999, *ApJl*, 527, L59
Ishihara, D., Takeuchi, N., Kobayashi, H., *et al.* 2016, arXiv:1608.04480
Ingalls, J. G., Chamberlin, R. A., Bania, T. M., *et al.* 1997, *ApJ*, 479, 296
Kama, M., Bruderer, S., Carney, M., *et al.* 2016, *A&A*, 588, A108
Kobayashi, H., Kimura, H., Watanabe, S.-i., Yamamoto, T., & Müller, S. 2011, Earth, Planets, and Space, 63, 1067
Kóspál, Á., Moór, A., Juhász, A., *et al.* 2013, *ApJ*, 776, 77
Kral, Q., Wyatt, M., Carswell, R. F., *et al.* 2016, *MNRAS*, 461, 845
Krips, M., Martín, S., Sakamoto, K., *et al.* 2016, *A&A*, 592, L3
Lagrange, A.-M., Beust, H., Mouillet, D., *et al.* 1998, *A&A*, 330, 1091
McElroy, D., Walsh, C., Markwick, A. J., *et al.* 2013, *A&A*, 550, A36
Moór, A., Henning, T., Juhász, A., *et al.* 2015, *ApJ*, 814, 42
Moór, A., Juhász, A., Kóspál, Á., *et al.* 2013, *ApJl*, 777, L25
Moór, A., Ábrahám, P., Juhász, A., *et al.* 2011, *ApJl*, 740, L7
Mumma, M. J. & Charnley, S. B. 2011, *ARAA*, 49, 471
Oka, T., Yamamoto, S., Iwata, M., *et al.* 2001, *ApJ*, 558, 176
Redfield, S. 2007, *ApJl*, 656, L97
Riviere-Marichalar, P., Barrado, D., Augereau, J.-C., *et al.* 2012, *A&A*, 546, L8
Riviere-Marichalar, P., Barrado, D., Montesinos, B., *et al.* 2014, *A&A*, 565, A68
Roberge, A., Kamp, I., Montesinos, B., *et al.* 2013, *ApJ*, 771, 69
Roberge, A., Feldman, P. D., Lagrange, A. M., *et al.* 2000, *ApJ*, 538, 904
Röllig, M., Abel, N. P., Bell, T., *et al.* 2007, *A&A*, 467, 187
Sakai, T., Oka, T., & Yamamoto, S. 2006, *ApJ*, 649, 268
Satou, N., Sekimoto, Y., Iizuka, Y., *et al.* 2008, *PASJ*, 60, 1199
Slettebak, A., Collins, G. W., II, Parkinson, T. D., Boyce, P. B., & White, N. M. 1975, *ApJs*, 29, 137
Shimajiri, Y., Sakai, T., Tsukagoshi, T., *et al.* 2013, *ApJl*, 774, L20
Tielens, A. G. G. M. & Hollenbach, D. 1985, *ApJ*, 291, 722
Tsukagoshi, T., Momose, M., Saito, M., *et al.* 2015, *ApJl*, 802, L7
Warin, S., Benayoun, J. J., & Viala, Y. P. 1996, *A&A*, 308, 535
White, J. A., Boley, A. C., Hughes, A. M., *et al.* 2016, *ApJ*, 829, 6
Wyatt, M. C. 2008, *ARAA*, 46, 339
Zuckerman, B. & Song, I. 2012, *ApJ*, 758, 77

Astrochemistry VII – Through the Cosmos from Galaxies to Planets
Proceedings IAU Symposium No. 332, 2017
M. Cunningham, T. Millar & Y. Aikawa, eds.

doi:10.1017/S1743921317006986

Imaging the water snowline in protostellar envelopes

Merel L. R. van 't Hoff

Leiden Observatory, Leiden University, PO box 9513, NL-2300 RA, Leiden, the Netherlands
email: vthoff@strw.leidenuniv.nl

Abstract. Determining the locations of the major snowlines in protostellar environments is crucial to fully understand the planet formation process and its outcome. Despite being located far enough from the central star to be spatially resolved with ALMA, the CO snowline remains difficult to detect directly in protoplanetary disks. Instead, its location can be derived from N_2H^+ emission, when chemical effects like photodissociation of CO and N_2 are taken into account. The water snowline is even harder to observe than that for CO, because in disks it is located only a few AU from the protostar, and from the ground only the less abundant isotopologue $H_2^{18}O$ can be observed. Therefore, using an indirect chemical tracer, as done for CO, may be the best way to locate the water snowline. A good candidate tracer is HCO^+, which is expected to be particularly abundant when its main destructor, H_2O, is frozen out. Comparison of $H_2^{18}O$ and $H^{13}CO^+$ emission toward the envelope of the Class 0 protostar IRAS2A shows that the emission from both molecules is spatially anticorrelated, providing a proof of concept that $H^{13}CO^+$ can indeed be used to trace the water snowline in systems where it cannot be imaged directly.

Keywords. stars: individual (TW Hydrae, NGC1333 IRAS2A), stars: formation, planetary systems: protoplanetary disks, astrochemistry

1. Introduction

The formation of low-mass stars begins with the collapse of a dense core in a molecular cloud. To conserve angular momentum, the infalling material forms a disk around the protostar. Due to the ongoing accretion of material from the surrounding envelope through the disk onto the star, together with the launching of an outflow, the envelope dissipates over time. What remains is a pre-main sequence star surrounded by a protoplanetary disk that contains the gas and dust from which a planetary system may be forming. The composition of the planets is thus determined by the chemical structure of the protostellar environment.

A key aspect of protostellar chemistry are snowlines. A snowline marks the midplane radius at which a molecular species freezes out from the gas phase onto dust grains. The location of a snowline depends both on the species-dependent sublimation temperature and the physical structure of the protostellar envelope or protoplanetary disk (e.g., temperature and density; see Fig. 1). The selective freeze out of major carbon or oxygen carrying species at different snowlines causes the elemental C/O-ratio of the planet forming material to vary with radial distance from the star (Öberg *et al.* 2011; Öberg & Bergin 2016). The bulk composition of planets may therefore be regulated by their formation location with respect to the major snowlines (e.g., Madhusudhan *et al.* 2014; Walsh *et al.* 2015; Eistrup *et al.* 2016). Moreover, the growth of dust particles, and thus the planet formation efficiency is thought to be enhanced in these freeze-out zones, for example due to the increase in solid density (e.g., Stevenson & Lunine 1988; Ros & Johansen 2013; Schoonenberg & Ormel 2017). Snowlines thus play a crucial role in the formation and

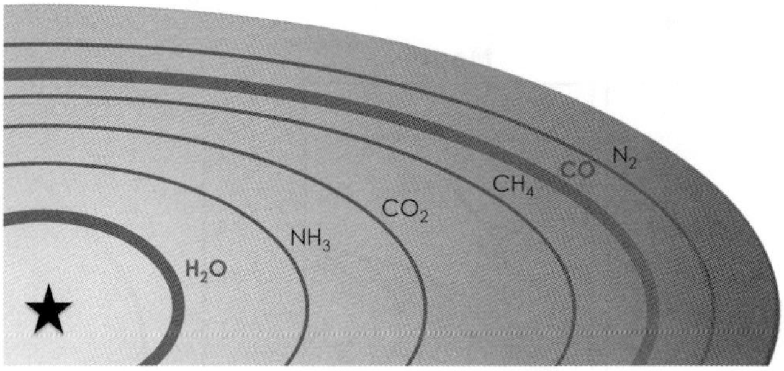

Figure 1. Schematic overview of the major snowlines. Their relative positions are set by the species-dependent freeze-out temperature, while the absolute positions depend on the physical structure of each individual object. The CO and H_2O snowlines discussed here are highlighted.

composition of planets. Determining their locations is therefore key to fully understand the planet formation process and its outcome.

2. The CO snowline

Of the major snowlines, the CO snowline is particularly interesting because CO ice is the starting point for the formation of many complex organic molecules (e.g., Herbst & van Dishoeck 2009). In addition, due to the low sublimation temperature of CO ($\sim$20 K), the CO snowline is located relatively far away from the central star (10s–100 AU in protoplanetary disks; see Fig. 1). Although we are now able to spatially resolve this with ALMA, locating the CO snowline directly remains difficult. CO line emission is generally optically thick and does not reveal the cold disk midplane. The logical step in such situation would be to use optically thin isotopologues. However, even $C^{18}O$ does not always offer a solution as is seen for TW Hya (Schwarz *et al.* 2016): the CO abundance remains 1–2 orders of magnitude below the 10^{-4} ISM abundance far within the radius where the CO snowline is expected based on the disk temperature. Furthermore, a recent ALMA survey of all disks in the Lupus star forming region suggest that CO abundances below the ISM abundance may be common (Ansdell *et al.* 2016; Miotello *et al.* 2017), indicating that other processes than freeze-out may contribute to lowering the gas-phase CO abundance. In contrast to $C^{18}O$, the double isotopologue $^{13}C^{18}O$ has recently been shown to reveal the snowline location in TW Hya (Zhang *et al.* 2017). However, the low $^{13}C^{18}O$ abundance restricts observations to the very few nearest and brightest disks.

An alternative approach is to use a molecule whose abundance is strongly affected by the freeze-out of CO. One such molecule is DCO^+, which forms via reaction of H_2D^+ and CO. Based on these chemical considerations, DCO^+ emission is expected to form a ring inside the CO snowline, where the CO abundance is low enough to enhance H_2D^+ formation, while there is still enough gaseous CO left to act as parent molecule for DCO^+. Mathews *et al.* (2013) indeed observed ring-shaped DCO^+ emission toward HD 163296, but the outer radius was later shown not to correspond exactly to the CO snowline (Qi *et al.* 2015); probably because DCO^+ can also form in warmer regions higher up in the disk from CH_2D^+ (Favre *et al.* 2015). ALMA observations of six protoplanetary disks by Huang *et al.* (2017) corroborate that DCO^+ emission, while still tracing relatively cold gas, does not have a direct relation with the CO snowline in disks.

Another molecule suggested to trace the CO snowline is N_2H^+. Formation of N_2H^+ occurs through proton transfer from H_3^+ to N_2,

$$N_2 + H_3^+ \rightarrow N_2H^+ + H_2, \tag{2.1}$$

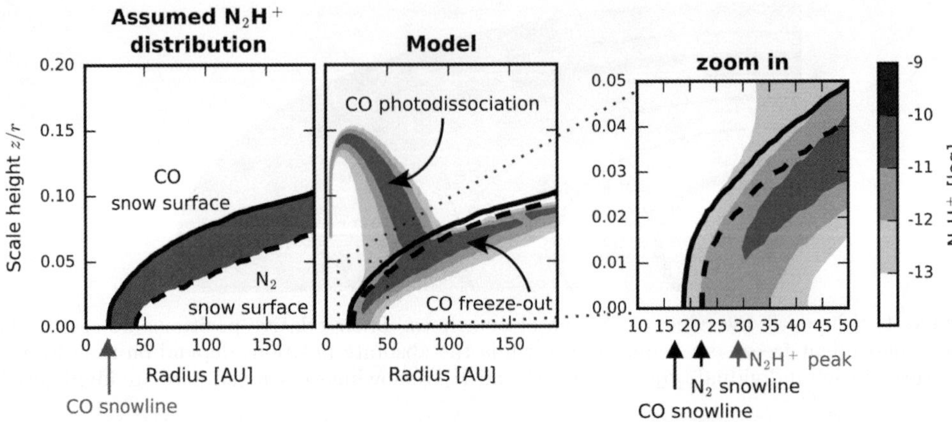

Figure 2. Generally assumed N_2H^+ distribution in protoplanetary disks, i.e., between the CO and N_2 snow surfaces (*left panel*) and the N_2H^+ distribution for TW Hya as predicted by our simple chemical model (*middle panel*). A zoom in on the region around the snowlines is shown in the *right panel.* The solid line indicates the CO snow surface and the dashed line the N_2 snow surface. The crucial differences between the expectations and the model predictions are the location of N_2H^+ peak (at the CO snowline versus beyond the CO and N_2 snowlines, resp.), and the formation of N_2H^+ above the CO snow surface in the model.

but is impeded when CO is present in the gas phase because CO competes with N_2 for reaction with H_3^+,

$$CO + H_3^+ \rightarrow HCO^+ + H_2. \tag{2.2}$$

In addition, reaction with gaseous CO is the dominant destruction route of N_2H^+:

$$N_2H^+ + CO \rightarrow HCO^+ + N_2. \tag{2.3}$$

N_2H^+ is therefore expected to be abundant only beyond the CO snowline, where CO is depleted from the gas phase. A CO snowline location has indeed been derived from N_2H^+ emission for the disks around TW Hya and HD 163296 (Qi *et al.* 2013, 2015).

However, a simple chemical model incorporating the main reactions for N_2H^+ (Eqs. 2.1–2.3) in addition to freeze out, thermal desorption and photodissociation of CO and N_2, combined with radiative transfer modeling, shows that the relationship between N_2H^+ and the CO snowline is more complicated (van 't Hoff *et al.* 2017). First, the N_2H^+ abundance peaks at temperatures slightly below the CO freeze-out temperature, instead of directly at the CO snowline (see Fig. 2), as also found by Aikawa *et al.* (2015). The snowline marks the radius where 50% of the CO is present in the gas phase and 50% is frozen out. Apparently, this reduction in gaseous CO is not yet enough to diminish the N_2H^+ destruction. Second, N_2H^+ can be formed higher up in the disk, above the layer where CO is frozen out, due to a small difference in the photodissociation rates for CO and N_2 (see Fig. 2). The slightly higher rate for CO creates another region where N_2 is still present in the gas phase, while CO is not. This "surface" layer of N_2H^+ can contribute significantly to the emission, shifting the N_2H^+ emission peak to larger radii, away from the CO snowline. N_2H^+ emission alone then merely provides an upper limit for the snowline location.

Applying our modeling approach to the TW Hya disk, using the physical model from Kama *et al.* (2016), we derive a CO snowline location of ~18 AU, instead of the ~30 AU suggested by Qi *et al.* (2013). The latter authors fitted a radial N_2H^+ column density profile to the emission with a steep rise at the CO snowline. Our outcome is consistent with the results from Schwarz *et al.* (2016) based on multiple ^{13}CO and $C^{18}O$ lines, and

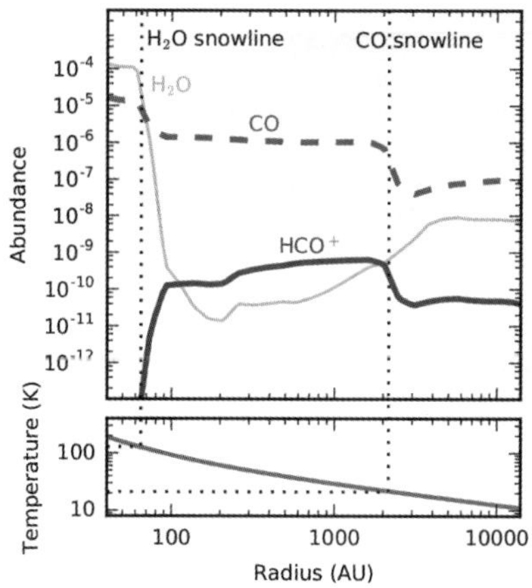

Figure 3. Temperature profile for the NGC1333 IRAS2A envelope from Kristensen *et al.* (2012) (*bottom panel*), and the corresponding H_2O (thin solid line), CO (dashed line) and HCO^+ (thick solid line) abundances predicted by the three-phase astrochemical model GRAINOBLE (*top panel*). The vertical dotted lines mark the H_2O and CO snowlines around 100 K and 20 K, resp.

the $^{13}C^{18}O$ analysis by Zhang *et al.* (2017). Deriving the CO snowline location from N_2H^+ emission is thus not as straightforward as was generally assumed, but, given a good physical model for the target disk, the location can be determined using simple chemistry in combination with radiative transfer modeling.

3. The H_2O snowline

The most important snowline is the water snowline, since the bulk of the oxygen budget and ice mass is in water ice. Unfortunately, this snowline is even more difficult to observe than that for CO. Because of the large binding energy of H_2O, water sublimates off the grains only at temperatures above ~100 K. This means that the snowline is located a few AU from the star in protoplanetary disks, that is, ~0.01″ in the nearest star-forming region Taurus. High angular resolution is thus required to observe it. Furthermore, except for the H_2O line at 183 GHz, the only thermal water lines observable from the ground are those from the less abundant isotopologue $H_2^{18}O$. As such, even ALMA will have great difficulty to detect the water snowline in protoplanetary disks. Cold water lines (<100 K) have been detected from space toward TW Hya with *Herschel* (Hogerheijde *et al.* 2011), but the *Herschel* beam is too large to resolve the snowline. So far, only for the disk around V883 Ori a snowline location has been reported, but this was inferred from a steep drop in the dust optical depth (Cieza *et al.* 2016).

The best way to locate the water snowline may therefore be by applying the same strategy as done for CO, that is, using a chemical tracer. The best candidate to trace the water snowline is HCO^+, because gaseous H_2O is its most abundant destroyer (Phillips *et al.* 1992; Bergin *et al.* 1998). A strong decline in HCO^+ abundance is thus expected when H_2O desorbs from the dust grains. This is corroborated by the results from chemical models. Figure 3 shows the outcome of the three-phase astrochemical code GRAINOBLE (Taquet *et al.* 2014) for the 1D temperature and density structure of the envelope around NGC1333 IRAS2A (Kristensen *et al.* 2012). Three-phase models consider reactions for gas-phase species, reactions for species on the ice surface, and reactions for bulk ice species. As expected, the HCO^+ and H_2O abundances show a strong anticorrelation. Hints of this anticorrelation are seen in observations towards the protostar

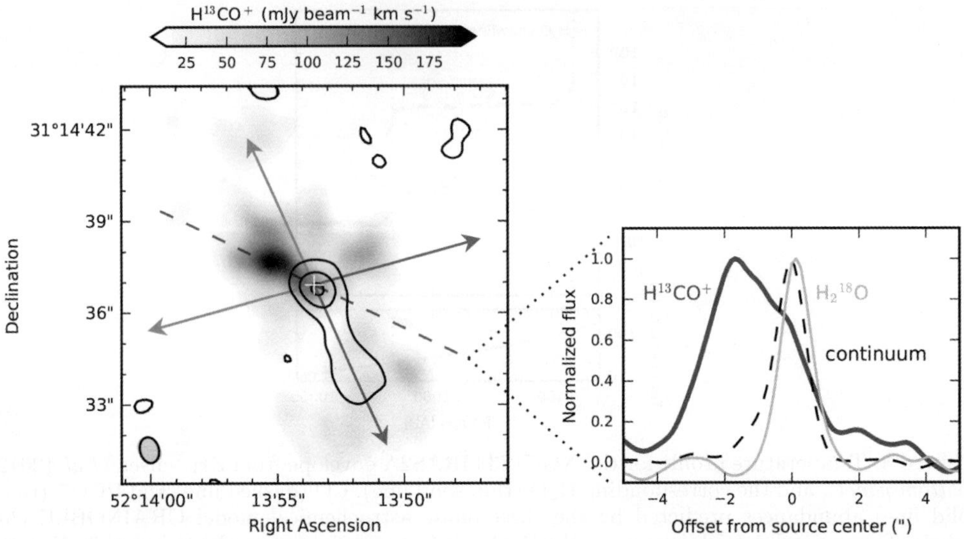

Figure 4. Integrated continuum subtracted intensity map for the $H^{13}CO^+$ $J = 3-2$ transition (grey scale) toward IRAS2A with the $H_2^{18}O$ $3_{1,3}-2_{2,0}$ transition overlaid in contours (*left panel*). The $H_2^{18}O$ contours are shown at 8.2 (1σ) $\times$ [3, 15, 35] mJy beam^{-1} km s^{-1}. The position of the continuum peak is marked by a white cross and the outflow axes by arrows. Both observations have a similar beam size (depicted in the lower left corner). The *right panel* shows the integrated intensities for $H^{13}CO^+$ (thick solid line), $H_2^{18}O$ (thin solid line) and the continuum (dashed line) along the dashed line, normalized to their maximum value.

IRAS 15398-3359, which show ring-shaped $H^{13}CO^+$ emission surrounding CH_3OH emission, another grain-surface molecule with a similar snowline location as water (Jørgensen *et al.* 2013). The non-detection of the high excitation $H_2^{18}O$ $4_{1,4}-3_{2,1}$ line prevented unambigious confirmation, but the detected HDO emission, although more complex, is consistent with the H_2O-HCO^+ anticorrelation scenario (Bjerkeli *et al.* 2016).

Protostellar envelopes, like IRAS 15398-3359, are the best sources to establish whether HCO^+ is a good snowline tracer: the water snowline is located further away from the star than in disks (10s–100s AU rather than a few AU; Harsono *et al.* 2015; Cieza *et al.* 2016), and compact warm water has already been observed toward four sources (Jørgensen & van Dishoeck 2010; Persson *et al.* 2012, 2013; Taquet *et al.* 2013). The only thing lacking are thus HCO^+ observations. We have therefore observed the optically thin isotopologue $H^{13}CO^+$ toward the Class 0 protostar NGC1333 IRAS2A using NOEMA (van 't Hoff *et al.* 2017). Comparison with the $H_2^{18}O$ emission presented by Persson *et al.* (2012) shows that while $H_2^{18}O$ peaks on source, $H^{13}CO^+$ has its emission peak $\sim 2''$ off source (see Fig. 4). As a first analysis, we performed a 1D radiative transfer calculation with Ratran (Hogerheijde & van der Tak 2000), using the 1D temperature and density structure derived by Kristensen *et al.* (2012) from DUSTY modeling (Ivezić & Elitzur 1997) of the continuum emission. A simple parametrized abundance profile for $H^{13}CO^+$ with sharp decreases inside the H_2O snowline and outside the CO snowline, as predicted by a full chemical model, can reproduce the observed location of the emission peak (see Fig. 5). The $H^{13}CO^+$ abundance drops outside the CO snowline because its parent molecule, CO, is frozen out. The temperature, however, has to be increased by a factor of $\sim$2 to reproduce the observed peak location with a drop in $H^{13}CO^+$ at 100 K. A constant $H^{13}CO^+$ abundance at all radii, on the other hand, produces emission peaking on source, unlike observed. These results suggest that water and HCO^+ are indeed anticorrelated,

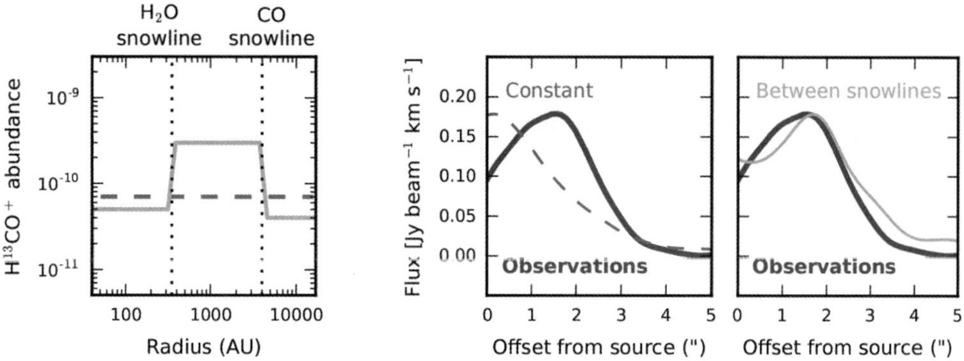

Figure 5. Different $H^{13}CO^+$ abundance profiles (*left panel*) and the resulting simulated integrated emission along the northeastern part of the radial cut shown in Fig. 4 (*right panels*). The vertical dotted lines in the *left panel* mark the H_2O and CO snowlines around 100 K and 20 K, resp. The observed $H^{13}CO^+$ emission is shown with the thick solid lines in the *right panels*.

and provide a proof of concept that the optically thin isotopologue $H^{13}CO^+$ can be used to trace the water snowline.

4. Summary and outlook

Due to its close proximity to the central star and the fact that only the less abundant isotopologue $H_2^{18}O$ can be observed from the ground, both high angular resolution and high sensitivity are required to observe the water snowline. Chemical imaging may therefore be the only way to locate this snowline in protoplanetary disks. This approach has proven useful for the CO snowline, although simple chemical considerations have to be taken into account when deriving the CO freeze-out radius from N_2H^+ emission. The best candidate to image the water snowline is HCO^+, because its main destructor is gaseous H_2O. The first observations of both $H^{13}CO^+$ and $H_2^{18}O$ toward the protostellar envelope of NGC1333 IRAS2A show that these molecules are indeed spatially anti-correlated. This suggest that $H^{13}CO^+$ may be used as a tracer of the water snowline in protoplanetary disks.

Acknowledgments

I would like to acknowledge and thank Catherine Walsh, Mihkel Kama, Stefano Facchini, Magnus Persson, Daniel Harsono, Vianney Taquet, Jes Jørgensen, Ruud Visser, Edwin Bergin and Ewine van Dishoeck for their contributions to this work.

References

Aikawa, Y., Furya, K., Nomura, H., & Qi, C. 2015, *ApJ*, 807, 120
Ansdell, M., Williams, J. P., & van der Marel, N. 2016, *ApJ*, 828, 46
Bergin, E. A., Melnick, G. J., & Neufeld, D. A. 1998, *ApJ*, 499, 777
Bjerkeli, P., Jørgensen, J. K., Bergin, E. A., *et al.* 2016, *A&A*, 595, A39
Cieza, L. A., Casassus, S., Tobin, J., *et al.* 2016, *Nature*, 535, 258
Eistrup, C., Walsh, C., & van Dishoeck, E. F. 2016, *A&A*, 595, A83
Favre, C., Bergin, E. A., Cleeves, L. I., Hersant, F., Qi, C., & Aikawa, Y. 2015, *ApJL*, 802, L23
Herbst, E. & van Dishoeck, E. F. 2009, *ARA&A*, 47, 427
Harsono, D., Bruderer, S., & van Dishoeck, E. F. 2015, *A&A*, 582, A41

Hogerheijde, M. R. & van der Tak, F. F. S. 2000, *A&A*, 362, 697
Hogerheijde, M. R., Bergin, E. A., Brinch, E. A., *et al.* 2011, *Science*, 334, 338
Huang, J., Öberg, K. I., Qi, C., *et al.* 2017, *ApJ*, 835, 231
Ivezić, Z. & Elitzur, M. 1997, *MNRAS*, 287, 799
Jørgensen, J. K. & van Dishoeck, E. F. 2010, *ApJL*, 710, L72
Jørgensen, J. K., Visser, R., Sakai, N., *et al.* 2013, *ApJL*, 779, L22
Kama, M., Bruderer, S., van Dishoeck, E. F., *et al.* 2016, *A&A*, 592, A83
Kristensen, L. E., van Dishoeck, E. F., Bergin, E. A., *et al.* 2012, *A&A*, 542, A8
Madhusudhan, N., Amin, M. A., & Kennedy, G. M. 2014, *ApJL*, 794, L12
Mathews, G. S., Klaassen, P. D., Juhász, A., *et al.* 2013, *A&A*, 557, A132
Miotello, A., van Dishoeck, E. F., Williams, J. P., *et al.* 2017, *A&A*, 599, A113
Öberg, K. I., Murray-Clay, R., & Bergin, E. A. 2011, *ApJL*, 743, L16
Öberg, K. I. & Bergin, E. A. 2016, *ApJL*, 831, L19
Persson, M. V., Jørgensen, J. K., & van Dishoeck, E. F. 2012, *A&A*, 541, A39
Persson, M. V., Jørgensen, J. K., & van Dishoeck, E. F. 2013, *A&A*, 549, L3
Phillips, T. G., van Dishoeck, E. F., & Keene, J. 1992, *ApJ*, 399, 533
Qi, C., Öberg, K. I., Wilner, D. J., *et al.* 2013, *Science*, 341, 630
Qi, C., Öberg, K. I., Andrews, S. M., *et al.* 2015, *ApJ*, 813, 128
Ros, K. & Johansen, A. 2013, *A&A*, 552, A137
Schoonenberg, D. & Ormel, C. W. 2017, *A&A*, 602, A21
Schwarz, K., Bergin, E. A., Cleeves, L. I., *et al.* 2016, *ApJ*, 823, 91
Stevenson, D. J. & Lunine, J. I. 1988, *Icarus*, 75, 146
Taquet, V., López-Sepulcre, A., Ceccarelli, C., *et al.* 2013, *ApJL*, 768, L29
Taquet, V., Charnley, S. B., & Sipilä, O. 2014, *ApJ*, 791, 1
van 't Hoff, M. L. R., Walsh, C., Kama, M., Facchini, S., & van Dishoeck, E. F. 2017, *A&A*, 599, A101
van 't Hoff, M. L. R., Persson, M. V., Harsono, D., *et al.* 2017, *A&A*, submitted
Walsh, C., Nomura, H., & van Dishoeck, E. F. 2015, *A&A*, 582, A88
Zhang, K., Bergin, E.A., Blake, G.A., Cleeves, L.I., & Schwarz K.R. 2017 *Nat. Astron.*, 1, 0130

Astrochemistry VII – Through the Cosmos from Galaxies to Planets
Proceedings IAU Symposium No. 332, 2017
M. Cunningham, T. Millar & Y. Aikawa, eds.

doi:10.1017/S1743921317007050

ALMA observations of Titan's atmospheric chemistry and seasonal variation

M. A. Cordiner[1,2], J. C. Lai[1,3], N. A. Teanby[4], C. A. Nixon[1], M. Y. Palmer[1,2], S. B. Charnley[1], A. E. Thelen[5], E. M. Molter[6], Z. Kisiel[7], V. Vuitton[8], P. G. J. Irwin[9] and M. J. Mumma[1]

[1]NASA Goddard Space Flight Center, 8800 Greenbelt Road, Greenbelt, MD 20771, USA
email: martin.cordiner@nasa.gov

[2]Department of Physics, Catholic University of America, Washington, DC 20064, USA

[3]McMaster University, Hamilton, Ontario, Canada

[4]School of Earth Sciences, University of Bristol, Wills Memorial Building, Queens Road, Bristol, UK

[5]New Mexico State University, 1780 E University Ave, Las Cruces, NM 88003, USA

[6]Astronomy Department, University of California, Berkeley, CA 94720, USA

[7]Institute of Physics, Polish Academy of Sciences, Al. Lotnikøw 32/46, 02-668 Warszawa, Poland

[8]Université Grenoble Alpes, CNRS, IPAG, F-38000 Grenoble, France

[9]Clarendon Laboratory, University of Oxford, Parks Road, Oxford, UK

Abstract. Results are presented from our ongoing studies of Titan using ALMA during the period 2012-2015, including a confirmation of the previous detection of vinyl cyanide (C_2H_3CN), as well as the first spatial map for this species on Titan. Simultaneous mapping of HC_3N, CH_3CN and C_2H_5CN reveal characteristic abundance patterns for each species that provide insight into their individual photochemical lifetimes, and help inform our understanding of Titan's unique, time-variable atmospheric chemistry and global circulation. A time-sequence of HC_3N maps covering 38 months reveals a dramatic change in the distribution of this gas consistent with high-altitude photochemical production followed by advection towards the southern (winter) pole, combined with rapid loss in the north after Titan's 2009 seasonal equinox. The 2015 C_2H_3CN and C_2H_5CN maps show abundance peaks in Titan's southern hemisphere, similar to those observed for the short-lived HC_3N molecule. The longer-lived CH_3CN, on the other hand, remains more concentrated in the north.

Keywords. Planets and satellites: individual (Titan), line: identification, techniques: interferometric, techniques: spectroscopic, astrochemistry, astrobiology

1. Introduction

Titan, Saturn's largest moon, has a carbon and nitrogen-rich atmosphere that hosts a complex photochemistry, resulting in the synthesis of a broad array of organic compounds including hydrocarbons, nitrogen-bearing and some oxygen-bearing species (see the review by Bézard *et al.* 2014). By virtue of their large dipole moments, several of Titan's nitriles possess strong rotational transitions in the wavelength range accessible to the Atacama Large Millimeter/submillimeter Array (ALMA). As a result of short photochemical lifetimes (on the order of a few years or less — see Wilson & Atreya 2004; Loison *et al.* 2015), some of these nitrile molecules can be used to probe the seasonally-variable chemistry and dynamics in Titan's upper atmosphere (e.g. Teanby *et al.* 2012; Cordiner *et al.* 2014; Vinatier *et al.* 2015). Understanding the complex chemistry at play in Titan's atmosphere is a work in progress, and several important problems remain unsolved, such as the origin and composition of the haze particles that envelop Titan, the source of the large negative ions found in the ionosphere, and the long-term behaviour of Titan's climate (Hörst 2017). The advent of ALMA provides a powerful new tool for the study of Titan's atmospheric composition and dynamics. With the end of the Cassini mission in September 2017, there is a strong need for continued, detailed molecular mapping to help test and refine atmospheric chemical/dynamical models for application to Titan and other carbon-rich planetary atmospheres. This will lead to improvements in our understanding of the possible chemical processes occurring on the early Earth, as well as other primitive bodies in our own Solar System and beyond.

The distribution of HC_3N has previously been studied by Cassini through its infrared (vibrational) emission, but CH_3CN, C_2H_3CN and C_2H_5CN have relatively weak IR bands, and have so far only been detected at radio/sub-mm wavelengths. Mapping all these species is now possible with ALMA, and each one provides a unique probe of dynamics and chemistry due to its individual production altitude and photochemical lifetime.

Vinyl cyanide (or acrylonitrile; C_2H_3CN), is currently of particular interest due to the work of Stevenson *et al.* (2015), who showed, through theoretical liquid-phase calculations, that this molecule may form thermodynamically stable membranes (known as azotosomes) in liquid methane at the surface temperature of Titan (approximately 94 K). Given that Titan possesses seas of liquid hydrocarbon, vinyl cyanide could be a strong candidate for forming membranes of potential astrobiological importance. Although the presence of C_2H_3CN on Titan was previously inferred from mass-spectrometric measurements of its protonated form and (neutral) dissociation fragments (Vuitton *et al.* 2007; Cui *et al.* 2009), the first spectroscopic detection was made only recently by Palmer *et al.* (2017), who reported observations of three mm-wave emission lines in ALMA Band 6. Here we report the detection and mapping of three additional lines of C_2H_3CN in Band 7, as well as data on C_2H_5CN, CH_3CN and HC_3N. Further details of this work are given by Lai *et al.* (2017).

2. Observations

The observations presented here come from short (2-3 min) integrations of Titan, obtained for the routine calibration of the ALMA flux scale. Raw interferometric data were obtained from the ALMA Science Archive covering five epochs in the range 2012-07-04 to 2015-08-29. Observed frequencies were in the range 221-349 GHz, with a spectral resolution of 976 kHz. Data were calibrated and processed as described by Cordiner *et al.* (2015). To produce spectral line maps, Titan's continuum was subtracted from the visibility amplitudes before imaging. The point-spread function was deconvolved using the Högbom algorithm, with natural visibility weighting and a threshold flux level of twice the RMS noise in each image. The resulting spatial resolution (FWHM of the Gaussian restoring beam) ranged from $0.48 \times 0.53''$ to $0.62'' \times 1.22''$. For comparison, the angular diameter of Titan's surface was $0.66''$ to $0.78''$ on the sky.

3. Results

An example spatially integrated spectrum (from 2015-04-29) is shown in Fig. 1, with multiple lines of C_2H_3CN, C_2H_5CN, CH_3CN, and vibrationally excited HC_3N identified and labeled. In addition, two lines of C_3H_8 (propane) were tentatively detected, as well as two unidentified (U) lines at 347635 MHz and 347719 MHz. For further details of

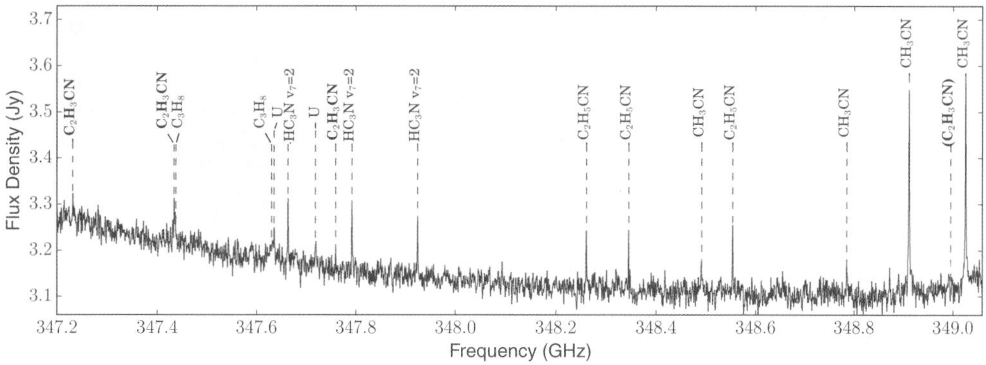

Figure 1. ALMA Band 7 spectrum of Titan observed 2015-04-29. Prominent emission lines are assigned by species; bold typeface highlights the C_2H_3CN emission features. The upwards slope towards lower frequencies is due to the pressure-broadened CO ($J = 3 - 2$) line wing.

the line assignments and radiative transfer modeling see Lai *et al.* (2017). The spatial distribution of flux from each species is shown in the contour maps in Fig. 2, in which the intensity is indicated with an orange color scale and the black contours correspond to increments of $n\sigma$ (where σ is the RMS noise in each map and n is a constant factor). The spatial scale is in sky-projected distances with respect to Titan's centre.

The three new C_2H_3CN emission lines detected improve the confidence level of our previous detection at lower frequency (Palmer *et al.* 2017). They also allow us to map the spatial distribution of this molecule for the first time. Our C_2H_3CN map (Figure 2; top left) shows a flux peak (at 5σ confidence) in Titan's southern hemisphere, consistent with enrichment of this gas over the south polar region. A similar, southerly enrichment is also evident for HC_3N and C_2H_5CN. Polar enrichments have previously been observed for a number of hydrocarbons and nitriles by Cassini and ALMA (Teanby *et al.* 2013; Cordiner *et al.* 2015; Vinatier *et al.* 2015). Titan's seasonal cycle is governed by the 29.5 yr orbital period of Saturn. It is theorized that as Titan transitions out of northern winter towards the summer solstice, the main atmospheric circulation cell, responsible for the redistribution of photochemical products from mid-latitudes towards the poles, should reverse its direction (e.g. Teanby *et al.* 2012). During the progrssion from the 2009 equinox to the southern winter solstice in 2017, the abundances of high-altitude photochemical products are thus expected to begin increasing in the southern polar region, while reactive species concentrated in the north would begin to be destroyed at a rate in accordance with the photochemical lifetime of each molecule.

The southern enhancement in C_2H_3CN abundance (about a factor of 3 stronger than in the north) is similar to that of HC_3N at this epoch, whereas the enhancement factor for C_2H_5CN is somewhat less (about a factor of 1.5), implying a shallower vertical gradient and a longer photochemical lifetime for C_2H_5CN compared to HC_3N and C_2H_3CN. Furthermore, we see a striking difference in the HC_3N distribution compared with the 2012 observations presented by Cordiner *et al.* (2015) — the HC_3N peak just three Earth years earlier was firmly in the north, so our ALMA data demonstrate rapid temporal variability of Titan's atmosphere. A factor of possible relevance is the difference in energy between the ground vibrational state transition ($v = 0$, $E_u = 150$ K) observed by Cordiner *et al.* (2015) and the excited-state ($v = 2$, $E_u = 970$ K) transition shown in Figure 2. We therefore need to consider the possibility that the previous HC_3N observation is more sensitive to colder gas, at lower altitudes than that probed by the vibrationally-excited HC_3N lines. To perform a more robust study of HC_3N temporal evolution, we need to

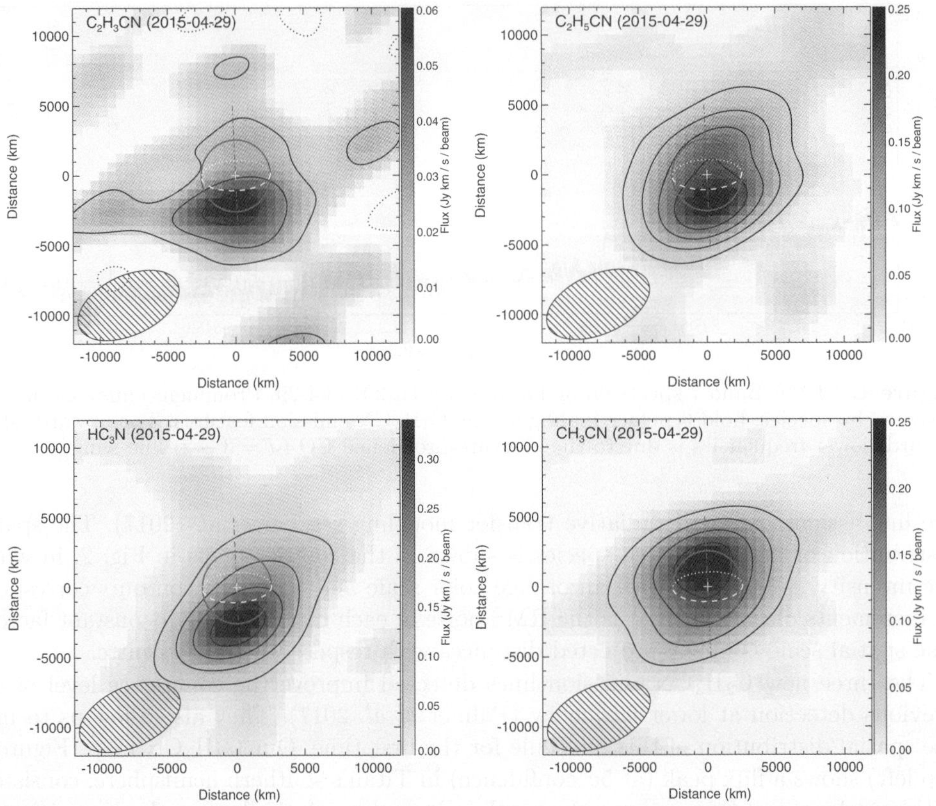

Figure 2. Flux contour maps for C_2H_3CN, C_2H_5CN, HC_3N, and CH_3CN. The coordinate scale is in Titan-projected distances. Titan's surface (blue circle) and polar axis (dot-dashed line) are marked. Contour intervals (in units of σ — the RMS noise level of each map), are as follows: C_2H_3CN: 1.5σ, C_2H_5CN: 3σ, HC_3N: 3σ, CH_3CN: 4σ. The FWHM of the Gaussian restoring beam ($1.22'' \times 0.62''$) and its orientation is shown with hatched ellipses.

compare transitions that all originate from similar (ground-state) vibrational energy levels. Ideally, the exact same ΔJ transition should be observed across multiple epochs, but for the present study, we have to work within the limitations of the serendipitous HC_3N detections available in the ALMA Science Archive.

Figure 3 shows a time series of HC_3N observations from 2012-07-04 to 2015-08-29. These observations include the transitions $J = 26 - 25, 40 - 39$ and $35 - 34$ in the vibrational ground state, and are expected to probe similar regions of the atmosphere. The $J = 26 - 25$ transition was observed on both the first and the final epoch, which helps rule out differences in thermal excitation as a cause for the observed variations. As explained by Lai *et al.* (2017), for the purpose of the present study, 5-10 K differences in temperature between Titan's north and south polar regions (as seen in Cassini CH_4 temperature data) are expected to have negligible impact on the relative emission intensity from these lower-energy HC_3N transitions observed by ALMA.

The panels on the right of Figure 3 show the total HC_3N flux for each epoch, extracted along a line from south to north across Titan's disk. These 'latitudinal profiles' provide a clearer view of the temporal variation, and highlight the tendency for HC_3N to be strongly peaked around both poles. The relative strength of the southern peak increases year-on-year, consistent with the buildup of HC_3N at the winter pole as a result of transport from its main production site at mid-latitudes. Between 2014 to 2015, the

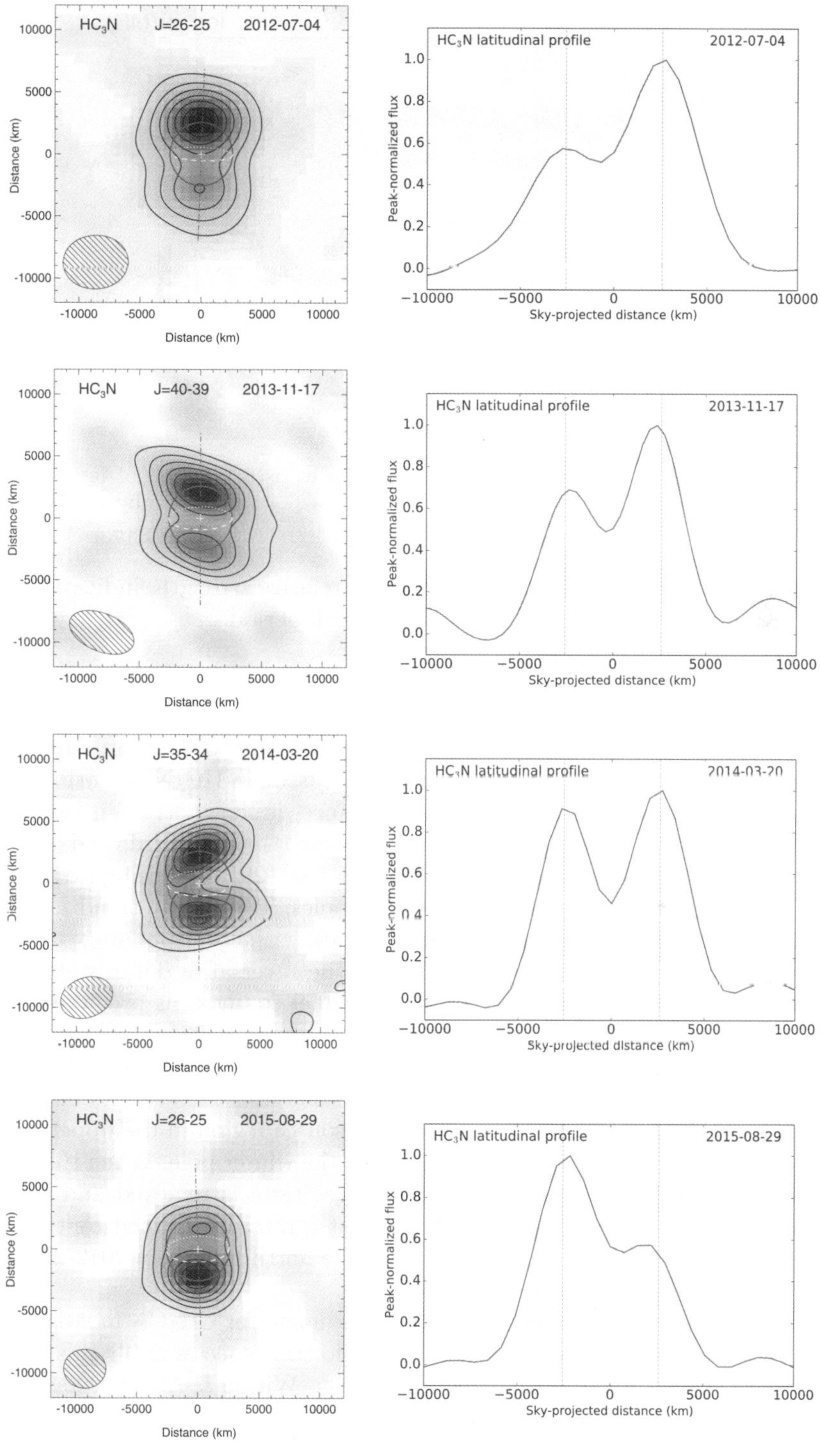

Figure 3. Titan HC_3N flux contour maps (left) and latitudinal profiles (right) for four epochs. Latitudinal profiles show the flux extracted along the polar axis from south to north (dashed line). Fluxes are normalized to the peak value (S_p) and the contour intervals are in units of $S_p/7$. Vertical lines in right-hand panels indicate the positions of Titan's S and N (polar) limbs.

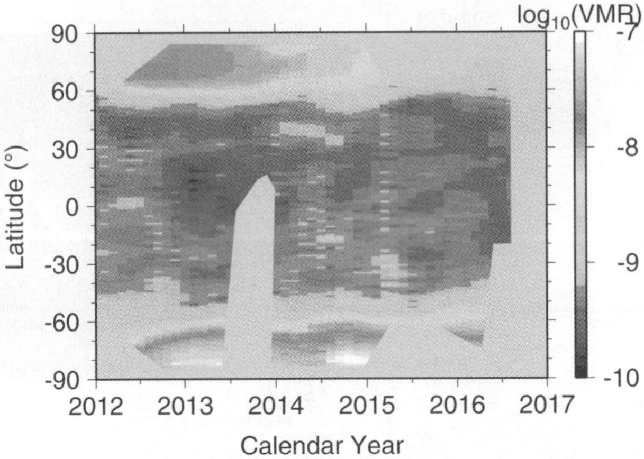

Figure 4. Cassini CIRS stratospheric HC_3N abundances as a function of latitude and time, using nadir observations of the vibronic emission band at 663 cm^{-1} Teanby *et al.* (2017). Grey indicates missing data.

northern peak underwent a dramatic reduction in relative strength, indicative of a rapid loss of HC_3N around the north pole over a 17 month period.

4. Discussion

The loss of photochemically-derived gases from the north and their buildup in the south since 2012 is consistent with the trend observed by Cassini in HC_3N and other short-lived molecules on Titan. A time series of HC_3N abundances is shown as a function of latitude in Figure 4. These data are based on observations using the Cassini Infrared Spectrometer (CIRS), following the method outlined by Teanby *et al.* (2010). This figure also confirms the degree to which HC_3N is confined about the poles. The steady trend for decreasing HC_3N abundances at the north pole ($> 60°$ latitude), with corresponding, rapid increase at the south pole ($< -60°$) since 2012 supports the reversal of Titan's global Hadley-type circulation cell within a few years after the 2009 equinox, as predicted by general circulation models (e.g. Newman *et al.* 2011; Lebonnois *et al.* 2012). The circulation cell occurs due to warm air rising over the summer hemisphere and sinking in the vicinity of the winter pole, and extends to altitudes above 500 km. This results in advection of gases (including the observed nitriles), from their source region in the upper atmosphere to the lower atmosphere, creating enrichment near the winter pole. When this circulation pattern reverses after equinox, the advective source term is removed and the enriched gases disperse and photochemically break down at a rate determined by their lifetime. The rapid (order of magnitude) loss of HC_3N at the north pole from 2012-2016 confirms the short chemical lifetime for this species.

The persistence of a northern-hemisphere enhancement for CH_3CN in 2015 (see Figure 2, bottom right), is indicative of a longer lifetime for this species in Titan's atmosphere, which agrees with prior theoretical expectations (e.g. Wilson & Atreya 2004; Krasnopolsky 2009). The more recent model of Loison *et al.* (2015) also found longer lifetimes for CH_3CN and C_2H_5CN than for HC_3N and C_2H_3CN. These differences in reactivity are related to molecular structure. For example, CH_3CN and C_2H_5CN are fully saturated, whereas HC_3N and C_2H_3CN are unsaturated. The saturated molecules are less likely to react with hydrogen-rich species in Titan's reducing atmosphere, which explains their longer lifetimes. Photolysis rates for CH_3CN and C_2H_5CN are also predicted to be lower than for HC_3N and C_2H_3CN as a result of UV line shielding by CH_4 (J. C. Loison 2017, private communication).

We find reasonable qualitative agreement between the photochemical lifetimes apparent from observations and those predicted by chemical models. This suggests that current theoretical understanding of Titan's complex nitrile chemistry is reasonably good. However, the chemistry of the observed nitriles is currently not fully constrained, which limits their use as probes of Titan's atmospheric dynamics. Additional laboratory experiments will be required to improve our knowledge of their chemical formation and loss pathways, which will be vital for a full understanding of the chemistry of these species in planetary atmospheres, as well as interstellar environments.

5. Conclusion

ALMA observations of Titan's atmospheric gases, including the first map of C_2H_3CN and a detailed time series of HC_3N measurements, are consistent with rapid photochemical production, transport and loss of these molecules over periods of a few Earth years. The abundance patterns for the different nitriles are explained as a result of their individual photochemical lifetimes, combined with the latitudinally-dependent, seasonal variations in atmospheric temperature and radiation input that give rise to changes in the direction of Titan's high-altitude global circulation cell.

The ability to obtain hemispheric maps of short-lived nitriles confirms the important role for ALMA in continuing Cassini's legacy for detailed studies of Titan's climate and chemistry. Ongoing observational, laboratory and theoretical studies are required to improve our understanding of Titan's atmospheric chemistry and dynamics, with applications to primitive planetary atmospheres in our Solar System and beyond.

Acknowledgements

This work was supported by The Goddard Center for Astrobiology, the National Science Foundation (under Grant No. AST-1616306) and the UK Science and Technology Facilities council. It makes use of ALMA data set ADS/JAO.ALMA#2013.1.00033.S. ALMA is a partnership of ESO, NSF (USA), NINS (Japan), NRC (Canada), and NSC and ASIAA (Taiwan), in cooperation with the Republic of Chile. The Joint ALMA Observatory is operated by ESO, AUI/NRAO, and NAOJ. The NRAO is a facility of the National Science Foundation operated under cooperative agreement by Associated Universities, Inc.

References

Bézard, B., Yelle, R., & Nixon, C. A. 2014, in Titan: Surface, Atmosphere and Magnetosphere, ed. I. Muller-Wodarg, C. Griffith, E. Lellouch, & T. Cravens (Cambridge, UK: Cambridge Univ. Press), 158

Cordiner, M. A., Nixon, C. A., Teanby, N. A., *et al.* 2014, *ApL*, 795, L30

Cordiner, M. A., Palmer, M. Y., Nixon, C. A., *et al.* 2015, *ApL*, 800, L14

Cui, J., Yelle, R. V., Vuitton, V., *et al.* 2009, *Icarus*, 200, 581

Hörst, S. M. 2017, *JGRE*, 122, 432

Lai, J. C., Cordiner, M. A., Nixon, C. A. *et al.* 2017, *AJ*, submitted

Lebonnois, S., Burgalat, J., Rannou, P., & Charnay, B., 2012, *Icarus*, 218, 707

Loison, J. C., Hébrard, E., Dobrijevic, M., *et al.* 2015, *Icarus*, 247, 218

Krasnopolsky, V. A. 2009, *Icarus*, 201, 226

Newman, C. E., Lee, C., Lian, Y. *et al.* 2011, *Icarus*, 213, 63

Palmer, M. Y., Cordiner, M. A., Nixon, C. A., *et al.* 2017, *Sci. Adv.*, 3, e1700022

Stevenson, J., Lunine, J., & Clancy, P. 2015, *SciA*, 1, e1400067
Teanby, N. A., Bezard, B., Vinatier, S. *et al.* 2017, *Nature Communications*, 8, 1586
Teanby, N. A., Irwin, P. G. J., de Kok, R., & Nixon, C. A. 2010, *ApJ*, 724, L84
Teanby, N. A., Irwin, P. G. J., Nixon, C. A., *et al.* 2012, *Natur*, 491, 732
Teanby, N. A., Irwin, P. G. J., Nixon, C. A., *et al.* 2013, *P&SS*, 75, 136
Vinatier, S., Bézard, B., Lebonnois, S., *et al.* 2015, *Icarus*, 250, 95
Vuitton, V., Yelle, R. V., & McEwan, M. J. 2007, *Icarus*, 191, 722
Wilson, E. H. & Atreya, S. K. 2004, *JGRE*, 109, 6002

Astrochemistry VII – Through the Cosmos from Galaxies to Planets
Proceedings IAU Symposium No. 332, 2017
M. Cunningham, T. Millar & Y. Aikawa, eds.

doi:10.1017/S1743921317007013

Unveiling the mid-plane temperature and mass distribution in the giant-planet formation zone

Ke Zhang[1], Edwin A. Bergin[1], Geoffrey A. Blake[2], L. Ilsedore Cleeves[3] and Kamber R. Schwarz[1]

[1]Department of Astronomy, University of Michigan, 311 West Hall, 1085 S University Ave., Ann Arbor, MI 48109, USA
email: kezhang@umich.edu

[2]Division of Geological & Planetary Sciences, California Institute of Technology, MC 150-21, Pasadena, CA 91125, USA

[3]Harvard-Smithsonian Center for Astrophysics, 60 Garden Street, Cambridge, MA 02138, USA

Abstract. Core-accretion theory predicts that the formation of giant planets predominantly occurs at the dense mid-plane of the inner ~50 AU of protoplanetary disks. However, due to observational limitation, this critical region remains to be the least charted area in protoplanetary disks. With its great sensitivity, ALMA recently started to image optically thin line emissions arisen from the mid-plane of the inner 50AU in nearby disks, which unlocks an exciting new path to directly constrain the physical properties of the giant planet formation zone through gas tracers. Here we present the first spatially resolved observations of the $^{13}C^{18}O$ J=3-2 line emission in the TW Hya disk. We show that this emission is optically thin even inside the CO mid-plane snowline. Combining it with the $C^{18}O$ J=3-2 images and the previously detected HD J=1-0 flux, we directly constrain the mid-plane temperature and optical depths of the CO gas and dust. We report a mid-plane CO snowline at 20.5 ± 1.3 AU, a mid-plane temperature distribution of $27^{+4}_{-3}\times(R/20.5AU)^{-0.47^{+0.06}_{-0.07}}$ K, and a gas mass distribution of $13^{+8}_{-5}\times(R/20.5AU)^{-0.9^{+0.4}_{-0.3}}$ g cm^{-2} between 5-20.5 AU in the TW Hya protoplanetary disk. We find a total gas/mm-sized dust mass ratio of 140 ± 40 in this region, suggesting that ~2.4 earth mass of dust aggregates have grown to > cm sizes (and perhaps much larger).

Keywords. planet formation, protoplanetary disks, astrochemistry

1. Introduction

Protoplanetary disks set the stage for planet formation. In particular, the equatorial plane of a disk, the so-called *mid-plane,* is the dominant location where planet formation occurs. The mid-plane region is the densest place along the vertical direction of a disk and it is the bulk mass reservoir for dust growth, drift, planetesimal formation, and ultimately the birth site of a planetary system. As a result, the basic properties of the mid-plane, such as temperature and density, are crucial initial inputs for planet formation models (Pollack *et al.* 1996; Benz *et al.* 2014).

However, even the most basic properties of the mid-plane is largely unconstrained. The most commonly used constraints of the surface mass distribution and the mid-plane temperature structure are based on spatially resolved observations of (sub)mm continuum emission of dust grains. At these wavelengths, the dust continuum emission is expected to be mostly optically thin. If the dust opacity and gas-to-dust mass ratio are constant throughout the disk, the mass distribution and temperature structure can be self-consistently constrained from spatially resolved images of dust continuum emission (e.g. Andrews & Williams 2007; Isella *et al.* 2009). However, the key assumption of constant gas-to-dust mass ratio and dust opacity is highly problematic. Growing observations show that the gas and millimeter-sized dust grains may have distinctive spatial distributions (Andrews *et al.* 2012; Rosenfeld *et al.* 2013b; Zhang *et al.* 2014), and the dust opacities vary significantly with radius in a disk (Pérez *et al.* 2012; ALMA Partnership *et al.* 2015).

Due to these large uncertainty in dust properties, an independent gas tracer is necessary to measure the mass distributions and temperature structures at the mid-plane of protoplanetary disks. But the dominant gas constituent H_2 lacks emissivity at the low temperatures present throughout most regions of the disk and thus cannot be directly measured. Instead, CO is typically used as a tracer of gaseous materials (Piétu *et al.* 2007; Ansdell *et al.* 2016). It has the advantages of being chemically stable in the absence of ionizing radiation and being resistant to freeze-out at the mid-plane out to a large radius (Qi *et al.* 2013).

The radius that CO starts to freeze-out at the mid-plane is called the mid-plane CO snowline. It sets an important boundary – outside the snowline, the CO gas condenses at different vertical heights and thus its lines provide no constraints on physical conditions at the mid-plane (Rosenfeld *et al.* 2013a). Inside the snowline, however, CO vapor has nearly constant abundance in both the radial and vertical directions (except for the very surface layer, where CO is photo-dissociated, but the surface layer only contains a negligible fraction of the total gas column density).

Using CO as a global tracer of mass distribution usually face two problems. First of all, the most abundant CO isotopologues ^{12}CO and ^{13}CO readily become optically thick near the disk surface (Beckwith & Sargent 1993; Piétu *et al.* 2007) and thus their line intensities cannot constrain the bulk mass reservoir at the mid-plane. Moreover, beyond the CO mid-plane snowline, CO gas freezes onto grains at different heights in the disk atmosphere (Schwarz *et al.* 2016). Thus the derived gas structure and, crucially, the radial surface density, is highly model dependent (Flaherty *et al.* 2015; Williams & McPartland 2016).

Here we aim to directly probe the mid-plane region inside the CO mid-plane snowline with an optically thin CO line. We targeted the ^{13}C^{18}O and C^{18}O J=3-2 transitions (on average 38433 and 577 times less abundant than ^{12}C^{16}O in the interstellar medium, respectively) in the TW Hya protoplanetary disks. It is one of the nearest disks (∼55 pc, van Leeuwen 2007), and also is one of the best-studied disk systems. This disk makes a useful analog to the solar nebula: the central star has a stellar mass about 0.6-0.8 $M_\odot$(Webb *et al.* 1999); the disk itself has a total mass (gas and dust) of ∼0.05 $M_\odot$(Bergin *et al.* 2013). This is comparable to the total mass of 0.01 $M_\odot$ in the Minimum Mass Solar Nebula (MMSN) which is the lowest possible mass needed to reproduce the Solar System (Hayashi 1981). The age of TW Hya is 8 ± 4Myr (Donaldson *et al.* 2016), within the expected window for giant planet formation (Chabrier *et al.* 2014).

2. Observations

The observations of the TW Hya protoplanetary disk were carried out with the Atacama Large Millimeter/submillimeter Array (ALMA) on 2015 March 8 (Cycle 3 project #2015.1.00308.S, P.I. Bergin). The lengths of baselines were between 15-460 m (16-492 kλ) and the total on-source integration time was 67.5 min. Three SPWs were used for line observations with a resolution of $\delta\nu$ = 122 kHz (a spectral resolution of 0.23 km s^{-1} after the default Hanning smoothing is applied), and the fourth one was dedicated for continuum observations with a total bandwidth of 1.875 GHz.

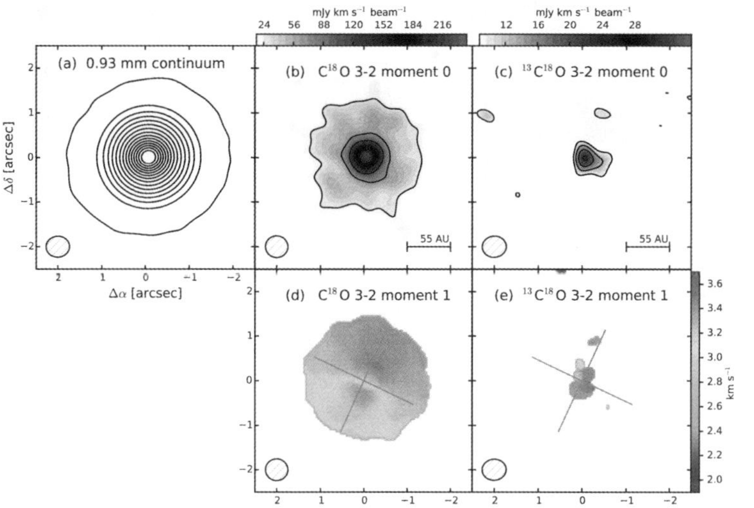

Figure 1. ALMA observations of the $C^{18}O$ and $^{13}C^{18}O$ (3-2) line emission in the TW Hya protoplanetary disk. (a) Dust continuum emission at 0.93 mm. The contours linearly span 10 to 3100σ. (b-c) Integrated emission from the $C^{18}O$ and $^{13}C^{18}O$ J=3-2 transitions. The $C^{18}O$ contour levels start at 3σ and increase in 5σ steps; the $^{13}C^{18}O$ contour levels are [2,4,5,6]σ. (d-e) Intensity weighted velocity maps of $C^{18}O$ and $^{13}C^{18}O$, showing that the emissions arise from a disk in Keplerian rotation.

The visibility data were calibrated by ALMA staff following standard procedures. We further performed self-calibration on the calibrated data in phase and amplitude in CASA 4.6.12, using the dedicated continuum observations. The self-calibrated phase and gain solutions were then applied to the three line SPWs. The data were subsequently continuum subtracted using line free channels. We generated synthesized images using a CLEAN deconvolution with a manual mask (Briggs weighting with a robust parameter of 0.5). For the 321 GHz continuum emission map, the synthesized beam is $0.54'' \times 0.47''$ (position angle = -87 degrees). The noise level of the $C^{18}O$ line images is 6.5 mJy beam^{-1} in each 0.23 km s^{-1} wide velocity bin, and 4 mJy beam^{-1} in the $^{13}C^{18}O$ line image cube.

Both $C^{18}O$ and $^{13}C^{18}O$ J=3-2 line emissions were robustly detected in the TW Hya disk, with a peak-to-noise ratio of 29 and 5, respectively. The synthesized images of line and continuum emission are displayed in Figure 1. Both line emissions show a bright central peak, and $C^{18}O$ line also shows broad diffuse emission in the outer disk. This morphology is consistent with the theoretical expectation that the mid-plane temperature gradually decreases with radius and that beyond the mid-plane CO snowline most of the CO condenses into ice. The central peaks of the $^{13}C^{18}O$ and $C^{18}O$ emission are 15-40% larger than the $0.5''$ ($\sim$27.5 AU) synthesized beam, suggesting that the region inside the CO snowline is partially resolved. In practice, the data provide constraints on smaller spatial scales than the synthesized beam, since it is also spectrally resolved (0.23 km s^{-1} resolution) and the velocity field is known to be Keplerian (see Figure 1 (d-e)). Inside the central bright peak emission, the peak flux of the $C^{18}O$ line emission is only 8 times stronger than the $^{13}C^{18}O$ emission, significantly less than the average isotopic ratio of $^{12}C/^{13}C$ = 69 in the ISM. Therefore, the $C^{18}O$ line emission is optically thick inside the mid-plane CO snowline.

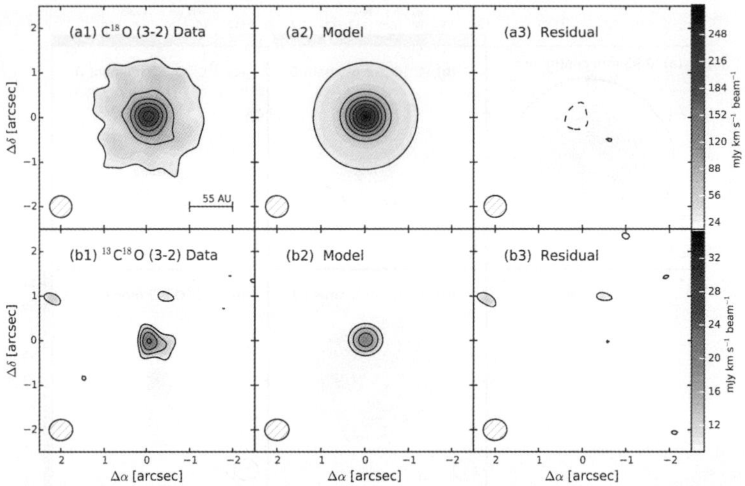

Figure 2. Comparisons of the $C^{18}O$ and $^{13}C^{18}O$ J=3-2 emission from the TW Hya disk with the best fitting model. The contour levels are the same as Figure 1(b-c).

3. Modeling

To retrieve the physical properties in the mid-plane, we use a parameterized disk model to reproduce the observations. Readers can find a full description of the model and the Markov chain Monte Carlo fitting algorithm in Zhang *et al.* (2017) and we briefly describe key elements of the modeling here.

We assume that the $^{13}C^{18}O$ and $C^{18}O$ emission arises predominately from the mid-plane or immediately adjacent regions. As the vertical temperature structures in protoplanetary disks are expected to be nearly isothermal within one scale height (Dullemond & Dominik 2004), both line emission can be characterized by the mid-plane temperature that is only a function of radius. This assumption is reasonable since $C^{18}O$ and $^{13}C^{18}O$ are 10^{-10}–10^{-6} times less abundant than H_2 (Wilson 1999).

To include continuum extinction from dust, we model the $C^{18}O$, $^{13}C^{18}O$ line emission and dust continuum simultaneously. For the dust emission, we use the radial profile of 345 GHz continuum emission of TW Hya derived from 1 AU resolution ALMA long baseline observations (Andrews *et al.* 2016). The advantage of using such high-resolution continuum emission is that we do not need to assume a function for the radial distribution of continuum optical depth. For the line emission, we use simple power-law functions to characterize the radial profiles of the temperature and integrated line strength. The best-fitting model is plotted in Figure 2 in comparison with the data.

4. Results

We report the first direct measurement of the mid-plane CO snowline at 20.5 ± 1.3 AU and a mid-plane temperature distribution of $27^{+4}_{-3}\times(\mathrm{R}/20.5\mathrm{AU})^{-0.47^{+0.06}_{-0.07}}$ K in the TW Hya disk. We find that the $^{13}C^{18}O$ J=3-2 emission is optically thin (τ <0.2) throughout the disk and the dust continuum emission is moderately optically thin (τ ~0.5) inside the CO mid-plane snowline. The two optical depths suggest that the $^{13}C^{18}O$ line emission arises from both sides of the disk and thus provides model-independent constraints on the gas mass distribution. Inside the mid-plane CO snowline, the $C^{18}O$ emission is optically thick (τ ~8) and thus provides direct constraints on the gas temperature instead of the mass distribution. Outside the mid-plane CO snowline, the $C^{18}O$ emission is optically thin. The observations provide constraints on physical properties of the mid-plane between 5-20.5 AU of the TW Hya disk: the outer limit is the mid-plane CO snowline location (beyond that CO condenses at the mid-plane); the inner boundary is limited by the radius of 10% cumulative flux at the highest velocity channel with a $^{13}C^{18}O$ emission detection.

With the temperature and optical depths from the best-fitting model, we can derive the column density distribution of $^{13}C^{18}O$ and then the gas mass distribution for a given $^{13}C^{18}O$ abundance ($x_{^{13}C^{18}O}$). Here we employ the spatially unresolved HD J=1-0 line flux of the TW Hya disk to estimate the $x_{^{13}C^{18}O}$ (Bergin *et al.* 2013). Combining the uncertainties in our model parameters and in the HD line flux, we find $x_{^{13}C^{18}O}=1.7^{+1.3}_{-0.8}\times10^{-10}$ and $\Sigma_{\rm gas}=13^{+8}_{-5}\times(R/20.5\,{\rm AU})^{-0.87^{+0.38}_{-0.26}}$ g cm^{-2} between 5-20.5 AU (for details, see Zhang *et al.* 2017).

Another constraint from our modeling is the optical depth of dust continuum emission at 0.93 mm between 5-20.5 AU. The optical depth is a combined result of dust mass and opacity. Recent observations have shown that the dust opacity may vary significantly with radius and as a result it is very challenging to measure the surface density profile of the dust mass. Due to these uncertainties, here we only derive a total gas-to-mass ratio for the region inside the CO mid-plane snowline. We adopt a constant dust opacity κ(0.93 mm) = 3.1 cm^{2}g^{-1}, a value similar to previous studies for the TW Hya disk (Andrews *et al.* 2012; Hogerheijde *et al.* 2016).

We measure the total gas-to-mm-sized dust ratio is 140^{+43}_{-31} for the 5-20.5 AU region, much higher than the typical ratio of 100 in molecular clouds. In order to recover a gas-to-dust ratio of 100, we find that $2.4^{+3.2}_{-2.0}\,M_\oplus$ solid masses are missing. This estimation is actually a lower limit, because the dust radial drift to the inner region should in general decrease the total gas-to-dust ratio in the inner disk (Birnstiel *et al.* 2012) and because the overall gas surface density also decreases with time. Therefore more than $2.4\,M_\oplus$ masses of solids may have grown to at least centimeter-sized and perhaps much larger bodies in the TW Hya disk.

Our observations demonstrate the possibility to use optically thin CO isotopologue lines to probe physical properties at the mid-plane of nearby protoplanetary disks. In the next a few years, ALMA will be able to carry out similar observations for more disks and provide critical constraints on the conditions of planet formation sites.

References

ALMA Partnership, Brogan, C. L., Pérez, L. M., *et al.* 2015, ApJL, 808, L3
Andrews, S. M. & Williams, J. P. 2007, ApJ, 671, 1800
Andrews, S. M., Wilner, D. J., Hughes, A. M., *et al.* 2012, ApJ, 744, 162
Andrews, S. M., Wilner, D. J., Zhu, Z., *et al.* 2016, ApJL, 820, L40
Ansdell, M., Williams, J. P., van der Marel, N., *et al.* 2016, ApJ, 828, 46
Beckwith, S. V. W. & Sargent, A. I. 1993, ApJ, 402, 280
Benz, W., Ida, S., Alibert, Y., Lin, D., & Mordasini, C. 2014, Protostars and Planets VI, 691
Bergin, E. A., Cleeves, L. I., Gorti, U., *et al.* 2013, Nature , 493, 644
Birnstiel, T., Klahr, H., & Ercolano, B. 2012, A&A, 539, A148
Chabrier, G., Johansen, A., Janson, M., & Rafikov, R. 2014, Protostars and Planets VI, 619
Donaldson, J. K., Weinberger, A. J., Gagné, J., *et al.* 2016, ApJ, 833, 95
Dullemond, C. P. & Dominik, C. 2004, A&A, 417, 159
Flaherty, K. M., Hughes, A. M., Rosenfeld, K. A., *et al.* 2015, ApJ, 813, 99
Hayashi, C. 1981, PTPS, 70, 35

Hogerheijde, M. R., Bekkers, D., Pinilla, P., *et al.* 2016, A&A, 586, A99
Isella, A., Carpenter, J. M., & Sargent, A. I. 2009, ApJ, 701, 260
Pérez, L. M., Carpenter, J. M., Chandler, C. J., *et al.* 2012, ApJL, 760, L17
Piétu, V., Dutrey, A., & Guilloteau, S. 2007, A&A, 467, 163
Pollack, J. B., Hubickyj, O., Bodenheimer, P., *et al.* 1996, *Icarus*, 124, 62
Qi, C., Öberg, K. I., Wilner, D. J., *et al.* 2013, *Science*, 341, 630
Rosenfeld, K. A., Andrews, S. M., Hughes, A. M., Wilner, D. J., & Qi, C. 2013a, ApJ, 774, 16
Rosenfeld, K. A., Andrews, S. M., Wilner, D. J., Kastner, J. H., & McClure, M. K. 2013b, ApJ, 775, 136
Schwarz, K. R., Bergin, E. A., Cleeves, L. I., *et al.* 2016, ApJ, 823, 91
van Leeuwen, F. 2007, A&A, 474, 653
Webb, R. A., Zuckerman, B., Platais, I., *et al.* 1999, ApJL, 512, L63
Williams, J. P. & McPartland, C. 2016, ApJ, 830, 32
Wilson, T. L. 1999, *Reports on Progress in Physics*, 62, 143
Zhang, K., Bergin, E. A., Blake, G. A., Cleeves, L. I., & Schwarz, K. R. 2017, Nature Astronomy, 1, 0130
Zhang, K., Isella, A., Carpenter, J. M., & Blake, G. A. 2014, ApJ, 791, 42

Astrochemistry VII – Through the Cosmos from Galaxies to Planets
Proceedings IAU Symposium No. 332, 2017
M. Cunningham, T. Millar & Y. Aikawa, eds.

doi:10.1017/S1743921317007773

Reinforcement of the link between stellar metallicity and the zero age orbit of gas giant planet

Rafael Pinotti, Heloisa M. Boechat-Roberty and Gustavo F. Porto de Mello

Observatorio do Valongo, Universidade Federal do Rio de Janeiro,
Ladeira Pedro Antonio 43, 20080-090, Rio de Janeiro, Brazil
email: rpinotti@astro.ufrj.br

Abstract. In 2005 we suggested a relation between the optimal locus of gas giant planet formation, prior to migration, and the metallicity of the host star, based on the core accretion model, and radial profiles of dust surface density and gas temperature. At that time, less than 200 extrasolar planets were known, limiting the scope of our analysis. Here, we take into account the expanded statistics allowed by new discoveries, in order to check the validity of some premises. We compare predictions with the present available data and results for different stellar mass ranges. We find that the zero age planetary orbit (ZAPO) hypothesis continues to hold after a two order of magnitude increase in discovered planets, as well as the prediction that planets around metal poor stars would have shorter orbits.

Keywords. planetary systems, gaseous planets, formation - planets and satellites

---

1. Introduction

Twelve years ago, less than two hundred extrasolar planets were known, and the most successful technique employed was the radial velocity, which was used for the discovery of the first one, 51 Peg (Mayor & Queloz 1995). After that, the transit method became increasingly more productive, and, when used by the Kepler Mission (Morton *et al.* 2016, D'Angelo, Durisen e Lissauer, 2010), surpassed the radial velocity method and helped to unveil thousands of candidate planets, pushing the statistics to the present level of more than 3600 confirmed planets (Schneider *et al.* 2011).

In 2005 we proposed a relation between the optimal locus of gas giant planet formation, prior to migration, and the metallicity of the host star, which can be considered as a proxy of the metallicity of the protoplanetary disc (Pinotti *et al.* 2005). In order to build the model, we assumed a number of premises, the most fundamental of which being that the planet formation mechanism is the core accretion (D'Angelo, Durisen & Lissauer 2010, Pollack *et al.* 1996), which requires that dust evolves to planetesimals, which in turn form a rocky nucleus a few times the Earths mass, which will then capture an appreciable amount of gas from the protoplanetary disc until its final mass reaches approximately 10^2 to 10^3 Earth mass. Radial profiles of disc temperature and dust surface density were obtained from the literature; considering also that the dust surface density profile is altered by change in the discs metallicity, we developed a quantitative relation, which dictates that the optimum region of planet formation shifts outward for higher metallicity, reaching asymptotic values for both high and low values of metallicity Z. This behavior seemed to explain why, with the statistics available at the time, metal poor stars tended to harbor planets with smaller orbital radius when compared with metal rich ones. When used in a plot of stellar metallicity versus planet semi-major axis, the relation forms an S shaped curve, which we called Zero Age Planetary Orbit (ZAPO).

The dearth of gas giant planets around metal poor stars is a well-known observational fact (Mortier *et al.* 2012, Schlaufman and Laughlin 2011, Sozzetti *et al.* 2009, Fisher & Valenti 2005, Gonzalez 1997) and a natural consequence of the core accretion mechanism, since the resulting lower dust surface density would affect the formation of the rocky core (Pollack *et al.* 1996), and because a low metallicity protoplanetary disc has possibly a shorter lifetime, affecting the probability of the formation of a fully mature gas giant planet (Yasui *et al.* 2010, Ercolano & Clarke 2010). Our hypothesis would add a new cause to the observed scarcity of planets around metal poor stars, that is, the smaller formation radius would increase the fraction of planets engulfed by their stars during the migration process. In this work, we used the new available statistics to assess the validity of some of our premises, check if the predictions are still valid, and compare results for different stellar mass ranges.

2. Development of the hypothesis

For more details about the development see Pinotti, Boechat-Roberty and Porto de Mello (2017) and Pinotti *et al.* (2005). Briefly, the probability P of gas giant planet formation as a function of the radius r is proportional to the dust surface density σ_s and the disk mid-plane temperature T,

$$P \propto \sigma_s / T \tag{2.1}$$

The radial profiles of σ_s and T are in the form of power laws:
$\sigma_s \propto \mathrm{r}^{-\alpha}$
$\mathrm{T} \propto \mathrm{r}^{-\beta} + \mathrm{t}$
$\alpha = \alpha(\mathrm{Z})$ and $\beta = \beta(\mathrm{Z})$, where Z is the metallicity of the disk.
The optimum formation radius $\mathrm{r}_{opt}(Z)$ is given by

$$r_{opt}(Z) = \left(\frac{(\beta_a - \alpha_a)}{t\alpha_a}\right)^{\frac{1}{\beta(Z)}} \tag{2.2}$$

where t, α_a and β_a are parameters for the calibration of the ZAPO curve.

The curve exhibits two asymptotic values, for very low and very high metallicity; for the population of migrated planets we assumed $r(Z, n) = nr_{opt}(Z)$, where $n < 1$. We used in our recent analysis the Extrasolar Planets Encyclopedia (Schneider *et al.* 2011) listed 3406 planets as of May 12th, 2016, a number that increases steadily as new discoveries are made in a weekly basis. The criteria for the selection of planets were: data on mass, semi-major axis SMA and eccentricity, maximum and minimum mass, maximum eccentricity, maximum SMA, most massive planet in the case of multiple systems and stars with data on metallicity.

3. Results

In 2005 we used 72 planets to calibrate the ZAPO curve (Fig.1), all of them around stars with 1 ± 0.2 M_{Sun}. In 2016 the available number of adequate planets around stars with the same mass range jumped to 282 (Fig.2). All of them, except two, are at or below the ZAPO curve, as predicted. The two exceptions are planets with eccentricities close to our upper limit of 0.4, indicating planet-planet interaction. Our prediction that

Table 1. Details of the sample of selected planets.

Stellar Mass range $M_\odot$	Average stellar mass $[M_\odot]$	Number of planets
< 0.8	0.65	38
$0.8 \leqslant M_\odot \leqslant 1.2$	1.02	282
$1.2 \leqslant M_\odot \leqslant 1.6$	1.36	138
< 1.6	1.91	46

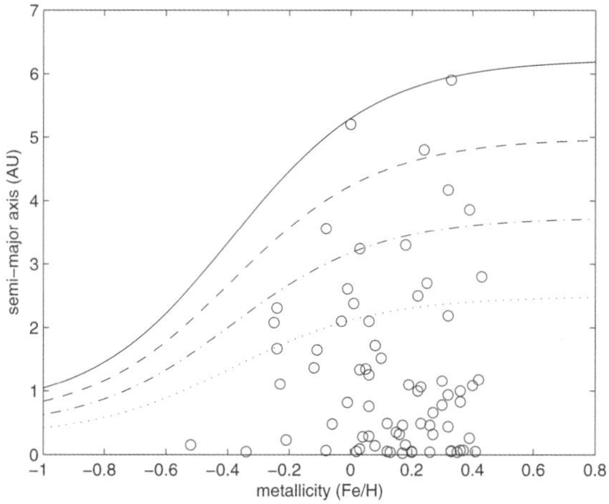

Figure 1. Semi-major axis as a function of metallicity (Fe/H). Reproduction of figure from Pinotti *et al.* 2005, showing the 72 planets used at the time, and ZAPO curve (n=1, full line), and curves for migrated planets: n=0.8 (dashed line), n=0.6 (dash-dot line) and n=0.4 (dotted line).

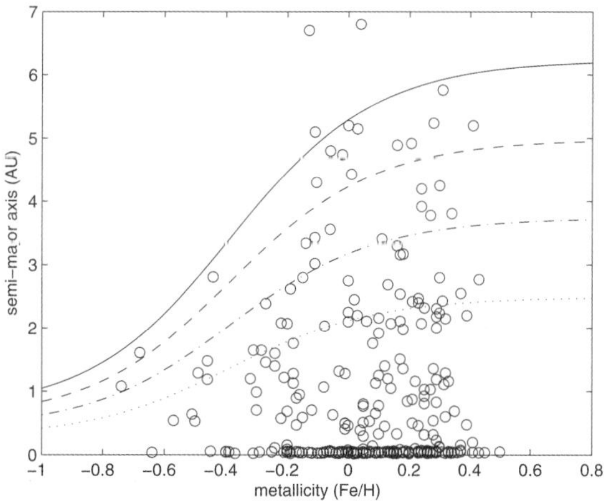

Figure 2. The same ZAPO curves of Figure 1, with the current group of 282 known planets around stars with mass range of $0.8 \leqslant M_{Sun} \leqslant 1.2$ (Pinotti *et al.* 2017).

planets around low metallicity stars will have small semi-major axes is also confirmed The ZAPO curve holds also for other stellar mass ranges.

References

D'Angelo, G., Durisen, R. H., & Lissauer, J. J., 2010, in S. Seager, ed., Exoplanets, University of Arizona Press, Tucson, p. 319

Ercolano, B. & Clarke, C. J. 2010 *MNRAS*, 402, 2735
Fisher, D. A. & Valenti, J. 2005, *ApJ* 622, 1102
Gonzalez, G., 1997, *MNRAS* 285, 403
Mayor, M. & Queloz, D., 1995, *Nature* 378, 355
Mortier A., Santos N. C., Sozzetti, A., Mayor M., Latham, D., Bonfils, X., & Udry, S., 2012, *A& A* 543, A45
Morton, T. D. *et al.*, 2016 *ApJ* 822, 86
Pinotti, R., Arany-Prado, L., Lyra, W., & Porto de Mello, G. F, 2005, *MNRAS* 364, 24
Pinotti, R., Boechat-Roberty, H. M., & Porto de Mello, G. F. 2017, *MNRAS* 28, 490
Pollack, J. B., Hubickyj, O., Bodenheimer, P., Lissauer, J. J., Podolak, M., & Greenzweig, Y., 1996 *Icarus* 124, 62
Schlaufman, K. C. & Laughlin, G., 2011 *ApJ* 738, id. 177
Schneider, J., Dedieu, C., Le Sidaner, P., Savalle R., & Zolotukhin, I., 2011, *A& A* 532, A79
Sozzetti, A., Torres, G., Latham, D. W., Stefanik, R. P., Korzennik, S. G., Boss, A. P., Carney, B. W., & Laird, J. B., 2009, *ApJ* 697, 544
Yasui, C., Kobayashi, N., Tokunaga, A. T., Saito, M., & Tokoku, C., 2010, *ApJ* 723, L113

Astrochemistry VII – Through the Cosmos from Galaxies to Planets
Proceedings IAU Symposium No. 332, 2017
M. Cunningham, T. Millar & Y. Aikawa, eds.

doi:10.1017/S1743921317007797

Possibility to locate the position of the H_2O snowline in protoplanetary disks through spectroscopic observations

Shota Notsu[1,2], Hideko Nomura[3], Catherine Walsh[4], Mitsuhiko Honda[5], Tomoya Hirota[6], Eiji Akiyama[6] and T. J. Millar[7]

[1]Department of Astronomy, Graduate School of Science, Kyoto University, Kitashirakawa-Oiwake-cho, Sakyo-ku, Kyoto 606-8502, Japan
e-mail: snotsu@kusastro.kyoto-u.ac.jp
[2]Research Fellow of Japan Society for the Promotion of Science (DC1)
[3]Department of Earth and Planetary Science, Tokyo Institute of Technology, 2-12-1 Ookayama, Meguro-ku, Tokyo 152-8551, Japan
[4]School of Physics and Astronomy, University of Leeds, Leeds, LS2 9JT, UK
[5]Department of Physics, School of Medicine, Kurume University, 67 Asahi-machi, Kurume, Fukuoka 830-0011, Japan
[6]National Astronomical Observatory of Japan, 2-21-1 Osawa, Mitaka, Tokyo 181-8588, Japan
[7]Astrophysics Research Centre, School of Mathematics and Physics, Queen's University Belfast, University Road, Belfast, BT7 1NN, UK

Abstract. Observationally measuring the location of the H_2O snowline is crucial for understanding the planetesimal and planet formation processes, and the origin of water on Earth. The velocity profiles of emission lines from protoplanetary disks are usually affected by Doppler shift due to Keplerian rotation and thermal broadening. Therefore, the velocity profiles are sensitive to the radial distribution of the line-emitting regions. In our work (Notsu *et al.* 2016, 2017), we found candidate water lines to locate the position of the H_2O snowline through future high-dispersion spectroscopic observations. First, we calculated the chemical composition of the disks around a T Tauri star and a Herbig Ae star using chemical kinetics. We confirmed that the abundance of H_2O gas is high not only in the hot midplane region inside the H_2O snowline but also in the hot surface layer and the photodesorption region of the outer disk. The position of the H_2O snowline in the Herbig Ae disk exists at a larger radius from the central star than that in the T Tauri disk. Second, we calculated the H_2O line profiles and identified that H_2O emission lines with small Einstein A coefficients ($\sim 10^{-6} - 10^{-3}$ s^{-1}) and relatively high upper state energies ($\sim$ 1000K) are dominated by emission from the hot midplane region inside the H_2O snowline, and therefore their profiles potentially contain information which can be used to locate the position of the H_2O snowline. The wavelengths of the H_2O lines which are the best candidates to locate the position of the H_2O snowline range from mid-infrared to sub-millimeter, and the total line fluxes tend to increase with decreasing wavelengths. We investigated the possibility of future observations using the ALMA and mid-infrared high-dispersion spectrographs (e.g., SPICA/SMI-HRS). Since the fluxes of those identified lines from a Herbig Ae disk are stronger than those of a T Tauri disk, the possibility of a successful detection is expected to increase for a Herbig Ae disk.

1. Introduction

Observationally locating the position of the H_2O snowline (Hayashi 1981; Hayashi *et al.* 1985) in a protoplanetary disk is important. It will provide information on the physical and chemical conditions in disks, such as the temperature structure, the dust-grain size distribution, and the water vapor distribution in the disk midplane (e.g., Oka *et al.*

2011; Piso *et al.* 2015), and will give constraints on the current formation theories of planetesimals and planets (e.g., Öberg *et al.* 2011; Okuzumi *et al.* 2012; Ros & Johansen 2013). It will help clarify the origin of water on rocky planets including the Earth (e.g., Morbidelli *et al.* 2000, 2012, 2016; Sato *et al.* 2016). Through recent space and ground infrared spectroscopic observations for protoplanetary disks, some infrared H_2O lines, which mainly trace the disk surface and the photodesorption region of the outer disk, have been detected (for more details, see e.g., Pontoppidan *et al.* 2010a,b; Hogerheijde *et al.* 2011; van Dishoeck *et al.* 2014; Blevins *et al.* 2016; Banzatti *et al.* 2017; Notsu *et al.* 2016, 2017).

The velocity profiles of emission lines from protoplanetary disks are usually affected by Doppler shift due to Keplerian rotation and thermal broadening. Therefore, the velocity profiles are sensitive to the radial distribution of the line-emitting regions. In our work (e.g., Notsu *et al.* 2016, 2017), we calculated the chemical composition and the H_2O line profiles in a T Tauri disk† and a Herbig Ae disk. We investigated the line properties in detail for candidate water lines to locate the position of the H_2O snowline over a wide wavelength range from mid-infrared to sub-millimeter, and discuss the possibility of detecting such lines with future observations. In this proceeding paper, we introduce the outline and key results of our recent work (Notsu *et al.* 2016, 2017).

2. Method & Result

2.1. *The physical models of the protoplanetary disks*

The physical structures of the protoplanetary disk models used here are calculated using the methods in Nomura & Millar (2005) including X-ray heating (Nomura *et al.* 2007). A more detailed description of the background theory and computation of this physical model is described in the original papers (Nomura & Millar 2005; Nomura *et al.* 2007) and our works (e.g., Notsu *et al.* 2015, 2016, 2017). Walsh *et al.* (2010, 2012, 2014, 2015), Heinzeller *et al.* (2011), Furuya *et al.* (2013) used the same physical models for a T Tauri disk and a Herbig Ae disk to study various chemical and physical effects, and they also describe the calculation of the physical structures in detail.

We adopted the physical models of a steady, axisymmetric Keplerian disks surrounding a T Tauri star with mass $M_* = 0.5M_\odot$, radius $R_* = 2.0R_\odot$, and effective temperature $T_* = 4000$K, and a Herbig Ae star with $M_* = 2.5M_\odot$, $R_* = 2.0R_\odot$, and $T_* = 10{,}000$K. We adopt the same compact and spherical dust-grain model of Nomura & Millar (2005). In Figure 1, we display the gas number density and the gas temperature of a Herbig Ae disk (Notsu *et al.* 2017).

2.2. *Chemical structures of the protoplanetary disks*

To investigate the chemical structure of disks, we use a large chemical network which includes gas-phase reactions and gas-grain interactions (freeze-out of gas molecules on dust grains, and thermal and non-thermal desorption from dust grains). The initial fractional abundances (relative to total hydrogen nuclei density) we use are atomic and oxygen-rich low-metallicity from Graedel *et al.* (1982), listed in Table 8 of Woodall *et al.* (2007).

Figure 2 shows the fractional abundances (relative to total gas hydrogen nuclei density, n_H) of H_2O gas and ice in a Herbig Ae disk (Notsu *et al.* 2017). The H_2O snowline of the Herbig Ae disk exists at a radius of $r \sim 14$ au in the midplane (~ 120K), which is significantly larger than that for the T Tauri disk model ($r \sim 1.6$ au, see Figure 2 of

† In the remainder of this paper, we define the protoplanetary disks around T Tauri/Herbig Ae stars as "T Tauri/Herbig Ae disks".

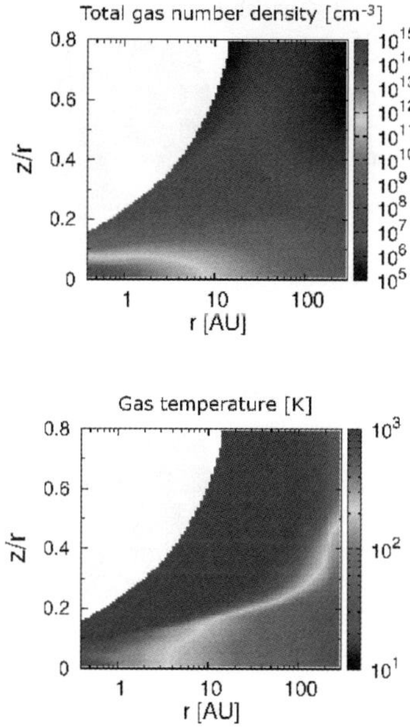

Figure 1. The total gas number density in cm^{-3} (top), the gas temperature in Kelvin (bottom) of a Herbig Ae disk as a function of the disk radius in au and height (scaled by the radius, z/r) up to maximum radius of r =300 au (Notsu *et al.* 2017).

Notsu *et al.* 2016). This is because the gas and dust temperatures, which are coupled in the midplane, are higher in the Herbig Ae disk than in T Tauri disk. We found that the abundance of H_2O is high (up to 10^{-4}) in the inner region with higher temperature ($\gtrsim$ 170K) within $\sim 7-8$ au, relatively high ($\sim 10^{-8}$) between $7-8$ au and 14 au (= the position of the H_2O snowline, ~120K) near the equatorial plane. In addition, it is relatively high ($\sim 10^{-8} - 10^{-7}$) in the hot surface layer and the photodesorbed region of the outer disk, compared to its value ($\sim 10^{-12}$) in the regions outside the H_2O snowline near the equatorial plane (for more details, see Notsu *et al.* 2017).

There is the radial difference between the exact H_2O snowline location (r ~14 au) and the outer edge of the hot water vapor area ($r \sim$ 8 au) in a Herbig Ae disk, although there is no significant difference between them in a T Tauri disk with the radial steeper temperature profile in the disk midplane. It is because the water formation rate by gas-phase reactions strongly depends on the gas temperature. Such water vapor distribution in the Herbig Ae disk midplane was discussed in Woitke *et al.* (2009).

2.3. *Calculations of* H_2O *emission lines from protoplanetary disks*

Using the H_2O gas abundance distribution obtained from our chemical calculation described in the previous paragraph, we calculate the H_2O emission line profiles ranging from near-infrared to sub-millimeter wavelengths from a T Tauri and a Herbig Ae disk assuming Keplerian rotation, and identify the lines which are the best candidates for probing emission from the inner thermally desorbed water reservoir, i.e., within the H_2O snowline. In our work (e.g., Notsu *et al.* 2016, 2017), we adopted the same calculation method to determine the H_2O emission line profiles from a T Tauri disk (based on Rybicki & Lightman 1986, Hogerheijde & van der Tak 2000, Nomura & Millar 2005, and Schöier *et al.* 2005), with the detailed model explained in Section 2.3 of Notsu *et al.* (2016). The code which we have built for calculating emission line profiles is a modification of

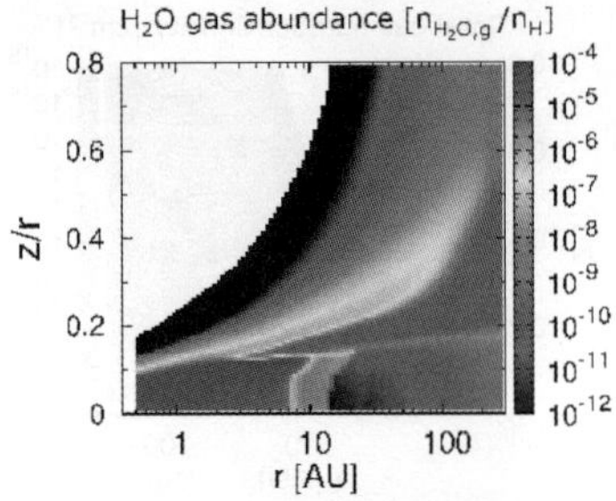

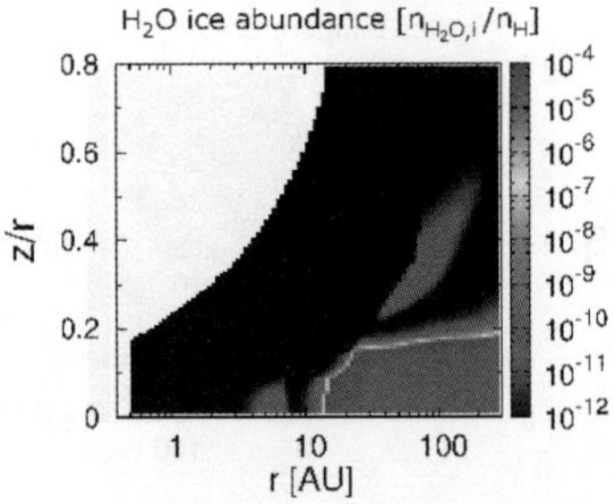

Figure 2. The fractional abundance (relative to total hydrogen nuclei density) distributions of H_2O gas (top) and H_2O ice (bottom) of a Herbig Ae disk as a function of disk radius and height (scaled by the radius, z/r) up to maximum radius of r =300au (Notsu *et al.* 2017).

the original 1D code called RATRAN‡ (Hogerheijde & van der Tak 2000). We adopt the data of line parameters in the Leiden Atomic and Molecular Database LAMDA¶ (Schöier *et al.* 2005). Here we note that in our method, we adopt the assumption of local thermal equilibrium (LTE) to obtain the level populations of the water molecule (n_u and n_l). In addition, we set the ortho to para ratio (OPR) of water to its high-temperature value of 3 throughout the disk.

According to our calculations (e.g., Notsu *et al.* 2016, 2017), we showed that H_2O emission lines with small Einstein A coefficients ($A_{ul} \sim 10^{-3} - 10^{-6}$ s^{-1}) and relatively high upper state energies ($E_{\rm up} \sim$ 1000K) are dominated by emission from the region inside the H_2O snowline, and therefore their profiles potentially contain information which can be used to locate the position of the snowline. This is because the water gas column density of the region inside the H_2O snowline is high enough that all lines are optically thick as long as $A_{ul} > 10^{-6}$ s^{-1}. On the other hand, the region outside the H_2O snowline has lower water gas column densities and lines with larger Einstein A coefficients have a more significant contribution to their fluxes since the lines are optically thin. In addition, we calculated the profiles of lines which have been detected by previous spectroscopic observations using *Herschel* (e.g., the ortho-H_2O 63.32μm and 538.29μm lines). These lines are less suited to locate the position of the H_2O snowline, because they are not dominated in flux by the region inside the H_2O snowline.

Figure 3 shows the emission profiles of ortho-H_2O lines at 682.66μm ($A_{\rm ul}$ $= 2.82\times10^{-5}$ s^{-1}, $E_{\rm u}$ = 1088.7K), 63.32μm ($A_{\rm ul}$ = 1.77 s^{-1}, $E_{\rm u}$ = 1070.6K) and 538.29μm ($A_{\rm ul}$ $= 3.50\times10^{-3}$ s^{-1}, $E_{\rm u}$ = 61.0K), for the Herbig Ae disk (Notsu *et al.* 2017). Figure 4 shows the line-of-sight emissivity (emissivity times extinction, $\eta_{ul}e^{-\tau_{ul}}$; see Equation (14) of Notsu *et al.* 2016) along the line-of sight direction (from $z = \infty$ to $-\infty$) of the

‡ http://home.strw.leidenuniv.nl/ michiel/ratran/
¶ http://home.strw.leidenuniv.nl/ moldata/

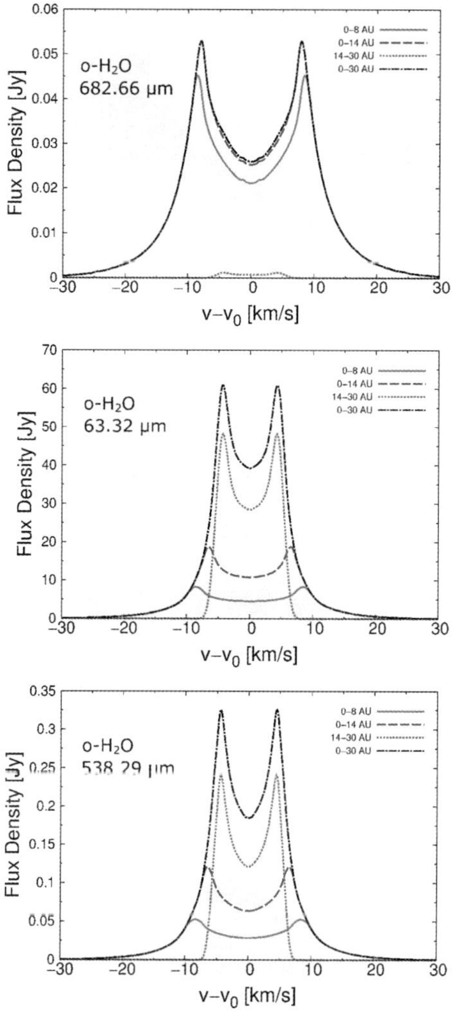

Figure 3. The emission profiles of ortho-H_2O lines at 682.66μm (top), 63.32μm (middle) and 538.29μm (bottom), for the Herbig Ae disk. We assume that the distance to the object d is 140pc ($\sim$ the distance of Taurus molecular cloud), and the inclination angle of the disk, i, is 30 degree (Notsu *et al.* 2017).

same three water lines for the Herbig Ae disk (Notsu *et al.* 2017). According to Figure 4, in the candidate 682.66μm line, the values of emissivity at $r \lesssim 14$ au (= the position of the H_2O snowline), $T_g \gtrsim$120K, and $z/r \sim 0.05 - 0.12$ are larger than those of the optically thin hot surface layer and the photodesorbed layer of the outer disk, and in particular those in the region with a higher H_2O gas abundance ($\sim$$10^{-4}$, $r < 7 - 8$ au, and $T_g \gtrsim$170K) and $z/r \sim 0.05 - 0.12$ are much larger.

Figure 5 shows the total fluxes of the various ortho-H_2O lines which are the candidates for tracing emission from hot water vapor within the H_2O snowline for a Herbig Ae disk (top panel) and a T Tauri disk (bottom panel) (Notsu *et al.* 2017). Since the fluxes of these lines from Herbig Ae disks are larger than those from T Tauri disks, the possibility of a successful detection is expected to increase for a Herbig Ae disk. The wavelengths of those lines which are the best candidates to locate the position of the H_2O snowline range from mid-infrared to sub-millimeter. The values of total fluxes tend to be larger as

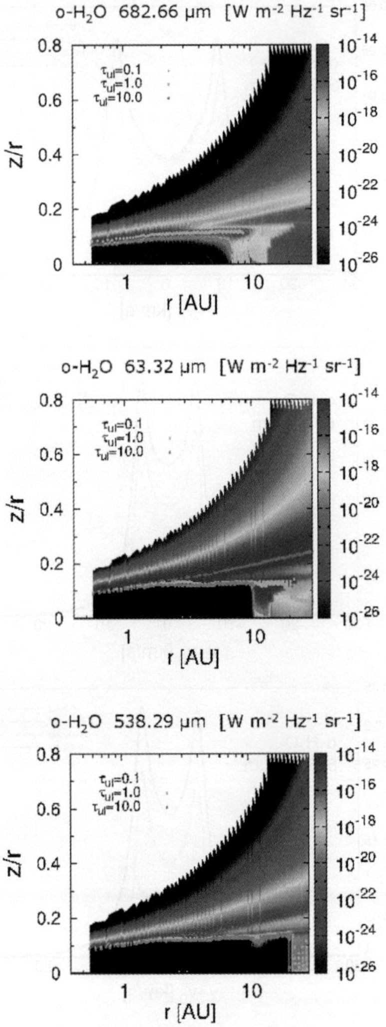

Figure 4. The line-of-sight emissivity of ortho-H_2O lines at 682.66μm (top), 63.32μm (middle) and 538.29μm (bottom), for the Herbig Ae disk. In these panels, we overplot the total optical depth contours (τ_{ul} = 0.1 (red cross points), 1 (cyan circle points), and 10 (orange square points)) on top of these line emissivity panels (Notsu *et al.* 2017).

the wavelengths of the H_2O lines become shorter. This is because the peak wavelength of the Planck function at the gas temperature around the H_2O snowline ($T_g \sim 100 - 200$ K) is in the mid-infrared region.

3. Discussion and Conclusion

In this work, we calculated the disk water vapor distribution and corresponding H_2O line profiles for a T Tauri disk and a Herbig Ae disk, and identified candidate water lines which can locate the position of the H_2O snowline across a wide wavelength range from mid-infrared to sub-millimeter.

The wavelengths of such candidate lines which trace emission from the hot water vapor within H_2O snowline overlap with the capabilities of ALMA and future mid-infrared high-

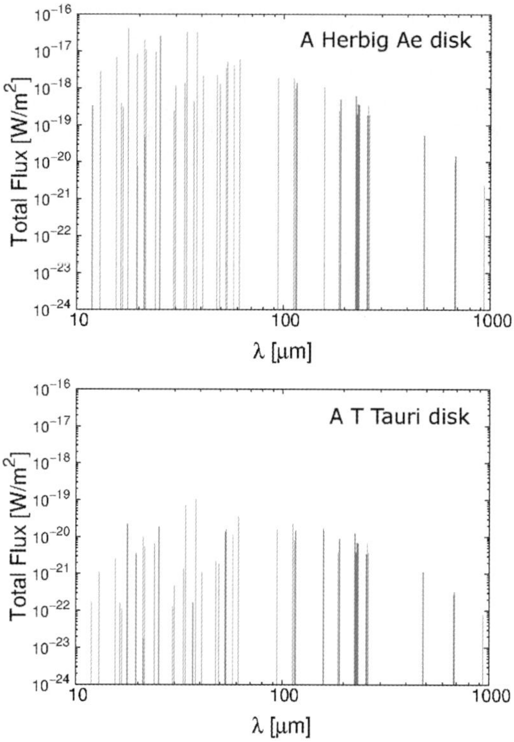

Figure 5. The total fluxes of the ortho H_2O lines which are best candidates to trace the emission from the water vapor within the H_2O snowline, for a Herbig Ae disk (top panel) and a T Tauri disk (bottom panel) (Notsu *et al.* 2017).

dispersion spectrographs (e.g., SPICA/SMI-HRS). The successful detection in a Herbig Ae disk could be achieved with current ALMA capabilities using several lines. Mid-infrared instruments such as HRS on SPICA/SMI would have a high sensitivity in the Q band (e.g., $\sim$ 16-18μm).

4. Acknowledgements

The numerical calculations in this study were carried out on SR16000 at Yukawa Institute for Theoretical Physics (YITP) and computer systems at Kwasan and Hida Observatory (KIPS) in Kyoto University, and PC cluster at Center for Computational Astrophysics, National Astronomical Observatory of Japan. This work is supported by JSPS (Japan Society for the Promotion of Science) Grants-in-Aid for Scientific Research (Grant Number; 23103005, 25108004, 25108005, 25400229, 15H03646), by Grants-in-Aid for JSPS fellows (Grant Number; 16J06887), and by the Astrobiology Center Program of National Institutes of Natural Sciences (NINS) (Grant Number; AB281013). S. N. is grateful for the support from the educational program organized by Unit of Synergetic Studies for Space, Kyoto University. C. W. acknowledges support from the Netherlands Organization for Scientific Research (NWO, program number 639.041.335). Astrophysics at Queen's University Belfast is supported by a grant from the STFC (ST/P000312/1).

References

Banzatti, A., Pontoppidan, K. M., Salyk, C., *et al.* 2017, *ApJ*, 834, 152
Blevins, S. M., Pontoppidan, K. M., Banzatti, A., *et al.* 2016, *ApJ*, 818, 22
Eistrup, C., Walsh, C., & van Dishoeck, E. F. 2016, *A&A*, 595, A83
Furuya, K., Aikawa, Y., Nomura, H., Hersant, F., & Wakelam, V. 2013, *ApJ*, 779, 11
Graedel, T. E., Langer, W. D., & Frerking, M. A. 1982, *ApJS*, 48, 321
Hayashi, C. 1981, *Progress of Theoretical Physics Supplement*, 70, 35
Hayashi, C., Nakazawa, K., & Nakagawa, Y. 1985, *Protostars and Planets II*, University of Arizona Press, 1100
Heinzeller, D., Nomura, H., Walsh, C., & Millar, T. J. 2011, *ApJ*, 731, 115
Hogerheijde, M. R., Bergin, E. A., Brinch, C., *et al.* 2011, *Science*, 334, 338
Hogerheijde, M. R. & van der Tak, F. F. S. 2000, *A&A*, 362, 697
Morbidelli, A., Bitsch, B., Crida, A., *et al.* 2016, *Icarus*, 267, 368
Morbidelli, A., Chambers, J., Lunine, J. I., *et al.* 2000, *Meteoritics and Planetary Science*, 35, 1309
Morbidelli, A., Lunine, J. I., O'Brien, D. P., Raymond, S. N., & Walsh, K. J. 2012, *Annual Review of Earth and Planetary Sciences*, 40, 251
Nomura, H. & Millar, T. J. 2005, *A&A*, 438, 923
Nomura, H., Aikawa, Y., Tsujimoto, M., Nakagawa, Y., & Millar, T. J. 2007, *ApJ*, 661, 334
Notsu, S., Nomura, H., Ishimoto, D., Walsh, C., Honda, M., Hirota, T., & Millar, T. J. 2017a, *ApJ*, 836, 118
Notsu, S., Nomura, H., Ishimoto, D., Walsh, C., Honda, M., Hirota, T., & Millar, T. J. 2016, *ApJ*, 827, 113
Notsu, S., Nomura, H., Ishimoto, D., *et al.* 2015, Revolution in Astronomy with ALMA: The Third Year, ASP Conference Series, 499, 289
Öberg, K. I., Murray-Clay, R., & Bergin, E. A. 2011, *ApJ*, 743, L16
Oka, A., Nakamoto, T., & Ida, S. 2011, *ApJ*, 738, 141
Okuzumi, S., Tanaka, H., Kobayashi, H., & Wada, K. 2012, *ApJ*, 752, 106
Piso, A.-M. A., Öberg, K. I., Birnstiel, T., & Murray-Clay, R. A. 2015, *ApJ*, 815, 109
Podio, L., Kamp, I., Codella, C., *et al.* 2013, *ApJ*, 766, L5
Pontoppidan, K. M., Salyk, C., Blake, G. A., *et al.* 2010a, *ApJ*, 720, 887
Pontoppidan, K. M., Salyk, C., Blake, G. A., K&aumlufl, H. U. 2010b, *ApJ*, 722, L173
Ros, K. & Johansen, A. 2013, *A&A*, 552, A137
Rybicki, G. B. & Lightman, A. P. 1986, Radiative Processes in Astrophysics, by George B. Rybicki, Alan P. Lightman, pp. 400. ISBN 0-471-82759-2. Wiley-VCH, June 1986
Sato, T., Okuzumi, S., & Ida, S. 2016, *A&A*, 589, A15
Schöier, F. L., van der Tak, F. F. S., van Dishoeck, E. F., & Black, J. H. 2005, *A&A*, 432, 369
van Dishoeck, E. F., Bergin, E. A., Lis, D. C., & Lunine, J. I. 2014, Protostars and Planets VI, University of Arizona Press, 835
Walsh, C., Millar, T. J., & Nomura, H. 2010, *ApJ*, 722, 1607
Walsh, C., Millar, T. J., Nomura, H., *et al.* 2014a, *A&A*, 563, AA33
Walsh, C., Nomura, H., Millar, T. J., & Aikawa, Y. 2012, *ApJ*, 747, 114
Walsh, C., Nomura, H., & van Dishoeck, E. 2015, *A&A*, 582, A88
Woitke, P., Thi, W.-F., Kamp, I., & Hogerheijde, M. R. 2009b, *A&A*, 501, L5
Woodall, J., Agúndez, M., Markwick-Kemper, A. J., & Millar, T. J. 2007, *A&A*, 466, 1197

Astrochemistry VII – Through the Cosmos from Galaxies to Planets
Proceedings IAU Symposium No. 332, 2017
M. Cunningham, T. Millar & Y. Aikawa, eds.

doi:10.1017/S1743921317007876

Unveiling the physical and chemical conditions in the young disk around L1527

M. L. R. van 't Hoff[1], J. J. Tobin[2], D. Harsono[1] and E. F. van Dishoeck[1,3]

[1]Leiden Observatory, Leiden University, PO box 9513, NL-2300 RA, Leiden, the Netherlands
email: vthoff@strw.leidenuniv.nl

[2]Homer L. Dodge Department of Physics and Astronomy, University of Oklahoma, 440 W. Brooks Street, Norman, OK 73019, USA

[3]Max-Planck-Institut für Extraterrestrische Physik, Giessenbachstrasse 1, 85748 Garching, Germany

Abstract. Planets form in disks around young stars. The planet formation process may start when the protostar and disk are still deeply embedded within their infalling envelope. However, unlike more evolved protoplanetary disks, the physical and chemical structure of these young embedded disks are still poorly constrained. We have analyzed ALMA data for ^{13}CO, $C^{18}O$ and N_2D^+ to constrain the temperature structure, one of the critical unknowns, in the disk around L1527. The spatial distribution of ^{13}CO and $C^{18}O$, together with the kinetic temperature derived from the optically thick ^{13}CO emission and the non-detection of N_2D^+, suggest that this disk is warm enough ($\gtrsim$ 20 K) to prevent CO freeze-out.

Keywords. stars: individual (L1527 IRS), stars: formation, astrochemistry

1. Introduction

Disks around young stars are the birthplace of planets. The chemical structure of these disks, and thus of the material that will build up the planets, determines planet compositions. In addition, the disk physical structure influences the planet formation process. Therefore, evolved protoplanetary disks (Class II sources) have been intensively studied and are becoming well characterized both physically (e.g. Andrews *et al.* 2010; Schwarz *et al.* 2016) and chemically (e.g. Thi *et al.* 2004; Dutrey *et al.* 2007; Öberg *et al.* 2010; Huang *et al.* 2017). However, grain growth already starts when the protostellar system is still deeply embedded in its natal molecular cloud (Kwon *et al.* 2009; Miotello *et al.* 2014) and the HL Tau images support the idea that planet formation begins during this class 0/I phase (ALMA Partnership 2015). Thus, young embedded disks may reflect the true initial conditions for planet formation.

Although several embedded disks are now known (e.g. Tobin *et al.* 2012; Murillo *et al.* 2013; Harsono *et al.* 2014; Aso *et al.* 2015), and ALMA enables us to spatially resolve molecular emission from these young disks, their physical and chemical conditions remain unconstrained. One of the critical unknowns is the temperature structure, since this directly influences the volatile composition of the planet-forming material. In regions where the temperature exceeds ~20 K, CO will be mainly present in the gas phase, while at lower temperatures CO is frozen out onto dust grains. In addition, this CO ice is the starting point for the formation of more complex molecules.

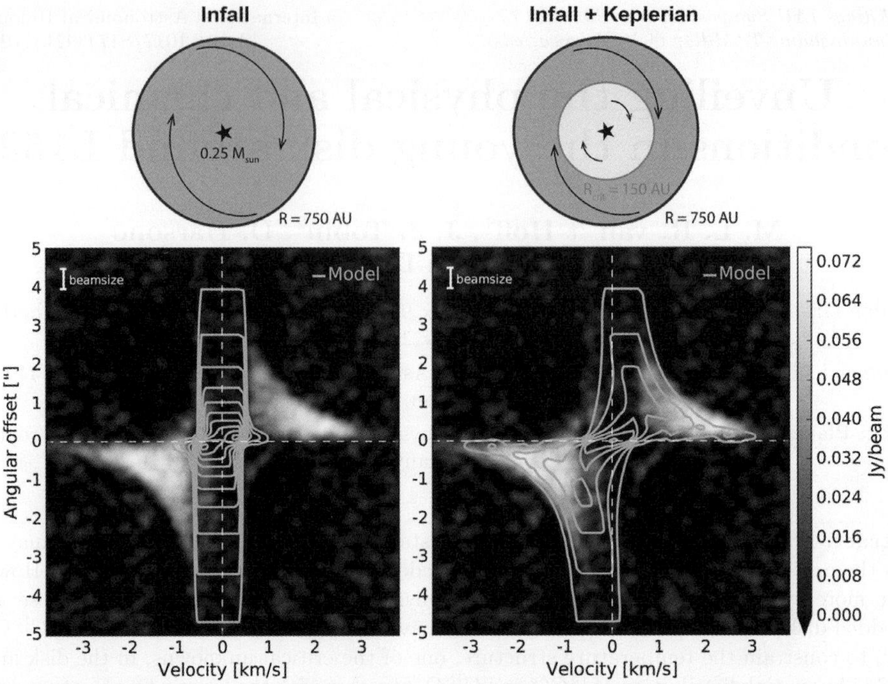

Figure 1. Position-velocity diagram for ^{13}CO observations (black-and-white scale) overlaid with contours for a thin disk model for a rotating infalling velocity profile (*left panel*) and for a rotating infalling velocity profile with Keplerian rotation in the inner 150 AU (*right panel*). The observations are taken with ALMA in Cycle 2 (PI: Koyamatsu).

2. Probing the disk temperature structure with ^{13}CO and N_2D^+

We have analyzed our own and archival ALMA data (PIs: Tobin, Koyamatsu, Ohashi, Sakai) and used ^{13}CO, C^{18}O and N_2D^+ observations to constrain the temperature in the embedded disk of L1527. This disk is particularly interesting because its almost edge-on configuration allows for direct probe of the vertical structure. Emission originating in the disk can be isolated from the envelope contribution based on velocity. Comparing the results from a thin disk model (Murillo *et al.* 2013) with an infall velocity profile to the results from a model including Keplerian rotation in the inner region shows that the emission in the highest velocity channels is solely due to the disk (Fig. 1). The spatial extent of the ^{13}CO and C^{18}O J = 2-1 emission in these channels suggest that CO is vertically present throughout the disk, including the disk midplane (Fig. 2). In addition, the ratio of the ^{13}CO and C^{18}O intensities shows that the ^{13}CO emission is optically thick ($\tau > 3$), and thus traces the kinetic temperature of the gas. The derived temperatures are ~30-40 K, above the CO freeze-out temperature of ~20 K (Fig. 2).

In contrast, N_2H^+ observations toward several protoplanetary disks have shown that the outer disk midplane becomes cold enough for CO to freeze out (Qi *et al.* 2013, 2015). The N_2H^+ ion traces CO freeze-out because its main destructor is gas-phase CO (Aikawa *et al.* 2015; van 't Hoff *et al.* 2017). The deuterated form of N_2H^+, N_2D^+, is not detected in the L1527 disk, corroborating the observation that CO is present in the gas phase throughout the disk.

Altogether, these preliminary results are in agreement with physical models of embedded disks (Harsono *et al.* 2014, 2015) and suggest that the young disk in L1527 is warm enough to prevent CO freeze-out.

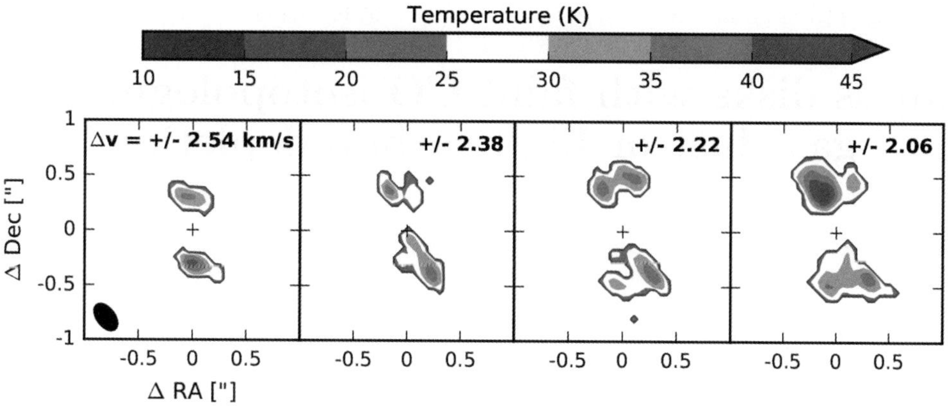

Figure 2. Channelmaps for ^{13}CO at the largest velocity offsets, that is, channels containing emission from the disk. Blue- and red-shifted channels with the same absolute velocity offset are shown in the same panel, and the velocity offset is shown in the top right corner. The beam is shown in the lower left corner of the first panel. CO freezes out at temperatures below ~20 K. The observations are taken with ALMA in Cycle 2 (PI: Koyamatsu).

Acknowledgments

Astrochemistry in Leiden is supported by the European Union A-ERC grant 291141 CHEMPLAN, by the Netherlands Research School for Astronomy (NOVA) and by a Royal Netherlands Academy of Arts and Sciences (KNAW) professor prize. M.L.R.H acknowledges support from a Huygens fellowship from Leiden University.

References

Aikawa, Y., Furya, K., Nomura, H., & Qi, C. 2015, *ApJ*, 807, 120

ALMA Partnership 2015, *ApJ*, 808, 3

Andrews, S. M., Wilner, D. J., Hughes, A. M., Qi, C., & Dullemond, C. P. 2010, *ApJ*, 723, 1241

Aso, Y., Ohashi, N., Saigo, K., *et al.* 2015, *ApJ*, 812, 27

Dutrey, A., Henning, T., Guilloteau, S., *et al.* 2007, *A&A*, 464, 615

Harsono, D., Jørgensen, J. K., van Dishoeck, E. F., Hogerheijde, M. R., Bruderer, S, Persson, M. V., & Mottram, J. C. 2014, *A&A*, 562, A77

Harsono, D., Bruderer, S., & van Dishoeck, E. F. 2015, *A&A*, 582, A41

Huang, J., Öberg, K. I., Qi, C., *et al.* 2017, *ApJ*, 835, 231

Kwon, W., Looney, L. W., Mundy, L. G., Chiang, H.-F., & Kemball, A. J. 2009, *ApJ*, 696, 841

Miotello, A., Testi, L., Lodato, G., *et al.* 2014, *A&A*, 567, A32

Murillo, N. M., Lai, S.-P., Bruderer, S., Harsono, D., & van Dishoeck, E. F. 2013, *A&A*, 560, A103

Öberg, K. I., Qi, C., Fogel, J. K. J., *et al.* 2010, *ApJ*, 720, 480

Öberg, K. I. & Bergin, E. A. 2016, *ApJL*, 831, L19

Qi, C., Öberg, K. I., Wilner, D. J., *et al.* 2013, *Science*, 341, 630

Qi, C., Öberg, K. I., Andrews, S. M., *et al.* 2015, *ApJ*, 813, 128

Schwarz, K., Bergin, E. A., Cleeves, L. I., *et al.* 2016, *ApJ*, 823, 91

Thi, W. F., van Zadelhoff, G. J., & van Dishoeck, E. F. 2004, *A&A*, 425, 955

Tobin, J.J., Hartmann, L., Chiang, H.-F., *et al.* 2012, *Nature*, 492, 83

van 't Hoff, M. L. R., Walsh, C., Kama, M., Facchini, S., & van Dishoeck, E. F. 2017, *A&A*, 599, A101

Astrochemistry VII – Through the Cosmos from Galaxies to Planets
Proceedings IAU Symposium No. 332, 2017
M. Cunningham, T. Millar & Y. Aikawa, eds.

doi:10.1017/S1743921317006603

Lupus disks with faint CO isotopologues: low gas/dust or high carbon depletion?

Anna Miotello

Leiden Observatory, Leiden University, Niels Bohrweg 2, NL-2333 CA Leiden, The Netherlands
email: miotello@strw.leidenuniv.nl

Abstract. With the advent of ALMA, complete surveys of gas and dust in protoplanetary disks are being carried out in different star forming regions. In particular, continuum emission is used to trace the large (mm-sized) dust grains and CO isotopologues are observed in order to trace the bulk of the gas. The attempt is to simultaneously constrain the gas and dust disk mass as well as the gas/dust mass ratio. In this work the observations from the *Lupus disk survey* have been analyzed with thermo-chemical disk models, including radiative transfer, CO isotope-selective processes and freeze-out. We find that CO-based gas masses are very low, often smaller than $1 M_{\rm J}$. Moreover, gas/dust mass ratios are much lower than value of 100 found in the ISM, being mainly between 1 and 10. This result can be interpreted either as rapid loss of gas, or as a chemical effect removing carbon from CO and locking it into more complex molecules or in larger bodies. Current data cannot distinguish between the two scenarios (except for sources with detected HD lines), but future observations of e.g. [CI] and hydrocarbon lines will help to calibrate CO-based gas masses and to constrain disk gas masses.

Keywords. astrochemistry – planetary systems: protoplanetary disks – planetary systems: planet-disk interactions – submillimeter: planetary systems

---

1. Introduction

There have been many attempts to study the bulk content of protoplanetary disks in gas and dust (see e.g., Thi *et al.* 2001; Pascucci *et al.* 2006; Panić *et al.* 2008, 2009; Andrews *et al.* 2012; Boneberg *et al.* 2016; Andrews 2015, as a review), but not in a statistically significant manner. Recently, the Atacama Large Millimeter/sub-millimeter Array (ALMA) has allowed the disk community to carry out complete surveys of gas and dust in protoplanetary disks in different star forming regions in a modest amount of time (Ansdell *et al.* 2016; Pascucci *et al.* 2016; Barenfeld *et al.* 2016). We focus on the first near-complete survey, that has been carried out in Lupus (Ansdell *et al.* 2016), where ~90 disks have been observed at a resolution of 0.3" (~ 20 – 30 au radius for a distance of 150 – 200 pc). A fraction of these disks were detected in the faint CO isotopologue lines (^{13}CO and C^{18}O). Such lines are commonly used to estimate the total disk gas mass, as they probe deeper into the disk than the optically thick ^{12}CO lines (van Zadelhoff *et al.* 2001; Dartois *et al.* 2003).

The main unknown is related to the conversion of the observed CO mass into total disk mass. A proper modeling of isotope-selective photodissociaton and freeze-out, which regulate CO isotopologues abundances, needs to be done in order to correctly interpret the observations (van Zadelhoff *et al.* 2001; Aikawa et al. 2002; Visser *et al.* 2009; Miotello et al. 2014; Miotello *et al.* 2016). Then, a value for the hydrogen to carbon elemental abundance has to be assumed, but this may not be the canonical value $X_{\rm C}$ =[C]/[H]=1.35×10^{-4} often used. In the case of TW Hya, where hydrogen deuteride was detected by the *Herschel Space Observatory* (Bergin *et al.* 2013), carbon is found to be depleted by 1 or 2 orders of magnitude (Favre *et al.* 2013; Kama *et al.* 2016; Trapman

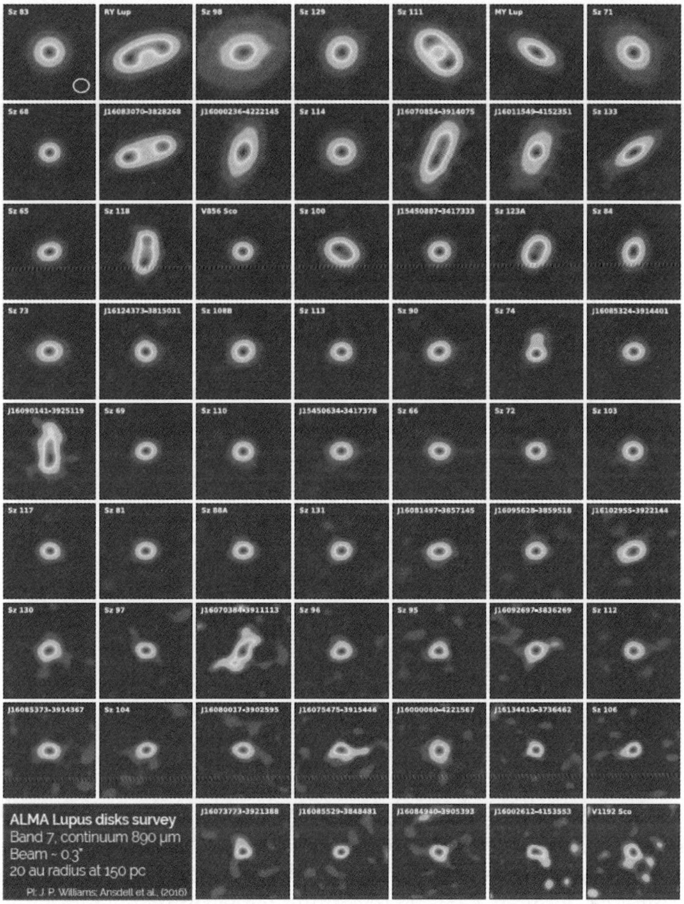

Figure 1. Collection of the disks detected in the continuum at 890 μm in the Lupus star forming region (each panel is $2'' \times 2''$). The resolution is $0.3''$, which corresponds to $\sim$ 20 au at 150 pc (adapted from Ansdell *et al.* 2016).

et al. 2017). This suggest that there may exist a class of disks where CO is not the main carbon reservoir.

2. Results and conclusions

In this proceeding we summarize some of the results obtained by Miotello *et al.* (2017), so we refer the reader to this paper for more detail. The modeling technique developed by Miotello *et al.* (2016) is applied to the ALMA continuum, ^{13}CO and C^{18}O Lupus observations (see Fig. 1). With our modeling procedure we are able to provide disk gas mass determinations also for the disks detected only in ^{13}CO, for a total of 34 objects. Previously, Ansdell *et al.* (2016) have presented an initial analysis for only 10 disks, for which both ^{13}CO and C^{18}O lines are available. This sample is extended and refined in Miotello *et al.* (2017). ^{13}CO alone can be used as a mass tracer only if its line emission are optically thin. We have used the grid of models presented in Miotello *et al.* (2016) to obtain the median of the ^{13}CO $J = 3 - 2$ simulated line luminosities. They can be

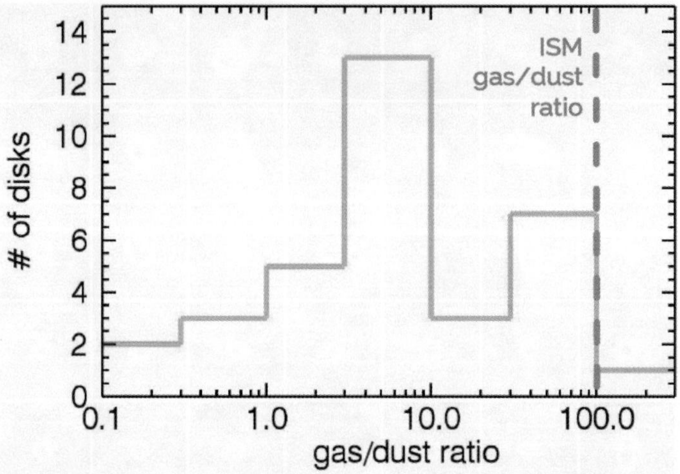

Figure 2. Histogram showing the number of disks presenting different levels of gas-to-dust ratio in the Lupus sample. Only disks detected both in continuum and ^{13}CO are considered (adapted from Miotello *et al.* 2017).

expressed by fit functions of the disk mass.

$$L_y = \begin{cases} A_y + B_y \cdot M_{\rm gas} & M_{\rm gas} \leqslant M_{\rm tr} \\ C_y + D_y \cdot \log_{10}(M_{\rm gas}) & M_{\rm gas} > M_{\rm tr}, \end{cases} \tag{2.1}$$

where $y = 13$ or 18, for ^{13}CO and C^{18}O respectively. The polynomial coefficients A_y, B_y, C_y, and D_y, as well as the transition masses $M_{\rm tr}$ are reported in Table 2 of Miotello *et al.* (2017). The dependence of the emission on the disk mass is linear for masses smaller than a transition point $M_{\rm tr}$, while it flattens becoming logarithmic for more massive disks. All the ^{13}CO-only detections are in the region of the emission-mass plane where $M_{\rm disk} < M_{\rm tr}$, so the dependence is linear. Therefore the fit function can be used to calculate the gas masses of these sources, as ^{13}CO is optically thin.

We derive very low gas masses that are generally lower than 1 $M_{\rm Jup}$, consistent with Ansdell *et al.* (2016). When compared with the derived dust masses from continuum emission, this translates to low global gas/dust ratios. Fig. 2 presents a histogram of the gas/dust ratio. Only objects for which both the continuum and ^{13}CO were detected are considered in the histogram. For most of the Lupus disks the global gas/dust ratio is between 1 and 10, much lower than the expected ISM-value of 100.

Low CO-based gas masses and gas/dust ratios are usually interpreted as the outcome of physical disk evolution that leads to a loss of gas. An alternative interpretation may be that the CO abundance is lower than what is usually assumed. Volatile carbon could be sequestered from CO to more complex less volatile molecules or being locked-up into larger bodies (Aikawa *et al.* 1996; Bergin *et al.* 2014; Du *et al.* 2015; Eistrup *et al.* 2016; Kama *et al.* 2016).

Current observations of continuum and ^{13}CO are not enough to distinguish between the two scenarios. Future ALMA observations of more complex tracers, such as hydrocarbons, or [CI] lines are needed to calibrate CO-based masses. An alternative method has been proposed by Manara *et al.* (2016) and Rosotti *et al.* (2017) to trace the gaseous disk component, independently from chemistry. This method is based on the simultaneous availability of dust disk mass and mass accretion rates measurement. Manara *et al.*

(2016) recently found for the first time a linear relation between the mass accretion rate using VLT/X-Shooter data (Alcalá *et al.* 2014, 2017) and the disk total mass inferred from the dust in Lupus. A linear correlation between $\dot{M}_{\rm acc}$ and $M_{\rm disk}$ has been theoretically predicted by viscous evolution theory (Hartmann *et al.* 1998). On the contrary, we have found no correlation between mass accretion and the disk masses derived from CO isotopologues lines for our 34 Lupus sources. Assuming that the evolution of all Lupus disks has occurred mainly viscously over the past few Myr, the observed correlation between the mass accretion rate and dust mass found by Manara *et al.* (2016) implies a constant gas/dust ratio of 100. This is a hint to a scenario in which carbon is depleted and this affects CO line emission.

Acknowledgements

The author thanks E. F. van Dishoeck, J. P. Williams, M. Ansdell, G. Guidi, M. Hogerheijde, C. F. Manara, M. Tazzari, L. Testi, N. van der Marel, S. van Terwisga, and the Lupus Disk Survey with ALMA team for the useful collaboration. This paper makes use of the following ALMA data: ADS/JAO.ALMA#2013.1.00220.S. ALMA is a partnership of ESO (representing its member states), NSF (USA) and NINS (Japan), together with NRC (Canada), NSC and SIAA (Taiwan), and KASI (Republic of Korea), in cooperation with the Republic of Chile. The Joint ALMA Observatory is operated by ESO, AUI/ NRAO, and NAOJ. Astrochemistry in Leiden is supported by the Netherlands Research School for Astronomy (NOVA), by a Royal Netherlands Academy of Arts and Sciences (KNAW) professor prize, and by the European Union A-ERC grant 291141 CHEMPLAN.

References

Aikawa, Y., Miyama, S. M., Nakano, T., & Umebayashi, T. 1996, *ApJ*, 467, 684
Aikawa, Y., van Zadelhoff, G. J., van Dishoeck, E. F., & Herbst, E. 2002, *A&A*, 386, 622
Alcalá, J. M., Natta, A., Manara, C. F., *et al.* 2014, *A&A*, 561, A2
Alcalá, J. M., Manara, C. F., Natta, A., *et al.* 2017, *A&A*, 600, A20
Andrews, S. M., Wilner, D. J., Hughes, A. M., *et al.* 2012, *ApJ*, 744, 162
Andrews, S. M. 2015, *PASP*, 127, 961
Ansdell, M., Williams, J. P., van der Marel, N., *et al.* 2016, *ApJ*, 828, 46
Barenfeld, S. A., Carpenter, J. M., Ricci, L., & Isella, A. 2016, *ApJ*, 827, 142
Bergin, E. A., Cleeves, L. I., Gorti, U., *et al.* 2013, *Nature*, 493, 644
Bergin, E. A., Cleeves, L. I., Crockett, N., & Blake, G. A. 2014, *Faraday Discussions*, 168, 61
Boneberg, D. M., Panić, O., Haworth, T. J., Clarke, C. J., & Min, M. 2016, *MNRAS*, 461, 385
Dartois, E., Dutrey, A., & Guilloteau, S. 2003, *A&A*, 399, 773
Du, F., Bergin, E. A., & Hogerheijde, M. R. 2015, *ApJL*, 807, L32
Eistrup, C., Walsh, C., & van Dishoeck, E. F. 2016, *A&A*, 595, A83
Favre, C., Cleeves, L. I., Bergin, E. A., Qi, C., & Blake, G. A. 2013, *ApJL*, 776, L38
Hartmann, L., Calvet, N., Gullbring, E., & D'Alessio, P. 1998, *ApJ*, 495, 385
Kama, M., Bruderer, S., van Dishoeck, E. F., *et al.* 2016, *A&A*, 592, A83
Manara, C. F., Rosotti, G., Testi, L., *et al.* 2016, *A&A*, 591, L3
Miotello, A., Bruderer, S., & van Dishoeck, E. F. 2014, *A&A*, 572, AA96
Miotello, A., van Dishoeck, E. F., Kama, M., & Bruderer, S. 2016, *A&A*, 594, A85
Miotello, A., van Dishoeck, E. F., Williams, J. P., *et al.* 2017, *A&A*, 599, A113
Panić, O., Hogerheijde, M. R., Wilner, D., & Qi, C. 2008, *A&A*, 491, 219
Pascucci, I., Gorti, U., Hollenbach, D., *et al.* 2006, *ApJ*, 651, 1177
Panić, O., Hogerheijde, M. R., Wilner, D., & Qi, C. 2009, *A&A*, 501, 269
Pascucci, I., Testi, L., Herczeg, G. J., *et al.* 2016, *ApJ*, 831, 125

Rosotti, G. P., Clarke, C. J., Manara, C. F., & Facchini, S. 2017, *MNRAS*, 468, 1631
Thi, W. F., van Dishoeck, E. F., Blake, G. A., *et al.* 2001, *ApJ*, 561, 1074
Trapman, L., Miotello, A., Kama, M., van Dishoeck, E. F., & Bruderer, S. 2017, *A&A*, 605, A69
van Zadelhoff, G.-J., van Dishoeck, E. F., Thi, W.-F., & Blake, G. A. 2001, *A&A*, 377, 566
Visser, R., van Dishoeck, E. F., & Black, J. H. 2009, *A&A*, 503, 323

Astrochemistry VII – Through the Cosmos from Galaxies to Planets
Proceedings IAU Symposium No. 332, 2017
M. Cunningham, T. Millar & Y. Aikawa, eds.

doi:10.1017/S1743921317007785

Gas vs dust radial extent in disks: the importance of their thermal interplay

Stefano Facchini
Max-Planck-Institut für Extraterrestrische Physik, Giessenbachstrasse 1, 85748 Garching, Germany
email: facchini@mpe.mpg.de

Abstract. A key parameter governing the secular evolution of protoplanetary disks is their outer radius. In this paper, the feedback of realistic dust grain size distributions onto the gas emission is investigated. Models predict that the difference of dust and gas extents as traced by CO is primarily caused by differences in the optical depth of lines vs continuum. The main effect of radial drift is the sharp decrease in the intensity profile at the outer edge. The gas radial extent can easily range within a factor of 2 for models with different turbulence. A combination of grain growth and vertical settling leads to thermal de-coupling between gas and dust at intermediate scale-heights. A proper treatment of the gas thermal structure within dust gaps will be fundamental to disentangle surface density gaps from gas temperature gaps.

Keywords. astrochemistry – planetary systems: protoplanetary disks – planetary systems: planet-disk interactions – submillimeter: planetary systems

1. Introduction

Spatially resolved observations of protoplanetary disks in both gas and dust are providing new estimates on their radial extents and surface density profiles (e.g. Piétu *et al.* 2014; Tripathi *et al.* 2017; Tazzari *et al.* 2017). Both quantities are fundamental parameters in any planet formation model (Mordasini *et al.* 2012), and together with the disk mass they define the planet formation potential of protoplanetary disks. In particular, the outer radius is a key quantity that could disentangle different evolutionary scenarios of disks, since it is expected to spread outwards in viscous evolution theory (Lynden-Bell & Pringle 1974), whereas spreading is suppressed if disk evolution is driven by magnetic winds (Bai 2016). The radial extent estimates mentioned above are all based on the (sub-)mm continuum brightness distribution, which traces the dust component. Observations of gas tracers, such as rotational transitions of ^{12}CO, are however showing that the gas seems to extend out to larger radii than the dust. Dutrey *et al.* (1998) and Guilloteau & Dutrey (1998) suggested that this difference may be due to optical thickness effects, with the gas lines being more optically thick than the (sub-)mm continuum counter-part. However, the steep radial decline in surface brightness at the outer edge of the dust disks has also been interpreted as the smoking gun of radial drift of large dust particles (e.g. Andrews *et al.* 2012; de Gregorio-Monsalvo *et al.* 2013; Piétu *et al.* 2014; Birnstiel & Andrews 2014).

In this paper, we address this potential difference between gas and dust radial extent taking into account the effects that the dust properties have on the gas temperature and chemistry. Precise estimates of the gas temperature and chemical abundances are needed to correctly predict the molecular emission in the outer regions of protoplanetary disks, which are used in observations to trace outer radii. The grain size distribution of dust particles affects the gas chemistry and temperature, and thus the CO abundance and excitation, in multiple ways. Among the most important effects, the grain size distribution

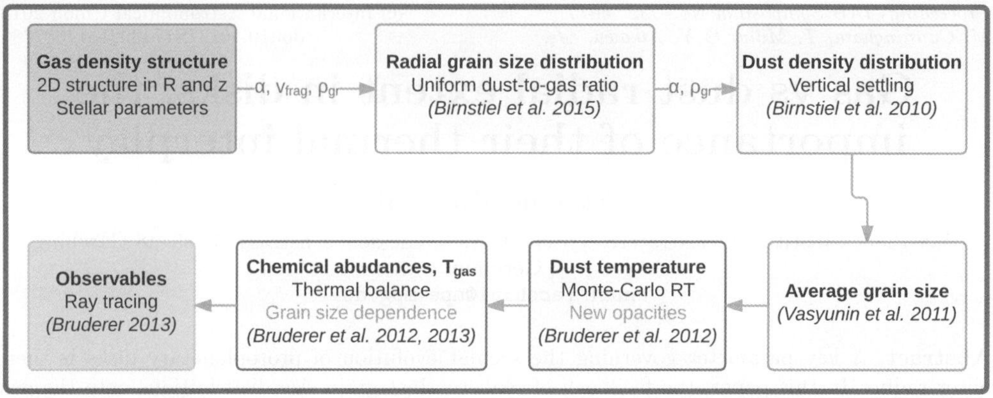

Figure 1. Diagram of the new DALI version used in this paper (taken from Facchini *et al.* 2017a). The blue box shows the input parameters of the models, in particular the disk gas density structure, and stellar parameters. Red parameters indicate physical quantities that are needed to compute the dust properties across the disk. Boxes with orange contours are DALI modules developed for this particular project. The final green box indicates the observables produced with the DALI ray-tracing.

resulting from grain growth, radial drift and vertical settling sets the opacity at far ultraviolet (FUV) wavelengths, and thus the penetration depth of photons capable of dissociating strongly bound molecules as CO. UV radiation and is important for the gas temperature through photoelectric heating. Then, the dust opacities at infrared (IR) wavelengths are important for the gas cooling, which in the upper layers is dominated by IR lines of simple chemical species. The dust surface area also sets the amount of thermal coupling between dust and gas components. Finally, the dust surface area affects the balance between desorption and adsorption of ices onto the surface of dust grains.

A simultaneous modelling of both dust and gas thermal structures in typical protoplanetary disks is performed in the following sections in order to compare the predictions of the dust and gas radial extents when the feedback of realistic grain size distributions of dust particles onto the CO abundance and emission is analysed. The dust properties are computed with state-of-the-art dust evolution codes (Birnstiel *et al.* 2015). In order to determine the gas temperatures and chemical abundances of both ice and refractory species, the thermo-chemical code DALI (Bruderer *et al.* 2012; Bruderer 2013) is used and generalised to consider the underlying dust properties.

2. Method

The method used in this project is thoroughly described in Facchini *et al.* (2017a). Here we present a short summary. The most important addition to previous thermo-chemical models is the calculation of realistic dust properties based on dust evolution models. The main input parameters of these calculations are the gas density structure, which in DALI is two dimensional in R and z, the stellar parameters, and the dimensionless turbulent parameter α (Shakura & Sunyaev 1973), which determines both the radial and the vertical dependence of the dust grain size distribution. The gas surface density is parametrized as an exponentially tapered power-law, with a gaussian vertical profile to mimic an isothermal steady-state. The radial grain size distribution is computed using the semi-analytical prescription of Birnstiel *et al.* (2015), where processes such as grain growth, fragmentation, erosion, and radial drift are taken into account.

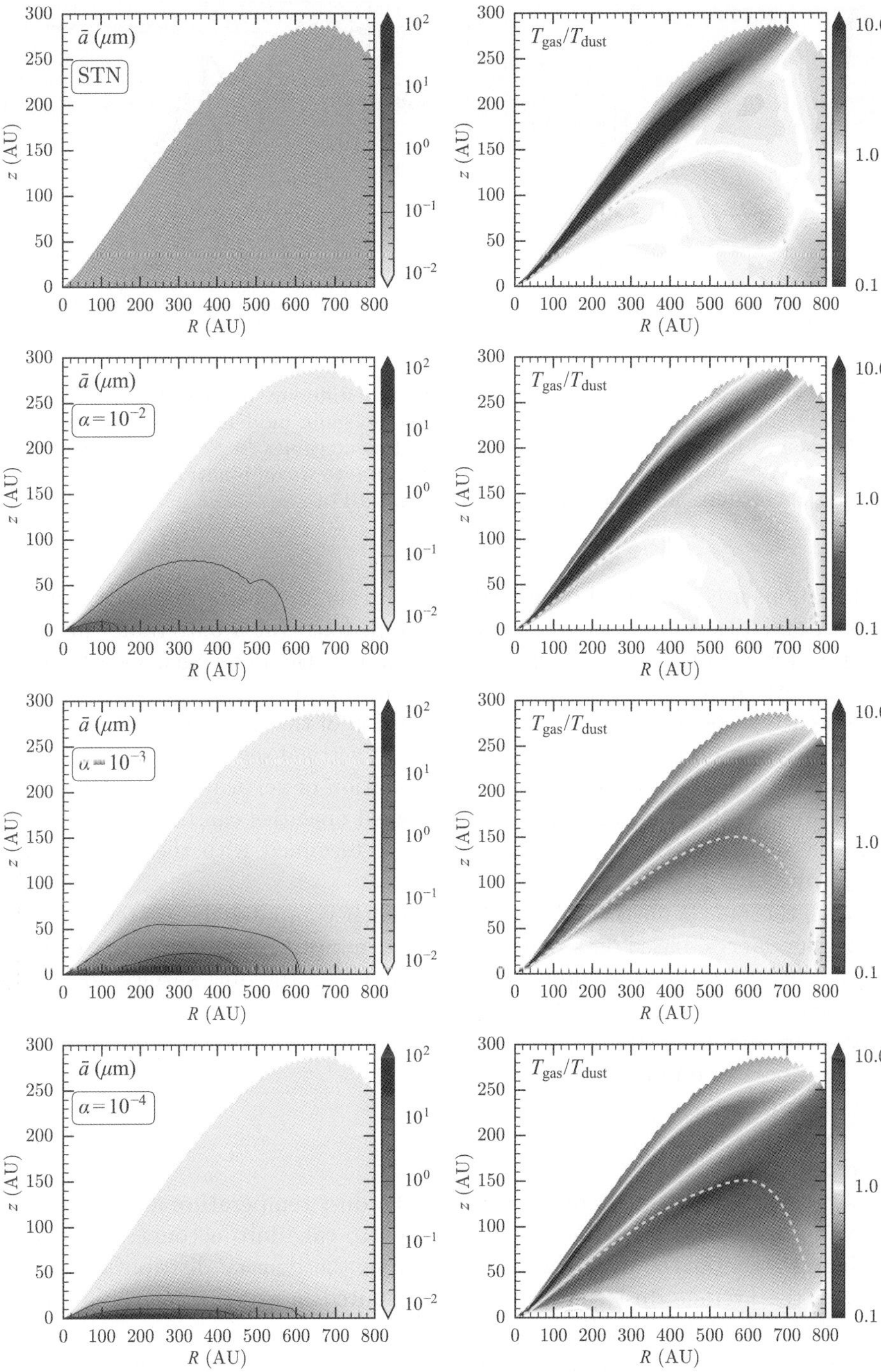

Figure 2. Left panels: average dust grain size used in the thermo-chemical calculations for a standard DALI model (STN), and three models with $\alpha = 10^{-2}$, 10^{-3} and 10^{-4}, from top to bottom. Note the growth of the average dust size to mm sizes in the disk midplane. Right panels: ratio of gas and dust temperatures for the same models. The yellow dashed lines indicate the $\tau = 1$ layer of the ^{12}CO J=3-2 transition. Note that $T_{\rm gas} < T_{\rm dust}$ in the ^{12}CO emitting layer for low α values. Figure adapted from Facchini *et al.* (2017a).

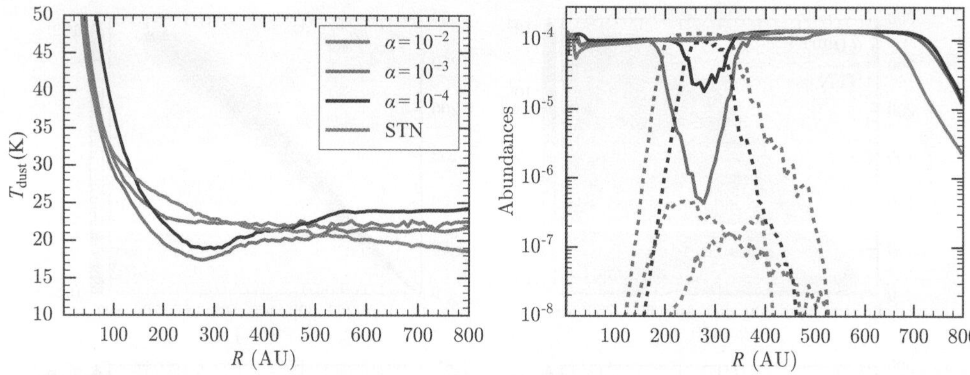

Figure 3. Left panel: dust midplane temperatures of different models. Right panel: volatile and solid CO abundances in the disk midplane for the same models. Solid lines represent gas phase CO, dashed lines indicate CO ice. For low turbulent values ($\alpha \lesssim 10^{-3}$) the disk midplane can witness a thermal inversion in its outer regions, due to a combination of grain growth and vertical settling. Figure adapted from Facchini *et al.* (2017a).

The calculations are performed using 250 bins in grain size, logarithmically sampled between 50 Å and 10 m. The dust-to-gas ratio is kept fixed (at a 0.01 value) in the whole radial grid. Thus, radial drift is taken into account in the determination of the largest particle size attainable in the disk, when fragmentation does not dominate, but it is not self-consistently considered to compute the evolution of the dust-to-gas ratio. The radial grain size distribution is then used to compute the vertical segregation of dust particles, based on the steady-state advection diffusion equation of vertical settling (e.g. Dubrulle *et al.* 1995). With these calculations, mass-averaged opacities can be calculated in every point in the disk, and the dust temperature is determined with the DALI continuum radiative transfer.

As for the thermo-chemistry, DALI self-consistently computes the gas temperature and chemical abundances, based on a reduced chemical network. Both in the thermal balance and in the chemistry there are processes that depend on the dust surface area available, in particular:

- freeze-out and desorption
- photo-processes on dust grains
- gas-grain collisions;
- H_2 formation
- hydrogenation

Obviously, there is also a clear dependence on the dust temperature and the continuum intensity spectrum. The code has been upgraded to take into account the dependence of all these effects on the dust surface area. From the density distribution of the 250 grain sizes, the average dust surface area is computed in every point of the disk using the prescription by Vasyunin *et al.* (2011).

The results of the models containing grain growth and vertical settling are compared against the output of a model where the dust properties are more simply parametrised. In the next sections this model will be referred to as "STN". All the models shown in this paper have the same gas density structure and stellar parameters, which are taken as the best representative values of the disk HD163296 from de Gregorio-Monsalvo *et al.* (2013). The differences that are observed between the models are due to the different dust grain size distributions across the disk.

3. Dust and gas temperatures

Grain growth and vertical settling have a significant effect both on the dust and on the gas temperatures. The left panels of Figure 2 show the average grain size that is obtained across disks where different turbulent parameters have been used, ranging between $\alpha = 10^{-2}$ and 10^{-4}. Comparison with the typical value of $0.1\,\mu$m used in the chemical calculations of disks is also portrayed. As turbulence decreases, the dust becomes more severely settled towards the disk midplane. Moreover, the maximum grain size reached in the disk midplane at a given radius is larger for lower turbulence, since turbulent velocities of dust particles become lower, and thus grains are less prone to fragment.

Severe vertical settling can lead to the peculiar effect of having a region where gas is colder than the dust particles at intermediate scale heights (see Figure 2, right panels). In the upper layers of the disk, the gas is usually much hotter than the dust, in particular in the region that is still optically thin to FUV wavelengths, since most of the gas heating in these regions comes from photoelectric effect on polyciclic aromatic hydrocarbons (PAHs). When significant vertical settling of dust particles occurs, there are two effects that come into play: there is a region in the intermediate layers that is optically thick to UV wavelengths, but optically thin to IR wavelengths, so most of the cooling lines can still escape. Moreover, in this same region the dust densities are not high enough to lead to thermal coupling between the gas and the dust components via gas-grain collisions. The combination of the two effects can make the gas colder than the dust component. Interestingly, on the same panels in Figure 2, the $\tau = 1$ layer of the ^{12}CO J=3-2 transition is overlaid, and it crosses exactly the region where the thermal de-coupling occurs.

In the disk midplane, when turbulence is low, there can be a radial range where the disk is self-shadowed, in particular when the radial gradient of the maximum grain size is steeper than the radial gradient of the gas surface density. However, in the disk outer regions, the dust particles are small enough that they are stirred up to high altitudes very easily. These two effects combined lead to a thermal inversion in the disk midplane (see Figure 3, left panel), with a dust temperature that can radially increase in the outer regions of the disk. For a warm disk, as the one analysed in this paper, this inversion can happen close to the sublimation temperature of CO, at $\sim 20-25$ K. The right panel of Figure 3 shows that for the $\alpha = 10^{-3}$ and 10^{-4} models, CO can freeze-out at the lowest temperatures of the respective models, but then is desorbed back into the gas phase in the outer regions of the disk, where the dust temperatures are higher than the CO sublimation temperature. In these calculations, a CO binding energy of 855 K was assumed, typical of CO pure ice (e.g. Collings *et al.* 2003). In the models presented here, this second desorption front is due to a temperature inversion, not to an increase penetration depth of UV photons, as in the models by Cleeves (2016). Interestingly, in both scenarios radial drift plays an important role in predicting this trend, which has been observed (or inferred) in recent ALMA images (e.g. Schwarz *et al.* 2016; Huang *et al.* 2016; Dutrey *et al.* 2017; Salinas *et al.* 2017).

4. Impact on outer radius estimates

It has been shown that the gas temperatures can be significantly affected by grain growth, vertical settling and radial drift. The gas temperatures are important in determining the radial intensity profiles of significantly optically thick and bright molecules, which are the more natural lines to use when trying to determine the radial extent of the gas component of a disk out to the very cold regions. If the radial extent of a disk is defined as the radius where the radial intensity profile reaches a specific value (that

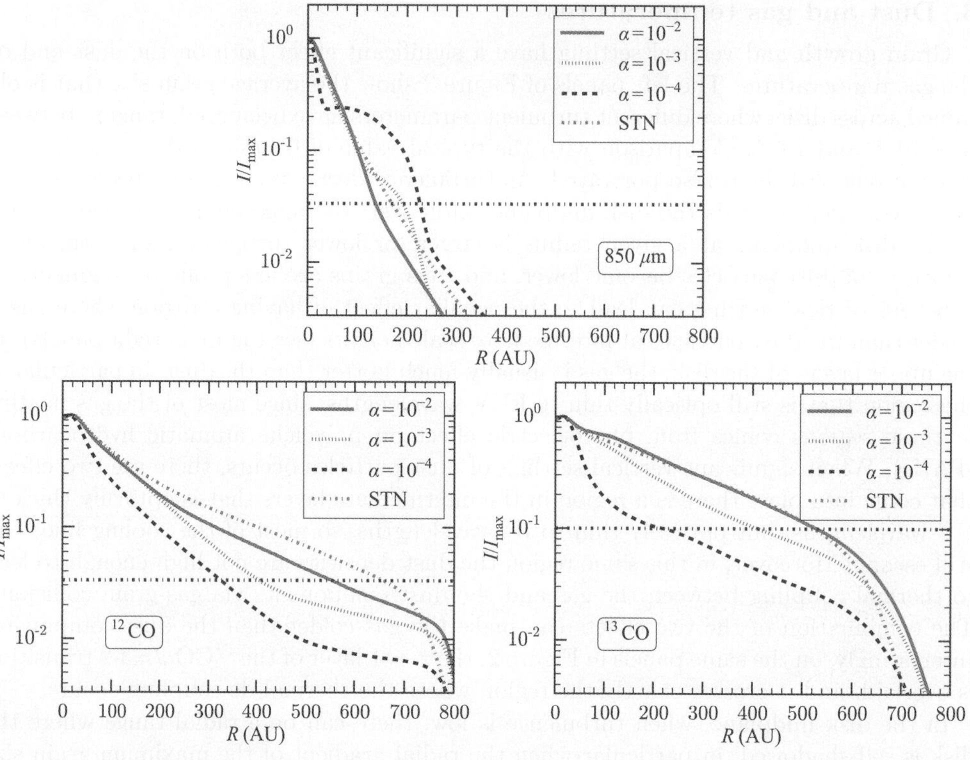

Figure 4. Normalised radial intensity profiles of the 850 μm continuum (top panel), ^{12}CO J=3-2 line, and ^{13}CO J=3-2 line. The dashed-dotted black line indicates a dynamic range of 30 for both the continuum and ^{12}CO, and a dynamic range of 10 for ^{13}CO. Note that the underlying gas surface density is the same in all models. Panels adapted from Facchini *et al.* (2017a).

can be set by the sensitivity of the observation, or as the radius at which the intensity is lower than the peak by some fixed amount), the thermal coupling between gas and dust can play a significant role. Figure 4 shows the radial intensity profiles of the models presented in Section 2, for sub-mm thermal continuum at 850 μm, for the ^{12}CO J=3-2 line, and for the ^{13}CO J=3-2 line. At a fixed peak intensity ratio, the estimated radial extent obtained from both ^{12}CO and ^{13}CO has a significant dependence on turbulence. Since the gas density structure is exactly the same for all the models, the difference is due to the radial gas temperature profile along the $\tau = 1$ layer of the lines being steeper for lower turbulence values. Thus, as the layers where the lines originate become colder, the line intensities become lower in the outer regions of the disk, and the disk looks more compact with a steeper surface density profile (see Figure 5).

The radial extent as probed by the submm continuum is not significantly affected by grain growth and radial drift. The top panel of Figure 4 shows the radial intensity profile at 850 μm, and the difference between the models including grain growth and radial drift and a model with a uniform grain size distribution (the STN model) is not very significant. However, for low turbulent values, the steep radial gradient in the maximum grain size attained in the disk midplane makes the outer regions of the intensity profile very steep. In these models, this is not due to a radial variation in the dust-to-gas ratio (which can have a significant effect, not considered here, as shown in Birnstiel & Andrews 2014), which is kept fixed, but to the radial gradient of the grain size distribution, a direct consequence of radial drift. Thus, the difference of the radial extent of the continuum and

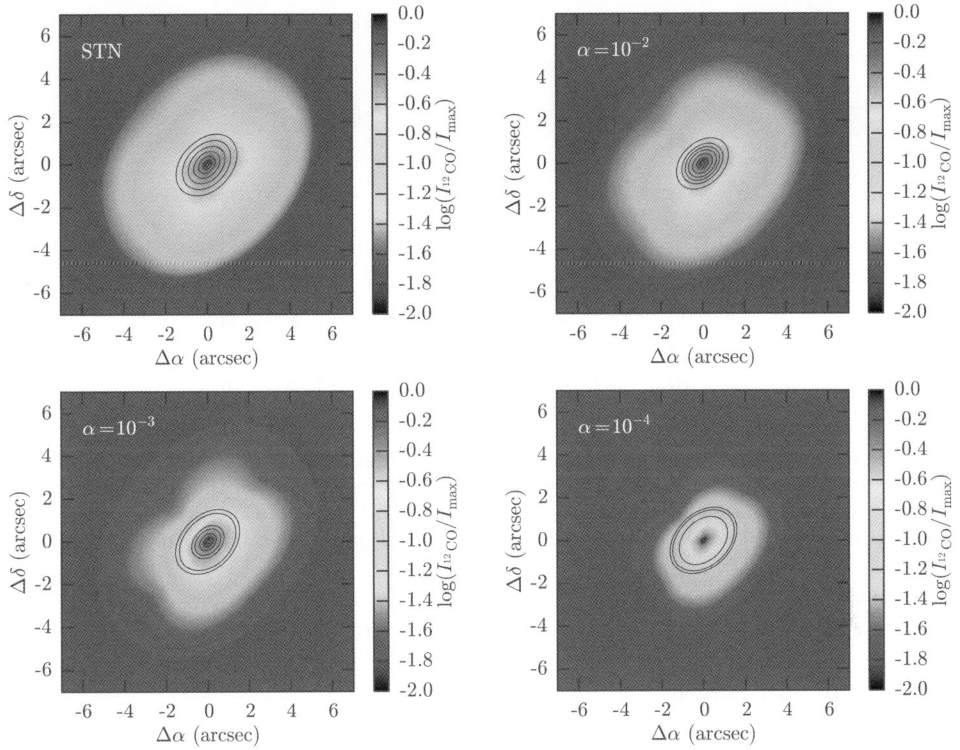

Figure 5. Normalized intensity maps of the ^{12}CO J=3-2 line for all models. Contours represent the peak normalized continuum levels at 850 μm between 1/2 and 1/128, logarithmically sampled. Panels from Facchini *et al.* (2017a).

line emission can be largely explained by an optical depth effect, with the dust continuum being optically thin and the CO lines being optically thick. The effect of radial drift can be seen in the sharpness of the continuum outer edge.

Figure 6 shows a summary plot of the ratio of disk outer radii as probed by submm continuum ($R^{\rm out}_{\rm mm}$) and CO isotopologue lines ($R^{\rm out}_{\rm CO}$), where we stress again that the underlying gas surface density is exactly the same for the three turbulent models. The ratio increases with turbulence, since $R^{\rm out}_{\rm mm}$ is not significantly affected by the value of α, whereas $R^{\rm out}_{\rm CO}$ shrinks for low turbulence values. This result emphasizes how important the simultaneous modeling of both gas and dust can be when interpreting observables as the disk outer radius, and in particular the significance of a proper implementation of realistic grain size distributions in the thermo-chemical models. This holds true for "simple" observables as the disk outer radius, but is at least as much important when trying to infer gas surface densities of disks, in particular when we know from observations that the dust properties are changing dramatically with radius, showing a plethora of substructures, as rings, gaps, spirals and azimuthal asymmetries (e.g. van der Marel *et al.* 2013; ALMA Partnership *et al.* 2015; Pérez *et al.* 2016). For example, determining the level of thermal coupling within the dust gaps will be important to correctly interpret the gaps that observations are starting to probe in gas tracers (Isella *et al.* 2016; Teague *et al.* 2017; Fedele *et al.* 2017). Good gas temperature estimates across the gaps, e.g. via line ratio determinations, will be important to disentangle between surface density and temperature gaps (Facchini *et al.* 2017b).

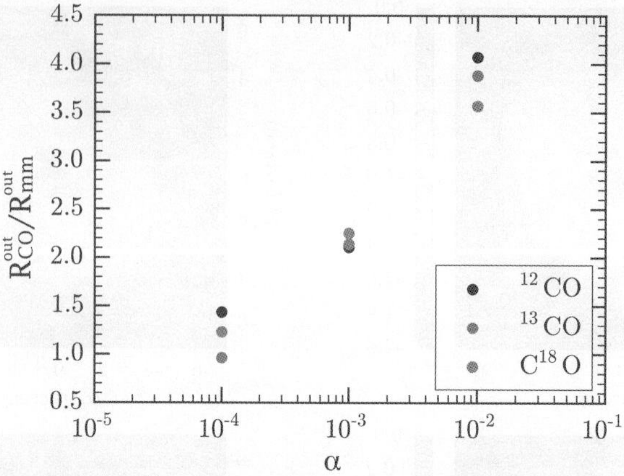

Figure 6. Ratio of the CO outer radius $R_{\rm CO}^{\rm out}$ and the mm outer radius $R_{\rm mm}^{\rm out}$ for the different turbulent models, where a dynamic range of 30 has been used for both the continuum and the gas outer radius estimates. Figure adapted from Facchini *et al.* (2017a). The color version of the figure is available in the electronic form or in the original paper (Facchini *et al.* 2017a).

References

ALMA Partnership, Brogan, C. L., Pérez, L. M., *et al.* 2015, *ApJ*, 808, L3

Andrews, S. M., Wilner, D. J., Hughes, A. M., *et al.* 2012, *ApJ*, 744, 162

Bai, X.-N. 2016, *ApJ*, 821, 80

Birnstiel, T. & Andrews, S. M. 2014, *ApJ*, 780, 153

Birnstiel, T., Andrews, S. M., Pinilla, P., & Kama, M. 2015, *ApJ*, 813, L14

Bruderer, S., van Dishoeck, E. F., Doty, S. D., & Herczeg, G. J. 2012, *A&A*, 541, A91

Bruderer, S. 2013, *A&A*, 559, A46

Cleeves, L. I. 2016, *ApJ*, 816, L21

Collings, M. P., Dever, J. W., Fraser, H. J., McCoustra, M. R. S., & Williams, D. A. 2003, *ApJ*, 583, 1058

de Gregorio-Monsalvo, I., Ménard, F., Dent, W., *et al.* 2013, *A&A*, 557, A133

Dubrulle, B., Morfill, G., & Sterzik, M. 1995, *Icarus*, 114, 237

Dutrey, A., Guilloteau, S., Prato, L., *et al.* 1998, *A&A*, 338, L63

Dutrey, A., Guilloteau, S., Piétu, V., *et al.* 2017, arXiv:1706.02608

Facchini, S., Birnstiel, T., Bruderer, S., & van Dishoeck, E. F. 2017a, *A&A* in press, arXiv:1705.06235

Facchini, S., Pinilla, P., van Dishoeck, E. F., & de Juan Ovelar, M. 2017b, *A&A* submitted

Fedele, D., Carney, M., Hogerheijde, M. R., *et al.* 2017, *A&A*, 600, A72

Guilloteau, S. & Dutrey, A. 1998, *A&A*, 339, 467

Huang, J., Öberg, K. I., & Andrews, S. M. 2016, *ApJ*, 823, L18

Isella, A., Guidi, G., Testi, L., *et al.* 2016, Physical Review Letters, 117, 251101

Lynden-Bell, D. & Pringle, J. E. 1974, *MNRAS*, 168, 603

Mordasini, C., Alibert, Y., Benz, W., Klahr, H., & Henning, T. 2012, *A&A*, 541, A97

Pérez, L. M., Carpenter, J. M., Andrews, S. M., *et al.* 2016, *Science*, 353, 1519

Piétu, V., Guilloteau, S., Di Folco, E., Dutrey, A., & Boehler, Y. 2014, *A&A*, 564, A95

Salinas, V. N., Hogerheijde, M. R., Mathews, G. S., *et al.* 2017, arXiv:1707.06475

Schwarz, K. R., Bergin, E. A., Cleeves, L. I., *et al.* 2016, *ApJ*, 823, 91

Shakura, N. I. & Sunyaev, R. A. 1973, *A&A*, 24, 337

Tazzari, M., Testi, L., Natta, A., *et al.* 2017, *A&A* in press, arXiv:1707.01499

Teague, R., Semenov, D., Gorti, U., *et al.* 2017, *ApJ*, 835, 228

Tripathi, A., Andrews, S. M., Birnstiel, T., & Wilner, D. J. 2017, *ApJ* in press, arXiv:1706.08977

van der Marel, N., van Dishoeck, E. F., Bruderer, S., *et al.* 2013, *Science*, 340, 1199

Vasyunin, A. I., Wiebe, D. S., Birnstiel, T., *et al.* 2011, *ApJ*, 727, 76

Interstellar Chemistry and the life cycles of stars

Astrochemistry VII – Through the Cosmos from Galaxies to Planets
Proceedings IAU Symposium No. 332, 2017
M. Cunningham, T. Millar & Y. Aikawa, eds.

doi:10.1017/S1743921317009784

Fire from Ice - Massive Star Birth from Infrared Dark Clouds

Jonathan C. Tan[1,2,3]

[1]Dept. of Space, Earth and Environment, Chalmers University, SE-412 96 Gothenburg, Sweden
[2]Dept. of Astronomy, University of Virginia, Charlottesville, VA 22904, USA
[3]Depts. of Astronomy and Physics, University of Florida, Gainesville, FL 32611, USA
email: jctan.astro@google.com

Abstract. I review massive star formation in our Galaxy, focussing on initial conditions in Infrared Dark Clouds (IRDCs), including the search for massive pre-stellar cores (PSCs), and modeling of later stages of massive protostars, i.e., hot molecular cores (HMCs). I highlight how developments in astrochemistry, coupled with rapidly improving theoretical/computational and observational capabilities are helping to improve our understanding of the complex process of massive star formation.

1. Introduction

Massive stars and their associated star clusters are important throughout astrophysics, yet there remain many open questions about how they form. These include: What is the accretion mechanism of massive star formation? What sets the initial mass function of stars, especially at the highest masses? What is the relation of massive star formation to star cluster formation? How do massive star and star cluster formation vary with galactic environment? The nature of the very first, metal-free, i.e., Population III, stars, often theorized to have been of high-mass, is an enduring open question.

In this short review, I focus on theoretical models and observational studies of local, i.e., Galactic, massive star and star cluster formation, concentrating on developments since the reviews of Tan *et al.* (2014, hereafter T14) and Tan (2016) and offering a somewhat different perspective compared to the review of Motte *et al.* (2017). I emphasize the importance of incorporating astrochemistry into theoretical models that try to predict the physical outcome of the massive star formation process, especially to predict the ionization fraction that couples the mainly neutral gas to magnetic fields. Astrochemistry is also crucial for providing diagnostics needed to test different theoretical models.

Following Williams *et al.* (2000) and T14, we define gas *clumps* as self-gravitating structures that fragment into star clusters, perhaps via a population of self-gravitating pre-stellar *cores* (PSCs). These cores are defined to be structures that collapse to a central, rotationally-supported disk leading to single or small-N multiple star formation. As massive stars tend to form in clusters (e.g., de Wit *et al.* 2005), massive star formation and cluster formation are connected processes that need to be understood together.

The observed range of masses of star-forming clumps, including early stage Infrared Dark Clouds (IRDCs), and young star clusters is from $M \sim 10$ to $\sim 10^6\ M_\odot$ (including the most extreme "super star clusters" seen in some relatively nearby galaxies, such as M82, NGC 1569 and NGC 5253), while mass surface densities, Σ, typically range from ~ 0.03 to $\sim 10\ \mathrm{g\,cm^{-2}}$ (i.e., ~ 200 to $\sim 5 \times 10^4\ M_\odot\ \mathrm{pc^{-2}}$) (T14). Thus, protocluster clumps have radii of ~ 1 to 10 pc and average H nuclei number densities of $n_\mathrm{H} \sim 10^3$ to $\sim 10^6\ \mathrm{cm^{-3}}$, although they may contain higher density substructures, including cores.

2. Theoretical Models of Massive Star Formation

2.1. *Overview of Formation Scenarios*

There is a long-standing debate about the basic formation mechanism of massive stars. Theories range from Core Accretion models (e.g., McLaughlin & Pudritz 1997; McKee & Tan 2003 [hereafter MT03], who presented the Turbulent Core Model [TCM]) that are scaled-up versions of standard low-mass star formation theories (Shu, Adams & Lizano 1987), to Competitive Accretion models at the crowded centers of forming star clusters (Bonnell *et al.* 2001; Wang *et al.* 2010) that do not involve massive starless cores, to Protostellar Collisions (Bonnell *et al.* 1998; Bally & Zinnecker 2005; Moeckel & Clarke 2011). Still, as Core Accretion models are the most conservative and simplest way to treat massive star formation and potentially offer a universal mechanism by which to understand all star formation, we will focus mostly on their predictions and comparison to observations, noting cases where these seem to be a relatively poor description.

2.2. *Initial Conditions of the Turbulent Core Model: Massive Pre-Stellar Cores*

The TCM assumes the initial conditions of massive stars are massive PSCs, approximated as polytropic spheres, that are in quasi virial equilibrium and pressure equilibrium with their surrounding clump environment. This implies that a core of mass M_c in a self-gravitating clump of mass surface density $\Sigma_{\rm cl}$, which sets ambient pressure, has radius

$$R_c \simeq 0.0574(M_c/60\,M_\odot)^{1/2}(\Sigma_{\rm cl}/1\,{\rm g\,cm^{-2}})^{-1/2}\ {\rm pc}, \tag{2.1}$$

(MT03), i.e., 12,000 AU, for fiducial parameter choices, including index of internal power law density profile, $k_\rho = 1.5$. The core has a mass-averaged 1D velocity dispersion of

$$\sigma_c \simeq 1.09(\phi_B/2.8)^{-1/2}(M_c/60\,M_\odot)^{1/4}(\Sigma_{\rm cl}/1\,{\rm g\,cm^{-2}})^{1/4}\ {\rm km\,s^{-1}}, \tag{2.2}$$

where $\phi_B = 1.3 + 1.5m_A^{-2}$ accounts for the effects of B-fields, taking a value of 2.8 in the fiducial case of an Alfvén Mach number, m_A, of unity. However, we note that MT03 approximated the magnetic pressure $B^2/(8\pi)$ as isotropic, which it is not. A random magnetic field has an isotropic pressure of $B^2/(24\pi)$, which would reduce ϕ_B, i.e., in this limit the fiducial value would become about 1.8 (C. McKee, private communication).

Note, these conditions apply to the gas structure at the moment just before protostar formation and little is yet assumed about how the core itself formed. By definition, it will have emerged from the clump gas, perhaps via top-down fragmentation, i.e., a massive starless core condensing out of the ambient clump medium, or via bottom-up growth of a smaller gravitationally bound starless core that gains mass by some combination of accretion from the clump or mergers with other cores (such core merging should not be confused with protostellar mergers). The timescale over which the core forms is also not specified: a range of possibilities from very fast, i.e., $t_{\rm c,form} \sim 1t_{\rm ff}$, where $t_{\rm ff} = (3\pi/[32G\rho])^{1/2} = 1.4 \times 10^5(n_{\rm H}/10^5\ {\rm cm^{-3}})^{-1/2}$ yr is the local free-fall time of the core given its mean density ρ or equivalently mean H nuclei number density, $n_{\rm H}$, to very slow, i.e., $t_{\rm c,form} \gtrsim 10t_{\rm ff}$ may be considered. If core formation is a relatively slow process, i.e., $\gtrsim$ a few $t_{\rm ff}$, then the conditions of quasi virial and pressure equilibrium are more likely to be achieved. Note also that while the boundary of a PSC has a precise theoretical definition, i.e., delimiting the material that is gravitationally bound to the core, observationally it can be very challenging to isolate core material by this criterion, especially since the mass surface densities of the core and the surrounding clump have similar values (MT03). We discuss observational definitions of PSCs in §3.1.

The mass scale of a core may potentially be set by its degree of magnetization. If it forms relatively slowly from a more strongly magnetized state as magnetic flux is gradually removed from the region via ambipolar (e.g., Mouschovias 1991) or reconnection (Lazarian & Vishniac 1999; Eyink *et al.* 2011) diffusion, then it will be close to the boundary between a sub- and super-critical state, i.e., where the B-field strength is just strong enough to impede collapse. Then the mass of the contracting core is effectively set by the magnetic critical mass (e.g., Bertoldi & McKee 1992)

$$M_{\rm c,B} = 51.8(y/0.5)^{-2}(B_c/200\,\mu{\rm G})^3(n_{\rm H,c}/10^5\,{\rm cm}^{-3})^{-2}\,M_\odot, \quad (2.3)$$

where $y = Z/R$ is the aspect ratio (R being the radius normal to axis of symmetry for an ellipsoidal core; $2Z$ being size of core along this axis) here normalized to a moderate degree of flattening along the field direction, B_c is the average magnetic flux in the core (here normalized to a typical value inferred from observations of molecular clouds at this density, Crutcher *et al.* 2010) and $n_{\rm H,c}$ is the mean density of the core. Kunz & Mouschovias (2009) have argued that the entire PSC mass function (PSCMF) may be set by modest variations in the degree of magnetization of gas within star-forming regions.

Eq. (2.2) implies massive PSCs are expected to have internal turbulent motions much greater than the isothermal sound speed at ~ 10 K, i.e., $c_{\rm th} = 0.19(T/10\,{\rm K})^{1/2}\,{\rm km\,s^{-1}}$. Thus they are likely be supersonically turbulent, which would induce internal substructure. The resulting localized shock heating, i.e., up to ~ 50 K for $\sim 1\,{\rm km\,s^{-1}}$ shock speeds, is expected to have astrochemical effects, e.g., liberation of species from grain ice mantles (CO freeze-out being efficient at $\lesssim 20$K) and reduction in the level of deuteration of key diagnostic species N_2D^+ (Kong *et al.* 2015, hereafter K15, see below). Dissipation of energy in these shocks leads to a reduction in the level of turbulence with decay times of about a few $t_{\rm ff}$ (e.g., McKee & Ostriker 2007), which may be partially offset by gravitational contraction of the core. If diffusion of B-field out of the core occurs more slowly, then one expects an evolution towards a state with a lower degree of turbulence, i.e., smaller values of m_A, and a stronger degree of support from large-scale B-fields.

Deuterated species, such as N_2D^+, are thought to be one of the key observational tracers of PSCs (§3.1). This expectation is based on studies of nearby, relatively low-mass PSCs, such as L1544 (e.g., Caselli & Ceccarelli 2012), with observations of such sources leading to the refinement of astrochemical models that aim to predict the abundance of these diagnostic species. High levels of deuteration arise because in the cold, dense conditions relevant to PSCs, CO molecules are mostly frozen-out onto dust grain ice mantles, which allows the abundance of H_2D^+ to build up via the exothermic reaction of $H_3^+ + HD \rightarrow H_2D^+ + o{-}H_2$, where we have indicated this applies only for the ortho form of H_2. The para form of H_2 has enough energy to drive the reverse reaction. As the ortho-to-para ratio of H_2 ($\mathrm{OPR}^{\mathrm{H2}}$) drops, e.g., mediated by either gas phase (e.g., Sipilä *et al.* 2015; K15) or grain surface phase (Bovino *et al.* 2017) reactions, this then leads to significant "deuteration fractions", $D_{\rm frac}$, i.e., abundance ratio to the non-deuterated species, of H_2D^+ and N_2D^+ (the latter formed via $H_2D^+ + N_2 \rightarrow H_2 + N_2D^+$).

The gas-phase astrochemistry associated with this deuteration, i.e., including ortho and para spin states, has been studied in the context of one-zone models by, e.g., Sipilä *et al.* (2010, 2015), Wirström *et al.* (2012) and K15, with the latter presenting an exploration of the parameter space relevant to IRDCs/protoclusters and massive PSCs (i.e., $n_{\rm H} = 10^3$ to $10^7\,{\rm cm}^{-3}$, $T = 10$ to 30K, cosmic ray ionization rates $\zeta = 10^{-18}$ to $10^{-15}\,{\rm s}^{-1}$, assumed fixed gas-phase heavy element depletion factors (a proxy for CO freeze-out) from $f_D = 1$ (no depletion) to 1000 (i.e., total abundance divided by gas phase abundance is 1000), and initial values of $\mathrm{OPR}^{\mathrm{H2}}$ from $\sim 10^{-3}$ to 3. Some of the main results of this study were the near-equilibrium values of $D_{\rm frac}^{\rm N_2D^+}$ under given conditions and the timescales needed to reach 90% of these near-equilibrium conditions. For example, for conditions of the fiducial

TCM, i.e., $n_{\rm H} \simeq 10^5\,{\rm cm}^{-3}$ and $T = 15\,{\rm K}$ (and $\zeta = 2.5 \times 10^{-17}$, $f_D = 10$), K15 found $D_{\rm frac,eq}^{\rm N_2D^+} = 0.181$ and the time to reach within 90% of this value to be $t_{\rm eq,90} = 1.25$ Myr (for initial ${\rm OPR^{H2}} = 0.1$). The fact that this is almost ten times longer than the local free-fall timescale for this density is the reason why the deuteration process may be a useful "chemical clock" to constrain the dynamics of core formation. However, this does depend on correct modeling of the timescale of conversion of ortho to para H_2 (see Bovino *et al.* 2017), as well as the assumption made for the "initial" value of ${\rm OPR^{H2}}$. Observational constraints on ${\rm OPR^{H2}}$ are important for breaking such model degeneracies (§3.1).

K15 also presented a series of chemodynamical models, including evolution of f_D by freeze-out and evolution of $n_{\rm H}$ at rates relative to that of free-fall collapse, i.e., following $dn_{\rm H}/dt = \alpha_{\rm ff} n_{\rm H}/t_{\rm ff}$. Again, since the time to reach high levels of deuteration of N_2D^+ can be significantly longer than a local free-fall time, these models can be used in conjunction with observed values of $D_{\rm frac}^{\rm N_2D^+}$ to constrain $\alpha_{\rm ff}$ (§3.1). Such chemodynamical models can be improved to include core radial structure; see, e.g., Pagani *et al.* (2009) for an example of a simple astrochemical model (no spin-state chemistry or N chemistry) applied to low-mass cores and the more general modeling of Gerner *et al.* (2015). Full 3D numerical (M)HD simulation that is fully coupled to an astrochemical network following deuteration is challenging. For example, Körtgen *et al.* (2017) carried out simulations with a reduced network that focussed on H_2D^+ (N_2D^+ was not modeled). Taking an intermediate approach, Goodson *et al.* (2016) analyzed the K15 models to derive analytic expressions for the growth rate of N_2H^+ and N_2D^+ abundances as a function of local density and starting ${\rm OPR^{H2}}$. These expressions were incorporated in look-up tables to estimate abundances of these species in MHD simulations of massive cores (Fig. 1). These results can be compared to observed candidate massive PSCs (§3.1).

In addition to PSC diagnostics, astrochemical modeling is also needed to estimate ionization fractions in these clouds. For example, for the fiducial parameters of the K15 gas phase astrochemical modeling described above, in particular with $n_{\rm H} = 10^5\,{\rm cm}^{-3}$ and $\zeta = 2.5 \times 10^{-17}\,{\rm s}^{-1}$, $n_e/n_{\rm H}$ is found to range from $\sim 10^{-8}$ when $f_D = 1$ (no depletion) up to $\sim 5 \times 10^{-8}$ in the limit of high f_D. In the case of $f_D = 10$, the positive charge carriers are dominated by about equal fractions of H_3^+ and HCO^+. The ambipolar diffusion time, $t_{\rm ad} = 2.5 \times 10^{13} (n_e/n_{\rm H})\,{\rm yr}$, then has a fiducial value of 5×10^5 yr, i.e., about three times greater than $t_{\rm ff}$. As discussed below, the ionization fraction also depends on assumptions about the dust grain size distribution, which introduces further uncertainties.

2.3. *Massive Protostellar Cores*

In Core Accretion models, a very low-mass ($\ll 1\,M_\odot$) protostar is first expected to form near core center. This acts as a seed for continued accretion from the infall envelope and the structure is now known as a protostellar core. In the fiducial TCM with collapse at rate similar to that of local free-fall, the accretion rate delivered by the infall envelope is

$$\dot{m}_{*d} = 9.3 \times 10^{-4} \epsilon_{*d} (M_c/60\,M_\odot)^{3/4} (\Sigma_{\rm cl}/1\,{\rm g\,cm^{-2}})^{3/4} (M_{*d}/M_c)^{1/2}\,M_\odot\,{\rm yr}^{-1}, \quad (2.4)$$

where $\dot{m}_{*d}$ is the rate of increase of the mass of the protostar and its disk, ϵ_{*d} is the current efficiency of the infall rate with respect to uninhibited collapse (values ~ 0.5 are expected to develope due to protostellar outflow feedback, e.g., Zhang, Tan & Hosokawa 2014 [ZTH14]) and M_{*d} is the idealized collapsed mass supplied to the central disk in the no-feedback limit. Here the ratio M_{*d}/M_c indicates the evolutionary stage of the collapse, i.e., advancing from 0 to 1. The increasing accretion rates during the collapse are a consequence of the assumed initial power law density profile of the PSC of $k_\rho = 1.5$. If an index of 2 is adopted, i.e., that of a singular isothermal sphere, then the accretion rate

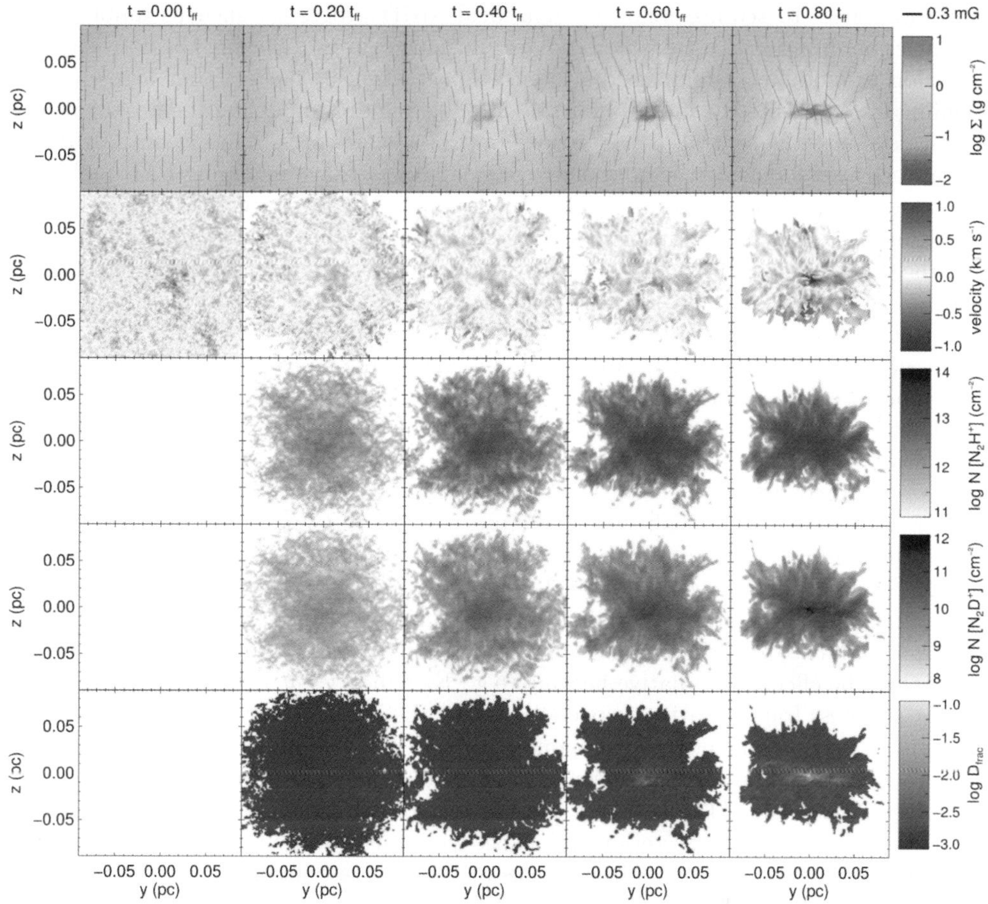

Figure 1. Evolution of a 60 $M_\odot$ PSC embedded in $\Sigma = 0.3$ g cm^{-2} clump (Goodson *et al.* 2016) with properties set by the TCM (MT03). Time evolution is left to right from 0 to 0.8 $t_{\rm ff}$, with initial mean local free-fall time, $t_{\rm ff} = 0.2$ Myr. 1st row: mass surface density, Σ. Total projected B-field strengths and orientations are shown by lines (scale in top right). 2nd row: mean velocity of N_2D^+. 3rd and 4th rows: N_2H^+ and N_2D^+ column densities, via analytic fits to the K15 model. 5th row: $D_{\rm frac}^{\rm N2H+} = [N_2D^+]/[N_2H^+]$. This particular simulated core undergoes relatively fast collapse, so there is little time to build-up very high values of $D_{\rm frac}^{\rm N2H+}$.

would be independent of M_{*d}/M_c. With the fiducial rate of eq. (2.4), the star formation timescale is $t_{*f} = 1.3 \times 10^5 (M_c/60\ M_\odot)^{1/4} (\Sigma_{\rm cl}/1\ {\rm g\,cm^{-2}})^{-3/4}$ yr, which has a weak dependence on M_c. Note, t_{*f} is similar to the clump mean free-fall time, as $t_{*f}/\bar{t}_{\rm ff,cl} \rightarrow 0.98 (M_c/60\ M_\odot)^{1/4} (M_{\rm cl}/4000\ M_\odot)^{-1/4}$. If cluster formation takes many clump free-fall times to complete, then this model allows stars of all masses to form contemporaneously, i.e., with individual formation times that are much shorter than that of the cluster.

Initially protostellar accretion may be quasi spherical, i.e., directly onto the surface of the protostar that is expected to have a radius of several $R_\odot$. Later, as material continues to accrete it is more likely to hit a centrifugal barrier and form an accretion disk, which then mediates accretion and angular momentum transfer, including by launching magneto-centrifugal disk winds and X-winds. For a review of these processes in the context of low-mass star formation, see, e.g., Inutsuka (2012). However, disk formation is a highly uncertain process given the expected role of magnetic braking for transferring angular momentum in the infall envelope (Li *et al.* 2014). Indeed, to fully model this process appears to require following non-ideal MHD processes, such as ambipolar or reconnection diffusion. To treat ambipolar diffusion requires following the astrochemical evolution to model the small residual ionization fraction in the gas that helps mediate coupling of B-fields to the mostly neutral gas. This requires modeling the propagation of cosmic rays into the central, dense regions of protostellar cores. Zhao *et al.* (2016) have also argued that following the abundance of small dust grains, including PAH molecules, that can act as charge carriers is also needed to make accurate predictions for disk formation. Given the uncertain nature of the initial conditions of massive star formation, e.g., the degree of magnetization of massive PSCs, and given the above complications of following the coupled chemistry and dynamics of the collapse, it is not yet possible to make accurate predictions for the sizes of disks around massive protostars.

The degree of fragmentation in the core will also depend on the magnetization of the PSC. For core formation mediated by magnetic fields, little fragmentation is expected below the magnetic critical mass scale given by eq. (2.3). This expectation is broadly confirmed by the results of numerical simulations. For example, Peters *et al.* (2011) simulated a core with $M_c = 100 M_\odot$, $R_c = 0.5$pc, $n_{\rm H,c} = 5,400{\rm cm}^{-3}$ and a weak magnetic field of $B_c = 10\,\mu$G. The result of the collapse was a small cluster of protostars. Seifried *et al.* (2011) simulated a core with $M_c = 100\,M_\odot$, $R_c = 0.25$ pc, $n_{\rm H,c} = 4.4 \times 10^4\ {\rm cm}^{-3}$ and $B_c \simeq 1$mG, while Myers *et al.* (2013) followed a core with $M_c = 300 M_\odot$, $R_c = 0.1$pc, $n_{\rm H,c} = 2.4 \times 10^6\ {\rm cm}^{-3}$ and $B_c \gtrsim 1$ mG. In these latter two simulations no fragmentation was seen, with collapse proceeding approximately monolithically towards a single, central protostar. The effects of radiative heating (e.g., Krumholz *et al.* 2007; Myers *et al.* 2013) further suppress fragmentation, especially during the later, higher-luminosity stages.

Semi-analytic treatments of protostar formation and evolution via the TCM in the limit of no fragmentation have been presented by Zhang & Tan (2011), Zhang, Tan & McKee (2013) and ZTH14. These models include treatments of the slowly rotating infall envelope, a viscous accretion disk, disk wind outflows and protostellar evolution. With the density structure specified by this modeling, the temperature structure is then calculated via Monte Carlo radiative transfer (RT) simulations, including both gas and dust opacities, along with emergent radiation, i.e., to predict multiwavelength images and spectral energy distributions (SEDs). Zhang & Tan (2017) present a suite of such models that cover the parameter space of $M_c = 10$ to $480\,M_\odot$ and $\Sigma_{\rm cl} = 0.1$ to $3\,{\rm g\,cm}^{-2}$. These models have been applied to observed sources by De Buizer *et al.* (2017) (§3.2).

The above models predict that the massive protostar will reach a H-burning phase while still accreting. The photosphere will have a high temperature and be a strong source of FUV and EUV radiation, creating photodissociation regions (PDRs) and HII regions, respectively. Ionizing feedback will first interact with the protostellar outflow, creating an "outflow-confined" HII region that will appear as elongated cm continuum emission (Tan & McKee 2003; Tanaka *et al.* 2016). A combination of radiative and mechanical (outflow) feedback is expected to eventually set the star formation efficiency from the core, ϵ_c. For $M_c = 60\,M_\odot$ and $\Sigma_{\rm cl} = 1\,{\rm g\,cm}^{-2}$, the models of Tanaka *et al.* (2017) have $\epsilon_c \simeq 0.4$. They also find ϵ_c decreases for larger M_c, as feedback from more massive stars is more powerful, and decreases as $\Sigma_{\rm cl}$ decreases, since lower density cores, which have lower accretion rates, are more easily disrupted. Axisymmetric, 2D numerical simulations of mechanical and radiative feedback from massive protostars have been presented by Kuiper *et al.* (2015, 2016) (see also the 3D radiative feedback only simulations of, e.g., Rosen *et al.* 2016 and Harries *et al.* 2017). One key property that is predicted by the above models and simulations is the opening angle of the outflow cavity at a given stage of the protostellar evolution, since this can be compared directly with observed systems.

The modeling of the astrochemistry of the "hot molecular cores" (HMCs) of high-mass protostars has been carried out by a number of groups (e.g., Charnley *et al.* 1992; Caselli *et al.* 1993; Viti *et al.* 2004; Doty *et al.* 2006; Garrod & Herbst 2006; Aikawa *et al.* 2012; Öberg *et al.* 2013; Gerner *et al.* 2014, 2015). Such modeling, typically done as post processing of a given physical model, requires specifying initial chemical conditions, e.g., of the PSC just before star formation, and then the evolution of densities and temperatures along streamlines.

Figure 2, from Drozdovskaya *et al.* (in prep.), shows an example of such modeling for a fiducial massive protostellar infall envelope of the TCM (ZTH14), utilizing a gas-grain network with ~700 species and 9,000 reactions, including surface chemistry and complex organic species (Drozdovskaya *et al.* 2014, 2015). After a static PSC phase of 3×10^5 yr, the streamlines of the infall envelope are followed for 8.2×10^4 yr, i.e., the time needed to form a 16 $M_\odot$ protostar. The protostar heats the infall envelope, driving astrochemical reactions in the gas and on grain surfaces, and liberating ice species. The figure shows examples of H_2CO gas and ice phase abundances, with the gas phase abundance rising strongly inside $\sim 3,000$ AU. Such models, with outputs then coupled to line radiative transfer codes, are important for making predictions that can be tested with sub-mm/mm observations. However, the full problem of coupled dynamical and chemical evolution, including non-ideal MHD effects and the interaction, via shocks, of protostellar outflows with the infalling core, is challenging to model accurately, given the many inherent uncertainties in both physical and chemical aspects.

2.4. *Massive Star Formation in the Context of Star Cluster Formation*

Aspects of star cluster formation have an impact on massive star formation. As with core formation times, similar uncertainty applies for clumps, i.e., their formation time from surrounding (giant) molecular cloud ([G]MC) gas, $t_{\rm cl,form}$, and the timescale to complete star cluster formation, $t_{*\rm cl,form} = (\epsilon/\epsilon_{\rm ff})\bar{t}_{\rm ff,cl}$, where ϵ is the final star formation efficiency and $\epsilon_{\rm ff}$ is the efficiency per free-fall time. Fast cluster formation in $\sim 1\bar{t}_{\rm ff,cl}$ has been proposed by Elmegreen (2000, 2007), Hartmann & Burkhart (2007) and Hartmann *et al.* (2012). Slower, quasi-equilibrium star cluster formation has been proposed by Tan, Krumholz & McKee (2006) and Nakamura & Li (2007, 2014), with turbulence maintained in the clump by protostellar outflows and/or by accretion (Klessen & Hennebelle 2010; Goldbaum *et al.* 2011). Turbulent gas is expected to have a low rate of star formation (Krumholz & McKee 2005; Padoan *et al.* 2011), i.e., $\epsilon_{\rm ff} \sim 0.02$. Such estimates are consistent with observed protoclusters (e.g., Krumholz & Tan 2007; Krumholz *et al.* 2012; Da Rio *et al.* 2014). Since ϵ needs to be $\gtrsim 0.3$ to form bound star clusters, then, at least in such systems, these estimates of $\epsilon_{\rm ff}$ imply $t_{*\rm cl,form} \sim 10\bar{t}_{\rm ff,cl}$.

The observational search for massive PSCs in protocluster clumps is described in §3.1, however, several points can be considered here to help guide expectations. First, since massive stars are rare, massive PSCs will also be rare objects. Most mass in a clump will not be part of a massive PSC. If we define massive PSCs as having $\gtrsim 16\,M_\odot$, i.e., able to form $\gtrsim 8\,M_\odot$ stars and if the PSCMF is described by a Salpeter (1955) power law of form $dN/d\log M_c \propto M_c^{-\alpha}$ with $\alpha = 1.35$ with lower limit of $M_c = 0.2$ or $1\,M_\odot$ (so that resulting stellar IMFs approximately bracket the characteristics of the observed IMF) and upper limit of 240 $M_\odot$, then the fraction of mass that is in massive cores is 0.144 or 0.272, respectively. Thus a typical mass fraction of cores that can form massive stars is $\simeq 0.2$. If the fraction of the total clump mass that forms PSCs is $\simeq 0.5$, i.e., so that total eventual star formation efficiency is $\simeq 0.25$, then only $\sim 10\%$ of the total clump mass is expected to be processed through massive PSCs. As discussed above, the protostellar phase is expected to take $t_{*f} \simeq 1\bar{t}_{\rm ff,cl}$ and also $t_{*f} \simeq 4.4\bar{t}_{\rm ff,c}$ (MT03). For a

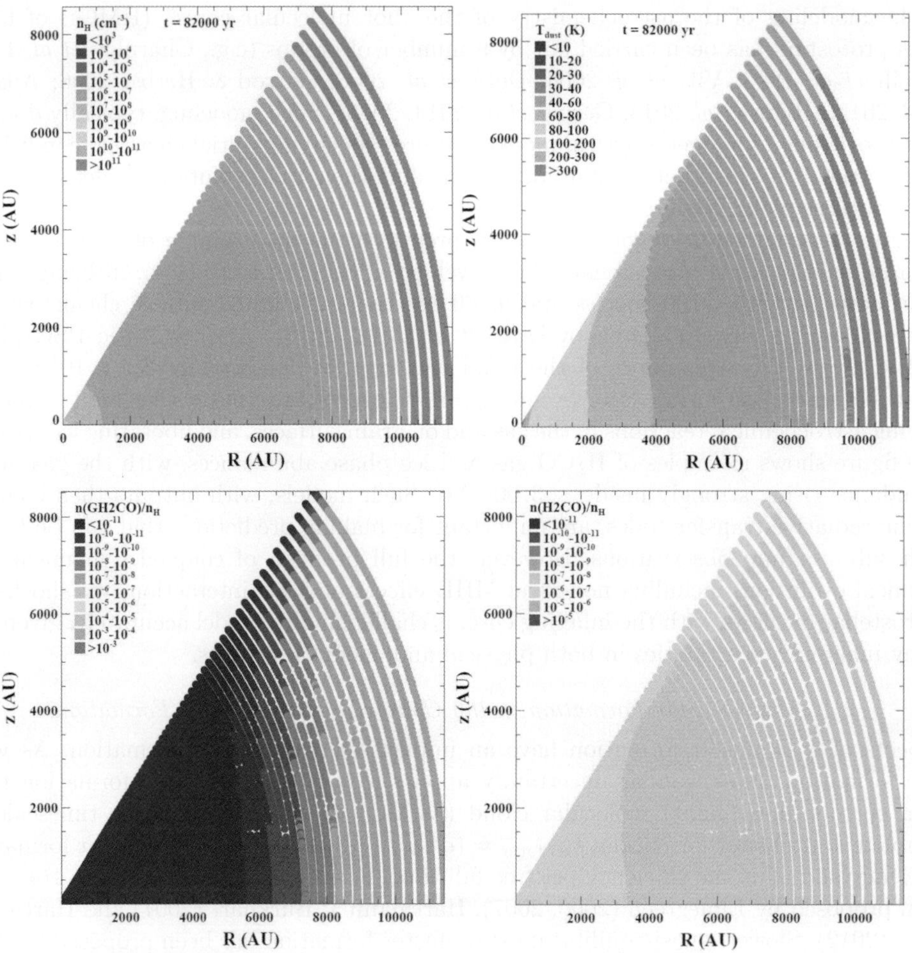

Figure 2. (From Drozdovskaya *et al.*, in prep.; see also Drozdovskaya *et al.* 2017) Astrochemical modeling of the infall envelope of a massive protostar for the case of a $60\,M_\odot$ core embedded in a $\Sigma_{\rm cl} = 1\,{\rm g\,cm^{-2}}$ clump with the protostar now at a mass of $16\,M_\odot$, located in lower left corner of each panel (ZTH14). By this point the outflow cavity has opened up to an angle of $\sim 40°$ (white sector: its density, temperature and chemical structure is not shown here). Note, the accretion disk is also not modeled here. *(a) Top Left:* Density structure ($n_{\rm H}$) at sampling points along streamlines of the infall envelope. *(b) Top right:* Temperature structure. *(c) Bottom Left:* Gas phase abundance of H_2CO relative to $n_{\rm H}$. *(d) Bottom right:* Ice phase abundance of H_2CO relative to $n_{\rm H}$. Note, H_2CO starts evaporating from the grain mantles inside $\sim 7,000$ AU from the protostar and becomes abundant in the gas phase inside $\sim 3,000$ AU.

steady star formation rate and, for simplicity considering closed box models that take $10\bar{t}_{\rm ff,cl}$ to form, then at any given time the observed massive protostellar core population will reflect only 10% of the total that ultimately forms and will only contain 1% of the initial clump mass. Similarly, if the same number of massive PSCs are observed as massive protostellar cores, then this would reflect those PSCs that are within $\simeq 4.4\bar{t}_{\rm ff,c}$ before they form a star (with $\bar{t}_{\rm ff,c}$ defined at this time). The implications of observed PSC and protostellar core demographics are described in §3.1.

Competitive Accretion and Protostellar Collision models both predict that massive protostars will be found near the centers of forming clusters in regions of high (proto)stellar densities. However, Core Accretion models may also predict a preference for massive

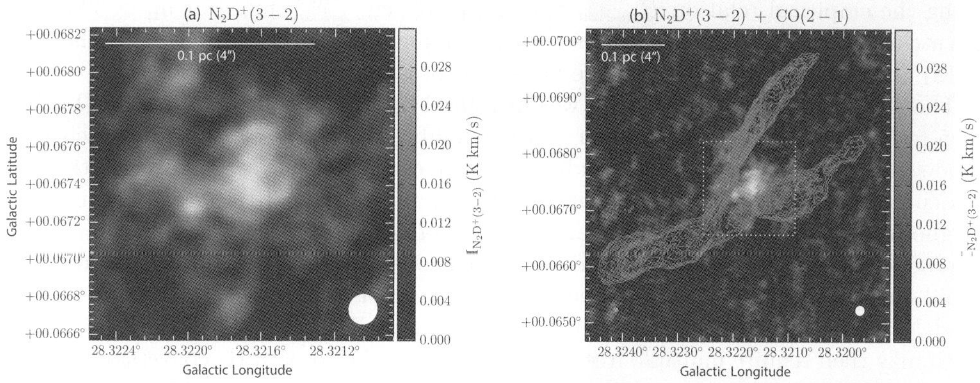

Figure 3. (From Kong *et al.* 2017b) *ALMA* observations of integrated intensity of N_2D^+ (3-2) emission from the G028.37+00.07 (C1-South) massive PSC candidate. The left panel shows a close-up view of the core, with the image smoothed to 0.5 arcsec resolution. The right panel shows an expanded view of the region, now including high velocity CO(2-1) emission revealing two collimated bipolar outflows from nearby, but separate, protostellar sources.

cores to form in denser regions near clump centers, e.g., if massive PSC formation occurs via an agglomeration of smaller PSCs. More isolated massive PSCs are also possible in Core Accretion models. There is a general expectation that massive protostars forming in crowded regions that suffer frequent tidal interactions with nearby stars will have smaller accretion disks and more disturbed accretion geometries. For example, the orientation of the disk and outflows would vary more in Competitive Accretion than Core Accretion models. Strong accretion variability, i.e., bursts, would also be expected to result from these interactions. However, accretion bursts due to disk instabilities and, more slowly, via infall variation in turbulent cores, are also possible in Core Accretion models.

3. Observational Studies of Massive Star Formation

3.1. *The Search for Massive Pre-Stellar Cores*

One of the best studied "low-mass" PSCs is L1544 (Caselli & Ceccarelli 2012). However, this core actually has $\sim 8\,M_\odot$, i.e., is quite massive, and its slow, subsonic infall ($\lesssim$10% of free-fall), suggests B-fields play a significant role in its dynamics (Keto *et al.* 2015).

Searches for more massive PSCs in higher $\Sigma_{\rm cl}$ environments have focussed on IRDCs. Tan *et al.* (2013), following up a sub-sample of the Butler & Tan (2012) MIR extinction map peaks, identified six cores via $N_2D^+(3-2)$ emission. Follow-up observations of the C1-S source identified protostellar outflows in its vicinity (Tan *et al.* 2016), but analysis of the highest resolution, highest sensitivity data (Kong *et al.* 2017b) indicates that C1-S, as defined by its $N_2D^+(3-2)$ emission, is spatially and kinematically distinct from these sources and is thus a promising massive PSC candidate (Fig. 3). The mass estimate, based on mm dust continuum emission, is about 50 $M_\odot$ inside a radius of 0.045 pc, depending on assumed dust temperature (fiducial value of 10 K). This implies a mean density of $n_{\rm H} \simeq 4\times10^6\,{\rm cm}^{-3}$. Note that due to systematically cooler temperatures, PSCs like C1-S do not stand out as strong continuum sources, especially compared to protostellar cores.

The velocity dispersion of C1-S is measured to be 0.28 km/s, which is about 1/3 of the level expected from virial equilibrium of the fiducial TCM core (eq. 2.2). For the core to be in virial equilibrium would require stronger large-scale B-fields, i.e., $\sim 3\,$mG, so that the Alfvén Mach number is about 0.2. Such B-field strengths are similar to those predicted using the empirical relation $B_{\rm median} \simeq 0.22(n_{\rm H}/10^5\,{\rm cm}^{-3})^{0.65}$ mG (for $n_{\rm H} > 300\,{\rm cm}^{-3}$) (Crutcher *et al.* 2010), to those observed on larger scales in some IRDCs (Pillai *et al.* 2015) and to values inferred in some massive protostars (§3.2).

The deuteration fraction of C1-S was measured by Kong *et al.* (2016) to be $D_{\rm frac}^{\rm N2D+} \sim$ 0.15 to 0.7. These values are similar to the equilibrium values of the K15 astrochemical models for the relevant physical conditions of the core. The timescales to reach this level of deuteration are $\gtrsim 10^5$ yr, significantly longer than the $\sim 2 \times 10^4$ yr free-fall time. Most chemodynamical models require relatively slow collapse compared to free-fall ($\alpha_{\rm ff} \lesssim 0.3$). Conversely, the example simulated core of Goodson *et al.* (2016), which undergoes more rapid collapse (Fig. 1), does not reach such high levels of $D_{\rm frac}^{\rm N2D+}$. Still, rapid collapse models can be made compatible if the starting value of ${\rm OPR}^{\rm H2}$ is very low (but which itself may then require multiple free-fall times) or if ortho to para H_2 conversion rates are dramatically sped up compared to gas phase estimates (Bovino *et al.* 2017). Observational contraints on ${\rm OPR}^{\rm H2}$ are needed to help break these degeneracies. Brüncken *et al.* (2014) achieved this for the low-mass protostellar core IRAS 16293-2422 A/B via observations of ortho- and para-H_2D^+, estimating it has a chemical age of > 1 Myr, i.e., $> 10t_{\rm ff}$. Similar studies are needed of more massive cores.

To increase the sample size of massive PSCs, Kong *et al.* (2017a) searched 30 IRDC clumps for N_2D^+(3-2) emission. Several promising candidates were detected. Dynamical analysis of the 6 strongest sources was carried out. Together with the 6 cores analyzed by Tan *et al.* (2013), overall this sample of 12 intermediate- and high-mass PSC candidates have observed velocity dispersions that are quite similar, within a factor of ∼0.8, compared to the fiducial virial equilibrium value of eq. (2.2).

Cyganowski *et al.* (2014) reported G11.920.61-MM2 as a massive PSC candidate. However, the non-detection of any molecular lines from this source is peculiar and makes it difficult to assess the reliability of the structure, e.g., via a dynamical mass measurement. The Cygnus X N53 MM2 core (Bontemps *et al.* 2010) and G11P6-SMA1 (Wang *et al.* 2014) are other potential massive PSCs based on the absence of obvious outflows (see also Motte *et al.* 2017). Sanhueza *et al.* (2017) searched IRDC G028.23-00.19 for massive PSCs. Given its current mass of 1,500 $M_\odot$, if it were to form a star cluster of $\sim 500\,M_\odot$, then the median expected mass of the most massive star would be about 26 $M_\odot$ (for Salpeter IMF from 0.1 to 120 $M_\odot$), so a $\sim 50\,M_\odot$ PSC is expected at some stage in the clump. However, Sanhueza *et al.* find five cores with masses up to only $\sim 15\,M_\odot$, though these do appear to be starless, i.e., lacking outflows. Such results may indicate that the most massive core has not yet formed. Note, if cluster formation is slow, then at any instant, only a small fraction $\sim \epsilon_{\rm ff}/\epsilon$ of the core population would be present.

Motte *et al.* (2007) and Russeil *et al.* (2010) (see also Motte *et al.* 2017) estimated massive PSC and starless clump lifetimes as short as $\lesssim 1$ to 3×10^4 yr in Cygnus X and NGC6334/NGC6357, by comparing to numbers of O to B3 stars and assigning a timescale of a few Myr to these stars. In addition to the already mentioned difficulty of identifying PSCs via dust continuum if they are systematically colder than protostellar cores and the ambient clump (e.g., Russeil *et al.* adopt a fiducial temperature of 20 K for their mass estimates), another potential problem with this analysis is that only core/clumps of $\geqslant 40 M_\odot$ and $\geqslant 200 M_\odot$ were counted in Cygnus X and NGC6334/NGC6357, respectively. For example, in the fiducial TCM, PSCs with masses $\sim 16\,M_\odot$ are expected to be able to produce $\sim 8\,M_\odot$ stars, i.e., B3 stars on the zero age main sequence.

3.2. *Massive Protostars, Accretion Disks, Outflows and Hot Molecular Cores*

Csengeri *et al.* (2017) studied mm dust continuum emission from 35 infrared quiet massive clumps, finding many examples of massive, protostellar cores that show limited fragmentation: most regions are dominated by just one or a few cores. The presence of strong B-fields is a plausible explanation for the limited degree of fragmentation in these sources, rather than, e.g., radiative heating. On the other hand, Cyganowski *et al.* (2017) have argued that there is a relatively high degree of fragmentation present in the G11.92-0.61 region. If massive stars are forming in an unbiased way within protoclusters, then, even in the context of Core Accretion models, one expects that many lower-mass protostellar cores will also be found in their vicinity.

Dynamically strong B-fields in massive protostellar cores have been inferred by Girart *et al.* (2009) and Zhang *et al.* (2014) from mm/sub-mm polarization observations. Vlemmings *et al.* (2010) measured even stronger B-field strengths of ~ 20 mG via 6.7 GHz methanol maser emission within $\sim 1,000$ AU of Cep A HW2.

Infall has been detected in 9 sources by Wyrowski *et al.* (2016), though it can be difficult to tell if this is at the clump or core scale. These authors find slow infall speeds: on average only $\sim$ 10% of the free-fall speed. Processes that may slow infall include support from B-fields and/or maintenance of clump turbulence by outflows and accretion.

The search for and characterization of rotationally supported disks remains challenging, which is not unexpected if diameters are $\lesssim 1,000$ AU, i.e., $\lesssim 0.5$" at 2 kpc. Since the review of T14, there have been several claims of detection of such disks (e.g., Ilee *et al.* 2016; Beuther *et al.* 2017a), with these studies utilizing emission of CH_3CN. In the latter work, the authors achieve 130 AU resolution, find limited fragmentation on the core scale and conclude the disk itself is also stable with respect to gravitational instability.

Collimated molecular outflows are often a feature of massive protostars (e.g., Beuther *et al.* 2002; Duarte-Cabral *et al.* 2013; see also Fig. 3, where the northern protostellar source is estimated to have a core mass of $\sim 30\,M_\odot$, Kong *et al.* 2017b). On small scales relevant to outflow launching, Hirota *et al.* (2017) have presented a high-resolution study of the closest example of a massive protostar, Orion source I, finding evidence of rotation near the base of the outflow, consistent with disk wind models. During the later stages of formation and for the more massive systems, the outflows are expected to become photoionized by the protostar. Ionized, collimated outflows traced as radio continuum "jets" have been seen in some massive protostars (e.g., Gibb *et al.* 2003; Guzmán *et al.* 2014), although the relative importance of shock- versus photo-ionization remains to be established. Cm continuum emission from ionized gas remains the most promising method to identify the precise locations of massive protostars over a range of evolutionary stages (e.g., Rosero *et al.* 2016), including detecting the presence of multiplicity (e.g., Beuther *et al.* 2017b). A growing sample of massive protostars, such as G35.20-0.74N, now have their IR to mm SEDs well-characterized and fit to predictions of the TCM (Zhang *et al.* 2013b; De Buizer *et al.* 2017). Elongation in the images from 10 to 40 μm is expected along the outflow cavity and this information helps to constrain the RT models. The goal of such studies is to determine to what extent simple, symmetric protostellar models can explain the observed dust continuum and, eventually, spectral line morphologies. The presence of order and symmetry in core and outflow features, especially when maintained over large scales, is not expected in Competitive Accretion and Protostellar Collision models. Core Accretion models may also exhibit spatial asymmetries, e.g., due to low-order multiplicity resulting from disk fragmentation and/or disk axis precession, as well as temporal variability, e.g., due to disk instabilities. Accretion bursts revealed by luminosity variations have been reported by Caratti o Garatti *et al.* (2017) and Hunter *et al.* (2017).

There are some examples of more disordered outflows, with the larger-scale outflow from the Orion KL region, potentially driven by source I, being a prime example (e.g., Bally *et al.* 2017). Dynamical interaction among protostellar and young stellar sources, some of which are now high proper motion runaway stars (e.g., Luhman *et al.* 2017) seems likely to have played a role in triggering the apparently "explosive" outflow, although the precise details of how this has occurred remain debated (e.g., Bally & Zinnecker 2005; Chatterjee & Tan 2012). It should be noted that the chemical complexity of the Orion Hot Core (e.g., Schilke *et al.* 2001; Crockett *et al.* 2010) is likely to have been affected by the strong shocks resulting from this enhanced outflow activity. Another potential example of an explosive outflow is the DR21 system (Zapata *et al.* 2013), but such systems appear to be relatively rare in the massive protostar population.

4. Summary and Outlook

Massive star formation involves many different complex physical and chemical processes that need to be followed over vast ranges of spatial and temporal scales. The initial conditions of the problem are poorly constrained and often poorly defined. However, there is progress driven by improving theoretical/computational modeling and improving observational capabilities. Astrochemical modeling has crucial roles to play in helping to carry out physical modeling, e.g., of ambipolar diffusion during the collapse of pre- and protostellar cores, and for interpretation of observational signatures of the various evolutionary phases of the massive star formation process. However, given the large uncertainties present in both physical and chemical models, great caution is needed when developing and interpreting model results. Careful testing of predictions against observations to then refine the models is essential.

Acknowledgements

We thank Chris McKee for helpful comments on the manuscript. We thank Maria Drozdovskaya, Matthew Goodson and Shuo Kong for providing figures from their papers.

References

Aikawa Y., Wakelam V. *et al.* 2012, *ApJ*, 760, 40
Bally, J., Ginsburg, A., Arce, H. *et al.* 2017, *ApJ*, 837, 60
Bally, J. & Zinnecker, H. 2005, *AJ*, 129, 2281
Beuther, H., Linz, H., Henning, Th. *et al.* 2017b, *ApJ*, 605, 61
Beuther, H., Schilke P. *et al.* 2002, *A&A*, 383, 892
Beuther, H., Walsh, A. J., Johnston, K. G. *et al.* 2017a, *A&A*, 603, 10
Bonnell, I. A., Bate, M., Clarke, C. J., & Pringle, J. E. 2001, *MNRAS*, 323, 785
Bonnell, I. A., Bate, M., & Zinnecker, H. 1998, *MNRAS*, 298, 93
Bontemps S., Motte F., Csengeri, T. & Schneider N. 2010, *A&A*, 524, 18
Bovino, S., Grassi, T., Schleicher, D. R., & Caselli, P. 2017, arXiv:1708.02046
Brüncken, S., Sipilä, O., Chambers, E. T. *et al.* 2014, *Nature*, 516, 219
Butler M. J. & Tan J. C., 2012, *ApJ*, 754, 5
Caratti o Garatti, A., Stecklum, B., Garcia Lopez, R. *et al.* 2017, *Nature Physics*, 13, 276
Caselli, P. & Ceccarelli, C. 2012, *A&AR*, 20, 56
Caselli P., Hasegawa T. *et al.* 1993, *ApJ*, 408, 548
Charnley S. B. *et al.* 1992, *ApJ*, 399, L71
Chatterjee, S. & Tan, J. C. 2012, *ApJ*, 754, 152
Crockett, N. R., Bergin, E. A., Wang, S. *et al.* 2010, *A&A*, 521, L21
Crutcher, R. M., Wandelt, B., Heiles, C., *et al.* 2010, *ApJ*, 725, 466
Csengeri, T. Bontemps, S., Wyrowski, F. *et al.* 2017, *A&A*, 600, L10
Cyganowski, C. J., Brogan, C. L., Hunter, T. R. *et al.* 2014, *ApJL*, 796, 2
Cyganowski, C. J., Brogan, C. L., Hunter, T. R. *et al.* 2017, *MNRAS*, 468, 3694
Da Rio, N., Tan, J. C., & Jaehnig, K. 2014, *ApJ*, 795, 55
De Buizer, J. M., Liu, M., Tan, J. C. *et al.* 2017, *ApJ*, 843, 33

de Wit, W. J., Testi, L., Palla, F., & Zinnecker, H. 2005, *A&A*, 437, 247
Doty, S. D., van Dishoeck, E. F., & Tan, J. C. 2006, *A&A*, 454, L5
Drozdovskaya, M., Tan, J. C. *et al.* 2017, in *Star Formation from Cores to Clusters*, 19
Drozdovskaya, M., Walsh, C., Visser, R. *et al.* 2014, *MNRAS*, 445, 913
Drozdovskaya, M., Walsh, C., Visser, R. *et al.* 2015, *MNRAS*, 451, 3836
Duarte-Cabral A. *et al.* 2013, *A&A*, 558, 125
Elmegreen, B. G. 2000, *ApJ*, 530, 277
Elmegreen, B. G. 2007, *ApJ*, 668, 1046
Eyink, G. L., Lazarian, A., & Vishniac, E. T. 2011, *ApJ*, 743, 51
Garrod R. T. & Herbst E. 2006 *A&A*, 457, 927
Gerner, T., Beuther, H., Semenov, D. *et al.* 2014, *A&A*, 563, 97
Gerner, T., Shirley, Y. L., Beuther, H. *et al.* 2015, *A&A*, 579, 80
Gibb, A. G. *et al.* 2003, *MNRAS*, 339, 198
Girart, J. M., Beltrán, M. T., Zhang, Q. *et al.* 2009, *Science*, 324, 1408
Goldbaum, N. J., Krumholz, M. R., Matzner, C. D., & McKee, C. F. 2011, *ApJ*, 738, 101
Goodson, M. D., Kong, S., Tan, J. C., Heitsch, F., & Caselli, P. 2016, *ApJ*, 833, 274
Guzmán, A. E., Garay, G., Rodríguez, L. F., *et al.* 2014, *ApJ*, 796, 117
Harries, T. J., Douglas, T. A., & Ali, A. 2017, *MNRAS*, 471, 4111
Hartmann, L., Ballesteros-Paredes, J., & Heitsch, F. 2012, *MNRAS*, 420, 1457
Hartmann, L. & Burkert, A. 2007, *ApJ*, 654, 988
Hirota, T., Machida, M. N., Matsushita, Y. *et al.* 2017, *Nature Astronomy*, 1, 146
Hunter, T. R., Brogan, C. L., MacLeod, G. *et al.* 2017, *ApJ*, 837, L29
Inutsuka, S. 2012, *PTEP*, 2012, 01A307
Keto, E., Caselli, P., & Rawlings, J. 2015, *ApJ*, 446, 3731
Klessen, R. S. & Hennebelle, P. 2010, *A&A*, 520, 17
Körtgen, B. Bovino, S., Schleicher, D. R. G. *et al.* 2017, *MNRAS*, 469, 2602
Kong, S., Caselli, P., Tan, J. C. *et al.* 2015, *ApJ*, 804, 98
Kong, S., Tan, J. C., Caselli, P. *et al.* 2016, *ApJ*, 821, 94
Kong, S., Tan, J. C., Caselli, P. *et al.* 2017a, *ApJ*, 834, 193
Kong, S., Tan, J. C., Caselli, P. *et al.* 2017b, *ApJ*, sub. (arXiv:1701.05953)
Krumholz, M. R., Dekel, A., & McKee, C. F. 2012, *ApJ*, 745, 69
Krumholz, M. R., Klein, R. I. *et al.* 2007, *ApJ*, 656, 959
Krumholz, M. R. & McKee, C. F. 2005, *ApJ*, 630, 250
Krumholz, M. R. & Tan, J. C. 2007, *ApJ*, 654, 304
Kuiper, R., Turner, N. J. & Yorke H. W. 2015, *ApJ*, 800, 86
Kuiper, R., Turner, N. J. & Yorke H. W. 2016, *ApJ*, 832, 40
Kunz, M. W. & Mouschovias, T. Ch. 2009, *MNRAS*, 399, L94
Lazarian, A. & Vishniac, E. T. 1999, *ApJ*, 517, 700
Li, Z.-Y., Banerjee, R., Pudritz, R. E. *et al.* 2014, *PPVI*, eds. Beuther *et al.*, U. Arizona, p173
Luhman, K. L., Robberto, M., Tan, J. C. *et al.* 2017, *ApJ*, 838, L3
McLaughlin, D. E & Pudritz, R. E. 1997, *ApJ*, 476, 750
McKee, C. F. & Ostriker, E. C. 2007, *ARA&A*, 45, 565
McKee, C. F. & Tan, J. C. 2003, *ApJ*, 585, 850 [MT03]
Moeckel, N. & Clarke, C. J. 2011, *MNRAS*, 410, 2799
Motte, F., Bontemps, S., Schilke, P., *et al.* 2007, *A&A*, 476, 1243
Motte, F., Bontemps, S., & Louvet, F. 2017, *AR&A*, in press (arXiv:1706.00118)
Mouschovias, T. Ch. 1991, *ApJ*, 373, 169
Myers A. T., McKee C. F. *et al.* 2013, *ApJ*, 766, 97
Nakamura, F. & Li, Z-Y. 2007, *ApJ*, 662, 395
Nakamura, F. & Li, Z-Y. 2014, *ApJ*, 783, 115
Öberg K. I., Boamah M. *et al.* 2013, *ApJ*, 771, 95
Padoan, P. & Nordlund, Å. 2011, *ApJ*, 730, 40
Pagani, L., Vastel, C., Hugo, E., *et al.* 2009, *A&A*, 494, 623
Peters, T., Banerjee, R., Klessen, R. S., & Mac Low, M.-M. 2011, *ApJ*, 729, 72

Pillai, T., Kauffmann, J., Tan, J. C. *et al.* 2015, *ApJ*, 799, 74
Rosen, A., Krumholz, M. R., McKee, C. F., & Klein, R. I. 2016, *MNRAS*, 63, 2553
Rosero, V., Hofner, P., Classen, M. *et al.* 2017, *ApJS*, 227, 25
Russeil D., Zavagno A., Motte F. *et al.* 2010 *A&A*, 515, 55
Salpeter, E. E. 1955, *ApJ*, 121, 161
Sanhueza, P., Jackson, J. M., Zhang, Q. *et al.* 2017, *ApJ*, 841, 97
Schilke, P. Benford, D. J., Hunter, T. R. *et al.* 2001, *ApJS*, 132, 281
Seifried, D., Banerjee, R., Klessen R. S. *et al.* 2011, *MNRAS*, 417, 1054
Sipilä, O., Caselli, P., & Harhu, J. 2015, *A&A*, 578, 55
Sipilä, O., Hugo, E., Harju, J. *et al.* 2010, *A&A*, 509, A98
Shu, F. H., Adams, F. C., & Lizano, S. 1987, *ARA&A*, 25, 23
Tan, J. C. 2016, *From Interstellar Clouds to Star-Forming Galaxies*, IAU Symp., 315, 154
Tan, J. C., Beltrán, M., Caselli, P. *et al.* 2014, *PPVI*, eds. Beuther *et al.*, U. Arizona, p149 [T14]
Tan, J. C., Kong, S., Butler, M. J., Caselli, P., & Fontani, F. 2013, *ApJ*, 779, 96
Tan, J. C., Kong, S. Caselli, P. *et al.* 2016, *ApJ*, 821, L3
Tan, J. C., Krumholz, M. R., & McKee, C. F. 2006, *ApJ*, 641, L121
Tan, J. C. & McKee, C. F. 2003, astro-ph/0309139
Tanaka, K. E. I., Tan, J. C., & Zhang, Y. 2016, *ApJ*, 818, 1
Tanaka, K. E. I., Tan, J. C., & Zhang, Y. 2017, *ApJ*, 835, 32
Viti S. *et al.* 2004, *MNRAS*, 354, 1141
Vlemmings W. *et al.* 2010, *MNRAS*, 404, 134.
Wang, P. Li, Z-Y., Abel, T., & Nakamura, F. 2010, *ApJ*, 709, 27
Wang K., Zhang Q., Testi L. *et al.* 2014, *MNRAS*, 439, 3275
Williams, J. P., Blitz, L., & McKee, C. F. 2000, *PPV*, eds. Mannings *et al.*, U. Arizona, p97
Wirström, E. S., Charnley, S. B., Cordiner, M. A., & Milam, S. N. 2012, *ApJL*, 757, L11
Wyrowski, F., Güsten, R., Menten, K. M. *et al.* 2016, *A&A*, 585, 149
Zapata, L. A., Schmid-Burgk, J., Pérez-Goytia, N. *et al.* 2013, *ApJ*, 765, L29
Zhang, Q., Qiu, K., Girart, J. M. *et al.* 2014, *ApJ*, 792, 116
Zhang, Y. & Tan, J. C. 2011, *ApJ*, 733, 55
Zhang, Y. & Tan, J. C. 2015, *ApJL*, 802, L15
Zhang, Y. & Tan, J. C. 2017, *ApJ*, submitted (arXiv:1708.08853)
Zhang, Y., Tan, J. C., & Hosokawa, T. 2014, *ApJ*, 788, 166 [ZTH14]
Zhang, Y., Tan, J. C., De Buizer *et al.* 2013b, *ApJ*, 767, 58
Zhang, Y., Tan, J. C. & McKee 2013a, *ApJ*, 766, 86
Zhao, B., Caselli, P., Li, Z.-Y. *et al.* 2016, *MNRAS*, 460, 2050

Astrochemistry VII – Through the Cosmos from Galaxies to Planets
Proceedings IAU Symposium No. 332, 2017
M. Cunningham, T. Millar & Y. Aikawa, eds.

doi:10.1017/S1743921317007852

Chemical highlights from the Rosetta mission

Kathrin Altwegg[1] **and the ROSINA Team**

[1]Physikalisches Institut, University of Bern
Sidlerstr. 5,CH-3012 Bern, Switzerland
email: altwegg@space.unibe.ch

Abstract. The overall goal of the ESA *Rosetta* mission was to help decipher the origin and evolution of our solar system. Looking at the chemical composition of comet 67P/Churyumov-Gerasimenko is one way of doing this. The amount of very volatile species found and the insight into their isotopic abundances show that at least some presolar ice has survived the formation of the solar system. It shows that the solar nebula was not homogenized in the region where comets formed. The D/H ratio in water furthermore indicates that Jupiter family comets and Oort cloud comets probably formed in the same regions and their difference is then purely due to their different dynamical history. The organics found in 67P are very diverse, with abundant CH- and CHO- bearing species. Sulphur bearing species like S_3 and S_4 and others show evidence of dust grain chemistry in molecular clouds.

Keywords. astrochemistry, comets: individual (67P/C-G), planetary systems: protoplanetary disks

1. Introduction

After a 10 year long cruise phase the European Rosetta mission finally reached its goal, comet 67P/Churyumov-Gerasimenko in August 2014. On board was a large payload consisting of 10 instruments on the Rosetta spacecraft and 8 instruments on the lander Philae (Glassmeier *et al.*, 2007). The scientific goal of Rosetta, and that is why the mission was called Rosetta, was to decipher the origin of our solar system, the Earth and finally the emergence of life from a tiny piece of well-preserved material from the time of solar system formation, namely a comet. If we want to follow the formation of the solar system, we have to look back to the interstellar medium. This interstellar medium is cold and very tenuous, but it already contains some characteristics which are the starting point of the composition of the solar system. The material is provided by the remnants of stars. Depending on what stars contributed we find isotopic differences which especially in dust can survive all the following stages. Once the interstellar medium forms a molecular cloud, more chemistry sets in. Depending on densities and temperature there is gas phase chemistry, but probably a lot of dust grain chemistry. The question now is how much of this material will be reprocessed during solar system formation and how much will survive the different stages from the molecular cloud through the star forming region and the early stages of our solar system until it is embedded in comets. Looking at isotopic ratios but also at molecules and their formation pathways help us to trace back this early solar system material to its origin. Comparing isotopic abundances with what we find in the Sun, in planets and especially on Earth gives us boundary conditions on how the solar system evolved and finally how life could emerge on Earth. In this review we will summarize the most important findings up to date on isotopic and molecular composition of comet 67P and draw some conclusions on what this means for the formation and evolution of the solar system.

2. Method

In order to be able to look at the chemical composition of the volatile material of comet 67P/Churyumov-Gerasimenko, Rosetta carried several mass spectrometers as well as optical spectrometers with different wavelength. Here we review some of the most important findings concerning cometary chemistry. Most of the measurements on cometary chemistry were done with the Rosetta Orbiter Sensor for Ion and Neutral Analysis (ROSINA) (Balsiger *et al.*, 2007). This instrument suite consists of two mass spectrometers, namely the Double Focusing Mass Spectrometer (DFMS) and the Reflectron Time of Flight spectrometer (RTOF). While DFMS has a very good mass resolution of 9000 (FWHM) at m/z = 28 and a high sensitivity with a dynamic range of 10^{10}, RTOF was measuring much faster giving the context to the DFMS measurements. Also part of ROSINA is the COmet Pressure Sensor (COPS), consisting of two ion gauges for the total density and the ram pressure from which gas dynamics could be deduced. All three sensors were measuring during most of the mission time which lasted from August 2014, when Rosetta was within 100 km of the comet to end of September 2016, when Rosetta softly crash-landed on the comet. By now by far not all data have been analysed and more results are to be expected. Details on the data analysis can be found in Le Roy *et al.* (2015) and Calmonte *et al.* (2016).

The trajectory of Rosetta around the comet was very variable. Early in the mission, as long as the comet was well outside 2 au from the Sun, Rosetta orbited the comet mostly in the terminator plane between 10 and 30 km from the comet. In spring 2015 two flybys with lower phase angles were done, whereby the second led to a safe mode of the spacecraft. This forced the spacecraft to move further from the comet. During the summer months of 2015 with perihelion on August 13, Rosetta spent most of the time several hundred kilometers from the comet. Only in 2016 lower altitudes could be flown again. In May 2016, the spacecraft spent an extended period between 10 and 7 km from the comet center. In August 2016 it finally went into elliptical orbits, gradually lowering the pericenter down to 4 km above surface. High dust impact forced it to stay above the 4 kilometers till September 30, when it softly crash-landed on the comet which was the end of the mission.

During all of these periods ROSINA was measuring the volatile part of the coma, dealing with large variations in coma density, but also composition. A full mass spectrum with DFMS between m/z = 13 to 100 amu/e lasts about 40 to 50 minutes. A comet rotation was between 12.2h when the spacecraft arrived at the comet and 12h when the mission ended. Every 6h the spacecraft performed a reaction wheel off-loading by burning some hydrazine, which forced the mass spectrometers to stop working for about an hour. Every few days there was a spacecraft maneuver lasting 6h during which time ROSINA was off. This means ROSINA-DFMS got on average 10 mass spectra per comet rotation, or 1.9 million single integer mass spectra over the mission, RTOF ~ 200000 full mass scale spectra and COPS ~ 20000 density measurements per month. By now, we are far from analysing all of them.

To get a quantitative assessment, species need to have been calibrated beforehand in the lab due to the individual fragmentation patterns which are especially in the case of DFMS different from e.g. the NIST database because DFMS uses a lower electron impact energy of 45 eV compared to 70 eV that is normally used in mass spectrometry. ROSINA measures local densities at the position of the spacecraft. These densities are affected by many parameters: heliocentric and cometocentric distances, sub-spacecraft latitudes and longitudes, seasons on the comet. To calculate production rates needs therefore a detailed shape model of the comet and a coma model. To derive bulk abundances is even more

Table 1. Measured minor isotopes in the coma of 67P

Isotope	Molecule	Isotopic ratio rel. to main isotope	Reference
D	H_2O	$(5.3 \pm 0.7) \times 10^{-4}$	Altwegg *et al.*, 2014
	HDO	$(1.80 \pm 0.9) \times 10^{-2}$	Altwegg *et al.*, 2017
	H_2S	$(1.2 \pm 0.3) \times 10^{-3}$	Altwegg *et al.*, 2017
^{13}C	CO	0.0116 ± 0.0012	Rubin *et al.*, 2017
^{13}C	CO_2	0.0119 ± 0.0006	Hässig *et al.*, 2017
^{18}O	H_2O	$(1.8 \pm 0.2) \times 10^{-3}$	Altwegg *et al.*, 2014
	CO_2	$(2.03 \pm 0.03) \times 10^{-3}$	Hässig *et al.*, 2017
^{29}Si	N/A	0.0434 ± 0.0050	Rubin *et al.*, 2017
^{30}Si	N/A	0.0263 ± 0.0038	Rubin *et al.*, 2017
^{33}S	H_2S	$(6.64 \pm 0.2) \times 10^{-3}$	Calmonte *et al.* (2017)
	CS_2	$(6.5 \pm 0.3) \times 10^{-3}$	Calmonte *et al.* (2017)
	OCS	$(6.06 \pm 0.45) \times 10^{-3}$	Calmonte *et al.*, 2017
^{34}S	H_2S	$(4.3 \pm 0.2) \times 10^{-2}$	Calmonte *et al.* (2017)
	CS_2	$(4.06 \pm 0.1) \times 10^{-2}$	Calmonte *et al.* (2017)
	OCS	$(4.2 \pm 0.15) \times 10^{-2}$	Calmonte *et al.* (2017)
^{38}Ar	N/A	0.185 ± 0.04	Balsiger *et al.*, 2015
^{128}Xe	N/A	0.076 ± 0.021	Marty *et al.*, 2017
^{129}Xe	N/A	1.398 ± 0.122	Marty *et al.*, 2017
^{130}Xe	N/A	0.184 ± 0.032	Marty *et al.*, 2017
^{131}Xe	N/A	0.864 ± 0.09	Marty *et al.*, 2017
^{134}Xe	N/A	0.238 ± 0.049	Marty *et al.*, 2017
^{136}Xe	N/A	0.166 ± 0.04	Marty *et al.*, 2017

difficult as species do not show the same desorption pattern with heliocentric distances (Gasc *et al.*, 2017). For relative isotopic abundances the problem is much simpler if the species are measured very close in time. For molecular abundances, species with similar volatilities often follow each other in sublimation, but not always. Details are given in the publications cited.

3. Isotopic ratios

Isotopic ratios play an important role when assessing on how much processing was done to the cometary material in the solar nebula and if / how much of the material is primordial coming unchanged from prestellar stages. A summary of all isotopic ratios measured so far in the volatile material of 67P is given in Table 1.

Deuterated water species Already early in the mission, when Rosetta was within 30km from the comet, D/H in cometary water could be determined (Altwegg *et al.*, 2014). Comet 67P shows an enrichment of more than a factor 3 relative to terrestrial water. The measured D/H is $(5.3\pm 0.7) \times 10^{-4}$. However, there remained some doubts if this ratio was somehow fractionated relative to the bulk, mainly because early in the mission the comet temperature was still very low, but the water ice measured from the northern, summer hemisphere was probably transported from the south during the previous perihelion passage. The measurements were therefore repeated later in the mission in December 2015 and March 2016, but the results were very similar (Altwegg *et al.*, 2017). At the same time also D_2O could be measured which gave a high relative value compared to HDO of 1.8%. There are several conclusions which can be drawn from these results.

Furuya *et al.* (2016) looked at the ratio D2/D1 = $[D_2O/HDO]$ / $[HDO/H_2O]$ and found that this ratio for reprocessed ice in the solar nebula would be close to the statistical value of 1/4. While HDO/H_2O ratios can remain largely unchanged even if water ice sublimates, reacts and then recondenses in the protosolar nebula, the D2/D1 ratio would not keep its primordial value. Measurements in molecular clouds give a D2/D1 ratio on the order of 10/1. D2/D1 is therefore indicative on how much water was reprocessed in the protosolar nebula. The value found in 67P (D2/D1 = 17, Altwegg *et al.*, 2017) clearly demonstrates that prestellar ice survived solar system formation until it was incorporated in comets. Comparing the D/H in 67P with other cometary values shows that the range of values of Oort cloud comets and Jupiter family comets span a range between almost terrestrial (Hartogh *et al.*, 2011) values up to 6×10^{-4} (e.g. Table 2 in Paganini *et al.*, 2017). This overlap demonstrates that Oort cloud comets and Jupiter family comets most probably formed in a wide range of heliocentric distances, but in regions which overlapped. The fate of the comet to either become an Oort cloud comet or to be pushed out into the Kuiper belt is therefore just due to its dynamical history and encounters with giant planets. The high D/H value in 67P brings the mean value of measurements in comets up to now to about twice the terrestrial value. Although the statistical sample is by now just about 10 comets this fact most probably excludes a mainly cometary origin of the terrestrial surface and atmospheric water reservoirs.

Deuterated H_2S ROSINA was able to measure the D/H in H_2S in the cometary coma. The value of $(1.2 \pm 0.3) \times 10^{-3}$ (Altwegg *et al.*, 2017) is again high. H_2S is most probably formed by dust grain chemistry in the presolar cloud because gas phase chemistry is very inefficient to form H_2S (Woods *et al.*, 2015). Calmonte *et al.* (2016) reported the presence of S_3 and S_4 in the coma of 67P which demonstrate that dust grain chemistry played a major role for the sulphur bearing species as these species cannot be formed in the gas phase at all. Dust grain chemistry can fractionate deuterated species as discussed in Hatchell *et al.* (1999). A high D/H ratio in H_2S is therefore fully compatible with a prestellar origin of H_2S.

Sulfur isotopes Sulfur bearing species are quite abundant in the coma of 67P. For some of them it was possible to derive the $^{33}S/^{32}S$ and the $^{34}S/^{32}S$ ratios, namely HS, CS_2 and OCS (Calmonte *et al.*, 2017). While $^{34}S/^{32}S$ is slightly depleted in most molecules considered, $^{33}S/^{32}S$ is much more depleted. The effect is neither due to mass fractionation nor to fractionation due to photoionisation.

Silicon Silicon is non-volatile and could most of the time therefore not be measured by ROSINA. However, early in the mission, solar wind could still hit the nucleus, especially on the winter hemisphere where outgassing was low (Wurz *et al.*, 2015). On the summer hemisphere, the solar wind was already attenuated by collisions with cometary gases. Silicon was then sputtered by the solar wind mostly from the winter hemisphere as neutral Si and could be detected by ROSINA. Three isotopes of silicon could clearly be identified: ^{28}Si, ^{29}Si and ^{30}Si (Rubin *et al.*, 2017). Comet 67P is slightly depleted in the heavy isotopes of silicon compared to the light ones. This is in line with isotopic anomalies found in presolar grains.

Carbon isotopes in CO and CO_2 Due to overlapping peaks at mass 29 amu, the $^{13}C/^{12}C$ ratio in CO has a large error bar. Within this error bar the isotopic ratio is compatible with solar (Rubin *et al.*, 2017). For CO_2, the overlap with other species is much smaller. The precision of the measurements did allow to deduce a $^{12}C/^{13}C$ ratio of 84 ± 3, which is within 1 sigma just outside of the normally considered value of 89 for the solar system (Hässig *et al.*, 2017).

Argon and Xenon isotopes Already early in the mission, in September 2014, two isotopes of argon, namely ^{36}Ar and ^{38}Ar were detected in the coma of 67P. The $^{38}Ar/^{36}Ar$ ratio derived from these measurements have quite a large uncertainty. Within this uncertainty, they seem to be solar. This is completely different for xenon. Only very late in the mission, when the spacecraft could approach the comet to within 7 km from its center, was also xenon detected. Seven out of the nine isotopes were found. The different isotopic abundances, normalized to ^{132}Xe and to solar abundance look quite unusual compared to what is known from the solar system. The cometary xenon is under abundant in ^{134}Xe and ^{136}Xe, but overabundant in ^{129}Xe (Marty *et al.*, 2017). While the depletion of the heavy ones can be explained by different nucleosynthetic processes, for ^{129}Xe there is most probably a contribution of the radioactive decay of ^{129}I responsible for the enhancement. This xenon signature resembles very much the U-xenon, postulated a long time ago by Pepin (2000), but never found.

Here are again several conclusions which can be drawn from the xenon isotopic ratios: 67P was formed somewhere where the composition of the solar nebula was not / not yet homogenized. The non-solar xenon isotopic ratios, combined with non-solar silicon and non-solar sulphur isotopes point to a protosolar nebula which still had some of its primordial heterogeneity due to contributions of different stars to the cloud like supernovae or AGBs.

The xenon isotopic ratios of the Earth atmosphere are neither solar nor chondritic. There is a mass dependent fractionation where heavier isotopes are less depleted than light ones. This fractionation was explained by loss from the Earth atmosphere. Corrected for this fractionation it turns out that ^{134}Xe and ^{136}Xe are heavily depleted in the Earth atmosphere. ^{129}Xe is enhanced and this was explained by contribution from ^{129}I from the Earth mantle. This for a long time served as a clock for the formation time of the Earth. If we now assume that 22% of the Earth atmospheric xenon was contributed by cometary impacts, we can easily explain the relative isotopic ratios of the atmospheric xenon, the depletion of the heavy isotopes as well as the enhancement of ^{129}Xe (Marty *et al.*, 2017). This also gives us a quantitative number for how much of the terrestrial water was contributed by comets. This value is low, a first estimate is below 1%, which is then fully compatible with the D/H difference in comets and the Earth water. What is more important is probably the contribution by comets of organics to the Earth which can be constrained by the xenon isotopes.

4. Molecular composition

The diversity of different species found by ROSINA was unexpected. Not only did ROSINA detect all the molecules found previously by in situ or remote sensing measurements in other comets (Le Roy *et al.*, 2015). ROSINA more than doubled the number of parent species known to exist in comets before Rosetta. Not all of the molecules were seen all the time in the cometary coma: some are highly volatile, but rare, like the noble gases and could only be found when the spacecraft was close to the nucleus. Others were associated to the dusty coma and were seen mainly during the perihelion passage in summer 2015. Because of the high numbers of unexpected species and therefore a lack of lab calibration data from the DFMS flight spare instruments, for many of the species, we still don't have quantitative abundances. In addition, the coma composition has been found to be highly heterogeneous and is a strong function of latitude and heliocentric distance (Hässig *et al.*, 2015) which makes a comparison e.g. to water not always very meaningful. To derive bulk abundances from coma measurements needs a good understanding of the kind of ice and the desorption patterns of the individual species and this work is still in progress. Here, we therefore do not give bulk abundances. For species, where there are numbers, which are mainly the sulfur species, we refer to the appropriate papers. Here we divide the molecules somewhat arbitrarily into highly volatiles where sublimation temperatures of the pure ices are well below CO_2, volatiles mostly with sublimation temperatures between CO_2 and water, which are present most of the time and semi-volatiles, which have higher sublimation temperatures than water, which

Table 2. Identified parent-species in the coma of 67P

Mass	Molecule	Classification	Reference
16	CH_4	highly volatile	Le Roy *et al.* (2015)
17	NH_3	volatile	Le Roy *et al.* (2015)
18	H_2O	volatile	Le Roy *et al.* (2015)
20	HF	volatile	Dhooghe *et al.*, 2017
26	C_2N_2	volatile	Le Roy *et al.* (2015)
27	HCN	volatile	Le Roy *et al.* (2015)
28	CO	highly volatile	Le Roy *et al.* (2015)
	N_2	highly volatile	Rubin *et al.*, 2015
30	C_2H_6	volatile	Le Roy *et al.* (2015)
31	P	parent unknown	Altwegg *et al.*, 2016
	CH_3NH_2	volatile	Altwegg *et al.*, 2016
32	CH_3OH	volatile	Le Roy *et al.* (2015)
	O_2	highly volatile	Bieler *et al.*, 2015
	S	volatile	Calmonte *et al.* (2016)
34	H_2S	volatile	Calmonte *et al.* (2016)
	H_2O_2	highly volatile	Bieler *et al.*, 2015
36	Ar	highly volatile	Balsiger *et al.*, 2015
	HCl	volatile	Dhooghe *et al.*, 2017
41	C_2H_3N	volatile	Le Roy *et al.* (2015)
44	CO_2	violatile	Le Roy *et al.* (2015)
	C_3H_8	volatile	
	C_2H_4O	volatile	Le Roy *et al.* (2015)
45	$C_2H_5NH_2$	volatile	Altwegg *et al.*, 2016
46	C_2H_5OH	volatile	
	CH_2S	volatile	Calmonte *et al.* (2016)
	CH_2O_2	volatile	Le Roy *et al.* (2015)
48	SO	volatile	Calmonte *et al.* (2016)
	CH_4S	volatile	Calmonte *et al.* (2016)
50	CH_3Cl	volatile	Fayolle *et al.*, 2017
52	C_2N_2	volatile	
58	C_4H_{10}	volatile	
60	C_3H_7OH	volatile	
	OCS	volatile	Calmonte *et al.* (2016)
62	C_2H_6S	volatile	Calmonte *et al.* (2016)
	$C_2H_6O_2$	volatile	
64	S_2	highly volatile	Calmonte *et al.* (2016)
	SO_2	volatile	Calmonte *et al.* (2016)
72	C_5H_{12}	volatile	
74	C_4H_9OH	semi-volatile	
75	$C_2H_5NO_2$	semi-volatile	Altwegg *et al.*, 2016
76	CS_2	volatile	Le Roy *et al.* (2015)
	$C_3H_8O_2$	semi-volatile	
78	C_6H_6	volatile	
81	HBr	volatile	Dhooghe *et al.*, 2017
84	Kr	highly volatile	
87	C_4H_9NO	semi-volatile	
92	C_7H_8	volatile	
96	S_3	volatile	Calmonte *et al.* (2016)
86	C_6H_{14}	volatile	
88	$C_5H_{11}OH$	semi-volatile	
100	C_7H_{16}	volatile	
106	C_8H_{10}	semi-volatile	
122	$C_7H_6O_2$	semi-volatile	
128	$C_{10}H_8$	semi-volatile	
	S_4	volatile	Calmonte *et al.* (2016)
132	Xe	highly volatile	Marty *et al.*, 2017

are present only in the dusty coma. A list of so far identified species is given in Table 2 together with their classification.

4.1. *Highly volatiles*

CO and CH_4 have been known to exist in cometary comae in rather large abundances, especially CO. These molecules were also detected in 67P, although rather depleted compared to Oort cloud comets (see Le Roy *et al.*, 2015). In addition to these well known molecules ROSINA has detected N_2, O_2, S_2, and noble gases Ar, Kr and Xe. The most surprising finding was probably O_2 (Bieler *et al.*, 2015). The O_2/H_2O ratio remained quite stable for the whole mission duration between 1-10%, O_2 following water outgassing remarkably well. O_2 is very scarce in the Universe. Up to now it was observed only in very few locations. Taquet *et al.*, 2016 explained the observed O_2 in comet 67P by grain surface chemistry in a moderately warm cloud whereas Mousis *et al.* (2016) found the O_2 abundance to be compatible with photolysis / radiolysis of water ice in presolar grains. Both scenarios are plausible. For the chemistry in molecular clouds the strong correlation between water and O_2 is not obvious, for the radiolysis scenario byproducts of this process are underabundant. While also H_2O_2 and HO_2 were found, the abundances are too low to be compatible with radiolysis. No O_3 was found. Other explanations like the one proposed by Yao and Giapis (2017) based on Eley-Rideal processes or by Dulieu *et al.* (2017) based on dismutation of H_2O_2 fail to reproduce the observed O_2 abundance without being in strong contradiction to the measured amounts of H_2O^+ or H_2O_2, respectively, by orders of magnitudes.

N_2 relative to CO was found by Rubin *et al.* (2015) to be $(5.70 \pm 0.66)\times10^{-3}$. This points to a low formation temperature of the cometary ice of 25-30K.

S_2 was found already very early in the mission (Calmonte *et al.*,2016). This makes the explanation (A'Hearn *et al.*, 1983), that S_2 is the product of coma chemistry unlikely because at that time the coma was very tenuous not allowing much chemistry to take place. S_2 has a very short lifetime in the gas phase under UV irradiation. The presence of S_2 in the coma of 67P is therefore an indication that the cometary ice at least partially survived the formation of the solar system as ice, because otherwise S_2 would have disappeared before incorporation into comets.

For the first time in a comet the three noble gases Ar, Kr, and Xe were detected. A quantitative analysis of their abundances is under way.

4.2. *Volatiles and semi-volatiles*

While the bulk of the volatile material of 67P is water with a relatively high abundance of CO_2, the number of carbon bearing species is very high. An analysis of the species detected early at around 3 au in the mission is given by Le Roy *et al.* (2015). The authors compared relative abundances on the northern (summer) hemisphere and on the winter hemisphere. It can be well seen, that abundance ratios vary quite a lot and that it is difficult to derive from these measurements bulk abundances. Later in the mission, after the equinox in May 2015, the heterogeneity in the coma was much less pronounced as by that time all species were predominantly outgassing from the south. However, the coma got dustier with a lot of outbursts in between July and September (Vincent *et al.*, 2016), which again made the coma much more variable. During this time, however, all species given in Table 2 could be measured except the noble gases, where densities were too low due to the large distance of the spacecraft from the comet.

One of the highlights in this period was the detection of the amino acid glycine together with its precursor molecule methylamine (Altwegg *et al.*, 2016). At the same time a clear signal of phosphorous was detected. However, it could not be determined which parent species is responsible because PH_3, PO. PN and CP all could not be resolved from neighboring peaks.Together with the isotopic measurements of xenon it should now become feasible to assess how much organic material was inserted into the Earth atmosphere and oceans and how likely it is that such impacts actually triggered the emergence of life.

Apart from the highly volatile S_2 more refractory sulphur species like S_3 and S_4 were also detected. These species are probably due to dust grain chemistry in molecular clouds and make the scenario by Woods *et al.* (2015), that much of the H_2S was destroyed on

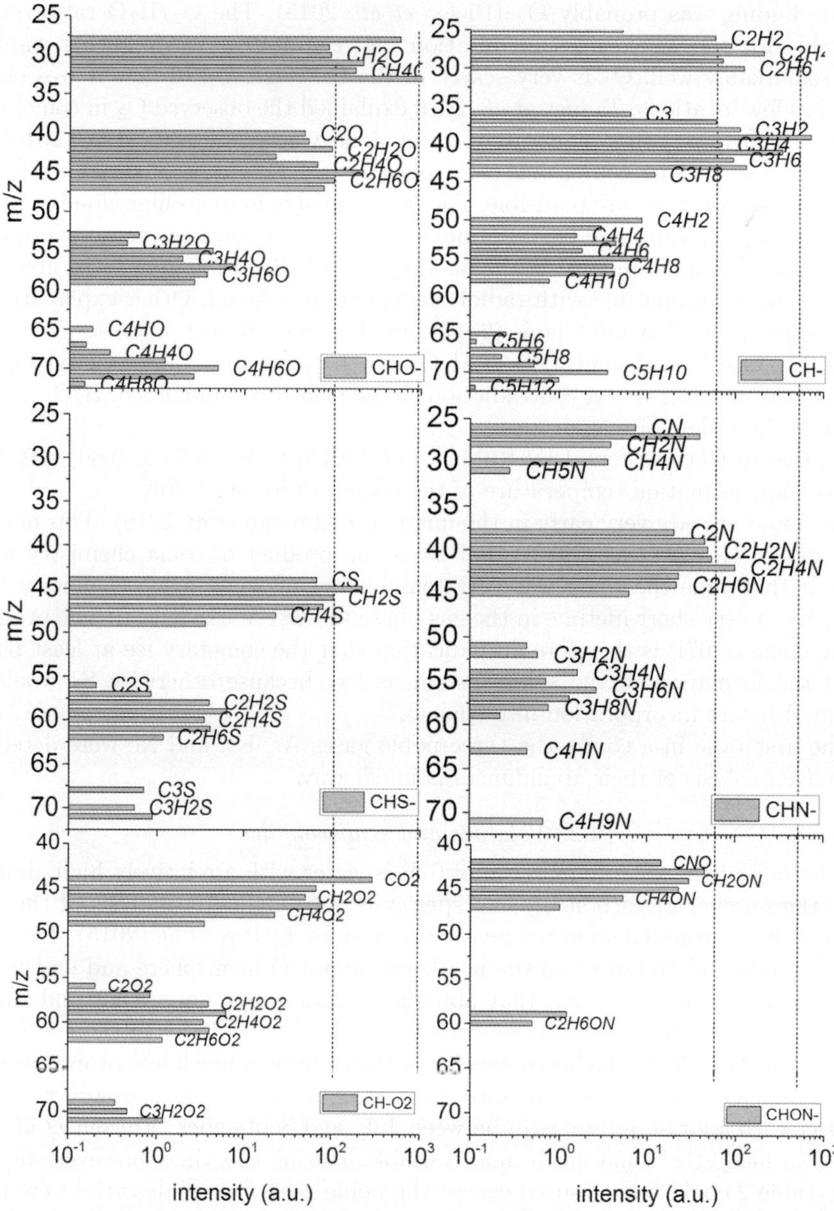

Figure 1. Fragments found in the mass spectra of DFMS during a dust impact on Sept. 5, 2016. For details see Altwegg *et al.* (2017a)

grains due to radiolysis and converted into more refractory sulphur, very likely. This would then also explain why sulphur is underabundant in molecular clouds relative to cosmic as these refractory species cannot be observed by remote sensing.

A direct hit by dust on September 5, 2016, when the spacecraft was 2 kilometers above the surface produced an enormous wealth of fragments in DFMS. Figure 1 shows the carbon bearing fragments detected in DFMS during this impact (Altwegg *et al.*, 2017a). DFMS got solid material into its ionization box which was at 1°C, allowing a relatively slow sublimation of the semi-volatile material. These fragments are due partly to the fragmentation inside of the electron impact ionization source of the instrument, partly to non-saturated molecules. Only a thorough calibration campaign in the lab, which gives exact fragmentation patterns, can therefore give quantitative results. ROSINA found many aliphatic hydrocarbons from methane to heptane. However, it also detected aromatic compounds with benzene, naphthalene, toluene and benzoic acid. Abundances generally decrease with higher number of C's. Apart from CH- bearing molecules CHO- bearing species are quite abundant.Even CHO_2- bearing fragments are rather abundant. Organo-sulphur species are surprisingly high, whereas nitrogen bearing species are depleted. Apart from organics, there are also of course inorganics, especially sulphur bearing species like SO_2, H_2O or NH_3.By far not all of the detected peaks can be explained by fragmentation of saturated molecules inside the instrument. There are clearly non-saturated species contributing to the fragments seen. The number of parent species is therefore very likely to increase in the future.

The data analysis from ROSINA/*Rosetta* is likely to go on for a few years to come and more, but especially more quantitative results can be expected. But it is also clear already now, that the amount of results from the *Rosetta* mission is unique. These results will indeed help to decipher the origin of our solar system, the evolution of the Earth and finally the emergence of life.

Acknowledgements

ROSINA would not give such outstanding results without the work of the many engineers, technicians, and scientists involved in the mission, in the *Rosetta* spacecraft, and in the ROSINA instrument team over the last 20 years, whose contributions are gratefully acknowledged. *Rosetta* is a European Space Agency (ESA) mission with contributions from its member states and NASA. We acknowledge herewith the work of the whole ESA *Rosetta* team.

References

A'Hearn, M. F., & Schleicher, D. G. and Feldman, P. D., 1983, *Astrophysical Journal*, 274, pp. L99-L103.

Altwegg, K., Balsiger, H., Bar-Nun, A., Berthelier, J. J., Bieler, A., Bochsler, P., Briois, C., Calmonte, U., Combi, M. R., Cottin, H., & De Keyser, J. 2016, *Science advances*, 2(5), p.e1600285.

Altwegg, K., Balsiger, H., Berthelier, J. J., Bieler, A., Calmonte, U., De Keyser, J., Fiethe, B., Fuselier, S. A., Gasc, S., Gombosi, T. I., & Owen, T. 2017, *Phil. Trans. R. Soc. A*, 375(2097), p.20160253.

Altwegg, K., Balsiger, H., Berthelier, J. J., Bieler, A., Calmonte, U., Fuselier, S. A., Goesmann, F., Gasc, S., Gombosi, T. I., Le Roy, L., & deKeyser, J., 2017a, *Mon. Not. R. Astron. Soc.*, p.stx1415.

Altwegg, K., Balsiger, H., Bar-Nun, A., Berthelier, J. J., Bieler, A., Bochsler, P., Briois, C., Calmonte, U., Combi, M., De Keyser, J., & Eberhardt, P. 2014, *Science*, p.1261952.

Balsiger, H., Altwegg, K., Bochsler, P., Eberhardt, P., Fischer, J., Graf, S., Jäckel, A., Kopp, E., Langer, U., Mildner, M., & Müller, J. 2007, *Space Science Reviews*, 128(1-4), pp. 745-801.

Balsiger, H., Altwegg, K., Bar-Nun, A., Berthelier, J. J., Bieler, A., Bochsler, P., Briois, C., Calmonte, U., Combi, M., De Keyser, J., & Eberhardt, P. 2015, *Science advances*, 1(8), p.e1500377.

Bieler, A., Altwegg, K., Balsiger, H., Bar-Nun, A., Berthelier, J. J., Bochsler, P., Briois, C., Calmonte, U., Combi, M., De Keyser, J., & van Dishoeck, E. F. 2015, *Nature*, 526(7575), pp. 678-681.

Calmonte, U., Altwegg, K., Balsiger, H., Berthelier, J. J., Bieler, A., Cessateur, G., Dhooghe, F., Van Dishoeck, E. F., Fiethe, B., Fuselier, S. A., & Gasc, S. 2016, *Mon. Not. R. Astron. Soc.*, 462, pp. S253-S273.

Calmonte, U., Altwegg, K., Balsiger, H., Berthelier, J. J., Bieler, A., Fiethe, B., Fuselier, S. A., Gasc, S., Gombosi T, Le Roy L., Rubin M., Semon T., Wampfler S., & Wurz P. 2017, *Mon. Not. R. Astron. Soc.*, submitted.

Furuya K, Van Dishoeck E.F., & Aikawa Y. 2016, *Astron. Astrophys.* 586, A127.

F. Dhooghe, J. De Keyser, K. Altwegg, Ch.Briois, H. Balsiger, J. Berthelier, U. Calmonte., G. Cessateur, M. Combi, E. Equeter, B. Fiethe, N. Fray, S. Fuselier, S. Gasc, A. Gibbons, T. Gombosi, H. Gunell, M. Hässig, M. Hilchenbach., L. Le Roy, R. Maggiolo, U. Mall, B. Marty, E. Neefs, H. RÃÍme, M. Rubin, T. Semon, C. Tzou, & P. Wurz 2017, *Mon. Not. R. Astron. Soc.*, in press

Dulieu, F., Minissale, M., & Bockelee-Morvan, D., 2017, *Astron. Astrophys.*, 597, p.A56.

E. C. Fayolle, K. I. Öberg, J. K. Jorgensen, K. Altwegg, H. Calcutt, H. S. P. Müller, M. Rubin, M. H. D. van der Wiel, P. Bjerkeli, T. L. Bourke, A. Coutens, E. F. van Dishoeck, M. N. Drozdovskaya, R. T. Garrod, N. F. W. Ligterink, M. V. Persson, S. F. Wampfler, & the ROSINA team, 2017, *Nature Astronomy*, in press

Gasc, S., Altwegg, K., Balsiger, H., Berthelier, J. J., Bieler, A., Calmonte, U., Fiethe, B., Fuselier, S., Galli, A., Gombosi, T., & Hoang, M., 2017, *Mon. Not. R. Astron. Soc.*, p.stx1412.

Glassmeier, K. H., Boehnhardt, H., Koschny, D., Kuhrt, E., & Richter, I. 2007, *Space Science Reviews*, 128(1-4), pp. 1-21.

Hartogh, P., Lis, D. C., BockelÃle-Morvan, D., de Val-Borro, M., Biver, N., KÃppers, M., Emprechtinger, M., Bergin, E. A., Crovisier, J., Rengel, M., & Moreno, R., 2011, *Nature*, 478(7368), pp. 218-220.

Hässig M., K. Altwegg, H. Balsiger, J. J. Berthelier, A. Bieler; U. Calmonte, F. Dhooghe, B. Fiethe, S. A. Fuselier; S. Gasc, T. I. Gombosi, L. Le Roy, A. Luspay-Kuti, K. Mandt, M. Rubin, C.-Y. Tzou, S. F. Wampfler, & P. Wurz, 2017, *Astron. Astrophys.* in press

Hässig, M., Altwegg, K., Balsiger, H., Bar-Nun, A., Berthelier, J. J., Bieler, A., Bochsler, P., Briois, C., Calmonte, U., Combi, M., & De Keyser, J. 2015, *Science*, 347(6220), p.aaa0276.

Hatchell J., Roberts H., & Millar T.J. 1999, *Astron. Astrophys.* 346, 227-232

Le Roy, L., Altwegg, K., Balsiger, H., Berthelier, J. J., Bieler, A., Briois, C., Calmonte, U., Combi, M. R., De Keyser, J., Dhooghe, F., & Fiethe, B., 2015, *Astron.Astrophys.*, 583, p.A1.

Marty, B., Altwegg, K., Balsiger, H., Bar-Nun, A., Bekaert, D. V., Berthelier, J. J., Bieler, A., Briois, C., Calmonte, U., Combi, M., & De Keyser, J. 2017, *Science*, 356(6342), pp. 1069-1072.

Mousis, O., Ronnet, T., Brugger, B., Ozgurel, O., Pauzat, F., Ellinger, Y., Maggiolo, R., Wurz, P., Vernazza, P., Lunine, J. I., & Luspay-Kuti, A. 2016, *Astrophys.J.Lett*, 823(2), p.L41.

Paganini, L., Mumma, M. J., Gibb, E. L., & Villanueva, G. L., 2017, *The Astrophysical Journal Letters*, 836(2), p.L25.

R. O. Pepin, 2000, *Space Sci. Rev.* 92, 371-39.

Rubin, M., Altwegg, K., Balsiger, H., Bar-Nun, A., Berthelier, J. J., Bieler, A., Bochsler, P., Briois, C., Calmonte, U., Combi, M., & De Keyser, J. 2015, *Science*, 348(6231), pp. 232-235.

Rubin, M., Altwegg, K., Balsiger, H., Berthelier, J. J., Bieler, A., Calmonte, U., Combi, M., De Keyser, J., Engrand, C., Fiethe, B., & Fuselier, S. A. 2017, *Astron.Astrophys.*, 601, p.A123

Taquet, V., Furuya, K., Walsh, C., & van Dishoeck, E. F. 2016, *Mon. Not. R. Astron. Soc.*, 462(Suppl1), pp. S99-S115.

Vincent, J. B., A'Hearn, M. F., Lin, Z. Y., El-Maarry, M. R., Pajola, M., Sierks, H., Barbieri, C., Lamy, P. L., Rodrigo, R., Koschny, D., & Rickman, H., 2016, *Mon. Not. R. Astron. Soc.*, 462(Suppl1), pp. S184-S194.

Woods P.M., Occhiogrosso A., Viti S., Kanuchova Z., Palumbo M.E., & Price S. D. 2015, *Mon. Not. R. Astron. Soc.* 450, 1256-1267.

Wurz, P., Rubin, M., Altwegg, K., Balsiger, H., Berthelier, J. J., Bieler, A., Calmonte, U., De Keyser, J., Fiethe, B., Fuselier, S. A., & Galli, A. 2015, *Astron. Astrophys.*, 583, p.A22.

Yao, Y. & Giapis, K. P., 2017, *Nature Communications* 8, 15298.

Astrochemistry VII – Through the Cosmos from Galaxies to Planets
Proceedings IAU Symposium No. 332, 2017
M. Cunningham, T. Millar & Y. Aikawa, eds.

doi:10.1017/S1743921317006810

Isotopic fractionation in interstellar molecules

Kenji Furuya

Center for Computational Sciences, University of Tsukuba, 1-1-1 Tennoudai, 305-8577 Tsukuba, Japan
email: furuya@ccs.tsukuba.ac.jp

Abstract. The level of isotopic fractionation in molecules provides insights into their formation environments and how they formed. In this article, we review hydrogen and nitrogen isotopic fractionation in low-mass star-forming regions. Interstellar molecules are significantly enriched in deuterium. The importance of the nuclear spin states of light species on deuterium fractionation and the usefulness of singly and doubly deuterated molecules as chemical tracers are discussed. Observations have revealed that molecules in prestellar cores are enriched in or depleted in ^{15}N depending on molecules. Compared with deuterium fractionation chemistry, our understanding of ^{15}N fractionation chemistry is not well established. We briefly discuss potential ^{15}N fractionation routes, i.e., isotopic-exchange reactions and isotopic selective photodissociation of N_2. In addition, the selective freeze-out of ^{15}N atoms onto dust grains around the transition between N atoms and N_2 is discussed as a potential mechanism that causes the depletion of ^{15}N in the gas phase.

Keywords. Astrochemistry, ISM: abundances, ISM: molecules, star: formation

1. Introduction

More than 150 different molecules have been detected in the interstellar medium (ISM). One of the significant features in interstellar molecules is that their isotope ratios for a given element, such as the D/H ratio, the ^{14}N/^{15}N ratio, and the ^{12}C/^{13}C ratio, significantly (up to several orders of magnitude for hydrogen and up to a factor of a few for nitrogen and carbon) deviate from their elemental abundance ratios. This is called isotopic fractionation. The level of isotopic fractionation in molecules depends on their formation environments and how they formed. Then isotope ratios in molecules have been widely used as tracers of physical and chemical conditions in the ISM. Primitive solar system materials, such as comets, also show isotopic fractionation (e.g., Mumma & Charnley 2011; Altwegg *et al.* 2015). By comparing isotope ratios in the interstellar materials and those in the solar system materials, their possible chemical link has been discussed (e.g., Caselli & Ceccarelli 2012; Cleeves *et al.* 2014; Furuya *et al.* 2017).

Isotopic fractionation in the ISM can be driven by a couple of mechanisms: isotopic-exchange reactions in the gas phase, isotope selective UV photodissociation, and grain surface reactions. In addition to the local processes, the elemental abundance ratios depend on the galactocentric distance and are not constant with time, because of stellar nucleosynthesis over time (e.g., Wilson 1999). Then one should be careful about differences in elemental abundances when comparing the isotope fractionation ratios in different objects, for example cometary material versus interstellar material.

In this paper, we briefly review deuterium fractionation and nitrogen isotope fractionation in low-mass star-forming regions. Due to limitations of space, we mostly focus on basic processes of deuterium fractionation, including the nuclear spin chemistry of H_2

(Section 2) and what information can be extracted from observationally derived deuterium fractionation ratios (Section 3). Nitrogen isotope fractionation is discussed in Section 4. More comprehensive reviews can be found in e.g., Caselli & Ceccarelli (2012), Bergin (2014), Ceccarelli *et al.* (2014), and Willacy *et al.* (2015).

2. Deuterium fractionation chemistry in star-forming regions

2.1. *Deuterium fractionation mechanisms*

Star formation takes place in molecular clouds. Molecular clouds are formed from diffuse atomic gas by gravitational collapse or by accretion caused by shock waves (e.g., Dobbs *et al.* 2014, Inutsuka *et al.* 2015). During the formation of molecular clouds, chemical transition from atomic hydrogen to H_2 occurs through recombination of H atoms on grain surfaces (e.g., Vidali 2013). The elemental abundance of deuterium with respect to hydrogen is $[D/H]_{elem} = 1.5 \times 10^{-5}$ in the local ISM (Linsky 2003). Chemical transition of atomic deuterium to HD occurs through the gas-phase reaction between D^+ and H_2 and the grain surface reaction between H and D atoms (Watson 1973). Because HD is much less abundant than H_2, HD needs a higher column density of the ISM gas for self-shielding. Then the transition from D atoms to HD resides deeper into the cloud, compared to the H I/H_2 transition (van Dishoeck & Black 1986).

In molecular clouds and prestellar cores, hydrogen and deuterium are primary in H_2 and HD, respectively. Deuterium fractionation can be understood as the process that distributes deuterium in HD to other species. Deuterium fractionation in the gas phase is mainly triggered by isotope exchange reactions (e.g., Watson *et al.* 1976);

$$H_3^+ + HD \rightleftharpoons H_2D^+ + H_2. \tag{2.1}$$

At a typical temperature of molecular clouds, ~10 K, the backward reaction is inefficient due to the endothermicity (but see below) and thus H_2D^+ becomes abundant with time with respect to H_3^+. Since H_3^+ plays a central role in the production of various gaseous molecules (Herbst & Klemperer 1973), the deuterium enrichment in H_3^+ is transfered to other gaseous molecules. For example, the N_2D^+/N_2H^+ abundance ratio is >0.01 in prestellar cores (e.g., Crapsi *et al.* 2005). Even doubly and triply deuterated molecules, such as D_2CO and ND_3, have been detected (e.g., Bacmann *et al.* 2003, Roueff *et al.* 2005). With increasing temperature, the rate of the backward reaction becomes significant, and thus deuterium fractionation becomes less efficient.

The enrichment of deuterium in H_3^+ leads to the increased atomic D/H ratio in the gas phase, since D atoms are formed via dissociative recombination of deuterated H_3^+. The main components of interstellar ices, such as water, ammonia, methanol, and their deuterated forms, are formed via sequential hydrogenation/deuteration of atoms and molecules on dust grain surfaces (Hama & Watanabe 2013; Linnartz *et al.* 2015 for recent reviews). Then deuteration levels of icy molecules reflect the increased atomic D/H ratio in the gas phase (e.g., Tielens 1983; Brown & Millar 1989). Summary of the deuterium fractionation chemistry is shown in Figure 1.

On the other hand, H_2D^+ can further react with HD to form D_2H^+ and D_3^+ (e.g., Roberts *et al.* 2003; Vastel *et al.* 2004);

$$H_2D^+ + HD \rightleftharpoons D_2H^+ + H_2, \tag{2.2}$$

$$D_2H^+ + HD \rightleftharpoons D_3^+ + H_2, \tag{2.3}$$

Since D_3^+ is the end product of a series of Reactions (2.1)-(2.3) and does not react with HD, D_3^+ can be more abundant than H_3^+ in extreme conditions (complete depletion of

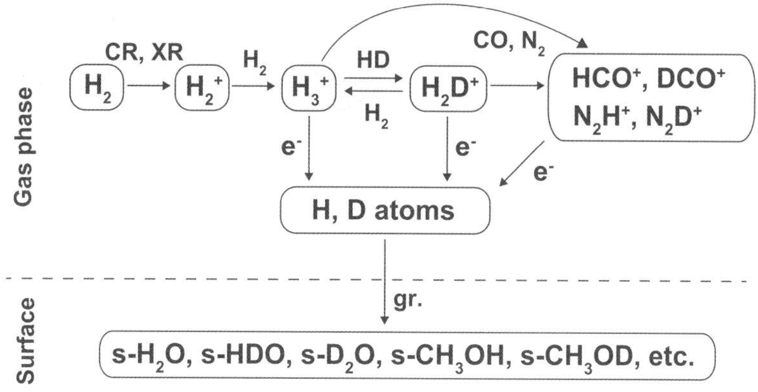

Figure 1. Summary of the main deuterium fractionation pathways of gas and icy molecules driven by Reaction (2.1). "gr." indicates adsorption of gas-phase species onto a grain surface followed by subsequent surface reactions. "s-X" indicates species X on a grain surface. In extreme condition, where most neutrals are frozen out and the H_2 OPR is lower than $\sim 6\times10^{-4}$, multiply deuterated H_3^+, D_2H^+ and D_3^+, become more abundant than H_3^+. Multiply deuterated H_3^+ is not shown in this figure for simplicity (see e.g., Walmsley *et al.* 2004 for more information).

heavy elements, see e.g., Walmsley *et al.* 2004 and Ceccarelli & Dominik 2005). Other deuterium isotopic exchange reactions, such as CH_3^+ + HD, become more important than the H_2D^+ chemistry at warm temperatures ($\gtrsim$50 K, e.g., Millar *et al.* 1989; Roueff *et al.* 2013), because of their higher exothermicity.

2.2. *Ortho-to-para ratio of H_2*

In addition to temperature, the level of deuterium fractionation depends on various parameters. Freeze-out of neutral species, such as CO and atomic oxygen, onto dust grains (i.e., ice formation) enhances deuterium fractionation, because the neutral species are the main destroyers of deuterated H_3^+ (e.g., Lepp *et al.* 1987). The rate of Reaction (2.1) in the backward direction strongly depends on the ortho-to-para ratios (OPRs) of H_2 and H_2D^+ (and also H_3^+). H_2 and H_2D^+ have two nuclear spin state isomers, the ortho-form and the para-form. For the ground states, o-H_2 and o-H_2D^+ have higher internal energy than their corresponding para forms by 170 K and 87 K, respectively, which are much higher than the typical temperature in molecular clouds (10-20 K). Since the high internal energy helps to overcome the endothermicity of the backward reaction, deuterium fractionation becomes less efficient as the ortho forms become abundant (e.g., Pagani *et al.* 1992; Gerlich *et al.* 2002; Flower *et al.* 2006; Hugo *et al.* 2009). Figure 2 shows the steady-state H_2D^+/H_3^+ ratio as functions of gas temperature, varying the H_2 OPR. The steady-state H_2D^+/H_3^+ ratio is roughly in inverse proportion to the H_2 OPR at given temperature. It should also be noted that at $\lesssim$20 K, H_2 rather than CO is the key regulator of the H_2D^+ chemistry as long as OPR(H_2) $\gtrsim 40x$(CO), where x(CO) is the CO abundance with respect to hydrogen nuclei (Furuya *et al.* 2015). Then detailed understanding of the H_2 OPR is required for understanding the deuterium chemistry.

The H_2 OPR in molecular clouds and cores is not well-constrained; it requires detailed physical and chemical modeling to estimate the H_2 OPR from observations of molecules other than H_2 in cold clouds and cores. Observational studies in literature consistently suggest the H_2 OPR is lower than unity. (e.g., Pagani *et al.* 2009, 2011; Brünken *et al.* 2014; Le Gal *et al.* 2014).

Theoretically, the H_2 OPR in the ISM is determined by the competition between the H_2 formation on grain surfaces and the ortho-para spin conversion in the gas phase via

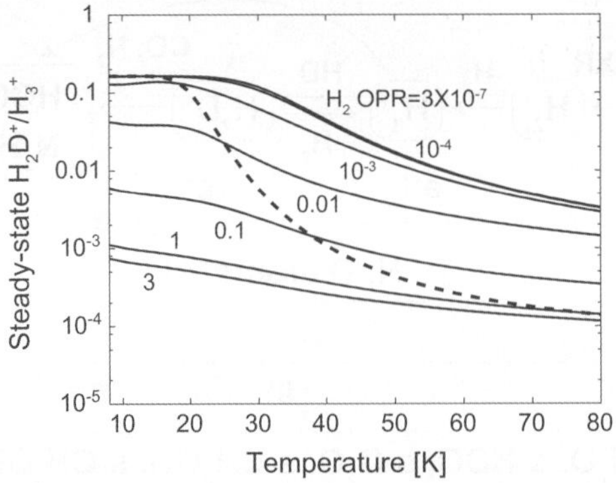

Figure 2. The steady-state H_2D^+/H_3^+ ratio at given H_2 OPR in the range of 3×10^{-7}-3 as functions of gas temperature (see Appendix of Furuya *et al.* 2015 for more information). In the black dotted line, H_2 OPR is assumed to be in thermal equilibrium. CO and electron abundances with respect to hydrogen nuclei are set to be 10^{-4} and 10^{-8}, respectively.

the proton exchange reactions between H_2 and H^+ (or H_3^+) (Gerlich 1990; Le Bourlot 1991; Honvault *et al.* 2011). H_2 molecules form on grain surfaces with the statistical OPR of three (Watanabe *et al.* 2010). The o-p conversion in the gas phase reduces the H_2 OPR to be the thermal equilibrium (TE) value of $9\exp(-170/T_{\rm gas})$ (Gerlich 1990). Laboratory experiments have found that the o-p spin conversion also occurs on grain surfaces in a laboratory timescale (Watanabe *et al.* 2010; Ueta *et al.* 2016). It is not trivial to quantify the effect of the o-p conversion on surfaces in the ISM, because of the uncertainties of the o-p conversion rates and the residence time of H_2 on surfaces (see e.g., Hincelin *et al.* 2015 for the latter). We do not consider the o-p conversion on a surface in the following discussions.

In astrochemical simulations, it has been generally assumed that hydrogen is already locked in H_2 at the beginning of the simulations, and the focus is placed on the subsequent molecular evolution. In this approach, the initial H_2 OPR is treated as a free parameter, due to the lack of observational constraints. The resultant deuteration ratio of molecules in the simulations strongly depends on the initial H_2 OPR, since the time it takes for the H_2 OPR to reach steady state is longer than the free-fall timescale and the freeze-out timescale (e.g., Flower *et al.* 2006; Kong *et al.* 2015). Furuya *et al.* (2015) studied the evolution of the H_2 OPR in forming molecular clouds from diffuse atomic gas. They found that at the H I/H_2 transition, the H_2 OPR is already lower than the statistical value of three by a factor of 10 or even more, because the timescale of the o-p conversion in the gas phase is shorter than the timescale of H_2 formation. The exact value, however, depends on physical environments, gas-phase abundances of heavy metals (e.g., sulfur), and the PAH chemistry. More studies of the transition from atomic gas to H_2 gas (i.e., molecular cloud formation) are required for better understanding of the initial H_2 OPR.

Whatever the initial H_2 OPR is, the H_2 OPR approaches the steady-state value with time. The steady-state H_2 OPR is given as follows (Le Bourlot 1991; Furuya *et al.* 2015):

$$\mathrm{OPR}(\mathrm{H}_2)_{\rm st} = \frac{9\exp(-170/T_{\rm gas}) + b_{\rm o}\tau_{\rm o\to p}/\tau_{\rm H_2}}{1 + (1 - b_{\rm o})\tau_{\rm o\to p}/\tau_{\rm H_2}}, \tag{2.4}$$

where b_o is the branching ratio to form o-H_2 for H_2 formation on grain surfaces (0.75, Watanabe *et al.* 2010). $\tau_{\rm o\to p}$ is the spin conversion timescale from o-H_2 to p-H_2 via the proton exchange reactions in the gas phase, while $\tau_{\rm H_2}$ is the formation timescale of H_2 on grain surfaces. The important point here is that there is a critical temperature under which the steady state H_2 OPR is larger than the TE value (see e.g., Fig. 1 of Faure *et al.* 2013); the critical temperature is 20 K for typical molecular clouds, where ξ_{CR}/n_{gas} is around 10^{-21} cm^3 s^{-1}, and it becomes lower with decreasing ξ_{CR}/n_{gas}. Below the critical temperature, the steady state H_2 OPR is given by $b_o\tau_{o\to p}/\tau_{H_2}$, that is a function of ξ_{CR}/n_{gas}. At $T_{gas} = 10$ K and $\xi_{CR}/n_{gas} = 10^{-21}$ cm^3 s^{-1}, the steady state H_2 OPR is $\sim 10^{-3}$, i.e., much higher than the TE value of $\sim 10^{-7}$ (e.g., Flower *et al.* 2006). Note that the enhanced H_2 OPR affects the H_2D^+ OPR thorough the non-reactive collision between H_2 and H_2D^+ (e.g., Pagani *et al.* 1992; Gerlich *et al.* 2002). This is the reason why the H_2D^+ OPR is a good tracer of the H_2 OPR (Brünken *et al.* 2014; Harju *et al.* 2017).

3. What can we learn from deuteration of icy molecules?

In this section, we discuss what information can be extracted from deuteration of icy molecules. The reason why we focus on ices is that bulk ice mantle compositions preserve the past physical and chemical conditions which the materials experienced. Deuteration measurements of the ISM ices rely heavily on the gas observations toward inner warm ($\gtrsim$150 K) regions in the deeply embedded protostars, where all ices have sublimated. High temperature gas-phase chemistry after ice sublimation could alter the original composition, but the chemical timescale is much longer ($\sim 10^5$ yr for the ionization rate of H_2 of 10^{-17} s^{-1}) than the infalling timescale of the warm gas onto the central star if the spherical collapse is assumed (e.g., Charnley *et al.* 1997; Aikawa *et al.* 2012; Taquet *et al.* 2014; Wakelam *et al.* 2014). Then the composition of the warm gas is most likely dominated by the ice composition, which was established in the cold prestellar stages.

3.1. *Singly deuterated molecules*

Ice formation in star-forming regions can be roughly divided into two stages (i.e., ice mantles have two layered structure in terms of their molecular compositions) as revealed by infrared ice observations (e.g., Pontoppidan 2006; Öberg *et al.* 2011); in the early stage, H_2O-dominated ice layers are formed. In the later stage, at higher extinction and density, the catastrophic CO freeze out happens, and ice layers which consist of CO and its hydrogenated species (H_2CO and CH_3OH) are formed.

Observations toward deeply embedded low mass protostars have revealed that the level of methanol deuteration, in particular the CH_3OD/CH_3OH ratio, is much higher than the HDO/H_2O ratio in the warm ($\gtrsim$150 K) gas around the protostars, where ices have sublimated ($\sim 10^{-2}$ versus $\sim 10^{-3}$; Parise *et al.* 2006; Coutens *et al.* 2014; Persson *et al.* 2013, 2014). Since the deuteration levels of icy molecules reflect the atomic D/H ratio in the gas phase, this trend indicates that deuterium fractionation processes are more efficient in the later stage of the ice formation, most likely due to CO freeze out and the drop of the H_2 OPR (Cazaux *et al.* 2011; Taquet *et al.* 2012). Then, generally speaking, one can constrain the relative timing of the formation of icy molecules by comparing their D/H ratios.

Care is required when one applies the above discussion to other icy molecules, such as formaldehyde; the D/H ratio of icy molecules can be altered after their formation. Once water and ammonia are formed, they do not efficiently react with D atoms to be deuterated (Nagaoka *et al.* 2005). On the other hand, the deuteration levels of formaldehyde and methanol can be enhanced by the substitution and abstraction reactions of H and D atoms at low temperatures ($\sim$10 K) after their formation (Watanabe & Kouchi 2008; Hidaka *et al.* 2009). As another example, methylamine (CH_3NH_2) is also subject to the H/D substitution reactions (Oba *et al.* 2014). Moreover the efficiency of the

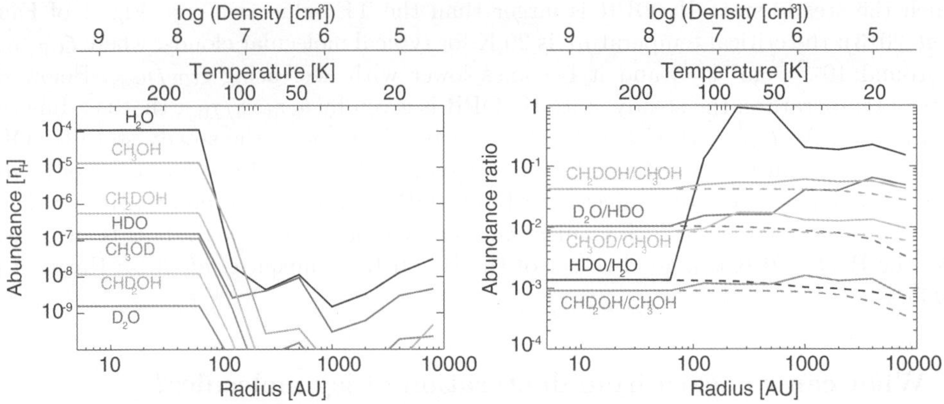

Figure 3. Radial profile of the molecular abundances (left panel) and the abundance ratios of the molecules (right panel) as functions of distance form the central low-mass protostar in the deeply embedded stage (Furuya *et al.* 2016 with some updates). The labels at the top represents the temperature and density structures. The solid lines represent gaseous molecules, while the dashed lines represent icy molecules in the bulk ice mantle.

substitution and abstraction reactions depend on a functional group in molecules, e.g., effective in the CH_3- group of methanol and not in its OH- group. Indeed, the observationally derived CH_2DOH/CH_3OD ratio is $\sim$10 in the warm gas of embedded low-mass protostars, contrasting with the expectation that the ratio is unity, when methanol ices are simply formed via addition reactions of H and D atoms on CO ice (Parise *et al.* 2006; Taquet *et al.* 2012). The CH_2DOH/CH_3OD ratio is close to unity in high-mass hot cores (Ratajczak *et al.* 2011). These observations likely indicate that the substitution and abstraction reactions are efficient in low-mass star-forming regions, but not the high-mass sources (Taquet *et al.* 2012).

State-of-the-art astrochemical models, which take into account the dynamical evolution of gravitationally collapsing cores, the deuterium and nuclear spin state chemistry, and a layered ice structure, have reproduced these observations at least qualitatively (Taquet *et al.* 2014; Furuya *et al.* 2016). Figure 3 shows radial profiles of the molecular abundances (left panel) and the abundance ratios of the molecules (right panel) in the deeply embedded protostellar stage as functions of distance from a central low-mass protostar. Inside $\sim$60 AU from the central star, dust temperature is high enough to sublimate all ices ($\gtrsim$150 K), and the gas composition is determined by the ice sublimation. At >150 K, the HDO/H_2O ratio is $\sim 10^{-3}$ and CH_3OD/CH_3OH ratio is $\sim 10^{-2}$, reproducing the observations. The CH_2DOH/CH_3OD ratio is higher than unity due to the efficient abstraction reactions in the model, but the model still underestimates the ratio compared to the observationally derived value of $\sim$10.

It is worth noting that the gaseous HDO/H_2O ratio is decreased from the cold outer envelopes to the inner warm ($\gtrsim$150 K) regions by more than one order of magnitude, which has been confirmed both by observations and astrochemical models (Fig. 3; Coutens *et al.* 2012, Persson *et al.* 2013). Then interferometer observations are more suitable for the deuteration measurements of warm water than single dish observations, which provide the integrated emission from much larger spacial scales than the warm gas (Jørgensen & van Dishoeck 2010). According to astrochemical models, unlike water, methanol does not show the decrease of the D/H ratio from the outer envelopes to the inner warm regions, because methanol efficiently forms only on grain surfaces and in the later stage of the ice formation (Fig. 3; see also Taquet *et al.* 2014).

3.2. *Doubly deuterated molecules*

In addition to singly deuterated molecules, doubly deuterated molecules in the inner warm regions around protostars can provide more constrains on the physical and chemical evolution in the cold prestellar stages. Recently, Coutens *et al.* (2014) found that the gaseous D_2O/HDO ratio is much higher than the HDO/H_2O ratio ($\sim 10^{-2}$ versus $\sim 10^{-3}$) in the inner warm regions of deeply embedded protostar NGC 1333-IRAS 2A. This finding contrasts with the expectation that the ratio is 1/4 (i.e., the statistical ratio), when water ices are formed via addition reactions of H and D atoms with atomic oxygen (Rodgers & Charnley 2002).

This higher D_2O/HDO than HDO/H_2O can be understood by taking into account chemical evolution (see Furuya *et al.* 2016 for details). Like the HDO/H_2O ratio, the D_2O/HDO ratio depends on the atomic D/H ratio. Then the high D_2O/HDO ratio indicates that the water ice formation still continues when deuteration becomes more efficient due to CO freeze out and the drop of the H_2 OPR. Indeed, the D_2O/HDO ratio ($\sim 10^{-2}$) is similar to the CH_3OD/CH_3OH ratio. But, the amount of water ice formed in this late stage should be limited to keep the HDO/H_2O ratio low. This requires that free atomic oxygen in the gas phase should be mostly (>90 %) consumed before the epoch of CO freeze out. The low abundance of atomic oxygen is consistent with the low abundance of O_2 gas in a protostellar envelope (Yıldız *et al.* 2013). This scenario was confirmed by the gas-ice astrochemical model of Furuya *et al.* (2016) (see also Fig. 3). Then, multiple deuteration of water is a strong tool to constrain the past history of water ice formation and the amount of free oxygen that is available for water ice formation.

A similar idea can be applied to atomic nitrogen and ammonia ice. The primary reservoir of nitrogen in dense star-forming clouds is not well constrained. One of the potentially important reservoirs is atomic nitrogen (Maret *et al.* 2006), but it is not directly observable in the cold gas. By conducting astrochemical modelings of star-forming cores, it was shown that if atomic nitrogen is the primary reservoir of nitrogen even when the catastrophic CO freeze out happens, the $[ND_2H/NH_2D]/[NH_2D/NH_3]$ ratio is close to the statistical value of 1/3, whereas if atomic nitrogen is largely converted into gaseous N_2 or icy molecules, the ratio should be significantly larger than 1/3 (Furuya & Persson 2018). The similar method could be applied to other elements, such as sulfur, by observing e.g., H_2S, H_2CS and their deuterated forms in the warm gas.

Schematic view of gas and ice evolution in cold prestellar stages is summarized in Figure 4. The chemical and isotopic (at least for deuterium) compositions of the ISM ice are characterized not only by the bulk ice composition, but also by the differentiation within the ice mantle, reflecting the physical and chemical evolution during the prestellar stages. There are several numerical methods to take into account a layered ice structure in astrochemical simulations, see e.g., Hasegawa & Herbst (1993), Taquet *et al.* (2012), Vasyunin & Herbst (2013), and Furuya *et al.* (2017).

3.3. *Caveat: H/D exchange reactions in warm ices*

Laboratory experiments have shown that thermally activated H-D exchanges between hydrogen-bonded molecules in mixed ices occur efficiently at warm temperatures of $\gtrsim 70$ K before sublimation of polar ices (Ratajczak *et al.* 2009; Faure *et al.* 2015b; Lamberts *et al.* 2015). The examples of this type of reaction are as follows:

$$H_2O + D_2O \rightleftharpoons 2\,HDO, \tag{3.1}$$

$$CH_3OD + H_2O \rightleftharpoons CH_3OH + HDO, \tag{3.2}$$

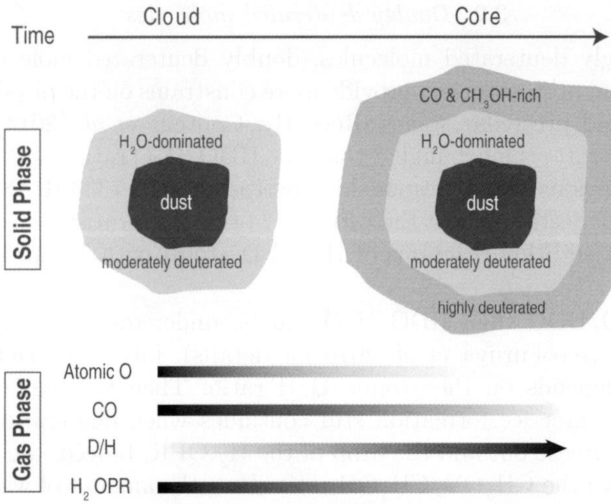

Figure 4. Schematic view of gas and ice evolution in cold prestellar stages, clouds and cores. Clouds) The main formation stage of H_2O ice. Deuteration in this stage is less efficient than in the later stage. The majority of oxygen is locked in O-bearing molecules in this stage. Cores) CO/CH_3OH-rich outer ice layers are formed, while the formation of water ice continues with much reduced efficiency, compared to the early stage. Deuteration is more efficient, because of CO freeze out and the drop of the ortho-to-para ratio of H_2.

If these reactions are efficient, the original deuteration ratios established during the cold prestellar stages would be largely lost before ice sublimation (Faure *et al.* 2015a). It should be noted, however, that the H/D exchange occurs only between closely interacting molecules in ice, and the efficiency of the H/D exchange in the inhomogenous ISM ice is unclear (Ratajczak *et al.* 2009; Faure *et al.* 2015a).

4. Nitrogen isotope fractionation in prestellar cores

Nitrogen has two isotopes, ^{14}N and ^{15}N, with the elemental abundance ratio $[^{14}N/^{15}N]_{elem}$ ~ 300 in the local ISM (Adande & Ziurys 2012; Ritchey *et al.* 2015). The dominant form of nitrogen in diffuse clouds is the atomic form with a small fraction of the molecular form (Viala *et al.* 1986; Knauth *et al.* 2004). In dense molecular clouds, atomic N is (at least in part) converted into gaseous and icy molecules (e.g., Herbst & Klemperer 1973; Hily-Blant *et al.* 2010), but the primary form of nitrogen is not well constrained. Gas-ice astrochemical models often predict most nitrogen is present as NH_3 ice and N_2 gas and ice (e.g., Daranlot *et al.* 2012; Chang & Herbst 2014).

Observations have revealed that N-bearing molecules in prestellar cores are enriched in or depleted in ^{15}N depending on molecules. Figure 5 shows the observationally derived $N/^{15}N$ ratio taken from literature. N_2H^+ and ammonia are either depleted in ^{15}N or not fractionated depending on source, while nitriles are either enriched in ^{15}N or not fractionated. Note that some care is required for the interpretation of the enrichment of ^{15}N in HCN and HNC as pointed out by Roueff *et al.* (2015); the $N/^{15}N$ ratios of these molecules were derived from ^{13}C species assuming the elemental $^{12}C/^{13}C$ ratio. HCN and HNC can be depleted in ^{13}C by a factor of up to a few (see Langer *et al.* 1984 for chemical fractionation of carbon isotope), and thus their $N/^{15}N$ ratios could be underestimated by the same factor.

It has been argued that this observed trend reflects the different gas-phase formation pathways of these molecules; gaseous ammonia and N_2H^+ are formed from N_2, while

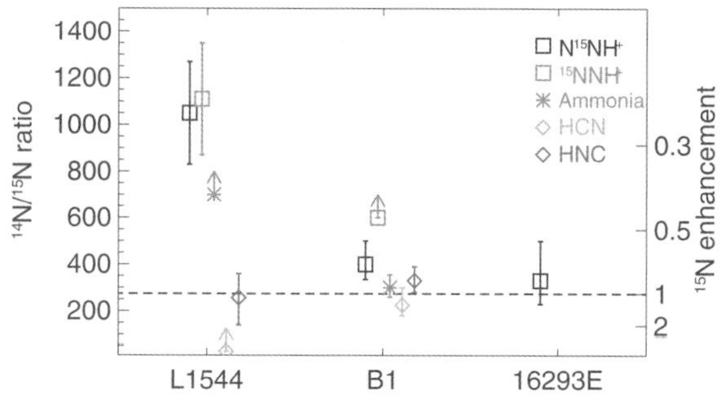

Figure 5. Observationally derived ^{15}N fractionation in various molecules in prestellar cores, where the ^{15}N isotopologues of N_2H^+ were observed (taken from Bizzocchi *et al.* 2013; Daniel *et al.* 2013, 2016; Gerin *et al.* 2009; Hily-Blant *et al.* 2013; Milam & Charnley 2012). The $N/^{15}N$ ratios in ammonia are the values for NH_2D in L1544 and for NH_3 in Barnard 1. The black dashed line indicates the elemental abundance ratio $[N/^{15}N]_{\rm elem}$ of ∼300 in the local ISM.

nitriles are formed from N atoms (Rodgers & Charnley 2008; Wirström *et al.* 2012; Hily-Blant *et al.* 2013). If so, then, the remaining question is what is the mechanism that causes the depletion of ^{15}N in N_2 and the enrichment of ^{15}N in N atoms.

There are several studies of astrochemical modeling of ^{15}N fractionation via isotope exchange reactions in prestellar cores (e.g., Terzieva & Herbst 2000; Rodgers & Charnley 2008; Wirström *et al.* 2012). The model of Wirström *et al.* (2012) can explain the observed ^{15}N fractionation in ammonia and nitriles, depending on the degree of CO freeze-out and the H_2 OPR, while their model predicts that N_2H^+ is enriched in ^{15}N, being inconsistent with the observations. Furthermore, recent quantum chemical calculations indicate the fractionation via isotope exchange reactions is not effective due to the presence of activation barriers for some key reactions (Roueff *et al.* 2015). See contribution by E. S. Wirström in this volume for more details.

Another mechanism that can cause ^{15}N fractionation is isotope selective photodissociation of N_2 (Liang *et al.* 2007; Heays *et al.* 2014). Photodissociation of N_2 is subject to self-shielding. Because $N^{15}N$ is much less abundant than N_2, $N^{15}N$ needs a higher column density of the ISM gas for self-shielding. As a result, in some regions, $N^{15}N$ is selectively photodissociated and then N_2 is depleted in ^{15}N, while atomic N is enriched in ^{15}N. Heays *et al.* (2014) developed a gas-phase astrochemical model of a molecular cloud, considering both isotope selective photodissociation of N_2 and isotope exchange reactions for ^{15}N fractionation. They found that the isotope selective photodissociation is at work only in the limited regions, where the interstellar UV radiation field is not significantly attenuated and the transition from atomic nitrogen to N_2 occurs. In these regions, N_2 and N_2H^+ are depleted in ^{15}N, while nitriles are enriched in ^{15}N. Note, however, that prestellar cores typically have higher density ($>10^5$ cm^{-3}) and higher A_V (>10 mag for a dust continuum peak) compared to their cloud model. Thus it seems difficult to explain the depletion of ^{15}N in N_2H^+ in the framework of the present gas-phase chemical network.

On the other hand, the combination of gas and surface chemistry may cause the ^{15}N fractionation. A potentially important mechanism is the freeze out of N atoms followed by the conversion into less volatile species (e.g., NH_3) in the regions where N atoms are enriched in ^{15}N, i.e., the N I/N_2 transition (Furuya & Aikawa 2018; see also Bizzocchi *et al.* 2013). In this case, ^{15}N is selectively depleted in the gas phase with respect to ^{14}N. After the N I/N_2 transition, overall gas would be depleted in ^{15}N compared to $[N/^{15}N]_{\rm elem}$, while ices would be enriched in ^{15}N. Once the differential $N/^{15}N$ ratio between the gas-

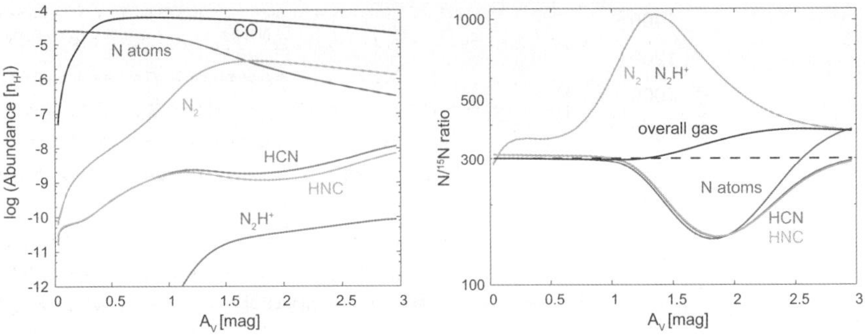

Figure 6. Molecular abundances of selected N-bearing molecules (left panel) and the $N/^{15}N$ abundance ratios (right) as functions of visual extinction, A_V (i.e., larger A_V corresponds to the deeper regions into the cloud center) taken from a gas-ice astrochemical model of molecular cloud formation via converging flows. The model includes gas phase chemistry, interaction between gas and grain, and grain surface chemistry in addition to isotope exchange reactions for ^{15}N fractionation chemistry and isotope selective photodissociation of N_2 (Furuya & Aikawa 2018). The black dashed line in the right panel indicates the elemental abundance ratio $[N/^{15}N]_{elem}$ of 300, while the black solid line indicates the $N/^{15}N$ ratio in the gas phase.

phase and the solid phase is established, it would remain as long as dust temperature is cold and ice sublimation is inefficient (until $\lesssim$150 K, if N atoms are mainly converted into ammonia ice). The $N/^{15}N$ ratios can vary among gaseous molecules, depending on the efficiency of ^{15}N fractionation by isotope exchange reactions.

In order to test this scenario, Furuya & Aikawa (2018) developed a gas-ice astrochemical model of a molecular cloud formation via converging flows of diffuse atomic gas, considering both the isotope selective photodissociation of N_2 (Heays *et al.* 2014) and isotope exchange reactions for ^{15}N fractionation (Roueff *et al.* 2015). Figure 6 shows molecular abundances of N-bearing molecules in the model (left panel) and the $N/^{15}N$ abundance ratios (right) as functions of visual extinction, A_V (i.e., larger A_V represents the deeper regions into the cloud center). It was found that the isotope selective photodissociation is efficient only around the N I/N_2 transition, consistent with Heays *et al.* (2014) (1 mag $\lesssim A_V \lesssim$ 2 mag in this model). Around the N I/N_2 transition, N_2H^+ is significantly depleted in ^{15}N, while nitriles are significantly enriched in ^{15}N. In deeper regions into the cloud, at $A_V \gtrsim$ 2 mag, the overall gas is somewhat depleted in ^{15}N due to the selective freeze-out of ^{15}N followed by the $^{15}NH_3$ formation on grain surfaces at 1 mag $\lesssim A_V \lesssim$ 2 mag. As a consequence, N_2H^+ is somewhat depleted in ^{15}N compared to $[N/^{15}N]_{elem}$ even in the regions where interstellar UV radiation is significantly attenuated. Nitriles have the similar $N/^{15}N$ ratios to $[N/^{15}N]_{elem}$ as a result of the local ^{15}N fractionation chemistry via isotope exchange reactions. The degree of ^{15}N depletion in the overall gas by this mechanism depends on the amount of ^{15}N atoms that are frozen out onto grain surfaces around the N I/N_2 transition.

References

Adande, G. R. & Ziurys, L. M. 2012, *ApJ*, 744, 194

Aikawa, Y., Wakelam, V., Hersant, F., Garrod, R., & Herbst, E. 2012, *ApJ*, 760, 40

Altwegg, K., Balsiger, H., Bar-Nun, A., *et al.* 2015, *Science*, 347, 1261952

Bacmann A., Lefloch B., Ceccarelli C., Steinacker J., Castets A., Loinard L., 2003, *ApJ*, 585, L55

Bergin, E. A. 2014. Astrobiology: An astronomer's perspective. American Institute of Physics Conference Series 1638, 5-34

Bizzocchi, L., Caselli, P., Lenardo, E., & Dore, L. 2013, *A&A*, 555, 109

Brown, P. D. & Millar, T. J. 1989, *MNRAS*, 237, 661

Brünken, S., Sipilä, O., Chambers, E. T. *et al.* 2014, *Nature*, 516, 219

Caselli, P. & Ceccarelli, C. 2012, *ARA&A*, 20, 56

Cazaux, S., Caselli, P., & Spaans, M. 2011, *ApJ*, 741, L34

Ceccarelli, C., Caselli, P., Bockelée-Morvan, D., Mousis, O., Pizzarello, S., Robert, F., & Semenov, D. 2014, Protostars & Planets VI (Tucson: Univ. Arizona Press), 859

Ceccarelli, C. & Dominik, C. 2005, *A&A*, 440, 583

Chang, Q. & Herbst, E. 2014, *ApJ*, 787, 135

Charnley, S. B, Tielens, A. G. G. M, & Rodgers, S. D. 1997, *ApJ*, 482, 203

Cleeves, L. I., Bergin, E. A., Alexander, C. M. O., *et al.* 2014, *Science*, 345, 1590

Coutens, A., Jørgensen, J. K., Persson, M., *et al.* 2014, *ApJL*, 792, L5

Coutens, A., Vastel, C., Caux, E., *et al.* 2012, *A&A*, 539, 132

Crapsi, A., Caselli, P., Walmsley, C. M., *et al.* 2005, *ApJ*, 619, 379

Dalgarno, A. 2006, *PNAS*, 103, 12269

Daniel, F., Faure, A., Pagani, L., *et al.* 2016, *A&A*, 592, 45

Daniel, F., Gérin, M., Roueff, E., *et al.* 2013, *A&A*, 560, 3

Daranlot, J., Hincelin, U., Bergeat, A., *et al.* 2012, *PNAS*, 109, 10233

Dobbs, C. L., Krumholz, M. R., Ballesteros-Paredes, J., *et al.* 2014, Protostars & Planets VI, 3

Faure, A., Faure, M., Theulé, P., Quirico, E., & Schmitt, B. 2015, *A&A*, 584, A98

Faure, M., Quirico, E., Faure, A., *et al.* 2015, *Icarus*, 261, 14

Faure, A., Hily-Blant, P., Le Gal, R., Rist, C., & Pineau des Forêts, G. 2013, *ApJ*, 770, L2

Fedoseev, G., Ioppolo, S., & Linnartz, H. 2015, *MNRAS*, 446, 449

Flower, D. R., Pineau Des Forêts, G., & Walmsley, C. M. 2006, *A&A*, 449, 621

Furuya, K., Aikawa, Y., Hincelin, U., Hassel, G. E., Bergin, E. A., Vasyunin, A. I., & Herbst, E. 2015, *A&A*, 584, 124

Furuya, K., van Dishoeck, E. F., & Aikawa, Y. 2016, *A&A*, 586, 127

Furuya, K., Drozdovskaya, M. N., Visser, R., van Dishoeck, E. F., Walsh, C., Harsono, D., Hincelin, U., & Taquet, V. 2017, *A&A*, 599, 40

Furuya, K., & Aikawa, Y. 2018, *ApJ*, 857, 105

Furuya, K., & Persson, M. V. 2018, *MNRAS*, 476, 4994

Gerin, M., Marcelino, N., Biver, N., *et al.* 2009, *A&A*, 498, L9

Gerlich, D. J. 1990, *Chem. Phys.*, 92, 2377

Gerlich, D., Herbst, E., & Roueff, E. 2002, *Plan. Sp. Sci.*, 50, 1275

Hama T., Watanabe N., 2013, *Chem. Rev.*, 113, 8783

Hama, T., Kuwahata, K., Watanabe, N., *et al.* 2012, *ApJ*, 757, 185

Harju, J., Sipilä, O., Brünken, S., *et al.* 2017, *ApJ*, 840, 63

Hasegawa, T. I. & Herbst, E. 1993, *MNRAS*, 263, 589

Heays, A. N., Visser, R., Gredel, R., *et al.* 2014, *A&A*, 562, 61

Herbst, E. & Klemperer, W. 1973, *ApJ*, 185, 505

Hidaka, H., Watanabe, M., Kouchi, A., & Watanabe, N. 2009, *ApJ*, 702, 291

Hily-Blant, P., Bonal, L., Faure, A., & Quirico, E. 2013, *Icarus*, 223, 582

Hily-Blant, P.,Walmsley, M., Pineau des Forts, G., & Flower, D. 2010, *A&A*, 513, 41

Hincelin, U., Chang, Q., & Herbst, E. 2015, *A&A*, 574, 24

Honvault, P., Jorfi, M., González-Lezana, T., Faure, A., & Pagani, L. 2011, *Phys. Rev. Lett.*, 107, 023201

Hugo, E., Asvany, O., & Schlemmer, S. 2009, *J. Chem. Phys.*, 130, 164302

Inutsuka, S.-I., Inoue, T., Iwasaki, K., & Hosokawa, T. 2015, *A&A*, 580, A49

Jørgensen, J. K. & van Dishoeck, E. F. 2010, *ApJ*, 725, L172

Kalvans, J., Shmeld, I., Kalnin, J. R., Hocuk, S. 2017 *MNRAS*, 467, 1763

Kong, S., Caselli, P., Tan, J. C., Wakelam, V., & Sipilä, O. 2015, *ApJ*, 804, 98

Knauth, D. C., Andersson, B.-G., McCandliss, S. R., & Warren Moos, H. 2004, *Nature*, 429, 636

Lamberts, T., Ioppolo, S., Cuppen, H. M., Fedoseev, G., & Linnartz, H. 2015, *MNRAS*, 448, 3820

Langer, W. D., Graedel, T. E., Frerking, M. A., & Armentrout, P. B. 1984, *ApJ*, 277, 581
Le Bourlot, J. 1991, *A&A*, 242, 235
Le Gal, R., Hily-Blant, P., Faure, A., *et al.* 2014, *A&A*, 562, 83
Lepp, S., Dalgarno, A., & Sternberg, A. 1987, *ApJ*, 321, 383
Liang, M.-C., Heays, A. N., Lewis, B. R., Gibson, S. T., & Yung, Y. L. 2007, *ApJ*, 664, L115
Linnartz H., Ioppolo S., Fedoseev G., 2015, *Int. Rev. Phys. Chem.*, 34, 205
Linsky, J. L. 2003, *Space Sci. Rev.*, 106, 49
Maret, S., Bergin, E. A., & Lada, C. J. 2006, *Nature*, 442, 425
Milam, S. N. & Charnley, S. B. 2012, in Lunar and Planetary Inst. Technical Report, Vol. 43, Lunar and Planetary Institute Science Conference Abstracts, 2618
Millar, T. J., Bennet, A., & Herbst, E. 1989, *ApJ*, 340, 906
Mumma, M. J. & Charnley, S. B. 2011, *ARA&A*, 49, 471
Nagaoka, A., Watanabe, N., & Kouchi, A. 2005, *ApJ*, 624, L29
Oba, Y., Watanabe, N., Hama, T., *et al.* 2012, *ApJ*, 749, 67
Oba Y., Chigai T., Osamura Y., Watanabe N., Kouchi A., 2014, *Meteor. Planet. Sci.*, 49, 117
Öberg, K. I., Boogert, A. C. A., Pontoppidan, K. M., *et al.* 2011, *ApJ*, 740, 109
Pagani, L., Wannier, P. G., Frerking, M. A., *et al.* 1992, *A&A*, 258, 472
Pagani, L., Vastel, C., Hugo, E., *et al.* 2009, *A&A*, 494, 623
Pagani, L., Roueff, E., & Lesaffre, P. 2011, *A&A*, 739, L35
Parise, B., Ceccarelli, C., Tielens, A. G. G. M., *et al.* 2006, *A&A*, 453, 949
Persson, M. V., Jørgensen, J. K., & van Dishoeck, E. F. 2013, *A&A*, 549, L3
Persson, M. V., Jørgensen, J. K., van Dishoeck, E., & Harsono, D. 2014, *ApJ*, 563, 74
Pontoppidan, K. M. 2006, *A&A*, 453, L47
Ratajczak, A., Quirico, E., Faure, A., Schmitt, B., & Ceccarelli, C. 2009, *A&A*, 496, L21
Ratajczak, A., Taquet, V., Kahane, C., *et al.* 2011, *A&A*, 528, L13
Ritchey, A. M., Federman, S. R., & Lambert, D. L. 2015, *ApJL*, 804, L3
Roberts, H., Herbst, E., & Millar, T. J. 2003, *ApJ*, 591, L41
Rodgers, S. D. & Charnley, S. B. 2002, *Planet. Space Sci.*, 50, 1125
Rodgers, S. D. & Charnley, S. B. 2008, *ApJ*, 689, 1448
Roueff, E., Gerin, M., Lis, D. C., *et al.* 2013, *Journal of Physical Chemistry A*, 117, 9959
Roueff, E., Loison, J. C., & Hickson, K. M. 2015, *A&A*, 576, 99
Roueff, E., Lis, D. C., van der Tak, F. F. S., Gerin, M., & Goldsmith, P. F. 2005, *A&A*, 438, 585
Taquet, V., Ceccarelli, C., & Kahane, C. 2012, *ApJ*, 748, L3
Taquet, V., Charnley, S. B., & Sipilä, O. 2014, *ApJ*, 791, 1
Taquet, V., Peters, P. S., Kahane, C., *et al.* 2013, *A&A*, 550, 127
Terzieva, R. & Herbst, E. 2000, *MNRAS*, 317, 563
Tielens, A. G. G. M. 1983, *A&A*, 119, 177
Ueta, H., Watanabe, N., Hama, T., & Kouchi, A., 2016, *Phys. Rev. Lett.*, 116, 253201
van Dishoeck, E. F. & Black, J. H. 1986, *ApJS*, 62, 109
Vastel, C., Phillips, T. G., & Yoshida, H. 2004, *ApJ*, 606, L127
Vasyunin, A. I. & Herbst, E. 2013, *ApJ*, 762, 86
Viala, Y. P. 1986, *A&AS*, 64, 391
Vidali, G. 2013, *ChRv*, 113, 8752
Wakelam, V., Vastel, C., Aikawa, Y., *et al.* 2014, *MNRAS*, 445, 2854
Walmsley, C. M., Flower, D. R., & Pineau Des Forêts, G. 2004, *A&A*, 418, 1035
Watanabe, N. & Kouchi, A. 2008, *Prog. Surf. Sci.*, 83, 439
Watanabe, N., Kimura, Y., Kouchi, A., Chigai, T., Hama, T., & Pirronello, V. 2010, *ApJ*, 714, L233
Watson, W. D. 1973, *ApJL*, 183, L17
Watson, W. D. 1976, *Reviews of Modern Physics*, 48, 513
Willacy, K., Alexander, C., Ali-Dib, M., *et al.* 2015, *SSRv*, 197, 151
Wilson, T. L. 1999, *Reports on Progress in Physics*, 62, 143
Wirström, E. S., Charnley, S. B., Cordiner, M. A., & Milam, S. N. 2012, *ApJ*, 757, L11
Yıldız, U. A., Acharyya, K., Goldsmith, P. F. *et al.* 2013, *A&A*, 558, 58

Astrochemistry VII – Through the Cosmos from Galaxies to Planets
Proceedings IAU Symposium No. 332, 2017
M. Cunningham, T. Millar & Y. Aikawa, eds.

doi:10.1017/S174392131700761X

Chemical Tracers of Dynamics in Low-Mass Protostellar Objects

Satoshi Yamamoto[1], **Yoko Oya**[1] **and Nami Sakai**[2]

[1]Department of Physics, The University of Tokyo, Bunkyo-ku, Tokyo, 113-0033, Japan
email: yamamoto@taurus.phys.s.u-tokyo.ac.jp, oya@taurus.phys.s.u-tokyo.ac.jp

[2]RIKEN, Wako, Saitama, 351-0198, Japan
email: nami.sakai@riken.jp

Abstract. Recent progress in astrochemistry of low-mass protostellar sources is reviewed. In particular, we focus on disk formation processes and associated chemical changes at a 50 au scale, which is extensively being studied with ALMA. A small scale chemical differentiation sensitively reflects changes in physical conditions, and hence, it provides us with unique opportunities of chemical diagnostics of disk-forming regions. Complex physical and chemical pictures of disk formation revealed by observations are summarized, and future prospects are discussed.

1. Historical Overview

One of important goals for astrochemistry and astrophysics is to understand chemical evolution from interstellar molecular clouds to planetary systems, which is directly related to the origin of the Solar System. So far, chemical compositions of molecular clouds and star-forming regions have extensively been studied, and their basic properties are reasonably understood. Likewise, chemical compositions of protoplanetary disks have also been explored, although the number of molecular species detected so far is not as many as that for molecular clouds. On the other hand, a link between these two stages, that is chemistry of disk-forming regions, has sparsely been investigated, because of its small size scale as well as physical and chemical complexities. However, ALMA is now changing this situation. Before moving on to the ALMA results on disk-forming regions, progress of astrochemistry of low-mass (solar-type) star-forming regions for the last three decades is overviewed with a particular focus on complex organic species.

1.1. *Early Days*

After discovery of various interstellar molecules in 1970s, chemical difference among observing objects was immediately recognized. Such information was used to choose an appropriate astronomical source for hunting new interstellar molecules. At the same time, chemistry was found to be sensitive to the physical condition of sources, and this characteristic suggested a possibility of chemical diagnostics of star formation to us.

In 1980s, chemistry of star-forming regions was almost equivalent to chemistry of high-mass star-forming regions. Spectral line observations in the millimeter-wave region were extensively conducted toward a few high-mass star-forming regions revealing bright molecular line emission (e.g., Cummins *et al.*1986; Blake *et al.* 1987; Masson & Mundy 1988; Turner 1989). Various large complex organic molecules (COMs) such as $HCOOCH_3$, $(CH_3)_2O$, and C_2H_5CN as well as sulfur-bearing species (SO and SO_2) were detected in a hot ($T > 100$ K) and dense region ($n(H_2) > 10^6$ cm^{-3}) around high-mass young stellar objects, which is called as *hot core* (Herbst & van Dishoeck 2009; Caselli and Ceccarelli

2012 for review). Since the large COMs were only found in hot cores, they were thought to be characteristic species of hot core chemistry.

In contrast, chemistry of low-mass star-forming regions was less studied in those days. Myers and his collaborators systematically studied physical properties of dense cores in dark clouds, and established them as birthplaces of low-mass stars with the aid of the infrared observations with the *IRAS* satellite (e.g. Myers & Benson 1983; Beichman *et al.* 1986; Benson & Myers 1989). However, their chemistry did not attract much attention of astrochemists, because the molecular emission is generally much fainter than that in high-mass star-forming regions. Low-mass version of hot cores was naturally expected, but the spectral lines of large COMs were not detected because of insufficient telescope sensitivity. Suzuki *et al.* (1992) revealed the systematic chemical difference between starless cores and star-forming cores: carbon-chain molecules are abundant in the former, while NH_3 are abundant in the latter. However, this is not the effect of star-formation activities, but mostly due to a combination of a chemical evolutionary effect and depletion of molecules onto dust grains (Suzuki *et al.* 1992; Aikawa *et al.* 2001).

1.2. *Chemistry of Low-Mass Star-Forming Regions*

In 1990s, pioneering works on chemistry of low-mass star-forming regions were reported by Blake *et al.* (1994) and van Dishoeck *et al.*(1995). They conducted a spectral line survey in the 230 and 345 GHz bands toward the low-mass star-forming region IRAS 16293–2422 in Ophiuchus with the CSO 10.4 m telescope and the JCMT 15 m telescope. In this observation, bright lines of sulfur-bearing molecules such as OCS, H_2CS, SO, and SO_2 were detected. In addition, high excitation lines of H_2CO and CH_3OH were also found. Excitation analyses of the observed lines strongly suggest the existence of a hot and dense region (hot core) near the protostar. A hot and dense region around the low-mass protostar was also suggested by the multi-line analysis of H_2CO and CH_3OH toward other protostars (Ceccarelli *et al.* 1998; 2000; Shoier *et al.* 2002). However, the large COMs, $HCOOCH_3$, $(CH_3)_2O$, and C_2H_5CN, which are characteristic to hot cores, were not detected. It should be noted that the distributions of the bright OCS and H_2CS emission in IRAS 16293-2422 were unclear in those days. Their distributions concentrated around the protostar have recently been revealed by Oya *et al.* (2016).

In 2003, Cazaux *et al.* detected the spectral lines of $HCOOCH_3$, $(CH_3)_2O$, and C_2H_5CN toward IRAS 16293–2422 with the IRAM 30 m telescope. This discovery was immediately followed up by interferometric observations, and the distribution of these molecules are found to be concentrated within a few 100 au region around each binary component (Source A and Source B) (Bottinelli *et al.* 2004a; Kuan *et al.* 2004). Thus, a hot-core like chemistry was found to be indeed occurring there. This discovery attracted many researchers in relation to organic materials in meteorites and comets.

It was thought that the above molecules are formed on dust grains and are sublimated in the hot region around the protostar (e.g., Garrod and Herbst, 2006; Aikawa *et al.* 2008; Herbst & van Dishoeck 2009). Later, it was also pointed out that they could be formed by gas-phase reactions starting from CH_3OH and H_2CO sublimated from dust grains (e.g. Balucani *et al.* 2015). The size of the hot region around a low-mass protostar is so tiny in comparison with a hot core around a high-mass protostar that it is named as *'hot corino'* (Bottinelli *et al.* 2004b). Extensive searches of hot corinos were conducted with large single dish telescopes, but only a few definitive hot corinos harboring the large COMs were found after IRAS 16293–2422. Those are NGC1333 IRAS 2A (Jørgensen *et al.* 2005), NGC1333 IRAS4A (Bottinelli *et al.* 2004b), NGC1333 IRAS4B (Sakai *et al.* 2006; Bottinelli *et al.* 2007), Serpens SMM1, and Serpens SMM4 (Öberg *et al.* 2011).

1.3. *Discovery of Chemical Diversity in Low-Mass Protostellar Sources*

In the course of a sensitive search for hot corinos, Sakai *et al.* (2008) discovered a protostellar source, whose chemical composition is completely different from the hot corino sources. It is L1527 in Taurus. In this source, carbon-chain molecules and their related species are extremely rich in a few 1000 au region around the protostar (Sakai *et al.* 2010). In general, carbon-chain molecules are only abundant in young starless cores, while they are deficient in protostellar cores (e.g., Suzuki *et al.* 1992; Aikawa *et al.* 2001). Hence, association of carbon-chain molecules around the protostar was very surprising. Sakai *et al.* (2008) proposed that carbon-chain molecules are *in situ* produced in the gas phase of a warm (> 25 K) region around the protostar by sublimation of CH_4 from grain mantle. This mechanism is named as *warm carbon-chain chemistry* (WCCC). An essential part of it was indeed confirmed by the chemical models (Hassel *et al.* 2008; Aikawa *et al.* 2008). Soon after the discovery of L1527 as the WCCC source, another protostar in Lupus, IRAS 15398–3359 was identified as the second WCCC source (Sakai *et al.* 2009).

In contrast to the rich existence of carbon-chain molecules, the large COMs are not detected in the WCCC sources, although L1527 harbors the Class 0-I protostar as in the case of the hot corino sources. This fact means that chemical compositions of protostellar cores have significant diversity even among the sources in similar evolutionary stages. The diversity can readily be seen as the different spectral appearance between IRAS 16293–2422 and L1527 (Figure 1). Since both hot corino chemistry and WCCC are triggered by evaporation of grain mantle, chemical compositions of grain mantle have to be different between these two cases. Sakai *et al.* (2009) proposed that a different duration time of the starless-core phase after shielding the interstellar UV radiation will make such a difference of grain mantle compositions. A shorter duration time favors WCCC, while a longer duration time results in hot corino chemistry. This situation is schematically shown in Figure 2. If this is the case, the chemical composition of protoatellar cores tells about the history of their starless-core phase which was memorized in the chemical composition of grain mantle. This would be a novel contribution of chemistry to star formation studies.

However, only a few definitive hot corino sources and a few definitive WCCC sources were recognized by single dish observations. This is partly due to insufficient sensitivity of the telescopes employed. In addition, the existence of intermediate character sources is suggested by survey observations of protostellar sources (Sakai *et al.* 2009; Sakai & Yamamoto 2013). Hence, it is still controversial what is the standard chemical composition for protostellar cores. It is also interesting how the chemical diversity in the protostellar cores is handed over to protoplanetary disks and eventually to planets. In 2011, ALMA started its early science operation, and is addressing these issues.

2. New Results with ALMA

2.1. *Detection of More Hot Corinos*

Thanks to high sensitivity and high angular resolution capabilities of ALMA, new hot corino sources have been detected. Codella *et al.* (2016) detected the CH_3CHO lines toward the low-mass protostellar source HH212 at a subarcsecond resolution with ALMA. This is a Class 0 protostar in Orion ($d = 400$ pc), showing a well collimated outflow (Codella *et al.* 2014; Lee *et al.* 2014; Podio *et al.* 2015; Lee *et al.* 2017). The distribution of CH_3CHO is concentrated around the protostar, and is similar to that of HDO. The line shape of CH_3CHO is also similar to that of HDO, whose line width is as broad as 7 km s^{-1}. Since HDO is thought to sublimated from grain mantle, the similarities between

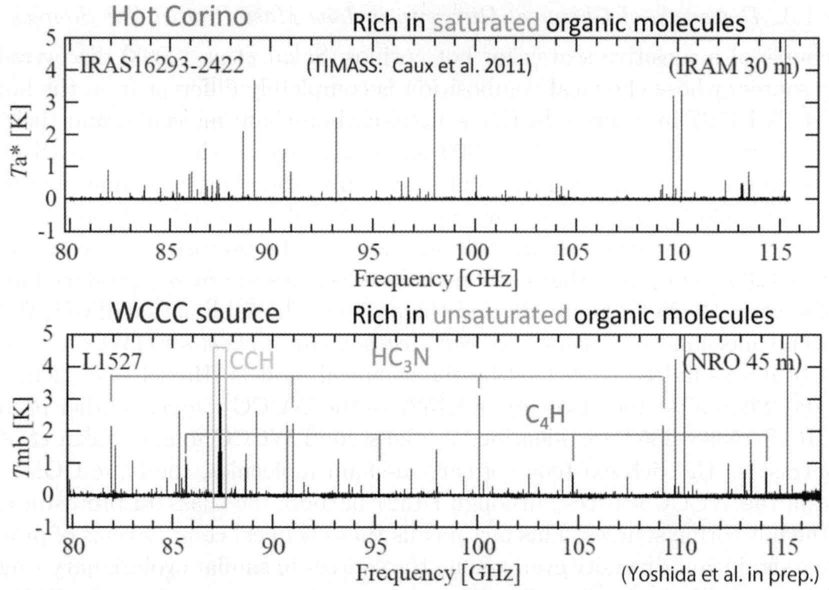

Figure 1. Spectral line surveys toward IRAS 16293–2422 (upper panel) and L1527 (lower panel). The spectrum patterns for the two sources are significantly different. Note that the lines appear to be weaker by a factor of about 2 due to insufficient spectral resolution of the spectrometer in the L1527 survey.

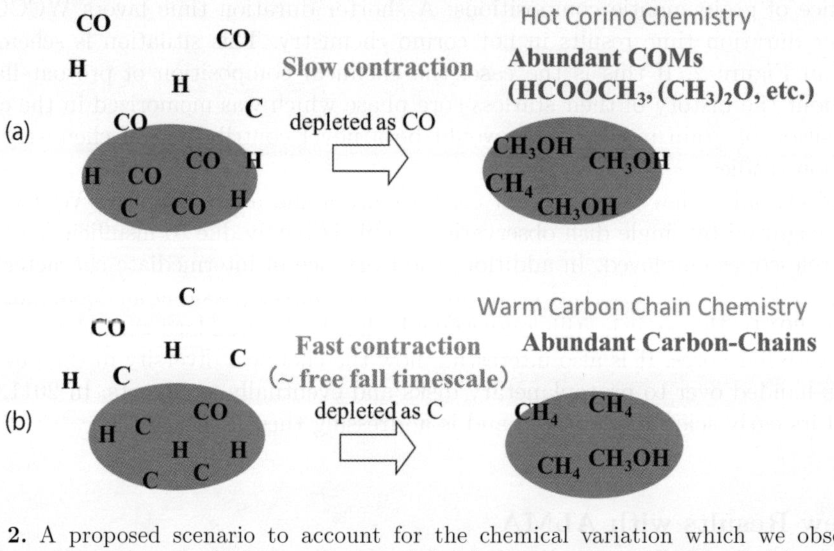

Figure 2. A proposed scenario to account for the chemical variation which we observe in low-mass star-forming cores (Sakai *et al.* 2009; Sakai & Yamamoto 2013). Sketch (a) represents the long starless-core phase which favors hot corino chemistry. In this case, most of the C atoms are converted to the CO molecules in the gas phase, which then deplete onto dust grains. They then form CH_3OH, H_2CO, and more complex organic molecules through hydrogenation reactions and radical reactions on the grain surfaces. Large COMs such as $HCOOCH_3$ may also be produced from CH_3OH and H_2CO in the gas phase after their evaporation. Sketch (b) represents a short starless-core phase in which most of the C atoms deplete directly onto the dust grains, and then form CH_4 through hydrogenation. This favors WCCC, with carbon-chain molecules produced in the gas phase after evaporation of CH_4 from dust grains.

CH_3CHO and HDO mean that CH_3CHO also exists in the hot and dense region near the porotstar. This is the first discovery of a hot corino in Orion. Recently, Lee *et al.* (2017) reported the detection of NH_2CHO in this source with ALMA.

Imai *et al.* (2016) also detected CH_3CHO, $HCOOCH_3$ and NH_2CHO toward the isolated low-mass protostar IRAS 19347+0727 in the Bok globule B335 (d = 100 pc) with ALMA. This source is well known as an ideal test-bed for star formation studies, and physical structures of its parent dense core and the associated outflow have extensively been studied (e.g., Hirano *et al.* 1988; Chandler and Sargent 1993; Evans *et al.* 2015; Yen *et al.* 2015). In contrast, studies on its chemical composition are relatively sparse, although the chemical composition at a few 1000 au scale was explored by observing fundamental molecules (Evans *et al.* 2005). The distribution of COMs is found to be concentrated around the protostar, which is unresolved with the 0.″5 (50 au) beam of ALMA. The outflow of this source is extending almost along the plane of sky, and hence, the envelope/disk structure likely has the edge-on configuration. Nevertheless, a rotation signature is not recognized. Thus, the broad spectral line profile of COMs suggests that the COM emission most likely comes from the infalling gas in the vicinity of the protostar. The fractional abundances relative to H_2 are derived to be 2.4×10^{-9}, 4.3×10^{-10}, and 4.6×10^{-9} for CH_3CHO, NH_2CHO, and $HCOOCH_3$, respectively. They are almost comparable to those found in the other hot corions (e.g., Taquet *et al.* 2015), and hence, B335 can be recognized as a hot corino source.

2.2. *Physical and Chemical Structures of Protostellar Cores: L1527*

Sakai *et al.* (2014a) observed the prototypical WCCC source L1527 with ALMA. This source has an almost edge-on disk/envelope system extending along the north-to-south direction and the extended outflow blowing the east and west directions (Ohashi *et al.* 1997; Hogerheijde *et al.* 1998; Tobin *et al.* 2013). Although carbon-chain molecules were known to be concentrated around the protostar at a 1000 au scale, its detailed distribution and kinematic structure at a smaller scale was unclear. The ALMA observation revealed that the distribution of the carbon-chain related species c-C_3H_2 is elongated along the inner part of the flattened envelope (Figure 3a). The kinematic structure clearly indicates the rotating motion of the envelope, as shown in the position-velocity diagram of Figure 3b. The rotation profile abruptly disappears inside the radius of 100 au from the protostar. In addition, the infalling gas is apparently visible toward the protostar. Hence, the kinematic structure traced by the c-C_3H_2 line cannot be explained by the Keplerian motion, but by an infalling-rotating gas conserving the angular momentum.

Figure 3c shows the simulation result for the infalling-rotating envelope model assuming the ballistic motion (Oya *et al.* 2014; 2015). The model result well reproduces the observed positon-velocity diagrams. The radius inside which the rotation profile disappears is thus found to be the radius of the centrifugal barrier of the infalling-rotating envelope. When the gas motion is approximated by the ballistic motion, the gas cannot fall inward of a certain radius because of the conservation of energy and angular momentum. This perihelion radius is called as the centrifugal barrier. It corresponds to a half of the centrifugal radius, where the centrifugal force balances to the gravity of the central star. The outgoing gas from the centrifugal barrier collides with the infalling gas, and hence, a week accretion shock will occur in front of the centrifugal barrier. As a result, the gas is stagnated there. If the angular momentum of the gas around the centrifugal barrier is extracted by some mechanisms such as disk winds and slow molecular outflows, a part of the gas can further fall toward the protostar to form a rotationally supported (Keplerian) disk inside the centrifugal barrier. The distribution of c-C_3H_2 clearly shows that this molecule is present in the infalling-rotating envelope, and is absent in the disk component. Sakai *et al.* (2014b) also revealed that CCH and CS shows the similar distribution. Carbon-chain related molecules and CS are thought to be frozen out onto dust grain in the mid-plane of the disk component, where the density is as high as 10^8 cm^{-3} and the temperature is as low as 30 K.

Figure 3d shows the position-velocity diagram of the SO line observed with ALMA (Sakai *et al.* 2014a). In contrast to c-C_3H_2, the distribution of SO is concentrated around the centrifugal barrier, as revealed by the bar-shaped feature in the position-velocity diagram. SO does not exist in the infalling-rotating envelope. Thus, a drastic chemical change is occurring around the centrifugal barrier. SO seems to be liberated from grain mantle by the weak accretion shock expected around the centrifugal barrier. It should be noted that a component having higher velocity than the maximum velocity at the centrifugal barrier can marginally be seen inside the centrifugal barrier. This may indicate the existence of SO in the disk component. Although both CS and SO are sulfur-bearing molecules, their distributions are found to be much different at a disk-formation scale in L1527. Note that a similar SO distribution was also reported by Ohashi *et al.* (2014).

On the other hand, the distribution of H_2CO is different from those of the above molecules (Sakai *et al.* 2014b). The position velocity diagram of H_2CO clearly shows the high velocity component associated with the disk component inside the centrifugal barrier in addition to the infalling-rotating envelope component and its centrifugal barrier.

The above results for L1527 shows the chemical differentiation around the low-mass protostars at a 50 au scale. More importantly, the chemical differentiation is related to the physical transition from the infalling-rotating envelope to the disk component. In other words, the disk formation around low-mass protostars can be studied by chemical diagnostics. For such studies, L1527 can be regarded as a *Rosetta Stone*, which allows us to translate the chemical compositions to the physical structures.

2.3. *The Other WCCC Sources*

Indeed, a similar chemical differentiation was found in the evolved WCCC source IRAS 04365+2535 in TMC-1A (d = 140 pc) (Sakai *et al.* 2016). This is a Class I protostar harboring rich carbon-chain molecules. The CS emission clearly traces the infalling-rotating envelope, while the SO emission selectively traces the centrifugal barrier. Although the distributions of CS and SO are asymmetric around the protostar, the essential feature of the chemical differentiation is similar to the L1527 case. With the aid of the ballistic model, the radius of the centrifugal barrier is determined to be 50 au. The gas certainly keeps falling beyond the centrifugal radius (100 au) in this source. The result contrasts the disk radius of 100 au inferred from the analysis of the $C^{18}O$ data (Aso *et al.* 2016). This contradiction would come from the complex physical structure in the transition zone (Section 3), which is ignored in their analysis.

Another WCCC source IRAS 15398–3359 shows a similar trend (Oya *et al.* 2014). In this source, the CCH and H_2CO distributions were studied. The both molecules trace the disk/envelope component as well as the outflow cavity. The integrated intensity distribution of the CCH emission shows a marginal dip toward the porotostar, while that of H_2CO has a clear peak at the protostar. This trend is consistent with that found in L1527, although the velocity gradient was not detected clearly at a resolution of 0.″5, because of the small protostar mass and the small radius of the centrifugal barrier.

The rotating ring traced by the SO emission is also found for L1489 by Yen *et al.* (2014). This source is a Class I protostar in Taurus, and the Keplerian disk structure is reported. The position velocity diagram of SO shows a bar-like feature, indicating the rotating ring with the radius of 300 au. Unfortunately, no carbon-chain molecules and CS were observed for this source, and the infalling-rotating envelope was not characterized.

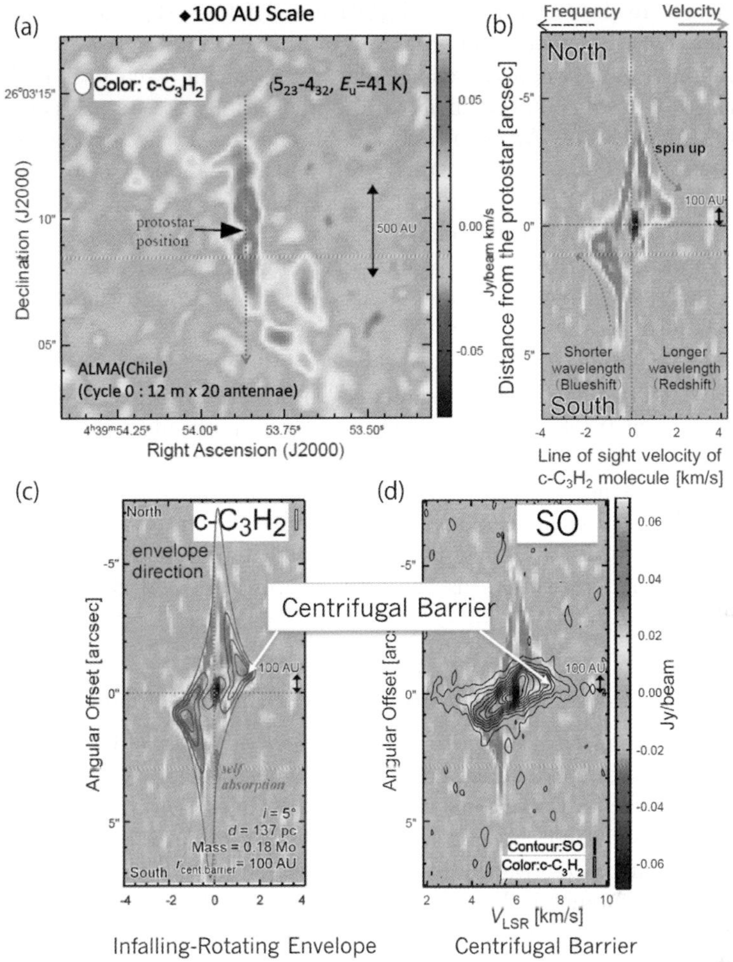

Figure 3. (a) Integrated intensity distribution of the c-C_3H_2 ($5_{2,3} - 4_{3,2}$; 249 GHz) line. (b) Position-velocity diagram of the c-C_3H_2 line along the north-to-south direction shown in panel (a) by the dashed arrow. (c) Position-velocity diagram of the infalling-rotating envelope model (contours) superposed on that of the c-C_3H_2 line (color). Contours are every 20% of the peak intensity. (d) Position-velocity diagram of the SO ($J_N = 7_8 - 6_7$; 341 GHz; contours) line superposed on that of the c-C_3H_2 line (color). Contours are every 4σ (28 mJy beam^{-1}), starting from 2σ. Taken from Sakai *et al.* (2014a; b).

Nevertheless, the SO ring may trace the centrifugal barrier, as in the case of L1527 and TMC-1A.

2.4. *Hot Corino Sources*

We now know the small scale chemical differentiation for the WCCC sources. Then what about hot corino sources? To address this question, Oya *et al.* (2016) studied the physical and chemical structures of IRAS 16293–2422 Source A in detail with ALMA. As mentioned in Section 1.2, IRAS 16293–2422 is a prototypical hot corino source, toward which various molecular-line observations have been carried out (e.g., Cazaux *et al.* 2003; Caux *et al.* 2011; Jørgensen *et al.* 2016). As presented in this symposium (Oya 2017), the kinematic structure of this source is well represented by an infalling rotating envelope and a Keplerian disk component inside it. The OCS line selectively traces the infalling-rotating envelope, while the CH_3OH and $HCOOCH_3$ lines highlight the centrifugal barrier. On the other hand, H_2CS exists in the infalling-rotating envelope and the disk component. Although the chemical species tracing each part is different from those for the WCCC sources, the chemical differentiation at a 50 au scale is evident in IRAS 16293–2422 Source A. An essentially similar chemical structure is also found for the companion Source B (Oya *et al.* 2018). CH_3OH and large COMs seem to be evaporated from grain mantle around the centrifugal barrier due to accretion shock of the infalling gas and/or direct heating by protostellar radiation. A faint high velocity component seen in these molecules implies that they would survive even in the disk component to some extent. Thus, a drastic chemical change is also occurring around the centrifugal barrier in the hot corino sources, as in the WCCC sources.

Very recently, Lee *et al.* (2017) revealed the distributions of HCO^+, CH_3OH, CH_2DOH, CH_3SH, and NH_2CHO in the hot corino source HH212 with ALMA. At a very high angular resolution of 0.″04, the distributions of these molecules including those of the vertical direction to the disk/envelope structure were beautifully resolved. The HCO^+ emission is found to trace the infalling-rotating envelope down to its centrifugal barrier, while CH_3OH and other organic species the atmosphere of the disk within the centrifugal barrier. This result is almost consistent with those found in IRAS 16293–2422 (Oya *et al.* 2016), and further reminds us the complex physical and chemical nature in the disk-forming region.

2.5. *Discovery of Hybrid Chemical Character Sources*

According to the proposed scenario of the chemical diversity (Figure 2), intermediate or hybrid character sources should naturally be expected. Although their presence is suggested by the survey observations with single-dish telescopes (Sakai *et al.* 2009), they were not identified at a high angular resolution. Oya *et al.* (2017) identified L483 as an example of such sources for the first time. L483 harbors the Class 0 protostar, IRAS 18148-0440, whose bolometric luminosity is as high as 13 $L_\odot$. It is located in Aquila Rift at the distance of 200 pc. This source was considered to be a WCCC candidate source, because carbon-chain molecules such as C_4H_2 and l-C_3H_2, which are usually undetected in protostellar cores, were detected (Hirota *et al.* 2009; Sakai *et al.* 2009). With ALMA, Oya *et al.* (2017) indeed confirmed the association of CCH extended over a few 100 au scale around the protostar, confirming the WCCC nature of this source. Its distribution has a hole toward the protostar position. Since the signal-to-noise ratio of the CCH emission is not enough to study its kinematic structure, we instead studied the kinematic structure of the CS line. Figure 4a shows the position velocity diagram of CS along the envelope/disk direction, while Figures 4b and c represent the simulation results of the infalling-rotating envelope and Keplerian models, respectively. Although CS traces the infalling-rotating envelope, it also traces the disk component inside the centrifugal barrier. In addition, SO traces the disk component inside the centrifugal barrier (Figure 4d). These features are in contrast to the L1527 and TMC-1A cases. This may be due to the high luminosity of this source, which prevents CS and SO from freezing out onto dust grains in the mid-plane of the disk component. Moreover, they detected the spatially-unresolved emission of $HCOOCH_3$, HNCO, and NH_2CHO toward the protostar. The detection of COMs clearly indicates the hot corino activity in this source. Hence, L483 can be regarded as the first source which definitively possesses both the WCCC and hot corino characters. Since WCCC is triggered by evaporation of CH_4, it is caused in the extended warm part ($T > 25$ K) of the protostellar core. On the other hand, evaporation of COMs, which characterizes the hot corino activity, only occurs in a hot part ($T > 100$

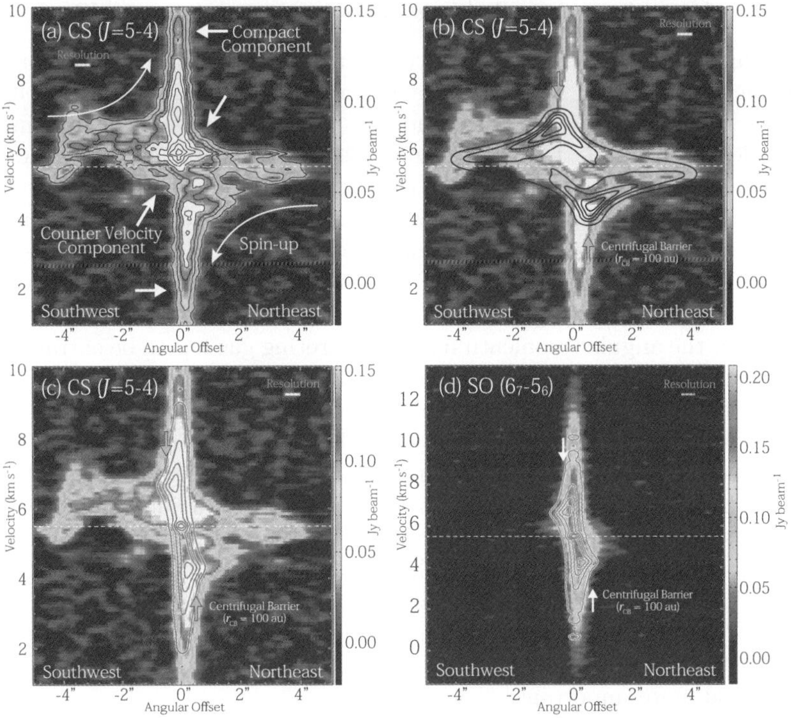

Figure 4. (a) Position-velocity diagram of the CS ($J = 5 - 4$) line along the envelope/disk direction. Contours are every 5σ (38 mJy beam^{-1}). (b) Position-velocity diagram of the infalling-rotating envelope model (contours) superposed on that of the CS line (color). Contours are every 20% from 5% of each peak intensity. (c) Position-velocity diagram of the Keplerian model (contours) superposed on that of the CS line (color). Contours are every 20% from 5% of each peak intensity. (d) Position-velocity diagram of the SO ($J_N = 6_7 - 5_6$; color) line and the Keplerian model (contours). Contours are every 30% from 3% of the peak intensity.

K) near the protostar. Such a difference of the distributions of CCH and COMs is indeed seen in L483.

A similar hybrid character has recently been recognized for B335. As mentioned above, this source harbors an unresolved hot corino. It also shows the relatively bright CCH emission concentrated at a 1000 au scale around the protostar. Since the CCH emission is almost absent or very faint in the vicinity of the other hot corino sources (IRAS 16293–2422 Source A, NGC 1333 IRAS 4A), B335 seems to have the WCCC character in addition to the hot corino character, as in the case of L483.

Such a hybrid character is predicted by the chemical model calculation by Aikawa *et al.* (2008). This model adopts the realistic physical model of a star-forming cloud, and simulates the temporal variation of the chemical structure. Their model indeed shows the enhancement of carbon-chain related species in the gas phase around the evaporation region of CH_4 ($T > 25$ K). This shows the WCCC character. On the other hand, CH_3OH and COMs are evaporated in the inner region at higher temperature ($T > 100$ K). Although the model does not consider the complex transition zone from the infalling-rotating envelope to the disk component, the result qualitatively explains the observational results for L483 and B335. The hybrid character source may be a more common occurrence in chemical compositions of protostellar cores. Further ALMA observations to characterize chemical properties of various protostellar cores are thus important.

3. What we have found and what we do not understand?

The infalling-rotating envelope and its centrifugal barrier are clearly demonstrated by molecular line observations. This fact means that the dynamic pressure of the infalling gas is higher than the static pressure and the magnetic pressure in infalling-rotating envelopes at least for L1527, TMC-1A, IRAS 16293–2422 Source A and B, and L483. For this reason, the gas keeps infalling through the centrifugal radius down to the radius of the centrifugal barrier. Then the gas is stagnated in front of the centrifugal barrier, and the accreting gas causes a weak shock there. The broadening of the envelope thickness around the centrifugal barrier is indeed found by the recent high resolution observation (Sakai *et al.* 2017). Inside the centrifugal barrier, the disk structure is expected to grow up. For the disk growth, the angular momentum of the accreting gas has to be extracted by some mechanisms such as disk wind or slow molecular outflows (e.g. Alves *et al.* 2017; Zhao *et al.* 2016; Codella *et al.* 2014). Thus, the transition from the infalling-rotating envelope and the disk component is occurring between the centrifugal radius to the centrifugal barrier, and seems to cause a complex physical structure and kinematics. Observations of the popular gas tracers such as the CO isotopologue lines at a limited angular resolution often tend to smear out the critical structural changes in the transition zone, because such tracers observe everything. On the other hand, chemical changes associated with the transition zone clearly highlight individual parts in the transition zone, just as chemical markers. Since the detailed understandings of the transition zone including its vertical structures are the central issue for the disk formation studies, such chemical approaches will be more and more important.

It should be noted, however, the infalling-rotating envelope model may not always be valid. In a very low mass protostar or a very young protostar, for instance, the central mass is so small that the dynamic pressure caused by the central gravity may not dominate the static pressure and the magnetic pressure. We need further considerations for these cases. In addition, the centrifugal barrier may appear at a much larger radius than a 50 au scale, if the angular momentum of the infalling gas is larger.

The transition zone is not only for physics but also for chemistry. Chemical compositions of the gas drastically change in the transition zone. So far, it has been thought that gas and dust in molecular clouds are smoothly delivered to protoplanetary disks. In fact, the chemical composition of molecular clouds is assumed as the initial condition for the chemical model simulation of protoplanetary disks (e.g., Walsh *et al.* 2014). However, the discovery of the centrifugal barrier and the associated chemical change clearly indicates that such a smooth transition picture is too simplified. In other words, the chemical composition at the centrifugal barrier both in the gas phase and on the surface provide us with the observational constraint to the initial condition of the model. Thus, we need to consider the chemical processes in the transition zone more seriously.

However, it is not well understood what chemical processes are actually occurring in the transition zone. Looking at the tracer molecules for the hot corino and WCCC sources, we recognize that the sulfur-bearing molecules reveals characteristic features. For instance, CS and SO behaves exclusively in the WCCC sources. The fact that OCS traces the infalling-rotating envelope in hot corino sources has not been expected in chemical models. Importance of the sulfur chemistry is often pointed out in various situations (e.g. Wakelam *et al.* 2011). It is also recognized in the analysis of the *Rosetta* data of 67P/Churyumov-Grasimenco (Calmonte *et al.* 2016). However, formation pathways of sulfur bearing species are quite uncertain both in the gas phase and on the grain surface. Thus, a thorough understanding of the sulfur chemistry is urgent.

Another remaining issue is how the isotope fractionation found in cold molecular clouds is transferred to the disk component. The isotope anomaly of molecules is widely recognized in cold interstellar medium, and is expected to be an important tracer which bridges interstellar chemistry and planetary chemistry. In particular, deuterium fractionation of water is extensively studied (e.g., Coutens *et al.* 2012). As for the other molecules, Murillo *et al.* (2015) studied the distribution of DCO^+ in the low-mass protostellar source VLA 1623. According to their result, DCO^+ is distributed in the outer envelope at a few 100 au scale, which shows a rotation signature. On the other hand, it is almost absent in the vicinity of the protostar. Yoshida *et al.* (2017) studied the D_2CO/H_2CO ratio in L1527. The D_2CO distribution peak is found to be offset from the protostar position, while the H_2CO is centrally peaked. Hence, the deuterium fractionation ratio seems to be lower in the disk-forming region. In contrast, Lee *et al.* (2017) very recently reported the detection of CH_2DOH in HH212, as mentioned above. The CH_2DOH distribution is almost the same as the CH_3OH distribution, and hence, substantial deuterium fractionation is seen even around the centrifugal barrier in this source. Hence, it is still controversial how deuterium fractionation changes in the disk formation process. Further observations with ALMA will answer this question.

4. Concluding Remarks

The ALMA observations briefly overviewed above reveal very complex physical and chemical processes occurring in disk-forming regions, most of which have not been anticipated before the ALMA era. More importantly, chemistry is further confirmed to be a powerful diagnostic tool of physical processes around the protostar at a 50 au scale. Such chemical approaches will be a major trend of star-formation studies in the next decade. In addition, we hope that the observational results will give rise to new research activities in molecular physics and chemistry.

5. Acknowledgements

This study is supported by KAKENHI (21224002,25400223, and 25108005) and the Japan France integrated action program by JSPS and MAEE.

References

Aikawa, Y., Ohashi, N., Inutsuka, S. *et al.* 2001, *ApJ*, 552, 639.
Aikawa Y., Wakelam, V., Garrod, R. T., & Herbst, E. 2008, *ApJ*, 674, 993.
Alves, F. O., Girart, J. M., Caselli, P. *et al.* 2017, *A&A*, 603, L3.
Aso, Y., Ohashi, N., Saigo, K. *et al.* 2015, *ApJ*, 812, 27.
Beichman, C. A., Myers, P. C., & Emerson, J. P. 1986, *ApJ*, 307, 337.
Benson, P. J. & Myers, P. C. 1989, *ApJS*, 71, 89.
Blake, G. A., Sutton, E. C., Masson, C. R., & Phillips, T. G. 1987, *ApJ*, 315, 621.
Blake, G. A., van Dishoeck, E. F., Jansen, D. J., *et al.* 1994, *ApJ*, 428, 680.
Bottinelli, S., Ceccarelli, C., Neri, R. *et al.* 2004a, *ApJ*, 617, L69.
Bottinelli, S., Ceccarelli, C., Lefloch, B. *et al.* 2004b, *ApJ*, 615, 354.
Bottinelli, S., Ceccarelli, C., Williams, J. P., & Lefloch, B., 2007, *A&A*, 463, 601.
Balucani, N., Ceccarelli, C., & Taquet, V. 2015, *MNRAS*, 449, L16.
Calmonte, U., Altwegg, K., Balsiger, H. *et al.* 2016, *MNRAS*, 462, S253.
Caselli, P. & Ceccarelli, C. 2012, *Astron. Ap. Rev.* 20, 56.
Caux, E., Kahane, C., Castets, A., *et al.* 2011, *A&A*, 532, A23.
Cazaux, S., Tielens, A. G. G. M., Ceccarelli, C. *et al.* 2003, *ApJ*, 593, L51.
Ceccarelli, C., Castets, A., Loinard, L. *et al.* 1998, *A&A*, 338, L43.

Ceccarelli, C., Loinard, L., Castets, A. *et al.* 2000, *A&A*, 357, L9.
Chandler, C. J. & Sargent, A. I. 1993, *ApJL*, 414, L29.
Codella, C., Cabrit, S., Gueth, F. *et al.* 2014, *A&A*, 568, L5.
Codella, C., Ceccarelli, C., Cabrit, S. *et al.* 2016, *A&A*, 586, L3.
Coutens, A., Vastel, C., Caux, E. *et al.* 2012, *A&A*, 539, A132.
Cummins, S. E., Linke, R. A., & Thaddeus, P. 1986, *ApJS*, 60, 819.
Evans, N. J., II, Lee, J.-E., Rawlings, J. M. C., & Choi, M. 2005, *ApJ*, 626, 919
Evans, N. J., II, Di Francesco, J., Lee, J., *et al.* 2015, *ApJ*, 814, 22
Garrod, R. T. & Herbst, E. 2006, *A&A*, 457, 927.
Hassel, G. E., Herbst, E., & Garrod, R. T. 2008, *ApJ*, 681, 1385.
Herbst, E. & van Dishoeck, E. F. 2009, *ARAAp*, 47, 427.
Hirano, N., Kameya, O., Nakayama, M., & Takakubo, K. 1988, *ApJL*, 327, L69.
Hirota, T., Ohishi, M., & Yamamoto, S. 2009, *ApJ*, 699, 585.
Hogerheijde, M. R., van Dishoeck, E. F., Blake, G. A. *et al.* 1998, *ApJ*, 502,315.
Imai, M., Sakai, N., Oya, Y. *et al.* 2016, *ApJL*, 830, L37.
Jørgensen, J. K., Bourke, T. L., Myers, P. C. *et al.* 2005, *ApJ*, 632, 973.
Jørgensen, J. K., van del Wiel, M. H. D., Coutens, A. *et al.* 2016, *A&A*, 595, A117.
Kuan, Y.-J., Huang, H.-C., Charnley, S. B. *et al.* 2004, *ApJ*, 616, L27.
Lee, C.-F., Hirano, N., Zhang, Q., *et al.* 2014, *ApJ*, 786, 114.
Lee, C.-F., Li, Z.-Y., Ho, P. T. P.. *et al.* 2017, *ApJ*, 843, 27.
Masson, C. R. & Mundy, L. G. 1988, *ApJ*, 324, 538.
Murillo, N. M., Bruderer, S., van Dishoeck, E. F. *et al.* 2014, *A&A*, 579, A114.
Myers, P. C. and Benson, P. J. 2003, *ApJ*, 266, 309.
Öberg, K., van der Marel, N., Kristensen, L. E., & van Dishoeck, E. F. 2011, *ApJ*, 740, 14.
Ohashi, N., Hayashi, M., Ho, P. T. P., & Momose, M. 1997, *ApJ*, 475, 211.
Ohashi, N., Saigo, K., Aso, Y., *et al.* 2014, *ApJ*, 796, 131.
Oya, Y., Sakai, N., Sakai, T. *et al.* 2014, *ApJ*, 795, 152.
Oya, Y., Sakai, N., Lefloch, B. *et al.* 2015, *ApJ*, 812, 59.
Oya, Y., Sakai, N., López-Sepulcre, A. *et al.* 2016, *ApJ*, 824, 88.
Oya, Y. 2017, *Proceedings of IAU Symposium* No. 332.
Oya, Y., Moriwaki, K., Onishi, S. *et al.* 2018, *ApJ*, 854, 96.
Oya, Y., Sakai, N., Watanabe, Y. *et al.* 2017, *ApJ*, 837, 174.
Podio, L., Codella, C., Gueth, F., *et al.* 2015, *A&A*, 581, 85.
Sakai, N., Sakai, T., & Yamamoto, S. 2006, *PASJ*, 58, L15.
Sakai, N., Sakai, T., Hitora, T., & Yamamoto, S. 2008, *ApJ*, 672, 371.
Sakai, N., Sakai, T., Hirota, T. *et al.* 2009, *ApJ*, 697, 769.
Sakai, N., Sakai, T., Hirota, T., & Yamamoto, S. 2010, *ApJ*, 722, 1633
Sakai, N. & Yamamoto, S. 2013, *ChRv*, 113, 8981.
Sakai, N., Sakai, T., Hirota, T., *et al.* 2014a, *Natur*, 507, 78.
Sakai, N., Oya, Y., Sakai, T., *et al.* 2014b, *ApJL*, 791, L38.
Sakai, N., Oya, Y., López-Sepulcre, A., *et al.* 2016, *ApJL*, 820, L34.
Sakai, N., Oya, Y., Higuchi, A. E. *et al.* 2017, *MNRAS*, 467, L76.
Schoier, F. L., Jørgensen, J. K., van Dishoeck, E. F., & Blake, G. A. 2002, *A&A*, 390, 1001.
Suzuki, H., Yamamoto, S., Ohishi, M. *et al.* 1992, *ApJ*, 392, 551.
Taquet, V., López-Sepulcre, A., Ceccarelli, C., *et al.* 2015, *ApJ*, 804, 81.
Tobin, J. J., Hartman, L., Chiang, H.-F. *et al.* 2013, *ApJ*, 771, 48.
Turner, B. E. 1989, *ApJS*, 70, 539.
van Dishoeck, E. F., Blake, G. A., Jansen, D. J., & Groesbeck, T. D. 1995, *ApJ*, 447, 760.
Wakelam, V., Hersant, F., & Herpin, F. 2011, *A&A*, 529, A112.
Walsh, C., Millar, T. J., Nomura, H. *et al.* 2014, *A&A*, 563, A33.
Yen, H. W., Takakuwa, S., Ohashi, N., *et al.* 2014, *ApJ*, 793, 1.
Yen, H. W., Takakuwa, S., Koch, P. M., *et al.* 2015, *ApJ*, 812, 129.
Yoshida, K., Sakai, N., Yamamoto, S. *et al.* 2017, in preparation.
Zhao, B., Caselli, P., Li, Z.-Y. *et al.* 2016, *MNRAS*, 460, 2050.

Astrochemistry VII – Through the Cosmos from Galaxies to Planets
Proceedings IAU Symposium No. 332, 2017
M. Cunningham, T. Millar & Y. Aikawa, eds.

doi:10.1017/S1743921317007414

On the origin of O_2 and other volatile species in comets

Vianney Taquet[1,2], Kenji Furuya[1,3], Catherine Walsh[1,4] and Ewine F. van Dishoeck[1,5]

[1]Leiden Observatory, Leiden University, PO Box 9513, 2300 RA, Leiden, The Netherlands
[2]INAF-Osservatorio Astrofisico di Arcetri, Largo E. Fermi 5, I-50125 Firenze, Italy
[3]Center for Computational Sciences, University of Tsukuba, 1-1-1 Tennoudai, 305-8577, Tsukuba, Japan
[4]School of Physics and Astronomy, University of Leeds, Leeds LS2 9JT, UK
[5]Max-Planck-Institut fur Extraterrestrische Physik, Giessenbachstrasse, 85741, Garching, Germany

Abstract. Molecular oxygen, O_2, was recently detected in comet 67P by the ROSINA instrument on board the Rosetta spacecraft with a surprisingly high abundance of 4% relative to H_2O, making O_2 the fourth most abundant in comet 67P. Other volatile species with similar volatility, such as molecular nitrogen N_2, were also detected by Rosetta, but with much lower abundances and much weaker correlations with water. Here, we investigate the chemical and physical origin of O_2 and other volatile species using the new constraints provided by Rosetta. We follow the chemical evolution during star formation with state-of-the-art astrochemical models applied to dynamical physical models by considering three origins: i) in dark clouds, ii) during forming protostellar disks, and iii) during luminosity outbursts in disks. The models presented here favour a dark cloud (or "primordial") grain surface chemistry origin for volatile species in comets, albeit for dark clouds which are slightly warmer and denser than those usually considered as solar system progenitors.

Keywords. astrochemistry, comets: individual: 67P/C-G, ISM: abundances, ISM: molecules, protoplanetary discs , stars: formation

1. Introduction

The *Rosetta* spacecraft analysed the Jupiter-family comet 67P/Churyumov-Gerasimenko (hereinafter comet 67P/C-G) in 2014 and 2015. The ROSINA instrument on board the *Rosetta* orbiter (Rosetta Orbiter Spectrometer for Ion and Neutral Analysis (Balsiger *et al.* 2007) detected a "zoo" of molecules in the coma of 67P/C-G from simple mono-atomic or di-atomic species to complex pre-biotic molecules, such as glycine, the simplest amino-acid (Le Roy *et al.* 2015, Altwegg *et al.* 2016).

One of the most surprising results provided by the ROSINA instrument is the *in-situ* detection of molecular oxygen, O_2, in the coma of comet 67P/C-G, resulting in the first detection of O_2 in a comet (Bieler *et al.* 2015). O_2 is strongly correlated with H_2O and is present at an average level of $3.80 \pm 0.85\%$ relative to H_2O, making it the fourth most abundant molecule in the comet, following H_2O, CO_2, and CO. The authors argue that O_2 does not originate from gas-phase chemistry in the coma but from direct sublimation from or within the comet surface. Moreover, the strong correlation with H_2O suggests that the O_2 is trapped within the bulk H_2O ice matrix of the comet and therefore that O_2 was present within the ice mantle on dust grains in the presolar nebula prior to comet formation since the surface of comet 67P/C-G revealed today is likely pristine. A reanalysis of data from the Neutral Mass Spectrometer on board the *Giotto* probe which did a fly-by of comet 1P/Halley in 1986, confirmed the presence of O_2 at a level similar to that seen in 67P/C-G (Rubin *et al.* 2015b). This suggests that O_2 is not only an abundant molecule in comets, but is also common to both Jupiter-family comets, such as 67P/C-G, and Oort Cloud comets, such as 1P/Halley, which have different dynamical behaviours and histories. These results raise the question whether O_2 was abundant in icy dust mantles entering the protoplanetary disk of the young Sun, or whether the conditions in the comet-forming zone of the early solar system were favourable for O_2 formation and survival.

O_2 is a diatomic homonuclear molecule; hence it does not possess electric dipole-allowed rotational transitions. Therefore, gas-phase O_2 has been particularly elusive in interstellar clouds. Recent high sensitivity observations with the *Herschel Space Observatory* allowed a deep search for O_2 towards sources considered as true solar system progenitors: low-mass protostars. A deep upper limit was determined towards the well-studied protostar, NGC 1333-IRAS 4A, ($O_2/H_2 \leqslant 6\times10^{-9}$ or $O_2/H_2O \leqslant 0.012\%$ with a H_2O abundance of $\sim 5\times10^{-5}$, Yildiz *et al.* 2013). This picture is consistent with laboratory experiments that have shown that O_2 ice is efficiently hydrogenated at low temperatures and converted into H_2O and H_2O_2 ices (Ioppolo *et al.* 2008, Miyauchi *et al.* 2008). This makes the close association of O_2 with H_2O in 67P/C-G an even stronger enigma.

Despite O_2 being a particularly elusive molecule in interstellar and circumstellar environments, there apparently do exist conditions which are favourable for the formation of O_2 and related species at abundance ratios similar to that observed in ices in comet 67P/C-G. Indeed, *Herschel* did reveal the presence of gas-phase O_2 in two sources: an active star forming region in Orion (Goldsmith *et al.* 2011) and in the dense core ρ Oph A located in the more quiescent ρ Oph molecular cloud complex, which stands out from other low-mass star-forming regions by exhibiting emission from relatively warm molecular gas (Larsson *et al.* 2007, Liseau *et al.* 2012). Subsequent observations of ρ Oph A have also determined the presence of related gas-phase species, HO_2 and H_2O_2, at abundance levels in reasonable agreement with those seen in 67P/C-G with ROSINA ($\sim 2\times10^{-3}$ that of O_2, see Bergman *et al.* 2011b, Parise *et al.* 2012). The chemically related species, O_3 (ozone), was not detected in the comet coma with a very low upper limit, $< 2.5\times10^{-5}$ with respect to O_2.

Other key di-atomic molecules of similar volatility, CO and N_2, have also been detected in 67P/C-G by *Rosetta* but with a much weaker correlation with H_2O (Rubin *et al.* 2015a, Bieler *et al.* 2015). Although CO shows a high abundance of 10 30% relative to H_2O, in good agreement with previous observations towards other comets, N_2 shows a much lower abundance of 0.57 ± 0.07 % relative to CO. The different correlations and abundances w.r.t. the H_2O clearly suggest a different chemical history for O_2, CO, and N_2.

Here we explore and discuss several different origins to explain the strong constraints provided by *Rosetta* on O_2 and other volatile species in comet 67P/C-G: i) in dark clouds ("primordial" origin), ii) during the journey from the protostellar envelope into the disk, iii) during luminosity outbursts within the protoplanetary disk.

2. Astrochemical models

The gas-grain astrochemical models by Taquet *et al.* (2014) and Furuya *et al.* (2015) have been used in this work to study the formation and survival of O_2 and other volatile species from dark clouds to the Solar System. These models couple the gas phase and ice chemistries with the approach developed by Hasegawa & Herbst (1993) to follow the multi-layer formation of interstellar ices and to determine the gas-ice balance. Several sets of differential equations, one for gas-phase species, one for surface ice-mantle species, and one (or several) for bulk ice-mantle species, are considered to follow the time-evolution of abundances. Following Vasyunin & Herbst (2013), the chemically-active surface is limited to the top four monolayers. The original three-phase model considered in the Taquet model assumes that the inert bulk ice mantle has a uniform molecular composition. In order to accurately follow the ice evolution in warm conditions, the Furuya model considers a depth-dependent molecular composition, through the division of the inert bulk ice mantle into five distinct phases (for details, see Furuya *et al.* 2016 and references therein).

The gas-phase chemical network used by the Taquet model is based on the 2013 version of the KIDA chemical database (Wakelam *et al.* 2012). It has been further updated to include warm gas-phase chemistry involving water and and ion-neutral reactions involving ozone. The network also includes the surface chemistry of all dominant ice components, as well as those important for water (e.g., O_2, O_3, and H_2O_2). Several new surface reactions were added involving O_3 and reactive species such as N, O, OH, NH_2, and CH_3, following the NIST gas-phase chemical database. The gas-ice chemical network of Garrod & Herbst (2006), based on the OSU 2006 network, is used in the Furuya model. The gas phase and surface networks in the Furuya model are more suited to the high density and warm temperatures conditions found in protostellar envelopes. It has therefore been supplemented with high-temperature gas-phase reactions from Harada *et al.* (2010) and includes the formation of many complex organic molecules. It is consequently more expansive than the network used in the Taquet model.

Elemental abundances of species used in the two models correspond to the set EA1 from Wakelam & Herbst (2008). Standard input parameters assumed for the two astrochemical models are: a cosmic ray ionisation rate ζ of 1×10^{-17} s^{-1}, a flux of secondary UV photons of 10^4 phot. cm^{-2} s^{-1}, a dust-to-gas mass ratio of 1%, a grain diameter of 0.2 μm, a volumic mass of grains of 3 g cm^{-3}, a grain surface density of 10^{15} cm^{-2}, a diffusion-to-binding energy ratio of 0.5, four chemically active monolayers, and a sticking coefficient of species heavier than H and H_2 of 1.

3. Interstellar chemistry of molecular oxygen

Two main processes have been invoked for the formation of molecular oxygen in the interstellar medium: i) gas-phase formation via neutral-neutral chemistry, and ii) formation via association reactions on/within icy mantles of dust grains. Gaseous O_2 is thought to form primarily via the barrierless neutral-neutral reaction between O and OH in cold and warm gas. Due to its importance, this reaction has been well studied both experimentally and theoretically. The formation of O_2 in cold dark clouds is initiated by the high initial abundance assumed for atomic oxygen, inducing an efficient ion-neutral chemistry that also forms OH. In warm environments ($T \gtrsim 100$ K), e.g., the inner regions of protostellar envelopes or the inner, warm layers of protoplanetary disks, OH and O are mostly produced through warm neutral-neutral chemistry driven by the photodissociation of water sublimated from interstellar ices. Solid O_2 in dark clouds is involved in the surface chemistry reaction network leading to the formation of water ice (Tielens & Hagen 1982, Miyauchi *et al.* 2008, Ioppolo *et al.* 2008). O_2 is formed through atomic O recombination on ices and efficiently reacts with either atomic O or atomic H to form O_3 or HO_2, respectively, eventually leading to the formation of water. The hydrogenation of O_3 also leads to the formation of O_2, in addition to dominating the destruction of O_3.

Radiolysis, i.e. the bombardment of (ionising) energetic particles depositing energy into the ice, and/or photolysis, i.e. the irradiation of ultraviolet photons breaking bonds, can trigger chemistry within the mantle of cold interstellar ices. We have investigated the impact of the UV photolysis induced by secondary UV-photons on the bulk ice chemistry and the formation and survival of O_2. We find that O_2 cannot be efficiently produced in the bulk through ice photolysis as the photodissocation of the main ice components not only produces O atoms, that recombine together to form O_2, but also H atoms that react with O_2 to reform water. Laboratory experiments show that O_2 can be efficiently formed through radiolysis of ices without overproducing H_2O_2 only if the radiolysis occurs as water is condensing onto a surface (see Teolis *et al.* 2006). However, in dark clouds water ice is mostly formed *in-situ* at the surface of interstellar grains through surface reactions involving hydrogen and oxygen atoms. This happens prior to the formation of the presolar nebula, i.e. the cloud out of which our solar system was formed, and it is possible that the comet-forming zone of the Sun's protoplanetary disk inherited much of its water ice from the interstellar phase (Visser *et al.* 2009, Cleeves *et al.* 2014).

4. Origin of cometary O_2

4.1. *Dark cloud origin?*

Impact of physical and chemical parameters. We first investigated whether the O_2 observed in 67P/C-G has a dark cloud origin. For this purpose, we used the Taquet astrochemical model presented in section 2. We carried out a first parameter study, in which several surface and chemical parameters are varied, in order to reproduce the low abundances of the chemically related species O_3, HO_2, and H_2O_2 with respect to O_2 seen in comet 67P/C-G. The low abundance of O_3 and HO_2 relative to O_2 ($\leqslant 2 \times 10^{-3}$) can be explained when a small activation barrier of $\sim$ 300 K is introduced for the reactions O + O_2 and H + O_2, in agreement with the Monte-Carlo modelling of laboratory experiments by Lamberts *et al.* (2013). However, the abundance of H_2O_2 is still overproduced by one order of magnitude, suggesting that other chemical processes might be at work.

A second parameter-space study was then conducted to determine the range of physical conditions over which O_2 ice and gas (and those for chemically-related species, O_3, HO_2, and H_2O_2) reach abundances (relative to water ice) similar to that seen in 67P/C-G. We ran a model grid in which four or five values for the total density of H nuclei, n_H, the gas and dust temperature, T (assumed to be equal), the cosmic ray ionisation rate, ζ, and the visual extinction, A_V are considered, following the methodology described in Taquet *et al.* (2012), resulting in 500 models in total. We explored the distribution of abundances of solid O_2, and the chemically related species, O_3, HO_2, and H_2O_2, relative to water ice, when the time reaches the free-fall time, t_{FF}. The results show that the formation and survival of solid O_2, and other reactive species, in interstellar ices, is strongly dependent upon the assumed physical conditions with abundance distributions ranging over several orders of magnitude. High O_2 abundances ($\geqslant$ 4% relative to water ice) are obtained only for the models with high densities ($n_H \gtrsim 10^5$ cm^{-3}). Higher gas densities result in a lower gas-phase H/O ratio, thereby increasing the rate of the association reaction between O atoms to form O_2 ice, and correspondingly decreasing the rate of the competing hydrogenation reactions, O + H and O_2 + H, which destroy O_2 ice once formed. An intermediate temperature of 20 K is also favoured because it enhances the mobility of oxygen atoms on the grain surfaces whilst at the same time allowing efficient sublimation of atomic H. This additionally enhances the rate of oxygen recombination forming O_2, with respect to the competing hydrogenation reactions. Moreover, because the density of gas-phase H atoms increases linearly with the cosmic-ray ionisation rate, ζ, a low value of ζ also tends to favour the survival of O_2 ice. On the other hand, the visual extinction does not have a strong impact on the abundance of solid O_2.

The ρ Oph A case. The parameter study presented above therefore suggests that the physical conditions of ρ Oph A, presenting a high density ($n_H \sim 10^6$ cm^{-3}), and a relatively warm temperature for a starless core ($T_{kin} = 24-30$ K and $T_{dust} \sim 20$ K; Bergman *et al.* 2011a) are consistent with those which facilitate the formation and survival of O_2 ice. This confirms that these properties offer optimal conditions for an efficient production of solid O_2 since ρ Oph A is the only interstellar source so far where gas-phase O_2, HO_2, and H_2O_2 have been detected. Figure 1 shows the chemical composition of the ice and gas obtained for the model using the physical conditions of ρ Oph A and that best reproduce the observations in comet 67P/C-G. The fractional composition in each ice monolayer is plotted as function of monolayer number, i.e. the ice thickness that grows with time. O_2 ice is mostly present in the innermost layers of the ice mantle and decreases in relative abundance towards the ice surface, reflecting the initial low ratio of H/O in the gas phase obtained at high densities, but tends to be well mixed with H_2O ice. CO_2 is also highly abundant because the higher temperature (21 K) enhances the mobility of heavier species, such as O or CO.

O_2, O_3, HO_2, and H_2O_2 are mostly, and potentially only, produced via surface chemistry; hence their gas-phase abundances depend on their formation efficiency in interstellar ices and on the probability of desorption upon formation through chemical desorption (thought to be the dominant desorption mechanism for these species in dark cloud conditions). The chemical desorption probabilities are highly uncertain and mainly depend on the ice substrate and the considered reaction. Figure 1 shows the temporal evolution of the gas phase abundances of O_2, O_3, HO_2, and H_2O_2 when the theoretical values by Minissale *et al.* (2016) relative to a bare grain substrate, and varying between 0 and 70%, are used. This model is almost able to simultaneously reproduce the gaseous abundances of O_2, HO_2, and H_2O_2 derived in ρ Oph A since the predicted O_2 abundance and the HO_2 and H_2O_2 abundances reach the observations at similar timescales (1.5×10^4 vs 2.2×10^4 yr). Using lower chemical desorption probabilities relevant to water ice substrates could improve the comparison with the observations.

Molecular nitrogen vs carbon monoxide and molecular oxygen. In contrast to O_2, it is seen that CO and N_2 are mostly formed in the outer part of the ices and would, therefore, undergo a more efficient sublimation, either thermally or through photo-evaporation, during their transport from dark clouds to forming disks in the subsequent protostellar collapse phase. The chemical heterogeneity predicted in ices can therefore naturally explain the high correlation between O_2 and H_2O signals together with the weak correlation between CO, N_2, and H_2O signals measured in comet 67P/C-G. However, it cannot explain the low N_2/CO abundance ratio of ~ 0.6 % observed in comet 67P/C-G since our dark cloud model predicts a N_2/CO of 50 %.

As shown by dynamical models of protoplanetary disk formation, volatile species that evaporated during their journey from dark clouds to upper disk layers can subsequently freeze-out onto ices again once they reach the colder disk midplane (see Drozdovskaya *et al.* 2014). N_2 is known to be slightly more volatile than CO with a binding energy lower by ~ 150 K for H_2O ice and by ~ 60 K for pure ices (Bisschop *et al.* 2006, Fayolle *et al.* 2016). We investigated the impact of these slightly different binding energies with a toy model on the recondensation of CO and N_2 during a cooling from 50 to 20 K that could occur during the transport of material from the upper disk layers to the disk midplane, assuming a constant density typical of a disk ($n_H = 10^8$ cm^{-3}) and that all CO and N_2 are initially in the gas phase. It is found that CO does freeze out more efficiently than N_2, inducing a low N_2/CO abundance ratio in ices down to 0.2% at 28 K before a re-increase to the initial abundance ratio at lower temperatures. A cooling of ices down to 26 - 28 K

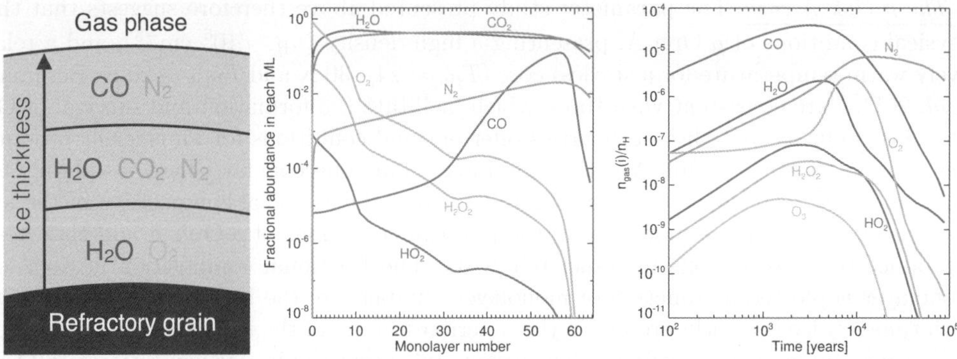

Figure 1. Left panel: Cartoon representation of the ice structure predicted in this work. Middle panel: Fractional composition of each ice monolayer as function of the monolayer number or ice thickness. Right panel: gas phase abundances as a function of time (right panel) of O_2 and its chemically related species predicted by the model using the ρ Oph A physical conditions.

near the cometary zone therefore induces a CO/H_2O abundance higher than 10% and a N_2/CO abundance lower than 0.6%, even if a high initial N_2/CO abundance is produced in the dark cloud phase. These numbers are close to those observed for 67P/C-G.

4.2. *Disk formation origin?*

Here, we discuss the role of chemistry during protostellar collapse and protoplanetary disk formation on the observed abundance of O_2 in 67P/C-G. To follow the chemical evolution from prestellar cores to forming disks, fluid parcels from the envelope to the disk are traced with the physical model initially developed by Visser *et al.* (2009). The Furuya astrochemical model is used to follow the gas-ice chemical evolution calculated along each individual trajectory. The physical model used here is an axisymmetric semi-analytical two-dimensional model that describes the temporal evolution of the density and velocity fields following inside-out collapse and the formation of an accretion disk described by the α-viscosity prescription (for details, see Harsono *et al.* 2013 and references therein). The model follows the physical evolution until the end of the main accretion phase when the gas accretion from the envelope onto the star-disk system is almost complete. A molecular cloud formation model is run to determine the composition of the gas and ice in the parent molecular cloud (Furuya *et al.* 2015). The chemistry is then evolved for an additional 3×10^5 yr under prestellar core conditions to compute the abundances at the onset of collapse. At the onset of collapse, models have a negligible O_2 ice abundance.

Figure 2 shows the spatial distributions of fluid parcels at the final time of the simulation in models in the so-called spread-dominated case (initial core rotation rate $\Omega = 10^{-13}$ s^{-1}). It is found that (i) some gaseous O_2 can form (up to $\sim 10^{-6}$) depending on the trajectory paths (left panels), and (ii) O_2 ice trapped within H_2O ice does not efficiently form en route into the disk (middle panels). Given that most elemental oxygen is in ices (H_2O and CO) at the onset of collapse, gaseous O_2 forms through photodissociation/desorption of H_2O ice by stellar UV photons in the warm (>20 K) protostellar envelope, followed by subsequent gas-phase reactions. The majority of parcels in each disk have a low final O_2/H_2O ice ratio, $\ll 10^{-2}$. However, the upper layers of the disk do have several parcels with a O_2/H_2O ice ratio higher than 10^{-2}. Analysis of the ice composition shows that the O_2 ice is associated with CO_2 ice rather than with H_2O. Upon water ice photodissociation, the warm temperatures encountered through the protostellar envelope mean that CO_2 ice (re)formation is more favorable than that for H_2O ice.

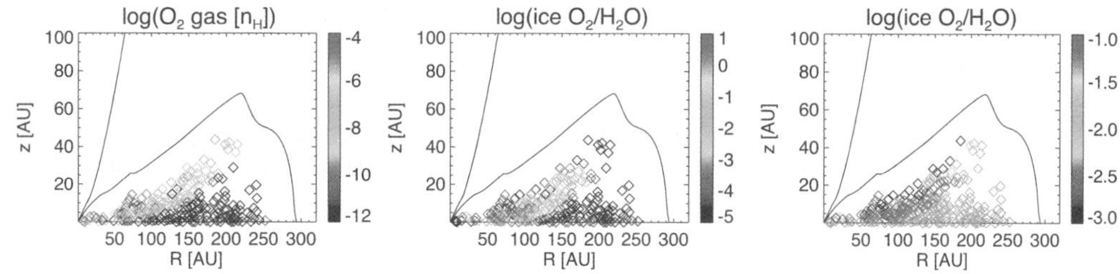

Figure 2. Spatial distributions of fluid parcels at the final time of the spreak-dominated disk formation simulation ($\Omega = 10^{-13}$ s^{-1}). The left panel shows the gaseous O_2 abundance with respect to hydrogen nuclei, the middle panel shows the abundance ratio between O_2 ice and H_2O ice. the right panel also shows the abundance ratio between O_2 ice and H_2O ice, but for the model where the initial ratio is artificially set to 5%. The solid lines represent the outflow cavity wall and the disk surface.

We also considered a case where the simulations begin with a similar fraction of O_2 ice to what is observed in comet 67P/C-G (5% relative to water ice). The O_2/H_2O ratio throughout both disks is largely preserved. Hence, O_2 which has a prestellar or molecular cloud origin, is able to survive the chemical processing en route into the comet-forming regions of protoplanetary disks. However, only little O_2 is formed during disk formation.

4.3. *Luminosity outburst origin?*

Observational and theoretical studies suggest that the luminosity evolution of low-mass stars is highly variable, with frequent and strong eruptive bursts, followed by long periods of relative quiescence (e.g. Hartmann & Kenyon 1985, Vorobyov *et al.* 2015). Such luminosity outbursts could have a strong impact on the morphology and the chemical composition of ices near the protoplanetary disk midplane. If the luminosity outburst is sufficiently strong, warm gas-phase formation of molecular oxygen could be triggered by the evaporation of water ice, if the peak temperature during the outburst is higher than $\sim$ 100 K. We explored the impact of a series of outbursts events in disks on the formation and recondensation of O_2, increasing the temperature from 20 to 100 K, every 10^4 yr for a total timescale of 10^5 yr. The Taquet astrochemical model described previously was used. Initial ice abundances are the median values derived by Öberg *et al.* (2011) from interstellar ice observations towards low-mass protostars. Thus it is assumed that the ice mantles are initially poor in O_2. The pre-outburst and post-outburst temperature is set to 20 K. Protoplanetary disk models suggest that the corresponding midplane density at this point is $\sim 10^8$ cm^{-3} (e.g. Walsh *et al.* 2014).

We varied several parameters that are thought to impact the gas phase formation of O_2 during the outburst and the efficiency of recondensiation during the cooling, such as the grain size, the cosmic ray ionisation rate, the cooling timescale after the outburst or the peak temperature. It is found that the maximum amount of O_2 formed during luminosity outbursts and then trapped within the ice mantle during the cooling does depend on the explored parameters but never exceeds $\sim$ 0.1% w.r.t. H_2O ice. This suggest that luminosity outbursts are too short to significantly produce O_2 with quantities similar to those observed in comet 67P/C-G. Assuming an initial O_2 abundance of 5% relative to water ice results in efficient trapping of O_2 within the water-ice mantle due to the fast cooling after the outburst. However, in that case also other volatile species, such as CO and N_2, become trapped, which is in contradiction with observations towards 67P/C-G.

5. Conclusions

The models presented here favour the scenario that molecular oxygen in 67P/C-G has a primordial origin (i.e., formed in the parent molecular cloud) and has survived transport through the protostellar envelope and into the comet-forming regions of protoplanetary disks. The "primordial" origin of O_2 is in good agreement with the conclusions of Mousis *et al.* (2016). Mousis *et al.* (2016) invoked radiolysis to efficiently convert water ice to O_2. However, we find here that the entrapment and strong association with water ice combined with low abundance of species like H_2O_2, HO_2, or O_3 can be explained by an efficient O_2 formation at the surface of interstellar ices through oxygen atom recombination in relatively warmer ($\sim$ 20 K) and denser ($n_{\rm H} \gtrsim 10^5$ cm^{-3}) conditions than usually expected in dark clouds. The weak correlation of CO and N_2 with water seen in 67P/C-G is explained by a later formation and freeze-out of these species in dark clouds with respect to O_2 and water. This picture would therefore be consistent with the physical and chemical properties of our Solar System, such as the presence of short-lived radio isotopes in meteorites or the orbits of Solar System planets, which suggests that our Solar System was born in a dense cluster of stars (see Adams 2010).

References

Adams, F. C. 2010, *ARAA*, 48, 47
Altwegg, K., Balsiger, H., Bar-Nun, A. *et al.* 2015, *Science*, 347, 27
Balsiger, H., Altwegg, K., Boschler, P., *et al.* 2007, *Sp. Sc. Rev.*, 128, 745
Bergman, P., Parise, B., Liseau, R., *et al.* 2011a, *A&A*, 527, A39
Bergman, P., Parise, B., Liseau, R., *et al.* 2011b, *A&A*, 531, L8
Bisschop, S. E., Fraser, H. J., Öberg, K. I. *et al.* 2006, *A&A*, 449, 1297
Cleeves, L. I., Bergin, E. A., Alexander, C. M. O. *et al.* 2014, *Science*, 345, 1590
Drozdovskaya, M. N., Walsh, C., Visser, R. *et al.* 2014, *MNRAS*, 445, 913
Fayolle, E. C., Balfe, J., Loomis, R. *et al.* 2016, *ApJ*, 816, L28
Furuya, K., Aikawa, Y, Hincelin, U. *et al.* 2015, *A&A*, 584, A124
Furuya, K., Drozdovskaya, M., Visser, R. *et al.* 2016, submitted to *A&A*
Garrod, R. T. & Herbst, E. 2006, *A&A*, 457, 927
Goldsmith, P. F., Liseau, R., Bell, T. A., *et al.* 2011, *ApJ*, 737, 96
Harada, N., Herbst, R., & Wakelam, V. 2010, *ApJ*, 721, 1570
Harsono D., Visser R., Bruderer S. *et al.* 2013, *A&A*, 555, A45
Hasegawa, T. I. and Herbst, E. 1993, *MNRAS*, 263, 589-606
Ioppolo, S., Cuppen, H. M., Romanzin, C. *et al.* 2008, *ApJ*, 686, 1474
Hartmann, L. & Kenyon, S. J. 1985, *ApJ*, 299, 462
Lamberts, T., Cuppen, H. M., & Ioppolo, S. and Linnartz, H. 2013, *PCCP*, 15, 8287
Larsson, B., Liseau, L., Pagani, L., *et al.* 2007, *A&A*, 466, 999
Liseau, R., Goldsmith, P. F., Larsson, B., *et al.* 2012, *A&A*, 541, A73
Minissale, M., Dulieu, F., Cazaux, S., & Hocuk, S. 2016, *A&A*, 585, A24
Mousis, O., Ronnet, T., Brugger, B. *et al.* 2016, accepted in ApJL
Miyauchi, N., Hidaka, H., Chigai, T. *et al.* 2008, Chem. Phys. Lett., 456, 27
Öberg, K. I., Boogert, A. C. A., Pontopiddan, K. M., *et al.* 2011, *ApJ*, 740, 109
Parise, B., Bergman, P., & Du, F. 2012, *A&A*, 541, L11
Rubin, M., Altwegg, K., Balsiger, H., *et al.* 2015a, *Science*, 348, 232
Rubin, M., Altwegg, K., van Dishoeck, E. F., & Schwehm, G. 2015b, *ApJl*, 815, L11
Taquet, V., Ceccarelli, C., & Kahane, C. 2012, *A&A*, 538, A42
Taquet, V., Charnley, S. B., & Sipilä, O. 2014, *ApJ*, 791, 1
Teolis, B. D., Loeffler, M. J., Raut, U., & Famá, M. and Baragiola, R. A. 2005, *ApJl*, 644, L141
Tielens, A. G. G. M., & Hagen, W., *A&A*, 114, 245
Vasyunin, A. I. and Herbst, E. 2013, *ApJ*, 762, 86

Visser, R., van Dishoeck, E. F., Doty, S. D., & Dullemond, C. P. 2009, *A&A*, 495, 881
Vorobyov, E. I. & Basu, S. 2015, *ApJ*, 805, 115
Wakelam, V. & Herbst, E. 2008 *ApJ*, 680, 371-383
Walsh, C., Millar, T. J., Nomura, H., *et al.* 2014, *A&A*, 563, 33
Yildiz, U. A., Acharyya, K., Goldsmith, P. F., *et al.* 2013, *A&A*, 558, A58

Discussion

Astrochemistry VII – Through the Cosmos from Galaxies to Planets
Proceedings IAU Symposium No. 332, 2017
M. Cunningham, T. Millar & Y. Aikawa, eds.

doi:10.1017/S1743921317009917

The chemical connection between 67P/C-G and IRAS 16293-2422

Maria Nikolayevna Drozdovskaya[1], Ewine F. van Dishoeck[2,3], Martin Rubin[4], Jes Kristian Jørgensen[5] and Kathrin Altwegg[4,1,6]

[1]Center for Space and Habitability (CSH), Universität Bern
Sidlerstrasse 5, CH-3012 Bern, Switzerland
email: maria.drozdovskaya@csh.unibe.ch

[2]Leiden Observatory, Leiden University
PO Box 9513, NL-2300 RA Leiden, The Netherlands
email: ewine@strw.leidenuniv.nl

[3]Max-Planck Institut für Extraterrestrische Physik (MPE)
Giessenbachstrasse 1, 85748 Garching, Germany

[4]Physikalisches Institut, Universität Bern
Sidlerstrasse 5, CH-3012 Bern, Switzerland
email: martin.rubin@space.unibe.ch

[5]Centre for Star and Planet Formation
Niels Bohr Institute & Natural History Museum of Denmark, University of Copenhagen
Øster Voldgade 57, DK-1350 Copenhagen K., Denmark
email: jeskj@nbi.ku.dk

[6]email: kathrin.altwegg@space.unibe.ch

Abstract. The chemical evolution of a star- and planet-forming system begins in the prestellar phase and proceeds across the subsequent evolutionary phases. The chemical trail from cores to protoplanetary disks to planetary embryos can be studied by comparing distant young protostars and comets in our Solar System. One particularly chemically rich system that is thought to be analogous to our own is the low-mass IRAS 16293-2422. ALMA-PILS observations have made the study of chemistry on the disk scales (< 100 AU) of this system possible. Under the assumption that comets are pristine tracers of the outer parts of the innate protosolar disk, it is possible to compare the composition of our infant Solar System to that of IRAS 16293-2422. The *Rosetta* mission has yielded a wealth of unique in situ measurements on comet 67P/C-G, making it the best probe to date. Herein, the initial comparisons in terms of the chemical composition and isotopic ratios are summarized. Much work is still to be carried out in the future as the analysis of both of these data sets is still ongoing.

Keywords. astrochemistry, comets: individual (67P/C-G), stars: individual (IRAS 16293-2422), (stars:) planetary systems: protoplanetary disks

1. The Assembly Line

Star and planet formation are two inseparable, intertwined processes that initiate upon the gravitational collapse of a prestellar core. The infalling gas and dust of the core are primarily partitioned between the young protostar and the protoplanetary disk. The leftover core materials remain in the form of an envelope around the young star-disk system during the embedded phase of star formation (e.g., Shu 1977). Some gas and dust are also expelled from the system via the action of outflows.

The deepest layer of a protoplanetary disk, the midplane, is the site of most-efficient build up of larger refractory bodies. It is still unclear exactly when grain growth from micron to millimeter and centimeter sizes occurs; however, meter- and kilometer-sized bodies will undoubtedly only be able to form in the midplanes, where the densities are high-enough for such masses to accumulate. This makes midplanes the birth places of comets and planetary embryos (e.g., Williams & Cieza 2011). In turn, it is collapsing core materials that feed the disk midplanes. What remains to be understood is the degree of processing or pristinity of the refractory and volatile components.

One potential means of tackling this question is by considering the chemical connection embedded in volatiles across the pre- and proto-stellar phases of star formation. It is possible to study the gaseous and solid volatiles in prestellar cores by means of observations in the submillimeter (e.g., ALMA) and infrared (e.g., *Spitzer*) frequency ranges. However, in the sub-surface layers of protoplanetary disks the densities are high and temperatures are low. This implies that most volatiles are in the solid phase and obscured from view by the thick dust disk. This makes direct midplane observations almost impossible (except less abundant isotopologes of highly volatile species such $^{13}C^{18}O$; Zhang *et al.* 2017) and imposes the need for indirect tracers, such as comets.

Comets are thought to form alongside planetary embryos in the disk midplane. It is possible that they remain relatively small, of km-size, due to being formed in the outer disk, where the densities are lower and thus there being less material available for growth of refractories. It is thought that Kuiper Belt comets remain in the distant parts of the disk until their orbit is perturbed by a giant planet's gravitational field, which can only occur in a mature planetary system, such as our own modern day Solar System. Oort Cloud comets have been suggested to have formed closer-in, e.g., near Jupiter, and then slung out to large distances during the early formation phases of our Solar System (e.g., A'Hearn *et al.* 2011, Mumma & Charnley 2011).

As cometary data are only available for our Solar System (the field of exo-comets still remains a future perspective) and direct observations of the pre- and proto-stellar phases are unavoidably only towards other forming stars, some assumptions need to be made. First of all, it is assumed that other low-mass protostellar systems are analogous to our innate protosolar nebula (PSN). Secondly, comets are seen as pristine tracers of the PSN, i.e., excluding any processing that may have occurred during their time in a mature planetary configuration.

2. *Rosetta* and 67P/C-G

Comets have primarily been studied via remote sensing and ground-based observations, and also dedicated *Stardust* and *Deep Impact* missions (as reviewed in A'Hearn *et al.* 2011). This picture changed in 2014 when the *Rosetta* mission approached, went into orbit and carried out a landing operation on the comet 67P/Churyumov-Gerasimenko (67P/C-G, hereafter). Between August 6th, 2014 and September 30th, 2016 the *Rosetta* orbiter remained in orbit around the comet and surveyed its surface and coma with an array of instruments. The lander *Philae* was in operation for about 3 days upon its triple landing on November 12th, 2014. The *Rosetta* mission has rendered the most unique and diligent dataset on cometary science yet with its in situ measurements.

Jupiter-family comets (JFCs) are short-period comets with low inclinations and aphelia near Jupiter's orbit, and are thought to originate from the Kuiper Belt. 67P/C-G is a JFC of around $\sim 4 \times 3 \times 2$ km in size with a period of 6.44 yr, an aphelion of 5.68 AU and a perihelion of 1.24 AU with the latest occurring on August 13th, 2015. Due to the inclination of its orbit relative to the orbital plane of the Solar System, it experiences seasonal changes. The Northern hemisphere experiences a long warm summer of ~ 5.5 yr, while the Southern hemisphere undergoes an intense short summer between its inbound equinox on May 5th, 2015 and its outbound equinox on April 9th, 2016. The *Rosetta* orbiter monitored its target during all three of these major events.

3. ALMA-PILS and IRAS 16293-2422 B

Low-mass protostars have been studied extensively with single-dish facilities, interferometers and space missions. The latest powerful successor is the Atacama Large Millimeter/submillimeter Array (ALMA). ALMA has been used to carry out the Protostellar Interferometric Line Survey (PILS; project-id: 2013.1.00278.S, PI: Jes K. Jørgensen) towards the low-mass binary system IRAS 16293-2422 in Band 7 (329 – 363 GHz). The survey is unique not only due to its wide frequency coverage, but also its high angular resolution of $\sim$ 60 AU, spectral resolution of $\sim$ 0.2 km s^{-1} and sensitivity of $4-5$ mJy beam^{-1} km s^{-1}. Of particular interest is source B, which has narrow lines with full width at half maximum (FWHM) of $\sim$ 1 km s^{-1} and its face-on disk orientation. This reduces line blending and facilitates line identification, thus revealing the chemistry of its disk-region (Jørgensen *et al.* 2016).

4. The chemical connection: composition

Jointly, the *Rosetta* data on 67P/C-G and the ALMA observations of IRAS 16293-2422 B, are some of the best data sets on a PSN midplane tracer and a PSN analogue. Hence, the two can be used to explore the chemical connection spanning the earliest stages of star formation and the protoplanetary building blocks, much better than was done in the past (e.g., Bockelée-Morvan *et al.* 2000). One comparative aspect is the chemical composition of the two targets.

The dominant volatiles of 67P/C-G have been shown to form two groups, those following H_2O and those following CO_2. The first group includes O_2 and NH_3, while the second group contains CO, H_2S, CH_4 and HCN. The partitioning is based on the density profiles pre- and post-perihelion (from January to August 2016), and the rate of decrease post-perihelion (Gasc *et al.* 2017). Water is the dominant volatile and the peak production rate was seen three weeks after perihelion (Hansen *et al.* 2016). All of these species have been detected towards IRAS 16293-2422 and are also well-known constituents of interstellar ice based on infrared observations against protostellar and background sources (Boogert *et al.* 2015).

The detection of O_2 at a level of $\sim$ 4% relative to H_2O remains a surprising result (Bieler *et al.* 2015), as it is a species rarely detected under interstellar conditions. The search for this molecule towards IRAS 16293-2422 is still ongoing with dedicated ALMA Cycle 4 data under analysis at the moment (Taquet *et al.*, 2018, in prep.). Astrochemical models are suggesting that such a high quantity of O_2 can only be formed through grain surface chemistry under prestellar core conditions, albeit at a lightly elevated $\sim$ 20 K temperature than normal. Chemistry within the protoplanetary disk is not efficient enough for the production of such a high abundance (Taquet *et al.* 2016). Other mechanisms such as radiolysis may also contribute to the prestellar O_2, although this mechanism overproduces HO_2, H_2O_2 and O_3 (Mousis *et al.* 2016).

A classical biomarker species, CH_3Cl, has been detected in the coma of 67P/C-G with the ROSINA instrument. The molecule has also been observed towards IRAS 16293-2422 A and B. The two detections, unfortunately, imply that CH_3Cl cannot be considered a biomarker. However, the Cl-bearing molecule remains the dominant organohalogen in the Earth's atmosphere and is seen at similar proportions towards 67P/C-G and

IRAS 16293-2422 B, $\sim 0.007 - 6 \times 10^{-4}$ and $\sim 7 \times 10^{-5}$ relative to CH_3OH, respectively (Fayolle *et al.* 2017).

Comet 67P/C-G has also been shown to carry glycine, atomic phosphorus and an entire zoo of complex organic molecules. It has been seen that it is abundant in S-bearing molecules, but poor in N-bearing molecules (Le Roy *et al.* 2015, Altwegg *et al.* 2016, Calmonte *et al.* 2016, Altwegg *et al.* 2017b, Wampfler *et al.*, 2018, in prep.). IRAS 16293-2422 is also rich in a variety of O-, N- and S-bearing complex organic species (e.g., Jørgensen *et al.* 2016); however, its complexity is more comparable to that of a petting farm rather than a zoo. It cannot be entirely excluded that the lack of detections of more complex molecules is not due to sensitivity issues as a result of complicated partition functions, which lead to weaker lines, and the larger distances between the observer and the target.

ROSINA measurements have estimated that $\sim 57\%$ of the volatile sulfur is incorporated into H_2S, $\sim 27\%$ into atomic S and then smaller portions into SO_2, SO, OCS, H_2CS and other molecules. Sulfur-chain species, including S_2, S_3, S_4, have been detected on 67P/C-G and are likely fragments of the semi-refractory S_8. S-bearing organics, CH_3SH and C_2H_6S have also been measured in the coma (Calmonte *et al.* 2016). All these cometary molecules containing a single S atom have been also detected towards IRAS 16293-2422, except for atomic S and C_2H_6S. Thus, in terms of diversity, there is an agreement. However, the latest analysis are showing that IRAS 16293-2422 B is much more abundant in OCS; and that H_2S is by-far not as much of a dominant S-carrier as is seen in the comet (Drozdovskaya *et al.*, 2017, under rev.).

5. The chemical connection: isotopic ratios

Another clue of the chemical trail across evolutionary stages lies in the isotopic ratios. The D/H ratio on 67P/C-G has now been measured in singly and doubly deuterated water, and in H_2S. The values are $(1.05 \pm 0.14) \times 10^{-3}$ for HDO/H_2O, $(1.8 \pm 0.9) \times 10^{-2}$ for D_2O/HDO and $(1.2 \pm 0.3) \times 10^{-3}$ for HDS/H_2S, respectively (Altwegg *et al.* 2017a). This implies that the ratio is molecule-dependent and may have been set at different evolutionary times corresponding to the time of formation of each molecule. The D/H as measured in terms of the singly deuterated water is higher than that of other JFCs, but closer to the range of protostellar envelope values. The best estimate in terms of singly deuterated water towards IRAS 16293-2422 B based on interferometric ALMA and SMA observations is $(9.2 \pm 2.6) \times 10^{-4}$ for HDO/H_2O(Persson *et al.* 2013). It is important to compare with observations at high spatial resolution, as water is present on large and small scales in such star-forming regions and the D/H ratio varies therein. Of relevance for the cometary comparisons, are the small disk-scales, which can only be reached with interferometers, where directly thermally desorbed ices are seen.

The higher D/H measured in the doubly deuterated water has been explained in astrochemical models via the process of deuteration occurring at a later stage of formation in the upper layers of icy grain mantles (Furuya *et al.* 2016, Furuya *et al.* 2017). This remains to be tested from the observational perspective towards IRAS 16293-2422. The D/H as measured in terms of H_2S and complex organics, for example, remains the topic of future work as the analysis of the ROSINA and PILS data is still ongoing. Further work is also required to compare isotopic ratios in volatiles of other species, including, C, O, N, S and Cl.

6. So far, what does this tell us?

With some of the best cometary and protostellar data at hand, the *Rosetta* measurements of 67P/C-G and ALMA-PILS observations of IRAS 16293-2422 B, it seems that the PSN, at least in part, contains prestellar materials. Qualitative comparisons in terms of the chemical composition of both targets are indicating that other low-mass systems, such as IRAS 16293-2422, may be analogous to our young Solar System. Quantitative comparisons and the detailed studies of isotopic ratios remain the subject of future research (Drozdovskaya *et al.*, 2018, in prep.).

References

A'Hearn, M. F., Belton, M. J. S., Delamere, W. A., Feaga, L. M., Hampton, D., Kissel, J., Klaasen, K. P., McFadden, L. A., Meech, K. J., Melosh, H. J., Schultz, P. H., Sunshine, J. M., Thomas, P. C., Veverka, J., Wellnitz, D. D., Yeomans, D. K., Besse, S., Bodewits, D., Bowling, T. J., Carcich, B. T., Collins, S. M., Farnham, T. L., Groussin, O., Hermalyn, B., Kelley, M. S., Kelley, M. S., Li, J.-Y., Lindler, D. J., Lisse, C. M., McLaughlin, S. A., Merlin, F., Protopapa, S., Richardson, J. E., & Williams, J. L. 2011, *Science*, 332, 1396

Altwegg, K., Balsiger, H., Bar-Nun, A., Berthelier, J.-J., Bieler, A., Bochsler, P., Briois, C., Calmonte, U., Combi, M. R., Cottin, H., De Keyser, J., Dhooghe, F., Fiethe, B., Fuselier, S. A., Gasc, S., Gombosi, T. I., Hansen, K. C., Haessig, M., Jäckel, A., Kopp, E., Korth, A., Le Roy, L., Mall, U., Marty, B., Mousis, O., Owen, T., Rème, H., Rubin, M., Sémon, T., Tzou, C.-Y., Waite, J. H., & Wurz, P. 2016, *SciA*, 2, 285

Altwegg, K., Balsiger, H., Berthelier, J. J., Bieler, A., Calmonte, U., De Keyser, J., Fiethe, B., Fuselier, S. A., Gasc, S., Gombosi, T. I., Owen, T., Le Roy, L., Rubin, M., Sémon, T., & Tzou, C.-Y. 2017, *RSPTA*, 375, 2097

Altwegg, K., Balsiger, H., Berthelier, J. J., Bieler, A., Calmonte, U., Fuselier, S. A., Goesmann, F., Gasc, S., Gombosi, T. I., Le Roy, L., de Keyser, J., Morse, A., Rubin, M., Schuhmann, M., Taylor, M. G. G. T., Tzou, C.-Y., & Wright, I. 2017, *MNRAS*, 469, 130

Bieler, A., Altwegg, K., Balsiger, H., Bar-Nun, A., Berthelier, J.-J., Bochsler, P., Briois, C., Calmonte, U., Combi, M., de Keyser, J., van Dishoeck, E. F., Fiethe, B., Fuselier, S. A., Gasc, S., Gombosi, T. I., Hansen, K. C., Hässig, M., Jäckel, A., Kopp, E., Korth, A., Le Roy, L., Mall, U., Maggiolo, R., Marty, B., Mousis, O., Owen, T., Rème, H., Rubin, M., Sémon, T., Tzou, C.-Y., Waite, J. H., Walsh, C., & Wurz, P. 2015, *Nature*, 526, 678

Bockelée-Morvan, D., Lis, D. C., Wink, J. E., Despois, D., Crovisier, J., Bachiller, R., Benford, D. J., Biver, N., Colom, P., Davies, J. K., Gérard, E., Germain, B., Houde, M., Mehringer, D., Moreno, R., Paubert, G., Phillips, T. G., & Rauer, H. 2000, *A&A*, 353, 1101

Boogert, A. C. A., Gerakines, P. A., & Whittet, D. C. B. 2015, *ARA&A*, 53, 541

Calmonte, U., Altwegg, K., Balsiger, H., Berthelier, J. J., Bieler, A., Cessateur, G., Dhooghe, F., van Dishoeck, E. F., Fiethe, B., Fuselier, S. A., Gasc, S., Gombosi, T. I., Hässig, M., Le Roy, L., Rubin, M., Sémon, T., Tzou, C.-Y., & Wampfler, S. F. 2016, *MNRAS*, 462, 253

Fayolle, E. C., Öberg, K. I., Jørgensen, J. K., Altwegg, K., Calcutt, H., Müller, H. S. P., Rubin, M., van der Wiel, M. H. D., Bjerkeli, P., Bourke, T. L., Coutens, A., van Dishoeck, E. F., Drozdovskaya, M. N., Garrod, R. T., Ligterink, N. F. W., Persson, M. V., & Wampfler, S. F., Rosina Team 2017, *NatAs*, 1, 703

Furuya, K., van Dishoeck, E. F., & Aikawa, Y. 2016, *A&A*, 586, A127

Furuya, K., Drozdovskaya, M. N., Visser, R., van Dishoeck, E. F., Walsh, C., Harsono, D., Hincelin, U., & Taquet, V. 2017, *A&A*, 599, 40

Gasc, S., Altwegg, K., Balsiger, H., Berthelier, J.-J.; Bieler, A., Calmonte, U., Fiethe, B., Fuselier, S., Galli, A., Gombosi, T., Hoang, M., De Keyser, J., Korth, A., Le Roy, L., Mall, U., Rème, H., Rubin, M., Sémon, T., Tzou, C.-Y., Waite, J. H., & Wurz, P. 2017, *MNRAS*, 469, 108

Hansen, K. C., Altwegg, K., Berthelier, J.-J., Bieler, A., Biver, N., Bockelée-Morvan, D., Calmonte, U., Capaccioni, F., Combi, M. R., de Keyser, J., Fiethe, B., Fougere, N., Fuselier, S. A., Gasc, S., Gombosi, T. I., Huang, Z., Le Roy, L., Lee, S., Nilsson, H., Rubin, M., Shou, Y., Snodgrass, C., Tenishev, V., Toth, G., Tzou, C.-Y., & Wedlund, C. S., *ROSINA* Team 2016, *MNRAS*, 462, 491

Jørgensen, J. K., van der Wiel, M. H. D., Coutens, A., Lykke, J. M., Müller, H. S. P., van Dishoeck, E. F., Calcutt, H., Bjerkeli, P., Bourke, T. L., Drozdovskaya, M. N., Favre, C., Fayolle, E. C., Garrod, R. T., Jacobsen, S. K., Öberg, K. I., Persson, M. V., & Wampfler, S. F. 2016, *A&A*, 595, A117

Le Roy, L., Altwegg, K., Balsiger, H., Berthelier, J.-J., Bieler, A., Briois, C., Calmonte, U., Combi, M. R., De Keyser, J., Dhooghe, F., Fiethe, B., Fuselier, S. A., Gasc, S., Gombosi, T. I., Hässig, M., Jäckel, A., Rubin, M., & Tzou, C.-Y. 2015, *A&A*, 583, A1

Mousis, O., Ronnet, T., Brugger, B., Ozgurel, O., Pauzat, F., Ellinger, Y., Maggiolo, R., Wurz, P., Vernazza, P., Lunine, J. I., Luspay-Kuti, A., Mandt, K. E., Altwegg, K., Bieler, A., Markovits, A., & Rubin, M. 2016, *APJL*, 823, 41

Mumma, M. J. & Charnley, S. B., 2011, *ARA&A*, 49, 471

Persson, M. V., Jørgensen, J. K., & van Dishoeck, E. F. 2013, *A&A*, 549, 3

Shu, F. H. 1977, *ApJ*, 214, 488

Taquet, V., Furuya, K., Walsh, C., & van Dishoeck, E. F. 2016, *MNRAS*, 462, 99

Williams, J. P. & Cieza, L. A. 2011, *ARA&A*, 49, 67

Zhang, K., Bergin, E. A., Blake, G. A., Cleeves, L. I., & Schwarz, K. R. 2017, *NatAs*, 1, 130

Astrochemistry VII – Through the Cosmos from Galaxies to Planets
Proceedings IAU Symposium No. 332, 2017
M. Cunningham, T. Millar & Y. Aikawa, eds.

doi:10.1017/S1743921317006779

Chemistry of the oldest white dwarf planetary systems

Mark A. Hollands[1], **Boris T. Gänsicke**[1], **Detlev Koester**[2], **Vadim Alekseev**[3] **and Emma L. Herbert**[1]

[1]Department of Physics, University of Warwick, Coventry CV4 7AL, UK
email: M.Hollands@warwick.ac.uk

[2]Institut für Theoretische Physik and Astrophysik, University of Kiel, 24098 Kiel, Germany

[3]St. Petersburg State University, 7/9 Universitetskaya Nab., 199034 St. Petersburg, Russia

Abstract. Almost all stars in the Milky Way, including the Sun, will end their lives as white dwarfs. Their relatively peaceful transition off of the main sequence implies that most of their planetary systems will survive engulfment during the deaths of their host stars. These remnant planetary systems remain detectable for many Gyr through the occasional metal-contamination of the white dwarf photospheres by tidally disrupted planetesimals. Spectral analysis of these "metal-polluted" white dwarfs therefore provides a direct method for measuring the chemical compositions of extrasolar material. Here we present our sample of 230 cool white dwarfs with metal-rich photospheres, explore the diverse range of compositions of the accreted matter, and discuss two extreme systems which have respectively accreted planetesimals consistent with crust-like and core-like planetary material.

Keywords. stars: planetary systems, white dwarfs, abundances

1. Introduction

1.1. *A review of white dwarf properties*

More than 95 % of all main-sequence stars will eventually become white dwarfs after exhausting their supplies of nuclear fuel. Immediately after leaving the AGB, these newly formed white dwarfs have effective temperatures ($T_{\rm eff}$) of several 100 000 K, but with no nuclear processes providing heat, these stellar embers radiate away their thermal energy over the course of billions of years. Cooling models provide a direct way to measure the age of a white dwarf from its mass and $T_{\rm eff}$.

Most of today's white dwarfs are the descendants of A- and F-type main sequence stars ($M_* \simeq 1.5$–$3.0\,{\rm M}_\odot$), resulting in final masses of $\simeq 0.6\,{\rm M}_\odot$, and due to their electron-degenerate structure small radii of $\simeq 0.01\,{\rm R}_\odot$. The moderate mass enclosed within such a small volume leads to immensely high surface gravities, typically of $\log g \simeq 8$ (cgs).

These extreme gravitational fields have an important consequence for the white dwarf structure, as they cause elements to stratify according to their atomic weights, with the lightest elements at the surface and heaviest in the core. Typically $\simeq 99$ % of the white dwarf mass is contained within the electron-degenerate carbon/oxygen core, shrouded by a thin envelope of helium. In most white dwarfs, the outermost layer, $\simeq 0.01$ % by mass, consists of hydrogen.

Because spectroscopy only probes the outermost layer, an optical spectrum of a "normal" white dwarf shows little more than a series of pressure-broadened Balmer lines (figure 1, top). In about one quarter of white dwarfs, all hydrogen is removed in a late shell flash, and hence their spectra consist only of helium lines (figure 1, bottom).

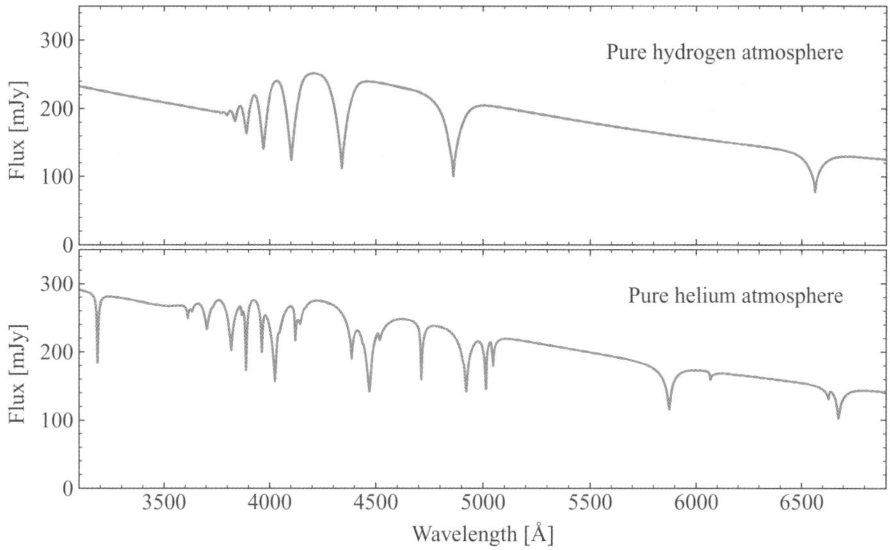

Figure 1. Spectral models for white dwarfs with atmospheres of pure hydrogen and helium. Both models are computed with effective temperatures of 20 000 K and log g of 8. The fluxes are scaled for white dwarfs observed from a distance of 10 pc. The naive expectation is that all white dwarfs should be observed with similar spectra.

In either case, one thing remains certain: if heavier elements are deposited onto the stellar surface, gravitational settling will cause them to sink below the photosphere on timescales many orders of magnitude shorter than the age of the white dwarf (Paquette *et al.* 1986). For a $T_{\rm eff} = 20\,000$ K white dwarf with a hydrogen atmosphere this process is as short as a few days (Koester 2009). Thus white dwarfs are expected to have pure hydrogen or helium atmospheres. Observationally this could not be further from the truth, with 25–50 % of white dwarfs having metal contaminated atmospheres (Zuckerman *et al.* 2010; Koester *et al.* 2014).

1.2. *Signatures of remnant planetary systems*

Over the last three decades, it has become clear that metals observed in the atmospheres of many white dwarfs, which have been known for a century (van Maanen 1917), signify the presence of remnant planetary systems. An understanding of this relation only began to emerge following observations of the white dwarf G29-38. In the infra-red, an excess of flux (figure 2) was found compared to what was expected if only a white dwarf was present (Zuckerman & Becklin 1987). This was later confirmed to originate from a compact circumstellar disk of dust (Graham *et al.* 1990). Because G29-38 also has a metal lines in its spectrum (Koester *et al.* 1997), it soon became apparent that these observables were connected, where eventually a planetary origin was suggested (Jura 2003).

The accepted scenario is that exoplanetesimals in such systems will sometimes have their orbits perturbed by more massive objects such as a planets (Debes & Sigurdsson 2002). If the perturbation increases the planetesimal eccentricity such that its periastron falls within about 1 $R_\odot$ of the white dwarf, tidal forces will unbind the asteroid (Jura 2003). The resulting dust will then circularise into a debris disk which over time will accrete onto the stellar surface, leading to the appearance of metallic absorption lines in the stellar spectrum.

Spectroscopic observations of G29-38 with Spitzer (Reach *et al.* 2005; Reach *et al.* 2009) have revealed strong 10 μm silicate emission (figure 2) which has been attributed

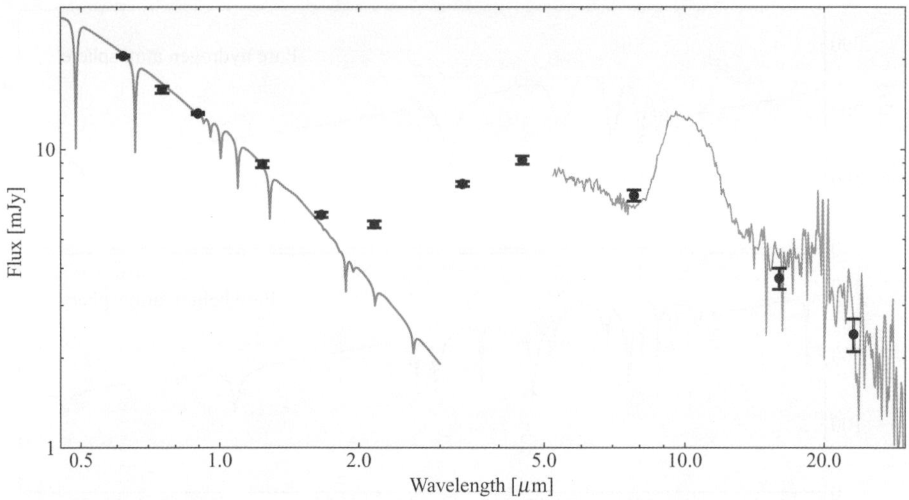

Figure 2. The white dwarf G29-38 exhibits a strong infra-red flux excess beyond about 1.5 μm, above the expected flux of the 12000 K white dwarf model. The Spitzer observations of Reach *et al.* (2005)/ Reach *et al.* (2009) demonstrate unambiguous 10 μm silicate emission. Photometry are from SDSS, APASS, 2MASS, WISE, and Spitzer.

to a mixture of enstatite and forsterite dust grains. While more than forty white dwarfs with circumstellar debris disks are now known from infra-red excesses (Rochetto *et al.* 2015), in addition to G29-38 only one other object (Jura *et al.* 2007) is sufficiently bright enough for spectroscopic follow-up with Spitzer. With JWST available in the near future, it will be possible to detect molecular emission at additional objects, and includes the prospect of carrying out detailed mineralogy in the brightest systems.

The most unambiguous evidence for remnant planetary systems was recently observed at WD 1145+017. This star shows highly variable, deep transits, demonstrating for the first time an asteroid disintegrating in real time (Vanderburg *et al.* 2015; Gaensicke *et al.* 2016). WD 1145+017 also exhibits both an infra-red excess and photospheric metals.

For detailed reviews to remnant planetary systems around white dwarfs, see Jura & Young (2014), Farihi (2016), and Veras (2016).

1.3. *White dwarfs as probes of exoplanetary material*

The otherwise pure hydrogen/helium atmospheres of white dwarfs make them perfect laboratories for analysing the bulk compositions of exoplanetesimals, by accurately measuring the atomic abundance ratios of the accreted material. Of course asteroids are chiefly comprised of molecular species, however the elemental abundance ratios obtained from spectral modeling of the white dwarf can be used to establish the most probable molecules/minerals in the accreted matter, preceding its violent disruption and vaporisation by the white dwarf.

Over the last decade, $\simeq$ 20 remnant planetary systems have been analysed in this way. The compositions of the disrupted planetesimals are found to be diverse (Gänsicke *et al.* 2012; Xu *et al.* 2014), indicating a significant fraction of these asteroids are differentiated. For instance the relatively high Ca and Al content found by Zuckerman *et al.* (2011) at NLTT 43806 points to lithospheric material, whereas the enhanced Fe and Ni abundances at Ton 345 found by Wilson *et al.* (2015) are better represented by core-material.

Analysis of two systems have also revealed the accretion of water-rich asteroids (Farihi *et al.* 2013; Raddi *et al.* 2015). In both cases, the oxygen abundances were found to be

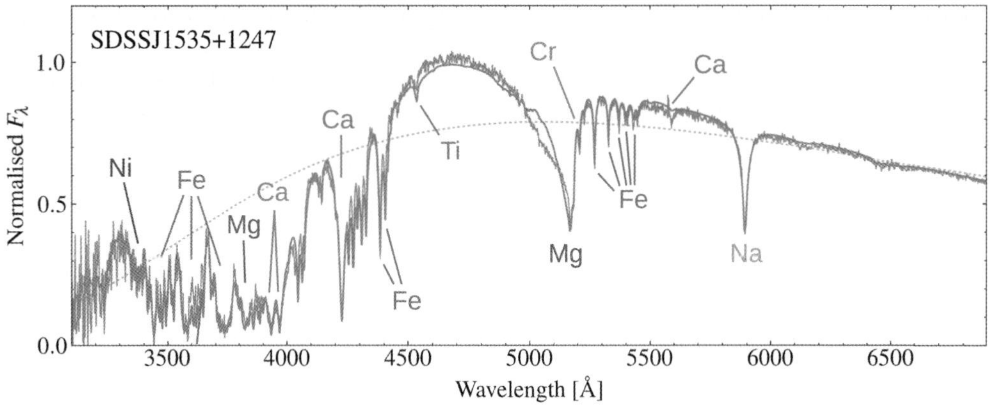

Figure 3. The optical spectrum of the DZ white dwarf SDSS J1535+1247, with its best fitting atmospheric model. The major metal transitions are labelled. A slight discrepancy is seen for the profile of the Mg-b blend (5171 Å), due to the challenge of modelling He-Mg collisions at the high atmospheric densities. The dotted line shows the corresponding model when no metals are present – essentially a blackbody, illustrating how SDSS J1535+1247 will look in a few Myr, once all metals have sunk below the photosphere.

vastly greater than expected from the accretion of minerals containing MgO, SiO_2, CaO etc. (bearing in mind that the dominant rock-forming elements Mg, Al, Si, Ca, and Fe are all easily detected). Because only a fraction of this oxygen could be explained through metal-oxide bearing minerals, and the non-detection of carbon rules out CO and CO_2, the excess oxygen strongly suggest a water-rich origin. In both systems the water mass-fractions of the asteroids were estimated to be a few 10 %, similar to the Solar System asteroid Ceres.

Thus far, all but one of the disrupted planetesimals detected at white dwarfs are consistent with asteroid-like parent bodies. The only exception is, WD 1425+540 which was recently found to have an atmosphere rich in volatile elements including C, N, and S, with an overall abundance pattern similar to the comet Halley, indicating that the white dwarf has accreted a large extrasolar Kuiper-Belt-Object (Xu *et al.* 2017).

2. A large sample of cool white dwarfs with metal-rich photospheres

2.1. *Cool DZ white dwarfs*

Below $T_{\rm eff} \simeq 11\,000$ K, white dwarfs with helium-dominated atmospheres are too cool to show helium transitions. Therefore, if metals have been accreted in sufficient quantities, their lines will be the only transitions observed in their spectra. Such white dwarfs are given the spectral class DZ.

As the $T_{\rm eff}$ is decreased further, helium becomes increasingly transparent and so relatively small traces of metals can have large observational signatures. Therefore, at moderate abundances, DZs often display absorption features from multiple elements, and thus spectroscopic analysis can reveal the composition of the accreted parent body with great accuracy.

Although not the focus of these proceedings, DZs are also old objects having already spent at least 1 Gyr cooling since first becoming a white dwarf. Hence DZs also allow us to study the properties of some of the oldest planetary material in the Milky Way.

Expanding upon the analysis of 26 cool, strongly metal-polluted DZs by Koester *et al.* (2011), we have enlarged this sample to 230 objects making use of spectroscopy available from the Sloan Digital Sky Survey (SDSS) which we first presented in Hollands *et al.* (2017). The stars in our sample, span a $T_{\rm eff}$ range of 4400–8800 K, with corresponding cooling ages of 1–8 Gyr (white dwarf cooling is non-linear in time), Including the main sequence life time of their progenitors, $\simeq$ 0.5–3.0 Gyr, implies some of these stars are nearly as old as the Galaxy.

The optical spectrum of the brightest DZ we found, SDSS J1535+1247, is shown in figure 3, which in all other regards is typical of the other stars in our sample, including its abundance pattern which is similar to that of the bulk Earth (McDonough 2000). Lines of Ca, Mg, Fe, and Na dominate the spectrum, with Cr, Ti, and Ni also detected owing to the high quality data available for this star. While, Si and O are expected to constitute approximately half of the accreted mass, these elements do not exhibit optical transitions at the low $T_{\rm eff}$ characteristic of DZs.

An important characteristic of the DZs in our sample are the deep convection zones in their outer helium envelopes, which typically comprise 10^{-6} to 10^{-5} of the total white dwarf mass (Hollands *et al.* 2017). Convection severely inhibits metals sinking into the white dwarf interior, with corresponding gravitational settling times of 1–3 Myr. While still significantly shorter than the white dwarf age, this is comparable to, if not longer than the duration of accretion an episode (Girven *et al.* 2012), and so for DZs, accretion-diffusion equilibrium is never reached. Instead the material is mixed throughout the convection zone, and therefore observations of DZs provide lower limits to the mass of the accreted planetesimals, which for our sample range from 10^{20} to 10^{24} g.

2.2. *Distribution of compositions*

Because of the large size of our sample, we are able to statistically explore the variety in exoplanetary material compositions, while also investigating the most interesting systems at a high level of detail (Hollands *et al.* in preparation). Model atmosphere fits to the entire sample yield abundances of Ca, Mg, and Fe for all 230 systems, and Na for around half of objects. Cr, Ti, and Ni are detected in only some systems subject to atmospheric abundances, spectral signal-to-noise, and wavelength coverage.

Considering the sample as a whole, we find that the Fe/Ca atomic abundance ratio provides a simple, but robust indicator of asteroid composition, with low and high Fe/Ca suggesting compositions similar to the Earth's crust and core, respectively. The full distribution of Fe/Ca, seen in Figure 4, spans two orders of magnitude from 1 to 100, where the mean of the distribution in seen to be close to the bulk Earth value of 1.11 (vertical dotted line).

2.3. *Systems exhibiting extreme extrasolar compositions*

In figure 5, we show two systems with high quality spectra from the extremes of the Fe/Ca distribution in figure 4. The atmospheric parameters of these two objects as well as SDSS J1535+1247 (figure 3) are listed in table 1. All three white dwarfs are of similar $T_{\rm eff}$ (5000–6000 K), but due to the variety in accreted material composition, their spectra are all remarkably different.

SDSS J0744+4649 shows an Fe/Ca atomic ratio only slightly above 1, signifying the accretion of a highly differentiated parent body, with a crust-like composition. This is further supported by the clear detections of Ti, and Na, which for the Earth, are both most prominent in the crust. Because crust-material material is largely composed of silicates the combined mass of metals detected from the spectrum likely only amounts to 20 % of the total, with the remaining 80 % contained within undetectable Si and O.

The Fe/Ca ratio for SDSS J0823+0546 is $\simeq$ 100 – despite the meagre abundance, Ca remains well measured from the strong H+K resonance lines. The additional detection of

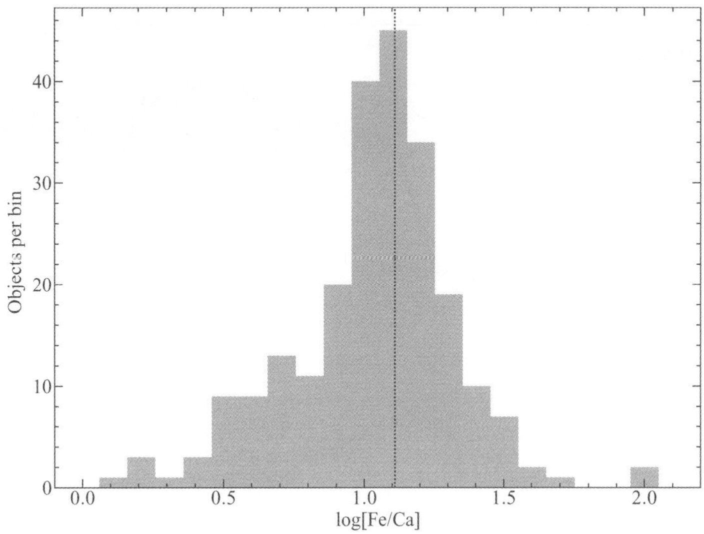

Figure 4. The Fe/Ca distribution for our sample spans two orders of magnitude, demonstrating the huge range in composition for the accreted material. The value for the bulk Earth, which many objects are clustered around, is shown by the dotted vertical line.

Table 1. Results from our fits for the three white dwarfs shown in figures 3 and 5 which we compare with the bulk Earth. In the final row we also list the white dwarf ages. Uncertainties on atomic abundance ratios are all approximately 0.1 dex.

	SDSS J			
	0744+4649	**0823+0546**	**1535+1247**	**Bulk Earth**
$T_{\rm eff}$ (K)	5030 ± 70	6020 ± 60	5770 ± 50	-
log[Ca/He]	−8.36	−9.34	−8.61	-
log[Fe/He]	−8.17	−7.36	−7.57	-
log[Fe/Ca]	+0.19	+1.98	+1.04	+1.11
log[Ti/Ca]	−1.02	-	−1.01	−1.37
log[Na/Ca]	−0.90	-	−0.11	−0.68
log[Ni/Fe]	-	−1.23	−1.33	−1.27
Age (Gyr)	$6.4^{+0.4}_{-1.1}$	3.2 ± 0.9	3.8 ± 0.9	-

Ni with Fe/Ni close to the bulk Earth ratio, strongly suggests that this white dwarf accreted a core fragment of a highly differentiated planetesimal. Almost the entire asteroid would have been composed of metallic Fe and Ni, where we estimate these two elements alone comprised at least 80 % of the total asteroid mass. Comparing SDSS J0823+0546 with analyses of similar objects (Wilson *et al.* 2015; Gänsicke *et al.* 2012; Kawka & Vennes 2014), we conclude that this is the most Fe-rich/core-like exoplanetary material detected to date. From the mass of material currently residing in the convection zone of SDSS J0823+0546, we derive a lower limit of $\simeq 4 \times 10^{21}$ g for the total asteroid mass, only one order of magnitude smaller than Psyche, the largest metallic Solar System asteroid (Carry 2012).

3. Conclusions and future work

We have identified a large sample of ancient, metal-rich cool DZ white dwarfs. Analysis of their spectra reveals the compositions of the accreted parent bodies to be extremely diverse, which in some cases are consistent with the accretion of either core or crust material from highly differentiated planetesimals.

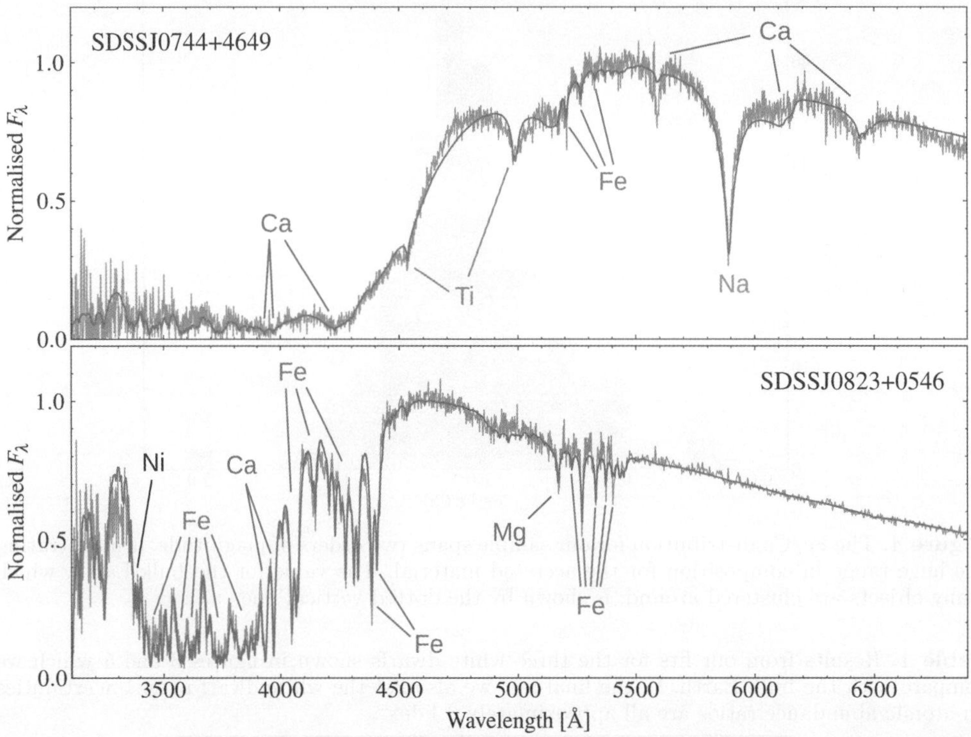

Figure 5. *Top*: SDSS J0744+4649 has a spectrum dominated by elements present in the Earth's crust, i.e. Ca, Ti, and Na. The intense absorption below 4500 Å is due to extreme pressure broadening of the H+K Ca II and 4226 Å Ca I resonance lines. *Bottom*: SDSS J0823+0546 exhibits a plethora of Fe transitions, particularly between 3400-3900 Å. Additionally it shows a blend of Ni lines near 3350 Å indicating the accreted body was rich in core-material.

Many of the objects identified in our sample are expected to appear in the 2nd Gaia data-release. Obtaining parallaxes to these white dwarfs will provide precise constraints on their radii, and in turn accurate estimates of their stellar masses. This will vastly improve the precisions of the white dwarf ages (see table 1) which are currently limited by the unknown stellar masses.

We will then be able to thoroughly investigate the chemistry of remnant planetary systems as a function of Galactic age.

Additional data releases from SDSS and other upcoming spectroscopic surveys will allow us to identify hundreds more metal-contaminated white dwarfs, increasing our understanding on the diversity of remnant planetary systems.

Acknowledgements

The research leading to these results has received funding from the European Research Council under the European Unions Seventh Framework Programme (FP/2007-2013) / ERC Grant Agreement n. 320964 (WDTracer).

References

Carry, B. 2012, *Planet. Space Sci.*, 73, 98

Debes, J. H. & Sigurdsson, S. 2002, *ApJ*, 572, 556

Farihi, J. 2016, *New Astron. Rev.*, 71, 9

Farihi, J., Gänsicke, B. T., & Koester, D. 2013, *Science*, 342, 218

Gänsicke, B. T., Aungwerojwit, A., Marsh, T. R., Dhillon, V. S., Sahman, D. I., Veras, D., Farihi, J., Chote, P., Ashley, R., Arjyotha, S., Rattanasoon, S., Littlefair, S. P., Pollacco, D., & Burleigh, M. R. 2016, *ApJ*, 818, L7

Gänsicke, B. T., Koester, D., Farihi, J., Girven, J., Parsons, S. G., & Breedt, E. 2012, *MNRAS*, 424, 333

Girven, J., Brinkworth, C. S., Farihi, J., Gänsicke, B. T., Hoard, D. W., Marsh, T. R., & Koester, D. 2012, *ApJ*, 749, 154

Graham, J. R., Matthews, K., Neugebauer, G., & Soifer, B. T. 1990, *ApJ*, 357, 216

Hollands, M. A., Koester, D., Alekseev, V., Herbert, E. L., & Gänsicke, B. T. 2017, *MNRAS*, 467, 4970

Jura, M. 2003, *ApJ Lett.*, 584, L91

Jura, M., Farihi, J., Zuckerman, B., & Becklin, E. E. 2007, *AJ*, 133, 1927

Jura, M. & Young, E. D. 2014, *Annual Review of Earth and Planetary Sciences*, 42, 45

Kawka, A. & Vennes, S. 2014, *MNRAS*, 439, L90

Koester, D. 2009, *A&A*, 498, 517

Koester, D., Gänsicke, B. T., & Farihi, J. 2014, *A&A*, 566, A34

Koester, D., Girven, J., Gänsicke, B. T., & Dufour, P. 2011, *A&A*, 530, A114

Koester, D., Provencal, J., & Shipman, H. L. 1997, *A&A*, 320, L57

McDonough, W. 2000, *Earthquake Thermodynamics and Phase Transformation in the Earth's Interior*, Elsevier Science Academic Press, pages 5–24

Paquette, C., Pelletier, C., Fontaine, G., & Michaud, G. 1986, 61, 197

Raddi, R., Gänsicke, B. T., Koester, D., Farihi, J., Hermes, J. J., Scaringi, S., Breedt, E., & Girven, J. 2015, *MNRAS*, 450, 2083

Reach, W. T., Kuchner, M. J., von Hippel, T., Burrows, A., Mullally, F., Kilic, M., & Winget, D. E. 2005, *ApJ Lett.*, 635, L161

Reach, W. T., Lisse, C., von Hippel, T., & Mullally, F. 2009, *ApJ*, 693, 697

Rocchetto, M., Farihi, J., Gänsicke, B. T., & Bergfors, C. 2015, *MNRAS*, 449, 574

van Maanen, A. 1917, *PASP*, 29, 258

Vanderburg, A., Johnson, J. A., Rappaport, S., Bieryla, A., Irwin, J., Lewis, J. A., Kipping, D., Brown, W. R., Dufour, P., Ciardi, D. R., Angus, R., Schaefer, L., Latham, D. W., Charbonneau, D., Beichman, C., Eastman, J., McCrady, N., Wittenmyer, R. A., & Wright, J. T. 2015, *Nat*, 526, 546

Veras, D. 2016, *Royal Society Open Science*, 3, 150571

Wilson, D. J., Gänsicke, B. T., Koester, D., Toloza, O., Pala, A. F., Breedt, E., & Parsons, S. G. 2015, *MNRAS*, 451, 3237

Xu, S., Jura, M., Koester, D., Klein, B., & Zuckerman, B. 2014, *ApJ*, 783, 79

Xu, S., Zuckerman, B., Dufour, P., Young, E. D., Klein, B., & Jura, M. 2017, *ApJ Lett.*, 836, L7

Zuckerman, B. & Becklin, E. E. 1987, *Nat*, 330, 138

Zuckerman, B., Koester, D., Dufour, P., Melis, C., Klein, B., & Jura, M. 2011, *ApJ*, 739, 101

Zuckerman, B., Melis, C., Klein, B., Koester, D., & Jura, M. 2010, *ApJ*, 722, 725

Astrochemistry VII – Through the Cosmos from Galaxies to Planets
Proceedings IAU Symposium No. 332, 2017
M. Cunningham, T. Millar & Y. Aikawa, eds.

doi:10.1017/S1743921317007761

The ALMA view of UV-irradiated cloud edges: unexpected structures and processes

Javier R. Goicoechea[1], S. Cuadrado[1], J. Pety[2,3], A. Aguado[4], J. H. Black[5], E. Bron[1], J. Cernicharo[1], E. Chapillon[2,6], A. Fuente[7], M. Gerin[3], C. Joblin[8], O. Roncero[9] and B. Tercero[7]

[1]ICMM-CSIC, Calle Sor Juana Ines de la Cruz 3, 28049 Cantoblanco, Madrid, Spain.
[2]IRAM, 300 rue de la Piscine, 38406 Saint-Martin-dHères, France.
[3]LERMA, Observatoire de Paris, PSL Research University, CNRS, Sorbonne Universités, UPMC Univ. Paris 06, ENS, 75005 Paris, France
[4]Facultad de Ciencias, Universidad Autónoma de Madrid, 28049 Madrid, Spain
[5]Chalmers University of Technology, Onsala Space Observatory, 43992 Onsala, Sweden
[6]OASU/LAB-UMR5804, CNRS, Université Bordeaux, 33615 Pessac, France
[7]Observatorio Astronómico Nacional (IGN). Apartado 112, 28803 Alcalá de Henares, Spain
[8]Université de Toulouse, UPS-OMP, CNRS, Institut de Recherche en Astrophysique et Planétologie , 9 Avenue du Colonel Roche, BP 44346, 31028 Toulouse, France.
[9]Instituto de Física Fundamental (IFF-CSIC), Calle Serrano 123, 28006 Madrid, Spain.

Abstract. Far-UV photons (FUV, $E < 13.6\,\mathrm{eV}$) from hot massive stars regulate, or at least influence, the heating, ionization, and chemistry of most of the neutral interstellar medium (H I and H_2 clouds). Investigating the interaction between FUV radiation and interstellar matter (molecules, atoms and grains) thus plays an important role in astrochemistry.

The Orion Bar, an interface region between the Orion A molecular cloud and the H II region around the Trapezium cluster, is a textbook example of a strongly illuminated dense PDR (photodissociation region). The Bar is illuminated by a FUV field of a few 10^4 times the mean interstellar radiation field. Because of its proximity and nearly edge-on orientation, it provides a very good template to investigate the chemical content, structure, and dynamics of a strongly irradiated molecular cloud edge. We have used ALMA to mosaic a small field of the Bar where the critical transition from atomic to molecular gas takes place. These observations provide an unprecedented sharp view of this transition layer ($\lesssim 1''$ resolution or $\lesssim 414\,\mathrm{AU}$). The resulting images (so far in the rotational emission of CO, HCO^+, $H^{13}CO^+$, SO^+, SO, and reactive ions SH^+ and HOC^+) show the small-scale structure in gas density and temperature, and the steep abundance gradients. The images reveal a pattern of high-density substructures, photo-ablative gas flows and instabilities at the edge of the molecular cloud. These first ALMA images thus show a more complex morphology than the classical clump/interclump static model of a PDR.

In order to quantify the chemical content in strongly FUV-irradiated gas, we have also used the IRAM-30 m telescope to carry out a complete line-survey of the illuminated edge of the Bar in the millimeter domain. Our observations reveal the presence of complex organic molecules (and precursors) that were not expected in such a harsh environment. In particular, we have reported the first detection of the unstable *cis* conformer of formic acid (HCOOH) in the ISM. The energy barrier to internal rotation (the conversion from *trans* to *cis*) is approximately $4827\,\mathrm{cm}^{-1}$ ($\approx 7000\,\mathrm{K}$). Hence, this detection is surprising. The low inferred *trans*-to-*cis* abundance ratio of 2.8 ± 1.0 supports a photoswitching mechanism: a given conformer absorbs a FUV stellar photon that radiatively excites the molecule to electronic states above the interconversion barrier. Subsequent fluorescent decay leaves the molecule in a different conformer form. This mechanism, which we have specifically studied with *ab initio* quantum calculations, was not considered so far in astrochemistry although it can affect the structure of a variety of molecules in PDRs.

Keywords. Astrochemistry – ISM: clouds – ISM: molecules – photodissociation regions (PDR)

1. Introduction: the role of far-UV stellar radiation on the ISM

Far-UV photons (FUV: $E < 13.6\,\text{eV}$) from hot massive O and B stars regulate, or at least greatly influence, the dynamics, heating, ionization, and chemistry of the neutral interstellar medium (ISM). From a global perspective, the photo-induced processes that take place in FUV-illuminated environments reflect the radiative feedback from such stars. This feedback takes place at very different spatial scales: in star-forming regions of the Milky Way, the ISM of starburst galaxies, and back in cosmic history: see the rapidly increasing number of ALMA detections of the [C II] 158 μm fine-structure line at high-redshifts (e.g., Capak *et al.* 2015, *Nature*). This far-IR emission line of ionized carbon (C^+) was first detected in the 80s (Russell *et al.* 1980) and is often the brightest and most important line coolant of the neutral ISM (hydrogen atoms in neutral form).

Investigating the interaction between FUV radiation and interstellar baryonic matter (atoms, molecules and dust grains) ultimately contributes to a better understanding of a varity of FUV-irradiated objects and environments of the galaxy: the diffuse interstellar clouds, the interfaces between H II regions near young massive stars and their natal clouds, reflection and planetary nebulae around evolved stars, and the externally irradiated protoplanetary disks (proplyds). All these environments in which the underlying physics and chemistry are driven by the presence of FUV photons are generically defined as photodissociation regions (PDRs; see the review paper by Hollenbach & Tielens 1999). FUV photons with wavelengths longer than 911 Å do not ionize H atoms (as in H II regions) but do dissociate molecules and ionize atoms with ionization potental below 13.6 eV (C, S, Si, Fe, P, etc.). Hence, PDRs are essentialy neutral, with ionization fractions as high as $[e^-]/[\text{H}]\simeq 10^{-4}$ when most of the electrons are produced by the ionization of carbon atoms, a key process in the ISM (see e.g., Dalgarno & McCray 1972).

PDRs are very relevant because they host the critical conversion from atomic to molecular ISM (the $H^+/H/H_2$ and $C^+/C/CO$ transition zones, see Figure 1). This stratification occurs as the column density of gas and dust increases deeper inside clouds and the FUV photon flux is attenuated (see Goicoechea & Le Bourlot 2007 and references therein). In addition, PDRs emit most of the IR radiation from the ISM of star-forming galaxies: mid-IR bands from polycyclic aromatic hydrocarbons (PAHs), IR H_2 ro-vibrational lines, [C II] 158 μm and [O I] 63 μm fine-structure lines, and warm dust continuum: after absorbing a FUV photon, grains are heated and radiation is reemitted in the mid- and far-IR. At the same time, photoelectrons ejected from grains and PAHs heat the gas.

The theoretical study and first thermo-chemical models of dense PDRs started to strongly develop more than 30 years ago. With the above broad definition, PDRs represent a significant fraction of the neutral atomic and molecular material in the Milky Way, perhaps more than 90%: all the gas and dust in the ISM between $A_V \approx 0.1$ and 7 mag (Hollenbach & Tielens 1999). At deeper cloud depths, the flux of stellar FUV photons is almost completely attenuated, gas-phase molecules and atoms freeze, and a much slower chemistry and weak ionization take place driven by cosmic-ray particles (see Figure 1).

In addition to being excellent tracers of the gas physical conditions (density, temperature, FUV field strenght, etc.), PDR observational diagnostics ([C II] 158 μm, PAH bands, H_2 lines, and far-IR continuum emission) can be used to estimate the star-formation rate in galaxies. More locally, PDR diagnostics can be used to determine the radiative feedback of massive stars on their natal clouds. Understanding how molecular clouds are evaporated by a strong FUV radiation field is important to determine their lifetimes, and how the star-formation process can be quenched in a given region. On the other hand, the dynamical effects induced by a strong stellar FUV field (e.g. radiation pressure), by

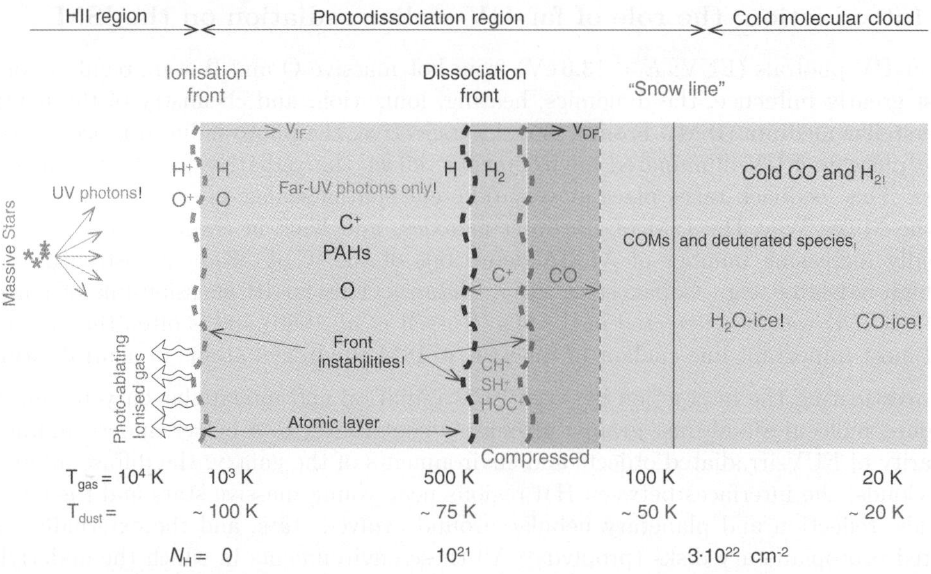

Figure 1. Structure of a strongly FUV-irradiated cloud edge (from Goicoechea *et al.* 2016).

the winds of massive stars, or by Supernova explosions could trigger, even regulate, the formation of a new generation of low-mass stars (Hollenbach & Tielens 1999).

In this contributed paper we summarize our latest results regarding the study of the Orion Bar PDR, an excellent astrochemical laboratory to investigate the interaction of interstellar matter with a strong flux of FUV photons from nearby massive stars.

The Orion Bar: Owing to its nearly edge-on orientation, "the Bar" (an interface region between the Orion A molecular cloud and the H II region around the Trapezium cluster, see Figure 2) is a prototypical strongly irradiated PDR with $G_0 \simeq 2\times10^4$ ($G_0=1.7$ is roughly the mean interstellar FUV field in Habing units, equivalent to a few 10^8 photons $cm^{-2}\,s^{-1}$ between 13.6 and 6 eV). Its closeness (~414 pc), high gas temperatures ($T_k \simeq$150-300 K) and thus bright molecular lines, makes it an ideal target for high signal-to-noise spectral-imaging at different wavelengths. In the last years, our team has led several efforts to constrain the molecular content of the Orion Bar by using the *Herschel Space Observatory* and the IRAM-30 m telescope, and more recently to study the small-scale density and temperature structures with ALMA. From these works **we highlight**:
(i) A line survey of the Bar edge in the 3 to 0.8 mm bands, complemented with $\sim2'\times2'$ maps at $\sim7''$ resolution with the IRAM-30m telescope (Cuadrado *et al.* 2015, 2016, 2017). Once thought to be a harsh environment to develop a rich chemistry, our observations show that PDRs host unique chemical features: reactive ions (SH^+ or HOC^+; Fuente *et al.* 2003, Goicoechea *et al.* 2017), hydrocarbon ions (C_3H^+; Pety *et al.* 2012; Cuadrado *et al.* 2015), and unstable isomers of organic species (*cis*-HCOOH, Cuadrado *et al.* 2016) that are not seen, or that are not abundant, in regions shielded from FUV radiation.
(ii) *Herschel*/HIFI large-scale ($7.5'\times11.5'$) velocity-resolved maps of Orion A (including the Bar) in FUV-radiation diagnostics such as [C II] 158 μm (Goicoechea *et al.* 2015).
(iii) Pilot ALMA mosaics of a small field ($\sim50''\times50''$) of the Bar in several molecular line diagnostics at $\sim1''$ resolution (Figures 3 and 4) (Goicoechea *et al.* 2016, 2017).

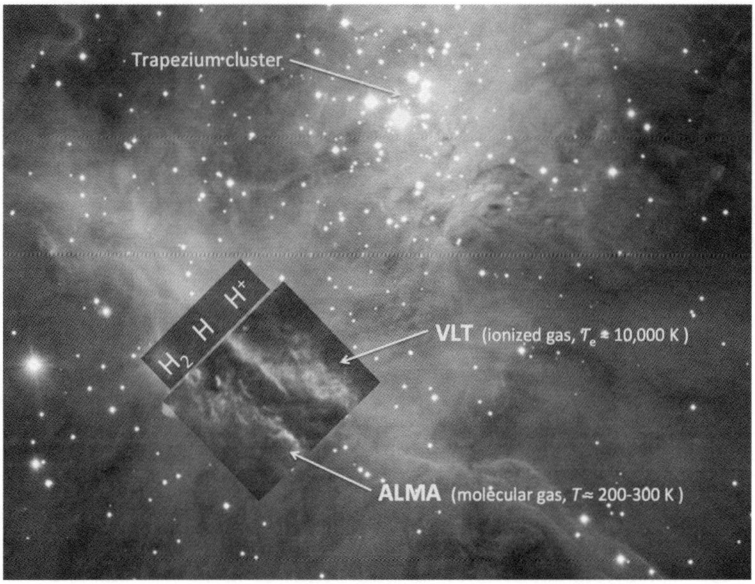

Figure 2. Part of the Orion nebula (ionized gas irradiated by UV photons from the Trapezium cluster) observed by the VLT, and a small field of the Orion Bar PDR observed by ALMA in molecular gas emission (adapted from Goicoechea *et al.* 2016, *Nature*).

2. Morphology of the PDR: small-scale structures and flows

Surprisingly, our pilot ALMA images show that there is no appreciable offset between the peak of the H_2 vibrational emission at 2.2 μm (delineating the H_2 dissociation front) and the edge of the observed CO and HCO^+ rotational emission (Figures 2 and 3). This implies that the H/H_2 and C^+/C/CO transition zones are very close, much closer than stationary PDR model predictions. In addition, our ALMA mosaics reveal a fragmented ridge of high-density substructures, photoablative gas flows and instabilities at the irradiated molecular cloud surface (Figure 3; see Goicoechea *et al.* 2016 for details). These results suggest that the cloud edge has been compressed by a high-pressure wave that is moving into the molecular cloud, demonstrating that dynamical effects are more important than previously thought. Whether the newly detected small-scale structures could be the seed of future star-forming clumps (e.g., by merging into more massive collapsing clumps) is uncertain and is subject of study. Gravitational collapse is not yet apparent from their density distribution nor from their low masses. However, our images resolve the HCO^+ J=4-3 emission from *proplyd 203-506* (Figure 3a and Champion *et al.* 2017). Interestingly, its location in the atomic PDR (A_V=0 to 1 mag, thus in-between the ionization and dissociation fronts) suggests that this proplyd (a low-mass protostar and protoplanetary disk) could have emerged from the molecular cloud after surviving the passage of the dissociation front. Even on its early cycles (the antena array was not yet completed at the time of our observations) ALMA clearly goes a step beyond in our understanding of the closest, and most iconic, massive star forming region. The results obtained from these pilot ALMA mosaics were unexpected and challenge many years of PDR modelling. However, they were obtained by imaging a small fraction of the Bar ($\approx$10-20% of its size; see Figure 2). In order to support and better understand our findings, it is mandatory to obtain ALMA images of the entire Bar, and of course, to observe other (more distant) massive star forming regions, i.e., at different stages of evolution and having different stellar populations, thus different number of ionizing massive stars.

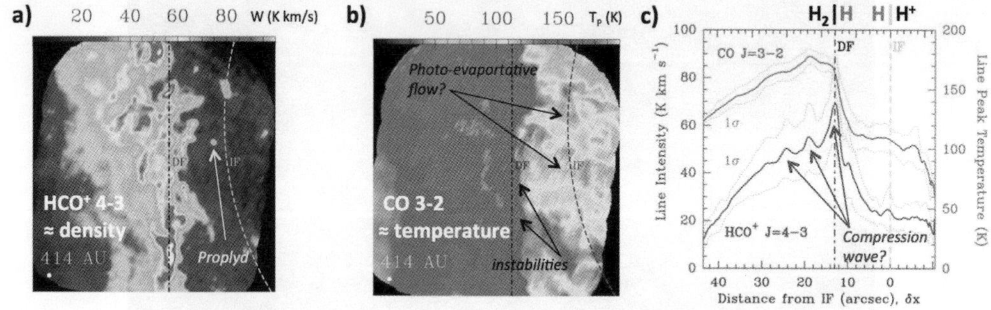

Figure 3. ALMA view of a small field of the Orion Bar in two tracers of the molecular gas: a) HCO^+ J=4-3 and b) CO J=3-2. c) Vertically averaged intensity cuts perpendicular to the Orion Bar. The flux of FUV photons decreases from right to left in these rotated images.

3. The short, but interesting, life of reactive ions

Reactive ions (CH^+, SH^+, CO^+, HOC^+, ...) are transient species for which the timescale of reactive collisions with H_2, H, or e^- (leading to a chemical reaction, and thus molecule destruction) is comparable to, or shorter than, that of inelastic collisions (Black 1998). The formation of reactive ions such as SH^+, for example, depends on the availability of S^+ ions (i.e., on the availability of FUV photons that can ionize sulphur atoms) and on the presence of excited H_2 (either UV-pumped or hot and thermally excited). These two ingredients allow overcoming the high endothermicity (and sometimes energy barrier) of some of the key initiating chemical reactions (Gerin *et al.* 2016 for a review). The reaction $S^+ + H_2(\mathrm{v}) \rightarrow SH^+ + H$, in particular, is endothermic by $\Delta E/k \simeq 9860\,\mathrm{K}$ if v=0, and only becomes exothermic when v$\geqslant$2 (Zanchet *et al.* 2013). Despite their short lifetimes, from a few years in diffuse gas to only a few hours (!) in dense PDR gas, reactive ions can be detected and probe energetic processes in FUV-irradiated warm molecular gas.

We have used the ALMA-ACA array to obtain high angular resolution ($5'' \times 3''$) images of some reactive ions toward the Bar (Goicoechea *et al.* 2017). The observed SH^+ and HOC^+ emission is restricted to a narrow layer of $2''$- to $10''$-width ($\approx$800 to 4000 AU depending on the assumed PDR geometry) that follows the vibrationally excited H_2^* emission (Figure 4). Both ions efficiently form very close to the H/H_2 transition zone, at a depth of $A_V \lesssim 1$ mag into the neutral cloud, where abundant C^+, S^+, and H_2^* coexist. The observed ions have low rotational temperatures ($T_{\mathrm{rot}}\approx$10-30 K$\ll T_{\mathrm{k}}$) and narrow line-widths ($\sim$2-3 km s^{-1}), a factor of $\simeq$2 narrower that those of the lighter reactive ion CH^+ (Nagy *et al.* 2013). In Goicoechea *et al.* (2017) we show that this is consistent with the higher reactivity† and faster radiative pumping rates of CH^+ compared to the heavier ions, which are driven relatively more quickly toward smaller velocity dispersion by elastic collisions and toward lower T_{rot} by inelastic collisions (see also Black 1998). Regardless of the excitation details, the ALMA-ACA images show that SH^+ and HOC^+ clearly trace the most exposed layers of the FUV-irradiated molecular cloud surface.

† For the physical conditions at the edge of the Bar, the destruction time scales (τ_{D}) of CH^+, HOC^+, and SH^+ are $\simeq$5 h, $\simeq$25 h and $\simeq$50 h respectively. Comparing with the timescales for collisional and radiative excitation shows that CH^+ molecules are excited by radiation many times during their short lifetime, but not by collisions. Hence, CH^+ can remain rotationally warm (owing to formation pumping; Nagy *et al.* 2013, Godard & Cernicharo 2013) while it emits. On the other hand, comparing $\tau_{\mathrm{D}}(HOC^+)$ with the timescales for collisional and radiative excitation shows that HOC^+ molecules (and likely the other heavier ions as well) are excited by collisions several times during their lifetime. Inelastic collisions thus can drive their rotational populations to lower T_{rot} relatively fast ($T_{\mathrm{rot}}(HOC^+) \ll T_{\mathrm{rot}}(CH^+) \approx 150\,\mathrm{K}$; Goicoechea *et al.* 2017).

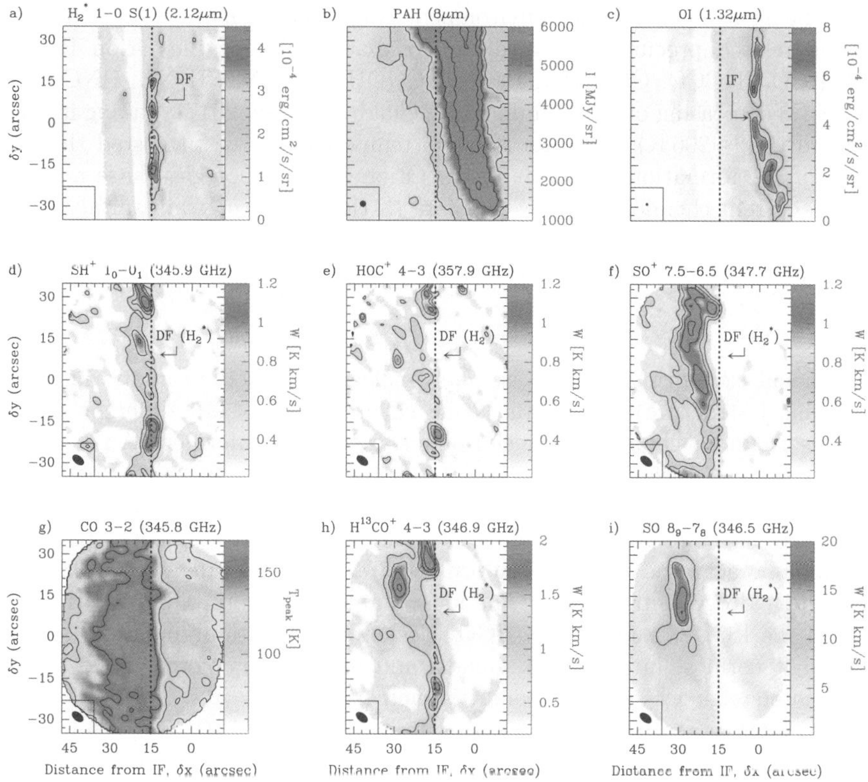

Figure 4. ALMA-ACA observations of several molecules, including the reactive ions SH^+ and HOC^+ and complementary images of the Bar (Goicoechea *et al.* 2017). All images have been rotated to bring the FUV illuminating direction in the horizontal direction (from the right). The upper row shows images of a) the H_2 v=1-0 $S(1)$ line at 2.12 μm, delineating the dissociation front (DF) (Walmsley *et al.* 2000); b) the Spitzer 8 μm emission produced mainly by PAHs, and c) the fluorescent O I 1.32μm line arising from the H II/PDR boundary (Walmsley *et al.* 2000).

4. Yes, COMs are seen in strongly FUV-irradiated gas

Observations towards UV-shielded cold cores (e.g. TMC 1, L1689B, or B1-b) have revealed molecules once considered to be present only in hot cores and corinos (e.g., Cernicharo *et al.* 2012). In these environments, complex organic molecules (COMs) are thought to form on the surface of grains and to be released through non-thermal desorption processes, chemical desorption, direct desorption by cosmic rays impacts, or by secondary FUV-photon induced processes. Until recently, observational studies of environments illuminated by stellar FUV photons were more scarce. Guzmán *et al.* (2014) presented the unexpected detection of *trans*-HCOOH, CH_2CO, CH_3CN, CH_3OH, CH_3CHO, and CH_3CCH at the edge of the Horsehead (a low FUV-illumination PDR with $G_0 \approx 100$). They found enhanced abundances compared to a nearby cold and dense core shieled from external FUV radiation. Guzmán *et al.* (2014) proposed that owing to the cold dust temperatures in this PDR ($T_d \lesssim 30$ K, i.e., grains should be coated by ices), ice-mantle photodesorption processes dominate the formation of gaseous COMs. In lower density translucent clouds ($G_0 \approx 1$), only H_2CO has been unambiguously detected (Liszt *et al.* 2006). These observations might suggest that the presence of COMs diminishes as G_0/n_H increases, and hence, that COMs might not be present in strongly FUV-irradiated gas.

Despite being a much harsher environment, we detected more than 250 lines from COMs and related precursors toward the irradiated edge of the Orion Bar: H_2CO, CH_3OH, HCO, H_2CCO, CH_3CHO, H_2CS, HCOOH, CH_3CN, CH_2NH, HNCO, $H_2^{13}CO$, and HC_3N (in decreasing order of abundance, Cuadrado *et al.* 2017). Taking into account the elevated gas (≈200 K) and dust (≈60 K) temperatures, we suggested the following scenarios for the formation of COMs in the Orion Bar: *(i)* hot gas-phase reactions not included in current chemical models; *(ii)* bare (little ice) warm grain surface chemistry; or *(iii)* the PDR dynamics is such that COMs previously formed in cold icy grains deeper inside the molecular cloud desorb and advect into the PDR (Cuadrado *et al.* 2017).

4.1. *Trans-cis photoisomerization of HCOOH*

Conformational isomerism refers to isomers (molecules with the same formula but different chemical structure) having the same chemical bonds but different geometrical orientations around a single bond. Such isomers are called conformers. An energy barrier often limits the isomerization. This barrier can be overcome by light. Photoisomerization (or photoswitching) has been studied in ice IR-irradiation experiments, in biological processes, and, for large polyatomic molecules, in gas-phase experiments. HCOOH is the simplest organic acid and has two conformers (*trans* and *cis*) depending on the orientation of the hydrogen single bond. The most stable *trans* conformer was the first acid detected in the ISM. Gaseous *trans*-HCOOH shows moderate abundances towards hot cores and hot corinos, in cold dark clouds, and in cometary coma. Solid HCOOH is present in interstellar ices and in chondritic meteorites.

The energy barrier to internal conversion from *trans*- to *cis*-HCOOH is much higher than the thermal energy available in molecular clouds†. Thus, only the most stable conformer (*trans*) is expected to exist in detectable amounts. In Cuadrado *et al.* (2016) we reported the first interstellar detection of *cis*-HCOOH. Its presence in the Orion Bar (but not in the hot core of Orion nor in the cold core of Barnard 1-b) with a very low *trans*-to-*cis* abundance ratio of 2.8 ± 1.0 (>100 and >60 in the Orion hot core and in B1-b respectively), supports a photoswitching mechanism: a given conformer absorbs a FUV photon with energies around 5 eV (precisely in the $\lambda \simeq 2300$-2800 Å range, higher energy photons dissociate the molecule) that radiatively excites the molecule to electronic states above the interconversion barrier. Subsequent fluorescent decay leaves the molecule in a different conformer form. This mechanism, which we specifically studied with *ab initio* quantum calculations (to compute the ∼5 eV-photon absorption cross–sections and the *trans*-to-*cis* and *cis*-to-*trans* photoswitching probabilities) was not considered in the ISM before, but likely induces structural changes of a variety of molecules in PDRs. In Cuadrado *et al.* (2016) we used the *Meudon* PDR code to estimate the flux of FUV photons at different positons of the Bar. The well-known “2175 Å bump” of the dust extinction curve (produced by the absorption of FUV photons by PAHs and small carbonaceous grains) reduces the number of HCOOH dissociating photons relative to those producing fluorescence (Figure 5 *right*). We determine that at a cloud depth of $A_V \approx 2 - 3$ mag (for the Orion Bar conditions), the *cis* conformer should be detectable with a *trans*-to-*cis* abundance ratio of 3.5-4.1. These ratios are remarkably close to the observed values. Therefore, both the detection of *cis*-HCOOH, and the low *trans*-to-*cis* abundance ratio in the Bar (much lower than in environments shielded from FUV radiation) would agree with the photoswitching scenario.

† The ground-vibrational state of *cis*-HCOOH is 1365 ± 30 cm^{-1} higher in energy than that of the *trans* conformer. The energy barrier to internal rotation (the conversion from *trans* to *cis*) is approximately 4827 cm^{-1} (≈ 7000 K in temperature units).

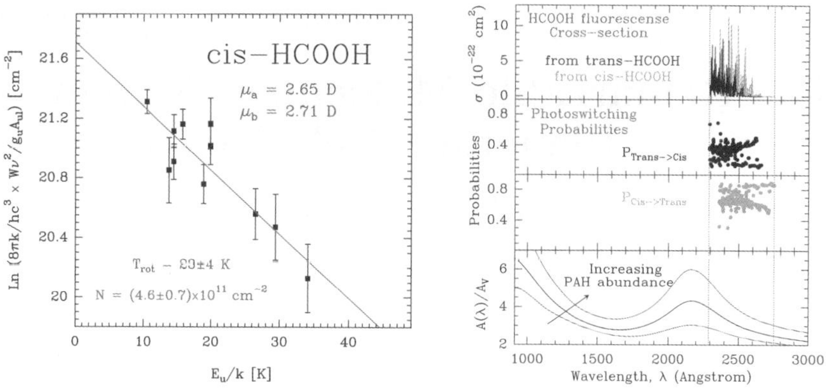

Figure 5. *Left*) Rotational diagram of the *cis*-HCOOH lines detected toward the Orion Bar. *Right*) Top panel: *Trans-* and *cis*-HCOOH absorption cross-sections for photons with $E \lesssim 5\,\mathrm{eV}$ (those leading to fluorescence) and photoisomerization probabilities. Bottom panel: standard dust extinction curve (blue) and effect of an increased PAH abundance (Cuadrado *et al.* 2016).

To summarize, the presence of transient reactive ions such as CH^+, SH^+ or HOC^+ is an unambiguous indication of FUV-irradiated warm molecular gas. The presence of COMs, however, is more widespread, and now includes harsh environments such as strongly FUV-irradiated gas (where conformers can photoisomerize). COM formation reflects the complicated interplay between gas and grain chemistry. Hence, more theory and laboratory experiments of the possible formation routes are needed. ALMA and soon JWST (for non-polar molecules, H_2 lines, PAHs and dust features) will open the study of PDRs to sub arcsecond resolution spectral-imaging. Exciting times ahead for PDR research.

References

Black, J. H. 1998, *Faraday Discussions*, 109, 257
Capak, P. L., Carilli, C., Jones, G., *et al.* 2015, *Nature*, 522, 455
Cernicharo, J., Marcelino, N., Roueff, E., *et al.* 2012, *ApJL*, 759, L43
Champion, J., Berné, O., Vicente, S., *et al.* 2017, accepted in *A&A*, arXiv:1702.00251
Cuadrado, S., Goicoechea, J. R., Pilleri, P., *et al.* 2015, *A&A*, 575, A82
Cuadrado, S., Goicoechea, J. R., Roncero, O., *et al.* 2016, *A&A*, 596, L1
Cuadrado, S., Goicoechea, J. R., Cernicharo, J., *et al.* 2017, *A&A*, 603, A124
Dalgarno, A. & McCray, R. A. 1972, *ARAA*, 10, 375
Fuente, A. *et al.* 2003, *A&A*, 406, 899
Gerin, M., Neufeld, D. A., & Goicoechea, J. R. 2016, *ARAA*, 54, 181
Godard, B. & Cernicharo, J. 2013, *A&A*, 550, A8
Goicoechea, J. R. & Le Bourlot, J. 2007, *A&A*, 467, 1
Goicoechea, J. R., Teyssier, D., Etxaluze, M., *et al.* 2015, *ApJ*, 812, 75
Goicoechea, J. R., Pety, J., Cuadrado, S., *et al.* 2016, *Nature*, 537, 207
Goicoechea, J. R., Cuadrado, S., Pety, J., *et al.* 2017, *A&A*, 601, L9
Guzmán, V. V., Pety, J., Gratier, P., *et al.* 2014, *Faraday Discussions*, 168, 103
Hollenbach, D. J. & Tielens, A. G. G. M. 1999, *Reviews of Modern Physics*, 71, 173
Liszt, H. S., Lucas, R., & Pety, J. 2006, *A&A*, 448, 253
Nagy, Z., Van der Tak, F. F. S., Ossenkopf, V., *et al.* 2013, *A&A*, 550, A96
Pety, J., Gratier, P., Guzmán, V., *et al.* 2012, *A&A*, 548, A68
Russell, R. W., Melnick, G., Gull, G. E., & Harwit, M. 1980, *ApJL*, 240, L99
Walmsley, C. M., Natta, A., Oliva, E., & Testi, L. 2000, *A&A*, 364, 301
Zanchet, A., Agúndez, M., Herrero, V. J., Aguado, A., & Roncero, O. 2013, *AJ*, 146, 125

Astrochemistry VII – Through the Cosmos from Galaxies to Planets
Proceedings IAU Symposium No. 332, 2017
M. Cunningham, T. Millar & Y. Aikawa, eds.

doi:10.1017/S1743921317011589

The Remarkable Molecular Content of Planetary Nebulae

D. R. Schmidt[1] **and L. M. Ziurys**[1,2]

[1]Department of Astronomy, University of Arizona,
933 North Cherry Avenue, Tucson, AZ 85719
email: drschmidt@email.arizona.edu

[2]Department of Chemistry and Biochemistry, University of Arizona
1306 E. University Blvd., Room 221, Tucson, AZ 85721
email: lziurys@email.arizona.edu

Abstract. In order to further define the molecular content of planetary nebulae (PNe), we have conducted searches for HCN, HCO^+, HNC, and CCH at millimeter wavelengths in a sample of seventeen PNe using the new 12 m and Sub-Millimeter Telescopes of the Arizona Radio Observatory (ARO). HCN and HCO^+ were identified in 75 % of the PNe, with corresponding fractional abundances of $f(HCN/H_2) \sim 0.1\text{-}9.1 \times 10^{-7}$ and $f(HCO^+/H_2) \sim 0.04\text{-}7.4 \times 10^{-7}$. HNC was subsequently identified in twelve PNe with $f(HNC/H_2) \sim 0.02\text{-}2.2 \times 10^{-7}$. The [HCN]/[HNC] ratio was found to be ~1-8 in nebulae observed. CCH was also detected in eight PNe. The abundances for all molecules were found to remain relatively constant with nebular age across 10,000 years, in contrast to model predictions. They are also 10-100 greater than those observed in diffuse clouds, and suggest that molecular material from PNe seed the diffuse ISM.

Keywords. ISM: molecules - planetary nebulae: general - radio lines: ISM

Planetary nebulae (PNe) represent the final stage of evolution for low- to intermediate-mass stars ($M_* \sim 0.5\text{-}8\ M_{sol}$), which make up the majority of stars in the Galaxy. During the asymptotic giant branch (AGB) phase, such stars lose copious amounts of atomic and molecular material (upwards of $10^{-4}\ M_{sol}yr^{-1}$) in the form of a superwind, from which a circumstellar envelope arises (e.g., Kwok 2000). The remnant stellar core, evolving toward a white dwarf, continues to increase in temperature until, at a temperature of 30,000 K, it is capable of ionizing the remnant material in the surrounding shell, thereby creating a PN. The powerful ultraviolet radiation field from this central white dwarf (10^5 times that of the interstellar medium, or ISM) is expected to photodissociate any remnant molecular content in the PN envelope within ~1000 years (e.g., Redman *et al.* 2003).

In spite of this, the presence of molecular material in PNe has been found to be a common occurrence, rather than the exception. Since the 1970s, CO and H_2 have been detected in numerous PNe, particularly through large-scale surveys (e.g., Huggins & Healy 1989; Huggins *et al.* 1996, 2005; Hora *et al.* 1999; Likkel *et al.* 2006). Further, targeted searches in a number of PNe have revealed a diverse array of polyatomic molecules. The Helix Nebula, for instance, one of the oldest and most well-studied PNe, has been identified in transitions of HCO^+, HCN, HNC, CCH, c-C_3H_2, and H_2CO (Bachiller *et al.* 1997; Tenenbaum *et al.* 2009). Subsequent mapping studies have shown that this molecular material is located throughout the PN (Zeigler *et al.* 2013; Zack and Ziurys 2013). The young PNe NGC 7027 and NGC 6537 have been detected in transitions of HCN, HNC, CCH, CS, SO, H_2CO, HCO^+, and N_2H^+ (Zhang *et al.* 2008; Edwards and Ziurys 2013), while the molecules SiO, SO, and SO_2 have been identified toward the middle-age

PN M2-48 (Edwards and Ziurys 2014). Remarkably enough, a molecule as complex as C_{60} has been observed in a handful of PNe (Cami *et al.* 2010; García-Hernández *et al.* 2010, 2011, 2012; Otsuka *et al.* 2013). Indeed, PNe have been found to be a veritable goldmine of polyatomic molecules; as the Helix proves, even the oldest nebulae can harbor significant quantities of molecular material. The survival of such material long past its expected destruction is thought to be due to shielding provided by high-density clumps of gas and dust. Such clumps, known as cometary globules, have been observed in a handful of PNe, such as NGC 6302 (Santander-García *et al.* 2017), NGC 7293, and NGC 6853 (O'Dell and Handron 1996; O'Dell *et al.* 2002).

Aside from a handful of targeted searches, however, most PNe have not had their molecular content explored in any detail beyond CO and H_2, if at all. Given that most stars will pass through the PN stage at the end of their lives, it is vital to understand the molecular composition of these nebulae, how it evolves over the course of the PN stage, and what role PNe play in seeding the ISM. In order to elucidate the answers to these questions, we undertook a search for a range of polyatomic molecules in a set of PNe which had previously been detected in CO. This work began in 2012, with a search for HCN and HCO^+ in seventeen nebulae (Schmidt and Ziurys 2016). HCN and HCO^+ were chosen due to their high dipole moments (2.98 D and 3.9 D, respectively), which allow them to be used as dense-gas tracers, and their chemical significance (HCO^+ is a tracer of ion-molecule chemistry, while HCN can be compared to its isomer HNC). The seventeen target PNe were selected from the CO surveys of Huggins and Healy (1989), Huggins *et al.* (1996), and Huggins *et al.* (2005). All sources had a measured CO $J=2\rightarrow1$ line temperature of at least 50 mK and did not have any previously known HCN or HCO^+ emission. Notably, these PNe span a range of ages, sizes, and morphologies, including young, small sources such as K3-17 and K3-45 and large, old nebulae like NGC 6772. As a result, molecular abundances could be examined with respect to such factors.

Rotational transitions of HCN and HCO^+ were searched for in these seventeen objects using the facilities of the Arizona Radio Observatory (ARO). The 3 mm, dual polarization receiver of the ARO 12 m Telescope at Kitt Peak (both the original as well as the recently installed former European ALMA prototype antenna) was used to observe the $J=1\rightarrow0$ line of both molecules; to study the $J=3\rightarrow2$ transitions, the 1.3 mm dual polarization receiver of the ARO Submillimeter Telescope (SMT) on Mt. Graham, Arizona was employed (see Schmidt and Ziurys (2016) for observational details). Of the seventeen PNe, eleven were detected in at least one transition of HCN, and ten in one transition of HCO^+ (with one nebula, K4-47, having been detected in HCO^+ prior to this survey by Edwards *et al.* (2014)). The HCN spectra for K4-47, in addition to those for many other molecules observed in this source, are given in Figure 1. Further, details on the sources detected in HCN and/or HCO^+ are provided in Table 1.

HCN and HCO^+ column densities were estimated from these data using the non-LTE radiative transfer code, RADEX. For details of this analysis, see Schmidt and Ziurys (2016). The resulting column densities varied from $N_{tot}(HCN) \sim 0.2\text{-}27 \times 10^{13}$ cm-2 and $N_{tot}(HCO^+) \sim 0.3\text{-}8.7 \times 10^{13}$ cm^{-2}. Corresponding abundances with respect to H_2 were calculated, resulting in ranges of $f(HCN/H_2) \sim 0.1\text{-}9.1 \times 10^{-7}$ and $f(HCO^+/H_2) \sim 0.04\text{-}7.4 \times 10^{-7}$.

In addition to reinforcing the idea that polyatomic molecules are common constituents of PNe, one remarkable outcome of this work is the apparent lack of evolution of molecular abundance over the nebular lifetime. Figure 2 demonstrates this conclusion with a plot of abundance versus age for the planetary nebulae observed, with circles representing the sources observed by Schmidt & Ziurys (2016; 2017a,b) and stars representing those previously measured. Here, red represents HCN while blue represents HCO^+. Also plotted

Table 1. Summary of Observed Planetary Nebulae

Source	Age[a] (years)	Morphology / C/O [a]	Molecules Detected by Schmidt and Ziurys (2016, 2017a,b)
Hb 5	1500	bipolar, C-rich	HCN, HCO^+, HNC, CCH
K3-17	970	bipolar	HCN, HCO^+, HNC, CCH
K3-45	830	bipolar	HCN, HNC, CCH
K3-58	11,070	bipolar	HCN, HCO^+, HNC, CCH
K3-83	4600	bipolar	HCO^+
M1-7	6000	bipolar, C/O=1.1	HCN, HCO^+, HNC, CCH
M2-9	2500	bipolar, C/O<0.5	HCO^+
M3-28	7700	quadrupolar	HCN, HCO^+, HNC, CCH
M3-55	5040	bipolar	HCN, HCO^+, HNC, CCH
M4-14	5550	quadrupolar	HCN, HCO^+, HNC, CCH
NGC 2440	3400	quadrupolar (?), C/O=1.1-1.9	HCN, HCO^+, HNC
NGC 6772	10,900	elliptical	HCN, HNC
K4-47	900	bipolar, C-rich (?)	HCN, HNC, CCH
NGC 7293	12,000	bipolar	HCN, HNC

[a] See Schmidt and Ziurys (2016, 2017a,b) for reference information.

are the predictions of Redman *et al.* (2003) (using triangles), whose model considered the evolution of a wide array of molecules in the envelopes of PNe up to an age of 10,050 years. As the plot demonstrates, while the model of Redman *et al.* (2003) predicts significant destruction of HCN between 2550 and 10,000 years and consistently low abundances of HCO^+ (between 10^{-10} and 10^{-12}), our measured abundances show little temporal variation and abundances that are much higher than those predicted by the model at all stages. Further, the PN abundances are seen to be on the order of 1-2 magnitudes greater than those reported for diffuse clouds (e.g., Liszt *et al.* 2006). This is consistent with the idea that dense globules harboring molecules in PNe disperse into the diffuse ISM and seed it with molecular material.

Upon completion of this survey, we commenced a search for HNC in the eleven PNe in which HCN had been identified, with the goal of measuring [HCN]/[HNC] ratios. While HCN and HNC are both predicted to be produced through dissociative recombination of HCNH+ in equal amounts, the ratio in AGB stars is skewed far from unity due to the hot, LTE chemistry in the inner envelope, which favors HCN. The ratio was though to decrease from the AGB stage through the protoplanetary nebula (PPN) and PN stages; however, few PNe had been detected in transitions of both molecules, and accurate measures of the [HCN]/[HNC] ratio were difficult to come by. The J = 1→0 and J = 3→2 transitions of HNC were measured, once again with the new ARO 12 m and Submillimeter Telescopes (see Schmidt and Ziurys (2017a)). In addition, eight positions of the Helix Nebula, which had previously been detected in H_2CO, HCO^+, and CO by Zack & Ziurys (2013), were also observed at the J=1→0 transition frequencies of HCN and HNC in order to analyze the ratio across the nebula.

HNC was detected in at least one transition in all the nebulae studied, and HCN and HNC were identified at all positions observed in the Helix. The HNC spectra obtained for K4-47 are displayed in Figure 1, while the HCN and HNC spectra for all positions of the Helix are shown in Figure 3. Once again, column densities were determined using RADEX in a similar fashion as that described for HCN and HCO^+; for details, see Schmidt & Ziurys (2017a). For the Schmidt & Ziurys (2016) survey sources, the HNC column densities were found to range from $\sim$(0.06-4.0) $\times$ 10^{13} cm^{-2}. Helix column densities ranged from $\sim$(0.2-2.4) $\times$ 10^{12} cm^{-2} for HCN and (0.07-1.6) $\times$ 10^{12} cm^{-2} for HNC. Once again, fractional abundances were calculated with respect to H_2, and were found to range

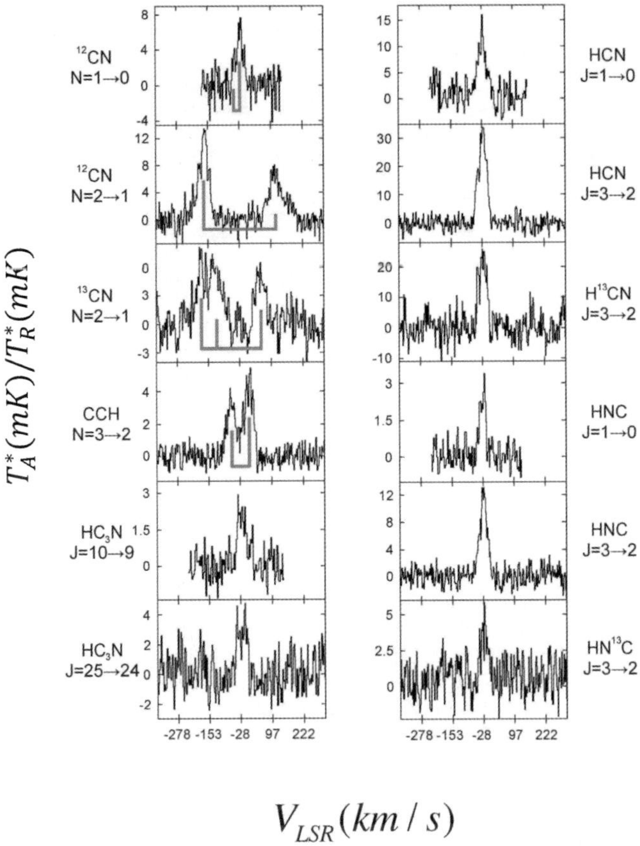

Figure 1. Spectra of the molecular transitions observed toward the planetary nebula K4-47 using the 12 m and Submillimeter Telescopes of the Arizona Radio Observatory. These include the ^{12}CN N=1→0 and N=2→1 transitions at 113 and 226 GHz, ^{13}CN: N=2→1 line at 217 GHz, CCH: N=3→2 line at 262 GHz, HC_3N: J=10→9 and J=25→24 transitions at 91 and 227 GHz, HCN: J=1→0 and J=3→2 lines at 88 GHz and 265 GHz, H^{13}CN: J=3→2 line at 259 GHz, HNC: J=1→0 and J=3→2 lines at 90 GHz and 271 GHz, HN^{13}C: J=3→2 line at 261 GHz. The strongest components of the CN hyperfine and CCH spin-rotation structures are indicated under the respective spectra.

from $f(\mathrm{HNC/H_2}) \sim (0.02\text{-}1.4) \times 10^{-7}$ for the survey sources. The fractional abundances for the Helix were found to be $f(\mathrm{HCN/H_2}) \sim (0.2\text{-}3.2) \times 10^{-7}$ and $f(\mathrm{HNC/H_2}) \sim (0.09\text{-}2.2) \times 10^{-7}$. The resulting [HCN]/[HNC] ratios were determined to vary between 2-8 for the survey sources and from 1-4 for the Helix.

The HNC abundances measured in this work showed no temporal variation, as opposed to the predictions of Redman *et al.* (2003) (see Figure 2; cyan symbols represent HNC measurements/expectations). While the model of Redman *et al.* (2003) predicted a sharp decrease of 10^4 by 6000 years, with a steady abundance of 10^{-12} thereafter, abundances calculated for our sources fell into the range $\sim 10^{-8} - 10^{-7}$, with no significant change between 1000 to 11,000 years. Further, the [HCN]/[HNC] ratios established in this work are notably constant with nebular age. These ratios are comparable to those measured for PPNe, and are significantly lower than those found for AGB stars (Woods *et al.* (2003), for instance, measured ratios of 40-300 for seven carbon stars). It is thought that the ratio significantly decreases in the PPN stage due to the increase in ionization, which drives reactions leading to the formation of $HCNH^+$, an intermediate in the formation of HNC

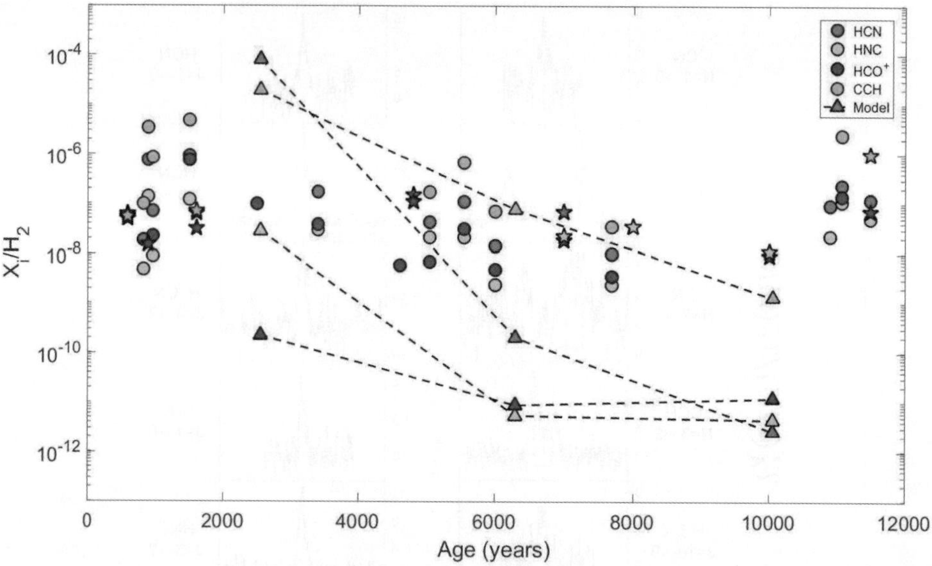

Figure 2. A plot of fractional abundances of HCN (red), HCO^+ (blue), HNC (cyan), and CCH (green) relative to H_2 in PNe vs. the ages of the PNe (in years). Circles represent data from the work of Schmidt and Ziurys (2016, 2017a,b), while stars indicate measurements for NGC 7027, NGC 6537, M2-48, NGC 7293, NGC 6853, M4-9, NGC 6781 and NGC 6720 from previous studies (see Schmidt and Ziurys 2016, 2017a; Zhang *et al.* 2008; Edwards and Ziurys 2013; Tenenbaum *et al.* 2009; Bachiller *et al.* 1997). The red, blue, cyan, and green triangles (with dashed lines) designate the fractional abundances for HCN, HCO^+, HNC, and CCH, respectively, from the model of Redman *et al.* (2003). The measured abundances of all molecules appear to stay relatively constant across the 10,000-year lifetime of PNe, while the model predicts a significant decrease during this time period.

through dissociative recombination. These and other reactions modify the [HCN]/[HNC] ratio, bringing it closer to unity with time. Based on our observations in PNe, we suggest that the chemistry which adjusts the abundances of AGB stars occurs primarily in the PPN stage, allowing the [HCN]/[HNC] ratio to freeze out in the PN phase. The fact that the ratio is remarkably consistent across the Helix Nebula provides further proof for the freeze-out hypothesis.

Additionally, comparison of the measurements presented in Schmidt and Ziurys (2017a) with those made in diffuse clouds by Liszt & Lucas (2001) and Liszt *et al.* (2006) once again show that the abundances measured in PNe are on the order of 1-2 orders of magnitude higher than those in diffuse clouds. Further, the average [HCN]/[HNC] ratio of 5 measured in diffuse clouds is comparable to those determined for PNe. This result is additional evidence that PNe seed the diffuse ISM with molecular material protected in dense clumps of gas and dust, which subsequently disperse into the surrounding medium. The gas is too diffuse for significant chemistry such that the [HCN]/[HNC] ratio remains fairly constant.

It is interesting that the ^{13}C isotopologues of HCN, HNC, and CN were often observed in these sources, when actually searched for. The carbon-13 species was observed in all three species in K4-47, for example, as shown in Figure 1. These three species suggest a $^{12}C/^{13}C$ ratio of $\sim$ 2-3. Considering the carbon-rich nature of this object, this very low ratio suggests that a J-type star was the progenitor for K4-47.

Due to the likelihood that the PNe containing both HCN and HNC are carbon-rich sources, we decided to initiate a search for CCH in the eleven PNe in which both molecules

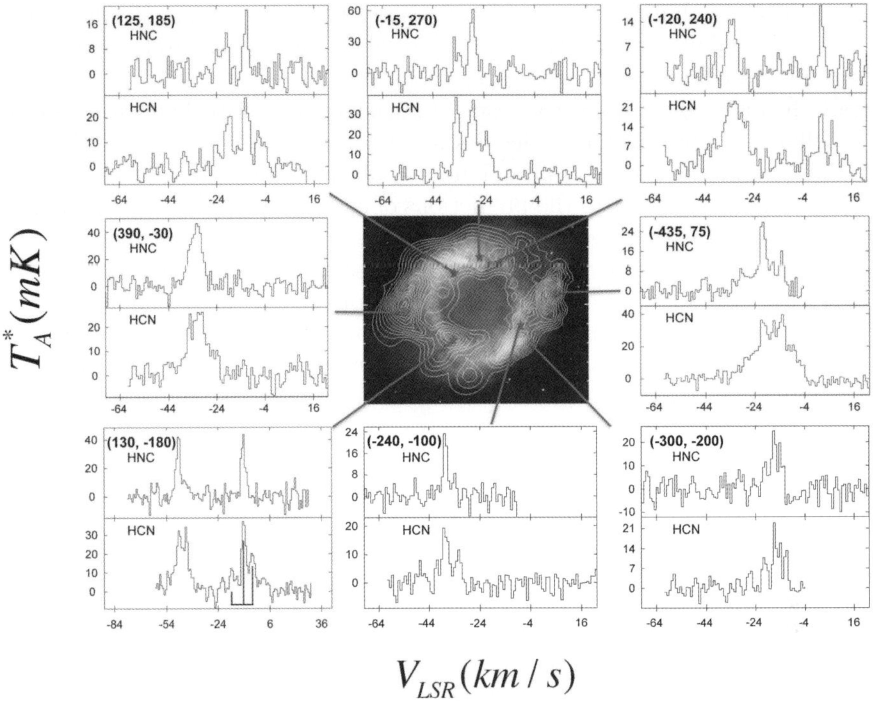

Figure 3. Spectra of the J = 1→0 transitions of HNC (upper panel) and HCN (lower panel) observed toward eight positions in the Helix Nebula, using the ARO 12 m telescope, the ALMA European prototype antenna. The R.A. and decl. offsets for each position (in arcseconds from the central star) are displayed in the top left corner of each panel. The HCN LTE hyperfine pattern is shown beneath the 10 km s^{1} velocity component of the (130, 180) spectrum. Positions are indicated on an optical image of the Helix (NASA *et al.* 2003). Overlaid contours on the image indicate the peak brightness temperature of the J = 1→0 transition of HCO^+ (from Zeigler *et al.* 2013). HCN and HNC are clearly visible in all positions, showing that these molecules have a widespread distribution in the nebula, and trace the complex, bipolar velocity structure of this object seen in HCO^+.

had been detected. Furthermore, two PNe in which C_{60} had previously been identified (M1-12 and M1-20; García-Hernández *et al.* 2010) were also observed. These targets were selected as C_2 is seen to be a building block of C_{60} in the laboratory, as well as a by-product of fragmentation of large carbon clusters. Given the inability for C_2 to be detected in the millimeter regime due to its lack of an electric dipole moment, CCH is a proxy, thereby allowing the exploration of the potential connection between C_{60} and C_2 in space. The results of this search will be published in a forthcoming work (Schmidt and Ziurys 2017b).

Using the facilities of the ARO, the N=1→0 and N=3→2 transitions of CCH were observed toward these target sources. Of the thirteen target sources, nine were detected in at least one transition of CCH, all from the original Schmidt and Ziurys surveys. Through a similar analysis to those for HCN, HCO^+, and HNC, column densities were determined, and were seen to range between 0.2-3.3 × 10^{15} cm^{-2}. Correspondingly, fractional abundances varied between ∼0.02-4.7 × 10^{-6}. One of the primary outcomes of this work was, once more, the lack of variation in abundance with nebular age (see Figure 2; green symbols represent CCH measurements/predictions). Redman *et al.* (2003) expect a decline in abundance of 10^4 over the course of the nebular lifetime; however, our measurements show that the ratio remains within the range of ∼10^{-7}-10^{-6} over this time. Further, these abundances are approximately 1-2 orders of magnitude greater than those measured in diffuse clouds (Lucas & Liszt 2000; Gerin *et al.* 2011), adding yet more evidence to the seeding hypothesis described earlier.

Overall, the main conclusions from each of the surveys described above can be summarized as follows: (1) polyatomic molecules are a common constituent of PNe regardless of age, size, and morphology, (2) abundances of all molecules studied show no variation with nebular age, in contrast to the predictions of chemical models, (3) abundances of all molecules compared with those measured for diffuse clouds are consistent with the idea that dense clumps of gas and dust in PNe disperse into the surrounding ISM, thereby seeding it with molecular material. More observations of a greater number of PNe and a wider variety of molecules will of course be necessary in order to provide a fuller understanding of the molecular content of PNe. However, observations of PNe like K4-47 (see Figure 1) already prove that such an undertaking can be incredibly fruitful.

References

Bachiller, R., Forveille, T., Huggins, P. J., & Cox, P. 1997, *A&A*, 324, 1123
Cami, J., Bernard-Salas, J., Peeters, E., & Malek, S. E. 2010, *Sci*, 329, 1180
Edwards, J. L. & Ziurys, L. M. 2013, *ApJL*, 770, L5
Edwards, J. L., Cox, E. G., & Ziurys, L. M. 2014, *ApJ*, 791, 79
Edwards, J. L. & Ziurys, L. M. 2014, *ApJL*, 794, L27
García-Hernández, D. A., Iglesias-Groth, S., & Acosta-Pulido, J. A., *et al.* 2011, *ApJL*, 737, L30
García-Hernández, D. A., Manchado, A., & Garca-Lario, P., *et al.* 2010, *ApJL*, 724, L39
García-Hernández, D. A., Villaver, E., & Garca-Lario, P., *et al.* 2012, *ApJ*, 760, 107
Gerin, M., Kamierczak, M., Jastrzebska, M., *et al.* 2011, *A&A*, 525, A116
Hora, J. L, Latter, W. B., & Deutsch, L. K. 1999, *ApJS*, 124, 195
Huggins, P. J., Bachiller, R., Cox, P., & Forveille, T. 1996, *A&A*, 315, 284
Huggins, P. J., Bachiller, R., Planesas, P., Forveille, T., & Cox, P. 2005, *ApJS*, 160, 272
Huggins, P. J. & Healy, A. P. 1989, *ApJ*, 346, 201
Kwok, S. 2000, The Origin and Evolution of Planetary Nebulae (Cambridge: Cambridge Univ. Press)
Likkel, L., Dinerstein, H. L., Lester, D. F., Kindt, A., & Bartig, K. 2006, *AJ*, 131, 1515
Liszt, H. & Lucas, R. 2001, *A&A*, 370, 576
Liszt, H. S., Lucas, R., & Pety, J. 2006, *A&A*, 448, 253
Lucas, R. & Liszt, H. S. 2000, *A&A*, 358, 1069
O′Dell, C. R. & Handron, K. D. 1996, *AJ*, 111, 1630
O′Dell, C. R., Balick, B., Hajian, A. R., Henney, W. J., & Burkert, A. 2002, *AJ*, 123, 332
Otsuka, M., Kemper, F., Hyung, S., *et al.* 2013, *ApJ*, 764, 77
Redman, M. P., Viti, S., Cau, P., & Williams, D. A. 2003, *MNRAS*, 345, 1291
Santander-García, M., Bujarrabal, V., Alcolea, J., *et al.* 2017, *A&A*, 597, A27
Schmidt, D. R. & Ziurys, L. M. 2016, *ApJ*, 817, 175
Schmidt, D. R. & Ziurys, L .M. 2017a, *ApJ*, 835, 79
Schmidt, D. R. & Ziurys, L. M. 2017b, submitted
Tenenbaum, E. D., Milam, S. N., Woolf, N. J., & Ziurys, L. M. 2009, *ApJL*, 704, L108
Woods, P. M., Schier, F. L., Nyman, L. Å., & Olofsson, H. 2003, *A&A*, 402, 617
Zack, L. N. & Ziurys, L. M. 2013, *ApJ*, 765, 112
Zeigler, N., Zack, L. N., Woolf, N. J., & Ziurys, L. M. 2013, *ApJ*, 778, 16
Zhang, Y., Kwok, S., & Trung, D.-V. 2008, *ApJ*, 678, 328

Astrochemistry VII – Through the Cosmos from Galaxies to Planets
Proceedings IAU Symposium No. 332, 2017
M. Cunningham, T. Millar & Y. Aikawa, eds.

doi:10.1017/S1743921317007608

Shocks and UV radiation around low-mass protostars: the Herschel-PACS legacy

Agata Karska[1,2,3], Michael J. Kaufman[4], Lars E. Kristensen[5], Ewine F. van Dishoeck[3,2] and WISH, DIGIT, and WILL teams

[1]Centre for Astronomy, Faculty of Physics, Astronomy and Informatics, Nicolaus Copernicus University, Grudziadzka 5, 87-100 Torun, Poland
email: agata.karska@umk.pl

[2]Max-Planck Institut für Extraterrestrische Physik (MPE), Giessenbachstr. 1, D-85748 Garching, Germany

[3]Leiden Observatory, Leiden University, P.O. Box 9513, 2300 RA Leiden, The Netherlands

[4]Department of Physics and Astronomy, San Jose State University, One Washington Square, San Jose, CA 95192-0106, USA

[5]Centre for Star and Planet Formation, Niels Bohr Institute, University of Copenhagen, Øster Voldgade 5-7, DK-1350 Copenhagen K, Denmark

Abstract. Far-infrared spectroscopy reveals gas cooling and its underlying heating due to physical processes taking place in the surroundings of protostars. These processes are reflected in both the chemistry and excitation of abundant molecular species. Here, we present the *Herschel*-PACS far-IR spectroscopy of 90 embedded low-mass protostars from the WISH (van Dishoeck *et al.* 2011), DIGIT (Green *et al.* 2013), and WILL surveys (Mottram *et al.* 2017). The 5 × 5 spectra covering the $\sim 50'' \times 50''$ field-of-view include rotational transitions of CO, H_2O, and OH lines, as well as fine-structure [O I] and [C II] in the ~50-200 μm range. The CO rotational temperatures are typically ~300 K, with some sources showing additional components with temperatures as high as ~1000 K. The H_2O / CO and H_2O / OH flux ratios are low compared to stationary shock models, suggesting that UV photons may dissociate some H_2O and decrease its abundance. Comparison to C shock models illuminated by UV photons show a good agreement between the line emission and the models for pre-shock densities of 10^5 cm^{-3} and UV fields 0.1-10 times the interstellar value. The far-infrared molecular and atomic lines are the unique diagnostic of shocks and UV fields in deeply-embedded sources.

Keywords. Young stellar objects, outflows, jets, shocks

1. Introduction

During the early formation of low-mass protostars, the mass accretion rates are high and thus, the feedback from the protostar on its surrounding is most spectacular. The launching of jets and winds creates outflow cavities and generates shock waves propagating in the envelope. Shocks compress and heat the gas but may also produce UV photons, which penetrate to large distances due to low densities and scattering in the outflow cavities. Characterising these processes requires spectroscopy in the far-infrared (IR), where cooling of the warm, dense gas is very efficient (Kaufman & Neufeld 1996). The question remains, how do these feedback processes influence the initial conditions (both physics and chemistry) of star and planet formation.

In this paper, we briefly summarize the survey of 90 deeply-embedded low-mass protostars observed as part of the three large *Herschel* programs: 'Water in Star forming regions with Herschel' (WISH, van Dishoeck *et al.* 2011), 'Dust, Ice, and Gas in Time'

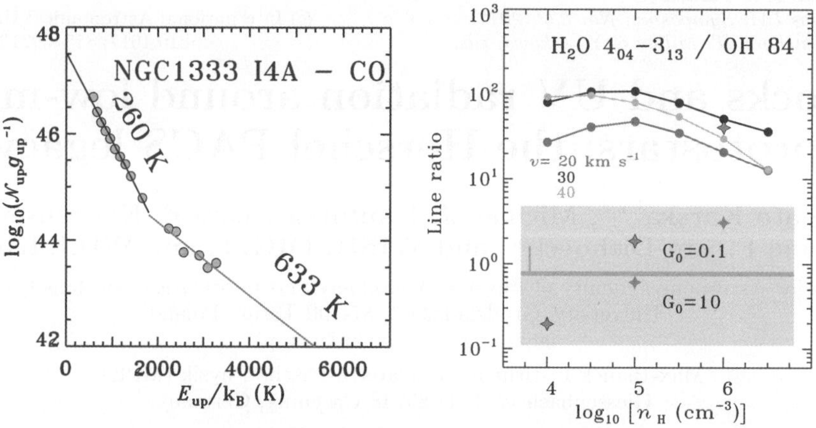

Figure 1. *Left:* CO rotational diagram of a Class 0 protostar NGC 1333 IRAS 4A. *Right:* Line ratio of H_2O $4_{04} - 3_{13}$ at 125 μm and OH at 84 μm as a function of logarithm of density of the pre-shock gas. The range observed in ~90 protostars is shown as filled, light-blue box, and the models of C shocks are shown as solid lines (models of shielded shocks, Kaufman & Neufeld 1996) and violet stars (models of UV illuminated shocks, Melnick & Kaufman 2015).

(DIGIT, Green *et al.* 2013), and 'William Herschel Line Legacy' (WILL, Mottram *et al.* 2017). All observations were obtained using the Photodetector Array Camera and Spectrometer (PACS, Poglitsch *et al.* 2010) that provided spectral maps consisting of 25 spatial pixels of ~ 9.4″ × 9.4″, corresponding to pixel sizes of ~ 2000 × 2000 AU for typical distances of 200 pc.

2. Results

Rich molecular line emission is commonly detected in Class 0/I protostars including highly-excited H_2O ($E_{\rm up}$ ~ 1000 K) and CO lines ($E_{\rm up}$ ~ 5000 K). The line fluxes of molecular species correlate strongly with each other and show similar extent of emission, typically following the outflow direction. Multiple transitions of CO lines are used to determine rotational temperatures of the gas (see left panel of Figure 1). A ~ 300 K component is universally seen in almost all protostars, whereas the hotter gas component show a broad distribution with a median at 720 K (detected in 27% of sources). These two components are identified in the spectrally-resolved CO 16-15 lines obtained with HIFI (Kristensen *et al.* 2017). The ~ 300 K component in the PACS CO ladder corresponds to the broad, Gaussian component in the line profiles originating in outflow cavity shocks or disk winds. The hotter component is associated with the offset component likely produced in the irradiated shocks (see Kristensen *et al.* 2017, for discussion).

The right panel of Figure 1 shows a comparison of the measured H_2O / OH line ratios and the predictions from stationary (1D plane-parallel) C shock models from Kaufman & Neufeld (1996). Clearly, these models overestimate the observed line ratios by 1-2 orders of magnitude. Low H_2O / OH line ratios require partial photodissociation of H_2O into OH and O. Predictions from C shock models illuminated by UV photons (Melnick & Kaufman 2015) reproduce the observed ranges of H_2O / OH for pre-shock densities of 10^5 cm^{-3} and UV fields 0.1-10 times the interstellar value. These preliminary comparisons show that UV radiation influences the chemistry in deeply-embedded protostars, as also evidenced by observations of light ionised hydrides (Kristensen *et al.* 2013, Benz *et al.* 2016).

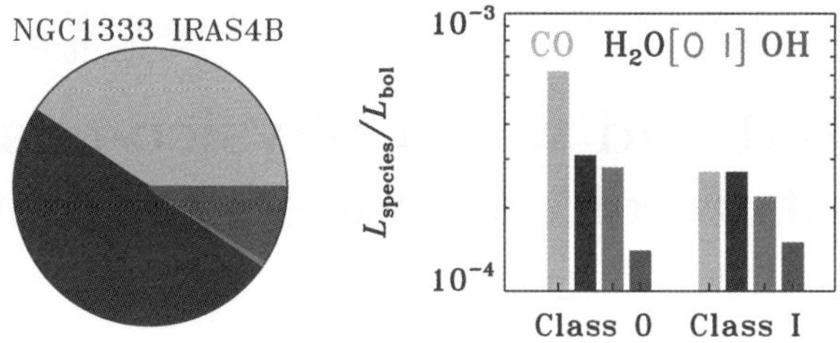

Figure 2. *Left:* Fractions of gas cooling contributed by CO (orange), H_2O (blue), OH (green), and [O I] (red), to the total far-infrared line cooling in Class 0 protostar NGC 1333 IRAS 4B. *Right:* Median line cooling in CO, H_2O, [O I], and OH over bolometric luminosities for Class 0 and Class I sources.

Low abundances of H_2O are also confirmed by relatively low total line cooling in H_2O versus other molecules in far-IR. Kaufman & Neufeld (1996) predict that H_2O should account for 70-90% of total molecular line cooling in the far-IR (H_2O+CO+OH, Karska *et al.* 2014). At the same time, the observed H_2O emission is typically below 30% and at most about 50% in the most H_2O-rich source NGC1333 IRAS4B (left panel of Figure 2). Clearly, some H_2O is missing and the UV photodissociation is a possible mechanism to decrease its abundance.

Far-IR molecular line cooling decreases from Class 0 to Class I (right panel of Figure 2), but the differences are significant only in CO and not in H_2O (Karska *et al.* 2018). Proper interpretation of these trends requires the analysis of velocity-resolved profiles, which reveal multiple kinematic components with varying contributions depending on the transition (e.g. Kristensen *et al.* 2017).

The far-IR data presented here demonstrate the necessity for excellent space-based facilities to quantify the chemistry and excitation of warm molecular and atomic gas. The next major step will occur when the *James Webb Space Telescope* (JWST) is launched in 2018; this telescope will provide crucial data on not only the fmid-IR cooling species, but most importantly on the dominant gas coolant: molecular hydrogen, H_2. Direct measurements of the atomic and molecular abundances will be performed and tested against sophisticated shock models.

Acknowledgement. The authors are grateful to G. Herczeg, J. Mottram, L. Tychoniec, J. Lindberg, N. Evans II, J. Green, Y.-L. Yang, A. Gusdorf, and N. Siodmiak for collaborations on this project. This work is supported by the Polish National Science Center grant 2013/11/N/ST9/00400 and 2016/21/D/ST9/01098.

References

Benz, A. O., Bruderer, S., van Dishoeck, E. F., *et al.* 2016, *A&A*, 590, A105
Green, J. D., Evans, II, N. J., Jrgensen, J. K., *et al.* 2013, *ApJ*, 770, 123
Karska, A., Kristensen, L. E., van Dishoeck *et al.* 2014, *A&A*, 572, A9
Karska, A., Kaufman, M., Kristensen, L.E., *et al.* 2018, *ApJS*, 235, 30
Kaufman, M. J. & Neufeld, D. A. 1996, *ApJ*, 456, 611
Kristensen, L. E., van Dishoeck, E. F., Benz, A. O. *et al.* 2013, *A&A*, 557, A23
Kristensen, L. E., van Dishoeck, E. F., Mottram, J. C. *et al.* 2017, arXiv No. 1705.10269
Melnick, G. J. & Kaufman, M. J. 2015, *ApJ*, 806, 227
Mottram, J. C., van Dishoeck, E. F., Kristensen, L. E., *et al.* 2017, *A&A*, 600, A99
Poglitsch, A., Waelkens, C., Geis, N., *et al.* 2017, *A&A*, 518, L2
van Dishoeck, E. F., Kristensen, L. E., Benz, A. O., *et al.* 2011, *PASP*, 123, 138

Astrochemistry VII – Through the Cosmos from Galaxies to Planets
Proceedings IAU Symposium No. 332, 2017
M. Cunningham, T. Millar & Y. Aikawa, eds.

doi:10.1017/S1743921317008080

The cold and warm physico-chemical structure of embedded protostars

N. M. Murillo

Leiden Observatory, Leiden University, P.O. Box 9513, 2300 RA, Leiden, the Netherlands
email: nmurillo@strw.leidenuniv.nl

Abstract. The physical evolution of low-mass protostars is relatively well-established, however, there are many open questions on the chemical structure of protostars. The chemical fingerprint generated in the early embedded phase of star formation may be transmitted to the later stages of star, planet and comet formation. The factors that influence the chemical fingerprint are then of interest to study, and determine whether the chemical structure is inherited from the parent cloud or product of the physical processes during star formation. Results of observations and modelling of molecules that trace the cold and warm extended structures of embedded protostars are briefly presented here. Two multiple protostellar systems are studied, IRAS 16293-2422 and VLA 1623-2417, both located in ρ Ophiuchus. We find that the physical structure of the protostars, that is the disk(-like) strucutres, outflow cavity and different luminosities, are important factors in determining the chemical structure of these embedded protostars.

Keywords. astrochemistry, instrumentation: interferometers, methods: data analysis, stars: formation, stars: individual (IRAS 16293-2422, VLA 1623-2417)

1. Introduction and observations

Two low-mass embedded Class 0 protostars, both located in ρ Ophiuchus (d $\sim$ 120 pc) only 2.xx pc apart, are studied and compared in order to explore the chemical structure of the earliest stages of star formation. IRAS 16293-2422, located in L1689N, is a well-studied multiple system with the main components A and B separated by 800 AU. Source A drives a complex outflow, has a large, close to edge-on, disk-like structure with a radius of 200 AU (Oya *et al.* 2016), and has a bolometric luminosity of 18 $L_\odot$ (Jacobsen *et al.* submitted). Source B has a bolometric luminosity of 3$L_\odot$ (Jacobsen *et al.* submitted) and is positioned face-on, thus the presence of outflow and disk-like structure is difficult to determine. VLA 1623-2417, located in L1688 (ρ Oph A), is a triple protostellar system with a prominent outflow. The three components have similar inclinations, but are at different evolutionary stages. Source A, with a bolometric luminosity of 1 $L_\odot$, has a confirmed rotationally supported disk with a radius of 150–180 AU (Murillo *et al.* 2013). VLA1623-2417 has been shown to be largely line poor, in contrast to IRAS 16293-2422 which shows complex organic molecules. This study does not focus on the compact "hot core" region (<1"), but on the more extended cold and warm gas.

The results and analysis of Atacama Large Millimeter/submillimeter Array (ALMA) observations of DCO^+, N_2D^+, $c{-}C_3H_2$ and C_2H toward both embedded protostellar systems are presented here. DCO^+ and N_2D^+ are good tracers of cold molecular gas where CO is frozen out onto the dust grains, and are found in the envelope of young embedded protostars. $c{-}C_3H_2$ and C_2H trace warmer gas due to being produced when material is UV-irradiated. The two molecules are generally found in photon-dominated regions (PDRs) and outflow cavities in protostellar disk-like structures.

IRAS 16293-2422 was observed in the "Protostellar Interferometric Line Survey" (PILS) program (Project-ID: 2013.1.00278.S; PI: Jes K. Jørgensen) in Cycle 2 with Band 7, using both the main 12-m array and the Atacama Compact Array (ACA). The combined observations provide a synthesized beam of 0.5". Detailed description of the observations and data reduction is given in Jørgensen *et al.* (2016).

VLA 1623-2417 was observed in Cycle 0 and 2, centered on source A. Cycle 2 includes the ACA. Band 6 observations of DCO^+ 3–2 provided a resolution of 0.85"×0.65" in Cycle 0. The Band 6 observations of $c-C_3H_2$ have a resolution of 1.68"×0.88". Band 7 observations of N_2D^+ have a spatial resolution of of 0.87"×0.54". Details on the data and reduction are given in Murillo *et al.* (2015; in prep.). Additional single-dish Atacama Pathfinder EXperiment (APEX) observations of C_2H toward VLA 1623-2417 were carried out using single-pointing mode with a beam of 28.7" (Murillo *et al.* in prep.).

2. Results and Analysis

DCO^+ 5–4 is detected in IRAS 16293-2422 around source A in a half-crescent shape, offset from the source position by 2.5" (Figure 1). No DCO^+ emission is found around source B. Toward VLA 1623-2417, DCO^+ 3–2 is also found around source A, but not the other two components of the system, B and W. DCO^+ 3–2 peaks about 2.5" away from source A to the south (Figure 1). In order to understand the observed DCO^+ emission, the simple chemical model from Murillo *et al.* (2015) is used, coupled with spherically symmetric density and temperature profiles for each source (IRAS 16293-2422: Crimier *et al.* 2010; VLA 1623-2417: Jørgensen *et al.* 2002). The model results show that the peak of the DCO^+ emission is not where expected from the spherically symmetric profiles (IRAS 16293-2422: ~10"; VLA 1623-2417: ~5"). Instead, a drop in the temperature profile is needed to reproduce the observed DCO^+ peak position. The model results indicate that the temperature at the DCO^+ peak position is 18 K and 14 K for IRAS 16293-2422 and VLA 1623-2417, respectively. This drop in temperature is most likely caused by the presence of the disk(-like) structure, which shadows the envelope material at its edge, generating an inward shift of cold molecules along the flattened structure but not elsewhere.

The position of the N_2D^+ 3–2 detected around IRAS 16293-2422 A (Jorgensen *et al.* 2011) is also well described by the temperature structure obtained from the DCO^+ observations. For VLA 1623-2417, N_2D^+ 3–2 is not detected with ALMA. Single-dish observations find N_2D^+ peaks to the north and away from VLA 1623-2417 by XX (Punanova *et al.* 2016). The lack of compact N_2D^+ emission in the inner envelope of VLA 1623-2417 can be explained as N_2, the precursor of N_2D^+, being frozen onto the dust grains due to low gas and dust temperatures. This is also in agreement with the temperature structure obtained from modeling the DCO^+ observations.

$c-C_3H_2$ is found to trace the outflow cavity of VLA 1623-2417 A, and one side of the south outflow cavity of IRAS 16293-2422 A (Figure 1). This is consistent with observations of other objects that find $c-C_3H_2$ in UV-irradiated regions. No $c-C_3H_2$ is found in the disk(-like) structures of either source, in contrast to what is observed in L1527 (Sakai *et al.* 2014). C_2H is detected on-source of VLA 1623-2417 A with single-dish observations, and is either located in the envelope or outflow cavity, indicating spatial correlation with the observed $c-C_3H_2$ emission around source A. In contrast, toward IRAS 16293-2422 A, C_2H does not show any spatial correlation with $c-C_3H_2$ (Figure 1). Instead, C_2H is located in a clumpy filament-like structure stretching from north to south, and apparently passing through IRAS 16293-2422 B. This lack of correlation is surprising, since both molecules have been found to have a correlation from low- and high-mass

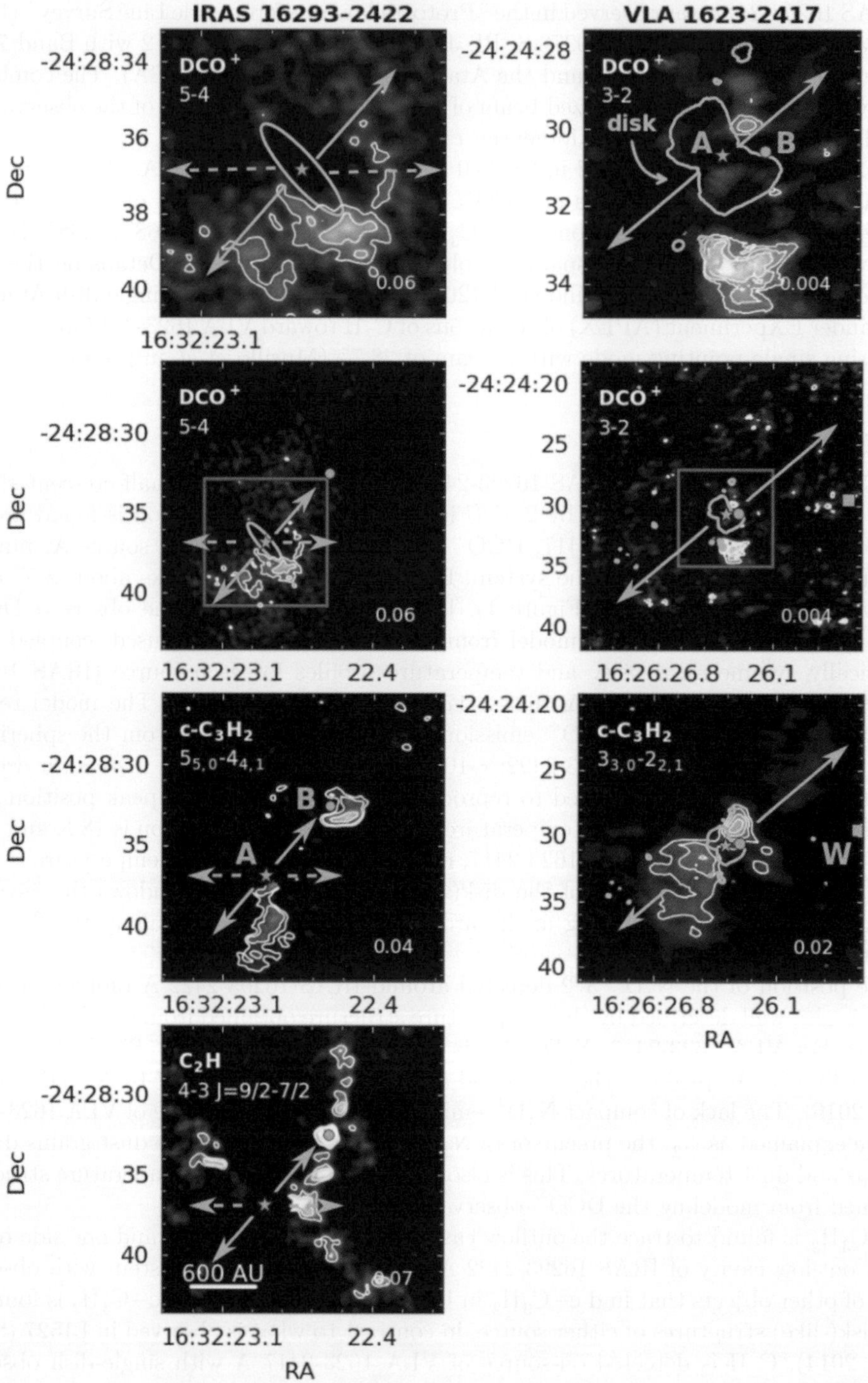

Figure 1. Intensity integrated maps of DCO$^+$, c$-$C$_3$H$_2$ and C$_2$H toward IRAS 16293-2422 (left column), and VLA 1623-2422 (right column). The green box shows the zoomed region shown in the top row. The gray arrows indicate the direction of the outflow(s). The cyan contour traces the extent of the observed disk(-like) structures in each system. Contours are in steps of 2, 3 and 5σ for IRAS 16293-2422. For VLA 1623-2417, DCO$^+$ is in steps of 4, 5 and 6σ, and c$-$C$_3$H$_2$ is in steps of 3, 4, 5 and 6σ, with σ (Jy beam^{-1} km s^{-1}) indicated in the lower right of each panel.

protostars, as well as in diffuse clouds and PDRs. A possible explanation for the lack of C_2H in the outflow cavity could be the reaction of C_2H with sulfur, nitrogen or carbon chains. However, molecules such as C_2S and C_2N are not found in the same region as $c-C_3H_2$. Another explanation could be that C_2H is reacting with carbon chains, producing even larger carbon chains and molecular hydrogen, and thus cannot be detected due to the lack of a dipole moment. The physical structure of IRAS 16293-2422 could also impact the spatial distribution of $c-C_3H_2$ and C_2H.

3. Conclusions

Using cold (DCO^+, N_2D^+) and warm ($c-C_3H_2$, C_2H) molecules, the chemical structure of two low-mass embedded protostars, IRAS 16293-2422 and VLA 1623-2417, is studied. In IRAS 16293-2422 and VLA 1623-2417, the spatial distribution of DCO^+, N_2D^+ and $c-C_3H_2$ can be explained with the same physical structures: drop in temperature for the cold molecules, and irradiated outflow cavities for $c-C_3H_2$. C_2H, on the other hand, is spatially correlated in VLA 1623-2417 but not in IRAS 16293-2422. While still a chemical puzzle, the anti-correlation of C_2H and $c-C_3H_2$ in IRAS 16293-2422 could be due to the complex structure of the system. Two conclusions can be drawn from the comparison of these two systems.

(*a*) Temperature is an important controlling factor in the chemical structure of embedded protostars.

(*b*) The presence of disk(-like) structures produces a drastic impact on the physical and chemical structure of the protostar, since the temperature structure becomes asymmetric and cold close to the source.

It is interesting to note that both systems studied here are multiple systems, and that the individual components have different chemical structures. This has also been observed in other embedded systems (e.g. NGC1333 SVS13: Chen *et al.* 2009; NGC1333 IRAS4A: Persson *et al.* 2012, López-Sepulcre *et al.* 2017). Thus, it is possible that there is no relation between multiplicity and chemical structure, at least in the embedded phase. In later stages, when the envelope has cleared, perhaps multiplicity plays a larger role in the chemical structure of stellar siblings.

Acknowledgements

The author thanks E. F. van Dishoeck, J. K. Jørgensen, M. H. D. van der Wiel, C. Walsh, S. Bruderer, D. Harsono, M. N. Drozdovskaya, S.-P. Lai, C. M. Fuchs and H. Calcutt for useful discussions on the observaitons and modelling that made this work possible.

References

Chen, X., Launhardt, R., & Henning, T, 2009, *ApJ*,a 691, 1729
Crimier, N., Ceccarelli, C., Maret, S., Bottinelli, S. *et al.* 2010, *A&A* 519, 65
Jørgensen, J. K., Schöier, F. L., & van Dishoeck, E. F., *A&A*, 389, 908
Jørgensen, J. K., Bourke, T. L., Nguyen Luong, Q., & Takakuwa, S., *A&A*, 534, 100
Jørgensen, J. K., van der Wiel, M. H. D., Coutens, A., Lykke, J. M. *et al.* 2016, *A&A*, 595, 117
López-Sepulcre, A., Sakai, N., Neri, R., Imai, M. *et al.* 2017, *arXiv:1707.03745*
Murillo, N. M., Lai, S.-P., Bruderer, S., Harsono, D., & van Dishoeck, E. F., 2013, *A&A*, 560, 103

Murillo, N. M., Bruderer, S., van Dishoeck, E. F., Walsh, C., *et al.* 2015, *A&A*, 579, 114
Persson, M. V., Jørgensen, J. K., & van Dishoeck, E. F., 2012, *A&A*, 541, 39
Punanova, A., Caselli, P., Pon, A., Belloche, A. & André, Ph., 2016 *A&A*, 587, 118
Oya, Y., Sakai, N., López-Sepulcre, A., Watanabe, Y., Ceccarelli, C. *et al.* 2016, *ApJ*, 824, 88
Sakai, Nami; S., T., Hirota, T., Watanabe, Y., Ceccarelli, C. *et al.* 2014, *Nature*, 507, 78

Astrochemistry VII – Through the Cosmos from Galaxies to Planets
Proceedings IAU Symposium No. 332, 2017
M. Cunningham, T. Millar & Y. Aikawa, eds.

doi:10.1017/S1743921317007839

Rings, rings, rings: what does CN tell us?

Paolo Cazzoletti

Max-Planck-Institut für Extraterrestrische Physik, Gieenbachstrae, 85741, Garching bei München, Germany
email: pcazzoletti@mpe.mpg.de

Abstract. CN emission lines are among the brightest, and have been observed in the last 20 years with single dish observations. With modern interferometers, we are now able to spatially resolve CN emission, which often shows ring-like structures. We investigate whether such structures trace the morphology of the disks, or if they have a chemical origin. By using the thermochemical code DALI, we conclude that CN formation is triggered by the existence of vibrationally excited H_2^*, produced by FUV pumping of H_2. Herbig stars therefore generally have larger rings and higher CN fluxes than TTauri. Disks with higher masses and flaring also show stronger CN emission and larger rings. CN observations could in the future provide important constraints on some important disk physical parameters. The results of the models are well consistent with the spatially resolved CN observations to-date available.

Keywords. astrochemistry – planetary systems: protoplanetary disks –methods: numerical – radiative transfer

1. Motivation

CN is one of the brightest molecules that have been observed in protoplanetary disks, and its lines can be as bright as those of ^{13}CO or even brighter. For this reason CN has been observed and studied both with single-dish (Thi *et al.* 2004; Kastner *et al.* 2008, 2014; Salter *et al.* 2011; Guilloteau *et al.* 2016) and interferometric observations (Dutrey *et al.* 1997; Öberg *et al.* 2010, 2011; Chapillon *et al.* 2012; Guilloteau *et al.* 2014)

However, only in recent years have spatially resolved observations of CN become available (Teague *et al.* 2016; van Terwisga *et al.* in prep.), showing ring shaped emission. The well studied disk around TW Hya, for example, shows a ring located at a radius of 47 AU from the star (Fig. 1), and the same is evident in two disks in the Lupus young star-forming region, namely Sz 78 and Sz 91.

We investigate whether such rings are due to the global morphology of the disks (e.g. transitional disks), or if they can originate from chemical processes depleting the inner disk of CN even in full disks. We also study which stellar and physical disk parameters mostly affect CN emission in order to understand what information can CN observation provide about protoplanetary systems.

2. Methods

We run a grid of full disk models using the thermochemical code DALI (Bruderer 2013) to self-consistently solve for the abundance of molecular species and the thermal balance of the gas. A chemical network accounting for the most relevant nitrogen chemistry reactions is adopted (Visser *et al.* in prep., this volume). The total mass of the disks M_{tot} is varied between 10^{-5} and 10^{-1} $M_{\odot}$, the flaring of the disk ψ between 0.1 and 0.3, and the disk aspect-ratio between 0.1 and 0.2. Two different stellar spectra are adopted,

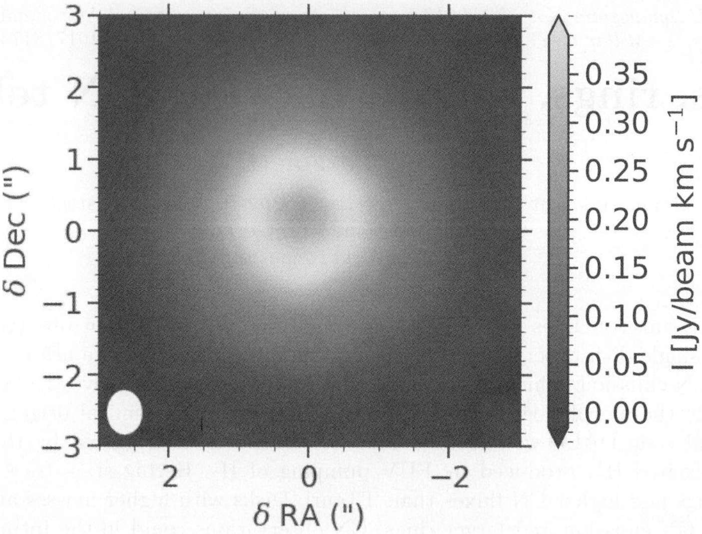

Figure 1. The CN ($N = 2 - 1$), 0.9" (47 AU) radius ring observed with ALMA in TW Hya (data from Teague *et al.* 2016).

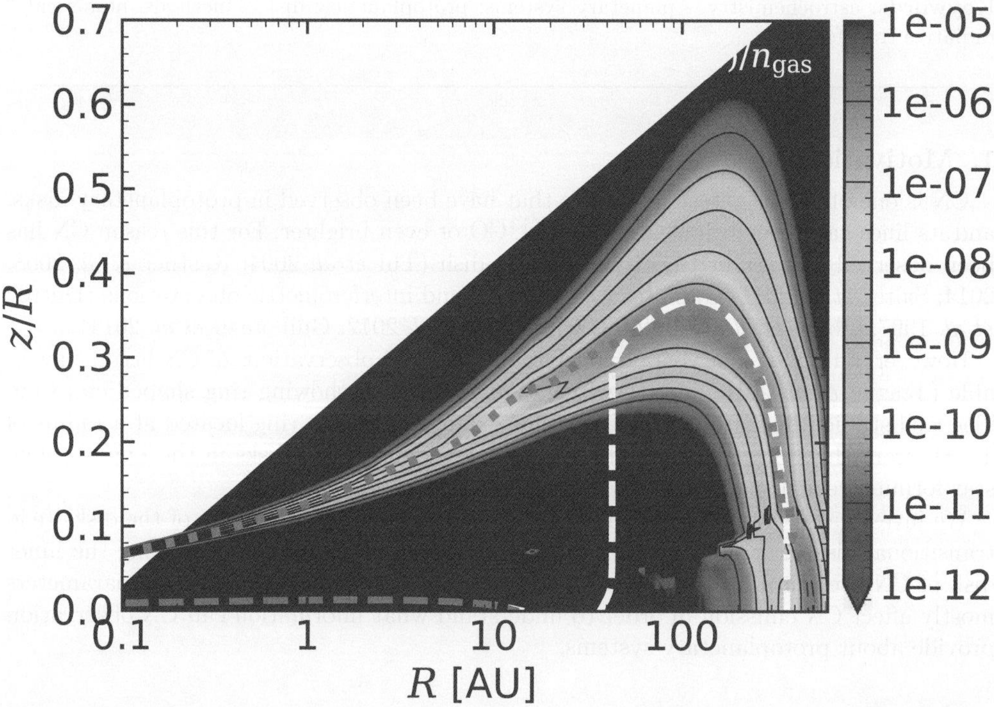

Figure 2. Abundance of CN in a representative disk around a T Tauri star. The green dot marks the location of the maximum, while the dotted green line marks the location of the vertical CN abundance maximum at each radius. The white dashed line marks the $\tau = 1$ layer of the CN ($N = 3 - 2$) line. Similarly, the red dashed line shows the $\tau = 1$ layer at the same wavelength for the dust. Such a profile is consistent with the observed ring-shaped emission. Adapted from Cazzoletti *et al.* (in press)

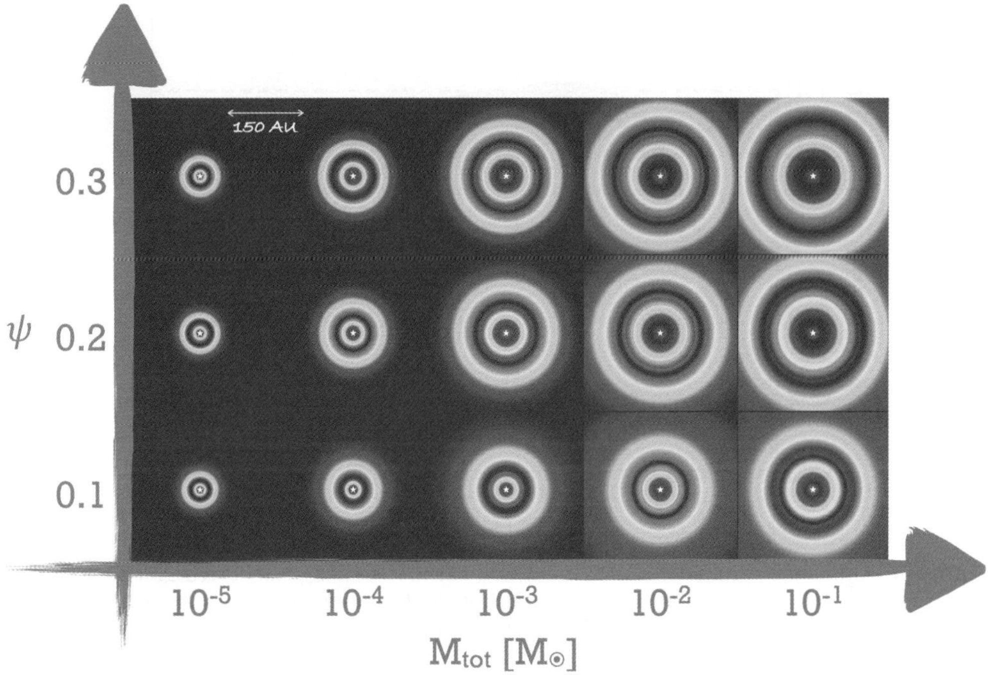

Figure 3. Model images of the CN (N,J)=(3-2, 7/2-5/2) transition emission from full disks around Herbig stars (normalized to peak). Higher masses and flaring show larger rings and higher fluxes. TTauri stars show similar trends, with smaller rings.

representing a Herbig star and a T Tauri star with a UV excess due to a $10^{-8}\,M_{\odot}$year^{-1} accretion rate ($L_{\rm UV} = 1.5 \times 10^{-4}\,L_{\odot}$).

3. Results

The CN abundance distribution is qualitatively similar for all of our models: CN is most abundant in the upper layers of the outer disk, and is less abundant in the inner disk (Fig. 2). Such a distribution naturally translates into a radial column density distribution peaking between a few tens and ~200 AUs, consistent with the observed ring-shaped emission. Indeed, ring-shaped emission is observed in all of our models, and the following additional results and trends are identified:

• T Tauri disks generally show smaller rings than Herbig disks

• CN fluxes in Herbig and T Tauri disks are comparable, ranging between $\sim 1\,{\rm Jy\,km\,s^{-1}}$ and $> 10\,{\rm Jy\,km\,s^{-1}}$

• The size of the rings increases with the mass and the flaring of the disks (Fig. 3). These trends are observed in both Herbig and T Tauri disks.

• CN emission is mostly optically thin, and the integrated flux increases with $M_{\rm tot}$. However, this dependence weakens for the most massive disks, where the emission becomes marginally optically thick.

• CN is not strongly affected by the overall abundance of volatile C and O, and even depleting these species up to a factor of 100 causes the CN flux only to vary by a factor of 2.

• The main formation route of CN is triggered by vibrationally excited H_2^* reacting with N to form NH, which in turn reacts with C and C^+ to form CN. This is evident

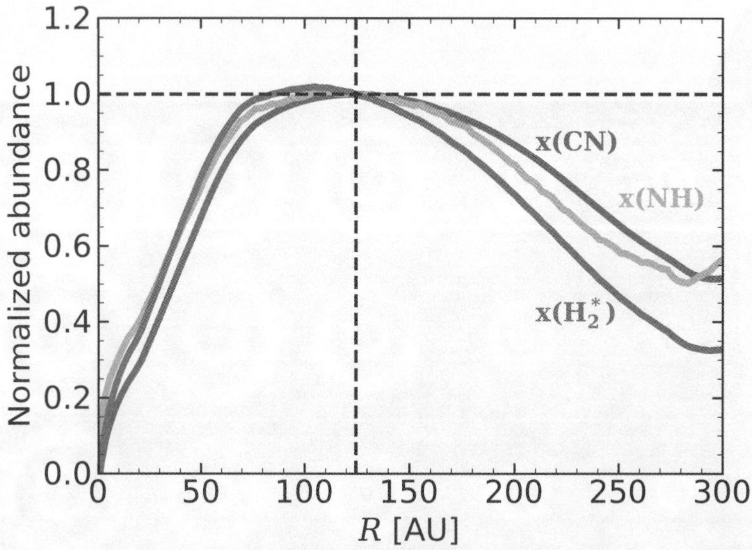

Figure 4. Abundance profile of H_2^*, NH and CN, calculated along the dotted green line in Fig. 2, and normalized at their value at the location of the CN abundance maximum (green dot in Fig. 2). The three species show very similar distributions. Adapted from Cazzoletti *et al.* (in press).

by comparing the abundance distributions of H_2^*, NH and CN (Fig. 4): all three species peak at the same radius.

CN measurements of flux and ring size can therefore be used to put constraints on some disk parameters such as the local UV flux, the flaring of the disks at large radii, and even the size of the disk. For a more details on the modelling and the results, we refer the reader to Cazzoletti *et al.* (in press).

Acknowledgements

I kindly thank E. F. van Dishoeck, R. Visser, S. Facchini and S. Bruderer with the help in this work.

References

Bruderer, S. 2013, *A&A*, 559, A46
Cazzoletti, P., van Dishoeck, E. F., Visser, R., Facchini, F., & Bruderer, S. in press on *A&A*
Chapillon, E., Guilloteau, S., Dutrey, A., Piétu, V., & Guélin, M. 2012, *A&A*, 537, A60
Dutrey, A., Guilloteau, S., & Guelin, M. 1997, *A&A*, 317, L55
Guilloteau, S., Simon, M., Piétu, V., *et al.* 2014, *A&A*, 567, A117
Guilloteau, S., Reboussin, L., Dutrey, A., *et al.* 2016, *A&A*, 592, A124
Kastner, J. H., Zuckerman, B., Hily-Blant, P., & Forveille, T. 2008, *A&A*, 492, 469
Kastner, J. H., Hily-Blant, P., Rodriguez, D. R., Punzi, K., & Forveille, T. 2014, *ApJ*, 793, 55
Öberg, K. I., Qi, C., Fogel, J. K. J., *et al.* 2010, *ApJ*, 720, 480
Öberg, K. I., Qi, C., Fogel, J. K. J., *et al.* 2011, *ApJ*, 734, 98
Salter, D. M., Hogerheijde, M. R., van der Burg, R. F. J., Kristensen, L. E., & Brinch, C. 2011, *A&A*, 536, A80
Teague, R., Guilloteau, S., Semenov, D., *et al.* 2016, *A&A*, 592, A49
Thi, W.-F., van Zadelhoff, G.-J., & van Dishoeck, E. F. 2004, *A&A*, 425, 955
van Terwisga, S., van Dishoeck, E. F., Facchini, F., Cazzoletti, P., *et al.* in prep.
Visser, R., Bruderer, S., Cazzoletti, P., Facchini, S., Heays, A. and van Dishoeck, E. F. in prep.

Astrochemistry VII – Through the Cosmos from Galaxies to Planets
Proceedings IAU Symposium No. 332, 2017
M. Cunningham, T. Millar & Y. Aikawa, eds.

doi:10.1017/S1743921317006871

On producers of cosmic organic compounds: exploring the boron abundance in lithium-rich K giant stars

N. A. Drake[1,2], R. de la Reza[1], V. V. Smith[3] and K. Cunha[1,3]

[1]Observatório Nacional-MCTIC, Rua Gen. José Cristino 77, 20921-400, Rio de Janeiro, Brazil,
email: drake@on.br, delareza@on.br, kcunha@on.br
[2]St. Petersburg State University, Laboratory of Observational Astrophysics,
Universitetski pr. 28, 198504, St. Petersburg, Russia
[3]National Optical Astronomy Observatory, 950 North Cherry Ave., Tucson, AZ 85719, USA,
email: vsmith@noao.edu, cunha@email.noao.edu

Abstract. The element boron belongs, together with lithium and beryllium, to a known trio of important elements for the study of evolutionary processes in low mass stars. Because B is the least fragile of this trio to be destroyed in the stellar interiors, it can be used to test if the Li enrichment is of planetary origin. Here, for the first time, boron lines are examined in the UV for four giants with different degrees of large Li enrichment by means of observations with the Hubble telescope. Two main results are found in our study. One is that to first approximation B abundances appear not to be in excess, invalidating the planet engulfment mechanism. The second one is that the two stars with very large Li abundances present emission lines indicating that quite strong active chromospheres are acting in these very Li-rich giants. These new results obtained from the UV complement our recent studies in the mid-IR (de la Reza *et al.* 2015) where strong emission-line features of organic material were found in the spectra of some Li-rich stars.

Keywords. stars: abundances – stars: chemically peculiar – stars: magnetic field – stars: mass loss – stars: individual: HD 787, HD 9746, HD 39853, HD 112127

1. Introduction

Following the standard evolution theory, the first ascent red giant stars (RGB) must destroy, by means of thermal reactions, almost all of their original lithium. However, since 1982, when Wallerstein & Sneden detected the first K giant with strong lithium lines, HD 112127, several Li-rich K giants have been detected having Li abundances, even as large as ten times the original Li abundance, indicating that new sources of ^{7}Li are in action. As an example, Fig. 1 shows the spectrum of the Li-rich giant HD 19475 in the region containing the Li I resonance line at λ6707.8 Å.

Li, Be, and B are relatively fragile elements which are destroyed by (p, α) reactions at relatively low temperatures at about $2.5 \cdot 10^6$, $3.5 \cdot 10^6$, and $5.0 \cdot 10^6$ K, respectively. Since 1982 the nature of these Li-rich red giant stars have been involved in a problem known as the "lithium puzzle".

Recently we have discovered in the near-IR (de la Reza *et al.* 2015) the presence of complex organic compounds in the shells of Li-rich K giant stars (Fig. 2). These stars present large mass loss transforming them into sources of organic material in the Galaxy.

Beryllium abundances in lithium-rich giants, including the star HD 787, were studied by Castilho *et al.* (1999) and Melo *et al.* (2005). The authors have found that Be abundances in these stars are low, suggesting that Be is very depleted (>90%) from the initial value of the Populaton I stars.

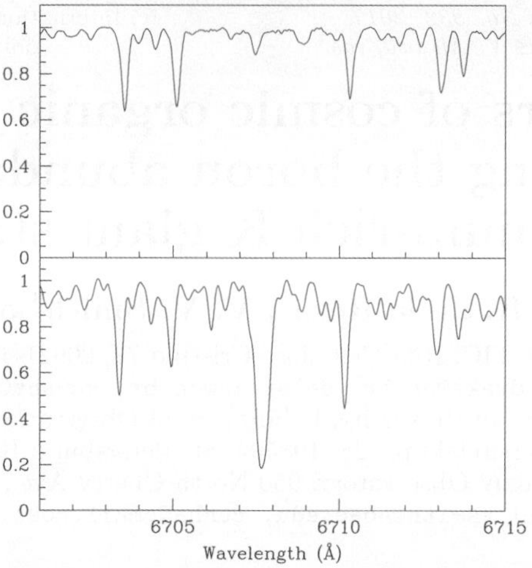

Figure 1. Spectrum of the Li-rich giant HD 787 (bottom panel) in comparison with a normal K giant star ϵ Vir (upper panel) in the region of the Li I line at 6707.8 Å.

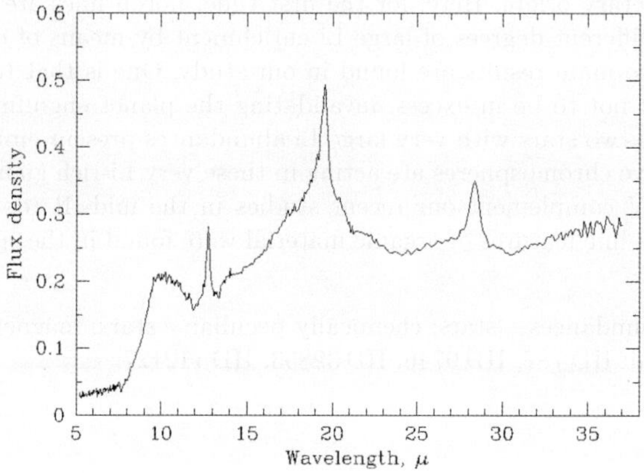

Figure 2. Spitzer mid-IR spectrum of the Li-rich K giant star PDS 68 between 5 and 38 μm presenting the UIE features superposed on a strong continuum emission (de la Reza *et al.* 2015). The spectrum was taken from the Cornell Atlas of *Spitzer*/IRS sources (CASSIS, Lebouteiller *et al.* 2011).

The nature of these stars is the object of recent studies. Here, it is explored for the first time the boron lines in the spectra of these objects in the UV spectral region with the aim to infer their main properties.

2. Observations and stellar parameters

We observed four Li-rich K giants HD 787, HD 9746, HD 3953, and HD 112127 with the Hubble Space Telescope and Space Telescope Imaging Spectrograph STIS with grating G230M having FWHM = 30 km s^{-1} and spectral region centered at the resonant B I lines at λ2496.8 and λ2497.7 Å. These four spectra are shown in Fig. 3.

The atmospheric parameters of the stars, effective temperature ($T_{\rm eff}$), surface gravity ($\log g$), and lithium abundance ($\log \varepsilon$(Li)) (where $\log \varepsilon(X) = \log(N_X/N_H) + 12.0$), are shown in Table 1.

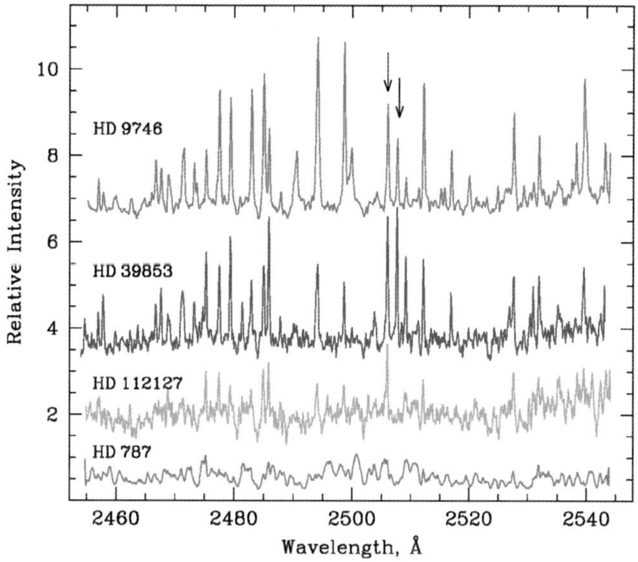

Figure 3. The HST/STIS UV spectra of four giant stars in the region of the B I lines at λ2496.8 and λ2497.7 Å. From top to bottom: HD 9746 ($\log\varepsilon$(Li) = 3.75), HD 39853 ($\log\varepsilon$(Li) = 2.85), HD 112127 ($\log\varepsilon$(Li) = 2.7), and HD 787 ($\log\varepsilon$(Li) = 1.8). The UV spectra of the Li-rich giants HD 9746 and HD 39853 are very complex and rich in strong emission lines, mainly of Fe II. As can be seen, the intensity of the emission lines correlates with the Li abundance. Very strong emission lines at λ2507 and λ2509 Å marked by arrows are observed in the spectra of HD 9746 and HD 39853. These two lines are the Fe II fluorescence lines pumped by H Lyα (Johanson & Hamann 1993) pointing to the existence of very strong Lyα emission. Star HD 787 shows a different spectrum probably due to its binary nature. The spectra are arbitrary shifted in the relative intensity axis.

Table 1. Atmospheric parameters of observed Li-rich K giant stars.

Star	$T_{\rm eff}$, K	$\log g$	$\log\varepsilon$(Li)	Reference
HD 787*	4220	1.50	1.8	(1)
HD 9746	4400	2.30	3.75	(2)
HD 39853	3900	1.16	2.85	(3)
HD 112127	4340	2.10	2.7	(4)

References: (1) Brown *et al.* (1989);
(2) Balachandran *et al.* (2000);
(3) Gratton & D'Antona (1989);
(4) Wallerstein & Sneden (1982)
Notes: *The star HD 787 is a proper-motion binary (Mason *et al.* 1999, Eggleton & Tokovinin 2008)

3. Results and conclusions: how to explain the Li puzzle?

• The boron abundances of the Li-rich giants appear to be low and in none of the cases presents an excess. This result is not compatible with any scenario of engulfment of planets in order to produce Li-rich K giant stars.

• In de la Reza *et al.* (2015) we propose a qualitative scenario to try to explain the "Li puzzle" in K giant stars. Here, a mass transfer produced by the diminishing of the stellar core rotation in the RGB phase, reaches the surface layers of the stars forming eventually the observed circumstellar shell. At the same time, this transport can, in some

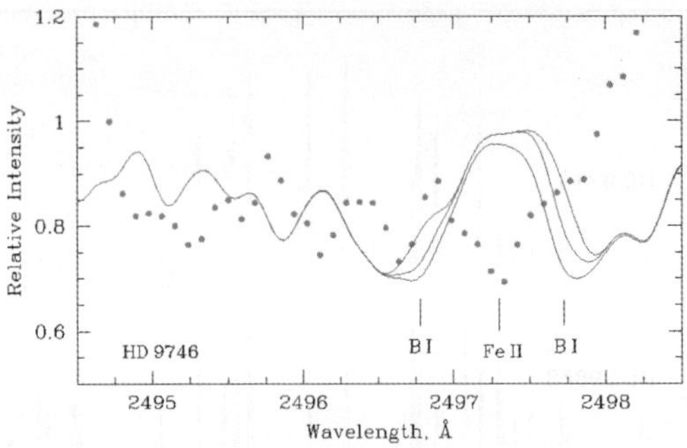

Figure 4. Fit of the observed spectrum of HD 9746 in the spectral region around the resonance doublet of B I at λ2496.8 and λ2497.7 Å in the spectrum of HD 9746. *Points*: observations. *Lines*: synthetic spectra computed using the recent version of MOOG code (Sneden 1973) for the boron abundances of $\log\varepsilon(\mathrm{B})$ = 0.60, 2.60, and 4.60 corresponding to the [B/Fe] = –2.0, 0.0, and +2.0. Strong contribution from the chromosphere causing veiling in this UV region, makes it difficult to determine the B abundance with high precision. Nevertheless, a direct comparison with theoretical B I lines shows that no expected B excess is present. The line lists were taken from the VALD Database, Duncan *et al.* (1998), and Cunha *et al.* (2000).

short time episodes, move the ^{7}Be produced in the internal H-burning region rapidly to the external convective zone where it is transformed into new ^{7}Li.

• Privitera *et al.* (2016) suggested that a strong magnetic field in a red giant star may be a sign of a past planet engulfment. It is interesting to note that our observed very Li-rich K giant HD 9746 has an important magnetic field detected by Aurière *et al.* (2015). We found in this work that this K giant star can be a descendant of an Ap star on the main sequence known to have very high Li abundance. Because HD 9746 has a relatively high rotation of $v \sin i = 8.7$ km s^{-1}, its magnetic field can have two origins: one fossil from the Ap ancestor and one produced by a rotationally induced dynamo. Goriely (2007) suggested that high lithium abundance in magnetic Ap stars may be produced by the spallation reactions in the regions of the magnetic poles, where accelerated protons and α-particles destroy CNO nuclei and produce lithium.

• Even if the direct relation of the Li enrichment phenomenon and existence of a magnetic field is not yet well understood, there exists a strong correlation between stellar activity and high Li abundances in red giant stars (Fekel & Balachandran 1993, Drake *et al.* 2002).

Acknowledgements. N.A.D. acknowledges IAU travel grant, FAPERJ, Rio de Janeiro, Brazil, for Visiting Researcher grant E-26/200.128/2015, and the Saint Petersburg State University for research grant 6.38.335.2015.

References

Aurière M., Konstantinova-Antova, R., Charbonnel, C., *et al.* 2015, *A&A*, 574, A90
Balachandran S. C., Fekel, F. C., Henry, G. W., & Uitenbroek, H. 2000, *ApJ*, 542, 978
Brown J. A., Sneden, C., Lambert, D. L., & Dutchover, E. 1989, *ApJS*, 71, 293
Castilho, B. V., Spite, F., Barbuy, B. *et al.* 1999, *A&A*, 345 249
Cunha K., Smith, V. V., Boesgaard, A. M., & Lambert, D. L. 2000, *ApJ*, 530, 939
de la Reza R., Drake, N. A., Oliveira, I., & Rengaswamy, S. 2015, *ApJ*, 806, 86
Drake N. A., de la Reza, R., da Silva, L., & Lambert, D. L. 2002, *AJ*, 123, 2703
Duncan, D. K., Peterson, R. C., Thorburn, J. A., & Pinsonneault, M. H. 1998, *ApJ*, 499, 871
Eggleton, P. & Tokovinin, A. 2008, *MNRAS*, 389, 869
Fekel F. C. & Balachandran S. 1993, *ApJ*, 403, 708

Goriely S. 2007, *A&A*, 466, 619
Gratton R. G. & D'Antona F. 1989, *A&A*, 215, 66
Johansson S. & Hamann F. 1993, *Phys. Scr.* T47, 157
Lebouteiller, V., Barry, D. J., Spoon, H. W. W., *et al.* 2011, *ApJS*, 196, 8
Mason, B. D., Martin, C., Hartkopf, W. I., *et al.* 1999, *AJ*, 117, 1890
Melo, C. H. F., de Laverny, P., Santos, N. C., *et al.* 2005, *A&A*, 439, 227
Privitera, G., Meynet, G., Eggenberger, P., *et al.* 2016, *A&A*, 593, L15
Sneden, C. 1973, PhD Thesis, Univ. of Texas at Austin
Wallerstein G. & Sneden C. 1982, *ApJ*, 255, 577

Astrochemistry VII – Through the Cosmos from Galaxies to Planets
Proceedings IAU Symposium No. 332, 2017
M. Cunningham, T. Millar & Y. Aikawa, eds.

doi:10.1017/S1743921317007748

Chemical Evolution of Turbulent Multiphase Molecular Clouds

Valeska Valdivia[1], Patrick Hennebelle[1], Benjamin Godard[2], Maryvonne Gerin[2], Pierre Lesaffre[2] and Jacques Le Bourlot[2]

[1]Université Paris Diderot, AIM, Sorbonne Paris Cité, CEA, CNRS, F-91191 Gif-sur-Yvette, France.
email: valeska.valdivia@cea.fr

[2]Laboratoire de radioastronomie, LERMA, Observatoire de Paris, École Normale Supérieure (UMR 8112 CNRS), 24 rue Lhomond, 75231 Paris Cedex 05, France

Abstract. Molecular clouds are essentially made up of atomic and molecular hydrogen, which in spite of being the simplest molecule in the ISM plays a key role in the chemical evolution of molecular clouds. Since its formation time is very long, the H_2 molecules can be transported by the turbulent motions within the cloud toward low density and warm regions, where its enhanced abundance can boost the abundances of molecules with high endothermicities.

We present high resolution simulations where we include the evolution of the molecular gas under the effect of the dynamics, and we analyze its impact on the abundance of CH^+.

Keywords. ISM: clouds, ISM: molecules, MHD, turbulence

1. Introduction

Turbulent molecular clouds present clumpy and filamentary structures, where warm and cold phases are interwoven in a very complex structure. The turbulent motions within molecular clouds can accelerate the conversion of atomic to molecular hydrogen in transient over-densities, and subsequently carry these molecules towards diffuse and warm regions (Valdivia *et al.* 2016), where they can participate in endothermic reactions, such as the formation of CH^+, which requires molecular hydrogen to be formed efficiently. In Valdivia *et al.* (2017) we have shown that the warm H_2 can enhance the CH^+ column densities by a factor 3 to 10.

Here we present a summary of these two works, as well as a study of the influence of the shielding by dust and the H_2 self-shielding on the evolution of the abundance and distribution of molecular hydrogen, as well as the influence of the magnetic field on the structure and distribution of H_2. For this study we used a modified version of the RAMSES code (Teyssier 2002) that includes the formation and destruction processes for H_2, its thermal feedback, the shielding by dust for the ultraviolet radiation, and the self-shielding due to other H_2 molecules. The implementation and the setup are fully described in Valdivia *et al.* (2016). For the shielding we have used our tree-based method, detailed in Valdivia & Hennebelle (2014).

2. Results

To understand the H_2 molecule formation process under the dynamical influence of a highly inhomogeneous structure, we have performed hydrodynamics and magneto-hydrodynamics (MHD) simulations of realistic molecular clouds formed through colliding streams of warm atomic gas in a 50 pc cubic box.

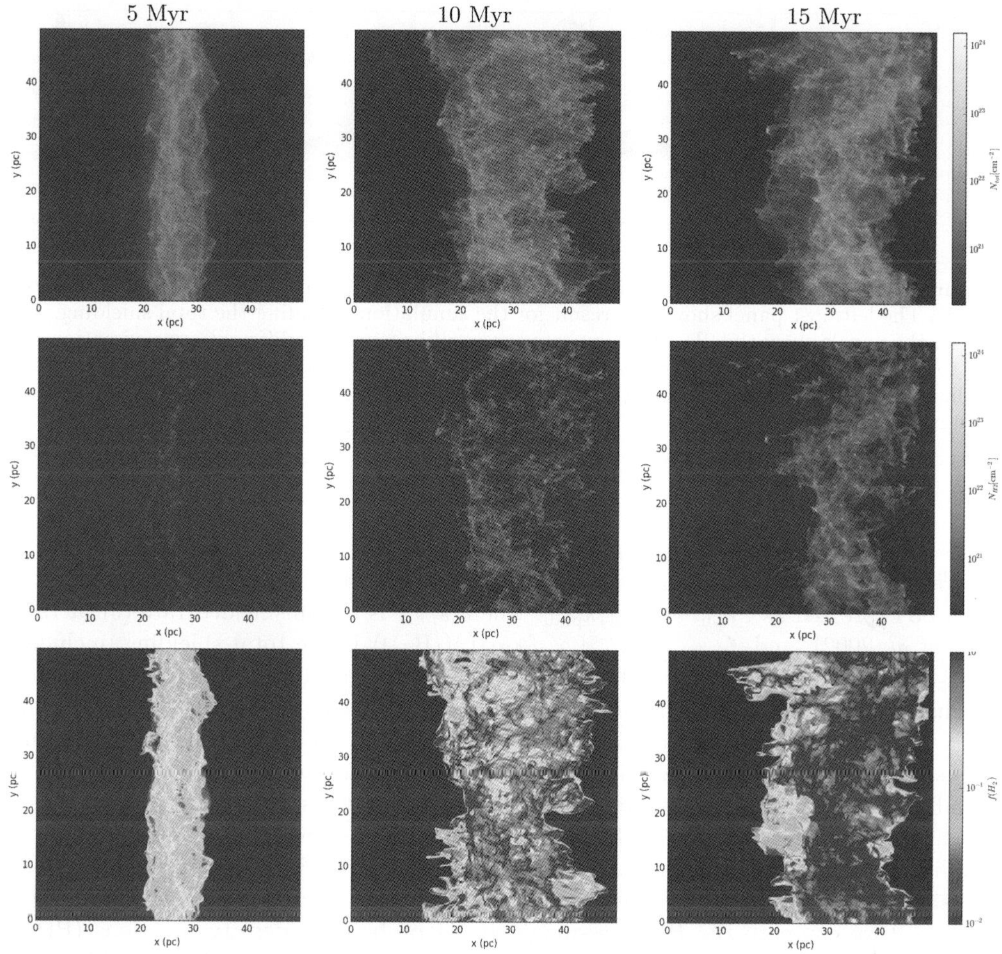

Figure 1. Evolution of a molecular cloud. From top to bottom: Total column density, H_2 column density, and the H_2 fraction at $t = 5$, 10, and 15 Myr (from left to right).

2.1. *General results*

In Valdivia *et al.* (2016) we have shown that H_2 molecular gas is formed faster than the usual estimates based on the mean density of the cloud, which can be interpreted as the result of the local density enhancements induced by turbulent compressions, causing a faster H_2 formation (Glover & Mac Low 2007a,b; Clark *et al.* 2012; Micic *et al.* 2012). Additionally, the relaxation of such over-densities along with the turbulent motions lead to a mixing between different phases and brings H_2 molecules to the warm phase, where they survive due to the shielding provided by the global structure, explaining the presence of warm H_2 in the conditions that prevail in molecular clouds. Figure 1 shows the general evolution of the molecular cloud, as well as the evolution of the molecular hydrogen.

2.2. *Influence of the total shielding*

Nowadays it is widely accepted that molecular hydrogen is formed on dust grain surfaces, and that it is mainly destroyed by UV fluorescent photo-dissociation. The strength of the UV field is attenuated by the presence of dust, and since UV photons are most likely absorbed by other H_2 molecules, both the dust and other H_2 molecules shield the

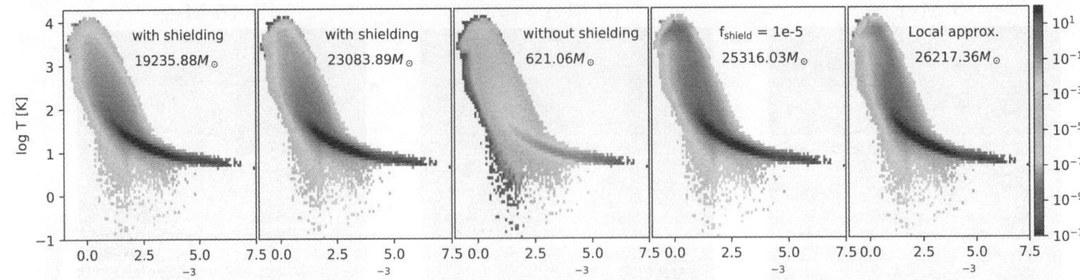

Figure 2. Comparison of the normalized $m(H_2)$-weighted 2D probability density functions (PDF).The leftmost panel shows the result for the simulation including the total shielding. The rest of the panels show the 2D-PDFs expected at equilibrium using different approximations for the total shielding. From left to right: shielding from the actual H_2 distribution in the simulation, without any shielding effect, using a constant shielding, and using the local approximation assuming 5% of H_2.

interior of the molecular cloud against photodissociation. The photo-dissociation rate can then be expressed as : $k_{\rm ph} = f_{\rm sh,H_2} k_{\rm ph,0}$, where $k_{\rm ph,0}$ is the unshielded rate (Jura 1974; Gry *et al.* 2002), and $f_{\rm sh,H_2} = \langle e^{-\tau_d} \times f_{\rm shield}(\mathcal{N}(H_2))\rangle$ is the total shielding, including dust and self shielding as described in Draine & Bertoldi (1996). Since both shielding mechanisms operate in a line-by-line basis, it is needed to compute them in all directions. This can be extremely expensive in terms of CPU time, so approached methods must be used. To compute total and H_2 column densities in our simulation we have used our tree-based method. To test different approximations for the shielding we post-processed a snapshot of our simulation at 15 Myr to calculate the expected abundances of H_2 at equilibrium using the actual shielding as well as other less rigorous approximations for the shielding factors. For all cases we used the temperature computed in the simulation. In Fig. 2 we compare the H_2 mass-weighted 2D probability density functions (PDF). The leftmost panel shows the actual distribution of the dynamically computed H_2 from our simulation, while the second panel show the distribution expected at equilibrium for the same shielding conditions. The third panel shows the distribution for the extreme unshielded case. The fourth panel corresponds to the case where the shielding is supposed uniform throughout the whole molecular cloud and equal to the mean shielding factor for gas in the density range between 3 to 3000 cm^{-3} from Fig. 7 in Valdivia *et al.* (2016). The rightmost panel corresponds to the local approximation, where column densities are approximated by $\mathcal{N} \sim n_{\rm local} \times L$, where $n_{\rm local}$ is the local total gas number density, and L is the size of the molecular cloud. For the self-shielding by H_2 we assume an abundance of 5%. The total mass of H_2 expected in each case is indicated in the panel in solar masses. This figure suggest that when the shielding effects are not included, only the gas denser than several 100 cm^{-3} is able to form H_2 efficiently enough to counterbalance the photodissociation. Another remark is that all the cases that included any treatment of the shielding produced comparable amounts of H_2. Nevertheless the treatments at equilibrium tend to overestimate the total abundance of H_2.

To better understand the influence of the shielding in the dynamical evolution of the cloud we performed two lower resolution simulations ($\sim$ 0.2 pc) using the same setup. Both simulations include the shielding for the UV radiation in the thermal balance, but only the first one includes the total shielding factor $f_{\rm sh,H_2}$ for the photodissociation of H_2.

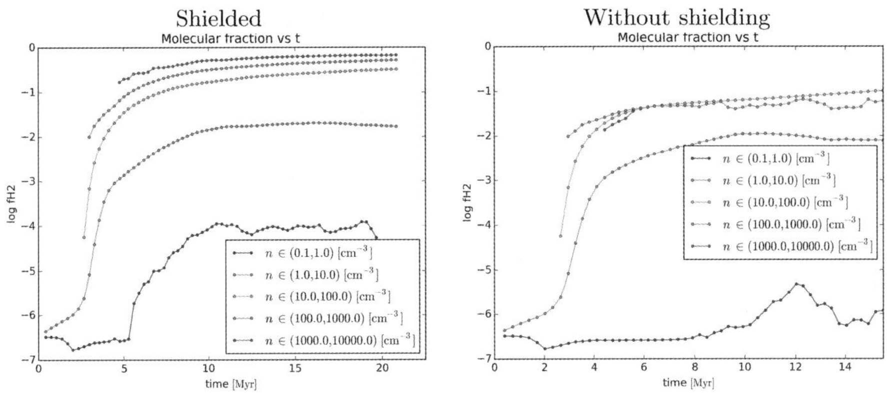

Figure 3. Evolution of the fraction of mass in H_2 form per density bin: including shielding effects (left) and without shielding (right).

Figure 3 shows the evolution of the molecular fraction $f(H_2)$ per density bin. When the shielding is not taken into account the presence of molecular gas is globally suppressed, and in a period of 15 Myrs the simulation converts less than 10% of total gas into molecular gas. This is true even for gas denser than several 100 cm^{-3}, suggesting that higher densities are not enough to counterbalance the fast destruction when the abundance of H_2 is treated dynamically.

We find that when the photodissociation rate includes the combined shielding effects, the amount of gas in H_2 form is about 5 times higher than when this effect is not included. An interesting result comes from the analysis of the most diffuse gas, for which the molecular fraction when the shielding is included is about 2 orders of magnitude higher than the case without shielding. This indicates that the shielding effect of dust and H_2 self-shielding is crucial to explain the presence H_2.

2.3. *Influence of warm H_2 on the abundance of CH^+*

Species with long evolution times are most likely influenced by the gas motions, and consequently can be found out-of-equilibrium, whereas species with shorter evolution times can react faster to variations in their environments. Since H_2 precedes the formation of other molecules, and since the formation time of H_2 is long compared to other molecules, its formation process is a bottleneck for the chemical evolution of the molecular cloud. This permits us to use a hybrid method, fully described in Valdivia *et al.* (2016b) and (Valdivia *et al.* 2017), where atomic and molecular hydrogen are the only species treated dynamically, while the abundances of the rest of the species are calculated at equilibrium. An interesting test case is the presence of CH^+ molecules in the diffuse ISM, where the only reaction efficient enough to produce CH^+ is highly endothermic ($\Delta E/k = 4300$ K, Agúndez *et al.* 2010), and involves H_2.

Figure 4 shows that the warm H_2 have an important influence on the abundance of CH^+ (Valdivia *et al.* 2017, 2016c). Most of CH^+ is formed in environments where molecular fraction is about a few percent and the gas temperature is higher than ~ 300 K. At the edges of clumps, where H_2 is found in excess with respect to the equilibrium value, the column densities of CH^+ are higher by a factor of 3 to 10 with respect to the values obtained when H_2 is also calculated at equilibrium.

2.4. *Influence of magnetic field*

It is well known that MHD simulations produce clouds with a different morphology than pure hydrodynamical simulations (Hennebelle *et al.* 2008). We have performed

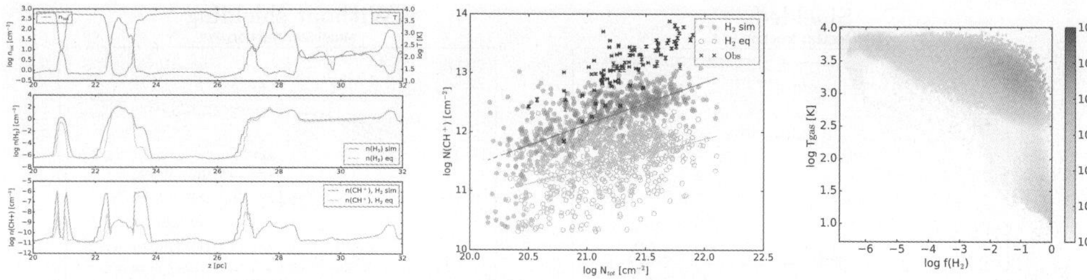

Figure 4. Influence of the warm H_2. Line-of-sight showing the increased abundance of CH^+ at the edge of clumps (left), CH^+ column densities (center), and the 2D PDF of CH^+ mass as a function of the molecular fraction and the gas temperature (right). The crosses in the central panel are the observational data from Crane *et al.* (1995); Gredel (1997) and Weselak *et al.* (2008). (Figures from Valdivia *et al.* 2017)

two simulations using the same set up described in Valdivia *et al.* (2016), at the same resolution. We use a base resolution of 256^3 and two additional refinement levels reaching an effective resolution of ~ 0.05 pc. The first one is a pure hydrodynamical simulation ($\vec{B} = 0$), and the second one is a MHD simulation, with a moderate magnetic field aligned with the gas inflow ($B \sim 2.5$ μG).

Figure 5 shows the general evolution of the cloud in terms of total column density (top panel) and the H_2 colum density (middle panel). The hydro case produces higher column densities at earlier times, as a consequence of the lack of magnetic support. In the MHD simulation, magnetic fields offer an additional support against gravity, delaying the collapse, as shown by Hennebelle *et al.* (2008). The cloud formed in the hydro simulations displays a more compact structure, reaching higher densities that favor an earlier formation of molecular gas (Fig. 6). Higher column densities favor higher molecular fractions, as shown in the bottom panel of Fig. 5. An interesting remark is that MHD simulations produce clumpier and more filamentary structures, allowing a more complex mixture of warm and cold gas with a wider range of abundances of H_2.

Even though the general evolution of the total amount of H_2, shown in Fig. 6, is very similar in both cases, the hydro case reaches 10% of molecular hydrogen about 1 Myr earlier than the MHD case. The difference between both cases is more pronounced at densities of a few $\sim 1 - 10$ cm^{-3}.

3. Conclusions

H_2 formation can proceed at early stages of the molecular cloud evolution due to compressive motions driven by the turbulence (Micic *et al.* 2012). The turbulent motions within the gas can mix the warm neutral and the cold neutral phases and transport long-lived molecules towards diffuse and warm environments.

The dust and the surrounding H_2 molecules shield cloud interior against the dissociating UV radiation, protecting the H_2 present in diffuse warm gas. The presence of magnetic fields offers a magnetic support against the gravitational collapse, delaying the formation of dense structure, and thus delaying the formation of H_2. Nevertheless, magnetized gas produces more complex and more filamentary structures displaying a wide variety of physical conditions.

The warm H_2 is an important ingredient in highly endothermic reactions, increasing the column densities of CH^+ by a factor 3 to 10. Locally this effect is even more striking, being about two orders of magnitude more abundant at the edge of clumps where there is an excess of H_2 with respect to the value expected at equilibrium. Nevertheless total

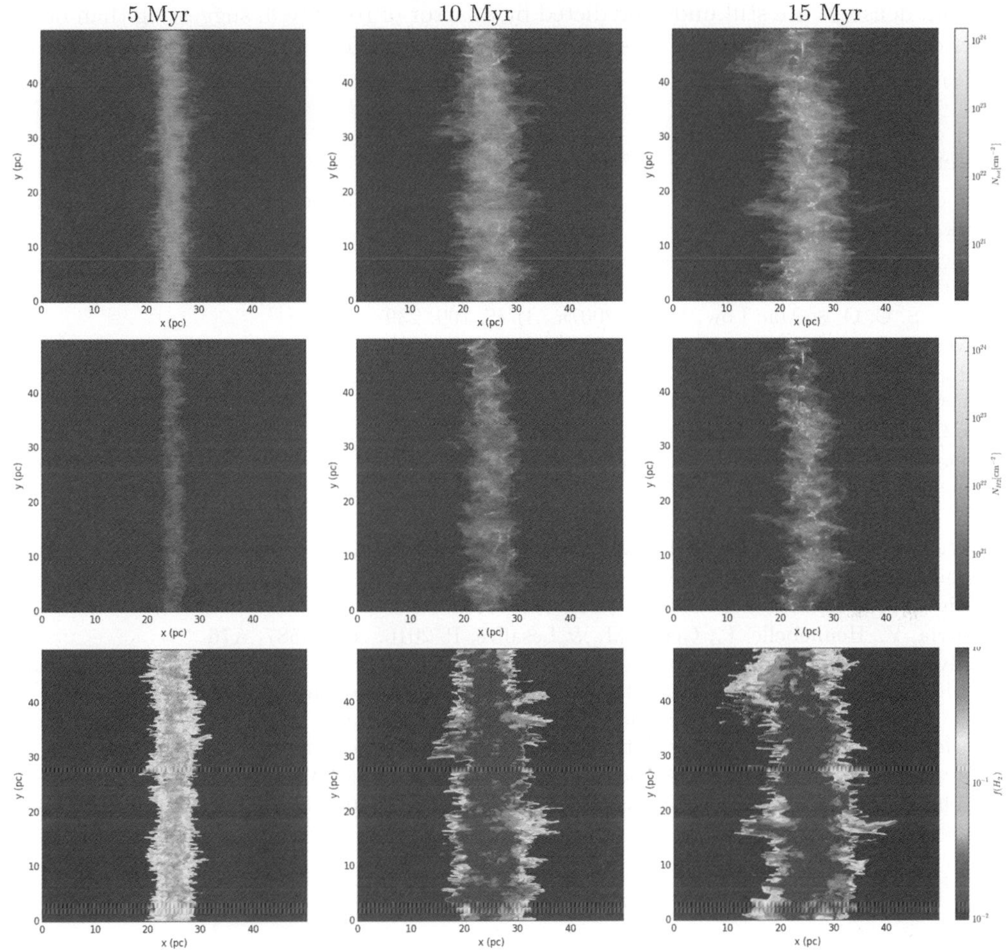

Figure 5. Evolution of a molecular cloud in absence of magnetic field. From top to bottom: Total column density, H_2 column density, and the H_2 fraction, at $t = 5$, 10, and 15 Myr.

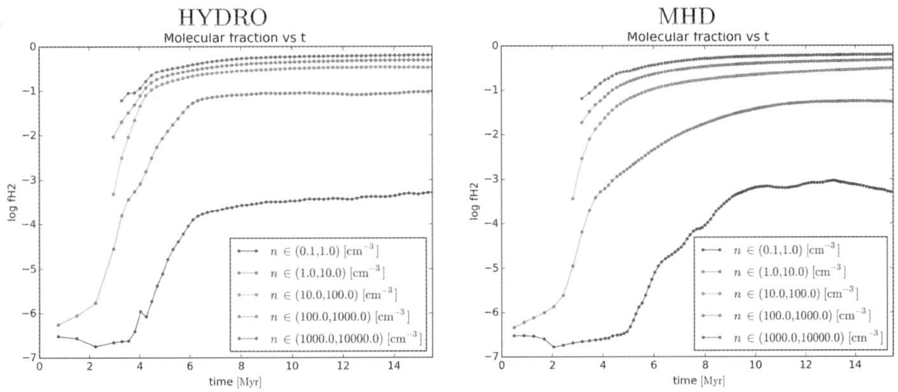

Figure 6. Evolution of the total molecular fraction and the molecular fraction per density bin. Hydro case (left panel), and the standard MHD simulation (right panel).

column densities are still under-predicted by a factor of roughly 6, suggesting than other physical process, such as the dissipation of turbulence, may be at play (Godard *et al.* 2014).

References

Agúndez, M., Goicoechea, J. R., Cernicharo, J., Faure, A., & Roueff, E. 2010, *ApJ*, 713, 662
Clark, P. C., Glover, S. C. O., Klessen, R. S., & Bonnell, I. A. 2012, *MNRAS*, 424, 2599
Crane, P., Lambert, D. L., & Sheffer, Y. 1995, *ApJS*, 99, 107
Draine, B. T. & Bertoldi, F. 1996, *ApJ*, 468, 269
Glover, S. C. O. & Mac Low, M.-M. 2007a, *ApJS*, 169, 239
Glover, S. C. O. & Mac Low, M.-M. 2007b, *ApJ*, 659, 1317
Godard, B., Falgarone, E., & Pineau des Forêts, G. 2014, *A&A*, 570, A27
Gredel, R. 1997, *A&A*, 320, 929
Gry, C., Boulanger, F., Nehmé, C., *et al.* 2002, *A&A*, 391, 675
Hennebelle, P., Banerjee, R., Vázquez-Semadeni, E., Klessen, R. S., & Audit, E. 2008, *A&A*, 486, L43
Jura, M. 1974, *ApJ*, 191, 375
Micic, M., Glover, S. C. O., Federrath, C., & Klessen, R. S. 2012, *MNRAS*, 421, 2531
Teyssier, R. 2002, *A&A*, 385, 337
Valdivia, V. & Hennebelle, P. 2014, *A&A*, 571, A46
Valdivia, V., Hennebelle, P., Gérin, M., & Lesaffre, P. 2016, *A&A*, 587, A76
Valdivia, V., Godard, B., Hennebelle, P., Gerin, M., & Lesaffre, P. 2016b, SF2A-2016: Proceedings of the Annual meeting of the French Society of Astronomy and Astrophysics, 163
Valdivia, V., Godard, B., Hennebelle, P., Gerin, M., & Lesaffre, P. 2016c, SF2A-2016: Proceedings of the Annual meeting of the French Society of Astronomy and Astrophysics, 309
Valdivia, V., Godard, B., Hennebelle, P., *et al.* 2017, *A&A*, 600, A114
Weselak, T., Galazutdinov, G., Musaev, F., & Krełowski, J. 2008, *A&A*, 479, 149

Astrochemistry VII – Through the Cosmos from Galaxies to Planets
Proceedings IAU Symposium No. 332, 2017
M. Cunningham, T. Millar & Y. Aikawa, eds.

doi:10.1017/S174392131700967X

Chemical and kinematic complexity of the very young star-forming region Serpens Main observed with ALMA

Łukasz Tychoniec[1], Charles L. H. Hull[2], John J. Tobin[3] and Ewine F. van Dishoeck[1,4]

[1]Leiden Observatory, Leiden University, PO Box 9513, 2300RA, Leiden, The Netherlands
email: tychoniec@strw.leidenuniv.nl

[2]Harvard-Smithsonian Center for Astrophysics, 60 Garden St., Cambridge, MA 02138, USA

[3]Homer L. Dodge Department of Physics and Astronomy, University of Oklahoma, 440 W. Brooks Street, Norman, OK 73019, USA

[4]Max Planck Institut für Extraterrestrische Physik, Giessenbachstrasse 1, 85748 Garching, Germany

Abstract. The youngest low-mass protostars are known to be chemically rich, accreting matter most vigorously, and producing the most powerful outflows. Molecules are unique tracers of these phenomena. We use ALMA to study several outflow sources in the Serpens Main region. The most luminous source, Ser-SMM1, shows the richest chemical composition, but some complex molecules are also present in S68N. No emission from complex organics is detected toward Ser-emb 8N, which is the least luminous in the sample. We discuss whether these differences reflect an evolutionary effect or whether they are due to different physical structures. We also analyze the outflow structure from these young protostars by comparing emission of CO and SiO. EHV molecular jets originating from SMM1-a,b and Ser-emb 8N contrast with no such activity from S68N, which on the other hand presents a complex outflow structure.

Keywords. astrochemistry, molecular data, techniques: interferometric, stars: formation, ISM: individual (Serpens), ISM: jets and outflows, ISM: molecules

1. Introduction

Deeply embedded Class 0 objects, the youngest low-mass protostars, present all kinds of chemical and kinematic complexity. The chemical richness is related to the physical processes associated with star formation within cold clouds of dust and gas. Deep inside, the gas is swept up and shocked by jets and outflows, while in parallel the material is heated, sublimates from the grains, and ultimately accretes onto the protostar through the disk. Class 0 sources accrete matter most vigorously, also producing the most powerful outflows (Bontemps *et al.* 1996), hence investigating this stage of protostellar evolution can provide a wealth of information about processes related to star formation.

Most of the outflowing molecular gas is present in relatively slow, massive and wide flows of entrained material. However, for some protostars an extremely high-velocity (EHV) component has been seen in e.g., CO, SiO, and SO, probably representing the original driving jet (e.g., Bachiller *et al.* 1990; Santiago-García *et al.* 2009; Hirano *et al.* 2010; Tafalla *et al.* 2010). Observing these very young jets could provide a tool to understand the launching mechanisms, interaction of the outflow with the envelope and chemical composition of the original jet.

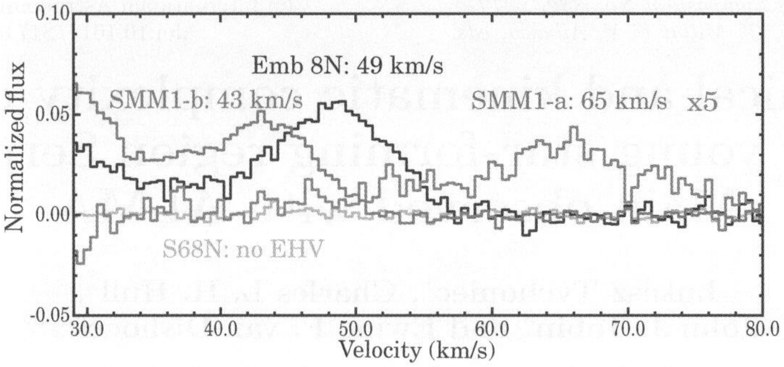

Figure 1. Comparison of redshifted EHV components in the line wings of all observed outflow spectra with labels marking the velocity at which the EHV component reaches peak intensity. Flux density was normalized to the maximum CO intensity for each source. For SMM1-a the normalized flux was multiplied by factor of 5.

Chemical complexity of the young protostars is particularly striking in the hot cores, where complex organic molecules can be detected. These molecules are of great importance in the context of understanding the origin of pre-biotic species, as many of those carbon-based molecules are building blocks of life as we know it. Since first detection of hot cores toward low-mass protostars (van Dishoeck *et al.* 1995; Cazaux *et al.* 2003), many more studies were dedicated to observe more sources that host complex organics (e.g., Bergner *et al.* 2017) or targeting one of the protostars and scanning its submillimeter spectrum to detect a wealth of complex molecules (Jørgensen *et al.* 2016).

We use the Atacama Large Millimeter/submillimeter Array (ALMA) to study four Class 0 objects in the Serpens Main region (Eiroa *et al.* 2008) located 436 pc away from the Sun (Ortiz-León *et al.* 2017). This small sample is attractive since it contains an intermediate mass protostar SMM1-a (100 $L_\odot$; Kristensen *et al.* 2012) as well as EHV molecular jets toward three protostars (SMM1-a, SMM1-b, and Ser-emb 8N). The high-resolution (0.5″) Band 6 (1.3 mm) ALMA observations yield spectra and maps showing various molecules. Here we present the analysis of CO and SiO emission in the outflows along with detections from complex organics molecules in the protostellar cores.

2. Molecular jets

Three outflows show a distinct high-velocity component in CO (2-1). Fig. 1 presents a comparison of those components. The luminous source SMM1-a has the highest velocity for the peak wing intensity. However, as no luminosity estimate is known for the SMM1-b and Ser-emb 8N, we cannot conclude whether or not there is a relation of the bolometric luminosity and EHV velocity. Inclination can also significantly change the observed radial velocity. We further discuss each source separately.

Serpens SMM1-a is the most massive protostar in the Serpens molecular cloud. A CO (2-1) jet with velocities up to 90 km/s was observed by Hull *et al.* (2016). The entrance of the outflow cavity wall is clearly delineated by the continuum dust emission, as well as in free-free emission (Hull *et al.* 2016). It appears that the molecular jet interacts heavily with the outflow cavity wall, as is indicated by enhanced ^{13}CS emission on the side of the outflow where the EHV CO jet is currently pointing (see Fig. 2, right).

SMM1-b is another protostellar component in the SMM1 system, discovered by Choi (2009). Hull *et al.* (2016, 2017) present the high-velocity molecular jet toward the SMM1-b in CO and SiO respectively. The latter work also reveals that SMM1-b is a binary

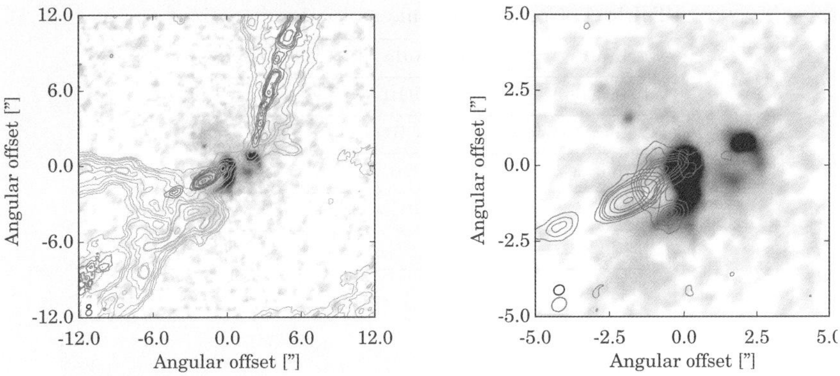

Figure 2. Serpens SMM1. Left: ALMA 1.3 mm continuum in grayscale with CO (2-1) integrated emission overlaid in contours. Red color represents EHV velocities from 44 to 70 km s^{-1} and orange low-velocity outflows from 10 to 15 km s^{-1}. Right: Blow-up of central part with red contours showing CO (2-1) EHV emission from 60 to 70 km s^{-1} and green contours show ^{13}CS (5-4) integrated emission from 9 to 14 km s^{-1}. See also Hull *et al.* (2016, 2017).

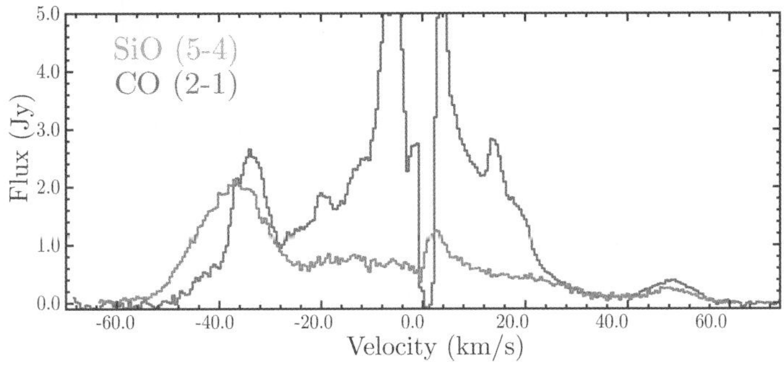

Figure 3. Integrated flux density over the total outflow area for Ser-emb 8N. SiO (5-4) is shown in green and CO (2-1) in purple.

system itself with the eastern component powering the EHV jet. This source shows a clear example of the evolution of the fast collimated jet into a slow molecular wind in CO emission (see Fig. 2, left).

Ser-emb 8 (S68N) is a protostar in the north-west part of the Serpens core. The outflow from S68N appears wide and evolved, with several spot shocks present in SiO, probably indicating interactions between the outflow and the surrounding cloud. This flow does not show signatures of EHV jets, consistent with the more evolved nature of the central source.

Protostar Ser-emb 8N is located about 10" north-east from S68N. This source shows a high degree of collimation, even at low velocities, suggesting that the outflow is very young (Arce & Sargent 2006). There are symmetric high-velocity bullets on both sides of the jet, probably related to the outbursts of the accretion activity. Although geometrically symmetric, the blueshifted component appears as more powerful with stronger SiO and CO emission (Fig. 3).

3. Complex organic molecules

With the high sensitivity of ALMA observations, a number of complex organic molecules are detected toward SMM1-a and S68N. Notably, no such emission is present in Ser-emb

Table 1. Complex organic molecules detections

Name	Formula	SMM1-a	S68N
Methanol	CH_2DOH	✓	✓
Methyl Formate	CH_3OCHO	✓	✓
Dimethyl Ether	CH_3OCH_3	✓	✓
Vinyl Cyanide	CH_2CHCN	✓	✗
Ethylene Glycol	$(CH_2OH)_2$	✓	✗

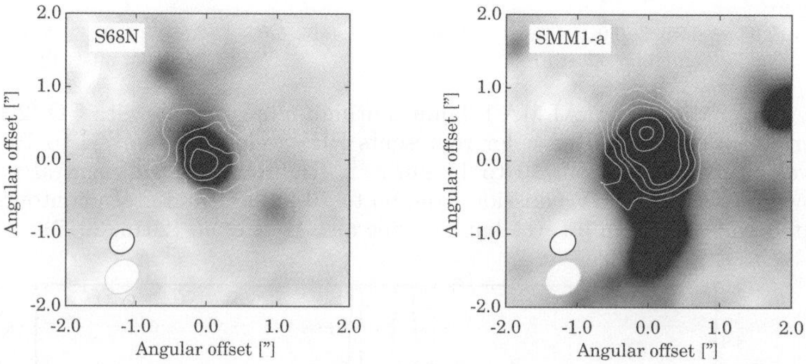

Figure 4. Left: 1.3 mm ALMA continuum emission in grayscale with an integrated emission of methyl formate in yellow contours toward S68N. Right: 1.3mm ALMA continuum emission in grayscale with an integrated emission of vinyl cyanide in yellow contours toward SMM1-a.

8N and SMM1-b. Detection of carbon based complex molecules was reported for SMM1-a (Öberg *et al.* 2011), while S68N is a new detection of a hot corino, hinted by the presence of methanol in the source position by McMullin *et al.* (1994). Table 1 shows the summary of detected molecules. Both SMM1-a and S68N have methanol, methyl formate, and dimethyl ether in their cores, while vinyl cyanide and ethylene glycol are present only toward SMM1-a. Emission from complex molecules is resolved, with peak of the emission located within the continuum source, with notable asymmetries for SMM1-a (Fig. 4) The luminosity of Ser-emb 8N and SMM1-b are poorly known, but based on their millimeter continuum they appear less massive than SMM1-a and S68N. Hence, the presence of the complex organic molecules may be dependent on the luminosity of the source, as the central protostar provides enough energy to heat up the dust in its immediate surroundings to the ice sublimation temperatures of $\sim$ 100 K out to a large enough distance. It appears that, as we detect EHV emission toward SMM1-a alongside with complex organic molecules, detection of hot core chemistry as a signature of a more evolved source (Santangelo *et al.* 2015; De Simone *et al.* 2017) seems to fall. However, the interpretation of the CO jet from SMM1-a is not straightforward. There is no corresponding EHV emission in SiO, which is the case for Ser-emb 8N and SMM1-b, and the slow CO outflow has an opening angle of $\sim$ 90°, which can be interpreted as an indicator of a more evolved source.

4. Summary

We detect the high-velocity jets toward 3 out of 4 Class 0 objects, which suggests that these phenomena may be more common but detectable only at high sensitivity observations. We also present detections and resolved images of complex organic molecules in the cores of two protostars. These results highlight new opportunities to study outflows from young protostars enabled by ALMA, as well as its unique place as the state-of-the-art astrochemistry tool, which brings us closer to understanding the path of complex molecules from interstellar clouds, through protostellar envelopes and disks, to planets.

Bibliography

Arce, H. G. & Sargent, A. I. 2006, Astrophysical Journal, 646, 1070

Bachiller, R., Martin-Pintado, J., Tafalla, M., Cernicharo, J., & Lazareff, B. 1990, *A&A*, 231, 174

Bergner, J. B., Öberg, K. I., Garrod, R. T., & Graninger, D. M. 2017, *ApJ*, 841, 120

Bontemps, S., Andre, P., Terebey, S., & Cabrit, S. 1996, *A&A*, 311, 858

Cazaux, S., Tielens, A. G. G. M., Ceccarelli, C., *et al.* 2003, *ApJ* (Letters), 593, L51

Choi, M. 2009, *ApJ*, 705, 1730

De Simone, M., Codella, C., Testi, L., *et al.* 2017, *A&A*, 599, A121

Eiroa, C., Djupvik, A. A., & Casali, M. M. 2008, The Serpens Molecular Cloud, ed. B. Reipurth, 693

Hirano, N., Ho, P. P. T., Liu, S.-Y., *et al.* 2010, *ApJ*, 717, 58

Hull, C. L. H., Girart, J. M., Kristensen, L. E., *et al.* 2016, *ApJ* (Letters), 823, L27

Hull, C. L. H., Girart, J. M., Tychoniec, Ł., *et al.* 2017, ArXiv e-prints, arXiv:1707.03827

Jørgensen, J. K., van der Wiel, M. H. D., Coutens, A., *et al.* 2016, *A&A*, 595, A117

Kristensen, L. E., van Dishoeck, E. F., Bergin, E. A., *et al.* 2012, *A&A*, 542, A8

McMullin, J. P., Mundy, L. G., Wilking, B. A., Hezel, T., & Blake, G. A. 1994, *ApJ*, 424, 222

Öberg, K. I., van der Marel, N., Kristensen, L. E., & van Dishoeck, E. F. 2011, *ApJ*, 740, 14

Ortiz-León, G. N., Loinard, L., Kounkel, M. A., *et al.* 2017, *ApJ*, 834, 141

Santangelo, G., Codella, C., Cabrit, S., *et al.* 2015, *A&A*, 584, A126

Santiago-García, J., Tafalla, M., Johnstone, D., & Bachiller, R. 2009, *A&A*, 495, 169

Tafalla, M., Santiago-García, J., Hacar, A., & Bachiller, R. 2010, *A&A*, 522, A91

van Dishoeck, E. F., Blake, G. A., Jansen, D. J., & Groesbeck, T. D. 1995, *ApJ*, 447, 760

Astrochemistry VII – Through the Cosmos from Galaxies to Planets
Proceedings IAU Symposium No. 332, 2017
M. Cunningham, T. Millar & Y. Aikawa, eds.

doi:10.1017/S1743921317007645

SuperMALT: Physical and Chemical Properties of Massive and Dense Clumps

Sudeep Neupane[1], Guido Garay[1], Yanett Contreras[2] and SuperMALT Team[3]

[1]Department of Astronomy, Universidad de Chile, Casilla 36D, Santiago, Chile
Email: sneupane@das.uchile.cl

[2]Leiden Observatory, Leiden University, PO Box 9513, NL-2300 RA Leiden

[3] http://home.strw.leidenuniv.nl/~ycontreras/supermalt.html

Abstract. The SuperMALT survey is observing 76 MALT90 clumps at different evolutionary stages (from pre-stellar or quiescent to HII) in high excitation molecular lines and key isotopomers using the Apex 12m telescope with an angular resolution of ~20" and a velocity resolution of ~0.1 km/s. The aim of this survey is to determine the physical, chemical, and kinematical properties of the gas within clumps as they evolve. Here we report some preliminary results based on observations of the J=3-2 & 4-3 lines of HNC, HCN, HCO^+, N_2H^+ and of the J=3-2 line of the isotopologue $H^{13}CO^+$. We find that the morphologies and line profiles vary with the evolutionary stage of the clumps. The average line width increases from quiescent to HII clumps while line ratios show hint of chemical differences among the various evolutionary stages.

Keywords. ISM: clouds, ISM: molecules, ISM: evolution, line: profiles

1. Introduction

Several observations have shown that high-mass stars form in massive dense clumps which have typically masses of $\sim 10^3$ $M_\odot$, densities of $\sim 10^5$ cm^{-3} and dust temperatures $\leqslant 20$ K. (eg., Faundez *et al.* 2004, Garay *et al.* 2004). Despite of the great role that high-mass stars play as energy input in the Galaxy, their formation mechanism is not cleary understood. Determining the properties of massive dense clumps and their fragmentation process, giving rise to cores, is crucial to understand the formation process of high-mass stars. Several molecular and continuum surveys have been made to investigate the characteristics and evolutionary stages of massive dense clumps (eg., ATLASGAL: Schuller *et al.* 2009, Hi-GAL: Molinari *et al.* 2010, GLIMPSE/MIPSGAL: Benjamin *et al.* 2003; Carey *et al.* 2009, MALT90: Foster *et al.* 2011; Jackson *et al.* 2013). In particular, the Millimeter Astronomy Legacy Team 90 GHz (MALT90) Survey observed ~3200 dense molecular clumps in different evolutionary states – from pre-stellar, to proto-stellar, and H II regions – in 16 low excitation molecular lines (mostly J=1→0), near 90 GHz. This survey, made using the 22m Mopra Telescope with an angular resolution ~40" and a velocity resolution of ~ 0.1 km/s, has provided important information concerning the morphology and kinematics of the massive dense clumps showing that clumps in different evolutionary stages indeed have different physical and kinematical properties (eg., Guzman *et al.* 2015, Rathborne *et al.* 2016, Contreras *et al.* 2017). However, since for a given molecular species only a single rotational transition (mostly J=1→0) was observed, the MALT90 data are insufficient to perform a robust derivation of the gas properties. Only with multi-level excitation analysis a robust determination of clump temperatures, densities, and column densities can be obtained, which are important to

Table 1. Molecular lines observed in SuperMALT. The OTF maps are 2'x2' in size.

Species	Frequency (MHz)	Observing Mode	Tracer
HCO^+(3-2)	267557.633	OTF	Density
HCN(3-2)	265886.431	OTF	Density
HNC(3-2)	271981.111	OTF	Density, Cold Chemistry
N_2H^+(3-2)	279511.760	OTF	Density, Chemically robust
HCO^+(4-3)	356734.242	OTF	Density
HCN(4-3)	354505.473	OTF	Density
HNC(4-3)	362630.303	OTF	Density
N_2H^+(4-3)	372672.480	OTF	Density
$H^{13}CO^+$(3-2)	260255.342	Point	Column Density
N_2D^+(3-2)	231321.830	Point	Deuterium Chemistry

understand the differences in physical and chemical properties of clumps at different evolutionary stages.

We conducted the SuperMALT survey to observe a subsample of MALT90 clumps in high excitation molecular lines and key isotopomers using the Apex 12m telescope with the aim of providing an important new legacy database to characterise the evolution of high-mass star forming clumps (Contreras *et al.* in preparation). In this work, we present some preliminary results from the survey.

2. Observations

We observed a subsample of 76 MALT90 clumps in different evolutionary states (pre-stellar or quiescent (Q), proto-stellar (P) and HII clumps (HII)) in high excitation molecular lines and key isotopomers (see Table 1). The sample was selected using the following criteria: (1) reliable detection ($\geqslant 5\sigma$) of key molecular transitions (i.e. N_2H^+, HCN, HNC, HCO^+(1-0)) in MALT90; (2) masses between 400 to 10^4 $M_\odot$; (3) distances between 3 to 5 kpc; and (4) representative of various evolutionary stages. We used the APEX telescope to observe the J=3-2 & 4-3 lines of HNC, HCN, HCO^+, N_2H^+ and the J=3-2 line of the isotopologue $H^{13}CO^+$ with angular resolution of ~23" at 270 GHz. As receivers we use the Swedish Heterodyne Facility Instrument (SHeFI) Apex-1 (213-275 GHZ) and Apex-2 (267-378 GHz), with a velocity resolution of ~0.1 km/s.

In the first phase of the project (2013-2016) we observed the J=3→2 transitions and in second phase, in progress, we are observing the J=4→3 transitions. We use the GILDAS software package to reduce and analyse the data. The CLASS and GREG package are utilised to produce the spectra, subtract baselines, fit profiles and make maps. The final spectra for analysis have a velocity resolution 0.2 km s^{-1} in 12 × 12 pixels with a pixel size of 10"x10". The noise in brightness temperature typically is ~0.03 K per channel for Apex1 observations and ~0.1-0.4 K per channel for Apex2 observations depending on the molecular lines. The peak spectra are obtained by averaging the data over a central circular region of size similar to the beam-size i.e. ~20". In the following section we report some early results on morphology, line profiles and line ratios mainly based on J 3→2 transition line data.

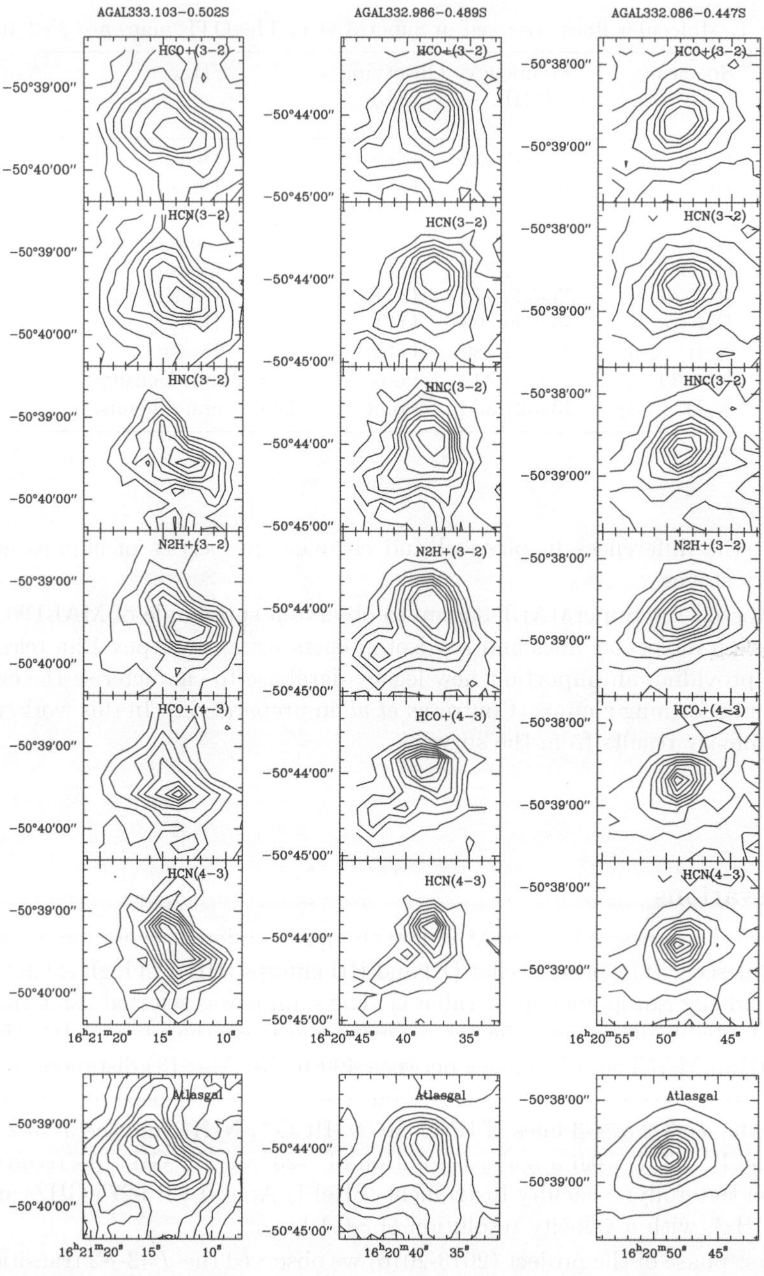

Figure 1. Contour maps of the emissions of 6 different molecular lines and Atlasgal (870μm) continuum map from three types of clumps (AGAL333.103-0.502S: pre-stellar (left), AGAL332.986-0.489S: proto-stellar (middle), and AGAL332.086-0.447S: HII clump (right)). Contour levels have steps of 10% and the peak contour is at 90% level of the peak emission. The molecular species and transition lines are shown inside each box.

3. Results

3.1. *Morphology*

We investigated the morphology of the clumps in the different lines by making maps of the velocity integrated emission using the same velocity interval for all lines. We find that

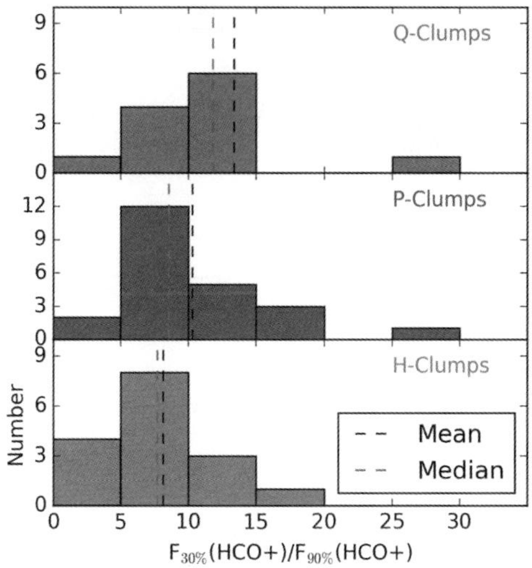

Figure 2. HCO^+ (3-2) integrated intensity ratios within contour region of 30% and 90% of the peak towards different type of clumps. The mean and median for each distribution is indicated by dotted color lines.

the morphologies of the emission in the J=3-2 lines of the different molecules are similar, although the HCO^+ (3-2) emission is usually more extended. This is illustrated in Figure 1 which shows maps of the emission from three clumps in different evolutionary stages in the J=3→2 lines of HCO^+, HCN, HNC and N_2H^+ and J=4→3 lines of HCO^+ and HCN. Also shown, for comparison, is a map of the dust continuum emission at 870 micron taken from ATLASGAL. At first order, the morphology of the molecular emissions is similar to that of the cold dust emission. The emission of J=4→3 lines of HCO^+ and HCN trace high density regions within the clump.

Using the HCO^+(3→2) data, we computed flux densities over two areas, enclosing, respectively, the 30% ($F_{30\%}$) and 90% ($F_{90\%}$) levels of the peak emission. We find that the average value of the flux density ratio ($F_{30\%}$)/($F_{90\%}$) is higher for quiescent clumps than for proto-stellar and HII clumps [see Fig. 2] suggesting that the prestellar clumps have shallower density profile than more evolved clumps.

3.2. *Line Profiles*

The line profiles of the emission from the clumps show a variety of shapes: gaussian, asymmetric, wings etc. The line profiles of Q-type clumps are typically nearly Gaussian, while the more evolved clumps (P and H type) show features such as self-absorption and line wings [see Fig. 3] in optically thick lines HCO^+(3-2) and HCN(3-2).

Line width: The width of optically thin lines provide information about the non-thermal or turbulent velocities in the clouds. The profiles of the $H^{13}CO^+$(3-2) and HNC(3-2) emission from most clumps are Gaussian, hence we fitted gaussian profiles to obtain FWHM widths. We used S/N $\geqslant 3\sigma$ as detection limit for the line widths analysis of different type of clumps. We find that the average line widths increases with the evolutionary stage of the clumps [see Fig.4].

Line ratios: The molecular line intensities and ratios reveal a wide range of chemical properties of clumps. Without going into complex chemistry of the molecules, we look for correlation of line intensity ratios of different species for clumps in all evolutionary stage. The preliminary analysis shows that the HCN(3-2) to HNC(3-2) intensity ratio

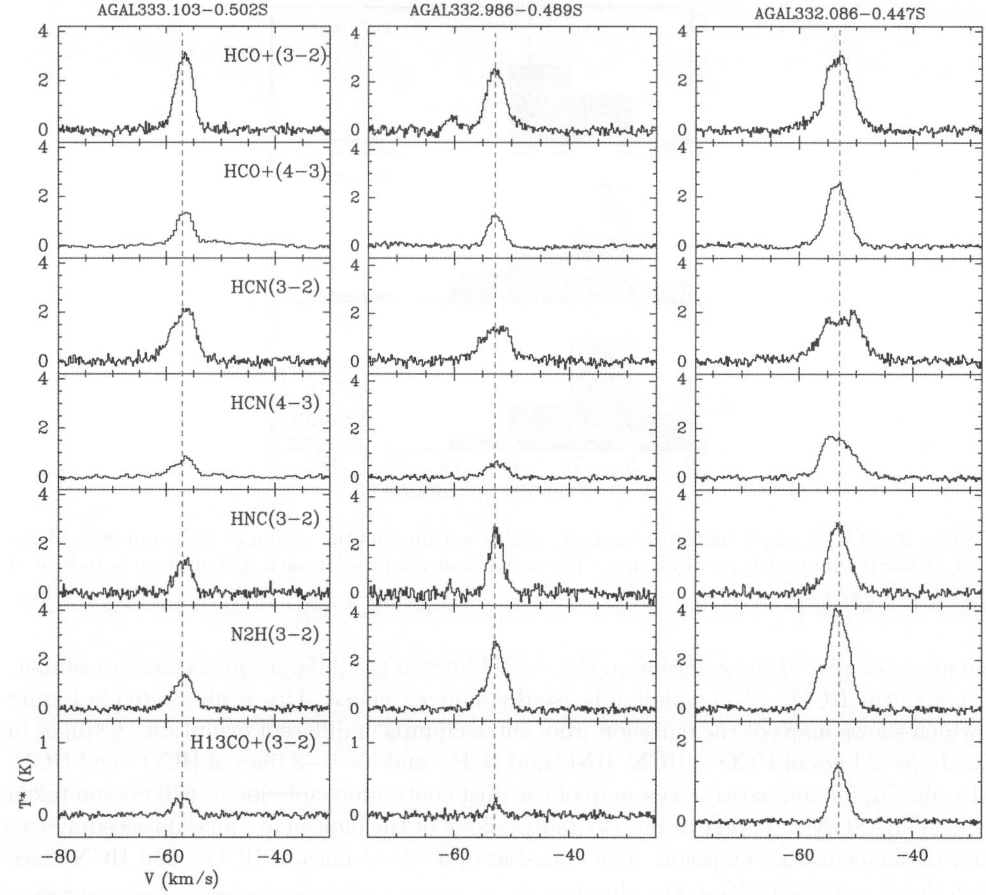

Figure 3. Line profiles of 6 different molecular lines towards three different types of clumps; pre-stellar: AGAL333.103-0.502S (left), proto-stellar: AGAL332.986-0.489S (middle), HII clump: AGAL332.086-0.447S (right). The molecular species, for each row, are shown inside each box on the left.

increases with evolutionary stage (Quiescent to HII clumps) suggesting a temperature dependency of the ratio [see Fig.5].

Physical properties: The multi J transition lines of molecular species can be used to constrain the physical parameters of the clumps. We use data from three rotational transitions ($J = 4 \rightarrow 3$, $3 \rightarrow 2$ from SuperMALT and $J = 1 \rightarrow 0$ from MALT90) in four molecular species and perform radiative transfer modelling, using RADEX (Schoier *et al.* 2005), to constrain the temperature and density of the clumps. This is ilustrated in Fig.6, which shows a relative chi-square map made using the observed data for clump AGAL333.103-0.502S and model integrated intensities. For this clump, the kinetic temperature is constrained to be in the range 10.4 to 12.4 K and the density in the range 2.5×10^6 and 1.6×10^7 cm^{-3}.

4. Summary and Conclusions

We report preliminary results from the SuperMALT survey. We find that the morphology of the emission observed in the J=3-2 lines of HNC, HCN, HCO^+, N_2H^+ are similar, although there are some differences. The HCO^+ emission is more extended than that seen

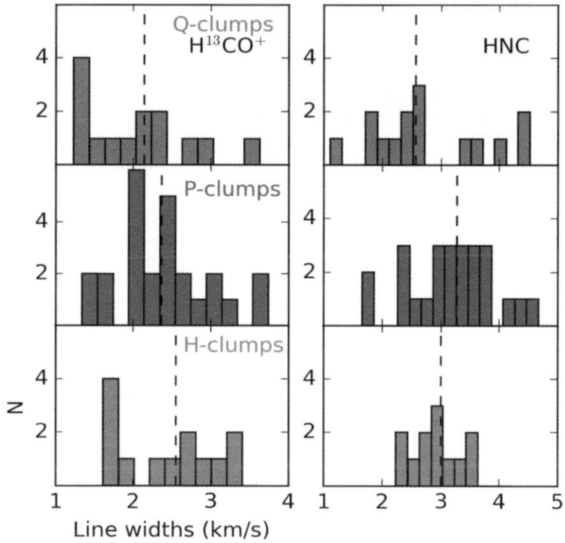

Figure 4. Line widths, of Q, P and HII clumps, determined from Gaussian fit to the peak spectra of optically thin lines $H^{13}CO^+$ (3-2) and HNC(3-2). The mean value for each distribution is indicated by dotted black line.

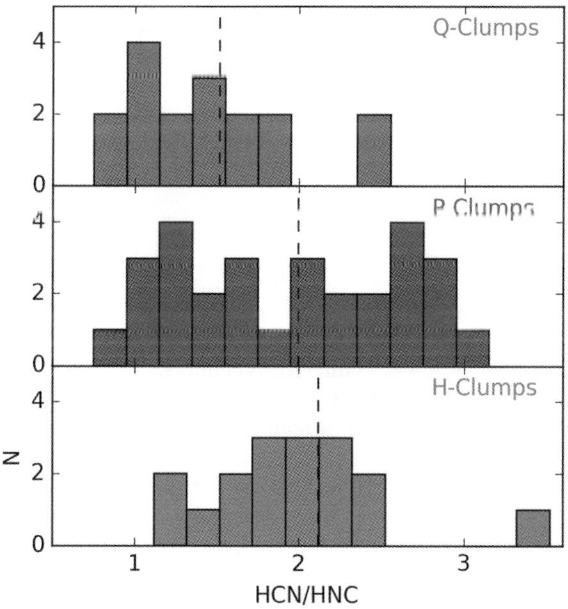

Figure 5. HCN(3-2) to HNC(3-2) integrated intensity ratios from the peak spectra towards Q, P and HII clumps. The average for each distribution is indicated by dotted black line.

in the N_2H^+ and HNC. The morphology of the molecular emissions are also similar to that of the dust emission at 870 micron (ATLASGAL). The line profiles appear different from quiescent clumps to HII clumps. Most quiescent clumps show nearly Gaussian line profiles while evolved clumps show features such as line wings and self absorption. Clumps in the late stages (HII) exhibit brighter lines than those of quiescent clumps

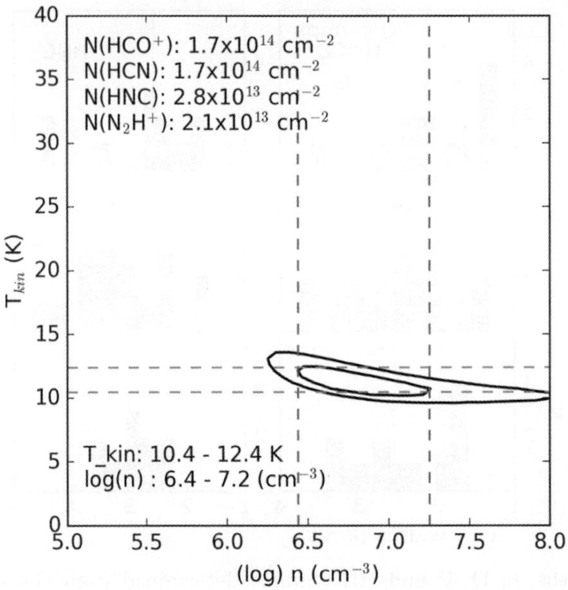

Figure 6. Chi-square minimization of observed and RADEX model intensities using four main molecular lines (HCO, HCN, HNC and N_2H^+) molecules for clump AGAL333.103-0.502S. Peak contour level shows the constraints in density and kinetic temperature.

and an increase in line widths or turbulence suggesting active star formation within the former clumps.

Acknowledgements

S.N. and G.G. greatfully acknowledge support from CONICYT project PFB-06.

References

Benjamin, Robert A., Churchwell, E., Babler, Brian L., *et al.*, 2003, *PASP*, 115, 953
Carey, S. J., Noriega-Crespo, A., Mizuno, D. R., *et al.*, 2009, *PASP*, 121, 76
Contreras, Y., Rathborne, Jill M., Guzman, A., *et al.*, 2017, *MNRAS*, 466, 340
Faúndez, S., Bronfman, L., Garay, G., Chini, R., Nyman, L., & May, J., 2004 *ApJ*, 426, 97
Foster, Jonathan B., Jackson, James M., Barnes, Peter J., *et al.*, 2011, *ApJ*, 197, 25
Garay, G., Faúndez, S., Mardones, D., Bronfman, L., Chini, R., & Nyman, L., 2004, *ApJ*, 610, 313
Guzmn, Andrs E., Sanhueza, P., Contreras, Y., *et al.*, 2015, *ApJ*, 815, 130
Jackson, J. M., Rathborne, J. M., Foster, J. B., *et al.*, 2013, *PASP*, 30, 57
Molinari, S., Swinyard, B., Bally, J., Barlow, M., *et al.*, 2010, *PASP*, 122, 314
Rathborne, J. M., Whitaker, J. S., & Jackson, J. M., 2016, *PASP*, 33, 30
Schuller, F., Menten, K. M., Contreras, Y., *et al.*, 2009, *A&A*, 504, 415
Schoier, F. L., van der Tak, F. F. S., *et al.*, 2005, *A&A*, 432, 369

Astrochemistry VII – Through the Cosmos from Galaxies to Planets
Proceedings IAU Symposium No. 332, 2017
M. Cunningham, T. Millar & Y. Aikawa, eds.

doi:10.1017/S1743921317007566

The Abundance of SiC_2 in Carbon Star Envelopes

Sarah Massalkhi, M. Agúndez, J. Cernicharo, J. P. Fonfría and M. Santander-García

Grupo de Astrofísica Molecular, Instituto de Ciencia de Materiales de Madrid, CSIC
C/ Sor Juana Inés de la Cruz 3, 28049 Cantoblanco, Spain

Abstract. Silicon carbide dust grains are ubiquitous in circumstellar envelopes around C-rich AGB stars. However, the main gas-phase precursors leading to the formation of SiC dust have not yet been identified. To date, only three molecules containing an Si–C bond have been identified to have significant abundances in C-rich AGB stars: SiC_2, SiC, and Si_2C. The ring molecule SiC_2 has been observed in a handful of evolved stars, while SiC and Si_2C have only been detected in the C-star envelope IRC +10216. We aim to study how widespread and abundant SiC_2, SiC, and Si_2C are in envelopes around C-rich AGB stars and whether or not these species play an active role as gas-phase precursors of silicon carbide dust in the ejecta of carbon stars.

1. Setting the Scene

During the late stages of their evolution, asymptotic giant branch (AGB) stars experience significant mass loss processes, which result in extended circumstellar envelopes (CSEs). These environments are efficient factories of molecules and dust grains. The main paradigm for the dust formation process involves a first step in which condensation nuclei of nanometer size are formed from some gas-phase precursors of highly refractory character and a second step in which the nuclei grow to micrometer sizes by accretion and coagulation as the material is pushed out by the stellar wind. The chemical nature of the molecules and dust grains formed depends to a large extent on the C/O elemental abundance ratio at the stellar surface. Although much has been advanced recently, there is still much to understand about how dust grains are formed and which are the main gas-phase precusors. This is the main driver of the ERC Synergy Project NANOCOSMOS.

2. SiC Dust: What is it about?

Silicon carbide (SiC) dust grains, which are detected through a solid-state band at $11.3\mu m$, are exclusively found in the envelopes around C-type stars ($C/O > 1$). Here, we explore which are the main precursors of SiC dust grains. Only three gas-phase molecules containing the Si-C bond have been observed in C-rich envelopes around AGB stars. The ring molecule SiC_2 has been observed towards a few AGB and post-AGB stars (Thaddeus *et al.* 1984; Bachiller *et al.* 1997; Zhang *et al.* 2009a,b), while SiC and Si_2C have been only observed in the C star envelope IRC +10216 (Cernicharo *et al.* 1989, 2015b). Much of the knowledge about the role of these three molecules as precursors of SiC dust grains comes from the study of IRC +10216, as in this source SiC_2 has been thoroughly identified across the mm and sub-mm ranges with ground based radio telescopes and with the Herschel Space Telescope (Lucas *et al.* 1995; Cernicharo *et al.* 2010; Müller *et al.* 2012, Fonfría *et al.* 2014; Velilla Prieto *et al.* 2015). The paradigm emerged from the studies of IRC +10216 suggests that only SiC_2 and Si_2C are present in the innermost regions, while SiC is probably a photodissociation product of these molecules, and it is thus restricted

Table 1. Sample of carbon stars

Name	V_{LSR} (km s^{-1})	D (pc)	$T_\star$ (K)	$L_\star$ (L$_\odot$)	$\dot{M}$ (M$_\odot$ yr^{-1})	V_{exp} (km s^{-1})	f_0
IRC +10216	−26.5	130 [a]	2330 [a]	8750 [a]	2.0×10^{-5} [a]	14.5	3.7×10^{-7}
CIT 6	−1	440 [g]	1800 [g]	10000 [g]	6.0×10^{-6} [g]	17	1.1×10^{-5}
CRL 3068	−31.5	1300 [g]	1800 [g]	10 900 [g]	2.5×10^{-5} [g]	14.5	1.9×10^{-6}
S Cep	−15.3	380 [k]	2200 [k]	7300 [k]	1.2×10^{-6} [k]	22.5	1.0×10^{-5}
IRC +30374	−12.5	1200 [j]	2000 [j]	9800 [j]	1.0×10^{-5} [j]	25	9.7×10^{-6}
Y CVn	+22	220 [k]	2200 [k]	4400 [k]	1.5×10^{-7} [k]	7	4.0×10^{-6}
LP And	−17	630 [g]	1900 [g]	9600 [g]	7.0×10^{-6} [g]	14.5	6.8×10^{-6}
V Cyg	+13.5	366 [g]	2300 [g]	6000 [g]	1.6×10^{-6} [g]	12	6.6×10^{-6}
UU Aur	+6.7	260 [k]	2800 [k]	6900 [k]	2.4×10^{-7} [k]	10.6	$< 1.0\times10^{-6}$
V384 Per	−16.8	560 [k]	2000 [k]	8100 [k]	2.3×10^{-6} [b]	15.5	1.3×10^{-5}
IRC +60144	−48.8	1030 [b]	2000 [b]	7800 [b]	3.7×10^{-6} [b]	19.5	6.0×10^{-6}
U Cam	+6	430 [i]	2695 [i]	7000 [i]	2.0×10^{-7} [i]	13	3.7×10^{-5}
V636 Mon	+10	880 [f]	2500 [n]	8472 [e]	5.8×10^{-6} [f]	20	1.7×10^{-6}
IRC +20370	−0.8	600 [k]	2200 [k]	7900 [k]	3.0×10^{-6} [b]	14	4.2×10^{-6}
R Lep	+11.5	432 [b]	2200 [b]	5500 [b]	8.7×10^{-7} [b]	17.5	$< 2.7\times10^{-6}$
W Ori	−1	220 [k]	2600 [k]	3500 [k]	7.0×10^{-8} [k]	11	1.4×10^{-5}
CRL 67	−27.5	1410 [d]	2500 [n]	9817 [d]	1.1×10^{-5} [d]	16	1.0×10^{-5}
CRL 190	−39.5	2790 [d]	2500 [c]	16 750 [d]	6.4×10^{-5} [d]	17.0	8.8×10^{-7}
S Aur	−17	300 [k]	3000 [k]	8900 [k]	4.0×10^{-7} [k]	24.5	7.3×10^{-6}
V Aql	+53.5	330 [k]	2800 [k]	6500 [k]	1.4×10^{-7} [k]	8	2.0×10^{-5}
CRL 2513	+17.5	1760 [d]	2500 [c]	8300 [c]	2.0×10^{-5} [d]	25.5	1.6×10^{-6}
CRL 2477	+5	3380 [m]	3000 [m]	13 200 [m]	1.1×10^{-4} [m]	20	6.0×10^{-7}
CRL 2494	+29	1480 [b]	2400 [b]	10 200 [b]	7.5×10^{-6} [b]	20	2.7×10^{-5}
RV Aqr	+0.5	670 [k]	2200 [k]	6800 [k]	2.3×10^{-6} [b]	15	3.0×10^{-6}
ST Cam	-13.6	360 [k]	2800 [k]	4400 [k]	1.3×10^{-7} [k]	8.9	$< 4.0\times10^{-6}$

The values f_0 in the last column are the derived SiC_2 abundances relative to H_2.
References: [a] Agúndez *et al.* (2012), [b] Danilovich *et al.* (2015), [c] Groenewegen *et al.* (1998), [d] Groenewegen *et al.* (2002), [e] Guandalini *et al.* (2013), [f] Guandalini *et al.* (2006), [g] Ramstedt *et al.* (2014), [i] parameters of the present-day wind of U Cam from Schöier *et al.* (2005),
[j] Schöier *et al.* (2006), [k] Schöier *et al.* (2013), [m] Speck *et al.* (2009),
[n] Assumed value for the stellar effective temperature $T_\star$ is 2500 K

to the outer envelope. The U-shaped line profiles of SiC (Cernicharo *et al.* 2000) and the upper limits derived to its abundance (Velilla Prieto *et al.* 2015) in the inner envelope support this idea. Moreover, chemical equilibrium calculations predict abundant SiC_2 and Si_2C, and little SiC in the hot and dense surroundings of the AGB star (Tejero & Cernicharo 1991; Yasuda & Kozasa 2012; Cernicharo *et al.* 2015b). It therefore seems that the gas-phase molecule SiC is not an important building block of SiC dust grains, while SiC_2 and Si_2C are probably the main gas-phase precursors.

However, little is still known on how widespread these molecules are in other carbon stars, what are their relative abundances, and what is their role on the formation of SiC dust. To explore the role of gas-phase SiC_2 molecules on the formation of silicon carbide dust, we carried out sensitive observations with the IRAM 30m telescope of a sample of 25 C-rich AGB stars (see Table 1) to observe emission lines of SiC_2, Si_2C and SiC in the λ 2mm band (see Table 2). The observations resulted in the detection of SiC_2 in 22 out of

Table 2. Rotational transitions of SiC_2, Si_2C, and SiC covered

Transition	Frequency (MHz)	A_{ul} (s^{-1})	E_u (K)
	SiC_2		
$6_{2,5} - 5_{2,4}$	140920.171	7.65×10^{-5}	31.5
$6_{4,3} - 5_{4,2}$	141751.492	4.87×10^{-5}	55.0
$6_{4,2} - 5_{4,1}$	141755.360	4.87×10^{-5}	55.0
$6_{2,4} - 5_{2,3}$	145325.875	8.39×10^{-5}	32.0
$7_{0,7} - 6_{0,6}$	158499.228	1.23×10^{-4}	31.0
	Si_2C		
$11_{1,11} - 10_{0,10}$	144033.475	9.57×10^{-6}	29.3
$22_{2,20} - 22_{1,21}$	155600.100	1.54×10^{-5}	115.0
$13_{1,13} - 12_{0,12}$	157768.156	1.28×10^{-5}	39.3
$20_{2,18} - 20_{1,19}$	157959.754	1.54×10^{-5}	97.4
$18_{2,16} - 18_{1,17}$	160644.790	1.55×10^{-5}	81.4
	SiC		
$^3\Pi_2$ $J = 4 - 3$	157494.101	3.98×10^{-5}	13.2

the 25 targeted sources. The lines corresponding to the rotational transitions $6_{4,3} - 5_{4,2}$ and $6_{4,2} - 5_{4,1}$ appear blended although in most sources each of these lines could be fitted individually. We also report the detection of SiC in 12 of the 25 targeted C-rich AGB envelopes (see Fig.1). We note that prior to this study, this radical had been detected only in the C-star envelope IRC +10216 (Cernicharo *et al.* 1989). The molecule Si_2C was not detected in any of the targeted sources, at the exception of IRC +10216, which at present is the only source in which this molecule has been identified (Cernicharo *et al.* 2015b). Hereafter, we focus on the analysis of SiC_2 and leave the interpretation of other molecules for future studies. We aim to model its emission and determine its abundance in each source to provide a global view of how abundant this molecule is in circumstellar envelopes around C-rich AGB stars.

3. Abundance of SiC_2: How we did it

To model all sources, we adopted a common physical scenario consisting of a spherically symmetric envelope of gas and dust expanding with a constant velocity and mass-loss rate around a central AGB star. The spherical envelope is described by the radial profile of various physical quantities, such as the gas density, the kinetic temperature of the gas, the dust temperature, the expansion velocity, and the microturbulence velocity. We also consider a dusty component of the envelope, although, we noticed that dust plays a minor role in the excitation of the rotational lines of SiC_2 observed.

We have performed non-LTE excitation and radiative transfer calculations to model the line emission of SiC_2 based on the multi-shell LVG (Large Velocity Gradient) method of the MADEX code. The circumstellar envelope is divided into a number of concentric shells, each of them with a characteristic set of physical properties and SiC_2 abundance, and statistical equilibrium equations are solved in each of them. In each shell, the contribution of the background radiation field (cosmic microwave background, stellar radiation, and thermal emission from surrounding dust) is included. In the excitation calculations, we consider rotational levels up to $J = 39$ (i.e., a total number of 620 energy levels) within

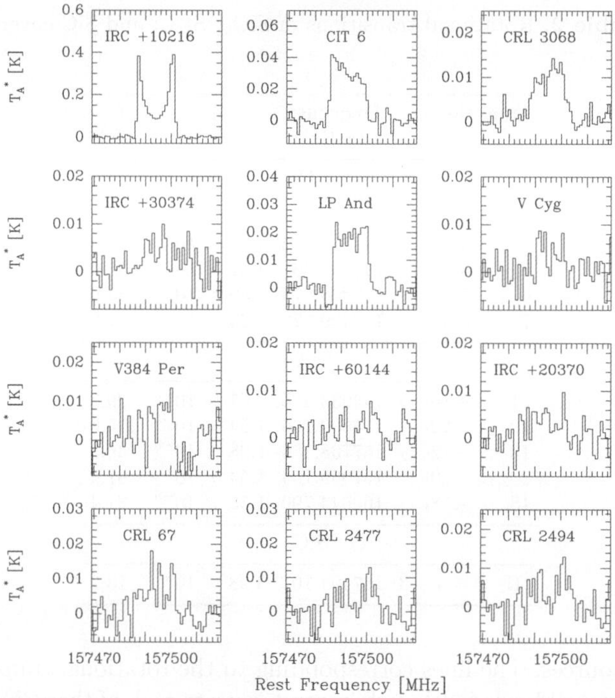

Figure 1. SiC $^3\Pi_2$ $J = 4 - 3$ line observed in 12 out of the 25 sources of our sample. The spectral resolution is 1 MHz.

the ground vibrational state of SiC_2. The rate coefficients through inelastic collisions with H_2 and He were taken from (Chandra, S. & Kegel, W. H. 2000). We consider that SiC_2 is formed close to the star with a given fractional abundance, which remains constant through the envelope up to the outer regions, where it is photodissociated by the ambient ultraviolet radiation field of the local Interstellar Medium. To model the emission lines of SiC_2 and determine its abundance, we varied the initial fractional abundance of SiC_2 relative to H_2, f_0, until the calculated line profiles matched the observed ones. We choose as best-fit model that which results in the best overall agreement between calculated and observed line profiles for the entire set of SiC_2 lines observed. In those cases in which no lines of SiC_2 are detected, we derive upper limits to the abundance of SiC_2 by choosing the maximum abundance that results in line intensities compatible with the noise level of the observations.

4. Results: What We Found

The calculated line profiles resulting from our best-fit model for each of the sources is compared with the observed line profiles in Fig. 2. In most sources, the shapes of the SiC_2 lines observed are nearly flat-topped, which is indicative of optically thin emission not resolved by the 15.0-17.5″ beam of the IRAM 30m telescope. One notable exception is IRC +10216, whose close proximity (130 pc) makes the emission to be spatially resolved by the telescope beam and line profiles show a marked double-peaked shape.

The fractional abundance of SiC_2 shows an interesting trend with the density in the envelope, evaluated through the quantity $\dot{M}/V_{\rm exp}$. As shown in Fig. 3, SiC_2 becomes less abundant as the density in the envelope increases. In a standard scenario describing the chemistry of an expanding envelope around an AGB star, the abundance with which

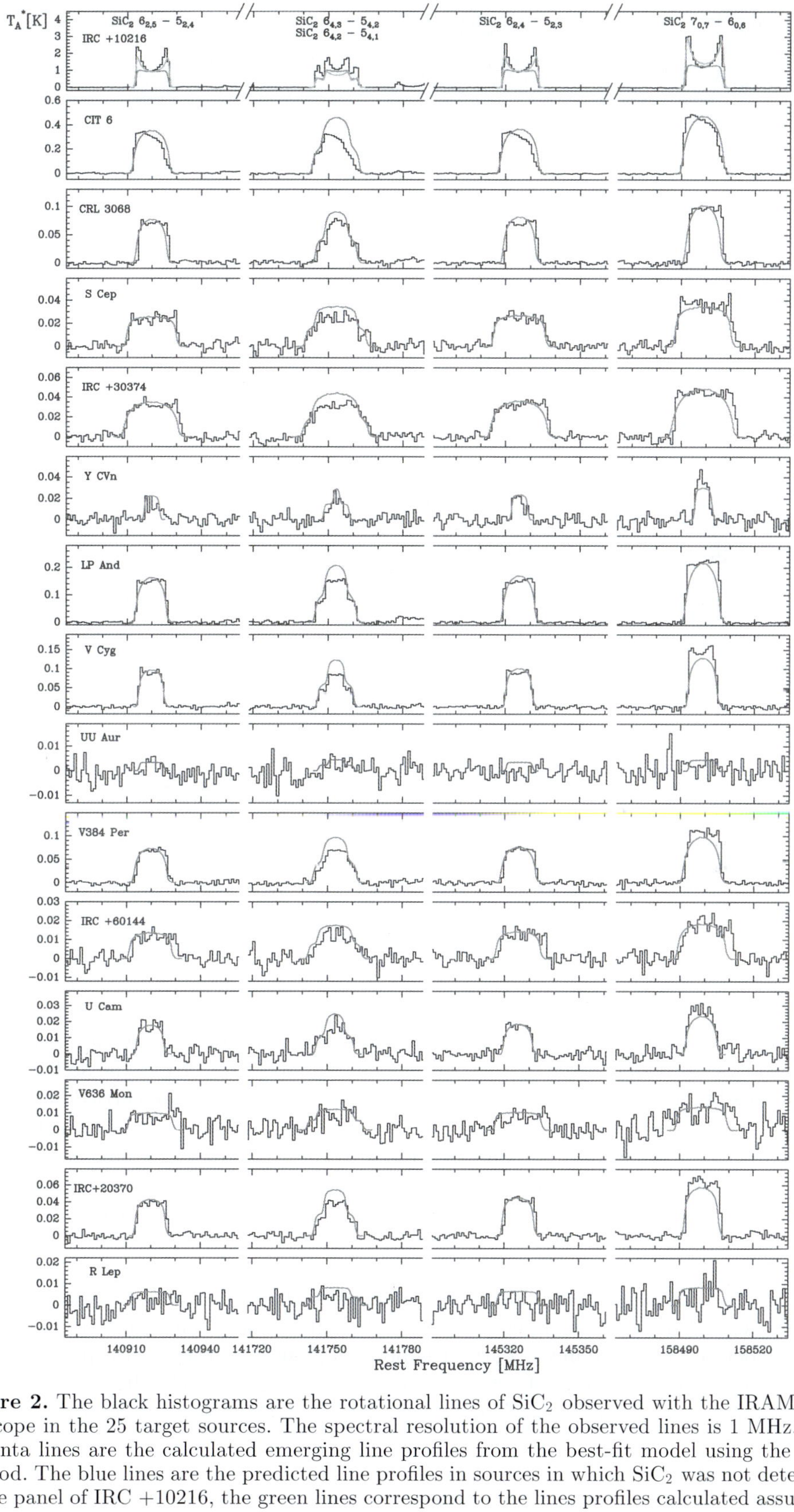

Figure 2. The black histograms are the rotational lines of SiC_2 observed with the IRAM 30m telescope in the 25 target sources. The spectral resolution of the observed lines is 1 MHz. The magenta lines are the calculated emerging line profiles from the best-fit model using the LVG method. The blue lines are the predicted line profiles in sources in which SiC_2 was not detected. In the panel of IRC +10216, the green lines correspond to the lines profiles calculated assuming LTE.

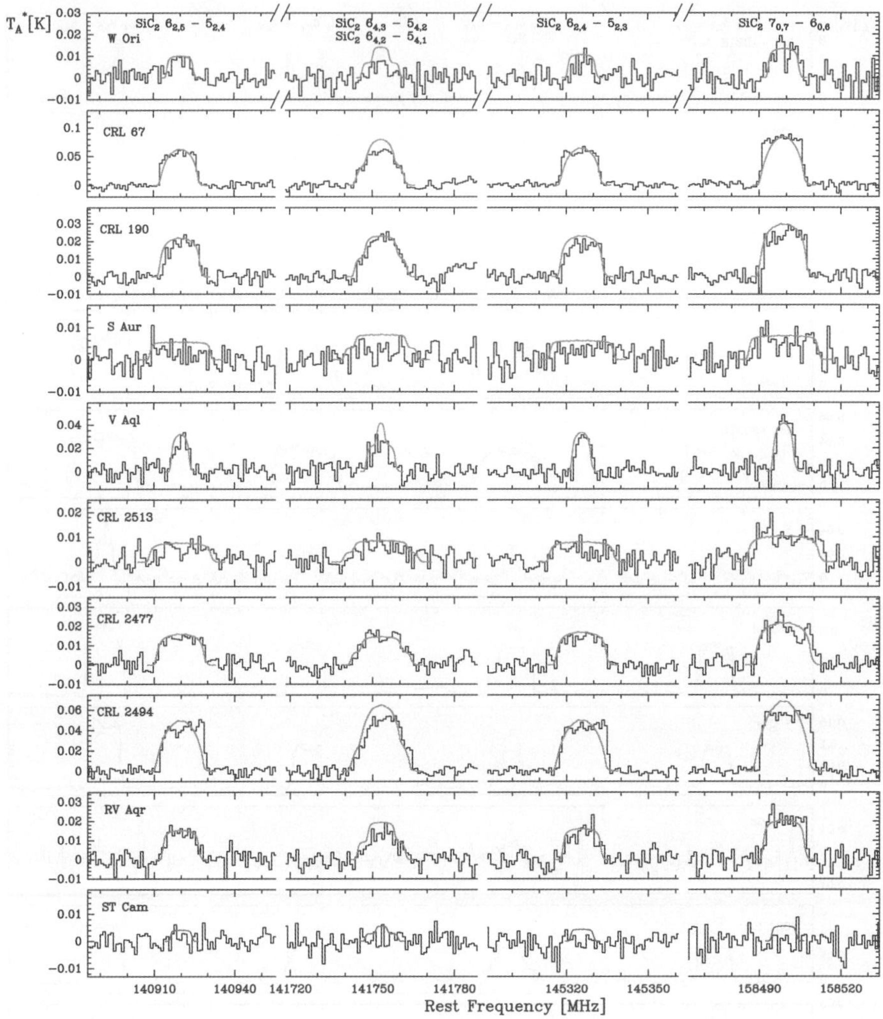

Figure 2. (cont.)

a molecule is injected into the intermediate and outer envelope is set by the processes occurring in the inner regions. In this sense, the abundance of SiC_2 in the intermediate regions of the envelope is set by thermochemical equilibrium (TE) at the stellar surface and is possibly modified by different processes during expansion. We carried out TE calculations which show that the SiC_2 abundances are insensitive to the gas density in the region where the molecular abundances quench to the TE values (see further, Massalkhi *et al.* 2017). We interpret this as that the SiC_2 molecules deplete from the gas phase to incorporate into solid dust grains, a process that is favored at higher densities owing to the higher rate at which collisions between particles occur. We note that the emission from the five SiC_2 lines observed in our sources arises typically from intermediate regions of the envelope, at radial distances in the range 10^{15}-10^{16} cm where dust formation has already taken place, and therefore the SiC_2 abundances have to be considered as post-condensation abundances. Further support for our scenario comes from mm-wave interferometric observations of SiC_2 in the C-star envelope IRC +10216 (Lucas *et al.* 1995; Fonfría *et al.* 2014; Velilla Prieto *et al.* 2015), which shows that SiC_2 is present in

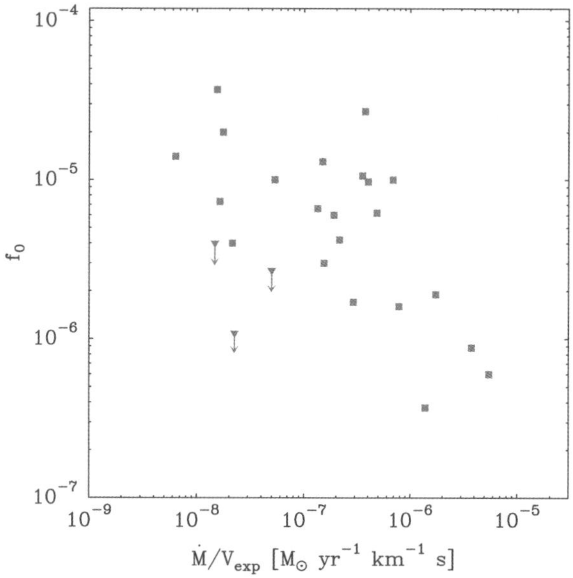

Figure 3. SiC_2 fractional abundance f_0 obtained from radiative transfer analysis, as a function of a density measure ($\dot{M}/V_{exp}$) for the C-rich AGB stars are represented with filled square. Upper limits for the non-detections are denoted with downward triangles.

regions close to the star, then experiences a marked abundance decline at 10-20 R_* (very likely due to condensation onto dust grains), and appears again in the outer envelope (probably as a result of the interaction between the UV radiation field and the envelope). The ring molecule SiC_2 thus emerges as a very likely gas-phase precursor in the process of formation of SiC dust in envelopes around C-rich AGB stars.

5. Similarly? C_2 in AGB stars

The dependence of the SiC_2 gas-phase abundance in the envelopes of different AGB stars on the mass-loss rate is similar to what we have found for C_2 in a sample of C-rich AGB stars (Fonfría *et al.* 2017). The observations of these stars were taken in the optical and near-IR wavelength ranges with the high spectral resolution spectrograph CARMENES ($R \simeq 85,000$) mounted on the 3.5 m telescope at the German-Spanish Astronomical Center at Calar Alto (CAHA). This project represents a big effort to understand the chemical evolution of C-rich AGB stars in their very complex, warm inner envelopes where the dust is formed and the gas expansion mechanism begins to work.

C_2 is a molecule found in C-rich AGB stars some decades ago (e.g., Bakker *et al.* 1995). The spectrum of this molecule shows two strong electronic systems in the near-IR and the optical comprising a large number of vibro-electronic bands: the Phillips system ($A^1\Pi_u - X^1\Sigma_g$ centered at 11900Å) and the Swan system ($d^3\Pi_g - a^3\Pi_u$ centered at 5200Å). No rotational or ro-vibrational electric dipolar transition is allowed in the ground electronic state of this homonuclear molecule, making its detection at wavelengths longer than $\simeq 2.5$ μm pretty difficult. The analysis of both systems is very useful to estimate the C_2 abundance with respect to H_2 in gas phase, the physical conditions, and the gas kinematics in the shells of the envelope where C_2 exists. Determining the fraction of C_2 in the $a^3\Pi_u$ electronic state is of particular interest since the chemical reactivity of this

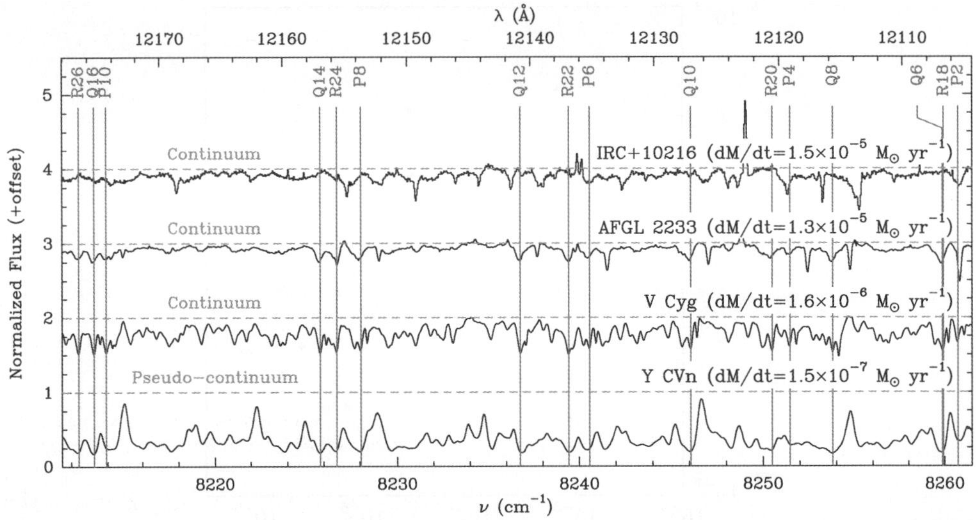

Figure 4. Comparison between the spectra of four C-rich stars with different mass-loss rates in the near-IR acquired with CAHA/CARMENES ($R \simeq 85,000$). The C_2 lines of the Phillips system (0, 0) band are identified.

molecule highly depends of the characteristics of its valence electrons, being significantly more reactive if it is electronically excited.

In our data, we have detected both systems. The obvious detection of the Swan system means that its lower electronic state $a^3\Pi_u$, which is at $\simeq 1000$ K, is significantly populated. This is an indication of warm C_2 that can be found only close to the photosphere of the central stars and can participate in the chemistry in high-temperature, high-density environments. It is believed that C_2 arises from carbon atoms recently ejected by the stellar pulsation that rapidly combine between them into carbon clusters (C_2, C_3, C_4,...), something that we are investigating with the Stardust machine, developed in the frame of the ERC project NANOCOSMOS. To date, it remains unknown if there is an additional contribution to the observed spectra from cold C_2 formed in the outer shells of the envelopes due to the active chemistry triggered by the external dissociating radiation field, as it happens with, e.g., C_3 and C_5 (Hinkle *et al.* 1988, Bernath *et al.* 1989). However, we expect this to be elucidated from our on-going research, which will allow us to describe the C_2 abundance profile throughout the whole envelope taking advantage of the kinematical effects on the line profiles (Fonfría *et al.* 2008).

The preliminary analysis of our data indicates that the gas-phase C_2 column density is different in several AGB stars, as can be seen in Fig. 4. The C_2 bands are clearly stronger in the AGB star with the lowest mass-loss rate (Y CVn) while they are barely noticeable in IRC+10216, which has a mass-loss rate two orders of magnitude higher. Thus, the higher the mass-loss rate of a C-rich AGB star, the less abundant is C_2 in gas-phase around the central object. Since the mass-loss rate is tightly related to the amount of dust grains available in the innermost envelope, the most plausible explanation of this dependence is that the gas density plays a very important role in the condensation of C_2 onto dust grains. The agreggation processes enhanced close to the central star due to the high gas density make of C_2 a very good candidate to form the seeds of the dust grains on which other refractory molecules such as SiC_2 will condense.

6. Conclusion

In this work, we found an evidence of efficient incorporation of SiC_2 and C_2 onto dust grains, a process that is more efficient at high densities because collisions between particles and coagulation processes become faster. The ring molecule SiC_2 emerges as a very likely gas-phase precursor in the process of formation of SiC dust while C_2 seems to be highly involved in the building of dust grain seeds in envelopes around C-rich AGB stars. The search for the gas-phase building blocks of dust grains in both carbon- and oxygen-rich AGB stars is a challenging scientific objective, for which astronomical observations still have to provide many answers. Observing other refractory molecules seems mandatory to understand which of the gas-phase molecules that are present in the atmospheres of AGB stars serve as precursor material to feed the process of dust grain formation.

References

Agúndez, M., Fonfría, J. P., Cernicharo, J., *et al.* 2012 *A&A*, 543, A48

Bachiller, R., Forveille, T., Huggins, P. J., & Cox, P. 1997, *A&A*, 324, 1123

Bakker E. J., Lamers H. J. G. L. M., Waters L. B. F. M. & Schoenmaker T., 1995, *Ap&SS*, 244, 335

Bernath P. F., Hinkle K. H. & Keady J. J., 1989, *Science*, 244, 562

Cernicharo, J., Gottlieb, C. A., Guelin, M., Thaddeus, P., & Vrtilek, J. M., 1989, *ApJ*, 341, L25

Cernicharo, J., Guélin, M., & Kahane, C., 2000, *A&AS*, 142, 181

Cernicharo, J., Waters, L. B. F. M., Decin, L., *et al.* 2010, *A&A*, 521, L8

Cernicharo, J., McCarthy, M. C., & Gottlieb, C. A., *et al.* 2015b, *ApJ*, 806, L3

Chandra, S. & Kegel, W. H. 2000 *A&AS*, 142, 113

Danilovich, T., Teyssier, D., Justtanont, K., *et al.* 2015 *A&A*, 581, A60

Fonfría J. P., Cernicharo J., Richter M. J. & Lacy J. H., 2008 *ApJ*, 673, 445

Fonfría, J. P., Fernández-López, M., Agúndez, M., *et al.* 2014 *MNRAS*, 445, 3289

Fonfría J. P., Santander-García M., Cernicharo J., & Amado, P. J., 2017 *in preparation*

Groenewegen, M. A. T., Whitelock, P. A., Smith, C. H., & Kerschbaum, F. 1998 *MNRAS*, 293, 18

Groenewegen, M. A. T., Sevenster, M., Spoon, H. W. W., & Pérez, I. 2002 *A&A*, 390, 511

Hinkle K. H., Keady J. J. & Bernath P. F., 1988 *Science*, 241, 1319

Lucas, R., Gulin, M., Kahane, C., Audinos, P., & Cernicharo, J., 1995, *Ap&SS*, 244, 293

Ramstedt, S. & Olofsson, H. 2014 *A&A*, 566, A145

Massalkhi, S., Agúndez, M., Cernicharo J., *et al.* 2017 *in preparation*

Müller, H. S. P., Cernicharo, J., Agúndez, M., *et al.* 2012 *Journal of Molecular Spectroscopy*, 271, 50

Schöier, F. L., Lindqvist, M., & Olofsson, H. 2005 *A&A*, 436, 633

Schöier, F. L., Olofsson, H., & Lundgren, A. A. 2006 *A&A*, 454, 247

Schöier, F. L., Ramstedt, S., Olofsson, H., *et al.* 2013 *A&A*, 550, A78

Speck, A. K., Corman, A. B., Wakeman, K., Wheeler, C. H., & Thompson, G. 2009 *ApJ*, 691, 1202

Tejero, J. & Cernicharo, J. *Anuario del Observatorio Astronómico de Madrid para, 1992,* 1992

Thaddeus, P., Cummins, S. E., & Linke, R. A., 1984, *ApJ*, 283, L45

Velilla Prieto, L., Cernicharo, J., Quintana-Lacaci, G., *et al.* 2015 *ApJ*, 805, L13

Yasuda, Y. & Kozasa, T. 2012 *ApJ*, 745, 159

Zhang, Y. & Kwok, S., Dinh-V-Trung, 2009a, *ApJ*, 691, 1660

Zhang, Y., Kwok, S., & Nakashima, J.-i., 2009b, *ApJ*, 700, 1262

Astrochemistry VII – Through the Cosmos from Galaxies to Planets
Proceedings IAU Symposium No. 332, 2017
M. Cunningham, T. Millar & Y. Aikawa, eds.

doi:10.1017/S1743921317007736

Multiline observations of S255IR with ALMA

Igor Zinchenko[1]**, Sheng-Yuan Liu**[2]**, Yu-Nung Su**[2] **and Petr Zemlyanukha**[1]

[1]Institute of Applied Physics of the Russian Academy of Sciences, 46 Uljanov str., 603950 Nizhny Novgorod, Russia
email: zin@appl.sci-nnov.ru

[2]Institute of Astronomy and Astrophysics, Academia Sinica, P.O. Box 23-141, Taipei 10617, Taiwan, R.O.C
email: syliu@asiaa.sinica.edu.tw, ynsu@asiaa.sinica.edu.tw

Abstract. We present preliminary results of the high resolution ($0.10'' \times 0.15''$) observations of the high mass star forming region S255IR with ALMA in several spectral windows from ~ 335 GHz to ~ 350 GHz. The main target lines were $C^{34}S$(7–6), $CH_3CN(19_K - 18_K)$, CO(3–2) and SiO(8–7), however many other lines of various molecules have been detected, too. We present sample spectra and maps, discuss briefly the source structure and kinematics. A new, never predicted methanol maser line has been discovered.

Keywords. ISM: clouds, ISM: molecules, ISM: abundances, ISM: individual objects (S255IR)

1. Introduction

Disks around young high mass stars attract an enhanced attention nowadays. Their investigations are essential for understanding the process of high mass star formation, in particular the role of the disk accretion.

Here we present preliminary results of our recent observations of the S255IR-SMA1 core with ALMA. Our previous SMA observations (Zinchenko *et al.* 2012, 2015) as well as the data by Wang *et al.* (2011) revealed a disk around the massive (20 $M_\odot$) YSO S255 NIRS3 associated with a spectacular outflow. A first disk-mediated accretion burst from the high mass YSO was detected recently here (Caratti O Garatti *et al.* 2017) following the methanol maser flare (Fujisawa *et al.* 2015).

The goals of our ALMA observations included an investigation of the structure, kinematics and physical properties of this disk and associated outflow at a much higher angular resolution and sensitivity in comparison with the previous studies.

2. Observations

Observations of the S255IR region were carried out with ALMA in Band 7 at two epochs during Cycle 4. The first observing epoch on 2016 April 21, carried out with baselines ranging between 15 m and 612 m. The second epoch of observations was on 2016 September 9, with baselines ranging between 15 and 3143 m. Four spectral windows centered at around 335.4 GHz, 337.3 GHz, 349.0 GHz, and 346.6 GHz, with bandwidths of 1875.0 MHz, 234.4 MHz, 937.5 MHz, and 1875.0 MHz, respectively, were observed. Our primary target lines were the $C^{34}S$(7–6), $CH_3CN(19_K - 18_K)$, CO(3–2) and SiO(8–7) transitions but many other lines are seen in the bands, too. The angular resolution of $0.10'' \times 0.15''$ and sensitivity are much better than in our previous SMA data.

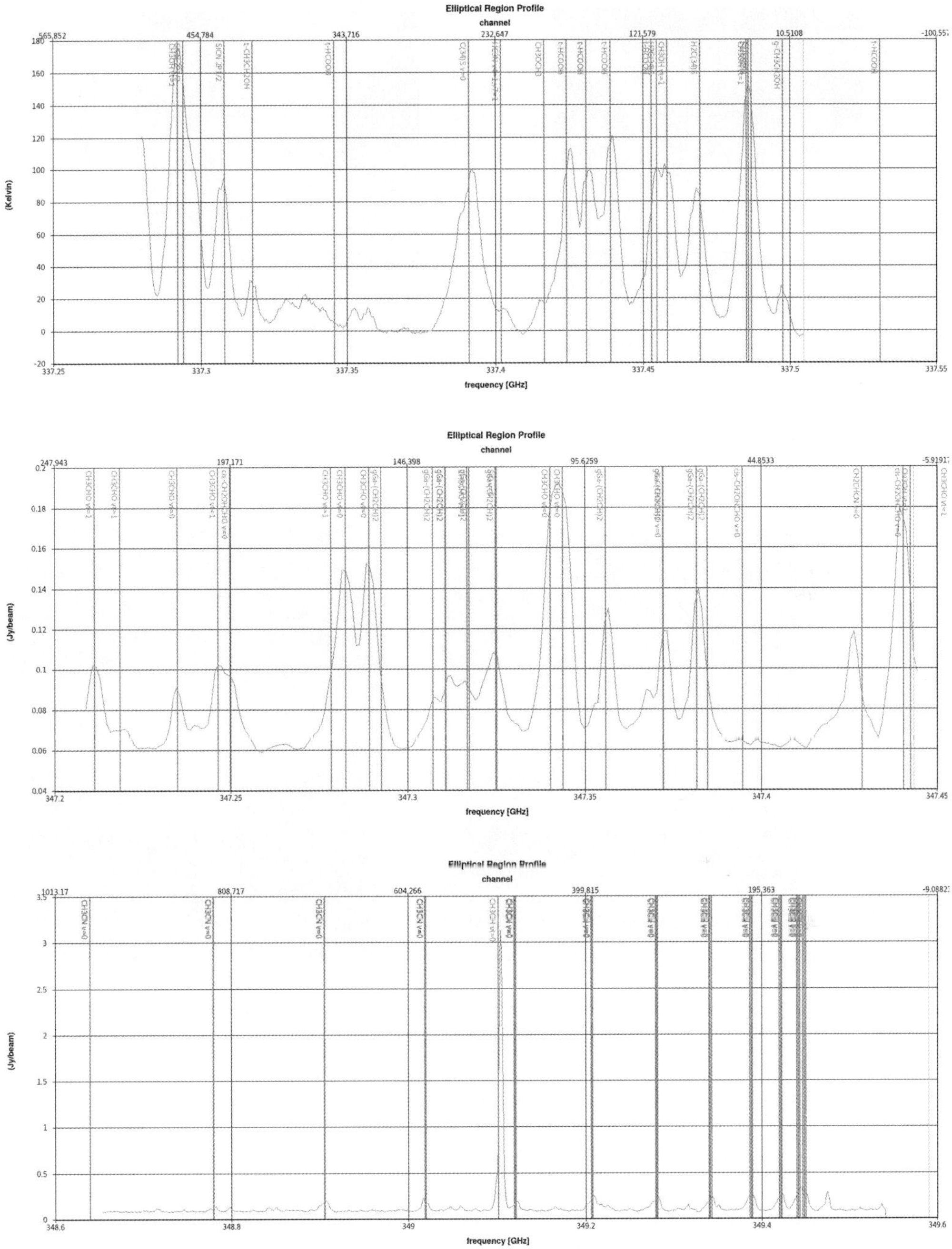

Figure 1. Examples of the spectra measured in S255IR.

3. Results

In Fig. 1 a part of the spectra is presented with possible line identifications (some of them need to be checked). The spectra are very rich, many complex molecules are observed, as expected for a hot core. A detailed analysis of the chemical composition will be given elsewhere. At the frequency of about 349.1 GHz a very bright apparently non-thermal maser line is detected. Most probably this is the methanol $14_1 - 14_0$ A^{-+}

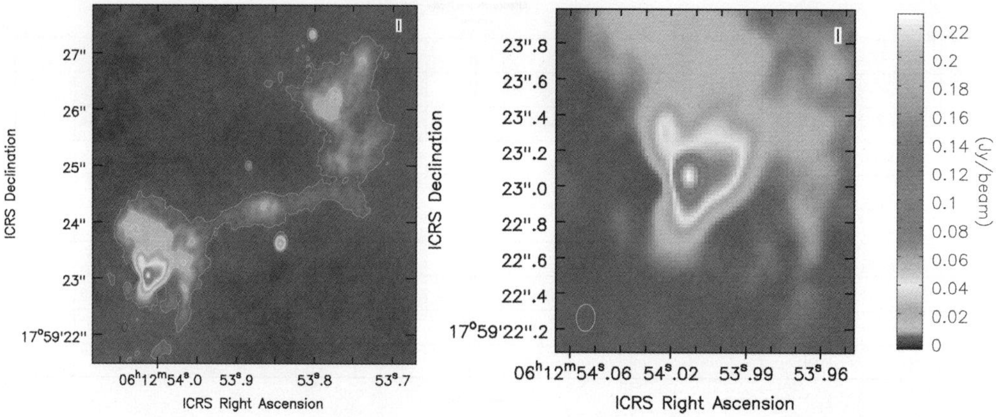

Figure 2. The continuum maps at 0.85 mm of the whole S255IR area (left panel) and of the S255IR-SMA1 clump (right panel).

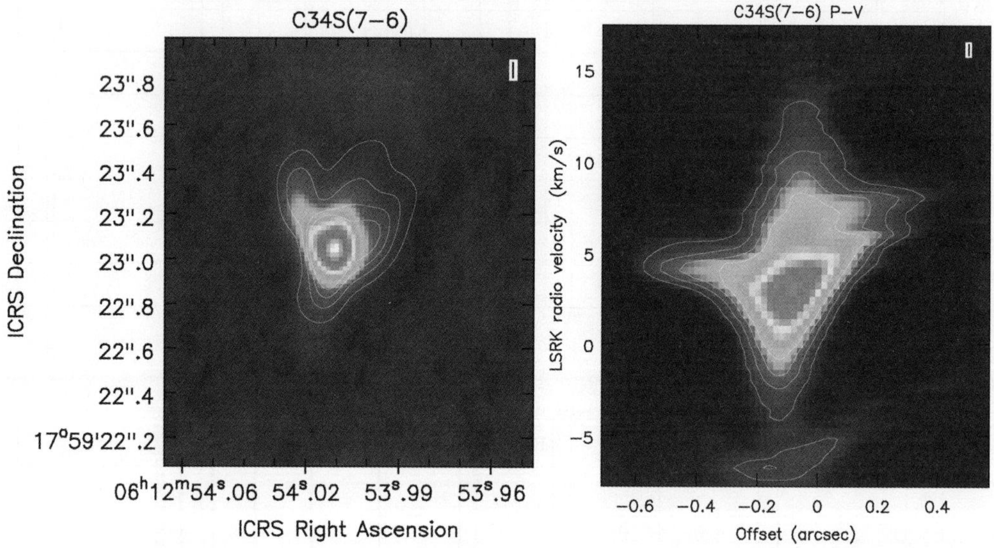

Figure 3. Left panel: the map of the integrated $C^{34}S$(7–6) emission in the S255IR-SMA1 clump overlaid with the contours of the 0.85 mm continuum emission. Right panel: the position velocity diagram of the $C^{34}S$(7–6) emission along the major axis of the disk identified here earlier.

transition. To the best of our knowledge this transition has never been observed or predicted to be masing. The maser emission is extended with a velocity gradient in same sense as the disk rotation. The peak of this maser emission is close to one of the clusters of the 6.7 GHz methanol masers observed by Moscadelli *et al.* (2017).

In Fig. 2 we present maps of the continuum emission of the S255IR area at 0.85 mm. They show in detail the main previously identified clumps: SMA1, SMA2 and SMA3 (Wang *et al.* 2011; Zinchenko *et al.* 2012, 2015). The SMA1 and SMA2 clumps appear to be physically related by a filamentary structure. The morphology of the SMA1 clump, which represents a disk around the massive YSO, looks rather complicated.

In Fig. 3 the map of the integrated intensity and the position-velocity diagram in the $C^{34}S$(7-6) line are presented. The latter indicates a Keplerian-like rotation (seen in the other lines, too).

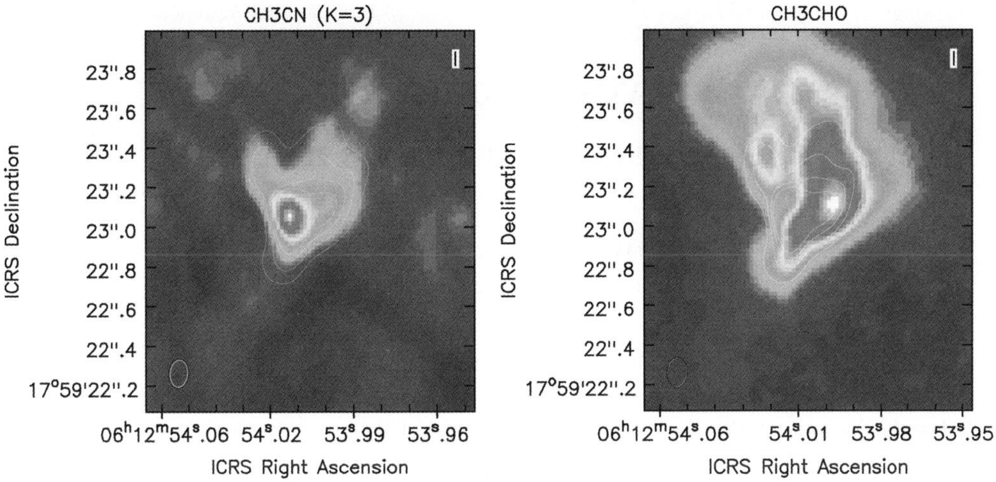

Figure 4. Maps of the integrated intensity in the $CH_3CN(19_3 - 18_3)$ line (left panel) and in the $CH_3CHO(18_5 - 17_5)$ E line (right panel) overlaid with the continuum maps at 0.85 mm toward the S255IR-SMA1 clump.

In Fig. 4 we present sample maps for two other species: CH_3CN and CH_3CHO.

4. Conclusions

The preliminary results of our observations of the S255IR area with ALMA indicate a physical relation of the two major clumps. The disk around the massive YSO S255 NIRS3 has a complicated morphology and close to Keplerian rotation. The chemistry of this object is very rich. A new methanol maser line is detected.

This research was supported by the Russian Foundation for Basic Research (grant No. 15-02-06098) in the part of the preparation of the observations and preliminary data reduction, and by the Russian Science Foundation (grant No. 17-12-01256) in the part of the data analysis.

References

Caratti O Garatti, A., Stecklum, B., Garcia Lopez, R., Eislöffel, J., Ray, T. P., Sanna, A., Cesaroni, R., Walmsley, C. M., Oudmaijer, R. D., de Wit, W. J., Moscadelli, L., Greiner, J., Krabbe, A., Fischer, C., Klein, R., & Ibañez, J. M. 2017, *Nature Physics*, 13, 276

Fujisawa, K., Yonekura, Y., Sugiyama, K., Horiuchi, H., Hayashi, T., Hachisuka, K., Matsumoto, N., & Niinuma, K. 2015, *The Astronomer's Telegram*, 8286

Moscadelli, L., Sanna, A., Goddi, C., Walmsley, M. C., Cesaroni, R., Caratti o Garatti, A., Stecklum, B., Menten, K. M., & Kraus, A. 2017, *A&A*, 600, L8

Wang, Y., Beuther, H., Bik, A., Vasyunina, T., Jiang, Z., Puga, E., Linz, H., Rodón, J. A., Henning, T., & Tamura, M. 2011, *A&A*, 527, A32

Zinchenko, I., Liu, S.-Y., Su, Y.-N., Kurtz, S., Ojha, D. K., Samal, M. R., & Ghosh, S. K. 2012, *ApJ*, 755, 177

Zinchenko, I., Liu, S.-Y., Su, Y.-N., Salii, S. V., Sobolev, A. M., Zemlyanukha, P., Beuther, H., Ojha, D. K., Samal, M. R., & Wang, Y. 2015, *ApJ*, 810, 10

Astrochemistry VII – Through the Cosmos from Galaxies to Planets
Proceedings IAU Symposium No. 332, 2017
M. Cunningham, T. Millar & Y. Aikawa, eds.

doi:10.1017/S1743921317007694

SO survey of massive cores

Igor Zinchenko[1] **and Christian Henkel**[2]

[1]Institute of Applied Physics of the Russian Academy of Sciences,
46 Uljanov str., 603950 Nizhny Novgorod, Russia
email: zin@appl.sci-nnov.ru

[2]Max-Planck-Institut für Radioastronomie,
Auf dem Hügel 69, D-53121 Bonn, Germany;
Astron. Dept., King Abdulaziz University, P.O. Box 80203, Jeddah 21589, Saudi Arabia
email: chenkel@mpifr-bonn.mpg.de

Abstract. We present the results of a survey of several tens of dense high mass star forming (HMSF) cores in three transitions of the SO molecule at 30 and 100 GHz with the 100-m Effelsberg and 20-m Onsala radio telescopes. The physical parameters of the cores are estimated from the line ratios and column densities. Relative abundances are derived as well.

Keywords. ISM: clouds, ISM: molecules, ISM: abundances

1. Introduction

The chemistry of sulfur-bearing molecules is still poorly explored, especially in high mass star forming regions. These molecules can be important tracers of various phenomena in star forming regions, including outflows. Relative abundances of sulfur-bearing molecules can serve as chemical clocks (e.g. Viti *et al.* 2004) and as an indicator of the initial C/O ratio (e.g. Bergin *et al.* 1997). However, there are some discrepancies between models and observations (e.g. Herpin *et al.* 2009)

2. Observations and data analysis

We have observed 3 transitions of the SO molecule: $J_N = 1_0 - 0_1$ at 30.0016 GHz (with the 100-m Effelsberg telescope), $J_N = 3_2 - 2_1$ at 99.2999 GHz and simultaneously $J_N = 4_5 - 4_4$ at 100.0296 GHz (with the 20-m Onsala telescope). The beam widths of the Effelsberg and Onsala antennas at these wavelengths are similar ($\sim 30''$ and $\sim 40''$, respectively).

We analyze the line ratios with RADEX (Van der Tak *et al.* 2007). The line ratios were modeled in wide ranges of density and temperature. Estimates of the density and temperature from the line ratios are used for estimates of the SO column densities. SO abundances are derived by comparison with the $C^{18}O$ column densities, found from our Onsala observations (Zinchenko *et al.* 2000). The relative $C^{18}O$ abundance is assumed to be $X(C^{18}O) = 1.7 \times 10^{-7}$ (Frerking *et al.* 1982).

3. Results

At Onsala we have observed 35 sources. The $J_N = 3_2 - 2_1$ transition has been detected in all of them. The $J_N = 4_5 - 4_4$ transition has been detected in 3 objects: W3 (OH), W51 M and ON2 N. In Effelsberg we have observed 22 sources from the same sample. The SO line has also been detected in all of them. The $J_N = 3_2 - 2_1$ and $J_N = 1_0 - 0_1$ spectra for these sources are presented in Fig. 1.

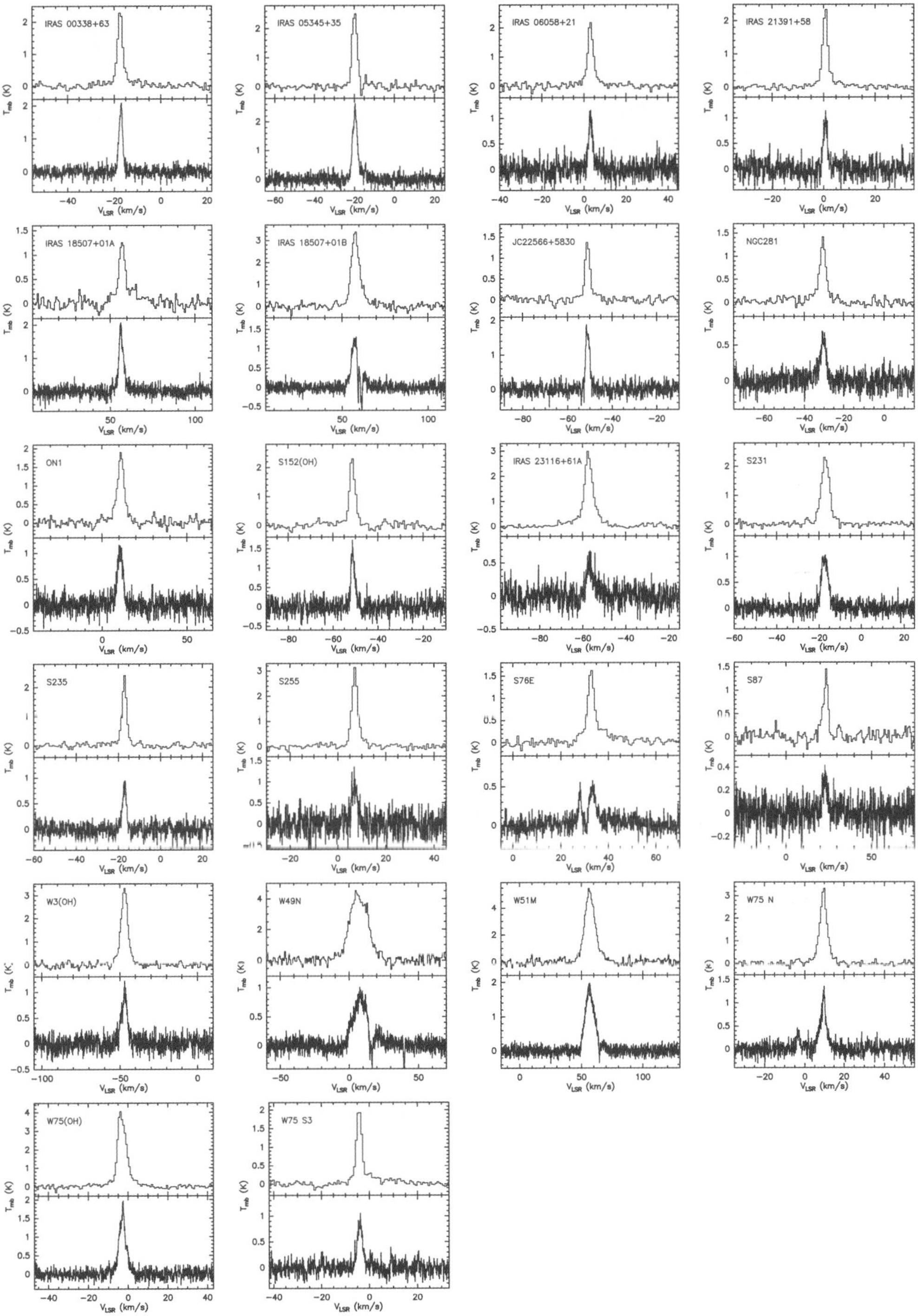

Figure 1. The $J_N = 3_2 - 2_1$ (upper panels) and $J_N = 1_0 - 0_1$ (lower panels) spectra for the 22 sources observed both in Effelsberg and in Onsala.

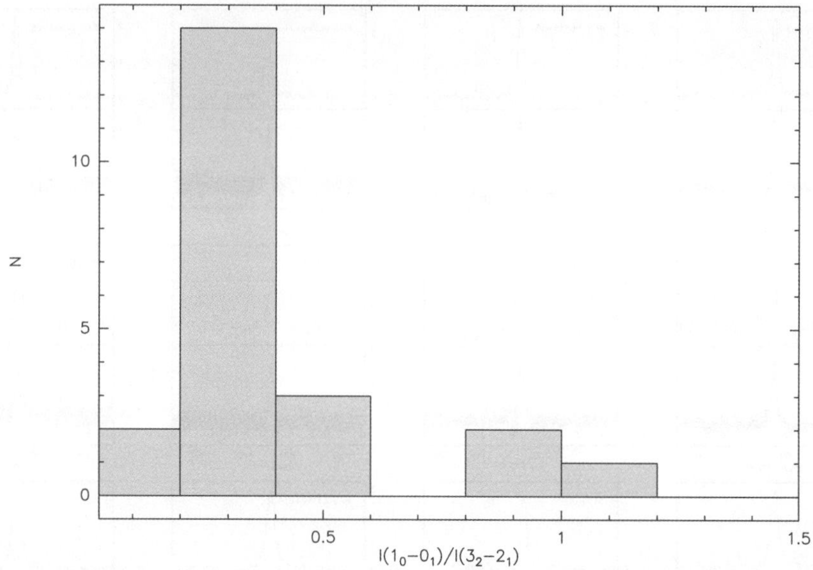

Figure 2. The histogram of the $(1_0 - 0_1)/(3_2 - 2_1)$ line intensity ratio.

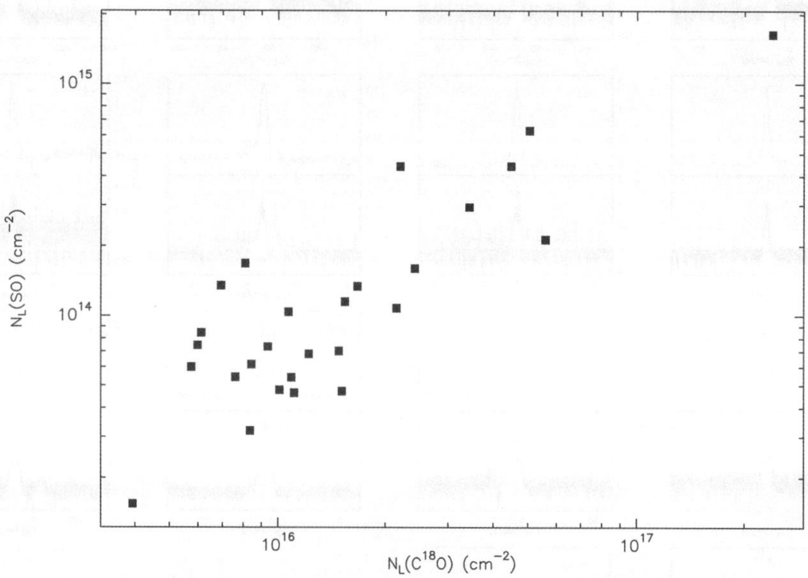

Figure 3. The SO column density in dependence on the $C^{18}O$ column density.

The $(1_0 - 0_1)/(3_2 - 2_1)$ intensity ratio lies in the range from ~ 0.15 to ~ 2 (Fig. 2). This corresponds to densities from $\sim 10^6$ cm^{-3} to $\sim 5 \times 10^3$ cm^{-3}. Most of the sources have a ratio of ~ 0.3, which corresponds to a density of $\sim 10^5$ cm^{-3}. The $(3_2 - 2_1)/(4_5 - 4_4)$ intensity ratio is ~ 20 for the detected sources. This detection indicates high temperature (> 50 K) and density ($> 10^6$ cm^{-3}).

There is a good correlation between the SO and $C^{18}O$ column densities (Fig. 3).

The SO relative abundance varies in the range $X(\mathrm{SO}) \sim (0.5 - 4) \times 10^{-9}$ with a median value of $\sim 1.3 \times 10^{-9}$. These values are rather close to the results of other recent studies

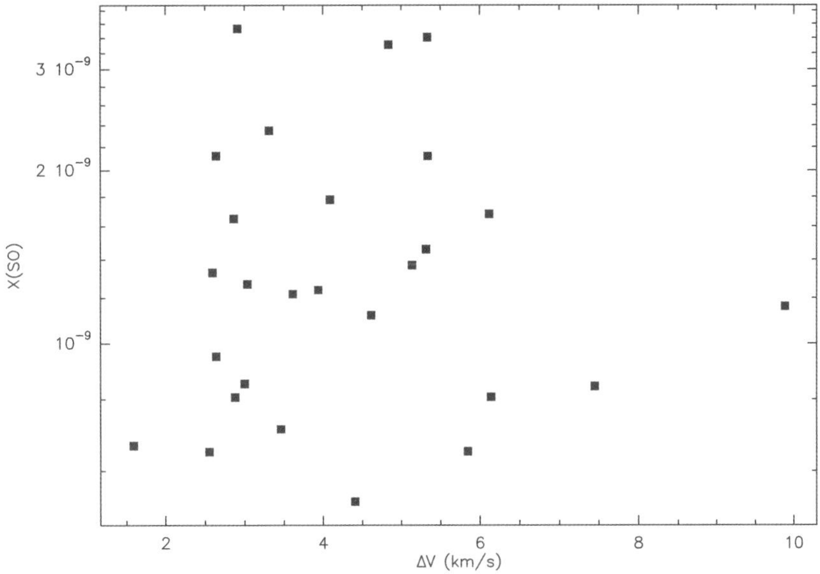

Figure 4. The SO relative abundance in dependence on the SO line width.

of similar objects (e.g. Li *et al.* 2015). We see no correlation between the SO abundance and the line width (Fig. 4).

4. Conclusions

We have surveyed 35 HMSF cores in the SO lines at 3 mm and 1 cm wavelengths. 22 objects have been observed (and detected) at both wavelengths. The line intensity ratios indicate a typical density of $\sim 10^5$ cm^{-3}. For several objects detected in the $J_N - 4_5 - 4_4$ line the density should be $> 10^6$ cm^{-3} and temperature > 50 K. There is a good correlation between the SO and C^{18}O column densities. The SO relative abundance varies in the range $X(\mathrm{SO}) \sim (0.5-4)\times 10^{-9}$ with a median value of $\sim 1.3\times 10^{-9}$. There is no correlation between the X(SO) and line width.

This research was supported by the Russian Foundation for Basic Research (recently by the grant No. 15-02-06098) and by the IAP RAS state program 0035-2014-0030.

References

Bergin, E. A., Goldsmith, P. F., Snell, R. L., & Langer, W. D. 1997, *ApJ*, 482, 285
Frerking, M. A., Langer, W. D., & Wilson, R. W. 1982, *ApJ*, 262, 590
Herpin, F., Marseille, M., Wakelam, V., Bontemps, S., & Lis, D. C. 2009, *A&A*, 853, 867
Li, J., Wang, J., Zhu, Q., Zhang, J., & Li, D. 2015, *ApJ*, 802, 40
Van der Tak, F. F. S., Black, J. H., Schier, F. L., Jansen, D. J., & van Dishoeck, E. F. 2007, *A&A*, 468, 627
Viti, S., Collings, M. P., Dever, J. W., McCoustra, M. R. S., & Williams, D. A. 2004, *MNRAS*, 354, 1141
Zinchenko, I., Henkel, C., & Mao R. Q. 2000, *A&A*, 361, 1079

Laboratory astrochemistry and modelling

Astrochemistry VII – Through the Cosmos from Galaxies to Planets
Proceedings IAU Symposium No. 332, 2017
M. Cunningham, T. Millar & Y. Aikawa, eds.

doi:10.1017/S1743921317007840

Processing of astrophysical ices by soft X-rays and swift ions

Sergio Pilling[1,2]

[1]Universidade do Vale do Paraíba (UNIVAP) / Laboratório de Astroquímica e Astrobiologia (LASA), Av. Shishima Hifumi, 2991, CEP: 12244-000, Sao Jose dos Campos, SP, Brazil.
email: sergiopilling@yahoo.com.br

[2]Instituto Tecnológico de Aeronáutica - ITA / DCTA, Praça Marechal Eduardo Gomes, 50, CEP:12228-900, São José dos Campos, SP, Brazil.

Abstract. The employment of soft X-rays and swift ions has been used in laboratory to simulate the physicochemical processing of astrophysical ice analogs by energetic photons and cosmic rays. This processing includes excitation, ionization and molecular dissociation, desorption, as well as triggers the formation of new compounds. Here we present some results from experiments employing infrared spectroscopy in two different laboratories: LNLS/CNPEM in Campinas/Brazil and GANIL/CIRIL/CIMAP in Caen/France. Among the results are the formation of alkenes and aromatic compounds during the irradiation of saturated hydrocarbon-containing ices by cosmic ray analogs, the production of the nucleobase adenine during soft X-ray photolysis of N_2:CH_4 ice, as well as the formation of peptide bonds during the bombardment of frozen glycine by cosmic ray analogs. The interaction between cosmic ray analogs and ionizing soft X-rays probed in the laboratory allows us to identify reaction routes that lead to chemistry enhancement of astrophysical ices and help us put constrains in prebiotic chemistry.

Keywords. Laboratory measurements, X-rays, Cosmic-rays, Astrophysical ices, FTIR, Astrochemistry, Astrobiology

1. Introduction

The birthplace of stars is the densest and coldest place of the interstellar medium (ISM) called dense molecular clouds. These regions have typical gas densities of 10^3-10^8 atoms cm^{-3} and temperatures in the order of 10-50 K. Because of the low temperature, dust particles (mainly silicates and carbonaceous compounds such as amorphous C and SiC) can accrete molecules from the gas phase and become coated with an ice mantle. The interstellar ice mantles (or simply interstellar ices) consist primarily of amorphous H_2O, but usually also contain a variety of other simple molecules, such as CO_2, CO, CH_3OH, and NH_3 (e.g., Ehrenfreund & Charnley 2000; Gibb *et al.* 2001; Boogert *et al.* 2004).

Laboratory studies and astronomical observations indicate that photolysis and radiolysis of such ices can create complex organic compounds, and even prebiotic molecules such as amino acids and nucleobases (e.g. Maloney *et al.* 1996; Bernstein *et al.* 2002; Muñoz Caro *et al.* 2002; Kobayashi *et al.* 2008; Pilling *et al.* 2009 Vasconcelos *et al.* 2017a, b, c). Figure 1 illustrates the scenario in which such cosmic rays, soft X-rays and fast electrons induce chemical changes in the icy environments.

Inside dense regions of the interstellar medium, where direct stellar UV is extremely attenuated, X-rays (despite some degree of attenuation) and cosmic rays are the main drivers of the gas phase and grain surface chemistry. For example, the penetration depth of 46 MeV Ni ions and equivelocity protons inside water ice is about 1 and 3 orders of magnitude higher than for 500 and 50 eV photons (considering the photoabsorption

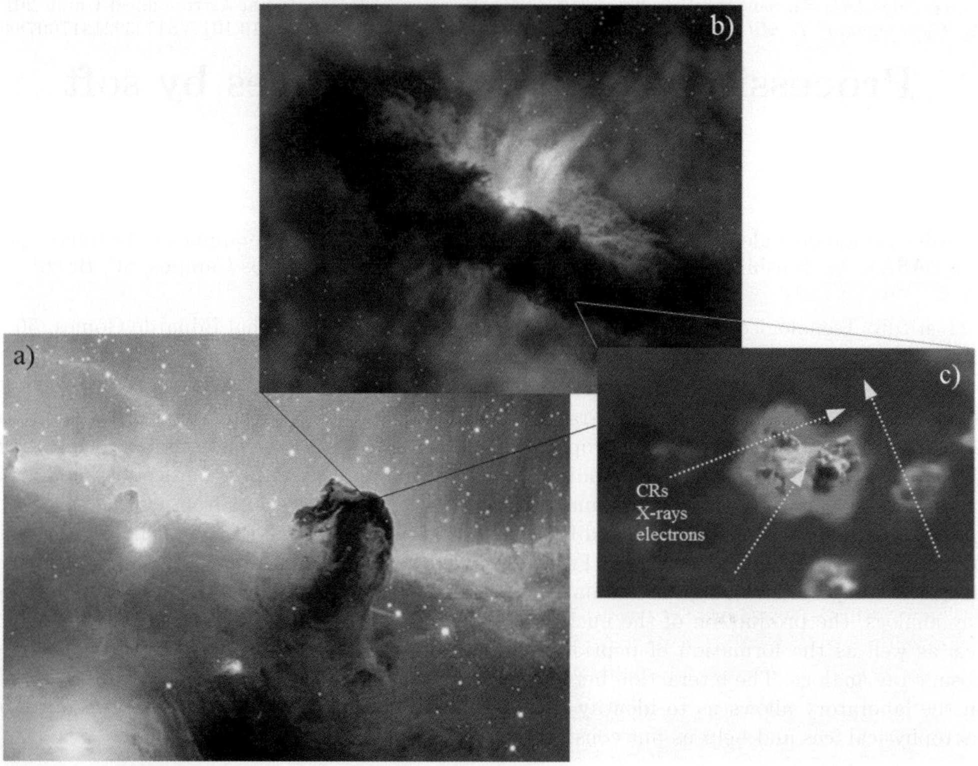

Figure 1. Mosaic with and schematic representation of the interaction between ionizing agents with interstellar grains inside dense clouds. A). Picture of the Horsehead Nebula, a dark molecular cloud, roughly 1,500 light years distant, embedded in the vast and complex Orion Nebula (Credit: Jean-Charles Cuillandre (CFHT), Hawaiian Starlight, CFHT. source: `http://www.nasa.gov`). b) Artist impression of a protostellar disk around a recently formed or forming star. (Credit: NASA/JPL-Caltech/R. Hurt -SSC). C) Illustration of interstellar grains covered with an ice cap being processed by ionizing radiation (Cosmic rays, X-rays and electrons).

cross-section from Chan *et al.* (1993) and McLaren *et al.* (1987)). In these environments, the dust grains can reach a size in the order of a micron, as a coagulation of sub-micron grains occurs with ice-rich mantles from tens of nanometers to tens of microns (Li & Greenberg 1997; Taylor *et al.* 1996; Witt *et al.* 2001; and references therein).

The desorption induced by energetic radiation is another important process that modifies the chemical abundances in the gas phase in space environments, mainly in cold and denser regions (see e.g. Prasad & Tarafdar 1983; Brown *et al.* 1984; Shen *et al.* 2004; Bonfim *et al.* 2017). As indicated by Shen *et al.* (2004), in the case of cosmic ray particles, the sputtering produced by direct impact and whole grain heating (classical sublimation) due to energy deposition in the bulk are also two efficient processes for releasing molecules from frozen surfaces to the gas phase.

Bonfim *et al.* (2017) pointed out, in a theoretical-experimental study employing X-rays on ices (broadband X-ray processing of SO_2 ice at 12 K), that the X-ray photodesorption yield (upper limit) is around 0.25 molecules/photon, which is between 100 and 1000 times higher than the one from Öberg *et al.* (2009), comparing with values obtained for CO and N_2 employing UV photons. The desortpion (or sputtering) induced by swift ions in ices is several orders of magnitude higher than the one produced by ionizing photons and can reach values around 10000 molecules/impact (Brown *et al.* 1984; Baragiola *et al.* 2003; Pilling *et al.* 2010a), and as discussed by Rocha *et al.* (2017), may be also dependent on ice polarity (ices composed of polar molecules are more resistant to destruction, whereas apolar molecule-containing ices are more easily destroyed). Additionally, it has been observed that the number of molecules ejected after ion impact scale with the square of the electronic projectile energy loss (Seperuelo-Duarte *et al.* 2010; Dartois *et al.* 2015; Mejía *et al.* 2015).

The processing of astrophysical ices by ionizing radiation such as soft X-rays and swift ions has been used in laboratory to simulate the effects of energetic photons (e.g. Pilling *et al.* 2009; Chen *et al.* 2013; Pilling *et al.*2011b; Pilling & Bergantini 2015; Jimenes-Escobar *et al.* 2016; Vasconcelos *et al.* 2017c; Bonfim *et al.* 2016; Rachid *et al.*2017; Almeida *et al.* 2014; and references therein) and cosmic rays (e.g. Strazzulla *et al.* 1983; Palumbo 2006; Strazzulla *et al.* 2007; Seperuelo-Duarte *et al.* 2009; Pilling *et al.* 2010a, 2010b; Hijazi *et al.* 2011; Andrade *et al.* 2013; Bergantini *et al.* 2014a; de Barros *et al.* 2014; Mejía *et al.* 2014; Almeida *et al.* 2017).

2. Experimental methodology

Over the years, several physicochemical techniques have been employed to study the surfaces and the bulk of astrophysical ices exposed to radiation (e.g. Thermal Programmed Desorption TPD, X-rays Spectroscopy XAS, Stimulations Ion Desorption SID, Plasma Desorption PD, Infrared Spectroscopy, and others). Here, we will focus on the study of the bulk of simulated ice by employing infrared (IR) spectroscopy in transmission mode (Fourier Transformed IR spectroscopy FTIR). Basically, the methodology employed requires two high-vacuum chambers (one to prepare the gas mixture of astrophysical interest and another to produce the icy film and perform the irradiation/bombardment), a helium cryostat, an ionization source and an infrared spectrometer that can operates in reflection or transmission mode. More details of such technique can be seen elsewhere (e.g. Seperuelo Duarte *et al.* 2009, 2010; Pilling *et al.* 2010a,b; Portugal *et al.* 2014; Pilling & Bergantini 2015, and references therein).

The highlight experiments described in this manuscript were performed employing two similar high-vacuum chambers at two different laboratories: the Brazilian Synchrotron light source (LNLS/CNPEM) in Campinas/Brazil, and the Grand Accélérateur d'Ion Lourdes (GANIL/CIRIL/CIMAP) in Caen/France. Briefly, the gaseous samples were inserted in the mixed chamber and the gas mixture was prepared in proportions similar to those observed in space environments (e.g. ices at protoplanetary disks PPDs, surface of Jovian moons, comets, etc.). For pure samples, only one type of compound is used. Basically, the pressure inside the mixed chamber is in the order of 10-100 mbars during sample preparation. The gas mixture (or pure sample) is usually inserted slowly into the main chamber trough an inlet needle with its end close to the IR-transparent cold substrate (e.g. ZNSe, KBr or CaF_2) resulting in ice film formation ($\sim$ 0.02-0.1 micron min^{-1}). The icy sample has thickness usually from 0.1-2 microns, but depending on the experiments a very thin or very thick sample can be produced. The base pressure of the main chamber is around 10^{-9} mbar and during the irradiation stage itself the chamber pressure can increase a little due to sputtering processes and reaches 10^{-8} mbar. After ice production, an infrared spectrum is obtained to characterize the sample and its thickness. The molecular column densities and ice thickness of samples were determined from the relation between the measured IR band area of a given vibrational mode in the IR spectrum and its respective and band strength, B (cm molecule-1) and average number density of ice, both taken from literature (see details at Pilling 2010a, 2010b, 2011a;

Portugal *et al.* 2014; Pilling & Bergantini 2015 and references therein). The in-situ IR spectra were recorded for ices irradiated at different fluences using FTIR spectrometer (e.g. Agilent Carry 630 or Nicolet Magna 550) from 4000 to 600 cm^{-1} with 1-2 cm^{-1} resolution. A background spectrum allowing the absorbance measurements was collected before gas deposition. As described, for example, by Pilling *et al.* (2010a) and Pilling & Bergantini (2015), in such experiments, the sample cryostat system can be rotated over 180 degrees and fixed at three different positions to allow gas deposition, FTIR measurement and ice exposure to the radiation beam.

From the analysis of the evolution of IR bands as a function of radiation fluence we derive molecular parameters such as effective destruction cross sections (for parent species) and effective formation cross section (for daughter species) as described in details elsewhere (e.g. Seperuelo Duarte *et al.* 2009; Pilling 2010a, Portugal *et al.* 2014; Pilling and Bergantini 2015 and references therein). Additionally, the desorption yield and molecular branching ratio at chemical equilibrium can be determined (e.g. Bonfim *et al.* 2017, Almeida *et al.* 2017; Rachid *et al.* 2017; Vasconcelos 2017a,b). Such chemical equilibrium occurs at a large radiation fluence and the composition of the sample, kept a constant temperature, stays constant even if the sample continues to be exposed to radiation. With such parameters, we can estimate the reaction rates and the half-life of parent species, as well as the yield of new compounds produced in the ices in astrophysical environments by knowing the radiation flux in space (e.g. Pilling 2010a; Pilling and Bergantini 2015; Rachid *et al.* 2017, and references therein).

X-ray induced processes in ices

In general, as discussed by Gullikson & Henke (1989) and Pilling & Bergantini (2015), X-ray absorption occurs via the photoelectric effect so that it results in the generation of a photoelectron with energy $E = h\nu - E_b$, where $h\nu$ is the soft X-ray photon energy (in this work 100 ¡ $h\nu$ (eV) ¡ 2000) and E_b is the binding energy of the electron. Electron binding energy is a generic term for the ionization energy that can be used for species with any charge state. For the species containing CHONS atoms, the highest electron binding energy within the available photon energy range is found for O K-edge electrons ($\sim$ 540 eV), followed by N K-edge electrons ($\sim$ 400 eV), C K-edge electrons ($\sim$ 290 eV), and S L-edge electrons ($\sim$160 eV). Additionally, the weakest electron binding energies come from the valence electrons of NH_3 ($\sim$ 10 eV), followed by valence electrons of SO_2 (12.3 eV), H_2O (12.6 eV), and CO_2 (13.7 eV). The effect of soft X-ray-induced secondary electrons has been studied by several groups in the literature (e.g., Henke *et al.* 1979; Gullikson & Henke 1989; Akkerman *et al.* 1993; Hüfner 1995; Cazaux 2006; Pilling *et al.* 2009; Pilling & Andrade 2012) and is a rather complex process.

As discussed by Pilling & Bergantini (2015), a simplified description for the interaction of soft X-ray with astrophysical ices can be divided into four steps: **i)** First, photoabsorption of X-ray and the generation of energetic primary electrons (see above). Such fast primary electrons lose energy by creating electron-hole pairs at a distance that is generally short compared with the penetration depth L of the incoming photon. As indicated by Bergantini *et al.* (2014b), the range of 1 keV electrons is around 60 nm inside such astrophysical analog samples and for low-energy electrons ($\sim$ 5 eV), the mean-free path inside such ices is around 1-2 nm; **ii)** Excitation of secondary electrons by fast primary electrons (photoelectrons). Once a secondary electron is created, it will undergo a short random walk in the material while losing energy due to the creation of phonons. Moreover, following Gullikson & Henke (1989), each soft X-ray may induce tens of secondary electrons inside matter, and this number increases as the temperature of the target increases; **iii)** Transport of the secondary electrons, including energy, to the lattice. Following Henke

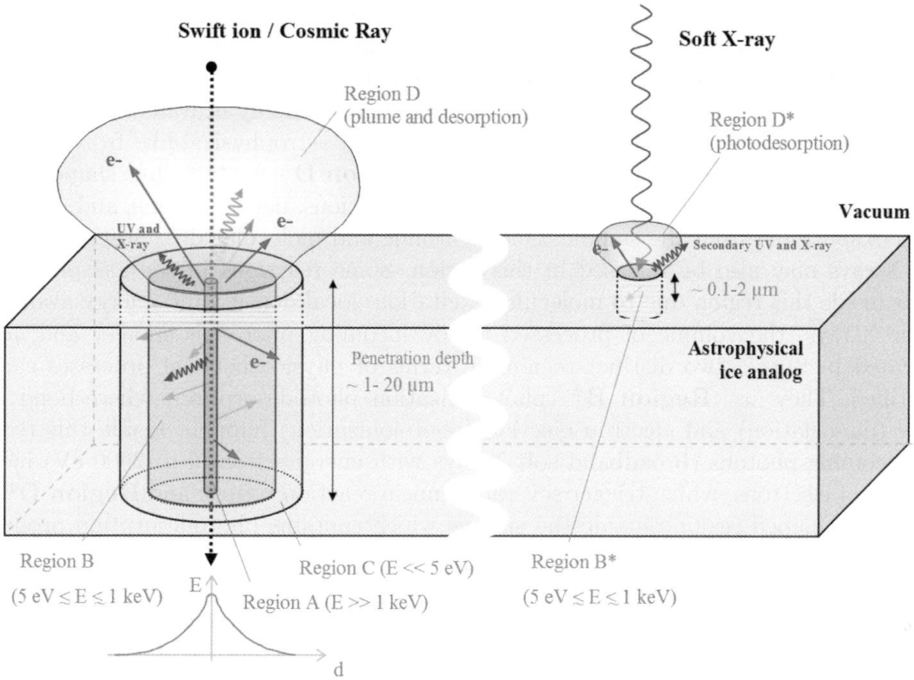

Figure 2. Schematic diagram of the different physicochemical regions surrounding the ion and photon track during processing of the typical astrophysical ice by cosmic rays and soft X-rays). Adapted from Vasconcelos *et al.* (2017c).

et al. (1979), most secondary electrons have energies below 5 eV. When captured by a molecular target inside the bulk, such electrons can induce further molecular dissociation, for example, via the dissociative electron attachment mechanism ($AB + e^- \rightarrow AB^- \rightarrow A- + B$). The mean-free path of such secondary electrons is much reduced in comparison to photoelectrons, and its yield can vary depending on sample density and temperature; and **iv)** Escape of secondary electrons reaching the surface with sufficient energy to overcome any potential barrier at the surface or be thermalyzed/captured by any molecular species or center in the bulk. In astrophysical water-rich ices (insulators), a secondary electron will undergo many collisions before it either escapes through the surface or loses enough energy so that it is unable to escape or becomes trapped by interacting with a center in the bulk.

Cosmic-ray induced processes in ices

The energy provided by the incoming swift oxygen ions and the energy delivered by the incoming soft photons induce different physicochemical processing inside the ice as illustrated in Figure 2. Its main difference comes from the radial energy gradient centered at the ion track compared to photons. For example, in the case of ion bombardment, we can identify four distinct regions (e.g. Portugal *et al.* 2014; Vasconcelos *et al.* 2017c): **Region A** (track core region) - A small cylindrical region, along the ion track in which the energy delivered is extremely high (E ¿¿ 1 keV); with complete ionization and thus destruction of molecules along the track; **Region B** (ionization/dissociation region) - A cylindrical surrounding region A, in which the energy is not enough (1keV ¡ E ¡ 5 eV) to atomize the target but induces bond ruptures (dissociation) and target electron ejection (ionization). Primary electrons and subsequent secondary electrons are a source of energy input for chemical reactions inside this region. As pointed out by Vasconcelos *et al.* (2017c), after ion bombardment on N_2:CH_4 ices, the molecules within this region can be converted to radicals that may react to produce new species such as (C_2H_2, C_2H_4, C_2H_6, HCN, HNC, NH_3, N_3, and others); **Region C** (region of morphological changes) - A region far from the ion track, in which not very much energy is available (E ¡¡ 5 keV), where only morphological changes are induced in the astrophysical-like frozen sample, such as change in the crystalline structure; and **Region D** - A plume-like shaped region outside the surface that contains ejected lost electron, ions, neutral atoms, and molecular clusters sputtered from the sample. Due to atomic and molecular deexcitation, UV and soft X-rays may also be released in this region. Some reactions in the gas phase may occur inside this region due to molecular excitation, local density and energy available.

For X-rays, the volume of processed ice by incoming photon is smaller and, as illustrated in Fig. 2, two distinct regions in terms of physicochemical processes can be identified. They are: **Region B*** (photoionization/photodesorption) where bond ruptures (dissociation) and electron ejection (photoionization) happen. Inside this region, the incoming photons (broadband soft X-rays with energies from 6 to 2000 eV) induce ejection of electrons, which trigger several chemical reaction routes; and **Region D*** - A plume-like shaped region outside the surface which contains photodesorption products, for example, photoelectrons, ejected ionic and atomic species, and molecules sputtered from the sample (less pronounced than in the case of ion bombardment). Contrary to ion bombardment experiments, no molecular clusters are expected to desorb from the surface. According to Vasconcelos *et al.* (2017c), both ionizing agents (swift ions and soft X-rays) produce the same kind of new species during ice processing with different production efficiencies and equilibrium branching ratio - EBR (reached after a longer radiation fluence). More details about EBR of irradiated ices is given elsewhere (e.g. Almeida *et al.* 2017; Rachid *et al.* 2017).

3. Experiments highlights

3.1. *Production of nucleobase adenine and nitriles during soft X-rays photolysis of N_2:CH_4 ices*

Pilling *et al.* (2009) presented, for the first time, the chemical alteration produced by the interaction of broad band soft X-rays (in the 0.5-3 keV range), and its induced primary and secondary electrons, on N_2:CH_4 (95:5) ice at 15 K in an attempt to simulate the photochemistry induced in the frozen surface of outer solar system bodies such as Pluto, Triton and other KBOs (Kuipier belt objects), as well as in the surface of aerosols in the upper atmosphere of the Saturn moon, Titan. The experiments have been performed in a high-vacuum chamber coupled to the SXS beamline SXS at LNLS/CNPEM. In-situ sample analyses were performed by an FTIR spectrometer.

Concerning the applicability of results in Titan chemistry, according to the authors, the experiments simulated roughly 7×10^6 years of solar soft X-ray exposure on Titan atmosphere. Besides the parent species N_2 and CH_4, the ice sample had also a small amount of water and CO ice (less than 1%), which simulates periods of heavy cometary bombardment on such solar system bodies. The in-situ IR analysis showed several organic molecules created and trapped in the ice at $\sim$ 15 K, including the reactive cyanate ion OCN^-, nitriles, and possibly amides and esters. The authors also observed that the thermal heating of a frozen sample drastically changes its chemistry, resulting in an organic residue rich in C-C and C-N aromatic structures. Ex-situ Gas Chromatography (GC-MS) and H-NMR analysis of the organic residue at room temperature (300 K) showed that among several nitrogen compounds, adenine ($C_5H_5N_5$), one of the

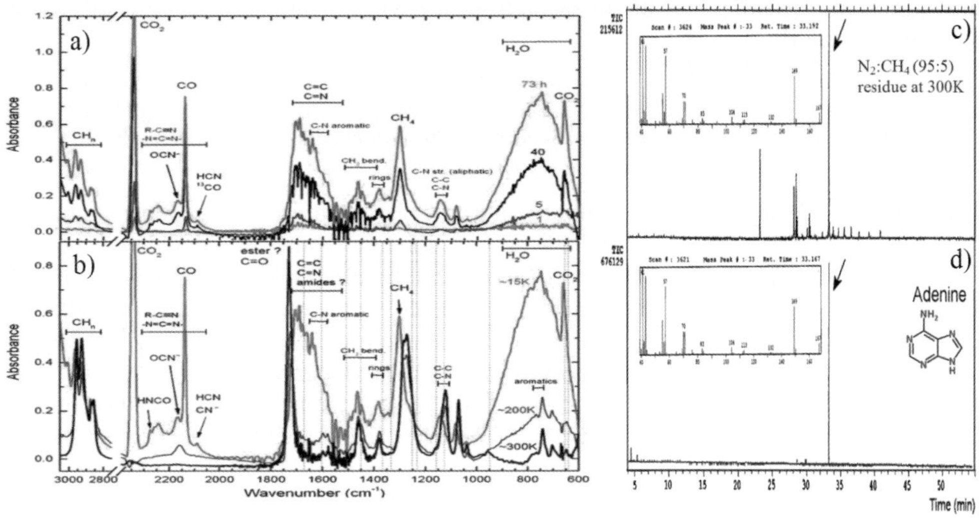

Figure 3. FTIR spectra of the irradiated N_2:CH_4 (95:5) ice at 15 K by soft X-rays: a) Evolution of spectral features as a function of irradiation time (1 h $\sim$ 3 $\times$ 10^{10} erg cm^2). b) Comparison between FTIR spectra of a 73 h irradiated sample at 15, 200, and 300 K. The vertical dashed lines indicate the frequency of some vibration modes of crystalline adenine. The total-ion current chromatogram obtained after the derivatization process of the ice residue at 300 K, and adenine standard are given in panels c) and b), respectively. The inset figures are the mass fragmentation of the sample and the adenine derivative at the retention time of around 33.2 min (arrows). Adapted from Pilling *et al.* (2009).

DNA-nucleobases, is one of the most abundant species produced due to irradiation by soft X-rays. The FTIR spectra of the irradiated N_2:CH_4 (95:5) ice at 15 K by soft X-rays, as well as, ex-situ analysis of the organic residue by GC-MC, are presented in Figure 3.

Vasconcelos *et al.* (2017c), in a similar experiment employing the SGM beamline of the LNLS/CNPEM, observed the production of several new organic molecules, including hydrocarbons and nitriles. Among the daughter species identified were C_2H_2, C_2H_4, C_2H_6, HCN, HNC, NH_3, and N_3 (the most abundant one). In this experiments, the broadband X-ray beam from 6 to 2000 eV was employed. The authors have also measured the effective dissociation cross-sections for CH_4 and N_2 at 12 K and the values obtained were 3.7 $\times$ 10^{-19} cm^2 and about 3.8 $\times$ 10^{-19} cm^2, respectively. A comparison with a similar experiment, but employing swift ions (15.7 MeV O^{+5}) as the ionizing agent shows that cross section is roughly 6 orders of magnitude higher ($\sim 10^{-13}$ cm^2) in case of ion bombardment. The authors also observe that molecular synthesis is strongly enhanced for swift ions in comparison to soft X-rays of the same energy fluence deposited and the results suggest that the modeling of the abundance of astrophysical ices over long periods of time may provide insights into the ion and photon chemistry present in the surface of space environments, such as in objects of the outer Solar System.

This investigation confirms previous studies suggesting that the organic chemistry in the outer solar system bodies such as KBO, Pluto and triton, as well as in Titan atmosphere (aerosols) and on its surface, should be highly complex, being rich in prebiotic molecules involving nitriles, adenine and other compounds. Molecules such as these on the early Earth have found a place that allows life (as we know it) to flourish, a place with liquid water. Additionally, it was observed that the stability of nucleobases in the presence of X-rays is very enhanced in comparison to the stability of amino acids allowing this species an extended half-life in astrophysical environments (Pilling *et al.* 2011b and references therein).

3.2. *Appearance of peptide bonds during the bombardment of frozen glycine by cosmic rays*

The study of glycine molecule stability in the crystalline zwitterion form, known as α-glycine ($^{+}NH_3CH_2COO^{-}$), under the action of heavy cosmic ray analogues, was performed by Portugal *et al.* (2014). The samples were bombarded at two temperatures (14 and 300 K) by $^{58}Ni^{11+}$ ions with energies of 46 MeV, up to a final fluence of 10^{13} ion cm^{-2}, inside a high-vacuum chamber at the heavy-ion accelerator GANIL. Cosmic ray processing of the sample shows the production of several daughter species, such as OCN^-, CO, CO_2 and CN^- and also suggests the appearance of peptide bonds (amide bonds at 1660, 1600 and 1320 cm^{-1}; Reis *et al.* 2006; Kaiser *et al.* 2013) during irradiation as presented in Figure 4. Selected reaction pathways for some daughter species, as well as the suggested peptide bond formation that can occur in region B and a schematic plot of the energy delivered within the sample as a function of the distance of the ion track, are also shown. The peptide bond occurs by the link between the carboxyl group of one amino acid and the amine group of another amino acid, forming an amide group with the release of a water molecule. Such intermolecular bonds between amino acids provide the basis for the production of longer molecular chains, including proteins.

The estimated half-life of glycine in the interstellar medium and in the Solar System in the presence of cosmic rays was around 3-8 $\times$ 10^3 yr and 0.8-4 $\times$ 10^3 yr, respectively. As also discussed by Pilling *et al.* (2014), in one experiment employing fast electrons in solid glycine, the survivability of glycine is highly increased if the samples are at 300 K when compared to low temperature experiments at 14 K due to the presence of free radicals (radiolysis products) in the ice which participate on further reaction pathways consuming the parent species.

Additionally, Portugal *et al.* (2014) suggested that glycine could be present in space environments that suffered aqueous changes, such as the interiors of comets, meteorites and planetesimals. This molecule is present in the proteins of all living beings. Therefore, studying its stability in these environments will provide further understanding of the role of this species in the prebiotic chemistry on Earth.

Pilling *et al.* (2013), in another experiment bombarding solid glycine with ions, have also observed a small feature around 1650-1700 cm^{-1}, tentatively attributed to an amide functional group, which was observed in the IR spectra of irradiated samples, suggesting that cosmic rays may induce peptide bond synthesis in glycine crystals. In such experiments, they have observed and quantified the selective destruction of glycine-bonds employing 1 MeV protons. They observed that the N-H bond was the most sensitive to incoming ions in two types of solid glycine at room temperature (α- and β-glycine). The authors estimated that the half-lives of α-glycine and β-glycine zwitterionic forms bombarded by 1 MeV protons at room temperature, extrapolated to the Earth orbit environment, were 9 $\times$ 10^5 and 4 $\times$ 10^6 years, respectively. In the diffuse interstellar medium, the estimated values are 1 order of magnitude lower. The results suggest that the pristine interstellar β-glycine form should be the most probable kind of glycine polymorph to survive hostile space radiation environments. Combining this result with the fact that this compound has the higher solubility among the other glycine polymorphs,the authors suggest that the β-glycine polymorph was relevant for the production of the first peptides on early Earth.

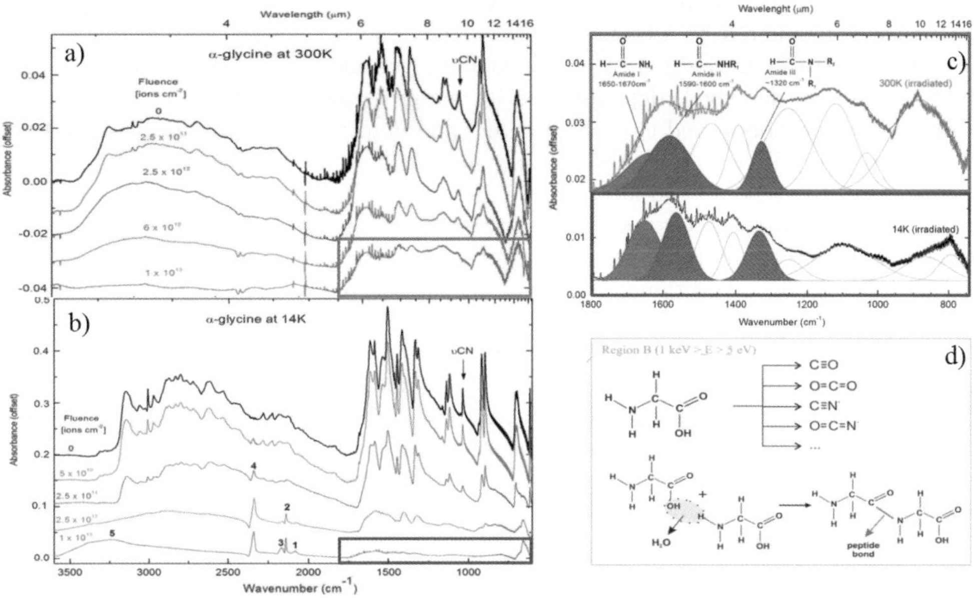

Figure 4. Processing of glycine by cosmic rays and formation of peptide bonds. Panels a and b show the IR spectra of α-glycine before (top dark line) and after different irradiation fluences. The peaks of daughter species formed are indicated by numbers: (1) CN^- (2080 cm^{-1}); (2) CO (2137 cm^{-1}); (3) OCN^- (2165 cm−1); (4) CO_2 (2336 cm^{-1}); (5) H_2O (3280 cm^{-1}). Panel c shows an expanded view of IR spectra of the residues produced after the bombardment employing the fluence of 10^{13} Ni ion cm^2. Filled curves present tentative assignments of amide bands obtained from a spectral deconvolution employing nine Gaussian profiles, in the range 1800-750 cm^{-1}. A scheme of peptide bond formation is presented in panel d. Adapted from Portugal *et al.* (2014).

3.3. *Formation of alkenes and aromatic compounds by cosmic rays*

The formation of C=C and C≡C bonds from the processing of pure c-C_6H_{12} (cyclohexane) and mixed H_2O:NH_3:c-C_6H_{12} (1:0.3:0.7) ices at 13 K by highly charged and energetic ions (219 MeV O^{7+} and 632 MeV Ni^{24+}) was experimentally studied by Pilling *et al.* (2012). The measurements were performed inside a high-vacuum chamber at GANIL and simulate the physical chemistry induced by medium-mass and heavy-ion cosmic rays in interstellar ice analogues. In situ analysis was performed by a Fourier transform infrared spectrometry at different ion fluences.

According to the authors, for the experiment involving the mixed ice bombarded by heavy-ion cosmic ray analogues the determined dissociation cross-sections of H_2O, NH_3 and c-C_6H_{12} are between 2-3 × 10^{-13} cm^2. The results indicate that the effective dissociation cross-section of cyclohexane is about one order of magnitude higher when swift heavy cosmic ray analogues were employed in comparison to the experiment employing medium-mass cosmic rays. Additionally, they observed that the radiolysis of simple alkanes such as cyclohexane in astrophysical ice analogues are consistent with the production of unsaturated molecules containing double bonds, triple bonds (e.g. 2334 cm^{-1}) and/or ring(s) (e.g. 3085 cm^{-1}). Extrapolating to an astrophysical scenario, the maximum production of alkenes is obtained after about 3-5 × 10^6 yr independent of the projectile type and the presence of polar species in the ice (e.g. water and ammonia). The authors also observed that effective formation cross-sections of alkenes, considering the C-H out-of-plane bending mode at 719 cm-1 (band strength, $B \sim 8 \times 10^{-18}$ cm molecule^{-1}), in the pure c-C_6H_{12} ice and mixed ice were about 10^{-15} and 10^{-14} cm^2, respectively. The maximum value was roughly the same for both experiments, 10^{-2} alkenes per cyclohexane molecule irradiated by a cosmic ray analogue. Another consequence of this induced unsaturation is the production of molecular hydrogen in the ice, which can be released to the gas phase (desorbed) depending on its energy or ice temperature.

Finally, the authors found that the 3.4 μm band detected in the IR spectrum of Planetary nebulae PN IRAS 05341+0852 is very similar to the band observed for the organic residue (at $\sim$ 130 K) obtained after ion bombardment of the H_2O:NH_3:c-C_6H_{12} ice. This suggests two different issues: first, it puts a constraint in the temperature of dust grains in the region, and secondly, it also indicates that a negligible amount of CO was initially present in the grains. Additionally, considering the sharp 3.37 μm absorption peak (C-H asymmetric stretching mode in -CH_3), they suggested that grains at the astronomical objects PPN CRL 618, DISM GC IRS 6E and at the Murchison carbonaceous meteorite were probably exposed to heavy cosmic rays with fluence higher than 3×10^{13} ions cm^{-2} or to some kind of ionizing agent that delivered an energy dose higher than 2×10^{8} J kg^{-1} ($\sim 10^{15}$ eV ng^{-1}). Additionally, a comparison between IR spectra from the laboratory and several YSOs suggests that the initial composition of grains in young stellar discs may include a mixture of H_2O, NH_3, CO (or CO_2), alkanes and CH_3OH. The 7.7 μm (1300 cm^{-1}) feature in YSOs, which is attributed to CH_4 (-CH deformation mode; Boogert *et al.* 1997), was best represented by the products coming from the radiolysis of ices initially containing alkanes, rather than by the ices where the primary source of carbon comes from CO. This suggests that saturated hydrocarbons, such as cyclohexane, may be a part of the initial chemical inventory of interstellar ices. However, the presence of broad features at 1450 and 1680 cm^{-1} in the astronomical observations, similar to the features observed in the IR spectra of the irradiated H_2O:NH_3:CO ice (Pilling *et al.* 2010a), requires the presence of CO (or/and CO_2) as one of the parental species in the interstellar medium.

4. Concluding remarks

In this work, we present some highlights of experiments employing swift ions broadband X-rays (plus fast electrons, low energy secondary electrons and a minor component of VUV photons) in astrophysical ices simulated in laboratory. The sample was analyzed by Fourier transform infrared spectroscopy FTIR at different radiation fluences. The experiments occurred at to laboratories in two countries: the Brazilian synchrotron light source (LNLS/CNPEM) in Campinas/Brazil and the Grand Accélérateur d'Ion Lourdes (GANIL/CIRIL/CIMAP) in Caen/France employing a similar methodology.

The selected results presented showed the interaction between cosmic ray analogs and ionizing soft X-rays probed in the laboratory and allow us to identify reaction routes that lead to chemical enhancement of astrophysical ices. For example, cosmic rays can induce the formation of alkenes and aromatic compounds (with eventual H_2 release to the gas phase) during the irradiation of saturated hydrocarbon-containing ices, as well as the formation of peptide bonds in ices containing amino acids. Such organic compound production was also observed during the processing of N_2:CH_4 ices by soft X-rays with the appearance of the nucleobase adenine, among the products of sample irradiation.

The results employing bombardment of astrophysical ice analogs by swift ions indicated that cosmic ray bombardment of saturated alkanes can be an alternative scenario for the production of unsaturated hydrocarbons and possibly aromatic rings (via dehydrogenation processes) in interstellar and protostellar ices. The results also indicate that a comparison between the laboratory spectra and IR observations of protostellar ices suggest that saturated hydrocarbons such as cyclohexane may be a part of the initial chemical inventory of interstellar ices.

The determination of the effective destruction and formation cross section in solid state, allows us to estimate the molecular half-life extrapolated for astrophysical environments and help us put constrains also in prebiotic chemistry that may have triggered life to arise in the Early Earth.

Acknowledgements

The author acknowledge the Brazilian agencies FAPESP (Projects JP 2009/18304-0, IC 2014/08643-0, DR 2012/17248-2 and MS 2016/11334-5) and CNPq (Research fellowship 304130/2012-5 and 306145/2015-4) as well as FVE/UNIVAP for the financial support. A special thank are given to the staff of UNIVAP and LNLS/CNPEM and GANIL/CIRIL/CIMAP for the technical support and help during the experiments.

References

Akkerman, A., Breskin, A., Chechik, R., & Gibrekhterman, A. 1993, *in Ionization of Solids by Heavy Particles, ed. R. A. Baragiola, (Berlin: Springer)*, Vol. 306, 359

Almeida, G. C., Pilling, S., Andrade, D. P. P., *et al.* 2014, *JPCC*, 118, 6193

Almeida, G. C., Pilling, S., de Barros, A. L. F., da Costa, C. P., Pereira, R. C., & da Silveira, E. F. 2017, *MNRAS*, 471, 1330

Andrade, D. P. P., de Barros, A. L. F., Pilling, S., Domaracka, A., Rothard, H., Boduch, P., & da Silveira, E. F. 2013, *MNRAS*, 787, 796

Baragiola, R. A., Vidal, R. A., Svendsen, A., Schou, J., Shi, M., Bahr, D. A., & Atteberrry, C. L. 2003, *Nuclear Instruments and Methods in Physics Research B*, 209, 294

Bergantini, A., Pilling, S., Rothard, H., Boduch, P., & Andrade, D. P. P. 2014a, *MNRAS*, 437, 2720

Bergantini, A., Pilling, S., Nair, B. G., Mason, N. J., & Fraser H. J. 2014b, *A&A*, 570, A120

Bernstein, M. P., Dworkin, J. P., Sandford, S. A., *et al.* 2002, *Nature*, 416, 401

Bonfim, V. S., Castilho, R. B., Baptista, L., & Pilling, S. 2017, *PCCP*, Accepted

Boogert, A. C. A., Schutte, W. A., Helmich, F. P., Tielens, A. G. G. M., & Wooden, D. H. 1997, *A&A*, 317, 929

Boogert, A. C. A., Pontoppidan, K. M., Lahuis, F., *et al.* 2004, *ApJS*, 154, 359

Brown, W. L., Augustyniak, W. M., Marcantonio, K. J., *et al.* 1984, *Nucl. Instr. Meth. B1*, IV, 307

Cazaux, J. 2006, *NIMPB*, 244, 307

Chan, W. F., Cooper, G., & Brion, C. E. 1993, *Chem. Phys.*, 178, 387

Chen, Y.-J., Ciaravella, A., Munõz Caro, G. M., *et al.* 2013, *ApJ*, 778, 162

de Barros, A. L. F., da Silveira, E. F., Pilling, S., Domaracka, A., Rothard, H., & Boduch, P. 2014, *MNRAS* 483, 2026

Dartois, E., Augé, B., Boduch, P., *et al.*, 2015, *A&A*, 576, A125

Ehrenfreund, P. & Charnley, S. B. 2000, *ARA&A*, 38, 427

Gibb, E. L., Whittet, D. C. B., & Chiar, J. E. 2001, *ApJ*, 558, 702

Gullikson, E. M. & Henke, B. L. 1989, *PhRvB*, 39, 1

Henke, B. L., Liesegang, J., & Smith, S. D. 1979, *PhRvB*, 19, 3004

Hijazi H., Rothard, H., Boduch, P., Alzaher, I., Ropars, F., *et al.* 2011, *Nucl. Instr. Meth. Phys. Res. B*, 269, 1003

Hüfner, S. 1995, *Photoelectron Spectroscopy: Principles and Applications (Berlin: Springer)*

Jimenez-Escobar A., Chen, Y.-J., Ciaravella, A., Huang, C.-H., Micela, G., & Cecchi-Pestellini, C. 2016, *ApJ*, 820, 25

Kaiser, R. I., Stockton, A. M., Kim, Y. S., Jensen, E. C., & Mathies, R. A. 2013, *AJ*, 765, 111

Kobayashi, K., Kaneko, T., Takano, Y., *et al.* 2008, *Proc. IAU Symp.* 251, 465

Li A., Greenberg J. M. 1997, *A&A*, 323, 566

Maloney, P. R., Hollenbach, D. J., & Tielens, A. G. G. M. 1996, *ApJ*, 466, 561

McLaren, R., Ishii, I., Hitchcock, A. P., *et al.* 1987, *J. Chem. Phys.*, 87, 4344
Mejía C., de Barros, A. L. F., Seperuelo Duarte, E., da Silveira, E. F., Dartois, E., Domaracka, A., Rothard, H., & Boduch, P. 2014, *Icarus*, 250, 222
Mejía C., Bender, M., Severin, D., *et al.* 2015, *Nucl. Instrum. Meth. B*, 353, 477
Muñoz Caro, G. M., Meierhenrich, U. J., Schutte, W. A., *et al.* 2002, *Nature*, 416, 403
Öberg, K. I., van Dishoeck, E. F., & Linnartz, H. 2009, *A&A*, 496, 281
Palumbo, M. E. 2006, *A&A*, 453, 903
Pilling, S., Andrade, D. P. P., Neto, A. C., Rittner, R., & Brito A. N. 2009, *JPCA*, 113, 11161
Pilling, S., Seperuelo Duarte, E., da Silveira, E. F., Balanzat, E., Rothard, H., Domaracka, A., & Boduch, P. 2010a, *A&A*, 509, A87
Pilling, S., Seperuelo Duarte, E., Domaracka, A., Rothard, H., Boduch, P., & da Silveira, E. F. 2010b, *A&A*, 523, A77
Pilling, S., Seperuelo-Duarte, E., Domaracka, A., Rothard, H., Boduch, P., & da Silveria E. F. 2011a, *PCCP*, 13, 15755
Pilling, S., Andrade, D. P. P., do Nascimento, E. M., Marinho, R. R. T., Boechat-Roberty, H. M., de Coutinho, L. H., de Souza, G. G. B., de Castilho, R. B., Cavasso-Filho, R. L., Lago, A. F., & de Brito A. 2011b, *MNRAS*, 411, 2214
Pilling, S. & Andrade, D. P. P. 2012, *in X-Ray Spectroscopy, ed. S. K. Sharma (Rijeka: InTech)*, 185
Pilling, S., Andrade, D. P. P., da Silveira, E. F., Rothard, H., Domaracka, A., & Boduch, P. 2012, *MNRAS*, 423, 2209
Pilling, S., Mendes, L. A. V., Bordalo, V., Guaman, C. F. M., Ponciano, C. R., & da Silveira, E. F. 2013, *Astrobiology*, 13, 79
Pilling, S., Nair, B. G., Escobar, A., Fraser, H., & Mason, N. 2014, *EPJD*, 68, 58
Pilling S. & Bergantini A. 2015, *ApJ*, 811, 151
Portugal, W., Pilling, S., Boduch, P., Rothard, H., & Andrade, D. P. P. 2014, *MNRAS*, 441, 3209
Prasad, S. S. & Tarafdar, S. P. 1983, *ApJ*, 267, 603
Rachid, G. M., Faquine, K., & Pilling, S. 2017, *PSS*, In press, DOI: 10.1016/j.pss.2017.05.003
Reis, E. F., Campos, F. S., Lage, A. P., Leite, R. C., Heneine, L. G., Vasconcelos, W. L., Lobato, Z. I. P., & Mansur, H. S. 2006, *Materials Res.*, 9, 185
Shen, C. J. & Greenberg, J. M., Schutte W. A. 2004, *A&A*, 415, 203
Seperuelo-Duarte, E., Boduch, P., Rothard, H., Been, T., Dartois, E., Farenzena, L. S., & da Silveira, E. F. 2009, *A&A*, 502, 599
Seperuelo-Duarte, E., Domaracka, A., Boduch, P., Rothard, H., Dartois, E., & da Silveira, E. F. 2010, *A&A*, 512, A71.
Strazzulla, G., Pirronello, V., & Foti, G. 1983, *ApJ*, 271, 255
Strazzulla, G., Leto, G., Spinella, F., & Gomis, O. 2007, *in Kuhs W. F., ed., Proc. 11th Int. Conf. on the Physics and Chemistry of Ice. Royal Society of Chemistry, Cambridge*, p. 561
Taylor, A. D., Baggaley, W. J., & Steel, D. I. 1996, *Nature*, 380, 323
Vasconcelos, F. A., Pilling, S., Rocha, W. R. M., Rothard, H., Boduch, P., & Ding, J. J. 2017a, *Phys. Chem. Chem. Phys.*, 19, 12845
Vasconcelos, F. A., Pilling, S., Rocha, W. R. M., Rothard, H., & Boduch, P. 2017b, *Phys. Chem. Chem. Phys*, 19, 24154
Vasconcelos, F. A., Pilling, S., Rocha, W. R. M., Rothard, H., & Boduch, P. 2017c, *ApJ*, submitted.
Witt, A. N., Smith, R. K., & Dwek E. 2001, *ApJ*, 550, L201

Astrochemistry VII – Through the Cosmos from Galaxies to Planets
Proceedings IAU Symposium No. 332, 2017
M. Cunningham, T. Millar & Y. Aikawa, eds.

doi:10.1017/S1743921317009929

Surface astrochemistry: a computational chemistry perspective

H. M. Cuppen[1], A. Fredon[1], T. Lamberts[2], E. M. Penteado[1], M. Simons[1] and C. Walsh[3,4]

[1]Radboud University, Institute for Molecules and Materials
email: h.cuppen@science.ru.nl

[2]Computational Chemistry Group, Institute for Theoretical Chemistry, University of Stuttgart, Pfaffenwaldring 55, 70569 Stuttgart, Germany

[3]School of Physics and Astronomy, University of Leeds, Leeds LS2 9JT, UK

[4]Leiden Observatory, Leiden University PO Box 9513, 2300 RA Leiden, The Netherlands

Abstract. Molecules in space are synthesized via a large variety of gas-phase reactions, and reactions on dust-grain surfaces, where the surface acts as a catalyst. Especially, saturated, hydrogen-rich molecules are formed through surface chemistry. Astrochemical models have developed over the decades to understand the molecular processes in the interstellar medium, taking into account grain surface chemistry. However, essential input information for gas-grain models, such as binding energies of molecules to the surface, have been derived experimentally only for a handful of species, leaving hundreds of species with highly uncertain estimates. Moreover, some fundamental processes are not well enough constrained to implement these into the models.

The proceedings gives three examples how computational chemistry techniques can help answer fundamental questions regarding grain surface chemistry.

Keywords. astrochemistry, molecular processes, methods: numerical, ISM: molecules

1. Introduction

Close to 200 molecules have been identified in interstellar space (see *http://www.astro.uni-koeln.de/cdms/molecules*). Among these, approximately 50 are classified as complex organic molecules (COMs), i.e., comprised of six atoms or more. COMs are generally observed in the gas phase, although it is widely accepted that saturated COMs are formed primarily through reactions on the surface of icy dust grains. This is where gas-phase species accrete, meet and react to form saturated molecules. The new generation of telescopes provides us with more information about that spatial extent, fractionation, and correlation with other COMs. The analysis of these data will require much more elaborate, and more diverse, gas-grain astrochemical models than have been developed so far (Cuppen *et al.* 2017). However, essential input information for gas-grain models, such as binding energies of molecules to the surface, have been derived experimentally only for a handful of species, leaving hundreds of species with highly uncertain estimates. Moreover, some fundamental processes are not well enough constrained to implement into the models.

For instance, the precise mechanism to form COMs with a C−C bond is unknown and several different mechanisms have been suggested over the past decades. Charnley & Rodgers (2009) applied a simple atom addition mechanism where species such as methanol can react with atomic carbon. Another mechanism was proposed by Garrod & Herbst (2006) and is more efficient at elevated temperatures. Cosmic-ray-induced photons can dissociate H_2CO and CH_3OH, creating functional-group radicals such as CH_3 and

CH_3O, on/within the ice mantle. At low temperatures, these radicals are hydrogenated again, but as the temperature increases to above $20-30$ K, the residence time of H atoms on the surface decreases substantially while at the same time radicals become mobile. Radical-radical association reactions become competitive with hydrogenation of radicals.

The detection of acetaldehyde CH_3CHO, dimethyl ether CH_3OCH_3, methyl formate CH_3OCHO, and ketene CH_2CO in the prestellar core L1689B suggest that COM synthesis has already started at the prestellar stage and suggests that there should be a viable grain surface reaction route that is possible at low temperatures and without the need for extensive irradiation (Bacmann *et al.* 2012). Woods *et al.* (2013) proposed a mechanism for the formation of COMs through recombination of two formyl radicals (HCO). Recently, Fedoseev *et al.* (2015) confirmed this picture by showing experimentally that the reaction between two HCO radicals yields glyoxal (HC(O)CHO) at 13 K; sequential hydrogenation with two or four hydrogen atoms produces glycoaldehyde and ethylene glycol, respectively. HCO radicals are formed in this case by CO hydrogenation and not by UV photolysis. Moreover, it showed that radical-radical recombination reactions are not only efficient at temperatures higher than 20–30 K, when radicals are commonly assumed to become mobile (e.g., Garrod 2013; Butscher *et al.* 2015).

In this proceedings, we will give three examples how computational chemistry techniques can help answer fundamental questions regarding grain surface chemistry. The low T COM formation route will be explored in more detail using a microscopic kinetic Monte Carlo (KMC) model. By applying a sensitivity analysis on binding energy data, more information on the relative importance of the different COM formation routes can be obtained. Finally, the role of exothermicity upon reaction for subsequent diffusion and desorption will be quantified using Molecular Dynamics simulations.

2. COM formation through HCO recombination without diffusion

The hypothesis of the low temperature COM formation route is that two formyl radicals (HCO) form in close proximity and recombine to from glyoxal (HC(O)CHO) (Fedoseev *et al.* 2015; Chuang *et al.* 2016). Through continued hydrogenation, glyoxal will consecutively be turned into glycoaldehyde ($HC(O)CH_2OH$) and ethylene glycol ($H_2C(OH)CH_2OH$). Finally, recombination of a formyl and a methoxy radical leads to the formation of methyl formate (CH_3OCHO). Formyl is formed through the well-known hydrogenation sequence of CO to methanol (Watanabe & Kouchi 2002; Fuchs *et al.* 2009) and hence without the need of UV photolysis or cosmic rays as external triggers. During CO freeze-out the gas phase H-atom abundance is similar to the CO gas phase abundance and there will be an insufficient number of H atoms to fully hydrogenate CO to H_2CO and CH_3OH, leaving a relatively high HCO surface abundance. Statistically these can be formed sufficiently close to recombine without the need of long scale diffusion.

This hypothesis has been shown to work in the laboratory (Fedoseev *et al.* 2015; Chuang *et al.* 2016), but here fluxes are necessarily high to enable molecule formation within the timescale of a few hours. To put the hypothesis really to the test, we simulate COM formation under interstellar conditions using a microscopic KMC model. We have previously been successful with this model to simulate both laboratory and ISM conditions for the formation of methanol (Fuchs *et al.* 2009; Cuppen *et al.* 2009). The main strength of microscopic simulation methods is that the location of individual species is traced throughout the simulation. Our study of hydrogenation of CO (Cuppen *et al.* 2009) showed that a gradient mantle composition builds up during catastrophic CO freeze-out. This is in agreement with ice observations of CO (Penteado *et al.* 2015).

In the past years, several codes have been developed that introduce multiple phases representing different layers within the ice (Taquet *et al.* 2012; Vasyunin & Herbst 2013;

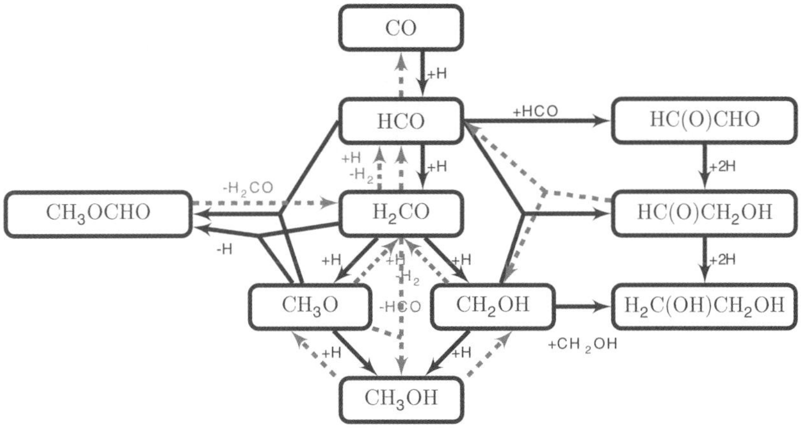

Figure 1. The new reaction network proposed in this paper. Solid lines indicate addition reactions whereas red dashed lines mark photo-dissociation reactions and blue dash-dotted lines are thermal dissociation reaction.

Furuya *et al.* 2016). However, the reaction rates are still calculated assuming a diffusive mechanism, by either rate equations or a macroscopic KMC routine. Crucial to the low temperature formation mechanism is that two HCO radicals form in close vicinity of each other and can react without the need for diffusion. A microscopic Kinetic Monte Carlo model has access to positional information of all species and does not make assumptions on the diffusive character of a reaction. It can therefore be used to estimated the relative importance of the different COM formation channels.

Figure 2 shows four simulation results for dark molecular clouds conditions with different hydrogen abundances. The simulated abundances are $n(\mathrm{H}) = 0.5, 2.5, 4.0$, and $9.0\ \mathrm{cm}^{-3}$, respectively. The grain surface temperature is 10 K and the initial CO abundance is $n_{\mathrm{initial}}(\mathrm{CO}) = 10\ \mathrm{cm}^{-3}$ in all cases. The gas phase abundance of atomic hydrogen is assumed to be in steady state in the gas phase and to remain constant throughout the simulation. The initial CO gas phase abundance is assumed to hold the total CO reservoir and is diminished throughout the simulation accounting for CO on the surface, and its products. The left-hand, vertical axes (in red) give the build-up in number particles per surface area whereas the right-hand axes (in blue) denote the grain surface abundance with respect to n_{H}. The total simulated time is 2×10^5 years. The upper left panel is a the result of simulations with a low H:CO ratio. The grain mantle is mainly composed of CO with H_2CO as second most abundant species. In these conditions the formation of HCO is relatively rare and two HCO radicals are hardly formed in close proximity. Indeed, the low amount of glycoaldehyde formed indicates that little HCO recombinations have occurred. However, enough hydrogen atoms are present to further hydrogenate HCO to H_2CO. A slow decrease in the mantle build-up (gray line) can be observed at approximately 10^5 years, due to CO depletion from the gas phase.

Results of simulations with a slightly higher gas-phase hydrogen abundance are shown in the upper right panel. The grain mantle now features less CO and H_2CO compared to the lower H:CO ratio. Because of the higher amount of hydrogen present in the gas phase, more HCO radicals are formed which can form HC(O)CHO instead of hydrogenating further to yield H_2CO. This is particularly true towards the end of the simulation, when the H:CO ratio increases further. Please notice that the overall grain mantle build-up in number of molecules is slightly lower than in the top left panel as indicated in the gray line. This is partly because it takes two molecules of CO to form HC(O)CHO and its

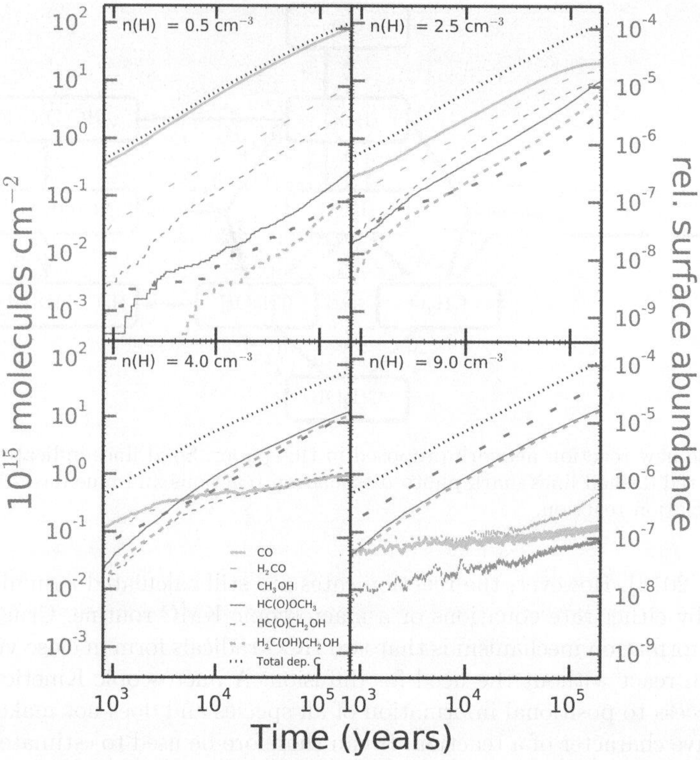

Figure 2. Time evolution of the grain surface composition for four different levels of initial hydrogen abundances: 0.5, 2.5, 4.0, and 9.0 cm^{-3}. The abundance of gas-phase carbon monoxide is 10.0 cm^{-3}. The temperature of the grain is 10 K. The relative surface abundance is given with respect to n_{H}.

hydrogenation products whereas it only takes one CO molecule to form H_2CO. The red dotted line accounts for this difference as it represents the number of equivalent CO molecules per unit array. The difference in mantle build-up can be fully explained by this effect.

Upon increasing the H:CO ratio further, HCO radical combination starts to compete with further hydrogenation. For the H:CO = 0.4 results in the bottom left bottom panel, the lifetime of the formation rates of CH_3OH by further hydrogenation, $H_2C(OH)CH_2OH$ through HCO radical formation, and CH_3OCHO through reaction of CH_3O are of comparable magnitude. This can be concluded from the abundance of these three species. The red dotted curve in the graph further shows that the mantle build-up is hampered. We will come back to this point later. For the highest H:CO ratio, hydrogenation wins from HCO radical recombination leading a CH_3OH dominated mantle with little $H_2C(OH)CH_2OH$ and CH_3OCHO.

Figure 3 shows cross sections of the grain mantles at $t = 2 \times 10^5$ years for the four simulations in Fig. 2. All panels show the grain in gray at the bottom of the image and the gas phase in black at the top. The different species that occupy the simulation lattice are represented by different colors. The same color coding is used as in Fig.2. The composition of the grain mantle reflects the earlier discussion on Fig. 2. What is immediately evident is the influence of the H:CO ratio on the surface morphology and the mantle height. Some of the reduction in height is because of the formation of the C–C bond which leads to a reduction in the number of molecules as mentioned earlier. The reduction for the

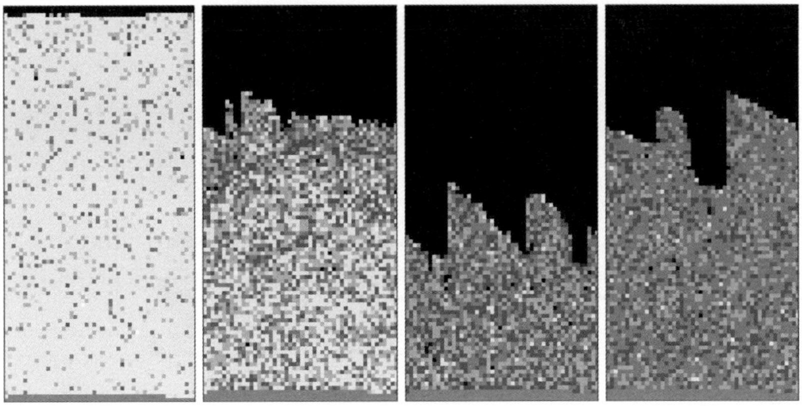

Figure 3. Cross-sections of the grain surface at $n(\mathrm{H})$ = 0.5, 2.5, 4.0, and 9.0 cm^{-3}. Colour indications are equal to Figure 2. Intermediates have slightly brighter colors than their immediate product. The grain is colored gray and the black parts are vacuum.

$n(\mathrm{H}) = 4.0\ \mathrm{cm}^{-3}$ simulation result is real. This appears be accompanied by a more rough surface morphology. We believe this because for more rough surfaces CO and H are more likely to land on top of a protrusion, which has less favorable binding since species are only interacting with a few neighboring species. CO and H are hence more likely to desorb from these protrusions before they can react. Moreover, these protrusions also hinder the diffusion to stronger binding sites.

The KMC simulations show that complex organic molecules with a C–C bond can be formed at temperatures as low as 10 K and without the need of external radiation. The H:CO gas phase abundance ratio is crucial in determining the amount of COMs and their relative composition. According to Goldsmith & Li (2005), the atomic hydrogen abundance should be in the range $n(\mathrm{H}) \approx 2$–$6\ \mathrm{cm}^{-3}$ which is in our intermediate H:CO range with optimal COM production circumstances. The amount of COMs formed is more than expected and for future simulations we need to consider more destruction channels.

Finally, all results discussed previously take into account the exothermicity of the surface reactions. The excess energy that is released during the reaction can be used for diffusion and desorption. In our scheme, a reaction product can move around a few hops before it is thermalized. This allows the reactants to find a site with a stronger binding energy and it makes the ice formed more compact. Simulations without exothermicity taken into account result in very fractal-like surfaces and show a substantial hampering in the ice build up. New species are less likely to accrete on these fractal-like surfaces and the binding energy in these structure is lower, enabling desorption of the reactants. This reinforces the conclusion that exothermicity is an important factor in the build up of ices (Lamberts *et al.* 2014) .

3. Sensitivity analysis: the relative importance of the different COM formation routes

Next, we present a sensitivity analysis study for the grain surface part of gas-grain codes by looking at the effect of uncertainties in the binding energy (E_{bind}) of species to the grain. We use a similar approach to Wakelam *et al.* (2006) and Wakelam *et al.* (2010). The binding energy of a species determines the upper value of the temperature range at which species are available for reactions on the grain surface. It also, indirectly, determines the onset temperature for surface reactions, since in many models the diffusion

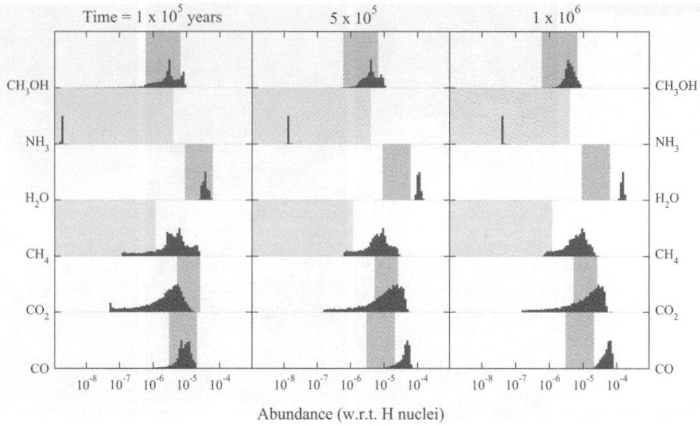

Figure 4. Distributions of simulated ice abundances (blue histograms) for three different times (10^5 in the left, 5×10^5 in the middle and 10^6 years in the right panels) compared to abundances derived from observations (yellow areas, light yellow for upper limits) of quiescent clouds and cores (see Boogert *et al.* (2015) for references). Figure reproduced from Penteado *et al.* (2016).

barrier is considered to be a fixed fraction of the binding energy (Hasegawa *et al.* 1992; Ruffle & Herbst 2000; Garrod & Herbst 2006; Cuppen *et al.* 2009). The initial goal of this work was to obtain constraints on the binding energies of different surface species by comparing the simulation results against ice observations. As will become clear, it is hard to constraint binding energies in this way, but crucial insight into the reaction network is obtained.

We start by compiling an updated list of binding energies by reviewing current experimental data. This resulted in recommended values for the binding energies and their uncertainties. These uncertainties reflect experimental uncertainties as well as intrinsic variations due to a range of different binding sites present on the surface. For radicals species, limited data is available and here we used uncertainties of half the binding energy with a maximum of 500 K. Next, 10,000 simulations of a standard homogeneous dark cloud with constant physical conditions are performed, where each simulation uses a set of binding energies that is randomly picked from a Gaussian distribution considering the recommended binding energies and their uncertainties. The rate equation model used is a two-phase chemistry model, since it treats the gas and solid phase; however, without using location information of ice species within the ice mantle. The chemical network used in the present work (Drozdovskaya *et al.* 2014, 2015; Garrod *et al.* 2008; Walsh *et al.* 2015, and references therein) contains 190 surface species. More details can be found in Penteado *et al.* (2016).

The simulations result in a very broad distributions of the obtained ice abundances. Only H_2O has a narrow distribution, other species can vary over orders of magnitude. Figure 4 shows the distributions for H_2O, CO, CH_3OH, NH_3, and CH_4 at three different times. Although the full distribution can cover several orders of magnitudes, the dispersion in terms of full-width-half-maximum is significantly narrower ($\lesssim 1$ order of magnitude) for most species. We compare the obtained model abundances with observational abundances taken from a recent review by Boogert *et al.* (2015) where we have taken the background star observations as representative of observations of quiescent clouds and cores. What is clear is the very large spread in observational abundances as indicated by the yellow areas in Figure 4, from upper limits to relatively large values compared with water ice observations. Inspection of Figure 4 shows that the best agreement is obtained at relatively early times of 10^5 years. This coincides with dark cloud

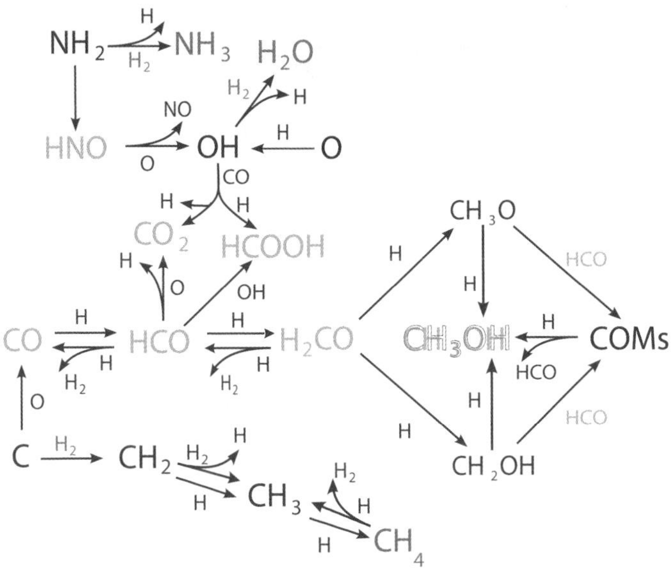

Figure 5. A limited surface network explaining the main results. Figure reproduced from Penteado *et al.* (2016).

model results focusing on the gas phase, where 10^5 years is considered as the early time of best overall agreement between models and observations for a large number of species. The observations of CH_4 and NH_3 show only upper limits. Our model results for NH_3 fall well below this limit, where as a large fraction of models heavily overproduce CH_4. We come back to this later.

A detailed analysis of the dependence on the different binding energies resulted in a limited surface network that can explain the main features of the results. This is presented in Figure 5. Crucial here is whether CO hydrogenation or CH_4 formation wins, which is critically determined by the amount of H_2 on the surface. For high H_2 abundance, the reactions and species in blue dominate, otherwise the yellow reactions and species dominate. Methanol is both yellow and blue and its abundance does not correlate with CO, H_2CO, and other CO hydrogenation products. The reason for this, is that for high HCO abundance, intermediate species CH_3O and CH_2OH rather react with HCO than hydrogenate to form CH_3OH. CH_3OH is, in this case, mainly formed through destruction of more complex species. For lower HCO abundances, CH_3OH is formed directly through hydrogenation of CH_3O and CH_2OH. Close inspection of the network shows that the HCO abundance is artificially high.

Using the standard network, its main formation reaction $H + H_2CO \longrightarrow HCO + H_2$ is much more efficient in the model (97 %) than $H + H_2CO \longrightarrow CH_3O$ or $H + H_2CO \longrightarrow CH_2OH$ which equally share the remainder. This is because the barriers for the different channels originate from different sources (gas phase data vs. surface experiments). Surface experiments do not support these extreme branching ratios on the surface. In a recent experimental study Chuang *et al.* (2016) investigated the hydrogenation of H_2CO. They indeed observe the addition reaction to CH_3O and the abstraction to $H_2 + HCO$, and they cannot exclude the channel to form CH_2OH. Here, we alter the network by adopting equal ratios for all three channels and perform 100 additional simulations using this network. The evolution of the ice abundance of the species involved in the hydrogenation of CO is shown in Figure 6 in yellow. A comparison is made between the original 100 runs of our standard network in blue.

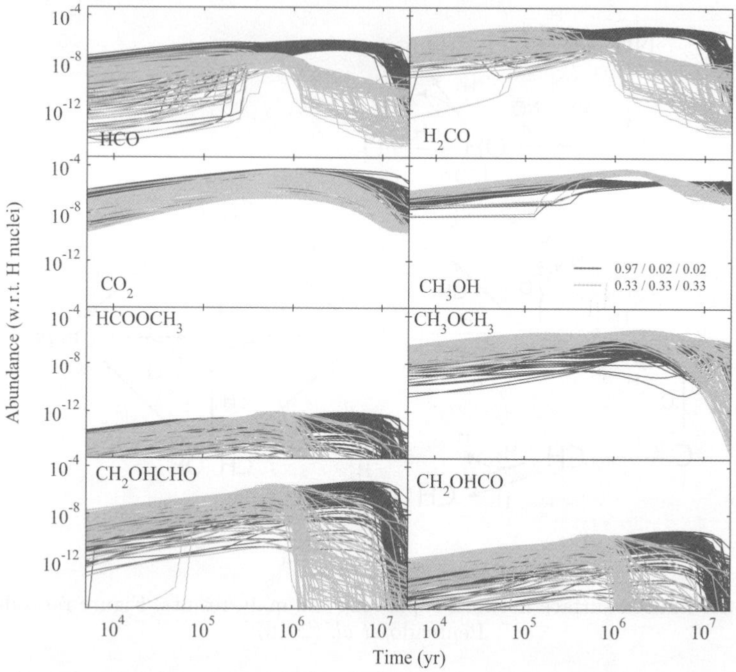

Figure 6. Evolution of ice abundances according to two different branching ratios: 0.97 / 0.02 / 0.02 (standard network) and 0.33 / 0.33 / 0.33 (updated network). Results using the standard network are shown in blue and using the updated network in yellow. Abundances much lower than 10^{-16} are negligible and are therefore not shown here. Figure reproduced from Penteado *et al.* (2016).

The timescale of conversion of HCO to H_2CO and CH_3OH is reduced by a few million years, since there are less back reactions. Moreover, as a consequence of the reduction in HCO, an increase in the methanol abundance can be observed, since CH_3O and CH_2OH will now predominantly react with H to form methanol. For the more complex species, both a reduction in the peak intensity and a change in the formation timescale can be observed.

In summary, the sensitivity analysis allows us to expose connections between different species in the total reaction network. It shows the importance of the $H_2CO + H$ reaction. By changing the branching ratios of this single reaction to more realistic values, the timescale and peak abundances of many COMs have changed. Several other reactions with branching ratios exist in the network and we advise a complete scrutinization of the whole network. The overproduction of CH_4 by the model with respect to observations could be due to too efficient destruction reactions of COMs, because of inaccuracy in branching ratios of reactions leading to CH_4 or its precursors.

4. Role of exothermicity in non-thermal desorption and diffusion

As a last example, we would like to present Molecular Dynamics simulations to examine the role of exothermicity in non-thermal desorption and diffusion. In our earlier example of low temperature COM formation, we already showed the importance of exothermicity in the build-up of ice mantles. Here we would like to quantify the outcome of exothermicity. Excess energy can be in the form of translational, rotational, or vibrational energy or electronic excitation. How this energy is spread is highly system dependent (Polanyi

1986); it does not only depend on the specific reaction but also on the reaction environment and the configuration in which it occurs. The excitation energy could be applied for desorption, but also for restructuring of the grain mantle as shown above, or diffusion of the species. Reactive or chemical desorption (Garrod *et al.* 2006, 2007; Dulieu *et al.* 2013) is hence only one possible outcome. In analogy, we introduce the concept of chemical or non-thermal diffusion where the energy released during the surface reaction allows the product to diffuse over the surface with the possibility to meet another reactant for subsequent reactions (Arasa *et al.* 2010; Lamberts *et al.* 2014). In this way, the excess energy of one reaction would enable subsequent reactions. We expect the type of excitation to have an effect on the outcome.

Molecular Dynamics (MD) simulations follow Newton's equations of motion and allow one to keep track of the energy, potential or kinetic energy, by giving a species some initial energy. KMC models, although microscopic, do not offer this opportunity. Here we aim to investigate the fate of kinetically excited species on the surface of an ice mantle. We consider an additional kinetic energy between 0.5 and 5 eV to cover the full range from reactions with a small exothermicity to photodissociation reactions where the excitation energy after bond breaking can be several eV. Crystalline water ice is used as a substrate and three different surface species are considered, CO_2, H_2O and CH_4, which are among the most common ice species (Boogert *et al.* 2015). Moreover, these species span a range in binding energies, number of internal degrees of freedom and molecular weight. In a later stage, we would also like to study the effect of other types of excitation, especially vibrational and rotational as well as the effect of the type of substrate (amorphous versus crystalline).

We roughly apply the following procedure: First the different binding sites on the substrate are identified by deposition of an admolecule at different positions on the water surface. Next, the strongest binding sites are selected, and from each site new simulations are issued where the admolecule is given a specific amount of additional translational kinetic energy in a random direction. Finally, the outcome of the simulation is detected by the software and stored for post-analysis. We classify each simulation into: desorption, adsorption, penetration, FAIL, and TIME LIMIT. For more details, we refer to Fredon *et al.* (2017).

For all three admolecules, desorption or adsorption are the most likely outcomes; penetration is found to occur in only a few simulations, all for high kicking energies. Figure 7 shows the fraction of simulations resulting in desorption as a function of the initial translational energy given to the three admolecules. Each fraction is based on 1,400 individual simulations. As expected, the desorption probability increases with increasing kick energies. There appears to be a threshold energy required for desorption, which coincides with the binding energy of the different species to the surface. The dashed and dotted curves represent models for chemical desorption that are generally used. Both the model by Garrod *et al.* (2007) and the model Minissale *et al.* (2016) cannot reproduce our simulation results with one universal expression. The model by Garrod *et al.* (2007) focuses on vibrational excitation instead of translational excitation which explains the discrepancy. The model by Minissale *et al.* (2016), on the other hand, does not reproduce the threshold value. We find our data to be best reproduced by

$$P_{\rm CD}(E_{\rm react}^{\rm trans}) = 1 - \exp\left(-\omega \frac{E_{\rm react}^{\rm trans} - |E_{\rm bind}|}{|E_{\rm bind}|}\right) \tag{4.1}$$

with $\omega = 0.3$. This is shown by the blue solid line in Figure 7.

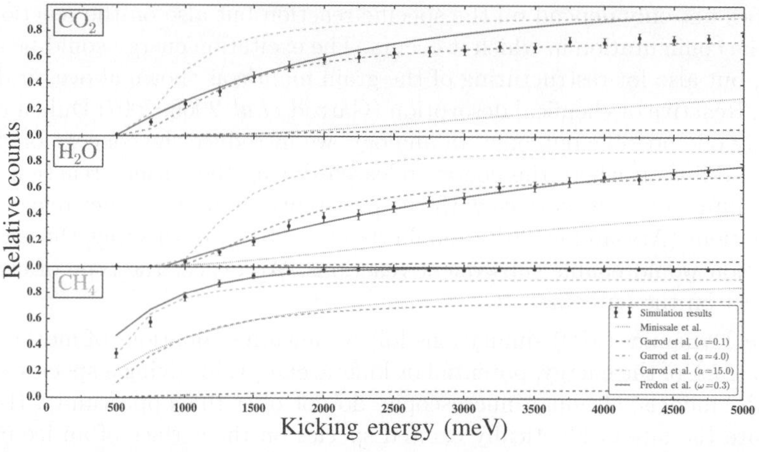

Figure 7. Desorption probability as a function of the initial translational energy given to the admolecule. The dots represent the data from our simulations with an error bar of 3σ. The blue lines, representing our model, show the closest agreement with the simulation data. Figure reproduced from Fredon *et al.* (2017).

As an example we will consider reactive desorption in two specific reactions. We start with

$$H + OH \longrightarrow H_2O. \tag{4.2}$$

This reaction has an enthalpy of reaction of 4.89 eV. If we assume an equipartition of the energy over all 9 degrees of freedom ($E_{\rm react}^{\rm trans} = 3 \times 1/9 \times 4.89 = 1.63$ eV), the desorption probability is

$$P_{\rm CD} = 1 - \exp\left(-0.3\frac{1.630 - 0.839}{0.839}\right) = 0.24. \tag{4.3}$$

Experiments on non-porous ASW give a desorption efficiency of 0.30 ± 0.10 (Minissale *et al.* 2016) and our values are hence in quite good agreement. We can make a similar calculation for $CH_3 + H \longrightarrow CH_4$ which has a heat of formation of 4.55 eV. The chemical desorption probability from a crystalline surface is now 0.75. This significantly higher efficiency of CH_4 desorption, versus the other molecules, is very interesting. Many chemical models of ice mantle production appear to overproduce solid-phase methane, as compared to astronomical measurements (as a fraction with respect to water)(Garrod & Herbst 2006; Öberg *et al.* 2008), including our simulations presented in Section 3.

Figure 8 indicates the traveled distance and direction of the admolecules for each "adsorption" trajectory. The figures shows multiple copies of the same simulation cell stacked in both x and y direction; the individual cells are indicated by the grid. All admolecules have started from the central simulation box and they traveled through the periodic boundary conditions (in some cases multiple times) to their final position indicated by the dots. The dots are color coded according to their initial kicking energy. The blue dots dominate the CH_4 panel because trajectories with higher initial translational energies will have resulted in desorption of the species and are hence not included in the plot. For many trajectories, the admolecule has traveled a substantial distance before coming to a stand still, up to 1157 Å for H_2O (not shown in the figure).

The CH_4 results in panel (c) in Fig. 8 show a strong angular dependence following the hexagonal pattern of the underlying crystalline substrate. Collisions of the CH_4 molecule with dangling protons lead either to redirection of the molecule through the "hexagonal channels" on the surface or to desorption of the molecule. In contrast, H_2O prefers to

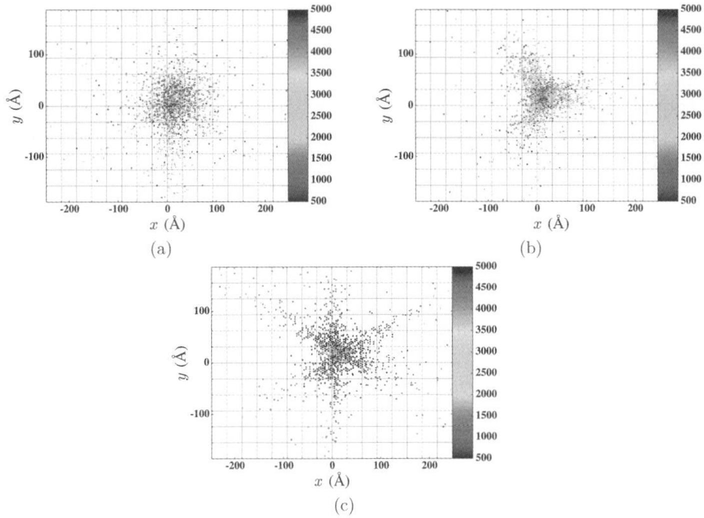

Figure 8. Top view of the simulation cells (original and periodic copies) indicating the traveled distance and direction of the admolecules for each "adsorption" trajectory. The original simulation cell is the one in the center and the copies are represented by the grid. The colored dots represent the positions of the CO_2 (a), H_2O (b), and CH_4 (c) admolecules at the end of the energy dissipation simulations. The colorbar represents the initial translational energy of the admolecules. Figure reproduced from Fredon *et al.* (2017).

interact with the dangling protons of the ice and three main directions in which H_2O travels on the surface are observed, following the dangling proton pattern. Indeed, as can be seen in Fig. 8, these directions do not coincide with the hexagonal channels of CH_4. CO_2 appears to be less affected by surface protrusions.

5. Conclusions

In conclusion, we have shown how different computational chemistry techniques can be applied to obtain more information about grain surface chemistry. We have used three different examples of different problems that required three very different computational techniques. To scrutinize a large gas-grain chemistry network through a sensitivity analysis requires a simulation method that can handle large reaction networks at very low computational costs to allow for 10,000 simulations. Currently, this is limited to the use of macroscopic methods, which we know not to capture all surface chemistry details. In this case, it exposed a weakness in the network: the uncertainty in the branching ratio of $H + H_2CO$. We hope that this will be followed up experimental groups.

The low temperature formation route of glycoaldehyde does not require diffusion or external UV. HCO radicals should simply form in close proximity. This required a microscopic, statistical approach. With our KMC model, we were indeed able to show that these complex molecules can form a temperatures as low as 10 K. It further showed the importance of exothermicity in the building of the ice layer. Again a microscopic method was required to obtain this type of information.

Finally, using Molecular Dynamics simulations we could quantify the result of exothermicity in the form of kinetic excitation. Species can indeed travel a substantial distance before thermalizing or kinetic excitation can result in non-thermal desorption. However, the picture may change when also including vibrational and rotational excitation and using an amorphous substrate.

References

Arasa, C., Andersson, S., Cuppen, H. M., van Dishoeck, E. F., & Kroes, G. 2010, J. Chem. Phys., 132, 184510

Bacmann, A., Taquet, V., Faure, A., Kahane, C., & Ceccarelli, C. 2012, A&A, 541, L12

Boogert, A. A., Gerakines, P. A., & Whittet, D. C. 2015, *Annual Review Astronomy and Astrophysics*, 53, 541

Butscher, T., Duvernay, F., Theule, P., Danger, G., Carissan, Y., Hagebaum-Reignier, D., & Chiavassa, T. 2015, MNRAS, 453, 1587

Charnley, S. B. & Rodgers, S. B. 2009, in *Astronomical Society of the Pacific Conference Series*, Vol. 420, Bioastronomy 2007: Molecules, Microbes and Extraterrestrial Life, ed. K. J. Meech, J. V. Keane, M. J. Mumma, J. L. Siefert, & D. J. Werthimer, Proceedings of a workshop held 16-20 July 2007 in San Juan, Puerto Rico., 29, aSP Conference Series, Vol 420

Chuang, K. J., Fedoseev, G., Ioppolo, S., van Dishoeck, E. F., & Linnartz, H. 2016, MNRAS, 455, 1702

Cuppen, H. M., van Dishoeck, E. F., Herbst, E., & Tielens, A. G. G. M. 2009, A&A, 508, 275

Cuppen, H. M., Walsh, C., Lamberts, T., Semenov, D., Garrod, R. T., Penteado, E. M., & Ioppolo, S. 2017, *Space Sci. Rev.*, in press

Drozdovskaya, M. N., Walsh, C., Visser, R., Harsono, D., & van Dishoeck, E. F. 2014, MNRAS, 445, 913

—. 2015, *Mon. Not. R. Astron. Soc.*, 451, 3836

Dulieu, F., Congiu, E., Noble, J., Baouche, S., Chaabouni, H., Moudens, A., Minissale, M., & Cazaux, S. 2013, *Sci. Reports*, 3, 1338

Fedoseev, G., Cuppen, H. M., Ioppolo, S., Lamberts, T., & Linnartz, H. 2015, MNRAS, 448, 1288

Fredon, A., Lamberts, T., & Cuppen, H. 2017, *ApJ*, submitted

Fuchs, G. W., Cuppen, H. M., Ioppolo, S., Romanzin, C., Bisschop, S. E., Andersson, S., van Dishoeck, E. F., & Linnartz, H. 2009, A&A, 505, 629

Furuya, K., Drozdovskaya, M. N., Visser, R., van Dishoeck, E. F., Walsh, C., Harsono, D., Hincelin, U., & Taquet, V. 2016, A&A, in press

Garrod, R., Park, I. H., Caselli, P., & Herbst, E. 2006, *Faraday Discussions*, 133, 51

Garrod, R. T. 2013, *ApJ*, 765, 60

Garrod, R. T. & Herbst, E. 2006, A&A, 457, 927

Garrod, R. T. & Herbst, E. 2006, A&A, 457, 927

Garrod, R. T., Wakelam, V., & Herbst, E. 2007, A&A, 467, 1103

Garrod, R. T., Weaver, S. L. W., & Herbst, E. 2008, *ApJ*, 682, 283

Goldsmith, P. F. & Li, D. 2005, *ApJ*, 622, 938

Hasegawa, T. I., Herbst, E., & Leung, C. M. 1992, ApJS , 82, 167

Lamberts, T., de Vries, X., & Cuppen, H. M. 2014, *Faraday Disc.*, 168, 327

Minissale, M., Dulieu, F., Cazaux, S., & Hocuk, S. 2016, A&A, 585, A24

Öberg, K. I., Boogert, A. C. A., Pontoppidan, K. M., Blake, G. A., Evans, N. J., Lahuis, F., & van Dishoeck, E. F. 2008, *ApJ*, 678, 1032

Penteado, E. M., Boogert, A. C. A., Pontoppidan, K. M., Ioppolo, S., Blake, G. A., & Cuppen, H. M. 2015, MNRAS, 454, 531

Penteado, E. M., Walsh, C. W., & Cuppen, H. M. 2016, *ApJ*, 2017, in press

Polanyi, J. C. 1986, *Noble Lecture*, Chemistry

Ruffle, D. P. & Herbst, E. 2000, MNRAS, 319, 837

Taquet, V., Ceccarelli, C., & Kahane, C. 2012, A&A, 538, A42

Vasyunin, A. I. & Herbst, E. 2013, *ApJ*, 762, 86

Wakelam, V., Herbst, E., Le Bourlot, J., Hersant, F., Selsis, F., & Guilloteau, S. 2010, A&A, 517, A21

Wakelam, V., Herbst, E., Selsis, F., & Massacrier, G. 2006, A&A, 459, 813

Walsh, C., Nomura, H., & van Dishoeck, E. 2015, A&A, 582, A88

Watanabe, N. & Kouchi, A. 2002, ApJ, 571, L173

Woods, P. M., Slater, B., Raza, Z., Viti, S., Brown, W. A., & Burke, D. J. 2013, *ApJ*, 777, 90

Astrochemistry VII – Through the Cosmos from Galaxies to Planets
Proceedings IAU Symposium No. 332, 2017
M. Cunningham, T. Millar & Y. Aikawa, eds.

doi:10.1017/S1743921317006408

Laboratory measurements of methanol photolysis branching ratios to guide astrochemical models

Susanna Widicus Weaver, Carson R. Powers, Morgan N. McCabe and Samuel Zinga

Department of Chemistry, Emory University,
1515 Dickey Drive, Atlanta, GA 30322, USA
email: swidicu@emory.edu

Abstract. Methanol is ubiquitous in star-forming regions, and has recently been detected in a protoplanetary disk. Astrochemical models have shown that methanol photolysis contributes to complex organic chemistry in interstellar ices. While some methanol photolysis branching ratios have been measured, infrared condensed-phase measurements rely on assumptions about the chemistry, and mass spectrometric measurements cannot distinguish structural isomers. To address these challenges, we are using pure rotational spectroscopy to quantitatively probe the methanol photolysis products. We use a VUV laser to dissociate methanol in the throat of a supersonic expansion, and probe the products downstream after cooling is complete. We then use a rotational diagram analysis to determine the relative density of each product relative to methanol. We have detected the methoxy, hydroxymethyl, and formaldehyde photolysis products. We present here the experimental setup and the initial results and discuss these results in the context of interstellar chemistry.

Keywords. astrochemistry, molecular data, molecular processes, ISM: molecules

1. Introduction

Complex organic molecules (COMs) are highly abundant in regions of star formation, but their formation pathways are not well understood. Gas-phase ion-molecule reactions have been shown to be inefficient for forming some of the most abundant COMs (Horn *et al.* (2004)). Therefore, recent astrochemical models (Garrod *et al.* (2008), Laas *et al.* (2011), Garrod (2013)) have explained COM formation via UV photolysis of icy grain mantles followed by radical-radical recombination reactions on the ice surface during star formation. Methanol is a good test of these proposed formation pathways because of its prevalence in interstellar ices. Methanol is commonly found in the condensed phase in dense molecular clouds, and can compose up to 20% of the ices in these regions (Gibb *et al.* (2004)). Methanol photolysis is therefore thought to be a major source of the organic functional groups that go on to form the backbones of COMs in the ISM. Despite its importance in astrochemistry, the methanol photodissociation mechanism has not been quantitatively determined in full. We therefore seek to provide laboratory measurements of the methanol photolysis branching ratios to guide astrochemical models.

The photolysis of methanol can proceed via multiple reaction pathways, which are overviewed in the reaction energy diagram shown in Figure 1. Two of the products of methanol photolysis, the radicals methoxy (CH_3O) and hydroxymethyl (CH_2OH), are of particular interest in astrochemistry. Further reaction of these radicals with hydroxyl radical (HCO), formed from photolysis of formaldehyde, can produce different structural

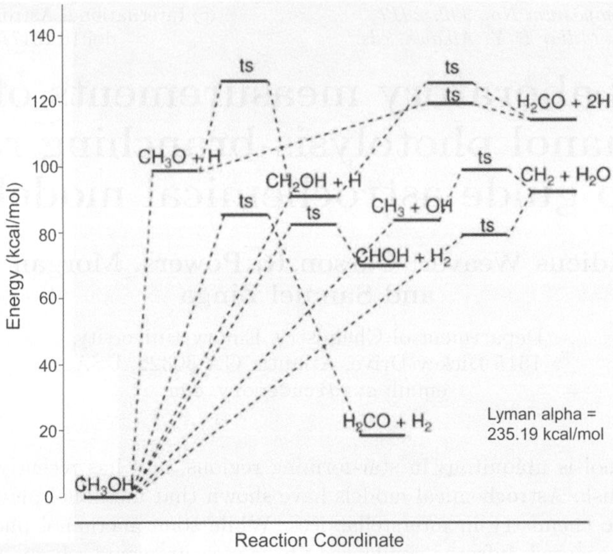

Figure 1. Reaction pathways for methanol photolysis.

isomers of molecules with the formula $C_2H_4O_2$. In the gas-grain model proposed by Garrod *et al.*, hydroxymethyl is linked to the production of glycolaldehyde, and methoxy to methyl formate; therefore, measuring the precise branching ratios of methanol photolysis is of interest to the astrochemical community at large.

Hagege *et al.* (1968) determined the gas-phase branching ratios for the main dissociation pathways of methanol at 184.9 nm via mass spectrometry. However, because the radicals hydroxymethyl and methoxy have the same mass, it was impossible to determine separate branching ratios for these two pathways; a total branching fraction for CH_3O + CH_2OH was determined to be ~75%. Öberg *et al.* (2009) performed studies in the condensed phase using reflection-absorption infrared spectroscopy (RAIRS). This study also determined that the two radicals of interest made up the majority of the photolysis products; more specifically, this study showed that hydroxymethyl was the main radical produced. However, this conclusion was made not by directly observing the production of radicals on the surface of the ice, but instead by observing the production of COMs from recombination reactions during warm-up. Based on their observations, this information was inferred from a chemical model which only assumed the production and recombination of first generation radicals produced from methanol photolysis.

Our experiment attempts to directly observe the products of methanol photolysis in the gas phase using direct absorption millimeter and submillimeter rotational spectroscopy. Through this technique, we aim to quantify the photolysis branching ratios for as many reaction pathways as possible, with a particular focus on distinguishing – for the first time – the hydroxymethyl and methoxy channels.

2. Experimental Design

The experimental setup can be seen in Figure 2. Argon was used as a carrier gas and was bubbled through liquid methanol. The gas mixture was flowed through a pulsed valve (Parker-Hannafin, Series 9, 1 mm pinhole) into a vacuum chamber held at a background pressure of 12 mTorr. The backing pressure for the valve was maintained at 32 psig, and the valve was operated with a 2 ms opening time at a rep rate of 44 Hz. The sample

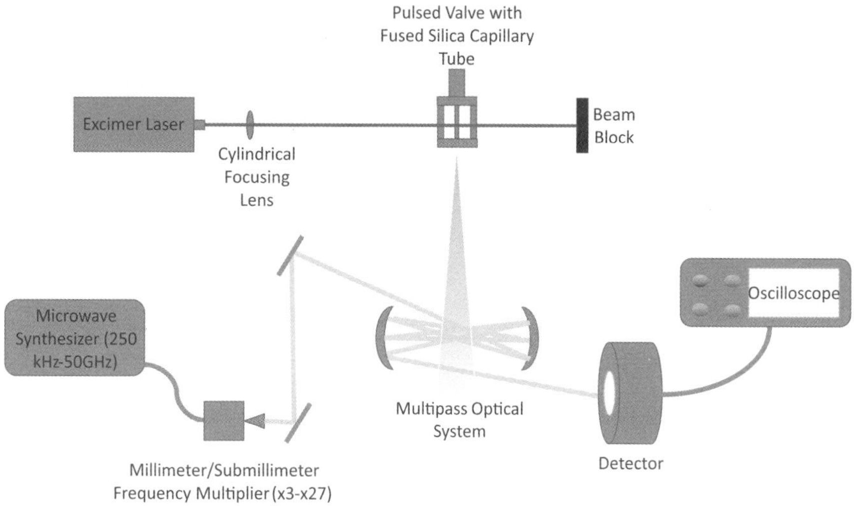

Figure 2. The design of the spectrometer used in conducting the methanol photolysis experiments.

passed through a fused silica tube affixed to the front of the pulsed valve. The output of a UV excimer laser (GAM LASER Inc., EX5 250 Hz) at 193 nm was focused onto the tube using a cylindrical lens. Photolysis occurred in the tube, and the photolysis products were rotationally and vibrationally cooled via supersonic expansion. The expansion was probed with millimeter/submillimeter light, produced from the output of a frequency synthesizer (Agilent Technologies, E8257D PSG with options 1EA, UNU, 550, and UNT) that was then multiplied to the range of 50 - 1000 GHz using frequency multipliers (Virginia Diodes Inc., AMC-S268). The millimeter/submillimeter beam path followed a multipass arrangement that enabled seven passes through the expansion. The output was detected using an InSb hot electron bolometer (QMC Ltd., QFI/XBI) and the resultant time-response of the detector was collected using a digital oscilloscope. From these data, 3D plots of spectral intensity versus frequency and time relative to the firing of the pulsed valve were constructed.

3. Results

To test the performance of the spectrometer and optimize experimental conditions, spectra of methanol were collected at the beginning of each set of measurements. This enabled all measurements to be referenced back to the parent methanol molecule, eliminating the effects of day-to-day fluctuations in valve conditions and sample delivery. Examples of typical methanol spectra are shown in Figure 3.

A Boltzmann diagram analysis was carried out for each molecule detected, and from this analysis the temperature and number density was determined. An example of this analysis for methanol can be seen in Figure 4.

Photolysis of the parent methanol is evidenced by the appearance of a depletion in the methanol absorption signal, shortly after the laser pulse. We time the submillimeter probe to coincide with this methanol depletion signal. A typical methanol depletion signal can be seen in Figure 5.

The first product targeted was formaldehyde; as shown by Hagege *et al.*, it is the most stable molecule produced via methanol photolysis, and was seen as a large percentage of

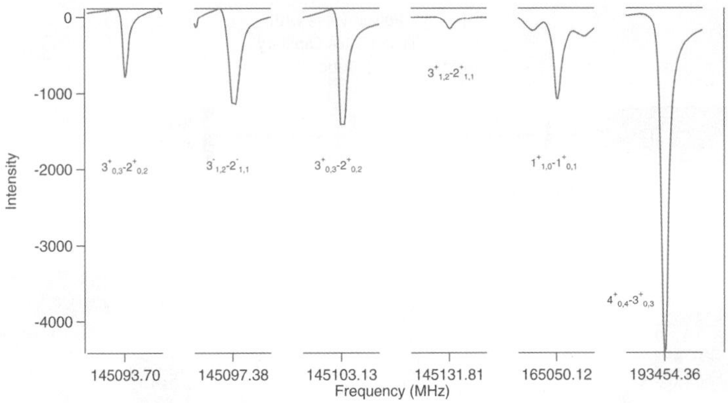

Figure 3. Several methanol lines observed in the experiment.

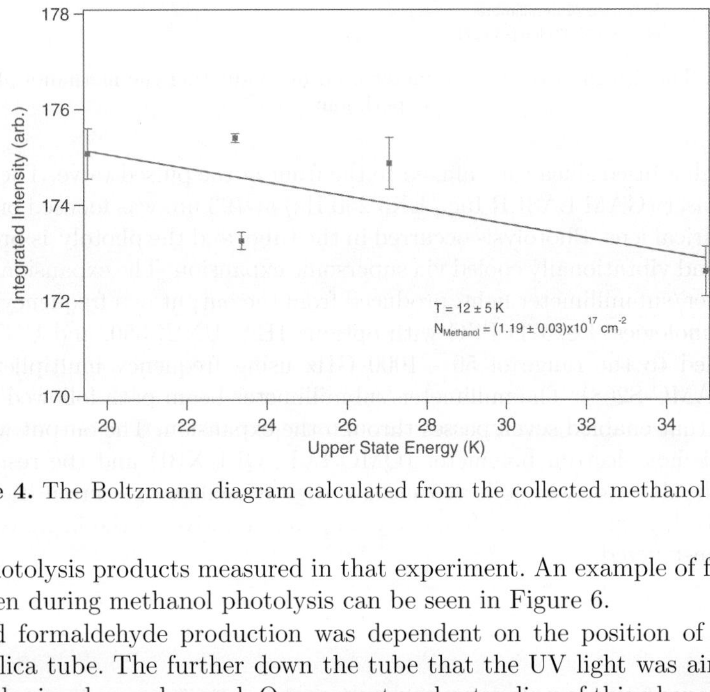

Figure 4. The Boltzmann diagram calculated from the collected methanol spectra.

the total photolysis products measured in that experiment. An example of formaldehyde spectra taken during methanol photolysis can be seen in Figure 6.

We found formaldehyde production was dependent on the position of the laser on the fused silica tube. The further down the tube that the UV light was aimed, the less formaldehyde signal was observed. Our current understanding of this phenomenon stems from the collisions happening in the fused silica tube. As shown in Figure 1, formaldehyde can either form directly from photolysis, or through further reactions of methoxy and hydroxymethyl. Therefore, formaldehyde is produced in greater concentrations when the laser is aimed at the beginning of the tube, allowing collisions between the radical products in the tube or with the walls of the tube that lead to further reactions. This idea is supported by the detection of an increase in observed methoxy radical signal when the laser is near the end of the tube (see Figure 7).

In addition to the detection of the methoxy radical, we have obtained a preliminary detection of the hydroxymethyl radical, shown in Figure 8. Bermudez *et al.* (2017) provided frequencies for rotational transitions of the hydroxymethyl radical that were used to guide our measurements. However, no line strength information was provided in this analysis, and our own fit of their experimental data did not converge. We can therefore only approximate the information needed for the Boltzmann analysis. Nonetheless, the

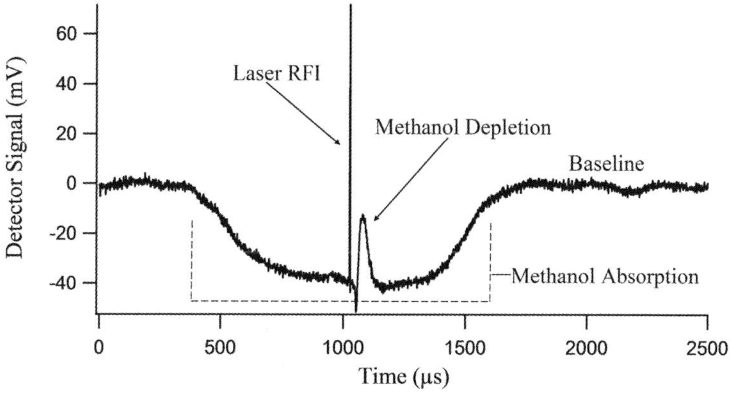

Figure 5. The time response of the methanol absorption feature across a valve pulse, with a depletion feature shortly after the UV laser pulse, taken from McCabe (2016).

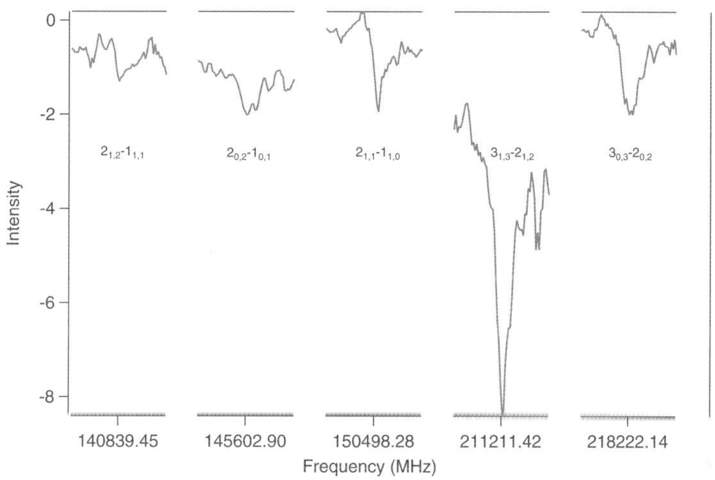

Figure 6. Spectra of formaldehyde produced from methanol photolysis.

lines are clearly present, and we are working to confirm this spectral assignment so that a full quantitative analysis is possible.

In our experiment, we can probe all methanol photolysis channels except OH, the transitions for which lie outside of the range of frequency coverage. For the sake of calculation, we therefore assumed the OH + CH_3 branching ratios determined by both Hagege *et al.* (1968) and Öberg *et al.* (2009). We used the results of the Boltzmann analysis for each product channel, referenced to the parent molecule signal, and calculated a tentative set of branching ratios for methanol photolysis. These ratios are given in Table 1. While these calculations make a number of assumptions, future experiments should help us to determine these branching ratios quantitatively.

Our experimental results can be compared to those of Hagege *et al.* (1968) and Öberg *et al.* (2009). While the results for the methoxy and hydroxymethyl channels are similar to these other studies, the results for formaldehyde are not. Hagege *et al.* (1968) found an order of magnitude larger branching fraction for formaldehyde than we see in our studies. On the other hand, Öberg *et al.* (2009) saw no production of Formaldehyde in their condensed phase work. In our work and that of Öberg *et al.* (2009), the hydroxymethyl channel is the dominant channel. This is in contrast to the results of the model by Laas

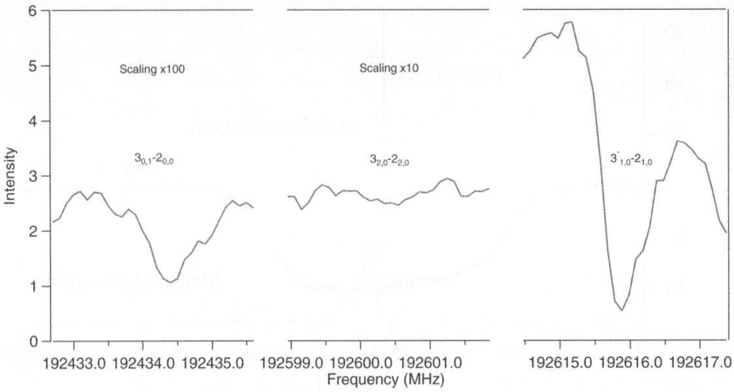

Figure 7. Spectra of methoxy radical produced from methanol photolysis.

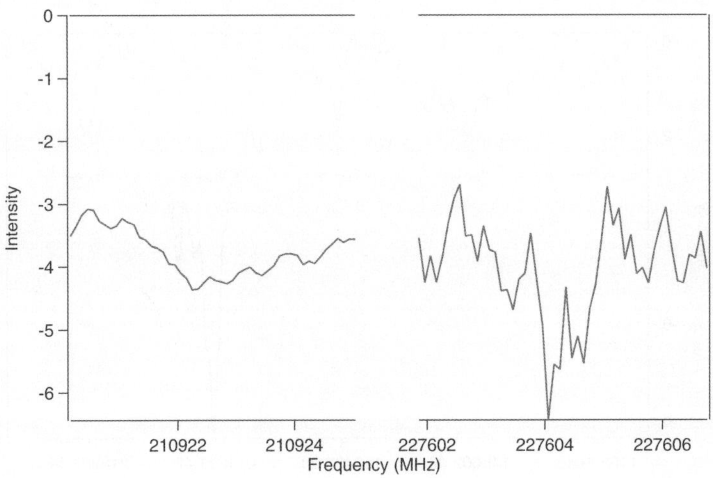

Figure 8. Spectra of hydroxymethyl radical produced from methanol photolysis.

Table 1. Tentative branching ratios calculated for methanol photolysis. The branching fraction for the CH_3 + OH channel are assumed from previous studies.

Photolysis Products	Hagege *et al.* (1968) (gas phase)	Öberg *et al.* (2009) (condensed phase)	Our Work (assuming CH_3=5%)	Our Work (assuming CH_3=12%)
CH_2OH	75% (+ CH_3O)	73%	53%	49%
CH_3O	75% (+ CH_2OH)	15%	39%	36%
CH_3	<5%	12%	5%	12%
H_2CO	20%	0%	2%	2%

et al. (2011), which indicates that the methoxy radical channel is predominant on the surface of ices in interstellar clouds. However, for some COMs the observed abundances indicate that hydroxymethyl may in fact be the dominant branching fraction. Additional studies that fully quantify this channel, as well as that of CH_3 + OH, are required before direct comparison with observations can be made.

4. Conclusions and Future Work

The branching ratios for methanol photolysis have been investigated. Tentative branching ratios have been extrapolated from experimental measurements presented here, and compared to previous literature values from Hagege *et al.* (1968) and Öberg *et al.* (2009). In the future, we will maximize radical production and find the optimal position for the laser photolysis to occur on the tube so that additional reactions are minimized. We will then finalize our measurements and determine quantitative branching ratios. To do so, we will further investigate the spectrum of hydroxymethyl radical and obtain the line strength information needed for a proper Boltzmann diagram analysis. We will also acquire a higher frequency multiplier and obtain the branching fraction for the CH_3 + OH channel. Once the experiment has been completed at 193 nm, we will measure the branching ratios using other UV wavelengths. These results can be used to refine models for chemistry on the surface of interstellar grains, and provide insight into the observations obtained by astronomers.

References

Horn, A., Møllendal, H., Sekiguchi, O., Uggerud, E., Herbst, E., Viggiano, A. A., & Fridgen, T. D. 2004, *ApJ*, 611, 605

Gibb, E. L., Whittet, D. C. B., Boogert, A. C. A., & Tielens, A. G. G. M. 2004, *ApJ* (Supplements), 151, 39

Garrod, R. T., Widicus Weaver, S. L., & Herbst, E. 2008, *ApJ*, 682, 283

Garrod, R. T. 2013, *ApJ*, 765, 60

Laas, J. C., Garrod, R. T., Herbst, E., & Widicus Weaver, S. L. 2011, *ApJ*, 728, 71

Hagege, J., Roberge, P. C., & Vermeil, C. 1968, *Trans. Faraday Soc.*, 64, 3288-3299

Öberg, K. I., Garrod, R. T., van Dishoeck, E. F., & Linnartz, H. 2009, *A&A*, 504, 891-913

McCabe M. N. 2016, *Ph.D. Thesis*, Emory University.

Bermudez, C., Bailleux, S., & Cernicharo, J. 2017, *A&A*, 598, A9

Astrochemistry VII – Through the Cosmos from Galaxies to Planets
Proceedings IAU Symposium No. 332, 2017
M. Cunningham, T. Millar & Y. Aikawa, eds.

doi:10.1017/S174392131700686X

Dust formation at cryogenic temperatures

Thomas Henning[1], Cornelia Jäger[2], Gaël Rouillé[2], Daniele Fulvio[3] and Serge A. Krasnokutski[2]

[1]Max Planck Institute for Astronomy, Königstuhl 17, 69117 Heidelberg, Germany
email: henning@mpia.de

[2]Laboratory Astrophysics Group of the Max Planck Institute for Astronomy at the Friedrich Schiller University Jena, Institute of Solid State Physics, Helmholtzweg 3, 07743 Jena, Germany

[3]Departamento de Física, Pontifícia Universidade Católica do Rio de Janeiro, Rua Marquês de São Vicente 225, 22451-900 Gávea, Rio de Janeiro, RJ Brazil

Abstract. The efficiency of dust formation in a variety of environments is an ongoing topic for discussions, especially if it comes to dust formation in the interstellar medium. Although this possibility is discussed in a wide range of numerical studies, experiments on the formation of dust at low densities and temperatures are mostly lacking. This contribution summarizes the main findings of our low-temperature condensation experiments including the formation of silica, complex silicates with pyroxene and olivine stoichiometry, and of carbonaceous refractory materials. Atomic and molecular species to be expected as products of supernovae shock fronts were produced by laser ablation of silicates and graphite. These species were deposited together with a rare gas on cold substrates representing the surfaces of surviving dust grains in the interstellar medium. After characterizing the precursor species, the rare gas matrix was annealed to induce diffusion and reactions between the initial components. We found the production of amorphous and homogeneous silica and magnesium iron silicates at temperatures of about 12 K in a barrierless reaction as monitored by infrared spectroscopy. The 10 μm band of the low-temperature siliceous condensates shows a striking similarity to the 10 μm band of interstellar silicates. Carbonaceous atoms and molecules can also react without a barrier and form an amorphous or hydrogenated amorphous carbon material. The refractory condensate has properties comparable to fullerene-like carbon grains formed at high temperatures.

Keywords. Cosmic Dust, Dust Formation, Cosmic Silicates, Laboratory Data

---

1. Introduction

Asymptotic giant branch (AGB) stars and supernovae (SNe) have been considered as the main dust factories in our and other galaxies. However, the freshly formed stardust is destroyed by SNe shocks. Models of the multi-phase interstellar medium (ISM) indicate that only a small fraction of the stardust survives the destructive processes and dust should be reformed efficiently in the denser phases of the ISM (Zhukovska, Gail & Trieloff 2008, Draine 2009, Zhukovska & Henning 2013). The importance of dust growth by the accretion of gas-phase metals was also demonstrated by recent cosmological simulations (McKinnon, Torrey & Vogelsberger 2016, Zhukovska *et al.* 2016). Aoyama *et al.* (2017) investigated the time evolution and spatial distribution of dust in an isolated galaxy and found that small grain production by shattering triggers accretion and coagulation. This process becomes significant at ages of about 1 Gyr after a large amount of small grains are produced by shattering. Michałowski (2015) derived dust yields for AGB stars and SNe required to explain dust masses of galaxies at z = 6.3–7.5 for which dust emission has been detected. The detected dust mass cannot be produced by AGB stars and SNe and suggests an efficient grain growth process in the ISM at these early epochs.

In the ISM, refractory atoms and molecules generated by the erosion of grains in SNe shocks can be re-accreted onto surviving grains. This process is thought to occur at low temperatures roughly between 10 and 20 K. Since all circumstellar and SNe-formed dust grains are mixed in the ISM, interstellar destruction processes such as sputtering and grain-grain shattering should produce carbonaceous and siliceous molecular and atomic precursor species. In order to provide an experimental basis for dust formation in the ISM, we performed dedicated laboratory experiments with the goal to study the condensation of major dust components including silicates and carbonaceous grains. We summarize the experimental setup and discuss the results of low-temperature condensation in differently doped cryogenic matrices.

2. Experimental methods

Laser ablation of cosmic dust analogs was used to simulate the destructive effect of SNe-induced shocks on siliceous and carbonaceous refractory dust grains. The resulting small molecules and atoms act as the precursors of dust formation. Polycrystalline graphite and amorphous silicates such as Mg_2SiO_4 (forsterite stoichiometry) and $Mg_{0.4}Fe_{0.6}SiO_3$ (pyroxene stoichiometry) synthesized by melting and quenching were used as targets for laser ablation. The experiment was carried out using a pulsed laser source emitting photons with a wavelength of 532 nm corresponding to a power density of 2 to 4 GW cm^{-2}. In a first step, the vaporized species were isolated in Ar or Ne matrices and deposited on cold KBr substrates (10 K) simulating the dust surface in space. Infrared (IR) and UV/VIS (ultraviolet/visible) spectra were recorded in situ providing the opportunity to identify the laser-vaporized species.

In a second step, the matrices were annealed by increasing the temperature ($\Delta T = 4$–6 K). The annealing process is performed in such a way that most of the matrix material remains intact. The annealing was applied to trigger the diffusion of species within the rare-gas matrix leading to collisions between the precursor species. After only a few minutes, the matrix was cooled down again to perform UV/VIS and IR spectroscopy and to monitor the formation of larger species or refractory grains. The final phase of the experiment consists of the heating of the substrate, the evaporation of the matrix, and the characterization of the material located on the substrate. The characterization of the produced refractory material was performed by high-resolution electron microscopy (HRTEM), and energy-dispersive X-ray (EDX) spectroscopy. The entire experiment is schematically described in Figure 1. Details of the experiments are summarized by Rouillé *et al.* (2013), Krasnokutski *et al.* (2014), Rouillé *et al.* (2014), and Fulvio *et al.* (2017).

Another experimental approach to study chemical reactions and the condensation of refractory materials at temperatures lower than 10 K is provided by helium droplets doped with a variety of species. The He droplet setup was described by (Krasnokutski, Rouillé & Huisken (2005)). The experiment consists of three differentially pumped chambers. In the source chamber, large He clusters are generated by supersonic expansion of He gas at $p = 20$ bar through a liquid He-cooled pinhole nozzle. In the second chamber, the He droplets can be doped with astrophysically relevant species. The gas-phase species are picked up by the He nanodroplets and carried to the third chamber, which is equipped with an electron bombardment ionizer and a quadrupole mass spectrometer. The temperature of the He droplets is 0.37 K and reactions inside occur under equilibrium conditions. The liquid He droplets serve as the cold reaction surface. The reaction energy between individual species can be experimentally determined using a calorimetric

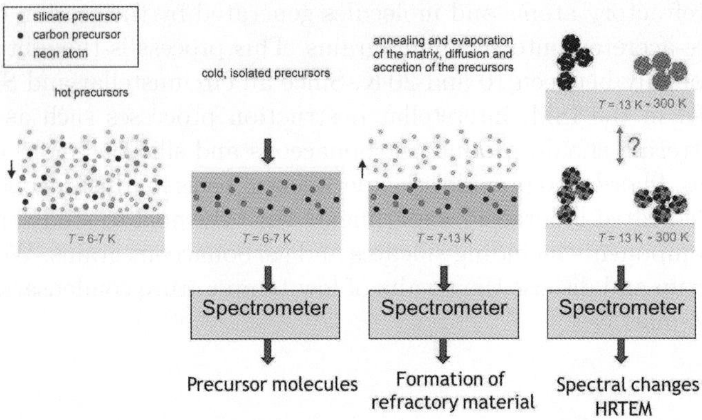

Figure 1. A sketch of the experimental procedure applied to study the cold condensation of silicates, carbonaceous grains, and mixed systems.

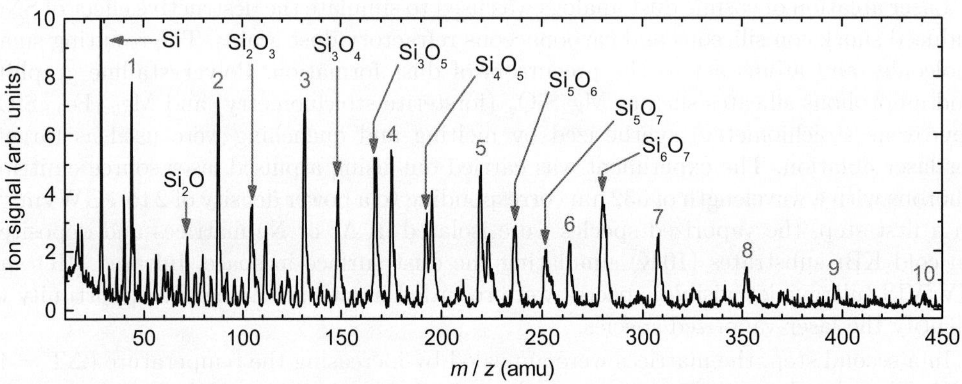

Figure 2. Mass spectrum of He droplets doped with SiO and H_2O molecules. The mass peaks corresponding to $(SiO)_k$ clusters are labeled with the corresponding index k. The data are from Krasnokutski *et al.* (2014).

approach in which the amount of evaporated He atoms can be converted into the reaction energy (Krasnokutski *et al.* 2014).

3. Low-temperature condensation of SiO_2 and complex silicates

SiO molecules are the main precursors for silicates in evolved stars (Gail & Sedlmayr 1999) and SiO_2 represents a major component of the solid silicates. Therefore, the experimental work primarily focused on the condensation of solid SiO_x materials at low temperature using the He droplet technique and the matrix approach. Figure 2 presents the mass spectrum recorded after doping He droplets with SiO and H_2O molecules. The mass peaks are attributed to various Si_xO_y clusters, including $(SiO)_k$ clusters. The energies of SiO polymerization reactions have been determined experimentally with the calorimetric method and compared to calculations based on the density functional theory. Both results confirmed the formation of cyclic $(SiO)_k$ ($k = 2-3$) clusters inside the He droplets. The SiO molecules react without an energy barrier and thus, oligomerization and formation of large molecules were found to be efficient. Such reactions can be expected on the surface of cold dust grains.

The results obtained in He droplets were confirmed by condensation studies in Ne matrices at around 10K. UV/VIS and IR spectroscopy were performed of a SiO-doped

Ne matrix at different annealing temperatures. $(SiO)_x$ oligomers ($x = 1$–3) identified in the cryogenic matrix disappeared during annealing of the Ne matrix up to 11 K and a broad band at about 9.5 μm characteristic of Si–O stretching vibrations in solid SiO_2 appeared. HRTEM and EDX spectroscopy revealed a homogeneous, amorphous structure and a SiO stoichiometry.

Apart from the simple SiO_x system, the successful condensation of complex silicates with pyroxene, olivine, and nonstoichiometric composition at low temperatures was also experimentally studied in cryogenic matrices. Two different silicates corresponding to Mg_2SiO_4 and $Mg_{0.4}Fe_{0.6}SiO_3$ (olivine and pyroxene) were used to produce refractory molecular and atomic species.

Spectroscopy in the UV/VIS and IR range was employed to identify the deposited species in the cryogenic matrices. Fe and Mg atoms were detected in the UV range demonstrating an efficient formation of Mg and Fe from the silicate targets. IR absorption bands point to the presence of O_3 (1039 cm^{-1}), O_4^+ (1164 cm^{-1}), and SiO (1228 cm^{-1}) in the matrix. Absorption features due to a complex possibly involving H_2 and H_3O^+ or $H_2O_5^+$ were also detected (Jacox & Thompson 2013). Based on comparisons with measurements on species isolated in Ar matrices, we assigned tentatively the bands at 1369 and 1424 cm^{-1} to SiO_3 and SiO_2, respectively (Tremblay *et al.* 1996). In addition, FeO (870 cm^{-1}) (Green, Reedy & Kay 1979) and OFeO (950 cm^{-1}) (Chertihin *et al.* 1996) could be identified.

IR spectroscopy is the most efficient method to monitor the condensation of silicates in the ice as a function of the temperature. Figure 3a shows the spectra of the material obtained by the condensation of species produced in the laser ablation of the Mg_2SiO_4 target. The spectrum was measured after the annealing of the doped Ne matrix to temperatures of 13 K and subsequent cooling to 6.5 K. Beside the features due to CO, CO_2, and H_2O, a band at 1000 cm^{-1} was recorded that is characteristic for a silicate of amorphous structure. The profile of this band is slightly affected by the presence of the water ice feature at around 780 cm^{-1} (Öberg *et al.* 2007). After warming up to room temperature, the position of the Si–O absorption band could be found at $\approx$1020 cm^{-1} (9.8 μm). Similar results were obtained with the $Mg_{0.4}Fe_{0.6}SiO_3$ target.

Figure 3b compares the IR spectra of the condensates obtained from the pure Mg- and the Mg-Fe-silicate targets to the interstellar silicate absorption feature. The Mg-Fe silicate shows a band at $\approx$990 cm^{-1} (10.1 μm). Similar band positions demonstrate that both condensates must have comparable stoichiometry even though different targets were used. The EDX analysis revealed the final non-stoichiometric compositions $Mg_{0.3}Fe_{0.9}SiO_{3.6}$ and $MgSiO_{2.8}$ for the condensates. This behavior was caused by a non-equal evaporation of Mg and Fe from the original silicate targets during the laser vaporization. HRTEM images of the $Mg_{0.3}Fe_{0.9}SiO_{3.6}$ condensate show fluffy aggregates composed of nanometer-sized, amorphous grains (see Fig. 4). This structure is also typical of the condensed $MgSiO_{2.8}$ material. No phase separations into individual oxides such as FeO, MgO, and SiO_2 are observed.

4. Low-temperature condensation of carbonaceous materials

The formation of a carbonaceous refractory condensate from molecular precursors was observed after deposition on a cold substrate at about 10 K (with or without Ar matrix). The isolation of the laser ablated carbonaceous species in an Ar matrix provided information on the precursor molecules that eventually form the refractory material. Beside the presence of species such as H_2O, CO, CO_2, C_3O, water clusters, various isolated odd- and even-numbered carbon chains were detected in the UV/VIS and IR ranging from

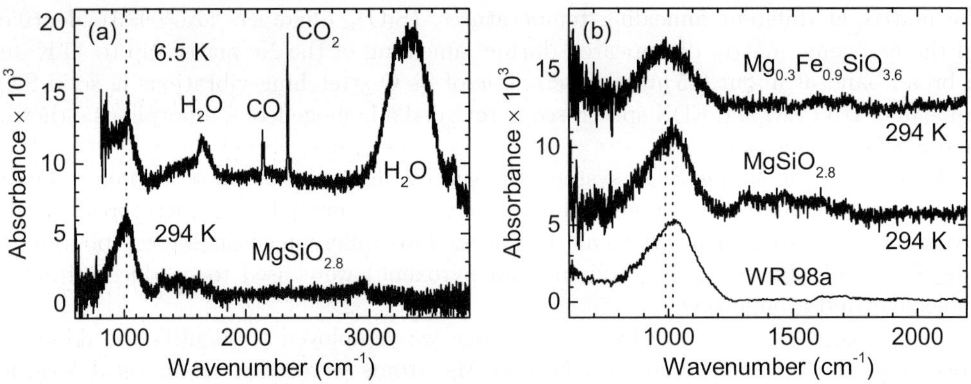

Figure 3. (a) IR spectra of the $MgSiO_{2.8}$ condensate. The top spectrum was measured at 6.5 K and the bottom spectrum was recorded at room temperature after sublimation of ices. (b) IR features of the silicate condensates obtained with the Mg_2SiO_4 and $Mg_{0.4}Fe_{0.6}SiO_3$ targets compared to the normalized silicate absorption feature toward WR 98a representing the local ISM (Chiar & Tielens 2006). The spectra were vertically shifted for clarity. The data are from Rouillé *et al.* (2014).

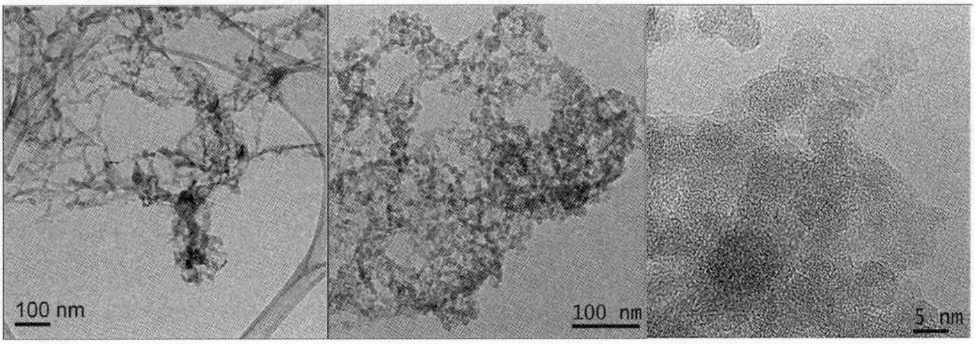

Figure 4. HRTEM image of the $Mg_{0.3}Fe_{0.9}SiO_{3.6}$ condensate prepared by the evaporation and recondensation of a magnesium iron silicate at about 10 K. The right image allows a direct view insight the amorphous grains whereas the two left images show the fluffy morphology of the condensate.

C_2 up to C_{13}. After the warming up of the cryogenic ice layer to room temperature, a refractory residue remained on the substrate.

Simultaneous deposition of carbonaceous species by laser ablation of graphite and molecular hydrogen in the Ar ice layer led to the formation of small hydrocarbon molecules such as C_2H_2, C_4H_2, $C_2H_2 \cdot H_2O$.

The carbonaceous species in pure and hydrogen-containing Ar matrices react between each other without an energy barrier resulting in the formation of a refractory carbonaceous material. IR spectra of the carbonaceous condensate at different temperatures are depicted in Fig. 5. Already at a low temperature of about 10 K, a decrease of the IR transmission with increasing ablation time can be observed. This is caused by the growth of a refractory carbonaceous film, which is characterized by free charge carriers responsible for a continuous absorption from the UV up to the MIR and weak scattering effects at small wavelengths. The increasing extinction is also accompanied by the occurrence of a broad plateau band between 1000 and 1400 cm^{-1}. The formation of such a plateau is a feature of amorphous and hydrogenated amorphous carbon samples prepared by gas-phase condensation techniques (Jäger, Mutschke & Henning 1998, Schnaiter *et al.* 1998,

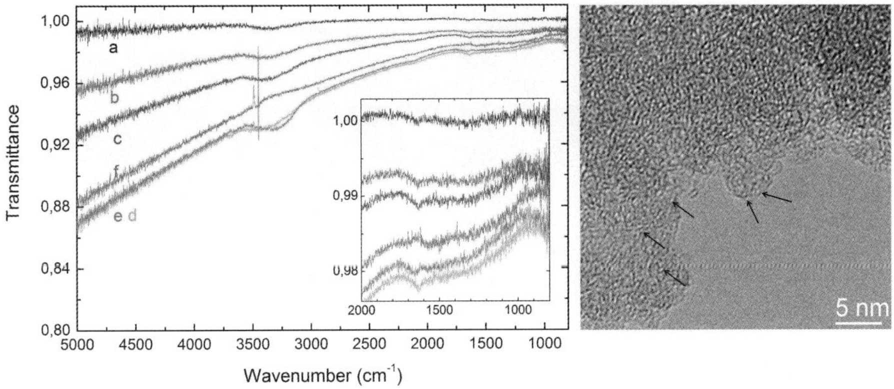

Figure 5. Left: Spectral evolution of the carbonaceous condensate formed at 10 K with increasing deposition time and after warming up to 200 K. Curves a, b, c, and d correspond to 5, 15, 25, and 35 min laser ablation at about 10 K. Curves e and f represent the warming up to 40 K and 200 K, respectively. Right: HRTEM micrograph of the internal structure of the condensed carbon material in Ar matrix. Fullerene-like carbon structures are visible such as small bucky onions and strongly bent graphene layers. Individual fullerene molecules of various sizes and shapes are identified and marked with arrows. The data are from Fulvio *et al.* (2017).

Llamas-Jansa *et al.* 2007). The broad absorption is a mixture of vibrational bands of various functional groups such as ν_{C-O-C}, ν_{C-C}, and δ_{C-H}. More details on the identification of individual species in the Ar ice matrix are provided in Fulvio *et al.* (2017).

The condensation of carbonaceous species at temperatures around 10 K provide similar fullerene-like carbon structures (see Fig. 5) as produced in high temperature condensation processes (Jäger *et al.* 2009). The presence of differently sized cage molecules, strongly bent graphene layers, and individual fullerenes can be observed in the images. The high-temperature formation pathway was initiated by the formation of chain-like structures that are formed as intermediates and finally convert into cage-like structures including fragments and completely closed cages. At low temperatures, the condensation process is also governed by carbon chains that grow and finally convert into more stable aromatic structures with the characteristic curved graphene layers.

5. Low-temperature condensation of of siliceous and carbonaceous materials

One of the challenges for interstellar dust models is the question if silicates and carbonaceous grains are two distinct dust populations or form Si-C-based compounds. Laboratory experiments to investigate the formation of refractory components in matrices containing both types of species were started. First experiments were dedicated to study possible reactions between carbonaceous and SiO_x species. Two different targets were used for the laser ablation process. SiO and carbon species such as C_2, C_3, C_n were deposited in a Ne matrix. After activation of the diffusion by moderate annealing to 9-11 K, we observed the rise of a band at 10 μm. This clearly indicates the appearance of a refractory SiO_x condensate in the remaining Ne ice. No hints on the formation of other components could be identified spectroscopically. After warming up, the condensate was studied by HRTEM and EDX analysis. A solid, refractory SiO component was found to be spatially separated from carbonaceous grains (see Fig. 6).

The results clearly point to a separate formation of carbonaceous and siliceous refractory materials. The formation of SiC was not observed in the condensate. Yet, small

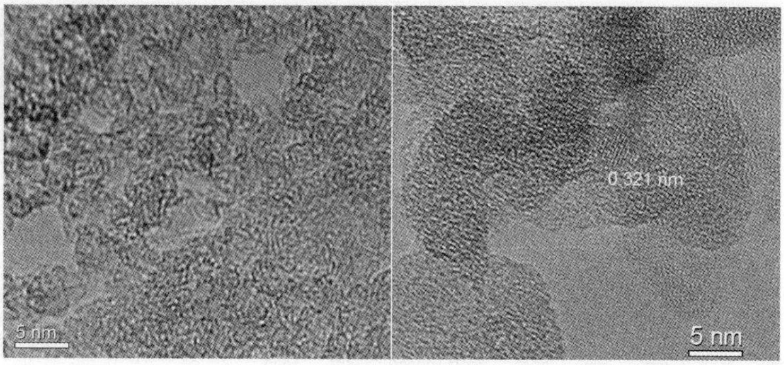

Figure 6. HRTEM images of carbonaceous (left) and SiO_x grains (right) condensed in a Ne ice containing siliceous and carbonaceous species prepared by simultaneous laser ablation of a graphite and a SiO target. Both materials are characterized by an amorphous structure. The silicate on the right sight contains a few small crystalline inclusions of about 2 nm that were identified as silicon based on the distances between their lattice fringes.

inclusions of silicon (~2 nm) were detected in the SiO_x grains. These silicon inclusions were probably formed by the reductive effect of the carbon species on SiO resulting in a oxygen-deficient SiO material. Further experiments on the simultaneous condensation of complex silicates and carbonaceous refractory material are still under progress but all results so far confirm the separation of silicates and carbonaceous grains.

6. Astrophysical implications and conclusions

In the denser ISM, gaseous species accrete on cold surfaces of pre-existing small particles. To simulate such processes in the laboratory, we have evaporated dust analogs that have successfully been used to model spectral energy distributions of many astrophysical sources (Gielen *et al.* 2011, Min *et al.* 2007). Important precursors for silicates such as Si, SiO, FeO, Mg, and Fe produced in our experiments were already observed in the ISM (Schilke *et al.* 1997, Walmsley *et al.* 2002, Furuya *et al.* 2003).

We detected a barrierless formation of amorphous and homogeneous silica and magnesium iron silicates at temperatures of about 12 K. The morphology and structure of the condensates are remarkably similar to those silicates produced by gas-phase condensation at high temperature. Both types of condensates are of amorphous structure and show very similar spectral properties comparable to the interstellar 10 μm absorption band. Distinct structural and spectral differences between high- and low-temperature condensates could not be identified.

Carbonaceous atoms and molecules also react without a barrier and form an amorphous or hydrogenated amorphous carbon material. One of the amazing findings of this study is the observation of fullerene-like carbon structures condensing at temperatures of about 10 K. The chemical evolution into fullerene-like carbon materials containing fullerene molecules of different sizes and shapes indicates a formation pathway via long and branched carbon chain molecules that eventually convert into aromatic structures even at temperatures around 15 K (Yamaguchi & Wakabayashi 2004). Such a process was predicted and observed for high-temperature condensation processes by Cherchneff *et al.* (2000) and by Jäger *et al.* (2009).

Experiments on the doping of cryogenic matrices with both carbonaceous and siliceous species show the condensation of distinct carbonaceous and silicate grains. No silicon carbide was detected. The formation of small crystalline Si inclusions could be observed in the SiO_x grains. These were formed by the reduction of silicon with carbon species leading to the oxidation of small carbon clusters or molecules.

References

Aoyama, S., Hou, K.-C., Shimizu, I., Hirashita, H., Todoroki, K., Choi, J.-H., & Nagamine, K. 2017, *MNRAS*, 466, 105

Draine, B. T. 2009, in: Th. Henning, E. Grün & J. Steinacker (eds.), *Cosmic Dust Near and Far*, ASP Conf. Ser. Vol. 414 (San Francisco: ASP), p. 453

Cherchneff, I., Le Teuff, Y. H., Williams, P. M., & Tielens, A. G. G. M. 2000, *A&A*, 357, 572

Chertihin, G. V., Saffel, W., Yustein, J. T., Andrews, L., Neurock, M., Ricca, A., & Bauschlicher, C. W. 1996, *J. Phys. Chem.*, 100, 5261

Chiar, J. E. & Tielens, A. G. G. M. 2006, *ApJ*, 637, 774

Fulvio, D., Gobi, S., Jäger, C., Kereszturi, A., & Henning, Th. 2017, *ApJS*, 233, 14

Furuya, R. S., Walmsley, C. M., Nakanishi, K., Schilke, P., & Bachiller, R. 2003, *A&A*, 409, L21

Gail, H.-P. & Sedlmayr, E. 1999, *A&A*, 347, 594

Gielen, C., Bouwman, J., Van Winckel, H., Lloyd Evans, T., Woods, P. M., Kemper, F., Marengo, M., Meixner, M., Sloan, G. C., & Tielens, A. G. G. M. 2011, *A&A*, 533, A99

Green, D. W., Reedy, G. T., & Kay, J. G. 1979, *J. Mol. Spectrosc.*, 78, 257

Jacox, M. E. & Thompson, W. E. 2013, *J. Phys. Chem. A*, 117, 9380

Jäger, C., Huisken, F., Mutschke, H., Llamas Jansa, I., & Henning, Th. 2009, *ApJ*, 696, 706

Jäger, C., Mutschke, H., & Henning, Th. 1998, *A&A*, 332, 291

Krasnokutski, S., Rouillé, G., & Huisken, F. 2005, *Chem. Phys. Lett.*, 406, 386

Krasnokutski, S. A., Rouillé, G., Jäger, C., Huisken, F., Zhukovska, S., & Henning, Th. 2014, *ApJ*, 782, 15

Llamas-Jansa, I., Jäger, C., Mutschke, H., & Henning, Th. 2007, *Carbon*, 45, 1542

McKinnon, R., Torrey, P., & Vogelsberger, M. 2016, *MNRAS*, 457, 3775

Michałowski, M. 2015, *A&A*, 577, A80

Min, M., Waters, L. B. F. M., de Koter, A., Hovenier, J. W., Keller, L. P., & Markwick-Kemper, F. 2007, *A&A*, 462, 667

Öberg, K. I., Fraser, H. J., Boogert, A. C. A., Bisschop, S. E., Fuchs, G. W., van Dishoeck, E. F., & Linnartz, H. 2007, *A&A*, 462, 1187

Rouillé, G., Jäger, C., Krasnokutski, S. A., Krebsz, M., & Henning, T. 2014, *Faraday Discuss.*, 168, 449

Rouillé, G., Krasnokutski, S. A., Krebsz, M., Jäger, C., Huisken, F., & Henning, T. 2013, in: A. Andersen, M. Baes, H. Gomez, C. Kemper & D. Watson (eds.), *The Life Cycle of Dust in the Universe: Observations, Theory, and Laboratory Experiments*, PoS(LCDU2013) (Proceedings of Science), p. 47

Schilke, P., Walmsley, C. M., Pineau des Forêts, G., & Flower, D. R. 1997, *A&A*, 321, 293

Schnaiter, M., Mutschke, H., Dorschner, J., Henning, Th., & Salama, F. 1998, *ApJ*, 498, 486

Tremblay, B., Roy, P., Manceron, L., Alikhani, M. E., & Roy, D. 1996, *J. Chem. Phys.*, 104, 2773

Walmsley, C. M., Bachiller, R., Pineau des Forêts, G., & Schilke, P. 2002, *ApJ*, 556, L109

Yamaguchi, Y. & Wakabayashi, T. 2004, *Chem. Phys. Lett.*, 388, 436

Zhukovska, S., Dobbs, C., Jenkins, E. B., & Klessen, R. S. 2016, *ApJ*, 831, 147

Zhukovska, S., Gail, H.-P., & Trieloff, M. 2008, *A&A*, 479, 453

Zhukovska, S. & Henning, T. 2013, *A&A*, 555, A99

Astrochemistry VII – Through the Cosmos from Galaxies to Planets
Proceedings IAU Symposium No. 332, 2017
M. Cunningham, T. Millar & Y. Aikawa, eds.

doi:10.1017/S1743921317006950

Laboratory astrochemistry: catalytic reactions of organic molecules over olivine-type silicates and SiC

Qian Li[1], B. S. Liu[2], P. J. Sarre[3] and A. S-C. Cheung[1]

[1]Department of Chemistry, The University of Hong Kong, Pokfulam Road, Hong Kong, China.
email: hrsccsc@hku.hk

[2]Department of Chemistry, Tianjin University, Tianjin 300072, China.
[3]School of Chemistry, The University of Nottingham, University Park, Nottingham NG7 2RD, United Kingdom

Abstract. A series of catalytic reactions has been performed in our laboratory using olivine-type silicates (OTS) and SiC as catalysts for the conversion of carbon-containing molecules (such as acetylene, CO and methanol) to small organic molecules (C_2H_4, C_3H_3, CH_3O) and also polycyclic aromatic hydrocarbons (PAHs). Experimentally, small-to-medium-sized gas-phase compounds such as PAHs, reaction intermediates and hydrocarbon compounds were detected in situ using the time-of-light mass-spectrometry technique. Solid deposition on the catalyst surface was examined by high-resolution transmission electron microscopy and thermo-gravimetric analysis techniques. Our laboratory results show that the conversion of acetylene to PAHs, the CO disproportionation reaction for producing CO_2 and carbon deposition (graphitic and carbon nanostructures), and also the transformation of methanol to hydrocarbon compounds can easily be achieved with OTS as a catalyst. Furthermore, the conversion of acetylene to PAHs could also be achieved by SiC as the catalyst. It is proposed that these catalytic reactions mimic similar chemical processes in circumstellar envelopes (CSEs).

Keywords. Catalytic reaction, surface chemistry, dust grains: silicon carbide, olivine-type silicates.

---

1. Introduction

Astrochemistry is a subject that blends astronomy and chemistry in which each area of research mutually enhances and enriches the other in the interaction. Chemical processes in interstellar and circumstellar regions may be classified into two categories: gas-phase reactions and surface reactions. For general gas-phase reactions, when atoms and molecules in the gas-phase collide, and can overcome any activation barrier, product species can be formed. In low temperature regions ($T \sim 10$ K), chemical reactions are triggered by cosmic-ray ionization and proceed mainly via ion-molecule reactions in the gas-phase (Öberg. 2016). It is essential to have fundamental chemical knowledge of molecular collisions, and photochemical and reaction dynamics processes to understand such reactions in the universe. For grain surface reactions, atoms and molecules that strike grains and are efficiently adsorbed onto the grain surface. As far as the adsorption processes are concerned, both physisorption and chemisorption can in principle occur. Physisorbed species can migrate on a grain surface and react when they meet. Chemisorbed species, on the other hand, stick tightly to the surface even at temperatures of several hundred Kelvin, and this suppresses any diffusion process. The strong interaction between adsorbed species and the surface can eventually lead to cleavage of chemical bonds of the adsorbed species and that proceeds as an initial step of a catalytic reaction. There has

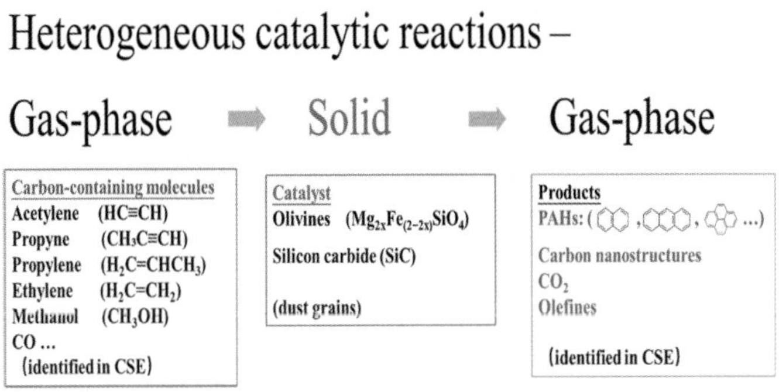

Figure 1. Schematic of heterogeneous catalytic processes of interest in our laboratory.

been some work concerning the study of physisorbed species on grain surfaces that are relevant to astrochemical processes (Pirronello *et al.* 1997, Herbst *et al.* 2005). However, only a handful of papers reported chemisorption processes or catalytic reactions that are of importance in astrophysical environments (Kress *et al.* 2001, Ferrante *et al.* 2000).

We have worked on heterogeneous catalytic reactions that are of significance in astrochemistry and explore new connections between circumstellar gas-phase molecules and grains. Recent studies of the circumstellar envelopes (CSEs) of evolved stars indicate that gaseous materials, the high near-zone temperature and high molecular density environment support a rich and complex chemistry (Ziurys. 2006). More than sixty different organic chemical compounds have been identified in the CSEs of stellar objects and many of them are exotic species with more than eight atoms. As far as circumstellar dust grains are concerned, there is evidence indicating the existence of olivine-type silicate (OTS) (forsterite, olivine, pyroxene) dust grains in oxygen-rich envelopes, and for the carbon-rich envelopes, silicon carbide (SiC) grains are an important component (Draine. 2003, Henning. 2010). The physical conditions in the circumstellar shell: the inner CSE is warm (the temperature can be up to 1500 K), and the number density of molecules is also high (around $n \approx 10^{10}$ cm^{-3}). With the presence of carbon-containing chemicals, solid grains as catalysts and the right physical conditions in a CSE, it is expected that catalytic reactions would be of significance in such astrophysical environments. Figure 1 shows schematically a summary of the catalytic processes studied in our laboratory with relevance to astrochemistry. Carbon-containing gas-phase molecules, catalysts and product species have all been identified in CSEs. In order to understand better the conversion reactions of organic molecules in circumstellar environments, this work is conducted in a hypothetical-deductive model of reasoning. The coexistence of the reagent, catalyst and product species suggests that there could indeed be a connection between them.

2. Experimental

2.1. *Characterization of catalysts*

The catalysts used in this work are solid SiC and olivine-type silicates. A range of characterization techniques were used to examine the catalysts before and after catalytic reaction (Ertl *et al.* 2008). The structure of the crystalline solid material and the size of the crystals were determined using X-Ray Diffraction (XRD). In addition, energy dispersive X-ray (EDX) spectroscopy was employed to measure the elemental composition and undertake chemical characterization of the catalyst. The organic molecules adsorbed on the catalytic surface were observed and analyzed using Fourier transform infrared (FT-IR) spectroscopy. In this work, various different forms of solid carbon deposition

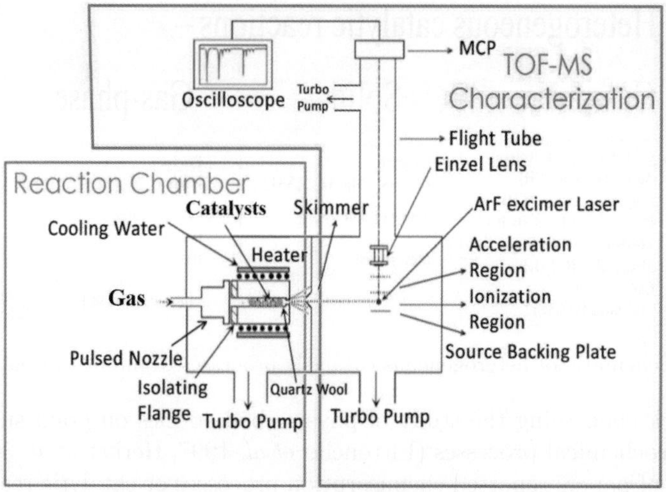

Figure 2. Schematics of the TOF-MS instrument used in this work.

on the catalytic surface were commonly observed. High-resolution transmission electron microscopy (HRTEM) was employed to examine the carbon deposits. The HRTEM images, with a spatial resolution below 2Å provided information on the structure of the deposited material. Furthermore, various thermal techniques such as thermal gravimetric analysis (TGA) and temperature-programmed desorption (TPD) proved to be useful in identifying the different forms of carbon deposition on the catalyst surfaces.

2.2. *Apparatus and experimental condition*

A schematic of the time-of-flight mass-spectrometer (TOF-MS) used for these studies is shown in Fig. 2 (Zhao *et al.* 2016). This homebuilt instrument consists of a source chamber with a quartz tube fixed-bed reactor, an ionization chamber, a TOF tube and a detection chamber. The feeding device is connected to the reactor and reactant gas is introduced via an electronically-controlled pulsed valve. A skimmer with a small aperture is used to separate the source chamber and ionization chamber, the background vacuum pressure of the chambers being typically 2×10^{-6} Torr. A two-stage TOF lens assembly (including source backing plate, ionization and acceleration region) and an ion-beam focusing Einzel lens setup are located inside the ionization chamber, where ionization of the gaseous species is achieved through multi-photon processes. The detection chamber is connected to the TOF tube and a multi-channel plate (MCP) detector is used to detect the arrival of ions. For each experiment, the same amount of catalyst is placed inside the quartz tube reactor. The reagent gas is introduced in pulsed mode to the reactor at a repetition rate of 10 Hz with a pulse duration of around 250-300 microsecond. The catalytic reaction was studied in the 100 to 750°C temperature range. The gaseous reagent, intermediates and products from the reaction are all injected directly into the ionization chamber through the skimmer and ionized by a 193 nm ArF excimer laser to generate molecular ions, which are then accelerated by high voltage and travel through a standard TOF field-free region. The signals recorded by the MCP detector are forwarded to a 300 MHz oscilloscope for averaging.

In order to further confirm the catalytic reactions and to obtain a sufficient amount of solid deposit, separate experiments using a continuous flow of reagent molecule under atmospheric pressure were conducted in a separate apparatus. A quartz tube reactor is housed in a furnace and the catalyst placed inside the quartz tube. After the catalyst

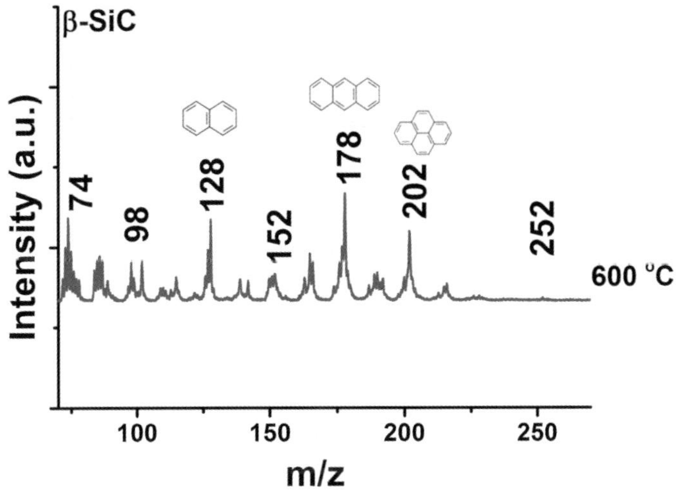

Figure 3. The formation of PAHs over SiC at 600°C.

is pretreated in argon, reagent gas is allowed to flow for a short period of time, and the effluent gas after passing through the catalyst is directed to go through a cold trap to condense any volatile compounds. The trapped species and the solid carbon deposits are, subsequently, analyzed using gas chromatography-mass spectrometry (GC-MS).

3. Results and discussion

We have studied catalytic reactions that are of significance in astrophysical environments, including acetylene on SiC, carbon monoxide disproportionation on OTS and methanol on OTS. The individual experiments conducted are summarized below:

3.1. *Catalytic conversion of acetylene to PAHs over SiC Grains*

Polycyclic aromatic hydrocarbons (PAHs) play an important role in astrochemistry. Despite the fact that the formation chemistry of PAHs has been widely discussed, a complete mechanism for cosmic PAH formation is still lacking or disputed. Our experimental work on catalytic conversion reactions of acetylene on a solid SiC grain surface provides an alternative or additional channel for the formation of PAHs. It is found that catalytic conversion reactions of acetylene on a solid SiC grain surface lead to the formation of polycyclic aromatic hydrocarbons (PAHs) and are expected to mimic chemical processes in some astrophysical environments. Gas-phase PAHs and intermediates were detected in situ using TOF-MS and their formation confirmed using GC-MS in separate experiments by flowing acetylene gas through a fixed-bed reactor. The formation of PAHs was found to be optimum around 600 °C. Figure 3 shows the formation of PAHs at this temperature (Zhao *et al.* 2016).

3.2. *Catalytic CO disproportionation reactions over olivine-type-silicates*

The catalytic carbon monoxide disproportionation (COD) reaction over OTS leads to formation of carbon dioxide gas and solid carbon allotropes on the surface of catalyst (Dai *et al.* 1996). Figure 4 illustrates the breaking down of CO and the formation of CO_2. The deposited carbon has nanotube, graphene and graphitic structure. These carbon nanotube and graphitic structures have important implications in astrochemistry, as cleavage can produce lower molecular mass PAHs and mimic chemical processes in

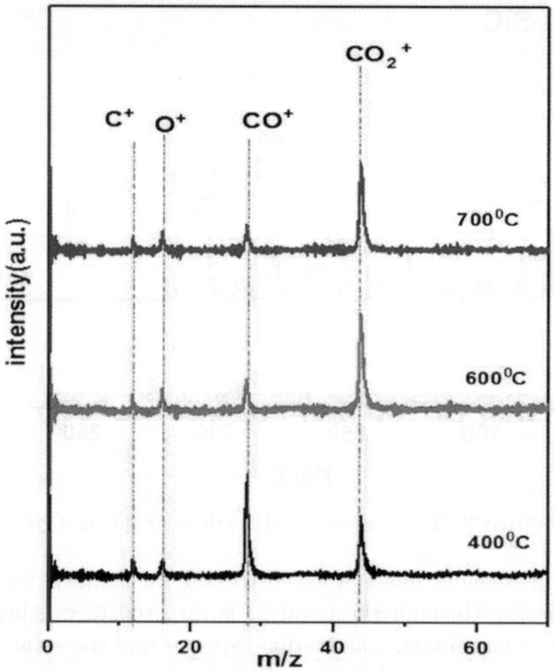

Figure 4. TOF mass spectrum of the COD reaction.

some evolved stellar environments. The catalytic COD reaction thus provides a top-down mechanism for forming smaller PAHs.

3.3. *Catalytic conversion of methanol to PAHs over olivine-type-silicates*

Catalytic reactions of methanol on olivine-type-silicates produce hydrocarbons and organic molecules, which indicates that there are important relationships among these organic molecules. The observed gas-phase molecules include C_2H_4, C_3H_3, CH_3O, CO, naphthalene, anthracene, pyrene etc., their mass-to-charge ratios were confirmed by using deuterated methanol. The reactions provide a link between dust particles and organic molecules, which may have implications for the chemistry in the CSE (Olah *et al.* 2016).

3.4. *Future experiments*

The catalytic reaction experiments performed thus far were with a single gas as reagent (acetylene, carbon monoxide, methanol etc.) on OTS and SiC catalysts. We are also interested in catalytic reactions using initial gas mixtures such as CO/H_2, C_2H_2/CO, CO/NH_3 etc. to examine whether other organic and nitrogen-containing molecules, and potentially prebiotic molecules, can be synthesized from these reactions.

3.5. *Astrophysical implications*

Interest in investigating chemistry involving olivine-type silicates and silicon carbide is due to the fact that the dust grains are ubiquitously present throughout interstellar and circumstellar media and because an important chemical role of cosmic dust particles can be as in terrestrial heterogeneous catalysis, to accelerate certain astrochemical reactions. For example, concerning the high abundance of large interstellar PAHs, a number of theories are based on bottom-up chemical processes initiated through gas-phase reactions or surface reactions (Zhao *et al.* 2016), in which there is a merging of small, possibly radical or ionized, PAH molecules into larger units as a result of collisions. A more recent theory is based on the completely opposite top-down principle, whereby PAHs form in interstellar space from decomposition of graphitic layers covering the surface of the particles leading to release of PAH molecules (Merino *et al.* 2014). In astrophysical environments, surface aromatic materials could indeed be generated initially from catalytic reactions such as those explored here experimentally, and eventually break down into smaller parts thus forming the PAHs that have been observed.

4. Acknowledgements

The work described here was supported by a grant from the University Research Committee of the University of Hong Kong. PJS thanks the Leverhulme Trust for award of a Research Fellowship.

References

Dai, H., Rinzler, G., Nikolaev, P., Thess, A., Colbert, D. T., & Smalley, R. E. 1996, *Chem. Phys. Lett.*, 260, 471

Draine, B. T. 2003 *ARAA*, 41, 241

Ertl, G., Knozinger,H., Schuth, F., & Weitkamp, J. 2008, *Handbook of Hetero. Cat.*, Vol.4

Ferrante, R. F., Moore, M. H., Nuth, J. A., & Smith, T. 2000, *Icarus*, 145, 297

Henning, T. 2010, *ARAA*, 48, 21

Herbst, E., Chang, Q., & Cuppen, H. M. 2005, *J. Phys. Conf. Sers.*, 6, 18

Kress, M. E. & Tielens, A. G. G. M. 2001, *Meteor. & Planet. Sci.*, 36, 75

Merino, P., Svec, M., Martinez, J. I., Jelinek, P., Lacovig, P., Dalmiglio, M., Lizzit, S., Soukiassian, P., Cernicharo, J., & Martin-Gago, J. A. 2014, *Nat. Comm.*,5, 3054

Öberg, K. I. 2016, *Chem. Rev.*, 116, 9631

Olah, G. A., Mathew, T., Prakash, G. K. S., & Rasul, G. 2016, *J. Am. Chem. Soc.*, 138, 1717

Pirronello, V., Biham, O., Liu, C., Shen, L., & Vidali, G. 1997, *ApJ*, 483, L131

Zhao, T. Q., Li, Q., Liu, B. S., Sarre, P. J., & Cheung, A. S.-C. 2016, *PCCP*,18, 3489

Ziurys, L. M. 2016, *Proc. Natl. Acad. Sci. U S A*,103, 12274

Astrochemistry VII – Through the Cosmos from Galaxies to Planets
Proceedings IAU Symposium No. 332, 2017
M. Cunningham, T. Millar & Y. Aikawa, eds.

doi:10.1017/S1743921317009401

Monte Carlo Modeling of Astrophysically-Relevant Temperature-Programmed Desorption Experiments

Aspen R. Clements,[1] Ilsa Cooke,[1] and Robin T. Garrod[1,2]

[1]Department of Chemistry, University of Virginia,
P.O. Box 400319, Charlottesville, VA 22904-4319
email: ac8hq@virginia.edu

[2]Department of Astronomy, University of Virginia,
P.O. Box 400325 Charlottesville, VA 22904-4325

Abstract. The formation of molecules in the interstellar medium is significantly driven by grain chemistry, ranging from simple (e.g. H_2) to relatively complex (e.g. CH_3OH) products. The movement of atoms and molecules on amorphous ice surfaces is not well constrained, and this is a quintessential component of surface chemistry. We show that ice structure created by utilizing an off-lattice Monte Carlo kinetics model is highly dependent on deposition parameters (i.e. angle, rate, and temperature). The model, thus far, successfully predicts the densities of deposition rate- and temperature-dependent laboratory experiments. The simulations indicate, when angle and deposition rate increase, the density decreases. On the other hand, temperature has the opposite effect and will increase the density. We can make ices with desired densities and monitor how molecules, like CO, percolate through H_2O ice pores. The strength of this model lies in the ability to replicate TPD-like experiments by monitoring molecules diffusing on and desorbing from user-defined surfaces.

Keywords. Keyword1, keyword2, keyword3, etc.

1. Introduction

Amorphous solid water (ASW) is a major component of astronomically-important structures, including mantles on interstellar grains, protoplanetary disks, planetary rings, satellites, and comets (e.g. Gibb *et al.* 2004 Oberg *et al.* 2011). Water ice in the ISM is found to be primarily amorphous, or in a disordered state, and can contain large, empty pockets called pores, which may trap species such as CO. The entrapment of CO enhances the production of molecules such as CH_3OH, a molecule potentially important for the formation of prebiotic molecules (Hidaka *et al.* 2008). The ability for molecules to incorporate into the ice is highly dependent on structure. For instance, porous ASW can absorb 20 to 50 times more gas than a compact, non-porous ASW (Ayotte *et al.* 2001). The formation of such an ice structure is dependent on many parameters, including the formation timescale, grain and gas temperatures, deposition angle, and deposition rate. Laboratory ice experiments are used by astrochemists to understand the behavior and formation of ices in the ISM; however, the deposition parameters often differ between the two, leading to cases in which the laboratory ice may not be truly representative of an ice in the ISM.

These ice experiments are essential for the understanding of astrochemistry because they provide insight into the identity of gas- and solid-phase species, the viability of proposed surface and gas reactions, and reaction rates and barriers. Given the utility of these experiments, ensuring they can be extrapolated to ISM ices is essential. For solid-phase chemistry, desorption energies set the temperature in which gas-phase species accrete and stick onto the grains. The method used to find the desorption, or binding, energies is called temperature-programmed desorption (TPD), in which molecules are thermally desorbed from an ice. In these experiments, an ice is deposited onto a surface at a specified temperature, determined by the desired structure (see below); the ice is then heated at a linear rate via an electrical current, and the desorption of molecules is monitored using a mass spectrometer. The TPD profiles, showing desorption rate as a function of temperature, will have unique profiles depending on the ice composition and structure, aiding in the identification of ice components (Collings *et al.* 2004). The interpretations of these results can be overly simplistic, not taking into account the ice structure or porosity and may not reflect processes in the ISM because of the discrepancies in conditions. Given the utility of laboratory experiments, ensuring their extrapolation to ISM ices and processes is a top priority. We attempt to control for various deposition parameters and mimic laboratory experiments with a kinetic Monte Carlo model.

2. Model

The model used in this study is the Model for Interstellar Monte-Carlo Ice Chemical Kinetics, or MIMICK. The predecessor of this model is discussed in more detail in Garrod (2013) . The code has been updated to use lateral periodic conditions to mimic a larger ice. All chemical species are treated as isotropic spheres with radii of 1.6 Å, for simplicity. MIMICK is an off-lattice model, allowing for amorphous surface structures with micro- and macro-pores, as opposed to the always-ordered structures of on-lattice models with pre-determined binding sites (Cuppen *et al.* 2005). Within the model, there are a large number of parameters that can be changed (e.g. deposition angle, rate, and temperature).

Figure 1. The CO diffuses into pores of H_2O until it begins to desorb around 25 K. By 130 K, H_2O has lost large pores and compacted. Experimental conditions from Fayolle *et al.* 2016.

3. Results

Full results and comparisons of models run with varying deposition parameters will be published shortly Clements *et al.* in prep). Preliminary research aimed to to replicate a TPD experiment from Fayolle *et al.* (2016a). The experiment involves depositing H_2O and, subsequently, CO; the chamber is slowly heated and desorbing molecules are monitored. The underlying H_2O ice structure will impact when and how CO molecules diffuse and, eventually, desorb into the gas phase. The laboratory deposition conditions are used in MIMICK to grow ices, with the aim of creating similar ice structures. Growing ices at 90 K and 11 K produces compact and porous ASW, respectively. The layered ices are then heated with a linear ramp rate around 1 K/min. Analogous to the purpose of a mass spectrometer, MIMICK monitors every molecule in the model, including when that leaves the ice surface. TPD profiles are generated and compared to the experimental data; figure 1 depicts the warm-up process within the model, following deposition. The model qualitatively demonstrates the expected changes in ice structure, during a warm-up process, and the diffusion and desorption of CO.

Generated TPD profiles of the described system can be seen in color in Figure 2. Figure 2(a) shows the compact H_2O and topped with CO ice. There is a two-peak structure, implying multiple CO binding energies. The first peak likely denotes CO-CO interactions because CO-CO interactions are expected to be weaker than CO-H_2O interactions.

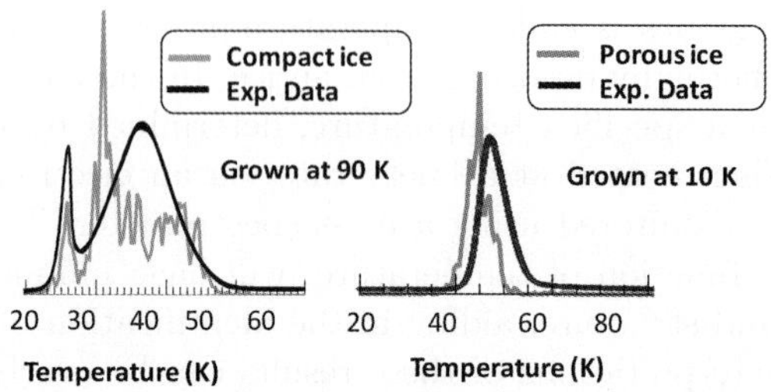

Figure 2. The TPD profiles from the model shown in color with the Fayolle data in black. The green shows the compact water ice and red the porous ice grown at 100 and 11 K, respectively.

Then, the second peak must correspond to the CO-H_2O interaction, likely showing the CO desorbing from within the ASW pores. Figure 2(b) (right) is the porous ice grown (11 K) and only has a one-peak structure, most likely the CO-H_2O interaction A compact ice grown at 90 K should have a significantly lower surface area, eliminating area for CO- H_2O interactions, and forcing more CO-CO interactions. When surface area is increased, like in the porous ice, CO-H_2O interactions should dominate, depending on the thickness of CO ice. The number of peaks in each panel matches the laboratory data, overlaid in black in the figure, and shows that MIMICK is already able to qualitatively replicate some experimental work.

4. Future Work

We have briefly shown, in these proceedings, the power of MIMICK and early results from initial experiments. The capabilities of MIMICK go far beyond TPD experiments, and results will be published shortly, as mentioned earlier, on experiments investigating the growth of ices under different deposition conditions (e.g. angle, temperature) and with incorporating a relaxation process. Clements *et al.* (in prep) The aforementioned TPD experiments will be optimized for use in quantitatively constraining the pair-wise interactions, rather than only qualitatively, as expressed here. As well, these experiments will be used to quantitatively characterize the diffusion of CO into and from H_2O. Furthermore, long-term goals include incorporating water ice formation reactions on the surface via depositing hydrogen and oxygen atoms, rather than direct deposition of water molecules, and running models with astronomical, rather than laboratory, deposition conditions to explore the formation of ices in the ISM. The strength of MIMICK lies in its vast number of utilities and applications, and we aim to make use of as many of them as possible.

5. Summary

We have introduced MIMICK as a tool to extend our understanding of surface processes within laboratory ice experiments. With the ability to control deposition parameters and resulting ice densities, observations and descriptions of how molecules percolate through ASW pores is possible, as shown in the case of CO in these proceedings. Noting the results that MIMICK produces, e.g. ASW porosity decreases during a warm-up process; pair-wise interaction energies are porosity dependent; and controlling for deposition parameters alters the resulting ice density, we have shown the model is able to successfully simulate laboratory ice experiments and can be used as a tool to study ice experiments. We then discussed the vast utility of MIMICK in a variety of applications.

References

Ayotte, P., Smith, R. S., Stevenson, K. P., *et al.* 2001, *JGR*, 106, 33, 387-33, 392
Boogert, A. C. A., Gerakines, P. A., & Whittet, D. C. B. 2010, *ARAA*, 53, 541-581
Collings, M. P., Dever, J. W., Fraser, H. J., *et al.* 2002, *ApJ*, 583, 1058?1062
Collings, M. P., Anderson, M. A., Chen, R., *et al.* 2004, *MNRAS,*, 354, 1133
Cuppen, H. M. & Herbst, E. 2005, *MNRAS,*, 361, 565-576
Fayolle, E. C., Balfe, J., Loomis, R., *et al.* 2016, *ApJL*, 816:L28, 6pp
Garrod, R. T. 2013, *ApJ*, 151, 158
Gibb, E. L., Whittet, D. C. B., Boogertand, A. C. A., *et al.* 2004, *ApJ*, 778, 35-74
Hidaka H., Miyauchi, N., Kouchi, A., & Watanabe, N. 2008, *CPL*, 456, 36-40
Oberg, K. I., Boogert, A. C. A., Pontoppidan, K. M., *et al.* 2011, *ApJ*, 740, 65-78

Astrochemistry VII – Through the Cosmos from Galaxies to Planets
Proceedings IAU Symposium No. 332, 2017
M. Cunningham, T. Millar & Y. Aikawa, eds.

doi:10.1017/S174392131700936X

Chemistry in velocity: Investigating gas chemical properties in 4D

Alvaro Hacar and Ewine van Dischoeck

Leiden Observatory, Leiden University, P.O. Box 9513, 2300-RA Leiden, The Netherlands
email: hacar@strw.leidenuniv.nl

Abstract. Gas chemistry is typically characterized comparing integrated intensities of different molecular species. With the advent of ALMA, chemistry is now resolved not only in space but also in velocity. When observed at high spectral resolution, most of the clouds, filaments, and cores in the solar neighbourhood show a complex internal structure with multiple velocity components. This intrinsic complexity should be disentangled in order to correctly interpret their chemical properties. We present the first results of the FIVE+ analysis technique combining both state-of-the-art structure reconstruction (2D), kinematic decomposition (+1D), and chemical analysis (+1D) in large molecular datasets.

Keywords. Astrochemistry, ISM: kinematics and dynamics, ISM: molecules

1. Introduction

The unprecedented spectral coverage of the new broad-band (sub-)millimeter observations is revolutionizing the analysis of the ISM chemistry. Due to sensitivity constraints, molecular surveys have classically focused their analysis on the gas chemistry in localized positions in star-forming regions (e.g., Blake *et al.* 1987). Similarly, mapping studies have surveyed large areas of the sky investigating the gas structure and kinematics using few (mainly CO isotopologues) of the brightest line tracers (e.g., Goldsmith *et al.* 2008). Now, ALMA routinely produces high-fidelity molecular line maps simultaneously observed in multiple tracers and transitions (e.g., Jørgensen *et al.* 2016). Resolved out in hundreds or thousands of beams, these new observations allow us to investigate the gas chemistry both in space and velocity. However, the intrinsic complexity of these multi-line spectral cubes challenges most of the traditional analysis techniques. In parallel to instrumental developments, innovative analysis tools are needed to characterize the rich gas chemistry revealed in these new molecular datasets.

2. Initial FIVE+ results: chemical properties of the NGC1333 fibers

In this poster we present the first results of the FIVE+ analysis technique. Based on the previous FIVE algorithm (Hacar *et al.* 2013), FIVE+ is designed to combine both observations and radiative transfer calculations in order to constrain both the physical and chemical gas properties from the simultaneously analysis of multiple line tracers. FIVE+ investigates the gas distribution and structure (2D), kinematics (+1D), and chemical composition (+1D) in complex molecular datasets. From the detailed study of the gas velocities in the Position-Position-Velocity (PPV) space, the 4D analysis carried out by FIVE+ is designed to reconstruct the emission of individual gas components independently disentangling their main physical properties (i.e., kinetic temperature, gas kinematics, molecular abundances...). The development of this new methodology is part

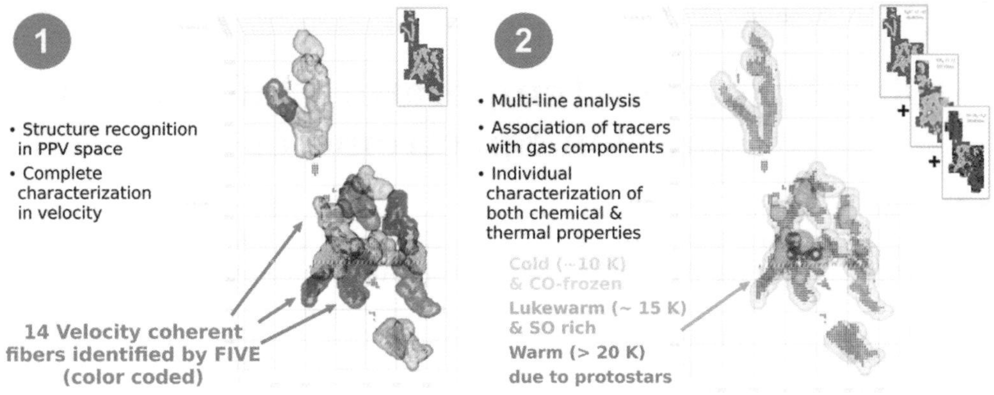

Figure 1. Preliminary FIVE+ results in the NGC1333 proto-cluster. **(Left)** 14 velocity–coherent fibers identified in the PPV space in N_2H^+ (color coded; Hacar *et al.* 2013). A full visualization of these results can be found in `https://sites.google.com/site/orion4dproject/visualization`. **(Right)** New multi-line FIVE+ analysis combining the information of all N_2H^+ (1-0), NH_3 (1,1) & (2,2), and SO (6_5-5_4) tracers within the same region. This joint analysis makes possible to systematically characterize the gas density (N_2H^+; grey structures), gas temperature (SO + NH_3; red and green regions), and chemical compositions of each individual structure both in position and velocity (see PPV iso-surfaces).

of the ORION-4D project (PI: Hacar, see also `https://sites.google.com/site/orion4dproject`).

We have carried out a pilot study of the chemical properties of the NGC 1333 protocluster using FIVE+. The dense gas substructure within NGC1333 has been extensively studied using large-scale N_2H^+ (1-0) observations (Hacar *et al.* 2017). As illustrated in Fig. 1 (Left), a complete characterization of the different gas components within this region reveals the presence of 14 independent velocity structures, the so-called fibers, forming a complex network. In Fig. 1 (Right), we have combined the above N_2H^+ emission with similar large-scale NH_3 (1,1) and (2,2), plus SO (6_5-5_4) observations. The combined analysis of these selective tracers allow us to identify, both in space and velocity, those regions presenting systematic CO-freeze out (aka rich in N_2H^+ and NH_3; grey and yellow PPV iso-surfaces), lukewarm regions affected by their internal stellar population (SO bright; green iso-surfaces), and warm material in the surroundings of the most massive protostars (via NH_3-derived temperatures) within this region. Our preliminary results indicate the potential of this multi-line analysis technique for the study of complex molecular datasets. Future developments of this new FIVE+ algorithm are expected to include the combination of additional line observations and radiative transfer codes (e.g., Brinch & Hogerheijde 2010) deriving better estimates for the excitation conditions, gas column density, and molecular abundances for each individual gas component.

References

Blake, G. A., Sutton, E. C., Masson, C. R., & Phillips, T. G. 1987, *ApJ*, 315, 621
Brinch, C. & Hogerheijde, M. R. 2010, *A&A*, 523, A25
Goldsmith, P. F., Heyer, M., Narayanan, G., *et al.* 2008, *ApJ*, 680, 428-445
Hacar, A., Tafalla, M., Kauffmann, J., & Kovács, A. 2013, *A&A*, 554, A55
Hacar, A., Tafalla, M., & Alves, J. 2017, arXiv:1703.07029
Jørgensen, J. K., van der Wiel, M. H. D., Coutens, A., *et al.* 2016, *A&A*, 595, A117

Astrochemistry VII – Through the Cosmos from Galaxies to Planets
Proceedings IAU Symposium No. 332, 2017
M. Cunningham, T. Millar & Y. Aikawa, eds.

doi:10.1017/S1743921317007803

100 GHz Room-Temperature Laboratory Emission Spectrometer

Nadine Wehres, Bettina Heyne, Frank Lewen, Marius Hermanns, Bernhard Schmidt, Christian Endres†, Urs U. Graf, Daniel R. Higgins and Stephan Schlemmer

I. Physics Institute, University of Cologne,
Zülpicher Str. 77, 50937 Cologne, Germany
email: wehres@ph1.uni-koeln.de; schlemmer@ph1.uni-koeln.de

Abstract. We present first results of a new heterodyne spectrometer dedicated to high-resolution spectroscopy of molecules of astrophysical importance. The spectrometer, based on a room-temperature heterodyne receiver, is sensitive to frequencies between 75 and 110 GHz with an instantaneous bandwidth of currently 2.5 GHz in a single sideband. The system performance, in particular the sensitivity and stability, is evaluated. Proof of concept of this spectrometer is demonstrated by recording the emission spectrum of methyl cyanide, CH_3CN. Compared to state-of-the-art radio telescope receivers the instrument is less sensitive by about one order of magnitude. Nevertheless, the capability for absolute intensity measurements can be exploited in various experiments, in particular for the interpretation of the ever richer spectra in the ALMA era. The ease of operation at room-temperature allows for long time integration, the fast response time for integration in chirped pulse instruments or for recording time dependent signals. Future prospects as well as limitations of the receiver for the spectroscopy of complex organic molecules (COMs) are discussed.

Keywords. astrochemistry, molecular data, methods: laboratory, techniques: spectroscopic

1. Introduction

High-resolution rotational spectroscopy in the laboratory enables us to obtain intrinsic molecular information, such as rotational constants, as well as (higher order) distortion constants with which precise transition frequencies for molecules can be predicted. This knowledge allows astronomers to model spectra of molecules under different physical conditions, i.e., different temperatures, which in turn enables identification in the observational data sets. This work becomes more and more demanding as the data sets include thousands of lines of many different species. The role of istopologes, vibrational excited states, isomers and even molecular conformers, but also the role of internal molecular excitation, are subjects of current research.

The field of molecular astrophysics strengthens our knowledge on the constituents of the interstellar and circumstellar medium (ISM/CSM) and deepens our understanding of the physical conditions in, for example, star forming regions. With the identification of molecules in the ISM/CSM we obtain information on temperatures, column densities, and abundances of such species. So called unbiased spectral line surveys can furthermore give insight into the details of chemical reaction networks, especially on the formation and destruction pathways of molecules within a certain astronomical environment. Molecules

† Present address: Max-Planck Institut für extraterrestrische Physik, Giessenbachstrasse 1, 85748 Garching, Germany

are ubiquitous in the ISM/CSM and the field of astrochemistry is closely linked to the progresses in the field of molecular physics in the laboratory and observational astronomy.

Millimetre and sub-millimetre (mm/submm) telescopes use very sophisticated receiver technology for the detection of the weak rotational emission spectra. The most recent developments in observational astronomy are large space based and ground based telescopes operating at frequencies of 80 GHz up and well into the THz regime, e.g., ALMA (the Atacama Large Millimeter/submillimeter Array), APEX (the Atacama PAthfinder Experiment), as well as SOFIA (the Stratospheric Observatory for Far-Infrared Astronomy), and the JWST (James Webb Space Telescope) that soon will be launched in October 2018. Observational Astronomy has therefore pushed forward the technological advances, leading to more sensitive receivers with higher spectral (and spatial) resolution covering up to 16 GHz bandwidth at present (Klein *et al.* (2012)). In this study we take advantage of the most recent developments in receiver technology that mainly has been pushed forward by the astronomy community. Room-temperature low-noise amplifiers operating at up to 300 GHz finally enable direct amplification of the weak molecular signal without lossy down-conversion of the high frequency signal first. Hence, the new advances in mm/submm technology become competitive with traditional frequency modulation (FM) absorption spectroscopy in the laboratory.

Here, we present the very first room-temperature heterodyne receiver dedicated to high-resolution gas phase spectroscopy of complex organic molecules in the laboratory. The advantage of using the new heterodyne spectrometer lies in the fact that receivers enable emission spectroscopy instead of using the traditional way of scanning frequency absorption spectroscopy. Heterodyne receiver detectors for the detection of weak emission signals of complex molecules offer a better understanding of intrinsic physical parameters such as lineshapes and absolute transition intensities. Such information can hardly be obtained by using the traditional way of FM absorption spectroscopy. Moreover, faster detection schemes are enabled by using emission spectroscopy, since spectra over several GHz of bandwidth can be detected instantly by using a sensitive receiver coupled to Fast Fourier Transform Spectrometers as backends.

Bringing these new technological advances to the laboratory will have the following advantages over traditional absorption technology using Schottky diodes or bolometers as detectors:

- absolute intensity calibration
- intensities reflecting LTE (local thermal equilibrium) conditions
- large instantaneous bandwidth of several GHz in one shot
- proper lineshapes/no derivatives
- faster detection schemes
- possible single sideband operation
- extension to quantum limited detectors is possible

All these advantages are now permitted by using radio astronomical receivers. Different experiments have been used in the past to also obtain broad instantaneous frequency coverage providing high spectral resolution. Especially test laboratories for the SWAS (Submillimeter Wave Astronomy Satellite), Odin and the Herschel satellite used a similar approach (Tolls *et al.* (2004), Teyssier *et al.* (2004), Frisk *et al.* (2003), Higgins *et al.* (2010)). However, all these experiments were solely used to understand and test the experimental set-ups and to obtain a pre-launch calibration for the onboard receivers. Due to the large complexity of the instruments none of these experiments used astronomical receivers with the sole purpose of focussing on laboratory studies. However, Tanarro *et al.* (2017) used the CNTRAG-IGN telescope (Guandalajara, Spain) in combination with a laboratory glass-cell to detect molecular emission lines for laboratory studies at

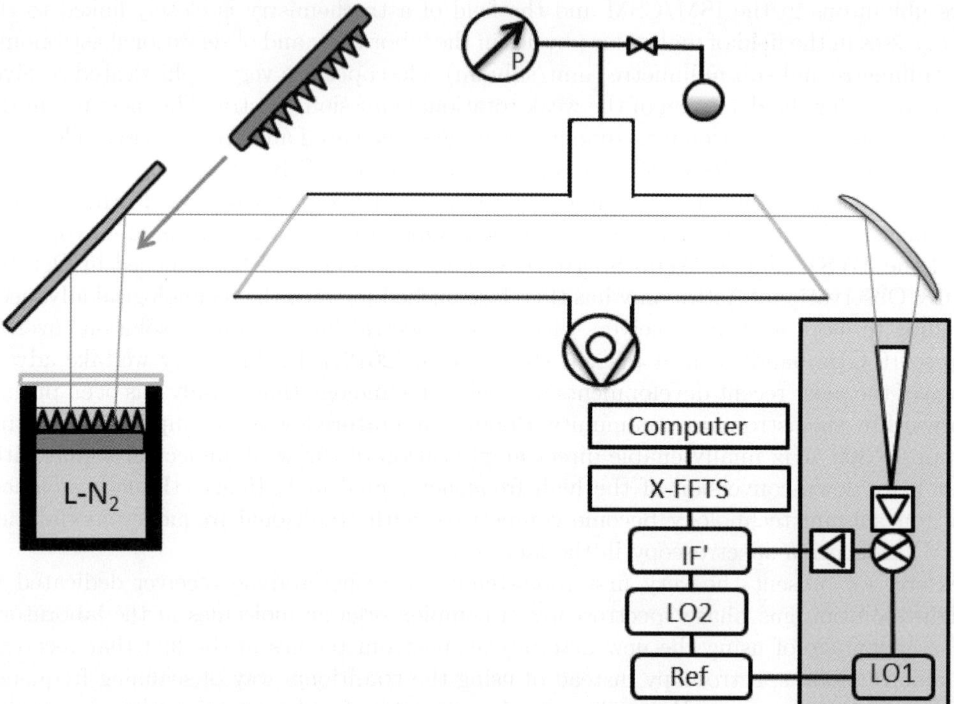

Figure 1. Schematic diagram of the 100 GHz room-temperature laboratory emission spectrometer. Light in the beam profile of the spectrometer (blue lines) is accepted by the receiver antenna. Light from molecules in the vacuum chamber is recorded against the liquid nitrogen continuum signal (cold load). The cold load consists of absorber foam immersed in liquid nitrogen acting as a blackbody emitter. For absolute intensity calibration a room-temperature absorber foam (hot load) is placed in between the chamber and the second mirror. A dual pumping stage allows for constant pressure of around 1×10^{-2} mbar during the experiment. The molecular sample, here methyl cyanide (CH_3CN), is located in a glass flask and admitted to the vacuum chamber through a needle valve. The grey shaded area underneath the receiver indicates cooling by a Peltier element of the copper block receiver mount installed for thermal stability of the sensitive detector. More details on the receiver electronics and the backend detector are given in the text.

frequencies between 41–49 GHz. Also other groups in Japan, Spain, the UK and the USA are pushing for the advancement of laboratory emission spectrometers.

In the following, first the instrument will be described, then the performance will be tested and checked against its design figures such as noise temperature, stability and sensitivity. In the Results section first spectra will be shown and compared to predictions based on previous knowledge. Finally, the laboratory emission spectrometer will be evaluated and future prospects will be discussed.

2. Experiment

2.1. *Laboratory Emission Spectrometer Set-Up*

The complete laboratory emission spectrometer set-up is shown in Fig. 1. The receiver is the heart of the instrument and shown on the right side of Fig. 1 and will be described in more detail in the following subsections. It detects light from the emitting molecules located in a pyrex vacuum cell against the cold background depicted in the left part of Fig. 1. The light is directed using a flat mirror (left part) and an elliptical mirror (right

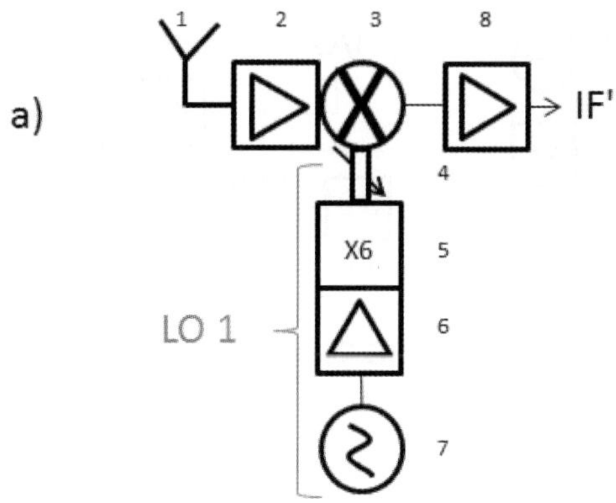

Figure 2. Electronics scheme of the room-temperature heterodyne receiver. The receiver consists of the corrugated horn antenna (1), followed by a W LNA (2), and a mixer (3), operational between 75-110 GHz. The mixer downconverts the molecular signal using the LO 1 signal (7). The LO 1 signal is based on a synthesizer signal which is amplified (6) and multiplied (x6) (5). To match the conversion loss a waveguide tuneable attenuator is used (4). The mixer output is amplified using an LNA (Miteq, +34 dB) (8) before entering the IF processor.

part). The cold load consists of a Dewar with a cold absorber foam at liquid nitrogen temperatures ($\simeq$ 77 K). For calibration purposes a hot load (room-temperature absorber foam) can be placed in between the cold load and glass cell (just in front of the flat mirror), as shown in Fig. 1.

A pressure gauge (Leybold, CERAVAC) is used to control the pressure inside the chamber upon gas injection. The molecules are located in a glass flask. Here, methyl cyanide (CH_3CN) is used. A needle valve allows for precise flow of the sample molecules into the cell. The total length of the glass chamber is around 2 m. On each side of the glass-cell HDPE (high density polyethylene) windows are mounted at Brewster angle. The diameter of the glass cell is about 10 cm. During the experiment the pressure is held constant. Typical pressures are around 1 x 10^{-2} mbar. For this, a dual pumping stage, consisting of a rotary vane pump (Leybold-Heraeus, 8 m^3 hr^{-1}) and a turbomolecular pump (Leybold-Heraeus 150 l s^{-1}) is used. After each data acquisition step, which usually lasts about 10 minutes, the glass chamber is pumped out and refilled again with "fresh" molecules. As will be discussed in the Results section temperature stability of the receiver is an issue of concern. Therefore we implemented a Peltier cooling element just underneath the receiver. This is indicated by the grey shaded area in Fig. 1. The Peltier element keeps the receiver at a constant temperature and prevents warming-up of the electronics during operation. The Peltier is usually kept at temperatures around 15°C.

2.2. *Frontend: The Heterodyne Receiver*

The new laboratory receiver depicted in Fig. 1 is shown in more detail in Fig. 2. The signal is received by a corrugated horn antenna (WR10, RPG company, 1) connected to a state-of-the-art low noise amplifier (W LNA 75–110, RPG company, 2) with a gain of 20 dB and a noise figure of 4 dB. The resulting signal is then fed into a fundamental balanced mixer (FBM, RPG company, 3) together with the signal from a local oscillator (7). The conversion loss of the mixer is 9.5 dB. The mixer multiplies the high frequency signal of the molecules (around 75–110 GHz) with the LO signal of a synthesizer which is synchronized to a 10 MHz Rubidium clock. The synthesizer is set to an output power of 6.5 dBm, and its output is guided into an amplifier and multiplier chain (AFM6, RPG

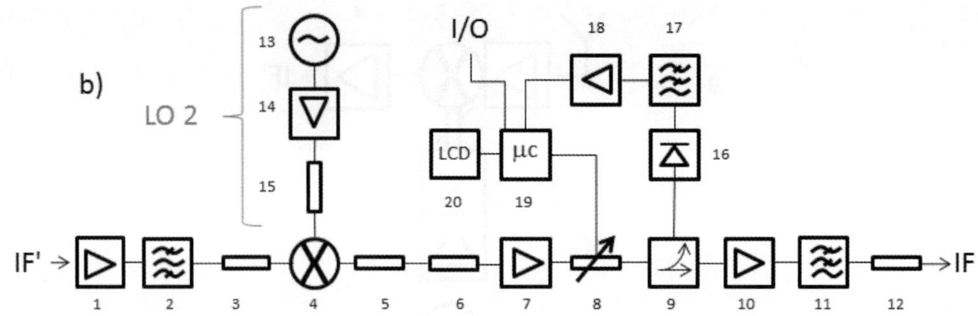

Figure 3. The intermediate Frequency Processor. The intermediate frequency signal (IF′) is amplified (25 dB, 1) and a bandpass filter (2) allows for frequencies of 6.35 ±1.25 GHz to pass. The signal is attenuated by 7 dB (3) before reaching the second mixing stage (4). Here, the IF′ is multiplied with the output of a 7.6 GHz direct digital syntheizser (DDS, 13), synchronized to a Rubidum clock. The LO (13) is then amplified by 24 dB (14) and attenuated by 3 dB (15). The mixing process results in frequencies of 7.6 GHz ± 6.35 GHz (4). The IF then gets attenuated (2 x 3 dB, 5 and 6), and amplified by +24 dB (7). A digital step attenuator (maximum 31 dB, 8) is controlled by a microcontroller (19). A small portion (about 1%) of the signal is coupled out by a directional coupler (9). Here, we obtain a continuum signal over the whole frequency range by using a continuum detector (tunnel diode, 16). A lowpass filter is applied for better signal stability (17). The counts are amplified (18) and a micro-controller (μC, 19) outputs the signal onto an LC-display (20). The remaining signal (about 99 %) passes the directional coupler (9), gets amplified (24 dB, 10), is send through a lowpass filter (loss 1 dB, 11), gets attenuated (7 dB, 12), and is fed into the XFFTS.

company, 6 and 5) resulting in about +10 dBm output power. A waveguide tuneable attenuator (WTA-110, RPG company, 4) with an attenuation range between 0.4 and 40 dB is used to match the conversion loss in the mixer. The intermediate frequency signal is then further amplified by a room-temperature low noise amplifier (LNA, Miteq company, +34 dB, 8) before entering the IF-processor as a signal indicated as IF′. The few components consisting of synthesizer (7), amplifier-multiplier chain (6, 5), as well as tuneable attenuator (4) are combined and called local oscillator 1 (LO 1) for future reference in the text.

2.3. *Backend: IF-Processor and XFFTS*

The intermediate frequency (IF′) is further processed by a home made IF-processor to match the input requirements of the XFFTS (eXtended Fast Fourier Transform Spectrometer) (Klein *et al.* (2012), Klein *et al.* (2006)). A schematic overview of the IF-processor is shown in Fig. 3. Here, the IF′ signal is once more amplified using a Miteq amplifier (25 dB, 1), and filtered through a bandpass filter (Reactel, Inc., loss -1 dB, 2). This will give signals at IF′ of 6.35 ± 1.25 GHz. We employ a 7 dB attenuator (3) before the signal enters a second heterodyne mixing stage (Marki, conversion loss -6.5 dB, 4). Here, the IF′ signal is multiplied with an LO frequency of 7.6 GHz (5 mW, 13), also synchronized to a 10 MHz Rubidium clock. The LO (13) is amplified (24 dB, 14), and attenuated (3 dB, 15), before entering the second mixer stage (4). This results in a center frequency of 1.25 GHz with a bandwidth of ± 1.25 GHz. The resulting frequency (IF) is then attenuated (2 x 3 dB, 5 and 6), and amplified (Mini Circuits, 24 dB, 7). A digital step attenuator (Mini Circuits, up to 31 dB, 8) mounted in front of a directional coupler (Mini Circuits, 9) matches the power level to the input requirements of the XFFT-spectrometer. The directional coupler takes about 1% off of the IF signal for continuum detection (16), the remaining ∼99% are amplified (Mini Circuits, 24 dB, 10), guided through a 2.5 GHz lowpass filter (-1 dB, 11), attenuated by 7 dB (12) and the resulting signal at 0.1–2.5 GHz is then detected by the XFFTS. The continuum signal that is coupled out in step (9) is detected by a continuum diode detector (16) with internal attenuation of 6 dB. An RC-lowpass filter (17) is applied for pre-integration, and an amplifier (18) is used before the signal is processed by a micro-controller (19). The output is shown on an LC-display (20), which also shows the applied digital attenuation (8), as well as the continuum signal. In total the molecular signal gets thus amplified by a factor of roughly 100 dB. Overall, we use two mixers. As described before, the first mixer is pumped by the so-called LO 1 stage included in the receiver itself (see Fig. 2). The second mixer is included in the IF-processor and pumped by the LO 2, which is labelled in Fig. 3, and includes parts 13, 14, and 15.

3. System Performance

The receiver described in the preceeding section is operated very similar to the receivers of a radio telescope, except the molecules are held in the vacuum cell. Our receiver is held at room-temperature while in contrast the ALMA band 3 receivers are held at some 4 K with a system noise temperature around 40 K. Therefore it is interesting how the warm detector might compete with the complex cryogenic temperature detector. For that reason the noise temperature as well as the system stability of the laboratory emission spectrometer will be tested in the following.

3.1. *Noise Temperature*

The receiver and system noise temperatures were evaluated using the so-called Y-factor method.

$$T_{\rm noise} = \frac{T_{\rm hot} - Y\, T_{\rm cold}}{Y - 1} \tag{3.1}$$

Y denotes the ratio of the power measured when detecting the hot and cold loads, respectively:

$$Y = \frac{P_{\rm hot}}{P_{\rm cold}} \simeq \frac{T_{\rm hot}}{T_{\rm cold}} \tag{3.2}$$

For absolute intensity calibration we use microwave absorber foam at room temperature (293-300 K) as well as at 77 K (liquid nitrogen temperature), respectively. For obtaining the receiver noise temperature the absorber is placed in front of the corrugated horn antenna, and the continuum output signal is recorded as a function of frequency. The result is shown in Fig. 4. The LO 1 frequency is varied over the bandwidth of the detector between 72 and 112 GHz, in steps of 1 GHz with input power levels of 6.5 dB in order to optimize the conversion gain. The black curve shows the receiver noise temperature (T_{rec}) as determined by the continuum diode detector described above when the absorber is placed in front of the antenna.

The blue curve shows the system noise temperature (T_{sys}) when the absorber is placed at the end of the glass cell (see also Fig. 1). In total, T_{sys} varies between 650 and 550 K over the whole frequency range (72 – 112 GHz). Here, three temperature spikes are visible at frequencies of 81, 91-92, and 101 GHz. Assumingly, the temperature spikes arise from intermodulation effects caused by non-linearities in the system. Furthermore, we see a clear slope in T_{sys} (blue curve, which includes the vacuum chamber). The slope is reproducible and most likely due to a non-optimal focussing mirror allowing the beam to spread too far and hit the glass-cell walls. This vignetting will then cause a slope in

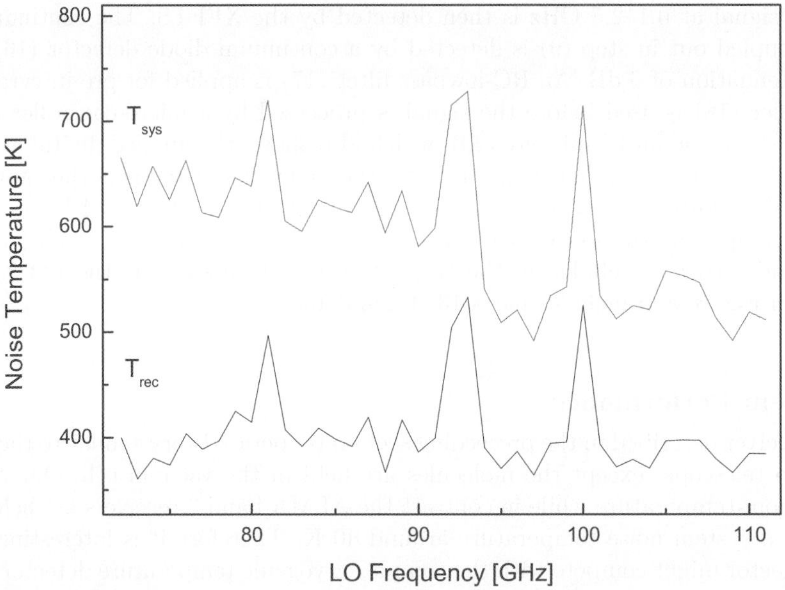

Figure 4. The receiver and system noise temperatures have been evaluated using the Y-factor method applied to the continuum signal detector. The figure shows the frequency dependence of the noise temperature of the receiver, T_{rec}, only (black curve), as well as the receiver plus optical path and vacuum chamber, T_{sys} (blue curve).

the detected noise temperature with an increase towards lower frequencies, as can be seen in the blue curve.

T_{rec} is significantly lower than T_{sys} and yields temperatures in the order of 400 K (black curve). Here, the same temperature spikes are detected independent of whether the glass-cell is used or not. T_{sys} (including vacuum chamber, HDPE windows, and mirrors, blue curve) is roughly 200 K higher. We determined that about 40 K are due to emission of the HDPE windows of which each contributes about 20 K to the total noise temperature.

For comparison, the theoretical value for the noise temperature of the receiver is calculated by taking into account the different noise figures given on the data sheets of the respective electronic components. According to the specifications, the main noise contribution is due to the low noise amplifier (LNA) with a noise figure of NF = 4 dB. The gain of the LNA is G = 20 dB. Hence, for the LNA by itself a noise temperature of 438 K, is expected, according to:

$$T_{\text{LNA}} = 290\ \text{K} \cdot (10^{\frac{\text{NF}}{10}} - 1) \tag{3.3}$$

The receiver noise temperature can then be calculated to be:

$$T_{\text{rec}} = T_{\text{LNA}} + \frac{T_{\text{Mixer}}}{G_{\text{LNA}}} \tag{3.4}$$

$$T_{\text{rec}} = 461\ \text{K} \tag{3.5}$$

The mixer losses are not included here, but are in the order of 9.5 dB and will likely increase T_{rec} even further. However, according to our measurements shown in Fig. 4 the receiver seems to perform slightly better than expected by the specifications. This result shows that the receiver noise temperature is about a factor of 10 worse than e.g. an

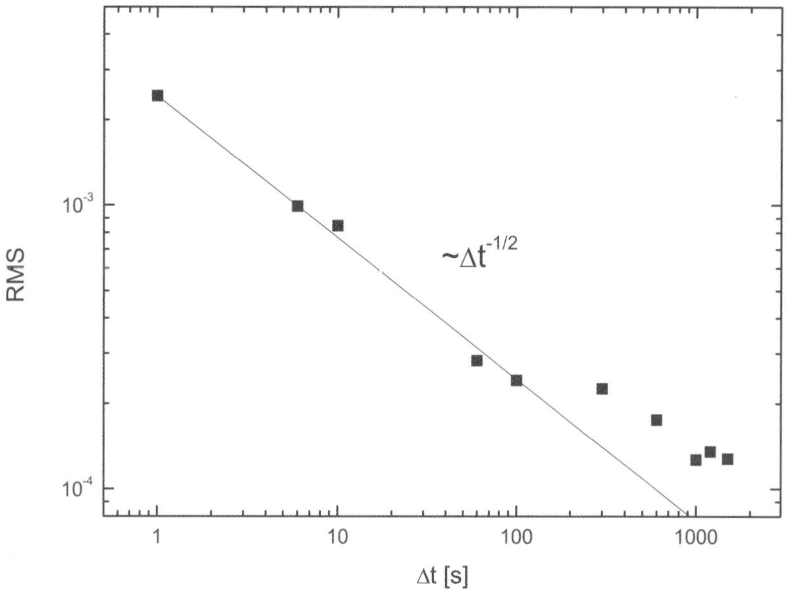

Figure 5. The stability of the spectrometer was evaluated from the background signal while taking spectra of methyl cyanide. The RMS value of the background signal was determined with respect to the integration time and a fixed channel width of 76 kHz (no binning). The integration times (Δt) were varied between 1 and 1500 s. The black line shows the expected improvement for statistical noise.

ALMA band 3 receiver. However, system operation at room temperature is much easier and allows even for student education as it was the case here.

3.2. *System Stability*

In order to obtain information on the system stability (so-called Allan stability) we recorded emission spectra of methyl cyanide for different integration times and fixed channel width of 76 kHz (no binning, about 88 kHz spectral resolution). In Fig. 5 the standard deviation of the baseline noise (RMS value) is plotted as function of the integration time. The integration time was varied between 1 and 1500 seconds. For integration times up to 100 s the improvement of the RMS with respect to the integration time is as expected and follows an inverse square-root dependence. The black line shows this expected improvement for statistical white noise. For integration times larger than 100 s the improvement deviates from the expected behavior and gets worse. Assumingly, this is due to drifts in the electronics during the experiments. This is not unexpected as the net amplification amounts to some 100 dB and is the result of many amplification and attenuation stages. Drifts occur naturally when the experimental set-up slowly warms up throughout the experiments. This causes variations in the continuum signal level. For this reason the receiver is kept at a constant temperature using a Peltier element. However, the intermediate frequency processor is not stabilized in our experiment. In order to avoid large temperature fluctuations throughout the measurements we kept integration times short (around 600 seconds at most), left the IF-processor constantly switched on, and stabilized the temperature in the laboratory (air conditioning). For better system stability it might be worthwhile to, e.g., introduce a temperature controlled water re-circulator to the IF processor. This could potentially result in longer Allan times.

3.3. *Spectrometer Sensitivity*

Once the system noise temperature and Allan stability are determined, we can calculate the sensitivity of the spectrometer by applying the radiometer formula:

$$\Delta T = \frac{T_{sys}}{\sqrt{\Delta\nu\Delta t}} \times 2 \tag{3.6}$$

Here, T_{sys} is the system noise temperature, Δt is the integration time, and $\Delta\nu$ is the channel width which is given by the XFFTS. The factor 2 arises from the fact that the receiver is operated in double sideband mode where the RF signal from both sidebands are combined for each IF frequency. Here we assume a 1:1 sideband ratio. Using a T_{sys} of around 600 K as a conservative value (see Fig. 5 with LO frequency of 86 GHz), and a 100 minute (6,000 s) integration time, as well as a channel width of 76 kHz (channel spacing of the XFFTS at a bandwidth of 2.5GHz), the resulting RMS noise value amounts to about 60 mK. Successive spectra have been obtained by integrating 10 spectra for about 600 s each. The spectra where then co-added to arrive at the noise equivalent temperature. This system noise equivalent temperature can be compared to the noise in calibrated spectra as will be described below.

In general, for each emission spectrum of 600 s duration, two separate spectra have to be recorded, i.e., the spectrum of the warm molecular gas (at room-temperature) in front of a cold background (liquid N_2), as well as a hot background reference spectrum for absolute intensity calibration, i.e., we record spectra at 77 K and 293 - 300 K, respectively. The signal S_{hot} is proportional to the power P_{hot} impinging on the receiver which is proportional to T_{hot} in case of the hot (293-300 K) background plus the signal from the receiver noise, i.e., proportional to T_{sys}. Thus,

$$S_{hot} \sim T_{hot} + T_{sys} \tag{3.7}$$

holds. Similarly, the signal S_{cold} for the cold (77 K) background measurement (without molecules) becomes

$$S_{cold} \sim T_{cold} + T_{sys}. \tag{3.8}$$

In fact from such measurements the system noise temperature for each IF frequency channel can be derived by solving the above equations for T_{sys}. When repeating these measurements (hot and cold) but with molecules in the cell the temperature calibrated signal becomes

$$\frac{S_{cold} - S_{hot}}{S_{hot}} \times (T_{hot} + T_{sys}). \tag{3.9}$$

Due to construction of this formula the positive molecular signal emerges on top of a negative background of size -(T_{hot} - T_{cold}). In these first measurements the signal was also subject to some baseline modulation issues based on standing waves. Therefore low frequency signals were corrected by twofold Fourier Transform with rejection of the very low frequency part including the constant background. As a result the negative background was rejected and the molecular signal was unaffected and could be further analysed based on Eq. 3.9. Consequent results are discussed in the following section.

A more thorough investigation of the signal shows that an analysis according to

$$\frac{S'_{cold} - S_{cold}}{S_{hot} - S_{cold}} \times (T_{hot} - T_{cold}) \tag{3.10}$$

where the signals S'_{cold} and S_{cold} refer to the signal with and without molecules in the cell are less prone to fluctuations and baseline effects. However, three independent measurements are needed for a complete molecular spectrum.

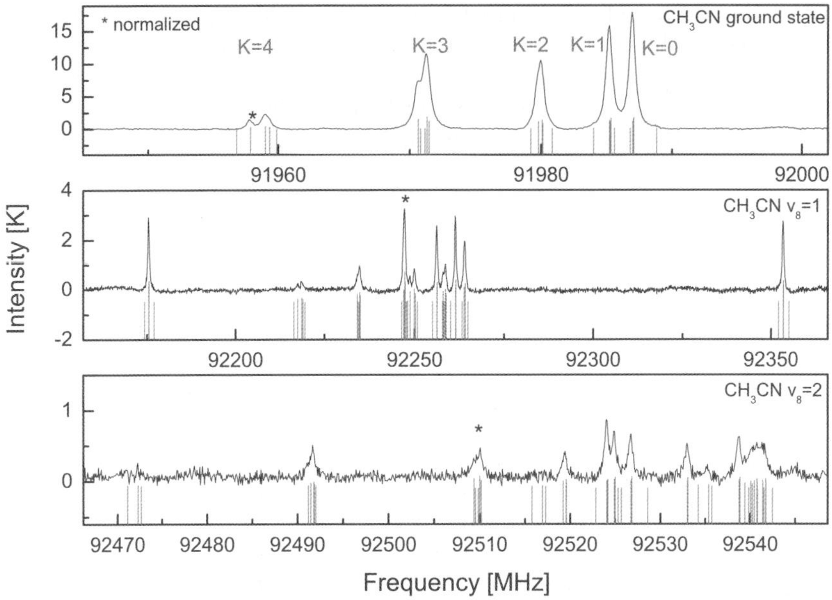

Figure 6. Emission spectrum of methyl cyanide. Upper panel: The emission spectrum of the ground vibrational state is clearly detected. Middle panel: The first vibrationally excited state of methyl cyanide is plotted ($v_8 = 1$). Lower panel: The second vibrationally excited state of methyl cyanide is detected ($v_8 = 2$). In order to overplot the CDMS entries for each vibrational state, we normalized the CDMS intensities to the respective band marked in the spectra with an asterisk. The total integration is 100 minutes and the pressure was kept constant at around 1 x 10^{-2} mbar. The RMS is in the order of 0.08 K.

4. Results and Discussion

As a proof of concept for the laboratory emission instrument spectra of methyl cyanide, CH_3CN, have been recorded. The molecule is per definitionem a so-called complex organic molecule. Following the convention of Herbst & van Dishoeck(2009) complex organic molecules are defined as a molecule containing 6 or more atoms including carbon and hydrogen . Rotational lines of many of those molecules dominate the rich spectra taken by ALMA and other telescopes. The dipole moment of CH_3CN is quite strong with about 3.9 Debye and makes this molecule a perfect test candidate for our new emission spectrometer. In terms of rotational spectroscopy this molecule is a symmetric top molecule, in which the rotational constant is about 9.2 GHz, giving rise to spacings of the major rotational features by 2 $B \sim$ 18.4 GHz. In the spectral range here, we trace the J=5→4 transition.

Fig. 6 shows the emission spectrum of methyl cyanide derived through measurements of the molecular emission against a cold and a hot reference background. The resulting spectra are divided into three panels in Fig. 6. In black the experimental spectrum is plotted, while the entries from the Cologne Database for Molecular spectroscopy (CDMS) are overplotted as red stick diagram (Endres *et al.* (2016), Müller *et al.* (2005), Müller *et al.* (2001)).

Each of the three panels shows lines associated with rotational transitions of CH_3CN in different vibrational levels. The y-axes are calibrated using the above mentioned procedure and give the intensities in Kelvin. The x-axes show the frequency in MHz. The rotational transitions within the ground vibrational state are shown in the upper panel, the rotational transitions within the vibrationally excited states, v_8=1 and v_8=2, are shown in the second and third panel, respectively. For all vibrational states the J=5→4 transitions are detected. It can clearly be seen that quite a fraction of the population is distributed over the vibrational manifold even in such a small molecule as methyl cyanide. Here, the first vibrationally excited state shows intensities in the order of one fourth of the intensities in the vibrational ground state. Even the second vibrational excited state is clearly detected and again about one fourth of the intensities is seen in this state. The vibrational states differ by about 350 cm^{-1} in energy each.

CH_3CN shows a characteristic splitting of the different K-levels for each J value ($|K| < J$). This K splitting is obvious in the ground vibrational state and labelled in the first panel. For some lines in the ground vibrational state a hyperfine splitting due to the nuclear spin of the nitrogen atom (I=1), is also resolved. This can be seen best for the peak at 91960 MHz, which is marked with an asterisk. Also, for the $K = 3$ transition, around 91970 MHz, a shoulder caused by the splitting of the individual hyperfine components can be detected. Here, a separation between the F=J and F=$J \pm 1$ components at 91970.7 and 91971.4 MHz can be seen.

Apart from these spectroscopic details Fig.6 shows that substantial parts of the spectra can be recorded simultaneously with the emission spectrometer on an hour time scale. Much weaker signals are detectable thanks to the good S/N ratio of the measured spectra. In fact, the RMS of the intensities are in the order of 0.08 K and hence the spectrometer sensitivity meets our expectation (RMS=0.06 K, according to system calibration). This RMS results in a S/N of about 210 for the strongest transition within the ground vibrational state (first panel), a S/N of 38 for the strongest transition within the first vibrationally excited state (second panel), and a S/N of around 10 for the second vibrationally excited state (third panel).

Each channel spacing of the XFFTS is around 76 kHz (2500 MHz divided by 32768 channels), while the spectral resolution is in the order of 88 kHz (equivalent noise bandwidth). While this is sufficient to compare spectra to observations, it is not sufficient to fully resolve the hyperfine structure of (complex) molecules. The hyperfine components are partially resolved in our spectrum, as can be seen by the red stick diagram. There are several reasons why we do not fully resolve the hyperfine components: First: Only if the spacing between individual hyperfine components is in the order of the spectral resolution of the XFFT-spectrometer this is possible. This is not always the case for methyl cyanide as can be seen by the red stick diagram. Second: The Doppler broadening for 300 K warm lines is in the order of 200 kHz, and hence larger than the spectral resolution of the XFFTS. Third: Pressure broadening is not negligible in this spectrum and broadens the lines in addition to the Doppler broadening.

In order to show the frequency accuracy as well as the intensity accuracy obtained with our spectrometer, we simulate the expected spectrum under LTE (local thermodynamic equilibrium) conditions. In Fig. 7 we plot the v_8=1 experimental spectrum (in black) together with a simulated spectrum (in red). The simulation has been obtained using PGOPHER Version 10.0 (Western(2017)). The program simulates the rotational structure by assuming LTE conditions. Here, we used the molecular parameter files as given from the website of the Jet Propulsion Laboratory (JPL, Pickett *et al.* (1998)). The parameter file was downloaded for the first vibrational excited state and includes the hyperfine splitting for the nitrogen nucleus ($I = 1$). The temperature in the simulation is set to 300 K. The only free parameter that was chosen in the simulation is the linewidth for a Voigt line profile.

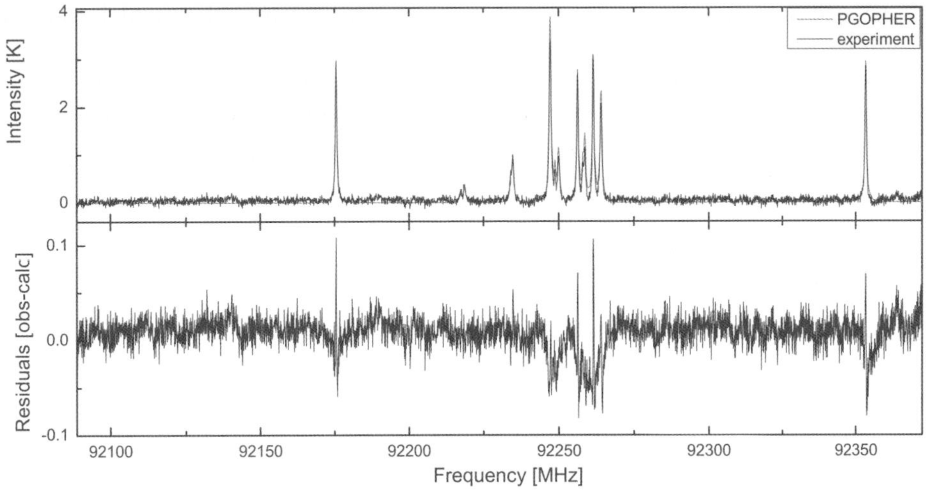

Figure 7. Emission spectrum of methyl cyanide. Upper panel: The emission spectrum of the $v_8 = 1$ is presented. The experimental spectrum is plotted in black. Overplotted is a PGOPHER simulation in red. Lower panel: The redisuals are shown when the simulation is subtracted from the experimental spectrum (obs-calc). The simulation has been obtained using PGOPHER under LTE conditions and for 300 K.

We assume Doppler broadening for our lines, which can be calculated to be in the order of FWHM=200 kHz at a center frequency of 92 GHz, applying the following formula:

$$\Delta\nu = 2\sqrt{\ln(2)} \times \nu_0 \ \mathrm{c}^{-1} \sqrt{\frac{2\mathrm{kT}}{\mathrm{m}}} \tag{4.1}$$

Here, ν_0 is the center frequency (here around 92 GHz), c is the speed of light, k indicates the Boltzmann constant, T is around room-temperature (300 K) and m is the molecular mass. Hence, we chose for a Gaussian component of 0.2 MHz in the simulation and subsequently added the Lorentzian lineshape to the simulation. The best results were thus obtained using 0.2 MHz Gaussian linewidths, in addition to 0.6 MHz Lorentzian linewidths.

The result can be seen in Fig. 7. The intensities are calibrated and given in Kelvin. In the simulation (red) the intensities are normalized to the intensity obtained in the emission experiment. Hence, for normalization purposes the emission line at around 92260 MHz was chosen. The frequency scale is given in MHz. The lower panel shows the resulting residuals when subtracting the simulated PGOPHER spectrum from the experimental spectrum (obs - calc). The intensity axis is still in Kelvin. Here, the residuals are in the order of a few percent (about 2.5 %) for the strongest lines. For weak emission lines (92220–92240 MHz), the residuals cannot be discriminated from the baseline noise.

5. Conclusion

In this paper we demonstrate the capabilities and limitations of a home-built 100 GHz room-temperature heterodyne receiver.

- The Allan stability was determined to be in the order of 100 s for 76 kHz channel width.
- The receiver noise temperature (at room-temperature) is around 400 K, and meets the expectations.

• The system noise temperature is in the order of 600 K and should be improved through better optics arrangement.

• Emission spectra of methyl cyanide show the J=5→ 4 rotational transitions in the ground vibrational state, as well as the first and second excited vibrational state v_8=1 and v_8=2, respectively.

• The spectral accuracy, i.e., the frequency and intensity accuracy was evaluated by comparing the methyl cyanide spectra obtained from the emission experiment to simulations obtained from PGOPHER. The residuals of the intensities were in the order of a few percent for the strongest deviation from the simulation.

We conclude that the new technology that can now be used in the laboratory environment has advantages with respect to the traditional frequency modulation (FM) absorption spectroscopy. Given the fact that broadband spectrometers up to 16 GHz are currently available, this will result in even more efficient ways of obtaining spectra of complex organic species over large frequency ranges. Relative intensity measurements show a reproducibility of the emission spectra much better than the few percent reported in our comparison between experiment and theoretical predictions. Thanks to this high quality intensity information it is foreseen that temperature dependent intensity measurements can be employed to derive molecular energy diagrams without associated rotational model Hamiltonians as in the traditional bootstrap approach (Medvedev & De Lucia(2007), De Lucia(2010)). This is thought to open the door to increased analysis speeds with many more molecular levels being included. This development is urgently needed in the era of ALMA where very complex spectra with unprecedented sensitivity and thus many molecular levels need to be understood. Since such species show a spectrum in a wide frequency regime with thousands of lines, it would be preferable to use more efficient detection methods as currently possible with FM technique. Coupling of laboratory emission receivers with other techniques, such as chirped-pulse excitation, would also result in an increased S/N ratio. First experiments along these lines demonstrate increased sensitivity with detection of even higher vibrational states of methyl-cyanide.

Currently, tests are underway to cool-down the low-noise amplifier to cryogenic temperatures, which also will decrease the system noise temperature and hence increase the spectrum acquisition efficiency. Furthermore, a new spectrometer making use of SIS-receivers (superconductor-insulator-superconductor) at 300 GHz is tested in our laboratory, which shows promising application for the acquisition of spectra of complex organic molecules with astrophysical importance.

To this date the fast response time of the emission detection has not been used except in first chirped-pulse experiments. However, these capabilities may also be used to identify transient species such as radicals and ions which are a challenge to laboratory work and astrophysics as well. All together the emission spectrometers developed in several laboratories around the globe will allow new experiments including the study of the influence of collisions or energy transfer on molecular spectra or the detection of molecules in non-standard environments such as desorbing molecular ices or products of chemical reactions.

Acknowledgements

The emission spectrometer development is carried out within the Collaborative Research Centre 956, "Conditions and Impact of Star Formation", sub-projects B3, B4 and S, funded by the Deutsche Forschungsgemeinschaft (DFG). M.H. acknowledges funding through the Bonn-Cologne-Graduate-School for physics and astronomy (BCGS). We would like to thank Gundolf Wieching at MPIfR Bonn for fruitful discussions.

References

De Lucia, F. C. 2010, *J. Molec. Spectrosc.*, 261, 1

Endres, C. P., Schlemmer, S., Schilke, P., Stutzki, J., & Müller, H. S. P. 2016, *J. Molec. Spectrosc.*, 327, 95

Frisk, U., Hagström, M., Ala-Laurinaho, J., *et al.* 2003, *A&A*, 402, L27

Herbst, E. & van Dishoeck, E. F. 2009, *ARAA*, 47, 427

Higgins, R. D., Teyssier, D., Pearson, J. C., Risacher, C., & Trappe, N. A. 2010, *Twenty-First International Symposium on Space Terahertz Technology*, 390

Klein, B., Hochgürtel, S., Krämer, I., *et al.* 2012, *A&A*, 542, L3

Klein, B., Philipp, S. D., Krämer, I., *et al.* 2006, *A&A*, 454, L29

Medvedev, I. R. & De Lucia, F. C. 2007, *ApJ*, 656, 621

Müller, H. S. P., Schlöder, F., Stutzki, J., & Winnewisser, G. 2005, *J. Mol. Struct.*, 742, 215

Müller, H. S. P., Thorwirth, S., Roth, D. A., & Winnewisser, G. 2001, *A&A*, 370, L49

Pickett, H. M., Poynter, R. L., Cohen, E. A., *et al.* 1998, *JQSRT*, 60, 883

Tanarro, I., Aleman, B., de Vicente, P., *et al.* 2017, submitted to *A&A*, private communication

Teyssier, D., Dartois, E., Deboffle, D., *et al.* 2004, *Fifteenth International Symposium on Space Terahertz Technology*, 306

Tolls, V., Melnick, G. J., Ashby, M. L. N., *et al.* 2004, *ApJS*, 152, 137

Western, C. M. 2017, *JQSRT*, 186, 221

Astrochemistry VII – Through the Cosmos from Galaxies to Planets
Proceedings IAU Symposium No. 332, 2017
M. Cunningham, T. Millar & Y. Aikawa, eds.

doi:10.1017/S1743921317008110

The influence of chemical environment on the infrared spectra of embedded molecules in astrophysical ices

Victor S. Bonfim[1] **and Sergio Pilling**[1,2]

[1]Instituto de Pesquisa e Desenvolvimento, Universidade do Vale do Paraíba - UNIVAP,
Av. Shishima Hifumi, 2911, Urbanova, São José dos Campos, SP - Brazil
email: victordsb@gmail.com

[2]Instituto Tecnológico de Aeronáutica, ITA - DCTA
Praça Marechal Eduardo Gomes, 50 - Vila das Acacias, São José dos Campos, SP - Brazil
email: sergiopilling@yahoo.com.br

Abstract. In this work, one intends to computationally simulate and investigate, via thermochemical calculations, how the chemical environment influences some molecular properties, such as IR spectra and absorption cross section, of individual species embedded in the solid phase employing the Polarized Continuum Model (PCM) approach. The trial molecules used here to check these effects are CO, CO_2 and H_2O. The solid phase (bulk ice) is simulated using different dielectric constant values representing different types of astrophysical ice at PCM approach. The effect of temperature is also investigated since it is known it affects the dielectric constant of the solvent medium.

Keywords. ISM: molecules, abundances, infrared: general, dense matter, astrochemistry

1. Introduction

Both on Earths surface and in astrophysical environments, various molecules are found embedded into solid water structure. In the case of interstellar medium, the astrophysical ice mantles over dust grains surface can vastly vary in composition. For some cases, the chemical species of interest are extremely diluted in the icy matrix containing them. Besides solid water, the main constituents of such matrices are also CO_2, CO, CH_3OH, or CH_4, among others, and a combination of them.

Astrophysical ices may also vary a lot in temperature, depending on various parameters of the surrounding environment. Nonetheless, they can be divided (at least) in two temperature regimes for the ice mantles (Van Dishoeck *et al.* 2013), and these regimes we call in the present study as cold ice and hot ice. This arbitrary classification is illustrated in Figure 1. for an astrophysical context where there is ice at both cold and hot regimes, depicted in the same figure in blue and in red colors, respectively.

As a first approximation, the net effect of the solvent medium (solid matrix) in a given chemical species which is there embedded can be accounted by simply considering the presence of an average electric field surrounding this species, this field being parameterized by the dielectric constant (ε) of the medium. This level of description can be achieved employing Polarized Continuum Model (PCM) approach (Miertuš *et al.* 1981), which is implemented in several computational Quantum Chemistry packages. Indeed, there are already some works in literature where the authors utilized PCM (or a similar continuum solvent model) to model the average icy matrix effect on the embedded molecule reactivity (Bonfim *et al.* 2015, 2017; Park & Woon 2004; Pilling *et al.* 2011a; Woon & Park 2004). However, none of them attempted to simulate the effects of one

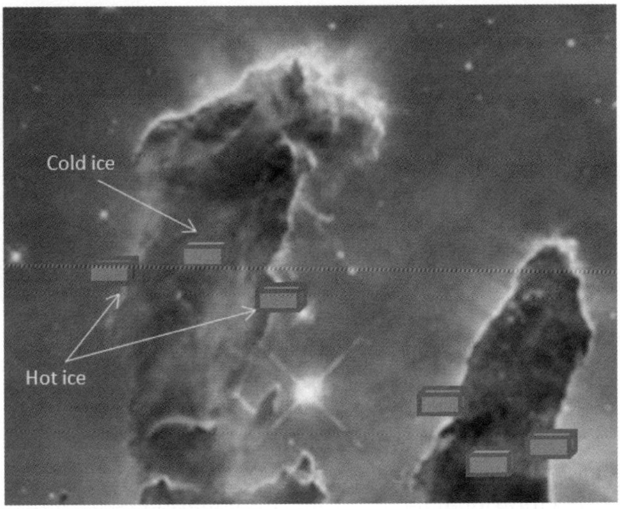

Figure 1. Representative illustration of an astrophysical scenario where ice is present at two regimes of temperature, here called cold ice (in blue) and hot ice (in red), comprising a star formation region in the Eagle Nebula (adapted from Constellation Guide 2013).

specific bulk ice property on the vibrational modes of each trapped molecule, which are closely related to the determination of molecular abundances in observational studies.

In this work we evaluate, through vibrational analysis from PCM ab initio calculations, how the chemical environment influences some features of molecular infrared (IR) spectrum profiles, specifically, the peak positions and their respective intensities, of three trial molecules. The chosen molecules are CO, CO_2 and H_2O, thus involving two polar molecules and one nonpolar species (CO_2). Very often these three molecules are the major components of the astrophysical ices matrix, from which H_2O is the dominant one for most cases. Very often these three molecules are the major components of the astrophysical ices matrix, from which H2O is the dominant one for most cases. Accordingly, we also intend to probe the role of the studied properties in determining molecular abundances.

2. Methods and Models

All Calculations were performed at Second-order Møller-Plesset perturbation theory (MP2) level, with the correlation consistent triple zeta basis set, or just cc-pVTZ (Dunning Jr 1989). The number of electrons in the system varies quite well among the molecules investigated here, and also for a larger set of molecules on which calculations are in progress. Thus, for reasons of computational convenience, we have used the MP2/cc-pVTZ calculation level for all the electronic structure calculations of the current work. The chemical species had the equilibrium geometry confirmed for each ε value by applying harmonic vibrational analysis, which has also provided the desired output data. The complete study was performed using the Gaussian 09 package (Frisch *et al.* 2010).

An interesting feature of vibrational intensity I changing is to allow one to know the influence felt by band strength B, since the ratio I/I_0 is equivalent to the quotient B/B_0. This latter one is a molecular property (rigorously, it varies with each vibrational mode) of major astrophysical interest as it is sensitive to changes in the chemical environment. This relationship between I and B can be seen from the equation:

$$B = \frac{2.3}{N} A, \; A \propto I \; [cm \; molecule^{-1}] \tag{2.1}$$

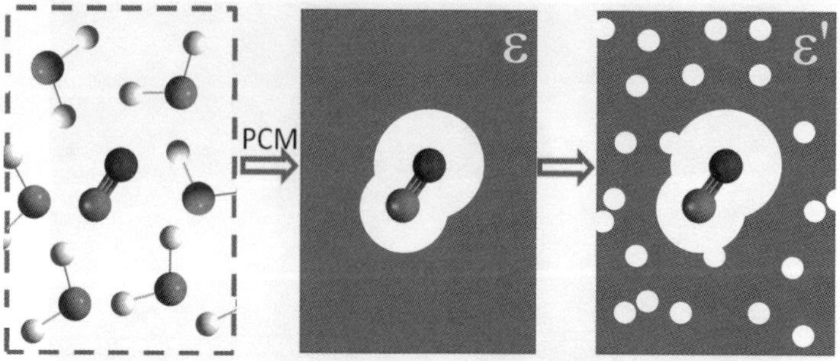

Figure 2. Illustration of the PCM approximation and the porous sample approximation (solid filled with holes containing vacuum). The molecular environment is represented by its dielectric constant ε and here ε' stands for the dielectric constant of the porous sample ($\varepsilon' < \varepsilon$). See details in the text.

where N is the column density of a given species, and A is the band area (in units of cm^{-1}) of a specific IR vibrational mode. From this equation (see e.g. Pilling *et al.* 2011b), we note that when comparing two samples containing the same N, variations in the ratio between two B values, namely B and B_0, will be equivalent to the ratio between I and I_0, assuming the change in A is very little affected by enlargement or narrowing of the peak. Once we are aware that our theory level is not accurate enough to predict each I value quantitatively, predicting the ratio of two band strengths gets us as close as possible from connections with observational and / or experimental data. Due to the exposed equivalence, Intensity variations are expressed in this work as B/B_0.

In the adopted effective-medium approach context, we can use the link between ε and other physicochemical properties of the bulk ice to probe how they influence the IR spectra of the embedded molecules. For practical means, it may be assumed that variations in such properties, such as ice temperature, are dominated by changes in the solid water matrix, and consequently in the corresponding ε values. In the current study, we propose an exponential function to fit ε as a function of temperature T, since ε (T) for $H_2O(s)$ may vary quite abruptly in the studied interval of 10-140 K (Tsekouras *et al.* 1998). The employed function is presented as follows:

$$\varepsilon = (9.0 \times 10^{-6})\, e^{0.12\,T} + 3.0 \tag{2.2}$$

Aiming not to get too far from experimental data, we have constrained the function during fitting process in order to reproduce ε (10 K) $\sim$ 3 and ε (140 K) = 180 (Johari & Whalley 1981; Tsekouras *et al.* 1998). Concerning porosity, an increase in the ice porosity can be simulated by decreasing the ice ε_{pi}, starting from the ε value for the compact ice, and adopting the proper relation between the two simulated quantities. Some variety of such relations have already been proposed and described in literature, and here we adapt the model from Sun & Goldberg (2005) for ternary mixtures:

$$Porosity = 100\% \times \frac{\sqrt{\varepsilon_{pi}} - \sqrt{\varepsilon}}{\sqrt{\varepsilon_{pi}} - 1} \tag{2.3}$$

In 2.3, ε_{pi} is the dielectric constant for the pure ice, assuming a structure without pores. In this work we have considered two temperature regimes, cold ices (close to 10 K) (Tsekouras et al., 1998), and hot ices, where ε_{pi} reaches about 180 for 140 K (Aragones et al., 2011; Johari & Whalley, 1981). A schematic illustration of the PCM approximation and how it approaches a porous ice is shown in Figure 2, in which the white circles represent holes containing vacuum (ε = 1).

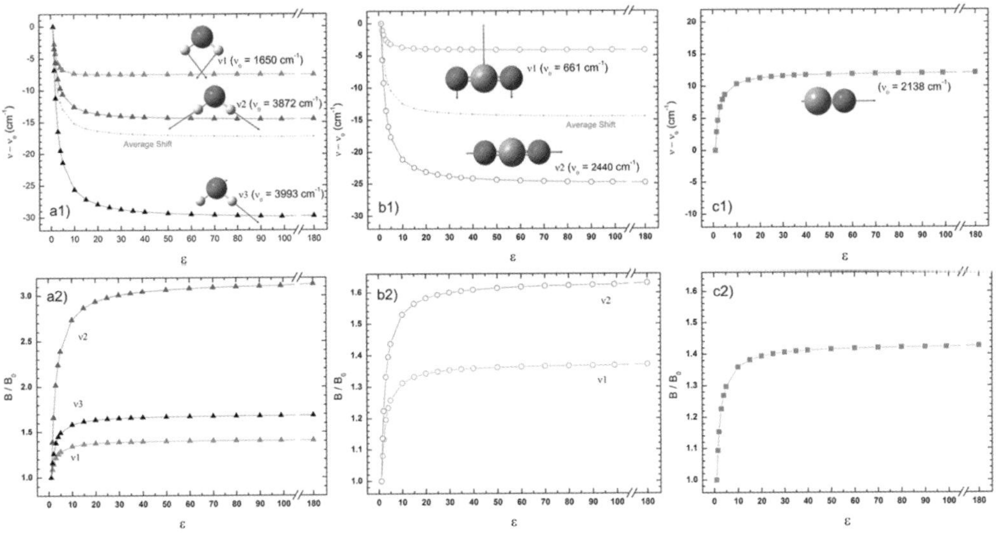

Figure 3. Changes in the IR spectral features of each vibrational mode of H_2O, CO_2 and CO molecules embedded in the icy matrix as a function of its ε, displayed in two different aspects: 1) frequency shifts; 2) band strength ratios. Results from vacuum calculations ($\varepsilon = 1$) are considered as reference values and are indicated by ν_0 and B_0 symbols. See details in the text.

3. Results and Discussion

The evaluated IR spectrum features have presented considerable variations with systematic ε changing. Figure 3 shows those variations for the three studied molecules, H_2O, CO_2 and CO in the panels a, b and c, respectively. The increase or decrease in the vibrational frequencies (in wavenumbers) is displayed in panels numbered as "1" and panels numbered as "2" show the variations on band strengths. In each case, the subscript "0" refers to the value obtained for the vacuum environment, where $\varepsilon = 1$ (by definition).

Regarding wavenumbers, the general trend was a decrease with ε increasing, indicating there would be a shift of the peaks to the right in their IR spectrum, which would also be worth if we expressed the results in wavelengths. The only exception to this trend was the CO vibrational mode, which in a spectrum would have shifted to the left. The maximum wavenumber shift occurred for the asymmetric stretch of the water molecule. This species also has the largest average shift, although the highest variation on CO_2 vibrational modes (also for an asymmetric stretch) is greater than the average shift for H_2O.

Differently from frequencies, band strengths have all varied in the same way for the investigated molecules, rapidly increasing with ε for the $1 < \varepsilon < 15$ interval, and showing a gradually slow increase for roughly $\varepsilon = 15$ onwards. Interestingly, the lowest-frequency modes all had a similar increase in magnitude, while the CO_2 asymmetric stretching (asym. str.) mode increased twice that amount. In turn, the water mode with highest B/B_0 values, the symmetric stretching mode, showed a maximum growing of around three times the original value, or about four times the maximum increase in B/B_0 for the CO_2 asym. str. mode. Hence, our results bring evidence that not taking in account at least the average effects of the chemical environment can lead to significant errors in molecular abundance determinations.

Figure 4 shows the total set of probable effects of changing ice temperature or ice porosity in the IR spectra for all the studied molecules. Here, these changes are presented

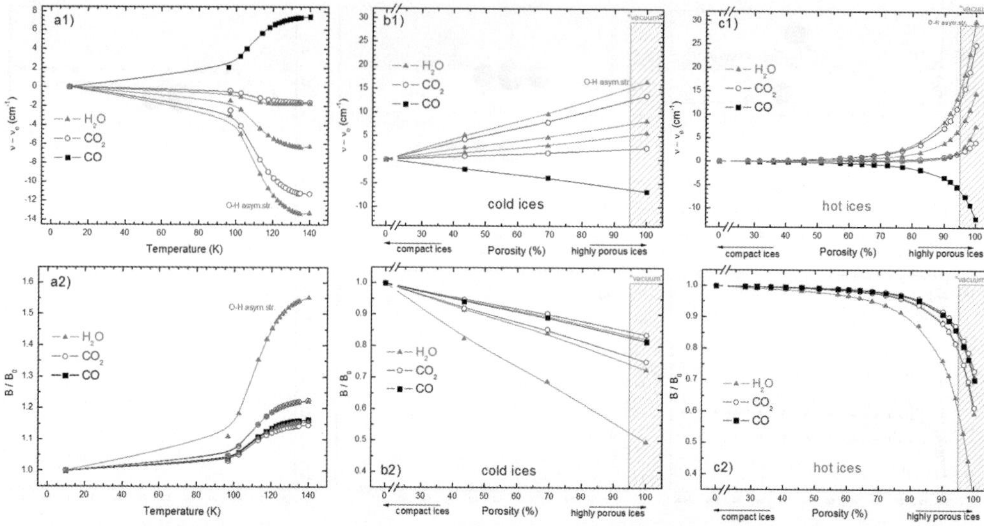

Figure 4. Changes in the IR spectral features of all studied species H_2O, CO_2 and CO embedded in a typical water icy matrix, as a function of temperature - from 10 K up to 140 K - (a) and porosity for both cold ice (b) and hot ice (c). Frequency shifts (ν - ν_0) and band strength ratios (B/B_0) are presented in top (1) and bottom (2) panels, respectively (ν_0 and B_0 are equivalent to the results for $\varepsilon = 3$, except for hot ice panels, where it was considered $\varepsilon_{pi} = 180$). The shaded area assumes that porosity above 95% would not allow an ice phase to exist.

as a function of temperature (from 10 K up) and porosity for both cold ice and hot ice. Here, the subscript "0" denotes $\varepsilon = 3$ results when probing temperature effects, and concerning results about porosity effects, it implies the ν_0 and I values obtained for ε_{pi}. Still related to porosity investigations, there are shaded areas in the corresponding panels, highlighting that a value equal to or very close to 100% would imply in the absence of a solid matrix. Nevertheless, we decided to keep these points in view in order to assist in visualizing the identified tendencies in vibrational frequencies and band strengths. Strictly saying, all the data displayed in this figure comes from the same calculation results already presented in Figure 3, but here the presentation format has changed to fit the evaluated properties.

The observed variations profile for both wavenumbers and band strength ratios are quite dependent from the relationship between ε and the linked property. For temperature, they are biased from the mathematical limitation in getting the inverse function of 2.2, which is a natural logarithm, and consequently is not defined for $\varepsilon = 3$ or lower. Also, the investigated vibrational modes have shown to be much more sensitive to the ice porosity while incorporated into cold ices, than into hot ices. Evidently, this comparison disregards data related to extremely high porosity (close to 95% and above). A recurring overall trend is that the "ν1" mode (lowest frequency) of each molecule has shown to be the least sensitive to variations in the surrounding chemical environment.

It is important to keep in mind that our vibrational modes analysis is based on an average field approach, which hinders direct comparisons with laboratory spectra. Even so, the qualitative relevance of the predicted deviations for astrophysical scenarios still remains. Especially in the case of amorphous ices, usually formed at low temperatures, which have been subject of Astrochemistry studies (Collings *et al.* 2004; Öberg *et al.* 2009; Pilling & Bergantini 2015).

4. Conclusions

For most vibrational modes employed in this study, band strengths have increased with ε, which implies they have decreased with respect to porosity. The frequency shifts (in cm^{-1}) showed opposite behavior in relation to the band strengths, with the exception of the CO mode, revealing that different vibrational modes are influenced by the solid matrix diversely. In order to achieve a more precise description of this influence, a larger set of molecules would be desirable. Besides that, all the trends here observed seem to be more pronounced in low temperature ices than in higher temperature ones.

The B/B_0 value for the H_2O symmetric stretching mode varied by more than 50% with the change from $\varepsilon = 3.0$ to $\varepsilon = 180$, which represents the difference in our models from a cold astrophysical ice to a hot ice. The determination of molecular abundances depends a lot on the band strength. Therefore, it becomes clear that direct transposition of a B value obtained in one experiment (or observation) to another with different physicochemical characteristics can lead to considerable errors in abundance determinations.

This study may also provide a better understanding of some astrophysical ices properties and their interaction with radiation. Based on the IR spectral profile of diluted species, one could be able to infer physical and chemical properties of the surrounding medium.

Acknowledgements

The authors would like to thank the Brazilian agencies FAPESP (JP2009 / 18304-0), CNPq and CAPES.

References

Aragones, J. L., MacDowell, L. G., & Vega, C. 2011, *J. Phys. Chem. A* 115, 5745-5758

Bonfim, V. S., Castilho, R. B., Baptista, L., & Pilling, S. 2017, *Phys. Chem. Chem. Phys.* submitted

Bonfim, V. S., Castilho, R. B., Baptista, L., & Pilling, S. 2015 *J. Phys. Conf. Ser.* 635, 102012. doi:10.1088/1742-6596/635/10/102012

Collings, M. P., Anderson, M. A., Chen, R., Dever, J. W., Viti, S., Williams, D. A., & McCoustra, M. R. S. 2004 *Mon. Not. R. Astron. Soc.* 354, 1133-1140

Dunning Jr, T. H. 1989, *J. Chem. Phys.* 90, 1007

Frisch, M. J., Trucks, G. W., Schlegel, H. B., Scuseria, G. E., Robb, M. A., Cheeseman, J. R., Scalmani, G., Barone, V., Mennucci, B., Petersson, G. A., Nakatsuji, H., Caricato, M., Li, X., Hratchian, H. P., Izmaylov, A. F., Bloino, J., Zheng, G., Sonnenberg, J. L., Hada, M., Ehara, M., Toyota, K., Fukuda, R., Hasegawa, J., Ishida, M., Nakajima, T., Honda, Y., Kitao, O., Nakai, H., Vreven, T., Montgomery Jr., J. A., Peralta, J. E., Ogliaro, F., Bearpark, M., Heyd, J. J., Brothers, E., Kudin, K. N., Staroverov, V. N., Kobayashi, R., Normand, J., Raghavachari, K., Rendell, A., Burant, J. C., Iyengar, S. S., Tomasi, J., Cossi, M., Rega, N., Millam, J. M., Klene, M., Knox, J. E., Cross, J. B., Bakken, V., Adamo, C., Jaramillo, J., Gomperts, R., Stratmann, R. E., Yazyev, O., Austin, A. J., Cammi, R., Pomelli, C., Ochterski, J. W., Martin, R. L., Morokuma, K., Zakrzewski, V. G., Voth, G. A., Salvador, P., Dannenberg, J. J., Dapprich, S., Daniels, A. D., Farkas, O., Foresman, J. B., Ortiz, J. V, Cioslowski, J., & Fox, D. J. 2010, Gaussian 09, Revision B.01

Constellation Guide 2013, Eagle Nebula - Messier 16 URL: http://www.constellation-guide.com/eagle-nebula-messier-16/ (accessed 4.2.17)

Johari, G. P., & Whalley, E. 1981, K. *J. Chem. Phys.* 75, 1333-1340

Miertuš, S., Scrocco, E., & Tomasi, J. 1981, *Chem. Phys.* 55, 117-129

Öberg, K. I., Linnartz, H., Visser, R., & van Dishoeck, E. F. 2009, *ApJ* 693, 1209-1218

Park, J. Y., & Woon, D. E. 2004, *J. Phys. Chem. A.* 108, 6589-6598

Pilling, S., Baptista, L., Boechat-Roberty, H. M., & Andrade, D. P. P. 2011a, *Astrobiology* 11, 883-893

Pilling, S. & Bergantini, A. 2015, *ApJ* 811, 151
Pilling, S., Duarte, E. S., Domaracka, A., Rothard, H., Boduch, P., & Silveira, E. F. Da 2011b *Phys. Chem. Chem. Phys.* 13, 15755-15765
Sun, Y. F., & Goldberg, D. 2005, *Geophys. Res. Lett.* 32, L04313
Tsekouras, A. A., Iedema, M. J., & Cowin, J. P. 1998, *Phys. Rev. Lett.* 80, 5798-5801
Van Dishoeck, E. F., Herbst, E., & Neufeld, D. A. 2013, *Chem. Rev.* 113, 9043-9085
Woon, D. E., & Park, J. 2004, *ApJ. J.* 607, 342-345

Astrochemistry VII – Through the Cosmos from Galaxies to Planets
Proceedings IAU Symposium No. 332, 2017
M. Cunningham, T. Millar & Y. Aikawa, eds.

doi:10.1017/S1743921317007864

Photodissociation processes of Bisanthenquinone cation

Tao Chen[1,2], Junfeng Zhen[2,3], Ying Wang[1], Harold Linnartz[3] and Alexander G. G. M. Tielens[2]

[1]Department of Theoretical Chemistry and Biology, School of Biotechnology, Royal Institute of Technology, 10691, Stockholm, Sweden
[2]Leiden University, Leiden Observatory, Niels Bohrweg 2, NL-2333 CA Leiden, Netherlands
[3]Sackler Laboratory for Astrophysics, Leiden Observatory, University of Leiden, P.O. Box 9513, 2300 RA Leiden, The Netherlands
email: chen@strw.leidenuniv.nl

Abstract. A systematic study, using ion trap time-of-flight mass spectrometry, is presented for the photo-dissociation processes of Bisanthenquinone (Bq) cations, $C_{28}H_{12}O_2{}^+$, a ketone substituted Polycyclic Aromatic Hydrocarbon (PAH). The Bq cation fragments through sequential loss of the two neutral carbonyl (CO) units upon laser (626nm) irradiation, resulting in a PAH-like derivative $C_{26}H_{12}{}^+$. Upon further irradiation, $C_{26}H_{12}{}^+$ exhibits both stepwise dehydrogenation and C_2/C_2H_2 loss fragmentation channels. Quantum chemistry calculations reveal a detailed picture for the first CO-loss, which involves a transition state with a barrier of ~ 3.4 eV, which is lower than the energy required for the lowest H-loss pathway (~ 5.0 eV). The barrier for the second CO-loss is higher (~ 4.9 eV). The subsequent loss of this unit changes the Bq geometry from a planar to a bent one. It is concluded that the photodissociation mechanism of the substituted PAH cations studied here is site selective in the substituted subunit. This work also shows that an acetone substituted PAH cation is not photo-stable upon irradiation.

Keywords. PAH, molecule, photodissociation, quinone, carbonyl, TOF

1. Introduction

Bisanthenquinone (Bq) molecules are commonly referred to as Polycyclic Aromatic Hydrocarbon (PAH) quinones. They form a subclass within the PAH family, molecules that are important from a physical chemical point of view and that also act as an important component of the interstellar medium (ISM) (Tielens (2008)). The vibrational signatures of PAHs dominate the mid-infrared spectra of most objects in space and they are key contributors to the energy and ionization balance of the gas. These PAHs are thought to form in stellar ejecta. Subsequently they are further processed for some hundred million years in the harsh environment of the ISM (Tielens (2005)). As part of this evolution, PAHs may acquire functional groups such as methyl, methoxy, and hydroxyl groups (Bernstein *et al.* (2002)). Little information is available on how photolysis affects substituted derivatives of PAH cations (Jochims *et al.* (1999), Rapacioli *et al.* (2015)). Such species contain highly active reaction centers that will affect the photodissociation efficiency and photoisomerization process (Lifshitz (1997), Bernstein *et al.* (2002), Walsh (2008)). This has been studied for methyl- and methoxy-substituted derivatives of hexa-peri-hexabenzocoronene (HBC) cations, i.e., $(OCH_3)_6HBC^+$ and $(CH_3)_4(OCH_3)_2HBC^+$ that dissociate through the loss of CH_3CO units upon laser (595 nm) irradiation, resulting in a PAH-like derivative $C_{36}H_{12}^+$ and a methyl-substituted PAH derivative $C_{44}H_{24}^+$,

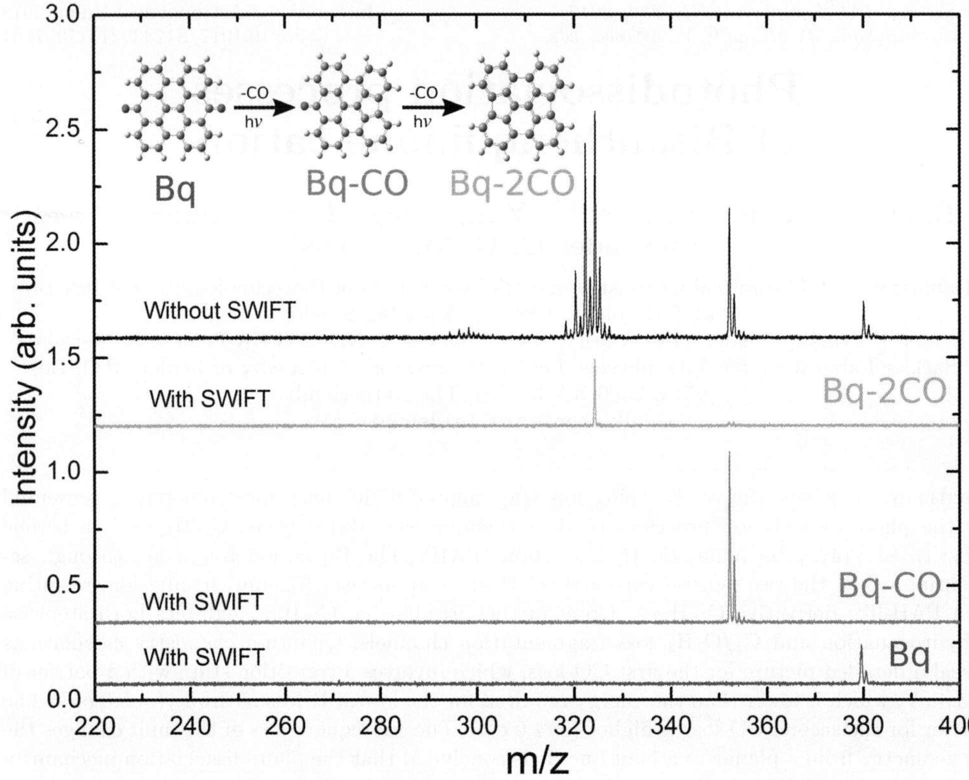

Figure 1. The mass spectrum of Bq cation, before (black curve) and after (green, red and blue curves) SWIFT signal isolation. The blue curve at mass 380.1 amu corresponds to the molecule as shown in the upper-left inset (Bq). The red curve at mass 352.1 amu corresponds to the molecule as shown in the upper-left inset (Bq-CO). The green curve at mass 324.1 amu corresponds to the molecule as shown in the upper-left inset (Bq-2CO). The small peaks on the right side of the main peak in graphs A and B are isotopic contributions from ^{13}C or ^{18}O. SWIFT isolation effectively excludes isotopes in the mass spectrum of Bq-2CO cation.

respectively (Zhen *et al.* (2016)). In this work, we present a study of Bq cations for understanding the effect of acetone site-substitution on photostability and reactivity behavior of PAHs and to elucidate the possible two step mechanism in the photolysis of methoxylated PAHs (e.g., loss of a CH_3 unit followed by CO elimination). The experiments are conducted by using quadrupole ion trap time-of-flight (QIT-TOF) mass spectrometry. Density functional theory (DFT) calculations are performed to explore the dissociation process of Bq cation.

2. Experimental setup

The experiments on photo-fragmentation processes of Bq cations are conducted with a quadrupole ion trap (QIT) time-of-flight (TOF) system: i-PoP (instrument for Photodissociation of PAHs) developed by Zhen *et al.* (2014). In general, it works in the following way: commercially available Bq powder (purity higher than 99.0 % from Kentax) is heated to ~ 500 K in an oven. Subsequently, the evaporated molecules are ionized by an electron gun and guided into the ion trap. Once the ions are trapped, a Stored Waveform Inverse Fourier Transform (SWIFT) excitation technique developed by Doroshenko & Cotter (1996) is applied to isolate a specific range of mass/charge (m/z) species or a

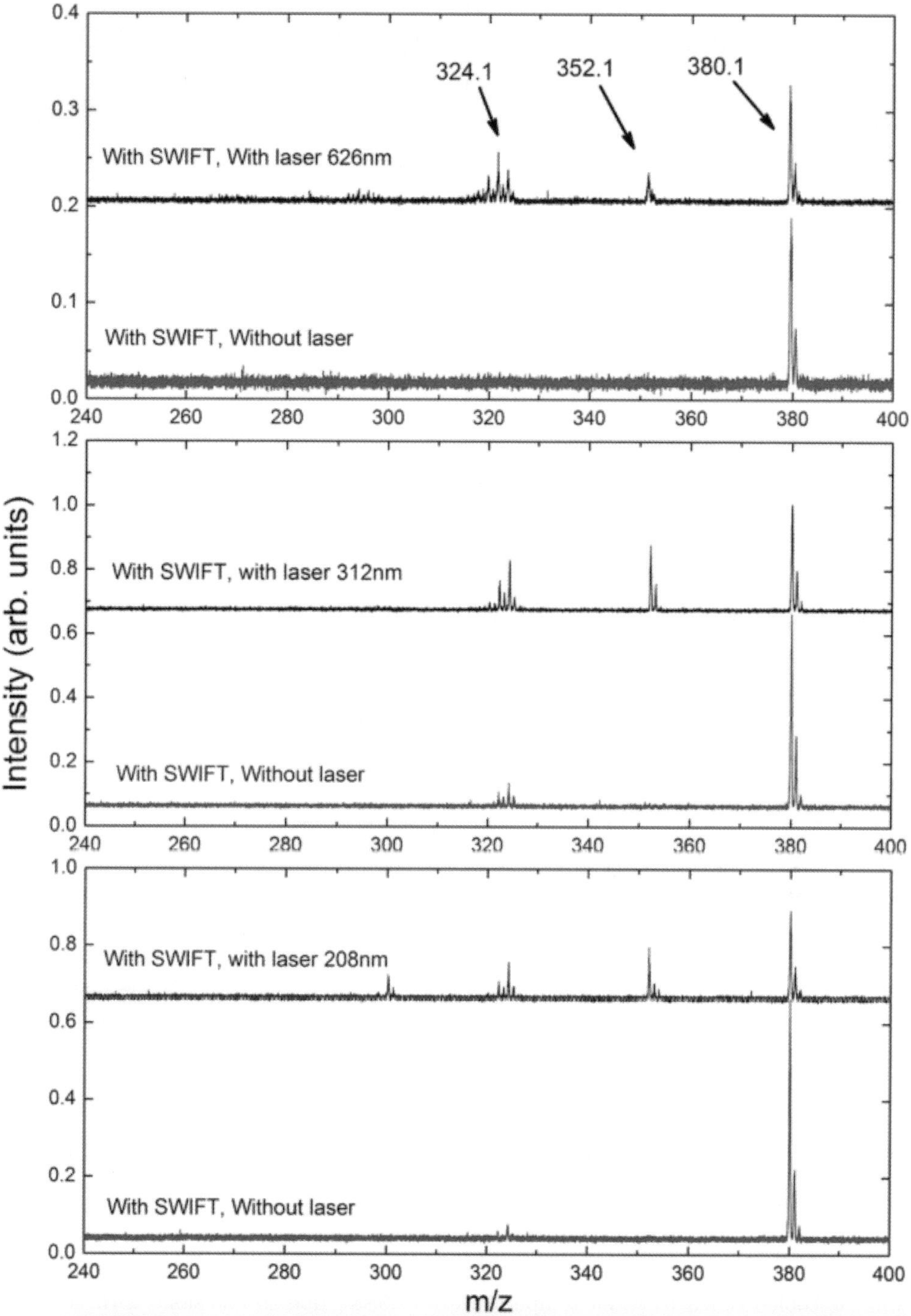

Figure 2. Mass-spectrum of Bq cation, before and after irradiation by 626 nm, 312 nm, and 208 nm laser light. Bq (380.1 amu) is dissociated to Bq-CO (352.1 amu) and Bq-2CO (324.1 amu) at all given wavelength. The black curves represent the selected molecules with SWIFT and with irradiation of laser. The blue curves are SWIFT selected Bq without laser irradiation. No other peaks are observed at Bq and Bq-CO except isotopes, meaning that CO is the dominant dissociation channel for Bq and Bq-CO.

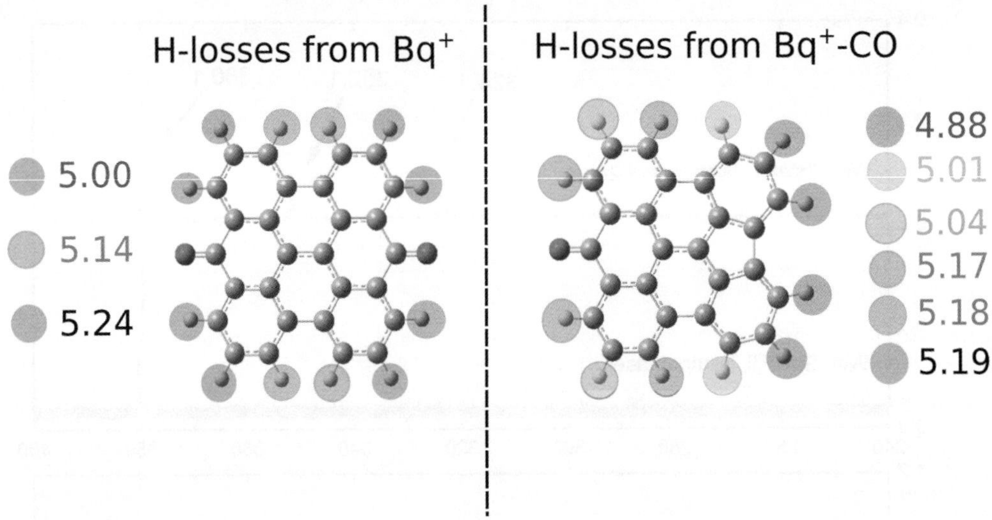

Figure 3. The dissociation energies for H-losses from Bq and Bq-CO cation. All values are given in eV.

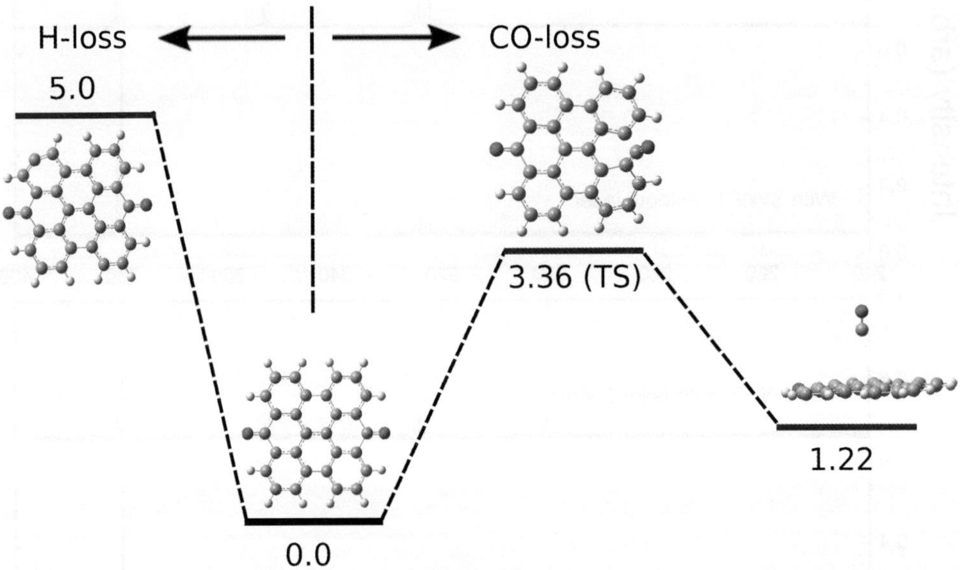

Figure 4. Calculated dissociation energies and reaction barriers for H-loss and first CO-loss from Bq cation. All values are given in eV.

single peak species (i.e. $^{12}C_{26}H_{12}{}^{+}$). After a short time delay (typically $\sim$ 0.2s), the ion cloud is cooled down to room temperature ($\sim$ 298K) due to collisions with He buffer gas that is continuously added to the trap. Subsequently, the ion cloud is irradiated by the light pulses generated by a tunable dye laser (LIOP-TEC, Quasar2-VN) pumped by a Nd:YAG laser (DCR-3, Spectra-Physics), operated at 10Hz. DCM is used to produce 626 nm light as well as 312 and 208 nm radiation through doubling and tripling. As no clear wavelenghth dependency is found, we only present the results obtained upon 626 nm irradiation. The fragments or intact molecules after irradiation are accelerated by a negative square pulse to transport them from the trap to the field-free TOF region. The signals are finally detected by a MicroChanel Plates (MCP) detector.

3. Computational methods

Our theoretical calculations are carried out using DFT. The dissociation energies, transition state energies and charge distributions presented in this work are calculated using the hybrid density functional B3LYP method as implemented in the Gaussian 09 program. All structures are optimized using the 6-31G(d) basis set. The vibrational frequencies are calculated to confirm that the structures correspond to minima (no imaginary frequencies) or transition states (one imaginary frequency). We have taken the Zero Point Vibrational Energy (ZPVE) into account. The ZPVE values are scaled by the empirical factor 0.9806 from Scott & Radom (1996) to correct for anharmonic effects. The Potential Energy Surface (PES) of all possible dissociation pathways is scanned for transition state calculations. Intrinsic reaction coordinate calculations are performed to confirm that the transition state structures are connected to their corresponding local minima.

4. Results and Discussion

Figure 1 shows the mass spectrum of the trapped Bq cations. The top mass spectrum is obtained without laser irradiation and without SWIFT signal isolation. It illustrates by Beynon *et al.* (1959) that the parent cations experience substantial fragmentation due to the electron impact ionization. The improvement in mass purity becomes clear from the lower three curves, where mass peaks at m/z=380.1 for the precursor species (Bq cation), at 352.1 for the first photofragment (Bq-CO cation) and at 324.1 for the second photofragment (Bq-2CO cation) are shown using the SWIFT pulse isolation. The mass spectra of Bq and Bq-CO cation show that the relatively small isotopic contributions from ^{13}C or ^{18}O, at m/z = 381.1 & 382.1 and 353.1 & 354.1 are suppressed effectively by the SWIFT pulse. Isolation of the parent mass peak, excludes ^{13}C contributions in the mass spectrum of Bq-2CO cation.

Figure 2 shows the mass-spectrum of Bq cation, before and after irradiation by 626 nm, 312 nm, and 208 nm laser light. There exist no clear wavelength dependency and therefore below only the 626 nm data are presented. As shown in Figure 1, the precursor molecule is located at m/z=380.1. Two main fragments are found with peaks at m/z = 352.1 and 324.1. The separation between precursor and first fragment peak as well as between first and second fragment peak amounts to 28 Da mass differences. This corresponds to one oxygen plus one carbon, suggesting an effective CO unit (m/z = 28) fragmentation channel. Also these two peaks are separated by a 28 Da mass difference. Moreover, no other peaks are observed at Bq and Bq-CO except isotopes, the loss of hydrogens cannot be detected in the experiments. The interpretation of these observations is that Bq photofragmentation is governed by the sequential loss of the two CO units, transfering Bq into Bq-CO and Bq-2CO. It is interesting to note that this CO-loss channel is clearly preferred above dehydrogenation, i.e., fragmentation through H-atom loss. For many regular PAHs, i.e., PAHs without side group additions, H-loss and 2H/H_2-loss are dominant dissociation channels (Chen *et al.* (2015)). As reported by Zhen *et al.* (2016) PAHs with side group additions, other dissociation pathways have been found, e.g. methoxy groups.

Figure 3 shows calculated dissociation energies for all possible H-loss pathways from Bq and Bq-CO cation. In the Bq case, with its high level of symmetry, only three different types of H-atoms can be distinguished. The lowest dissociation energy for H-loss is found for H-atoms dissociating from the corner locations and amounts to 5.0 eV. For the Bq-CO cation the PAH geometry looses its symmetry and the lowest dissociation energy (4.88 eV) is now found for the two H-atoms that were closest to the lost CO-unit.

Figure 4 describes the CO and H dissociation pathways from Bq. The lowest dissociation pathway for H-loss is a one-step process without any transition state involved, and the calculated barrier is 5 eV. The CO-loss starts with the dissociation of one of the two C-C bonds (between Bq carbon and carbonyl). A transition state, in which the CO is connected with only one carbon atom of Bq, is formed by overcoming a barrier of 3.36 eV. The partially free CO in this intermediate can rotate along the binding carbon and upon breaking the bond, move out from the Bq-plane, resulting in the formation of a pentagon. Such process is similar to the photo-dissociation of methoxy-substituted derivatives of HBC cations Zhen *et al.* (2016), in which pentagons are formed after loss of a CH_3CO unit. The carbonyl keeps moving towards the center of the Bq-plane and it finally leaves vertically from the middle of the Bq-plane. Although the CO-loss pathway contains a transition state, the barrier is only 3.36 eV, which is much lower than the barrier of the H-loss (5 eV) from Bq. The loss of one CO-unit, therefore, is favoured above that of an H-atom, fully consistent with the experimental findings.

5. Conclusion

Experiments and quantum chemistry calculations are performed for understanding the photo-dissociation processes of Bq cations. The mass spectrum of Bq cation upon laser (626nm) irradiation presents clear evidence for pure neutral CO-losses, i.e., no other fragments are detected as a first or second dissociation product. Our quantum chemistry calculation reveals that for Bq the transition barrier for CO-loss is only $\sim$ 3.4 eV, which is much lower than alternative routes. We expect that CO loss from larger PAH quinones will also lead to pentagon formation and we speculate that this will facilitate the isomerization of large PAHs to cages and fullerenes in the interstellar medium (Berné & Tielens(2012)).

The results reported here are interesting also from a different point of view. Pentagon formation is important to understand how PAHs can transfer into carbon cages and fullerenes, as proposed by Berné & Tielens(2012) and experimentally investigated by Zhen *et al.* (2014). Several pathways have been discussed, and a recent study by de Haas *et al.* (2017) show the facile pentagon formation upon irradiation of PAHs. Here it is shown that pentagon formation can also be initialized through selective side group dissociation.

6. Acknowledgments

This work is supported by Swedish Research Council (Contract No. 2015-06501). Facility is supported by the Swedish National Infrastructure for Computing (SNIC). We acknowledge the European Union (EU) and Horizon 2020 funding awarded under the Marie Skłodowska Curie action to the EUROPAH consortium, grant number 722346. Studies of interstellar PAHs at Leiden Observatory are supported through a Spinoza award.

References

G. G. M. Tielens, *Annu. Rev. Astron. Astrophys.*, 2008, **46**, 289.

A. G. G. M. Tielens, The Physics and Chemistry of the Interstellar Medium, *Cambridge Univ Press, Cambridge, UK*, 2005.

M. P. Bernstein, J. E. Elsila, J. P. Dworkin, S. A. Sandford, L. J. Allamandola & R. N. Zare, *Astrophys. J.*, 2002, **576**, 1115.

A. I. S. Holm, H. A. B. Johansson, H. Cederquist, & H. Zettergren, *J. Chem. Phys.*, 2011, **134**, 044301.

T. Chen, M. Gatchell, M. H. Stockett, R. Delaunay, A. Domaracka, E. R. Micelotta, A. G. G. M. Tielens, P. Rousseau, L. Adoui, B. A. Huber, H. T. Schmidt, H. Cederquist, & H. Zettergren, *J. Chem. Phys.*, 2015, **142**, 144305.

O. Berné, & A. G. G. M. Tielens, *Proc. Natl. Acad. Sci., & USA*, 2012, **109**, 401.

J. Zhen, P. Castellanos, D. M. Paardekooper, H. Linnartz, & A. G. G. M. Tielens, *Astrophys. J. Lett.*, 2014, **797**, L30.

R. Walsh, *Chem. Soc. Rev.*, 2008, **37**, 686.

H. W. Jochims, H. Baumgartel, & S. Leach, *Astrophys. J.*, 1999, **512**, 500.

M. Rapacioli, A. Simon, C. C. M. Marshall, J. Cuny, D. Kokkin, F. Spiegelman, & C. Joblin, *J. Phys. Chem. A*, 2015, **119**, 12845.

J. Zhen, P. Castellanos, H. Linnartz, & A. G. G. M. Tielens, *Mol. Astrophys.*, 2016, **5**, 1.

C. Lifshitz, *Int. Rev. Phys. Chem.*, 1997, **16**, 113.

J. Zhen, D. M. Paardekooper, A. Candian, H. Linnartz, & A. G. G. M. Tielens, *Chem. Phys. Lett.*, 2014, **592**, 211.

M. V. Doroshenko, & R. J. Cotter, *Rapid Commun. Mass Spectrum.*, 1996, **10**, 65.

A. P. Scott, & L. Radom, *J. Phys. Chem.*, 1996, **100**, 16502.

J. H. Beynon, G. R. Lester, & A. E. Williams, *J. Phys. Chem.*, 1959, **63**, pp 1861

A. J. de Haas, J. Oomens, & J. Bouwmana, *Phys. Chem. Chem. Phys.*, 2017, in press, **DOI: 10.1039/C6CP08349H**.

Astrochemistry VII – Through the Cosmos
from Galaxies to Planets
Proceedings IAU Symposium No. 332, 2017
M. Cunningham, T. Millar & Y. Aikawa, eds.

doi:10.1017/S1743921317007827

CH_3NCO detections in observations and the laboratory

N. F. W. Ligterink and the PILS team

Sackler Laboratory, Leiden Observatory, Leiden University,
PO Box 9513, 2300 RA Leiden, The Netherlands

Abstract. Methyl isocyanate (CH_3NCO) belongs to a select group of peptide-like prebiotic molecules. In this paper we present its first detection toward the solar type low-mass protostar IRAS16293-2422 (hereafter IRAS16293). CH_3NCO is detected towards IRAS16293 as a warm component with $T_{ex} > 100$ K and HNCO/CH_3NCO ∼4-12. Also, its grain surface formation route is investigated in the laboratory. VUV processing of CH_4:HNCO mixtures, investigated by IR spectroscopy and mass spectrometry, revealed that it can be formed by reactions of CH_3 and with (H)NCO. Observations and experiments strongly hint that methyl isocyanate is formed on interstellar dust grains.

---

1. INTRODUCTION

Within astrochemistry prebiotic complex organic molecules hold a special place, due to their link to life-bearing molecules such as amino acids. Methyl isocyanate, CH_3NCO, is a molecule that is structurally similar to a peptide bond, the bond that links amino acids together. It belongs to a larger family of molecules that include isocyanic acid (HNCO) and formamide (NH_2CHO). These last two molecules have been observed in a variety of sources, including the sun-like low mass protostar IRAS16293-2422 (Bisschop *et al.* 2007; López-Sepulcre *et al.* 2015; Coutens *et al.* 2016). IRAS16293 (d = 120 pc) is considered as a protostellar template for low-mass sources and is particularly rich in organic molecules (van Dishoeck *et al.* 1995; Cazaux *et al.* 2003; Bottinelli *et al.* 2004; Kuan *et al.* 2004; Bisschop *et al.* 2008; Jaber *et al.* 2014). However, methyl isocyanate only started to gather interest in the last two years, after a detection was claimed on the comet 67P/Churyumov-Gerasimenko (Goesmann *et al.* 2015; but see for a rebuttal Altwegg *et al.* 2017). This triggered the interstellar observation of CH_3NCO toward Sgr B2(N) and Orion KL (Halfen *et al.* 2015; Cernicharo *et al.* 2016).

The astrochemical origin of methyl isocyanate is not yet understood and this is partly due to the limited number of laboratory studies that have been performed (Ruzi & Anderson 2012; Henderson & Gudipati 2015). A number of formation routes have been hypothesized by astrochemists, with one of the more prominent ones being the solid-state methylation (CH_3 addition) to (H)NCO proposed by Cernicharo *et al.* (2016); Belloche *et al.* (2017):

$$\mathrm{(H)NCO(s) + CH_3(g,s) \rightarrow CH_3NCO(s) + H(s)} \tag{1.1}$$

Table 1. Comparison of molecular abundance ratios toward different sources

Source	$HNCO/CH_3NCO$	CH_3CN/CH_3NCO	CH_3OH/CH_3NCO
IRAS16293 B	12	5	3333
IRAS16293 A	4	11	3200
Orion KL A[a]	15	10	400
Orion KL B[a]	15	25	225
Sgr B2(N)[a]	40	50	40
Comet 67P/C-G[b]	>0.2	>0.2	–

[a] *Cernicharo et al. (2016),*[b] *Goesmann et al. (2015); Altwegg et al. (2017)*

2. Observations

We searched for methyl isocyanate in the Protostellar Interferometric Line Survey (PILS) data, an unbiased spectral survey of the low-mass protostellar binary IRAS16293-2422 with ALMA (Jørgensen *et al.* 2016). Methyl isocyanate is detected toward both A and B component. In source B we have been able to identify 43 unblended lines of CH_3NCO, while in source A 11 unblended lines are found. Excitation temperatures of 100 and 300 K give good agreement with the observations. Our data are not sensitive to any cold CH_3NCO component since lines with low E_{up} values are missing in the spectral range of the PILS survey. The CH_3NCO column density is found to be 3-4 $\times 10^{15}$ cm^{-2} in source B and 6-9 $\times 10^{15}$ cm^{-2} in source A, for $T_{ex} = 300$ and 100 K, respectively in a 0.5" beam. An integrated intensity map of one of the brightest CH_3NCO lines is presented in Figure 1. The two isomers of methyl isocyanate - acetonitrile oxide (CH_3CNO) and methyl cyanate (CH_3OCN) - are not found and are less abundant than methyl isocyanate by at least two and one orders of magnitude, respectively.

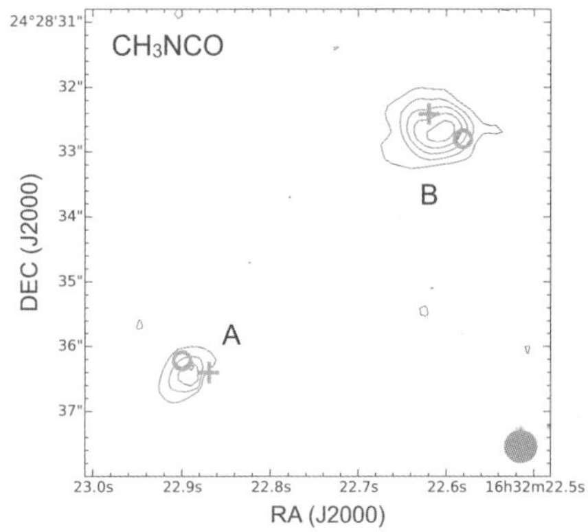

Figure 1. ALMA integrated intensity map of the CH_3NCO $39_{0,39,0}-38_{0,38,0}$ transition at 336339.9 MHz. The positions of the continuum of source A (South East source) and source B (North West source) are indicated with red crosses, while the positions studied in this paper are indicated with red circles.

The column densities of CH_3NCO have been compared with those of HNCO, CH_3CN and CH_3OH (see Table 1, Coutens *et al.* 2016, Calcutt *et al.* in prep., Jørgensen *et al.* subm.). Within IRAS16293 no significant differences are seen between source A and B. In general the abundance ratios are comparable to the values found in Orion KL (Cernicharo *et al.* 2016) and slightly lower than toward Sgr B2. Methanol is, however, clearly more abundant than methyl isocyanate toward IRAS16293 than toward Orion KL and Sgr B2.

3. Laboratory experiments

Methylation of HNCO was tested in the laboratory by VUV irradiating $^{12/13}CH_4$:HNCO mixtures at a cryogenic temperature of 20 K under ultra high vacuum (UHV) conditions. VUV radiation is used as a tool to form CH_3 radicals, which can further react with HNCO. Chemical changes in the ice are traced by IR spectroscopy and Temperature Programmed Desorption (TPD) mass spectrometry.

The strongest vibrational features of solid methyl isocyanate are found between 2320 and 2280 cm^{-1} for the -N=C=O antisymmetric stretching vibration and overtone $2\nu_7$ CH_3 rocking mode (Sullivan *et al.* 1994; Zhou & Durig 2009; Reva *et al.* 2010). The region around these bands is used to monitor CH_3NCO formation in the ice. In order to identify methyl isocyanate in the gas-phase using TPD, the mass fragmentation pattern available in the NIST database is used. The fragmentation pattern at 70 eV comprises unique peaks at $m/z = 57$ and 56 and these will be used as main TPD mass tracers. When using $^{13}CH_4$, both IR bands and the mass tracers will shift.

Figure 2 presents the IR spectra of VUV irradiated $^{12/13}CH_4$:HNCO samples. Three known spectroscopic features of CO_2, OCN^- and CO (blue) show up during irradiation. However, two new features are seen around 2300 cm^{-1} (red), which do not show up while processing samples of pure HNCO or CH_4. Also a clear redshift of about 10 cm^{-1} of the two features is seen between the sample of $^{12}CH_4$ and $^{13}CH_4$. These spectroscopic features are therefore the result of a reaction between methane and isocyanic acid, and, since they are found close to the positions of known CH_3NCO features, are plausibly identified with methyl isocyanate.

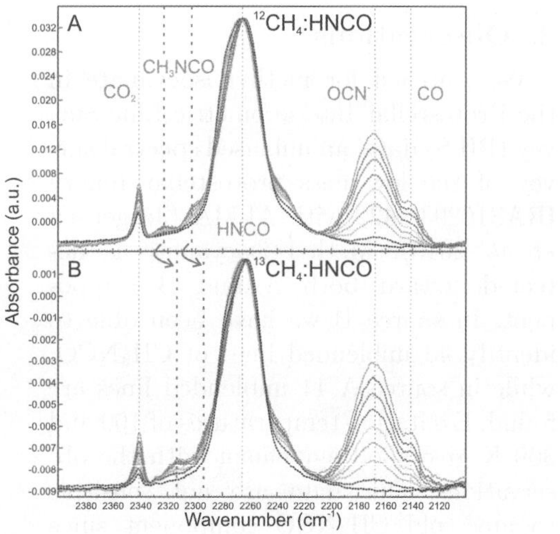

Figure 2. IR spectra taken over time for the $^{12/13}CH_4$:HNCO mixture. Spectroscopic features that coincide with CH_3NCO are found at the red lines and show a clear shift with the $^{13}CH_4$ isotopologue.

In the TPD trace m/z 57 and 56 are seen to desorb from the ice after irradiation, while the same desorption peak is not seen in control experiments. It is not possible to match the database fragmentation pattern of methyl isocyanate to the observed peaks, due to interference with other species in m/z 56, but the assignment to methyl isocyanate is plausible.

4. Conclusion

Methyl isocyanate is detected for the first time toward a low mass protostar, IRAS16293, on solar system scales (emission radius of 60 au). The abundance ratios of CH_3NCO relative to the N-bearing species HNCO and CH_3CN are similar to those found toward Orion KL and deviate by at most an order of magnitude from Sgr B2, making a common pathway in these sources likely. Laboratory experiments strongly suggest that reactions between CH_3 radicals and (H)NCO form CH_3NCO. Both observations and laboratory results hint that the interstellar formation route of methyl isocyanate should be found in icy grains.

The detection of CH_3NCO adds to the growing list of complex molecules known to be present around solar mass protostars, showing that the ingredients for prebiotic molecules are abundant. Future deeper searches for even more complex molecules relevant for the origin of life are warranted.

For more details on this work see Ligterink *et al.* (2017).

Acknowledgement

This work has been made possible by valuable contributions from A. Coutens, V. Kofman, H. S. P. Müller, R. T. Garrod, H. Calcutt, S. F. Wampfler, J. K. Jørgensen, H. Linnartz and E. F. van Dishoeck.

References

Altwegg, K., Balsiger, H., Berthelier, J. J., *et al.* 2017, *MNRAS*, 469, S130
Belloche, A., Meshcheryakov, A. A., Garrod, R. T., *et al.* 2017, *A&A*, 601, A49
Bisschop, S. E., Jørgensen, J. K., Bourke, T. L., Bottinelli, S., & van Dishoeck, E. F. 2008, *A&A*, 488, 959
Bisschop, S. E., Jørgensen, J. K., van Dishoeck, E. F., & de Wachter, E. B. M. 2007, *A&A*, 465, 913
Bottinelli, S., Ceccarelli, C., Neri, R., *et al.* 2004, *ApJ*, 617, L69
Cazaux, S., Tielens, A. G. G. M., Ceccarelli, C., *et al.* 2003, *ApJ*, 593, L51
Cernicharo, J., Kisiel, Z., Tercero, B., *et al.* 2016, *A&A*, 587, L4
Coutens, A., Jørgensen, J. K., van der Wiel, M. H. D., *et al.* 2016, *A&A*, 590, L6
Goesmann, F., Rosenbauer, H., Bredehöft, J. H., *et al.* 2015, *Science*, 349
Halfen, D. T., Ilyushin, V. V., & Ziurys, L. M. 2015, *ApJ*, 812, L5
Henderson, B. L. & Gudipati, M. S. 2015, *ApJ*, 800, 66
Jaber, A. A., Ceccarelli, C., Kahane, C., & Caux, E. 2014, *ApJ*, 791, 29
Jørgensen, J. K., van der Wiel, M. H. D., Coutens, A., *et al.* 2016, *A&A*, 595, A117
Kuan, Y.-J., Huang, H.-C., Charnley, S. B., *et al.* 2004, *ApJ*, 616, L27
Ligterink, N. F. W., Coutens, A., Kofman, V., *et al.* 2017, *MNRAS*, 469, 2219
López-Sepulcre, A., Jaber, A. A., Mendoza, E., *et al.* 2015, *MNRAS*, 449, 2438
Reva, I., Lapinski, L., & Fausto, R. 2010, *J. Mol. Struc.*, 976, 333–341
Ruzi, M. & Anderson, D. 2012, *J. Chem. Phys.*, 137, 194313
Sullivan, J., Heusel, H., Zunic, W., & Durig, J. 1994, *Spectrochimica Acta*, 50
van Dishoeck, E. F., Blake, G. A., Jansen, D. J., & Groesbeck, T. D. 1995, *ApJ*, 447, 760
Zhou, S. & Durig, J. 2009, *J. Mol. Struc.*, 924

Astrochemistry VII – Through the Cosmos
from Galaxies to Planets
Proceedings IAU Symposium No. 332, 2017
M. Cunningham, T. Millar & Y. Aikawa, eds.

doi:10.1017/S1743921317011619

Recent Progress in Laboratory Astrophysics Achieved with NASA Ames' COSmIC Facility

Farid Salama[1], Ella Sciamma-O'Brien [1,2], Cesar S. Contreras [1,2] and Salma Bejaoui[1]

[1]NASA-Ames Research Center, Space Science & Astrobiology Division, Moffett Field, California, USA

[2]Bay Area Environmental Research Institute, Petaluma, California, USA

Abstract. We describe the characteristics and the capabilities of the laboratory facility, COSmIC, that was developed at NASA Ames to generate, process and analyze interstellar, circumstellar and planetary analogs in the laboratory. COSmIC stands for 'Cosmic Simulation Chamber' and is dedicated to the study of neutral and ionized molecules and nanoparticles under the low temperature and high vacuum conditions that are required to simulate various space environments such as diffuse interstellar clouds, circumstellar outflows and planetary atmospheres. Recent results obtained using COSmIC will be highlighted. In particular, the progress that has been achieved in the domain of the diffuse interstellar bands (DIBs) and in monitoring, in the laboratory, the formation of circumstellar dust grains and planetary atmosphere aerosols from their gas-phase molecular precursors. Plans for future laboratory experiments on interstellar and planetary molecules and grains will also be addressed, as well as the implications of the studies underway for astronomical observations and past and future space mission data analysis.

Introduction

The Cosmic Simulation Chamber (COSmIC) was developed to generate, process, and analyze interstellar, circumstellar and planetary analogs in the laboratory (Salama (2008)). COSmIC is dedicated to the study of neutral and ionized molecules and nanoparticles under the low temperature and high vacuum conditions that are required to simulate space environments. The COSmIC experimental setup (Figures 1 and 2) is composed of a Pulsed Discharge Nozzle (PDN) expansion, that generates a plasma in the stream of a free supersonic jet expansion, coupled to high-sensitivity, complementary in situ diagnostics: cavity ring down spectroscopy (CRDS) and laser induced fluorescence (LIF) systems for photonic detection, and Reflectron Time-Of-Flight Mass Spectrometer (ReTOF-MS) for mass detection in real time (Rickets, *et al.* (2011)). *read more at:* `https://www.nasa.gov/feature/ames/cosmic`

1. INTERSTELLAR APPLICATIONS: Astronomical Surveys: Searching for the Signature of Neutral and Ionized PAHs in Absorption and in Emission

1.1. *Absorption Spectroscopy*

Cavity ring-down spectra of neutral and ionized PAH molecules isolated in a free cold jet expansion with Ar buffer gas have been measured in COSmIC, and then compared

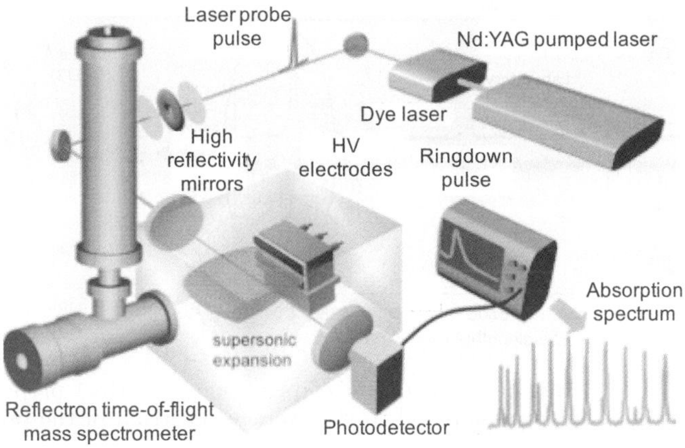

Figure 1. Schematic of the COSmIC experimental setup.

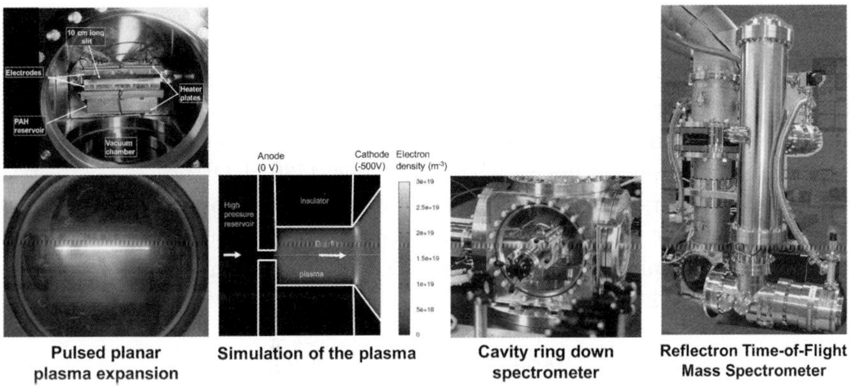

Figure 2. (Left) Picture of the pulsed discharge nozzle with and without plasma discharge. (Middle left) Simulation of the plasma discharge (Broks *et al.* (2015)). (Middle right) Picture of the cavity ringdown spectrometer (CRDS). (Right) Picture of the reflectron time-of-flight mass spectrometer.

to the average spectrum (S/N) of the program reddened stars along lines of sight probing translucent interstellar clouds (Salama, *et al.* (2011)). Upper limits for the column densities of the PAH molecular carriers were calculated at the wavelengths associated with specific vibronic transitions of the PAH molecules according to the formula (Frisch (1972)):

$N(cm^{-2}) = 1.13 \cdot 10^{20} EW(\AA)/\lambda(\AA)^2 \cdot f$ where EW is the equivalent width and f is the oscillator strength or the 'f value' of the vibronic transition that is being considered.

The comparison of these unique laboratory data with high-resolution, high signal-to-noise ratio spectra (Figure 3) led to the following findings:

- The abundance of the specific neutral and ionized PAHs examined in this study must be very low in the individual translucent interstellar clouds probed in the survey (PAH features remain below the level of detection).
- The derived column densities for neutral PAHs in these lines-of-sight range from 10^{11} to $10^{13} cm^{-2}$.

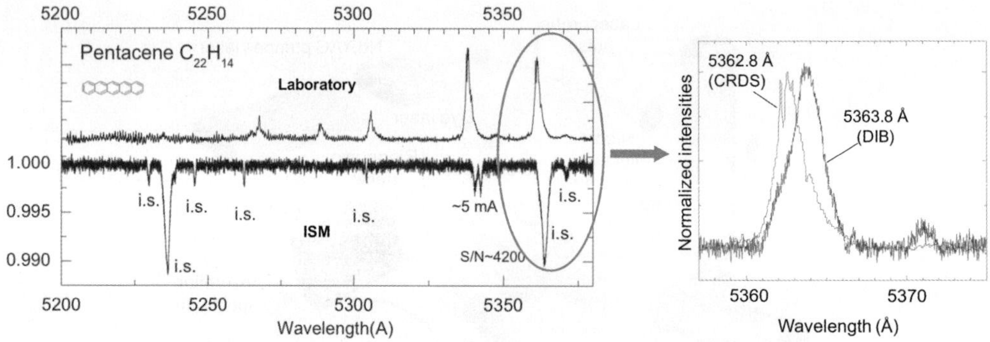

Figure 3. (Left) Cavity ring-down spectrum of pentacene ($C_{22}H_{14}$) prepared in a cold jet expansion with Ar buffer gas is compared to the average spectrum of the program stars (S/N 4200) (Salama, *et al.* (2011)). (Right) Comparison of the 5362.8 Å absorption band of jet-cooled pentacene ($C_{22}H_{14}$) measured in the laboratory with cavity ring-down spectroscopy with a FWHM of 1.9Å and the 5363.8 Å DIB, FWHM = 2.1Å extracted from the average spectrum of the program stars (S/N 2100).

- Neutral PAHs exhibit intrinsic band profiles that are similar to the profile of the narrow DIBs indicating that the carriers of the narrow DIBs must have close molecular structure and characteristics.
- The absorption bands of PAH ions are intrinsically broad (very fast relaxation time). The bands exhibit a close similarity with the strong 4430 Å DIB.

This study was the first quantitative survey of neutral and ionized PAHs in the optical range and it opened the way for unambiguous quantitative searches of PAHs in a variety of interstellar and circumstellar environments. *(read more at:* `nasa.gov/centers/ames/news/releases/2011/11-35AR.html`*)*

1.2. *Emission Spectroscopy*

In preparation for the development of a Laser Induced Fluorescence (LIF) diagnostic on COSmIC, a preliminary study was conducted on a matrix isolation experiment. LIF spectra of neutral PAHs measured in 10 K inert-gas matrices were investigated and compared with the luminescence spectra measured in the Red Rectangle Nebula, a protoplanetary nebula that is one of the brightest known sources of extended red emission as well as of ubiquitous infrared (UIR) band emissions (Bejaoui *et al.* (2005)). These measurements indicate that small neutral PAHs can contribute to the blue fluorescence observed in the Red Rectangle.

2. CIRCUMSTELLAR APPLICATIONS: Formation of carbon dust grains in circumstellar outflows of late carbon stars - From PAHs to carbon nanoparticles

Condensation experiments (Kimura & Saito (2010)) have concluded that 'Evolved stars supply organic materials into space. The organic materials evolve and alter to form more complex organic molecules before incorporation into the star formation stage (Figure 4). PAHs are the most plausible molecules abundantly produced within the gas ejected from evolved stars. During the growth of PAH molecules, and as they move away from the central star, larger PAHs coagulate to form carbon grains.'

In COSmIC, simple hydrocarbons (CH_4, C_2H_2,...) and PAHs seeded in Ar gas can be used as precursors to study grain formation in the gas phase. These Ar-based

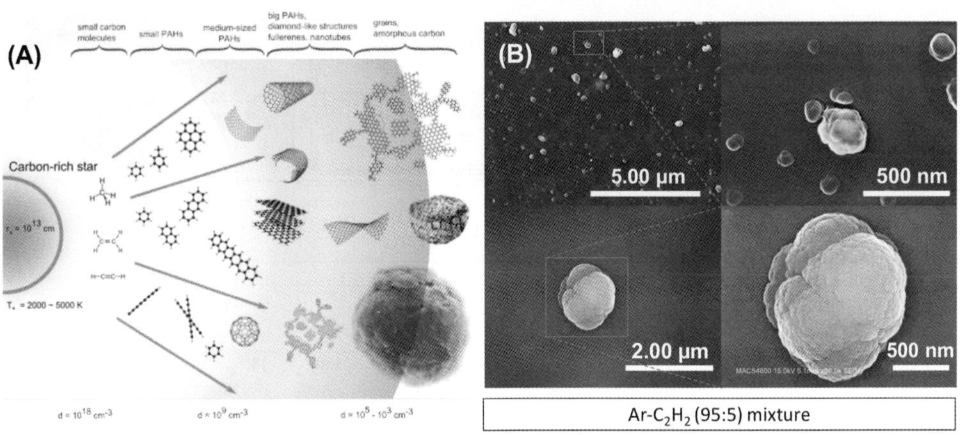

Figure 4. ((Left - A) Cartoon of carbon grain formation in the atmosphere of late-type carbon-rich stars (Contreras & Salama (2013)). (Right - B) SEM images of cosmic dust analogs formed in an Ar-based mixture in COSmIC.

mixtures are injected in the Pulsed Discharge Nozzle, generating a supersonically-cooled expansion in which a plasma discharge dissociates and ionizes the precursors and induces chemistry, producing hydrocarbon ions that can then be extracted and directly detected by Reflectron Time-of-Flight Mass Spectrometry, without fragmentation (Contreras & Salama (2013)).

The chemistry induced by plasma in these hydrocarbon-seeded argon jet expansions results in the formation of soot nanoparticles (nm-sized grains and μm-sized aggregates) that can be deposited onto different substrates for further ex situ analysis like Scanning Electron Microscopy (SEM) as shown in Figure 4. *read more at:* `nasa.gov/press/2014/may/nasa-simulator-successfully-recreates-space-dust`

These Ar-hydrocarbon plasma experiments have shown that:

• The use of cold plasma sources as tools for exploring the formation (and destruction) processes of carbon-bearing molecules is relevant to astrophysical environments.

• The combined techniques of PDN-CRDS-ReTOFMS are useful to study the dynamic behavior of soot particles formed in the plasma.

• Deposition of soot material and further characterization with scanning electron microscopy can provide insight into how the grains form in the gas phase.

• Three distinct types of materials are observed in the deposition of plasma formed carbon grains: bedding (10-50 nm nanoparticles covering the substrate), grains (100-300 nm) and aggregates (500-1000 nm).

• All the material exhibits a cauliflower-like morphology, similar to that of amorphous carbon solid particles.

3. PLANETARY APPLICATIONS: Simulating Titan's atmospheric chemistry at low temperature

Titan, Saturn's largest moon, has a dense atmosphere, mainly composed of nitrogen (N_2 at 95-98%) and methane (CH_4 at 2-5%). In Titan's atmosphere, a complex organic chemistry is induced by solar UV radiation and electron bombardment from Saturn's magnetosphere and leads to the production of heavy neutral, positively and negatively charged molecules and subsequently solid aerosols that form the orange haze surrounding Titan. We use the Titan Haze Simulation (THS) experiment on COSmIC to study Titan's

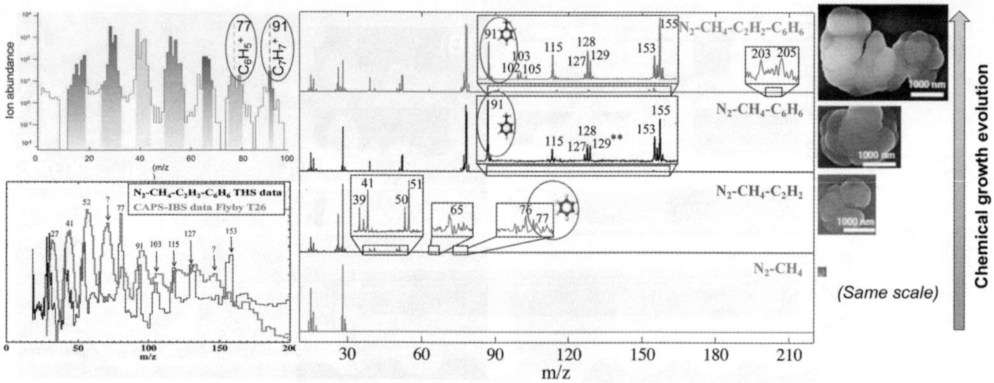

Figure 5. (Top left) INMS positive ion spectrum showing the peaks at 77 and 91 also observed in the THS mass spectra. (Top bottom) Comparison between the Cassini CAPS-IBS spectrum from Flyby T26 (red) (adapted from Crary *et al.* (2009)) and the THS experimental spectrum obtained in the N_2-CH_4-C_2H_2-C_6H_6 plasma (black) degraded to CAPS-IBS' resolution. (Courtesy of J. Westlake.) (Middle) THS mass spectra of the positive ions produced in N_2-CH_4, N_2-CH_4-C_2H_2, N_2-CH_4-C_6H_6, and N_2-CH_4- C_2H_2 -C_6H_6 gas mixtures, showing the chemical growth evolution in the THS experiment when gradually injecting heavier hydrocarbons in the initial gas mixture. (Right) SEM images of aggregates formed in the same four gas mixtures.

atmospheric chemistry and aerosol evolution at low temperature. In the THS, the gas is cooled to Titan-like temperature (~150 K) in the supersonic expansion of the PDN before inducing the chemistry by pulsed plasma, and remains at low temperature in the plasma discharge (~200 K). The pulsed nature of the plasma results in a truncated chemistry. Different N_2-CH_4-based gas mixtures can be injected in the plasma, with or without the addition of heavier precursors present as trace elements on Titan, in order to monitor the evolution of the chemical growth from molecular reactions to the formation of aerosols (Figure 5). Both the gas- and solid phase products resulting from the plasma-induced chemistry can be analyzed (Sciamma-O'Brien, *et al.* (2014), (2017)).

A recent mass spectrometry study (Sciamma-O'Brien, *et al.* (2014)) has shown the unique capability of the THS to monitor the first and intermediate steps as well as specific chemical pathways in Titan's N_2-CH_4 atmospheric chemistry. The mass spectra obtained in various N_2-CH_4-based gas mixtures in the THS experiment show that the N_2-CH_4 chemistry induced in the THS results in the production of species that match certain regions of the observational spectra returned by the Cassini Ion Neutral Mass Spectrometer (INMS) and Cassini Plasma Spectrometer (CAPS) instruments, as shown in Figure 5. The Titan aerosol analogs produced in the THS can be deposited on substrates and characterized ex situ. A scanning electron microscopy analysis of the THS solid aerosols has shown that heavier precursors lead to more complex chemistry and hence to larger aggregates, in agreement with the results of the mass spectometry gas phase analysis, as shown in Figure 5.

The gas and solid phase analyses conducted on the THS experiment have demonstrated that:

• COSmIC can be used to study for the first time, the first an intermediate steps of Titan's chemistry and the early stages of aerosol formation, at Titan like temperature (150 K) with the THS experiments.

• The solid phase analysis is consistent with the gas phase analysis and has shown that when larger molecules are present in the initial mixture, a more complex chemistry has time to occur in the pulsed discharge and produce heavier products and larger aerosols

(Sciamma-O'Brien, *et al.* (2014), (2017)). The THS can therefore be used to study the early stages of aerosol growth on Titan

• The gas phase analysis can be compared to CAPS-IBS data but some regions are missing. By changing the composition of the initial mixture, we will be able to determine what specific chemical pathways lead to the production of the species still missing in our spectra.

Acknowledgments

This work is supported by the NASA Astronomy and Physics Research and Analysis (APRA) and the Solar System Workings Program Elements of the Science Mission Directorate. The authors acknowledge the technical support of R. Walker and E. Quigley. E.S.O. and C.S.C are Research Associates with BAER Institute. S.B. is a NASA Postdoctoral Fellow (NPP) at NASA-Ames Research Center.

References

Bejaoui, S., Salama, F., & Sciamma-O'Brien, E. 2015, *Proceedings of the International Astronomical Union*, 11(A29A).

Broks, B. H. P., Brok, W. J. M., Remy, J., van der Mullen, J. J. A. M., Benidar, A., Biennier, L., & Salama, F. 2005, *Phys. Rev. E*, 71, 036409.

Crary, F.J., Magee, B.A., Mandt, K., Waite Jr., J.H., Westlake, J., Young, D.T., 2009 *Planet. Space Sci*, 57, 1847.

Frisch, P. 1972, *The Astrophys. J.*, 173, 301.

Contreras, C. S., & Salama, F., 2013, *Astrophys J Suppl Ser*, 208, 1,6.

Kimura, Y & Saito, M.,2010, *AAS BAAS # 215, id.301.08* 42, 298.

Ricketts C., Contreras C., Walker, R., Salama F. 2011, *Int. J. Mass Spec*, 300, 26.

&Salama, F., 2008, and references therein *In Organic Matter in Space*, Proceedings of the International Astronomical Union, 4 (S251), 357.

Salama F., Galazutdinov G., Krelowski J., Biennier L., Beletsky Y., In-Ok Song, 2011 *The Astrophys. J.*, 728, 154.

Sciamma-O'Brien, E. Ricketts, C. L., & Salama, F., 2014, *Icarus*, 243, 325-336.

Sciamma-O'Brien, E., Upton, K., & Salama, F., 2017, *Icarus*, 289, 214.

Astrochemistry VII – Through the Cosmos from Galaxies to Planets
Proceedings IAU Symposium No. 332, 2017
M. Cunningham, T. Millar & Y. Aikawa, eds.

doi:10.1017/S1743921317007633

Kinetic Monte Carlo simulations of the grain-surface back-diffusion effect

Eric R. Willis[1] **and Robin T. Garrod**[1,2]

[1]Department of Chemistry, University of Virginia
Charlottesville, VA 22904-4319
email: ew2zb@virginia.edu

[2]Department of Astronomy, University of Virginia
Charlottesville, VA 22904-4325

Abstract. Back-diffusion is the phenomenon by which random walkers revisit binding sites on a lattice. This phenomenon must occur on interstellar dust particles, slowing down dust-grain reactions, but it is not accounted for by standard rate-equation models. Microscopic kinetic Monte Carlo models have been used to investigate the effect of back-diffusion on reaction rates on interstellar dust grains. Grain morphology, size, and grain-surface coverage were varied and the effects of these variations on the magnitude of the back-diffusion effect were studied for the simple H+H reaction system. This back-diffusion effect is seen to reduce reaction rates by a maximum factor of ∼5 for the canonical grain of 10^6 binding sites. The resulting data were fit to logarithmic functions that can be used to reproduce the effects of back-diffusion in rate-equation models.

Keywords. astrochemistry, ISM: molecules, methods: numerical

1. Introduction

Astrochemical rate-equation models are widely-used to investigate the chemistry in interstellar environments, from molecular clouds to protoplanetary disks. Despite their accuracy in reproducing gas-phase chemistry, these models have some documented deficiencies with regards to grain-surface chemistry (e.g., Garrod 2008). While they can not be used to simulate large systems like rate-equation models, microscopic Monte Carlo models can supply accurate and detailed information about grain-surface chemistry (Garrod 2013, Pederson *et al.* 2014). This information can then be used to improve rate-equation models by, for example, incorporating physical effects (such as back-diffusion) that are otherwise difficult to analytically solve.

2. Methods

In our study, we have utilized two Monte Carlo kinetics models. The simpler of the two models consists of a 2-dimensional square lattice with periodic boundary conditions. Each binding site on this lattice has rectangular geometry. For simplicity, only one particle is allowed to diffuse for each model run. This is analogous to having one mobile H atom and numerous relatively immobile O atoms. The size of the surface was varied to test surface size effects, as was the number of stationary reactive particles.

The other model used in this investigation was MIMICK (Model for Interstellar Monte Carlo Ice Chemical Kinetics, Garrod 2013). MIMICK is a fully 3-dimensional code, and as such is capable of incorporating more complex grain geometries. For the purposes of this study, two geometries were used: a cubic grain and a bucky-ball. These different

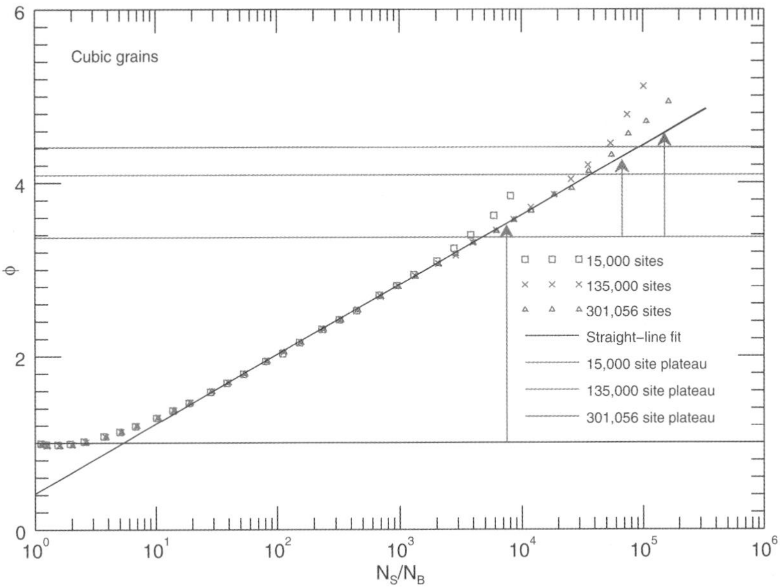

Figure 1. Results for cubic grains.. The arrows are color-coded to correspond to the data points for the same grain size, and indicate the points on the line corresponding to an average of two particles on each grain.

geometries were chosen in order to study the effect of large-scale and small-scale geometric changes on the back-diffusion effect. To this end, the cubic grain has rectangular binding site geometry (as the simple 2-dimensional model does), whereas the bucky-ball grain has hexagonal binding site geometry. In all MIMICK models, a single diffusion barrier is used for all binding sites (510.6 K, taken from Lohmar & Krug 2009), and a simplified chemical network including only atomic and molecular hydrogen is included. When two H atoms meet on the surface, they are assumed to immediately react to form H_2, which then desorbs from the surface.

3. Results

The primary results of this study are the data obtained with MIMICK for the cubic and bucky-ball 3-dimensional grains. These data are shown in Figures 1 and 2, and plot the back-diffusion factor (ϕ) varying with grain surface coverage. The back-diffusion factor is simply the factor by which grain-surface reactions are slowed compared to a standard rate-equation treatment.

Grain-surface coverage has the largest effect on the back-diffusion factor. The behavior can be divided into 3 basic regimes. At high coverage, ϕ approaches 1, indicating that there is a high probability of reaction within a few hops. At intermediate surface coverage, ϕ can be described by a logarithmic function dependent upon surface coverage. At very low surface coverage, the back-diffusion factor reaches a maximum at 2 particles on the grain, with a plateau value that can be fit using a logarithmic function dependent on surface size.

Smaller effects are observed for grain morphology and size. For each grain morphology, the different sizes are observed to follow the same logarithmic trend until a certain point of low coverage is achieved, where the grains deviate from the logarithmic fit. This deviation is due to "dead time" spent in the N=0 and N=1 population states for these

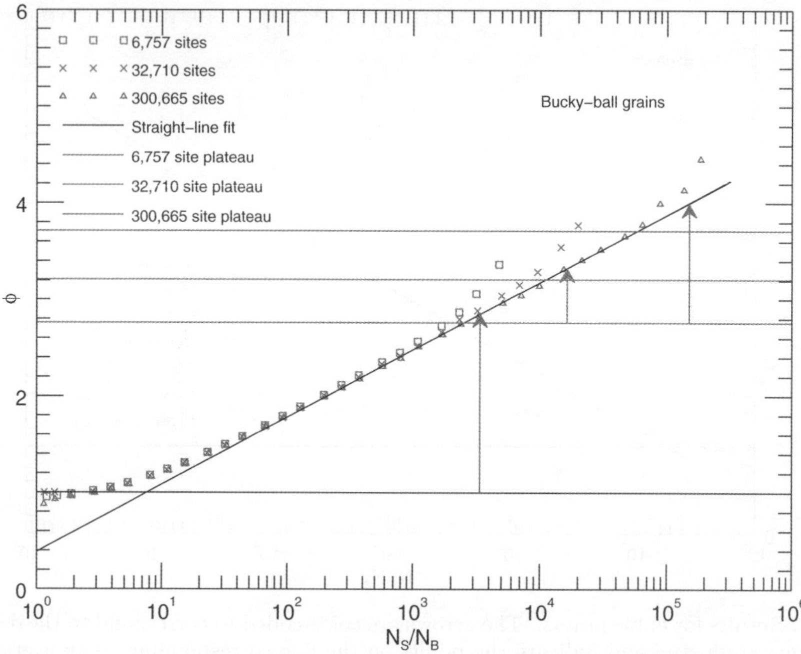

Figure 2. Results for bucky-ball grains, as per Fig. 1.

Table 1. Fits for each grain morphology. Straight-line fit corresponds to the main logarithmic portion of the data, while plateau fit corresponds to the maximum value for each grain size.

Morphology	Straight-line fit	Plateau fit
Flat (single mobile particle)	$0.315 \ln \frac{N_S}{N_B} + 0.950$	$0.315 \ln N_S + 0.200$
Cubic	$0.349 \ln \frac{N_S}{N_B} + 0.415$	$0.342 \ln N_S + 0.072$
Bucky-ball	$0.303 \ln \frac{N_S}{N_B} + 0.386$	$0.250 \ln N_S + 0.580$

low-coverage models. To account for this, simulations were run limiting the number of particles on the grain to 2. The back-diffusion factor calculated from these simulations is assumed to be a maximum, and is displayed as the color-coded horizontal lines in the figures.

The cubic and bucky-ball grain morphologies exhibit slight differences in their back-diffusion factors with coverage, corresponding to the aforementioned differences in local binding site geometry. It is observed that the bucky-ball grain with hexagonal binding site geometry produces lower values of ϕ for identical coverages. This effect is in agreement with previous simulations (Chang, Cuppen & Herbst 2005).

The logarithmic fits for ϕ were tested in a simple rate-equation model. These fits were seen to reproduce the back-diffusion behavior observed in the Monte Carlo models quite accurately, with >95% agreement at almost all coverages. The fits are simple to incorporate into rate-equation models, and should be inserted into the computation of the rate coefficients on the grain surface as follows, where ϕ is the back-diffusion factor:

$$k_{reac} = \frac{k_{hop,i} + k_{hop,j}}{\phi N_s} \tag{1}$$

The fits for all surface morphologies are shown in Table 1. If the calculated back-diffusion factor is less than 1 or greater than the plateau for a particular grain size, it should be corrected to these values.

The authors would like to acknowledge Eric Herbst, Aspen R. Clements, and Ilsa R. Cooke for useful discussions. This work was supported by the NASA Laboratory Astrophysics Program (Grant NNX15AG07G).

References

Chang, Q., Cuppen, H. M., & Herbst, E. 2005, *A&A*, 434, 599

Garrod, R. T. 2008, *A&A*, 491, 239

Garrod, R. T. 2013, *ApJ*, 778, 158

Lohmar, I. & Krug, J. 2009, *J. Stat. Phys.*, 134, 307

Pedersen, A., Wikfeldt, K. T., Karssemeijer, L., Cuppen, H., & Jónsson, H. 2014, *J. Chem. Phys.*, 141, 234706

Astrochemistry VII – Through the Cosmos
from Galaxies to Planets
Proceedings IAU Symposium No. 332, 2017
M. Cunningham, T. Millar & Y. Aikawa, eds.

doi:10.1017/S1743921317006706

Modeling the chemistry in the icy mantles of interstellar grains

Juris Kalvāns

Engineering Research Institute "Ventspils International Radio Astronomy Centre" of Ventspils University College,
Inženieru 101, Ventspils, LV-3601, Latvia
email: juris.kalvans@venta.lv

Abstract. The diffusion and photoprocessing of molecules within interstellar ices has been verified experimentally but often not fully included in astrochemical models. With models that consider photodissociation, binary reactions, and diffusion for molecules on the surface and in bulk ice, we explored the chemistry of interstellar and circumstellar ices in gravitationally contracting low-mass starless and prestellar cores, and a protostellar envelope.

Results. Photoprocessing gradually converts mixed H_2O and CO ices into CO_2 and allows for a late synthesis of icy organic species. Different layers within a single icy mantle favor the synthesis different species. Deuterium-rich molecules are concentrated on the outer surface of ice. Formation of organic molecules in bulk ice lowers their average deuterium content. The abundances of major icy species can be changed by about 25-50 % because of ice photoprocessing. Inter-layer diffusion of icy species allows sequential evaporation in protostellar envelopes, which occurs over a prolonged period.

Keywords. astrochemistry, molecular processes, methods: numerical, stars: formation, ISM: clouds, molecules

1. Introduction

Ices in star-forming regions are a likely environment for the production of various chemical species, such as water, ammonia, carbon dioxide, complex organic molecules, hydrogen sulfide. The relevant synthesis pathways and efficiency are affected by external factors, such as irradiation, and factors specifically characteristic to ices, such as diffusion of molecules. The aim of this paper (poster) is to present the effects on ice composition caused by photoprocessing by ultraviolet photons and molecule diffusion between ice layers.

We review our recent modeling work on chemistry within the icy mantles of interstellar grains. Three recent articles are discussed: Kalvāns (2015b, hereafter Paper I), Kalvāns (2015c, Paper II), and Kalvāns *et al.* (2017, Paper III). We also show new results concerning ice photoprocessing. One of the novelties of this research is that processes in the whole volume of the icy mantles are explicitly considered, not just the surface.

The objects in consideration are gravitationally contracting cloud cores (Paper I). As the core becomes denser and darker, ice starts to accumulate onto the dust grains. Depending on their parameters, such cores can develop into starless or star-forming cores. In the former, modeled in Paper II, ices are slowly processed by ultraviolet photons within the core. In the envelopes of star-forming cores (Paper III), any ices are heated until they eventually evaporate. In this paper we focus on the aspects concerning major ice components that define the properties and possible chemical transformations of any other molecules in the icy mantles.

2. Physical models

We employed the astrochemical model Alchemic-Venta, which is based on the ALCHEMIC code (Semenov *et al.* 2010). The center part of a collapsing cloud core was considered. The increase in gas density was calculated following Nejad *et al.* (1990). The adopted initial density is $\approx 3 \times 10^3$ cm^{-3}. This rises to 10^7 cm^{-3} for assumed prestellar cores (Papers I and III) and up to 4×10^5 cm^{-3} in the case of starless cores (Paper II). Additionally, in Paper III we explored the chemical evolution in envelope that is heated by the protostar according to the temperature profile T2 of Garrod & Herbst (2006). Papers II and III considered a Plummer-like density profile, which is more adequate than the flat density profile in our previous works.

A standard hydrogen ionization rate by cosmic rays of 1.3×10^{-17} s^{-1} was assumed. The flux of cosmic-ray induced secondary photons was taken to be 4875 s^{-1}cm^{-2} (Roberts *et al.* 2007).

3. Chemical model

The numerical codes of Papers I and II employed updated versions of the gas-grain chemical network by Garrod *et al.* (2008) with many organic compounds. Paper III used another derivation of the OSU chemical database, one that considers the chemistry of deuterium (Albertsson *et al.* 2013). Surface and bulk-ice chemical reactions were described with the modified rate-equations method with rates adjusted for reaction-diffusion competition according to Garrod & Pauly (2011).

For molecules in the outer layer, the whole grain surface is available for diffusion until they encounter a reaction partner (Hasegawa *et al.* 1992). However, bulk-ice molecules were assumed to be locked in cells with only a small number of possible reactants available at a given instant (Paper I). This approach evolved from a model that considered reactions on the surface of pores within ice (Kalvāns & Shmeld 2010).

Neutral gas-phase species were allowed to stick on grain surfaces. In the model, icy mantles are accumulated on grains with a radius of 0.1 μm and can reach thickness up to 180 monolayers (MLs). Adsorbed molecules can be released to the gas by several mechanisms:

- thermal evaporation;
- photodesorption by interstellar and secondary UV radiation;
- desorption by cosmic-ray induced whole-grain heating with iron cosmic-ray flux taken to be 2.06×10^{-3} s^{-1}cm^{-2} (Roberts *et al.* 2007);
- reactive desorption with the method outlined by Garrod *et al.* (2007);
- indirect reactive desorption, as described in Paper I.

Observations and laboratory studies have shown that interstellar ices have a polar component with water as dominating species and a non-polar component, dominated by solid CO (Sandford *et al.* 1988, Tielens *et al.* 1991. Thus, to describe properly the basic processes occurring in bulk-ice layers, these two components must be accounted for. We describe the icy mantles as consisting of a total of four layers – the surface (typical thickness 1-2 MLs) and three subsurface bulk-ice layers ('sublayers', up to 60 MLs). The separate MLs within these four layers are assumed to be completely mixed. This description allows to include polar, non-polar and mixed ice components, in addition to the reactive surface. A similar approach – the '(2+n)-phase model' – has been used by Furuya *et al.* (2017).

In the following sections we focus on the diffusion of icy species between the ice mantle layers and photoprocessing of bulk-ice molecules. This includes discussing the employed

Table 1. Calculated relative abundances of solid water and carbon oxides in ices. The ordinate shows integration time t prior to the supposed starbirth in the modelled cloud core.

Ice species	**Abundance** $\times 10^{-5}$ **with** $\epsilon_{\rm pd} = 0.0$	**Abundance** $\times 10^{-5}$ **with** $\epsilon_{\rm pd} = 1.0$
H_2O	11.9	8.89
CO	8.29	5.62
CO_2	5.67	8.35

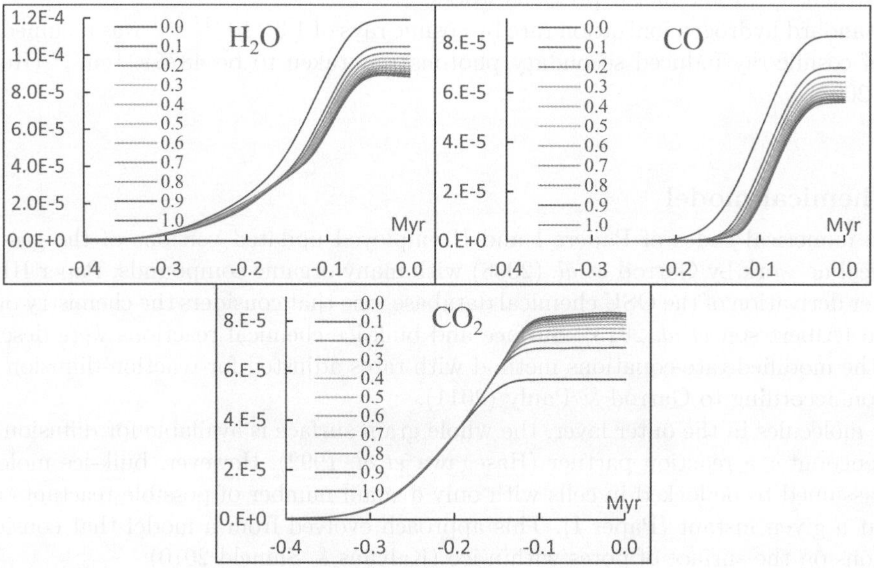

Figure 1. Calculated relative abundances of major icy species for models with various values for $\epsilon_{\rm pd}$ (ratio of photodissociation rates for icy species versus gas-phase species), as indicated in the plots.

methods and the resulting implications on the composition of interstellar and circumstellar ices.

4. Ice photoprocessing

Photoprocessing of icy molecules involves two main steps: the dissociation of a molecule and the subsequent reactions by the fragments, which often are active chemical radicals. The efficiency of photoprocessing is affected by the rate of dissociation. For adsorbed species, photodissociation was first used in astrochemical numerical simulations by Ruffle & Herbst (2001). They assumed that the photodissociation coefficient $k_{\rm pd,i}$ of icy species on grain surface is similar to the coefficient for these species in the gas phase $k_{\rm pd,g}$.

For the aims of this paper, the photodissociation rate of icy molecules can be loosely described with the help of an efficiency parameter $\epsilon_{\rm pd}$:

$$\epsilon_{\rm pd} = k_{\rm pd,g}/k_{\rm pd,i}. \tag{4.1}$$

The expected values for $\epsilon_{\rm pd}$ are in the range of 0...1, i.e., photodissociation rate of icy molecules is likely to be similar or lower than that in the gas phase. This is because the photodissociation fragments may immediately recombine to form the parent molecule.

Additionally, dissociation yield can be lower thanks to stabilizing bonds to a neighbor molecules in a solid. Current astrochemical models that explicitly consider bulk-ice typically assume either $\epsilon_{\rm pd} = 0$ (e.g., Furuya *et al.* 2017, Ruaud *et al.* 2016, and Taquet *et al.* 2014) or $\epsilon_{\rm pd} = 1$ (Chang & Herbst 2016, Furuya *et al.* 2017, Garrod *et al.* 2017, Vasyunin *et al.* 2017, and Papers I-III). Perhaps the only models that considered ice molecule photodissociation at a lower rate were those of Kalvāns & Shmeld (2010, 2013).

Photoprocessing most intensively affects ice composition during the period when a cloud core slowly contracts, darkens, and cools. Water ice appears at an extinction $A_{\rm V}$ along the line of sight of ≈ 3.2 mag (Whittet *et al.* 2001), CO_2 ice at ≈ 4.3 mag (Bergin *et al.* 2005), and CO ice at ≈ 6.8 mag (Whittet *et al.* 2007). These values correspond to translucent cloud cores. Ice accumulation in translucent cores is reproduced by the above-mentioned numerical simulations.

The above means that a part of the ice is formed under irradiation by interstellar photons. This affects the overall composition of ice because molecules can be dissociated and re-formed on the surface. The observed approximate proportions of solid H_2O:CO:CO_2 of 100:31:38 (Öberg *et al.* 2011) in cold cores are the result of molecule accretion onto grains, ice photoprocessing, and desorption. Paper I considered active bulk-ice chemistry with $\epsilon_{pd} = 1$. It was found that selective desorption of surface species is required to explain the observed proportions of H_2O, CO, and CO_2.

Photoprocessing continues to affect ice composition in dark, dense cores even when the interstellar photons are attenuated. In this case, cosmic-ray induced photons are the main dissociation agents (Prasad & Tarafdar 1983). Kalvāns & Shmeld (2010) found that photoprocessing in dark cores slowly converts CO and H_2O icy mixtures into CO_2, while the excess hydrogen diffuses out and is lost to the gas phase. This result was confirmed by the more detailed starless core model of Paper II.

The CO_2:H_2O ice abundance ratio can be used to estimate the age of a quiescent cloud core. This ratio likely does not exceed 50 % in the interstellar medium (Boogert *et al.* 2011). In Paper II we estimated the maximum age of quiescent cores to be 1 Myr. This time can be higher if ϵ_{pd} is taken to be lower than unity. A second interesting conclusion was that photoprocessing promotes the synthesis of different organic an inorganic compounds in different layers of ice. The production and destruction of a particular minor species depend on the abundances of the major ice components H_2O, CO, CO_2, NH_3, and N_2 in the considered layer. For example, formaldehyde formation is most effective in CO-rich ice, while methanol largely is formed in the water-rich layers.

In addition to UV photons, interstellar ices are chemically processed also by cosmic-ray particles. Kalvāns (2015a) estimated that such processing may change the abundances of the above-mentioned major ice components by no more than a few per cent. For minor species, e.g., carbon-chain compounds, the abundances can change much more significantly, by up to two orders of magnitude.

Photoprocessing has implications on the deuterium chemistry in the icy mantles. Deuterated species are concentrated on ice surface (Kalvans & Shmeld 2013, Taquet *et al.* 2013). For complex organic molecules (COMs) and many other minor species deuterium content decreases over time. This is because new molecules, synthesized primarily in the bulk-ice, are deuterium-poor (Paper III). However, the total deuterium proportion in ice mantles (i.e., D content in solid H_2O and NH_3) increases for time-scales of 100 kyr and longer (for both, starless cores and circumstellar envelopes). This is because the light hydrogen isotope diffuses out of the mantle more easily and its amount in photoprocessed ice decreases faster (Kalvāns 2014).

In addition to the findings of previous papers, described above, we present a small theoretical study of ice photoprocessing in a low-mass collapsing core. This is a preliminary study for a subsequent more detailed investigation. Our aim was to quantify the influence of photodissociation rate on the calculated composition of interstellar ices. For this aim, we calculated the abundances of major icy species (water and carbon oxides) in a collapsing prestellar core, i.e., during the ice formation epoch. The dissociation rate was modified by assuming different values of ϵ_{pd}.

Eleven simulations were performed with adopted $\epsilon_{\rm pd}$ values in the range of 0...1.0. We employed the exact model of Paper III with a single change: removed indirect reactive desorption. This mechanism has unknown efficiency, which has been adjusted in order to obtain a better agreement of calculated ice composition and observations of interstellar ices (see Paper I). Because ice composition *is the subject* of our limited study, this desorption mechanism was dropped. Here we considered only the prestellar phase of the model described in Paper III, i.e., up to integration time $t = 0$ Myr.

Figure 1 shows that photoprocessing has the effect of increasing the abundance of CO_2 ice at the expense of CO and H_2O. Table 1 presents a quantitative comparison – the relative abundances of H_2O, CO, and CO_2 ices at the densest stage of the prestellar core at $t = 0.0$ Myr. Comparing cases with no ice photodissociation versus full photoprocessing, the relative abundances of CO and H_2O have decreased by 2.7×10^{-5} and 3.0×10^{-5}, respectively. At the same time, the relative abundance of CO_2 ice has increased by 2.7×10^{-5} due to photoprocessing. These numbers represent abundance changes in the range of 25-50 %. Thus, while ice photoprocessing is not a decisive process in ice accumulation, its influence is significant, and probably can be quantified with the help the available observational data (e.g., Whittet *et al.* 2007).

5. Inter-layer diffusion of molecules in icy mantles

The inter-layer mobility of icy molecules was probably first considered by Cuppen & Herbst (2007). The rate of diffusion for each molecule on ice surface is governed by its binding, or diffusion, energy, $E_{\rm b}$. The latter is often obtained from the desorption energy $E_{\rm D}$ by assuming a fixed $E_{\rm b}/E_{\rm D}$ ratio. In Papers I-III this ratio was taken to be in the range 0.35-0.50 (Garrod & Herbst 2006, Garrod & Pauly 2011). Species in the bulk are surrounded by ice molecules and thus are more strongly bound in the ice lattice. We assumed that their *absorption energy* $E_{\rm B}$ is three times higher than the desorption energy of surface species (Paper I), $E_{\rm B}/E_{\rm D} \approx 3$. The binding energy for mantle species was modified proportionally.

It was assumed that radial diffusion of molecules in bulk ice between the four layers occurs thanks to irregularities and cracks in the amorphous ice. The diffusion rate of a molecule depends on its own binding energy. This is different from, e.g., Garrod (2013), who assumed that diffusion occurs via swapping of water molecules. The latter approach is not suited, for example, a case of CO-dominated ice.

The above means that each molecule diffuses through ice at its own rate. This is approximately consistent with temperature-programmed desorption experiments (e.g., Fayolle *et al.* 2011). In circumstellar envelopes, where the icy mantles are heated up, chemical radicals with increasing binding energies become mobile, activating various reaction chains that affect the abundances of minor species (see, e.g., the discussion by Kalvāns 2015a). The major ice components evaporate in three steps: first, CO and N_2, then CO_2, and then H_2O and NH_3 (Paper III). With the evaporation of CO, the production of new COMs in bulk-ice becomes insignificant. It is completely shut down when CH_4, H_2CO, and CO_2 disappear from the icy mantles.

The evaporation of icy species with our model is a prolonged process taking at least several kyrs. First, molecules on the outer surface evaporate. This produces an initial spike for the gas-phase abundance for some species (see figures 4, 5, A1, and A2 of Paper III). Because E_b for a molecule in bulk ice layers is higher than E_D for a molecule on the surface, radial diffusion occurs at higher temperatures than evaporation. This produces a close second peak in species' gas-phase abundance. The second peak often is broader because molecule diffusion from the inner ice layers takes time. These two evaporation stages must be temporally and spatially close, and are unlikely to be discerned by observations.

6. Summary

We have reviewed our recent astrochemical models of interstellar cloud cores and circumstellar envelopes. Results on photodissociation of icy species and diffusion between molecule layers in icy mantles were discussed. A modeling study showed that photodissociation can alter the abundances of major icy species, an effect that can be studied by numerical simulations calibrated by observational results.

7. Acknowledgments

I thank Ventspils City Council for support. This work was financed by ERDF project Physical and chemical processes in the interstellar medium, No. 1.1.1.1/16/A/213, being implemented in Ventspils University College. This research has made use of NASA's Astrophysics Data System.

References

Albertsson, T., Semenov, D. A., Vasyunin, A. I., Henning, T., & Herbst, E. 2013, *ApJS*, 207, 27
Andersson, S. & van Dishoeck, E. F. 2008, *A&A*, 491, 907
Bergin, E. A., Melnick, G. J., Gerakines, P. A., Neufeld, D. A., & Whittet, D. C. B. 2005, *ApJL*, 627, L33
Boogert, A. C. A., Huard, T. L., Cook, A. M., Chiar, J. E., Knez, C., Decin, L., Blake, G. A., Tielens, A. G. G. M., & van Dishoeck, E. F. 2011, *ApJ*, 729, 92
Chang, Q. & Herbst, E. 2016, *ApJ*, 819, 145
Cuppen, H. & Herbst, E. 2007, *ApJ*, 668, 294
Fayolle, E. C., Öberg, K. I., Cuppen, H. M., Visser, R., & Linnartz, H. 2011, *A&A*, 529, A74
Furuya, K., Drozdovskaya, M. N., Visser, R., van Dishoeck, E. F., Walsh, C., Harsono, D., Hincelin, U., & Taquet, V. 2017, *A&A*, 599, A40
Garrod, R. T. 2013, *ApJ*, 765, 60
Garrod, R. T., Belloche, A., Mller, H. S. P., & Menten, K. M. 2017, *A&A*, 601, A48
Garrod, R. T. & Herbst, E. 2006, *A&A*, 457, 927
Garrod, R. T. & Pauly, T. 2011, *ApJ*, 735, 15
Garrod, R. T., Wakelam, V., & Herbst, E. 2007, *A&A*, 467, 1103
Garrod, R. T., Weaver, S. L. W., & Herbst, E. 2008, *ApJ*, 682, 283
Hasegawa, T. I., Herbst, E., & Leung, C. M. 1992, *ApJS*, 82, 167
Hincelin, U., Chang, Q., & Herbst, E. 2015, *A&A*, 574, A24
Kalvāns, J. & Shmeld, I. 2010, *A&A*, 521, A37
Kalvāns, J. & Shmeld, I. 2013, *A&A*, 554, A111
Kalvāns, J. 2014, *PASP*, 126, 811
Kalvāns, J. 2015a, *A&A*, 573, A38
Kalvāns, J. 2015b, *ApJ*, 803, 52 (Paper I)
Kalvāns, J. 2015c, *ApJ*, 806, 196 (Paper II)
Kalvāns, J., Shmeld, I., Kalnin, J. R., & Hocuk, S. 2017, *MNRAS*, 467, 1763 (Paper III)
Nejad, L. A., Williams, D. A., & Charnley, S. B. 1990, *MNRAS*, 246, 183

Öberg, K. I., Boogert, A. C. A., Pontoppidan, K. M., van den Broek, S., van Dishoeck, E. F., Bottinelli, S., Blake, G. A., & Evans, N. J. II 2011 *ApJ*, 740, 109

Prasad, S. S. & Tarafdar, S. P. 1983, *ApJ*, 267, 603

Roberts, J. F., Rawlings, J. M. C., Viti, S., & Williams, D. A. 2007, *MNRAS*, 382, 733

Ruaud, M., Wakelam, V., & Hersant, F. 2016, *MNRAS*, 459, 3756

Ruffle, D. P. & Herbst, E. 2001, *MNRAS*, 322, 770

Sandford, S. A., Allamandola, L. J., Tielens, A. G. G. M., & Valero, G. J. 1988, *ApJ*, 329, 498

Semenov, D., Hersant, F., Wakelam, V., Dutrey, A., Chapillon, E., Guilloteau, St., Henning, Th., Launhardt, R., Pietu, V., & Schreyer, K. 2010, *A&A*, 467, 1763

Taquet, V., Charnley, S. B., & Sipilä, O. 2014, *ApJ*, 791, 1

Taquet, V., Peters, P. S., Kahane, C., Ceccarelli, C., López-Sepulcre, A., Toubin, C., Duflot, D., & Wiesenfeld, L. 2013, *A&A*, 550, A127

Tielens, A. G. G. M., Tokunaga, A. T., Geballe, T. R., & Baas, F. 1991, *ApJ*, 381, 181

Vasyunin, A. I., Caselli, P., Dulieu, F., & Jimnez-Serra, I. 2017, *ApJ*, 842, 33

Whittet, D. C. B., Gerakines, P. A., Hough, J. H., & Shenoy, S. S. 2001, *ApJ*, 547, 872

Whittet, D. C. B., Shenoy, S. S., Bergin, E. A., Chiar, J. E., Gerakines, P. A., Gibb, E. L., Melnick, G. J., & Neufeld, D. A. 2007, *ApJ*, 655, 332

Complex Organic Molecules

Astrochemistry VII – Through the Cosmos from Galaxies to Planets
Proceedings IAU Symposium No. 332, 2017
M. Cunningham, T. Millar & Y. Aikawa, eds.

doi:10.1017/S1743921317007815

Exploring molecular complexity in the Galactic Center with ALMA

Arnaud Belloche

Max-Planck-Institut für Radioastronomie, 53121 Bonn, Germany
email: belloche@mpifr-bonn.mpg.de

Abstract. The search for complex organic molecules (COMs) in the ISM has revealed chemical species of ever greater complexity. This search relies heavily on the progress made in the laboratory to characterize the rotational spectra of these molecules. Observationally, the advent of ALMA with its high angular resolution and sensitivity has allowed to reduce the spectral confusion and detect low-abundance molecules that could not be probed before. We present results of the EMoCA survey conducted with ALMA toward the star-forming region Sgr B2(N). This spectral line survey aims at deciphering the molecular content of Sgr B2(N) in order to test the predictions of astrochemical models and gain insight into the chemical processes at work in the ISM. We report on the tentative detection of N-methylformamide, on deuterated COMs, and on the detection of a branched alkyl molecule. Prospects for probing molecular complexity in the ISM even further are discussed at the end.

Keywords. ISM: molecules, astrochemistry, ISM: individual objects: Sagittarius B2(N), stars: formation

1. Complex organic molecules in the interstellar medium

The in-situ exploration of comet 67P/Churyumov-Gerasimenko by the Rosetta mission has led to the detection of more than twenty organic molecules in the coma of the comet and in material excavated from the surface by the impact of Rosetta's lander Philae (Goesmann *et al.* (2015), Altwegg *et al.* (2016), Altwegg *et al.* (2017)). Most of these molecules, such as methyl cyanide, CH_3CN, formamide, NH_2CHO, or acetone, $CH_3C(O)CH_3$, were already known to exist in the interstellar medium (ISM) but some, like methyl isocyanate, CH_3NCO, reported by Goesmann *et al.* (2015)†, were not. Shortly after, as soon as spectroscopic predictions of the rotational spectrum of the molecule became available, methyl isocyanate was detected in the ISM (Halfen *et al.* (2015), Cernicharo *et al.* (2016)). Another molecule that was not known to exist in the ISM, and has still not been found, is the amino acid glycine, NH_2CH_2COOH. Altwegg *et al.* (2016) reported its identification in the coma of comet 67P, confirming an earlier claim by Elsila *et al.* (2009) who analyzed samples returned from comet 81P/Wild 2 to Earth by the Stardust mission. In addition to these cometary detections, more than 80 different types of amino acids were identified in meteorites on Earth, with an isotopic composition and a racemic distribution suggesting that their presence is not due to contamination by Earth material and that they have instead a true extraterrestrial origin (see, e.g., Botta & Bada (2002)). The degree of molecular complexity found in these small solar-system bodies, in particular in meteorites, raises the following questions: is this chemical complexity a widespread outcome of interstellar chemistry in our Galaxy, and possibly even in other

† The identification of methyl isocyanate in the mass spectrum of the cometary material analyzed by Goesmann *et al.* (2015) has been recently disputed (Ligterink *et al.* (2017), Altwegg *et al.* (2017)).

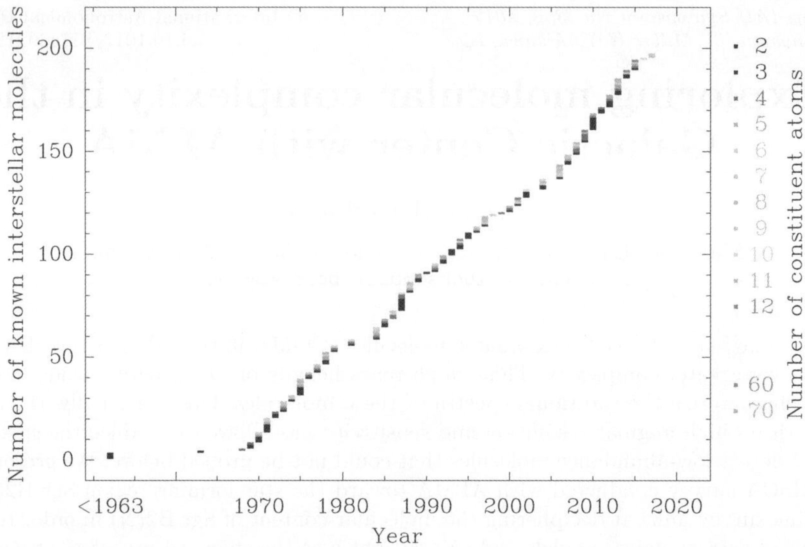

Figure 1. Number of known interstellar molecules as a function of time as of July 2017. The three molecules identified before 1963 are indicated on the left. The number of atoms each molecule contains is color-coded following the color scheme shown on the right. This diagram uses the information provided by H. S. P. Müller in the Cologne Database for Molecular Spectroscopy (CDMS, http://www.astro.uni-koeln.de/cdms/molecules).

galaxies? What degree of complexity is interstellar chemistry capable of achieving, in particular in star (and planet) forming regions?

As of today, 197 molecules have been identified in the interstellar medium or in circumstellar envelopes of evolved stars. This number does not include isotopologs, that is molecules containing less abundant isotopes such as D, ^{13}C, ^{18}O, ^{34}S, or ^{15}N. The first identifications of interstellar molecules, CH, CH$^+$, and CN, go back nearly eight decades ago and were made in absorption in the UV domain (McKellar (1940), Adams (1941), Douglas & Herzberg (1941)). However, the search for molecules in the ISM started to become efficient more than two decades later, with the advent of radio astronomy. Figure 1 shows the number of known interstellar molecules as a function of time, with the number of their constituent atoms color-coded. Since the first detection of a molecule in the radio domain, the hydroxyl radical OH by Weinreb *et al.* (1963), about seven molecules have been newly identified in the ISM every two years on average. While the rate of detection has been to first order constant over the past five decades, it is worth mentioning that the detection rate over the past decade ($\sim$5 new molecules per year) exceeds the rate achieved during the first decade when radio astronomy started to flourish ($\sim$4 new molecules per year between 1968 and 1977). This shows the current vitality of this field of research and suggests that the inventory of interstellar molecules should continue to grow significantly in the near future.

Following the review by Herbst & van Dishoeck (2009), it has become common in the astrochemical community to define a molecule as complex when it contains six atoms or more. With this definition, about one third of the known interstellar molecules are complex (Fig. 2). All these complex molecules contain at least one carbon atom and they are thus called complex organic molecules (COMs). The presence of such complex interstellar molecules raises the question of their origin. What are the chemical processes that lead to chemical complexity in the ISM? What are the respective roles of gas-phase and grain-surface/mantles processes? What are the relevant reactions? What

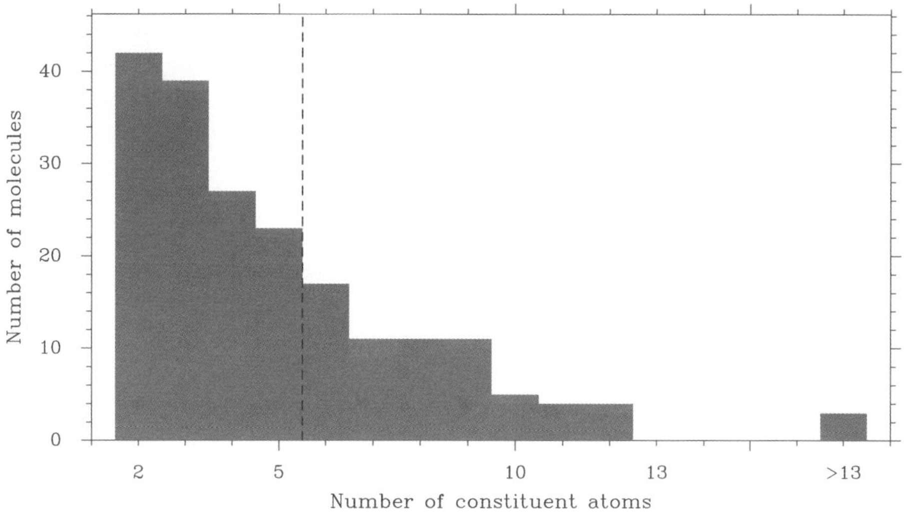

Figure 2. Number of known interstellar molecules as a function of the number of their constituent atoms as of July 2017. Detections published in 2015–2017 are emphasized in red. The dashed line indicates the threshold above which a molecule is defined as complex.

are their reaction rates and branching ratios? To answer all these questions, a close interplay between laboratory experiments, theoretical calculations, numerical simulations, and observations is necessary (see, e.g., Herbst & van Dishoeck (2009), Garrod & Widicus Weaver (2013), Öberg (2016)). Here, we focus on the latter with the aim of confronting results of observations to predictions of astrochemical models.

2. Sgr B2 and the EMoCA survey

One of the most prominent star-forming regions in our Galaxy is the giant molecular cloud Sagittarius (Sgr) B2, with a mass of $\sim 10^7$ $M_\odot$ in a diameter of $\sim$40 pc (Lis & Goldsmith(1990)). Sgr B2 is located about 100 pc from the Galactic Center in projection, at a distance of 8.3 kpc from the Sun (Reid *et al.* (2014)). It contains two main sites of on-going formation of high-mass stars, Sgr B2(N) and Sgr B2(M), that both host clusters of ultra-compact H II regions (see, e.g., Gaume *et al.* (1995), Schmiedeke *et al.* (2016)). Sgr B2(N), in particular, contains several hot molecular cores (Fig. 3). The main ones, Sgr B2(N1) and Sgr B2(N2), have been known for a long time (e.g., Snyder *et al.* (1994), Belloche *et al.* (2008), Qin *et al.* (2011), Belloche *et al.* (2016)), but several new, fainter ones were recently detected with the Atacama Large Millimeter/submillimeter Array (ALMA) (Bonfand *et al.* (2017), Sánchez-Monge *et al.*(2017)). The H_2 column densities of these hot cores are so high (10^{24}–10^{25} cm^{-2} over few arcsec) that the detection of molecules with low abundances in Sgr B2 is easier than toward other sources. This has been a key advantage for the detection of new COMs in the past few decades. Indeed, many COMs were first detected toward Sgr B2.

Following up a spectral line survey of Sgr B2(N) and Sgr B2(M) carried out with the IRAM 30 m single-dish telescope a decade ago, which led to the detection of several new COMs (Belloche *et al.* (2008), Belloche *et al.* (2009), Belloche *et al.* (2013)), we performed the EMoCA project with ALMA in Cycles 0 and 1. EMoCA stands for Exploring Molecular Complexity with ALMA. This interferometric project is a spectral line survey of Sgr B2(N) that covers the 3 mm atmospheric window between 84.1 GHz and 114.4 GHz, at an angular resolution of $\sim 1.6''$ and a spectral resolution of 488 kHz,

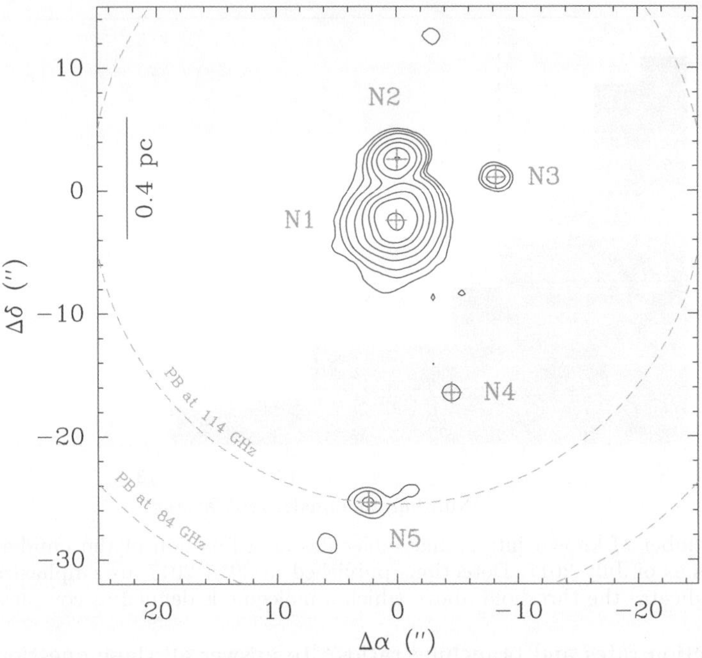

Figure 3. Contour map of Sgr B2(N) representing the number of channels between 84.1 GHz and 114.4 GHz with emission detected above the 7σ level in the continuum-subtracted spectra of the EMoCA survey. The contour levels are 500, 1000, 2000, 5000, 10000, and then increase by steps of 10000. The dashed circles represent the primary beam of the ALMA 12 m antennas at 84 GHz and 114 GHz. The equatorial offsets are given with respect to the phase center located at $(\alpha, \delta)_{\mathrm{J2000}} = (17^{\mathrm{h}}47^{\mathrm{m}}19.87^{\mathrm{s}}, -28^{\circ}22'16'')$. Sources identified as hot cores by Bonfand *et al.* (2017) are indicated with blue crosses and labels. Figure adapted from Bonfand *et al.* (2017).

that is a velocity resolution between 1.3 and 1.7 km s^{-1}. We achieved a sensitivity of ~3 mJy beam^{-1} (1σ), that is ~0.15 K in brightness temperature scale. EMoCA aims at detecting new COMs in the ISM and, thereby, testing the predictions delivered by numerical simulations of interstellar chemistry.

In addition to delivering maps of thousands of spectral lines, the EMoCA survey represents an improvement by more than one order of magnitude both in angular resolution and sensitivity compared to our previous single-dish survey. In particular, its angular resolution of 1.6″ (13000 au in projection) is sufficient to separate the two main hot cores, Sgr B2(N1) and Sgr B2(N2), that are separated by ~5″ in the north-south direction. A nice surprise of this survey is that the emission lines of the northern hot core, Sgr B2(N2), have widths (FWHM) of 5 km s^{-1} only, which reduces considerably the level of line confusion compared to previous single-dish data. As shown in Fig. 4, the lines detected toward the main hot core, Sgr B2(N1), are broader (~ 7 km s^{-1}) and some of them have stronger wings. With the aim to look for weak lines of low-abundance species, we have focused our analysis on Sgr B2(N2) so far. We model its emission spectrum using Weeds (Maret *et al.* (2011)) under the assumption of local thermodynamic equilibrium (LTE), which is reasonable given the high densities probed with ALMA. Our spectroscopic database, which is needed to model the spectra, contains all entries available in the JPL and CDMS databases (Pickett *et al.* (1998), Müller *et al.* (2001), Endres *et al.* (2016)), as well as additional predictions sent to us by spectroscopists. Our current model of the emission spectrum of Sgr B2(N2) is shown in red in Fig. 4. In the next three

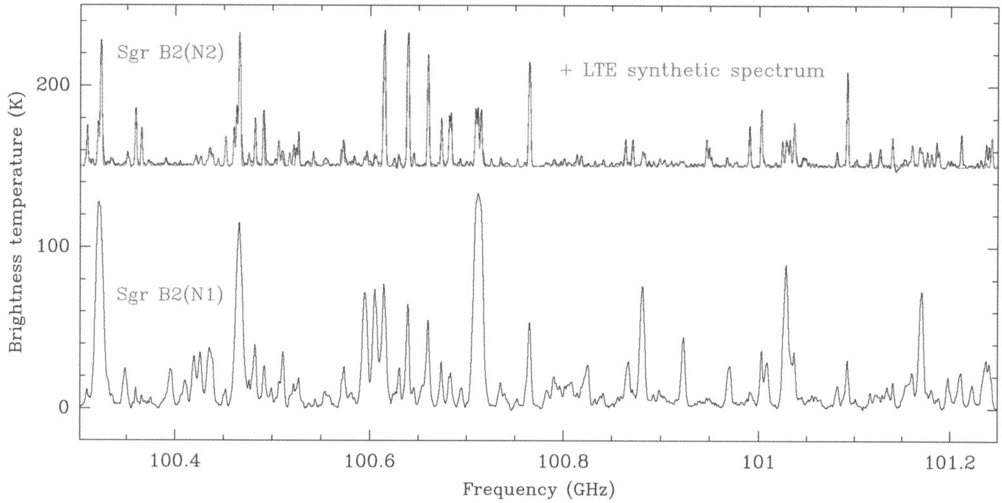

Figure 4. Short portion of the EMoCA spectrum observed toward the hot molecular cores Sgr B2(N1) and Sgr B2(N2) with ALMA at an angular resolution of $\sim 1.6''$. The red spectrum shows the LTE model that contains the contribution of all the molecules identified so far toward Sgr B2(N2).

sections, we report on some of the results obtained with the EMoCA survey based on this model.

3. Tentative detection of N-methylformamide

N-methylformamide, CH_3NHCHO, is a structural isomer of the known interstellar molecule acetamide, $CH_3C(O)NH_2$, that was detected in Sgr B2(N) by Hollis *et al.* (2006) (see also Halfen *et al.* (2011)). Both molecules contain a peptide bond which is the characteristic chemical bond linking amino acids in proteins. N-methylformamide can also be seen as a partially hydrogenated form of methyl isocyanate, CH_3NCO, which was recently detected in the ISM as mentioned in Sect. 1. Given that both acetamide and methyl isocyanate were known interstellar molecules, N-methylformamide appeared as a good candidate for an interstellar detection. New measurements of the rotational spectrum of the *trans* conformer of this molecule were performed in Lille (France) and Kharkiv (Ukraine) over the frequency range 45–630 GHz, and accurate spectroscopic predictions were then obtained with the RAM36 program (Ilyushin *et al.* (2010)) for its ground state and its first two torsionally excited states (Belloche *et al.* (2017)).

With these spectroscopic predictions, we searched for emission of N-methylformamide in the EMoCA spectrum of Sgr B2(N2). Five of the lines that we assigned to CH_3NHCHO suffer little from contamination by emission lines of other species. Four of these five lines are shown in Fig. 5. Given that our full model of Sgr B2(N2) includes the contribution of all species identified so far (red spectrum in Fig. 4 and green spectrum in Fig. 5), we selected seven additional lines of N-methylformamide that could be "decontaminated" with the full model and we used them together with the five well detected lines to build a population diagram for this molecule. A fit to this diagram yields a temperature of ~ 150 K, which is typical of the rotational temperatures derived for other COMs in Sgr B2(N2). Since only five lines are detected without contamination, the detection of CH_3NHCHO is considered as tentative only (Belloche *et al.* (2017)). However, the fact that our model takes into account the emission of all species identified so far and that

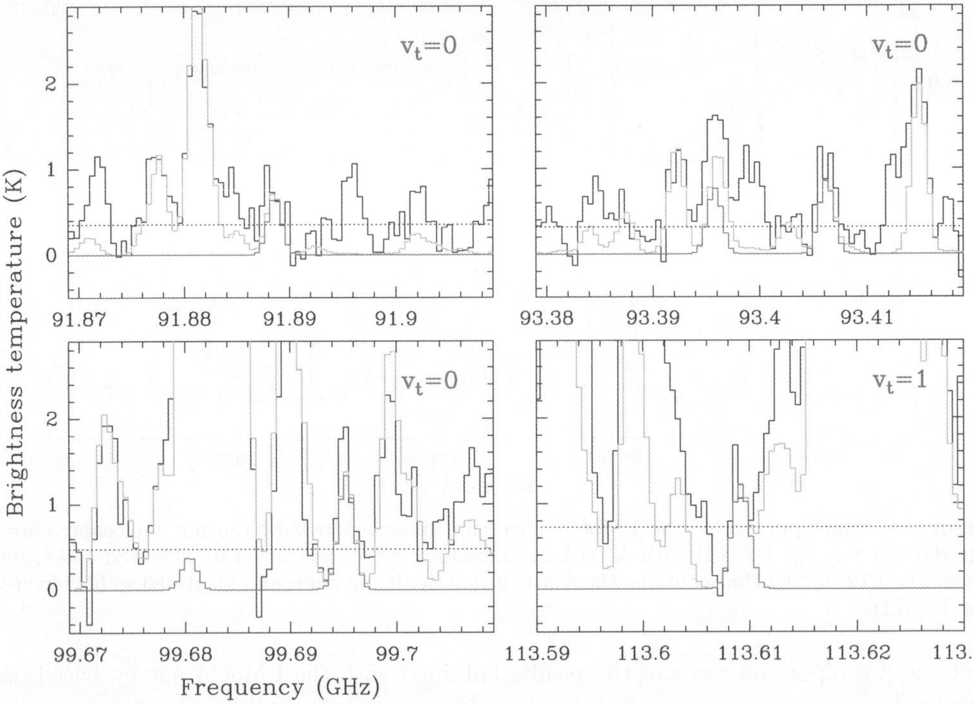

Figure 5. Spectra of Sgr B2(N2) showing four of the five transitions of N-methylformamide that do not suffer much from contamination by emission lines of other species in the EMoCA survey (Belloche *et al.* (2017)). In each panel, the observed ALMA spectrum is shown in black, the synthetic spectrum of N-methylformamide in red, and the model containing the contributions of all molecules identified so far in green. The dotted line indicates the 3σ detection level. The line in the bottom right panel belongs to the first torsionally excited state of N-methylformamide, while the other ones are ground-state lines.

the derived rotational temperature of N-methylformamide is typical of COM emission in Sgr B2(N2), we are confident that the identification of this molecule in this source is correct.

We used the EMoCA survey to compare the abundance of N-methylformamide to the ones of related species. CH_3NHCHO is, within a factor of two, nearly as abundant as CH_3NCO and $CH_3C(O)NH_2$, and more than one order of magnitude less abundant than formamide, NH_2CHO, and isocyanic acid, HNCO. We compared these abundance ratios to predictions obtained with the chemical kinetics model MAGICKAL (Garrod (2013)), the network of which was expanded to include N-methylformamide as well as related species. The first conclusion of this comparison is that an efficient formation of HNCO via NH+CO on grains with an activation energy below 1500 K is necessary to achieve the observed abundance of CH_3NCO. The second conclusion is that the production of CH_3NHCHO on grain surfaces is plausible via two formation routes: the successive addition of H to CH_3NCO and radical-radical reactions (CH_3+HNCHO or HCO+$HNCH_3$) (Belloche *et al.* (2017)). The good match between the predicted and measured abundance ratios is an additional support to our identification of N-methylformamide in the EMoCA spectrum of Sgr B2(N2). However, the chemical model produces an overabundance of acetamide, which should be investigated in more details in the future.

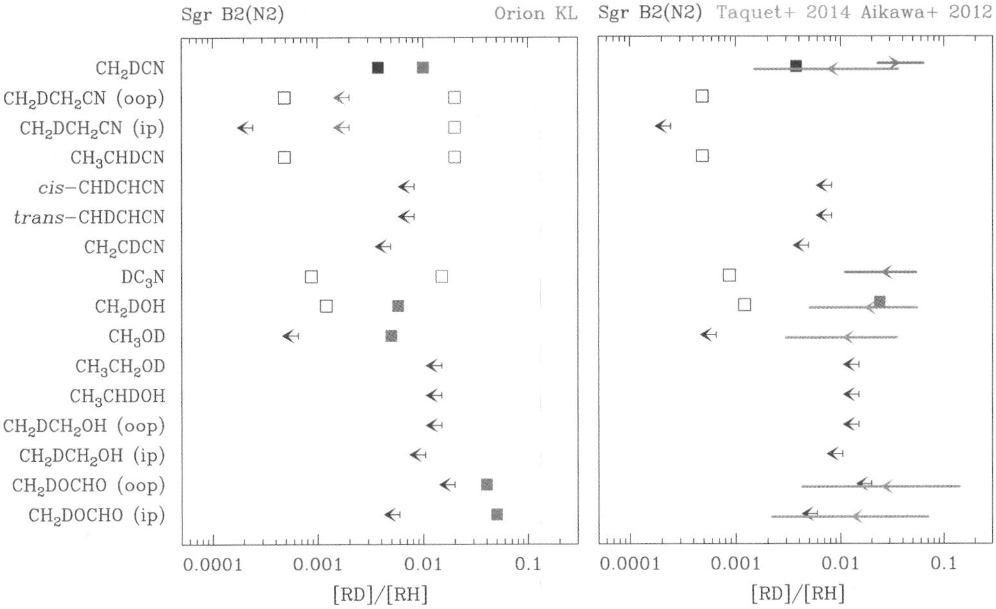

Figure 6. *Left panel:* Level of deuterium fractionation measured for a sample of (complex) organic molecules toward Sgr B2(N2) (Belloche *et al.* (2016)). Filled and empty squares indicate robust and tentative detections, respectively. Upper limits are indicated with arrows. The symbols in red show the level of deuteration reported for Orion KL by Gerin *et al.* (1992), Margulès *et al.* (2009), Daly *et al.* (2013), Esplugues *et al.* (2013), Neill *et al.* (2013), and Coudert *et al.* (2013). *Right panel*: Same measurements toward Sgr B2(N2) compared to predictions of chemical models of collapsing protostellar envelopes. The purple bars and arrows show the evolution of the level of deuterium fractionation as a function of time during the main accretion phase in the model of Taquet *et al.* (2014) while the blue ones and the blue square show the model predictions of Aikawa *et al.* (2012).

4. Deuterated complex organic molecules

The cosmic abundance of deuterium with respect to hydrogen is low ($\sim 1.5 \times 10^{-5}$, Linsky (2003)). Still, molecules in star-forming regions are often found to be enriched in deuterium (e.g., Parise *et al.* (2006)). This mainly results from the exothermicity of the reaction of H_3^+ with HD, the reservoir of deuterium in molecular clouds, that produces H_2D^+, and the depletion of CO, the main destroyer of H_3^+, onto the surface of dust grains that occurs in the dense and cold regions of molecular clouds. Given these properties, the degree of deuterium fractionation of molecules in star-forming regions is a powerful tool to trace the history of their cold prestellar phase (see, e.g., Caselli & Ceccarelli (2012), Ceccarelli *et al.* (2014)). Although the level of deuteration is known to be low in the Galactic Center region (e.g., Gerin *et al.* (1992)), we took advantage of the high sensitivity of the EMoCA survey to search for deuterated COMs in Sgr B2(N2).

We firmly detected CH_2DCN and could also report tentative detections of CH_2DCH_2CN, CH_3CHDCN, DC_3N, and CH_2DOH (Belloche *et al.* (2016)). The level of deuteration measured for these species, as well as upper limits derived for other deuterated COMs, are summarized in the left panel of Fig. 6, along with measurements or upper limits reported in the literature for Orion KL. Overall, the degree of deuteration of COMs in Sgr B2(N2) is on the order of 0.1%, a factor three to ten lower than in Orion KL. The right panel of Fig. 6 compares the Sgr B2(N2) results to predictions of chemical models of (low-mass) protostellar envelopes (Aikawa *et al.* (2012), Taquet *et al.* (2014)). Apart for a possible agreement for CH_2DCN in one model, the models produce overabundances of deuterated COMs by at least a factor of four.

The low level of deuteration of COMs in Sgr B2(N2) has at least two possible origins. First, it may reflect a lower elemental abundance of deuterium in the Galactic Center region compared to the local ISM, owing to stellar processing of previous generations of stars, as has been claimed, albeit with large uncertainties, by previous studies that were based on small molecules (Jacq *et al.* (1999), Lubowich *et al.* (2000), Polehampton *et al.* (2002)). A second possibility could be the thermal history of star formation in Sgr B2. Both the gas kinetic temperature (30–50 K, Ott *et al.* (2014)) and the dust temperatures (20–28 K, Guzmán *et al.* (2015)) are higher in Sgr B2 than in nearby star forming regions. In the model of Taquet *et al.* (2014), the level of deuteration of methanol is reduced by a factor three at the end of the prestellar phase when the initial dust temperature is increased by 3 K only, from 17 K to 20 K. The deuteration of COMs is thus very sensitive to the temperature during the prestellar phase, and the higher temperatures prevailing in the Galactic Center region could have contributed to the low level of COM deuteration found in Sgr B2(N2). In addition, it is worth mentioning that the factor five difference found in Sgr B2(N2) between the deuteration level of methyl cyanide, CH_3CN, and ethyl cyanide, C_2H_5CN, may point to different formation pathways, in the gas phase for the former and in the ice mantles for the latter. Dedicated numerical simulations of deuteration designed for Sgr B2 will be needed in the future to investigate its thermal history and the chemical pathways of its deuterium chemistry in detail.

5. Branched molecules in the interstellar medium

One of the first results of the EMoCA survey was the interstellar detection of a branched alkyl molecule (Belloche *et al.* (2014)). The rotational spectrum of this molecule, *iso*-propyl cyanide, i-C_3H_7CN, was studied in the laboratory by Müller *et al.* (2011). It is a structural isomer of the straight-chain form *normal*-propyl cyanide, n-C_3H_7CN, that we detected earlier in our single-dish survey of Sgr B2(N) (Belloche *et al.* (2009)). Both isomers are clearly detected in the EMoCA spectrum, with about 50 transitions for the branched form and 120 transitions for the chain-like form (Fig. 7). The population diagram of n-C_3H_7CN yields a well constrained rotational temperature of $\sim$150 K, typical of the COM emission in Sgr B2(N2). From the LTE modelling, we derived a surprisingly high abundance ratio $[i\text{-}C_3H_7CN]/[n\text{-}C_3H_7CN]$ of 0.4, with the branched form nearly as abundant as the chain-like one.

We compared the measured abundance ratio of the two isomers of propyl cyanide again to predictions of the chemical kinetics model MAGICKAL (Garrod (2013)). The reaction network of this model was expanded to include, for the first time, the formation of branched molecules (Garrod *et al.* (2017), Belloche *et al.* (2014)). The measured abundance ratio of the two isomers of propyl cyanide is well reproduced by the model. The model indicates that both molecules can be formed on grain surfaces, but via different formation routes. The dominant routes for n-C_3H_7CN are the addition reactions $CH_3\dot{C}H_2+\dot{C}H_2CN$ and $\dot{C}N+CH_2{=}CHCH_3$ while the formation of i-C_3H_7CN is dominated by the radical-radical reaction $\dot{C}H_3+CH_3\dot{C}HCN$. In the model, the key parameter influencing the i/n ratio is the barrier of the $\dot{C}N$ addition to C_2H_2 and C_2H_4 (Garrod *et al.* (2017)).

As mentioned in Sect. 1, amino acids of extraterrestrial origin were detected in meteorites on Earth. Both branched and chain-like isomers of these amino acids were identified, and the branched isomers were even found to dominate over the chain-like ones (e.g., Cronin & Pizzarello (1983)). The detection of a branched alkyl molecule in the

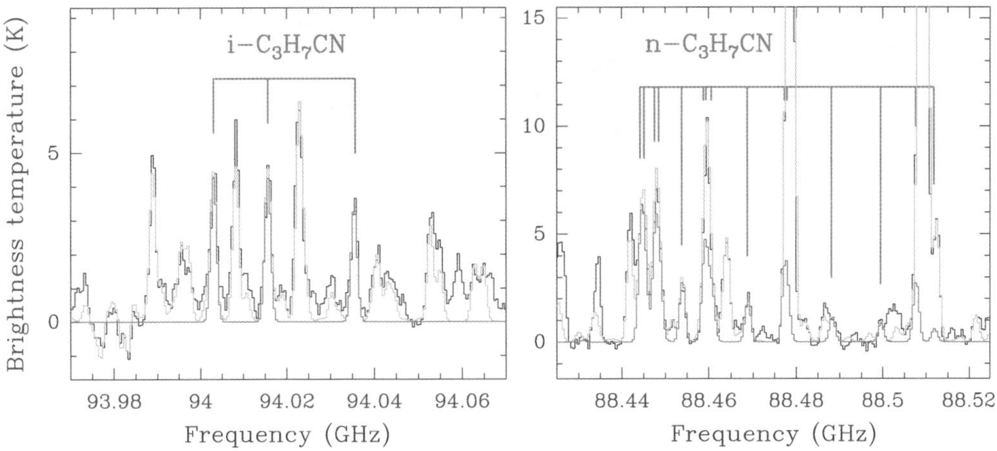

Figure 7. *Left panel:* Example of transitions of *iso*-propyl cyanide detected toward Sgr B2(N2) with ALMA (Belloche *et al.* (2014)). The observed spectrum is shown in black, the synthetic spectrum of *iso*-propyl cyanide in red, and the model containing the contributions of all molecules identified so far in green. *Right panel*: Example of transitions of *normal*-propyl cyanide detected toward Sgr B2(N2) with ALMA. The red spectrum is the synthetic spectrum of *normal*-propyl cyanide.

ISM thus establishes a further link between the chemical composition of meteorites and interstellar chemistry.

6. Outlook

The EMoCA project is just one example illustrating the new window opened by ALMA in terms of angular resolution and sensitivity to probe the complexity of interstellar chemistry and confront the derived chemical composition of astronomical sources to the predictions of astrochemical models. While Sgr B2 is a place of choice to detect molecules of low abundance thanks to its high column densities, spectral line surveys of other sources with ALMA, such as the PILS survey toward a nearby low-mass protostar (The ALMA Protostellar Interferometric Line Survey, Jørgensen *et al.* (2016)), are exploring the outcome of interstellar chemistry in other environments and will also contribute to improving our understanding of the chemical processes at work in the ISM.

Our study of deuterated COMs in Sgr B2(N2) has shown that a sensitive, unbiased spectral line survey at high angular resolution is an exquisite tool to measure the level of deuterium fractionation of molecules with high accuracy. This is made possible by the conditions prevailing in this hot core: the densities are so high that collisions dominate over radiative processes and the LTE approximation is valid. The spectra of COMs, which can have several dozens of lines detected, can thus be fitted with great accuracy. In this respect, Sgr B2 is an excellent laboratory to test the predictions of chemical models quantitatively. Numerical simulations of deuterium chemistry designed for Sgr B2(N) should help us in the future set valuable constraints on the D/H elemental abundance in the Galactic Center region.

After the detection of acetamide, $CH_3C(O)NH_2$, and the tentative detection of N-methylformamide, CH_3NHCHO (Sect. 3), it is tempting to search for the next structural isomer of these two molecules, acetimidic acid, $CH_3C(OH)NH$, which is the third most stable isomer in the C_2H_5NO family (Lattelais *et al.* (2010)). However, the rotational spectrum of this molecule has not been characterized in the laboratory yet, and its dipole

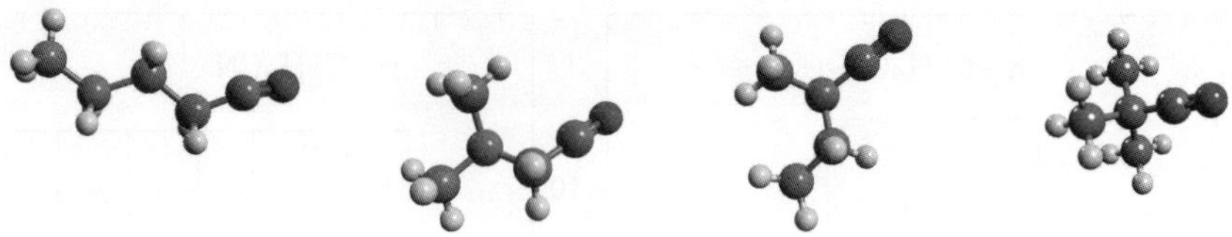

Figure 8. Ball-and-stick representations of the chain-like and branched forms of butyl cyanide. From left to right: n-C_4H_9CN, i-C_4H_9CN, s-C_4H_9CN, and t-C_4H_9CN. Figure adapted from Garrod *et al.* (2017), image credit: E. Willis.

moment is a factor of two smaller than the ones of acetamide and N-methylformamide (Lattelais *et al.* (2009)). Therefore, its detection in the ISM will be challenging.

The detection of a branched alkyl molecule (Sect. 5) has opened a new domain in the structures available to the chemistry of star forming regions. The next step in complexity in the alkyl cyanide family is butyl cyanide, C_4H_9CN. This molecule has one straight-chain isomer and three branched ones (Fig. 8). The straight-chain form, *normal*-butyl cyanide, was searched for in the ISM but it has not been detected so far, neither in our previous IRAM 30 m survey of Sgr B2(N) nor in the EMoCA survey (Ordu *et al.* (2012), Garrod *et al.* (2017)). The chemical model MAGICKAL predicts that one of the branched forms, s-C_4H_9CN, should in fact dominate over the straight-chain form in Sgr B2(N2) and that we may have been only short of a detection with the EMoCA survey (Garrod *et al.* (2017)). This motivated us to start a follow-up survey of Sgr B2(N) at 3 mm with ALMA in its Cycle 4. This new project will represent an improvement by a factor of three both in angular resolution and sensitivity with respect to the EMoCA survey. The characterization of the rotational spectrum of s-C_4H_9CN was performed in Cologne (Müller *et al.* (2017)) and the Cycle 4 observations have been completed recently. s-C_4H_9CN will be one of the first molecules we will search for in this new, sensitive spectral line survey that may reveal the presence of other interesting COMs as well.

As mentioned in Sect. 2, the EMoCA survey revealed that the hot core Sgr B2(N2) has linewidths as narrow as 5 km s^{-1} at arcsecond scale. This helped reduce the spectral confusion compared to previous single-dish surveys. EMoCA has indeed not reached the confusion limit on this source yet, while the spectrum of Sgr B2(N1) is almost confusion-limited. There is thus still room for improvement on Sgr B2(N2), and also on the other fainter hot cores detected in Sgr B2(N) (Bonfand *et al.* (2017), Sánchez-Monge *et al.*(2017)), which is one of the goals of our new Cycle 4 project. However, it is likely that the confusion limit will almost be reached in this next survey and the following question will arise: how will we beat the confusion limit in order to explore even further the chemical complexity of the ISM in the future? At least two possibilities can be investigated. As spectral confusion increases with frequency in Sgr B2(N) (Belloche *et al.* (2013)), the first option is to go to lower frequencies, as has already been explored at low angular resolution with the PRIMOS spectral survey of Sgr B2(N) performed with the Green Bank Telescope (e.g., Neill *et al.* (2012)). Higher angular resolution will however be necessary to probe the hot cores in an optimal way: this will be possible with ALMA once it is equipped with receivers operating below 80 GHz (bands 1 and 2 or 2+3, Di Francesco *et al.* (2013), Fuller *et al.* (2016)), and on a longer term with the Next Generation Very Large Array (ngVLA, Hughes *et al.*(2015)). The second direction will be to perform very deep integrations with ALMA on sources that have narrower linewidths, such as the Class 0 protostar IRAS 16293–2422 which is the target of the PILS survey (Jørgensen *et al.* (2016)).

Acknowledgements

I would like to thank Holger Müller, Rob Garrod, Karl Menten, Mélisse Bonfand, Vivien Thiel, and Eric Willis for their precious contributions to the EMoCA project. I am also thankful to the whole spectroscopic community for tirelessly providing spectroscopic predictions of molecules of ever greater complexity to the astronomical community. This paper makes use of the following ALMA data:
ADS/JAO.ALMA#2011.0.00017.S, ADS/JAO.ALMA#2012.1.00012.S. ALMA is a partnership of ESO (representing its member states), NSF (USA), and NINS (Japan), together with NRC (Canada), NSC and ASIAA (Taiwan), and KASI (Republic of Korea), in cooperation with the Republic of Chile. The Joint ALMA Observatory is operated by ESO, AUI/NRAO, and NAOJ. This work has been in part supported by the Deutsche Forschungsgemeinschaft (DFG) through the collaborative research grant SFB 956 "Conditions and Impact of Star Formation", project area B3.

References

Adams, W. S. 1941, *ApJ*, 93, 11
Aikawa, Y., Wakelam, V., Hersant, F., Garrod, R. T., & Herbst, E. 2012, *ApJ*, 760, 40
Altwegg, K., Balsiger, H., Bar-Nun, A., *et al.* 2016, *Science Advances*, 2, e1600285
Altwegg, K., Balsiger, H., Berthelier, J. J., *et al.* 2017, *MNRAS*, 469, S130
Belloche, A., Menten, K. M., Comito, C., *et al.* 2008, *A&A*, 482, 179
Belloche, A., Garrod, R. T., Müller, H. S. P., *et al.* 2009, *A&A*, 499, 215
Belloche, A., Müller, H. S. P., Menten, K. M., Schilke, P., & Comito, C. 2013, *A&A*, 559, A47
Belloche, A., Garrod, R. T., Müller, H. S. P., & Menten, K. M. 2014, *Science*, 345, 1584
Belloche, A., Müller, H. S. P., Garrod, R. T., & Menten, K. M. 2016, *A&A*, 587, A91
Belloche, A., Meshcheryakov, A. A., Garrod, R. T., *et al.* 2017, *A&A*, 601, A49
Bonfand, M., Belloche, A., Menten, K. M., Garrod, R. T., & Müller, H. S. P. 2017, *A&A*, in press, arXiv:1703.09544
Botta, O. & Bada, J. L. 2002, *Surveys in Geophysics*, 23, 411
Caselli, P. & Ceccarelli, C. 2012, *A&AR*, 20, 56
Ceccarelli, C., Caselli, P., Bockelée-Morvan, D., *et al.* 2014, Protostars and Planets VI, 859
Cernicharo, J., Kisiel, Z., Tercero, B., *et al.* 2016, *A&A*, 587, L4
Coudert, L. H., Drouin, B. J., Tercero, B., *et al.* 2013, *ApJ*, 779, 119
Cronin, J. R. & Pizzarello, S. 1983, *Advances in Space Research*, 3, 5
Daly, A. M., Bermúdez, C., López, A., *et al.* 2013, *ApJ*, 768, 81
Di Francesco, J., Johnstone, D., Matthews, B. C., *et al.* 2013, arXiv:1310.1604
Douglas, A. E. & Herzberg, G. 1941, *ApJ*, 94, 381
Elsila, J. E., Glavin, D. P., & Dworkin, J. P. 2009, *Meteorit. Planet. Sci.*, 44, 1323
Endres, C. P., Schlemmer, S., Schilke, P., Stutzki, J., & Müller, H. S. P. 2016, *J. Mol. Spectr.*, 327, 95
Esplugues, G. B., Cernicharo, J., Viti, S., *et al.* 2013, *A&A*, 559, A51
Fuller, G. A., Avison, A., Beltran, M., *et al.* 2016, arXiv:1602.02414
Garrod, R. T. 2013, *ApJ*, 765, 60
Garrod, R. T. & Widicus Weaver, S. L. 2013, *Chemical Reviews*, 113, 8939
Garrod, R. T., Belloche, A., Müller, H. S. P., & Menten, K. M. 2017, *A&A*, 601, A48
Gaume, R. A., Claussen, M. J., de Pree, C. G., Goss, W. M., & Mehringer, D. M. 1995, *ApJ*, 449, 663
Gerin, M., Combes, F., Wlodarczak, G., *et al.* 1992, *A&A*, 259, L35
Goesmann, F., Rosenbauer, H., Bredehöft, J. H., *et al.* 2015, *Science*, 349, aab0689
Guzmán, A., Sanhueza, P., Contreras, Y., *et al.* 2015, *ApJ*, 815, 130

Halfen, D. T., Ilyushin, V. V., & Ziurys, L. M. 2011, *ApJ*, 743, 60
Halfen, D. T., Ilyushin, V. V., & Ziurys, L. M. 2015, *ApJ*, 812, L5
Herbst, E. & van Dishoeck, E. F. 2009, *ARA&A*, 47, 427
Hollis, J. M., Lovas, F. J., Remijan, A. J., *et al.* 2006, *ApJ*, 643, L25
Hughes, A. M., Beasley, A., & Carilli, C. 2015, IAU General Assembly, 22, 2255106
Ilyushin, V. V., Kisiel, Z., Pszczólkowski, L., Mäder, H., & Hougen, J. T. 2010, *J. Mol. Spectr.*, 259, 26
Jacq, T., Baudry, A., Walmsley, C. M., & Caselli, P. 1999, *A&A*, 347, 957
Jørgensen, J. K., van der Wiel, M. H. D., Coutens, A., *et al.* 2016, *A&A*, 595, A117
Lattelais, M., Pauzat, F., Ellinger, Y., & Ceccarelli, C. 2009, *ApJ*, 696, L133
Lattelais, M., Pauzat, F., Ellinger, Y., & Ceccarelli, C. 2010, *A&A*, 519, A30
Ligterink, N. F. W., Coutens, A., Kofman, V., *et al.* 2017, *MNRAS*, 469, 2219
Linsky, J. L. 2003, *Space Sci. Revs*, 106, 49
Lis, D. C. & Goldsmith, P. F. 1990, *ApJ*, 356, 195
Lubowich, D. A., Pasachoff, J. M., Balonek, T. J., *et al.* 2000, *Nature*, 405, 1025
Maret, S., Hily-Blant, P., Pety, J., Bardeau, S., & Reynier, E. 2011, *A&A*, 526, A47
Margulès, L., Motiyenko, R., Demyk, K., *et al.* 2009, *A&A*, 493, 565
McKellar, A. 1940, *PASP*, 52, 187
Müller, H. S. P., Thorwirth, S., Roth, D. A., & Winnewisser, G. 2001, *A&A*, 370, L49
Müller, H. S. P., Coutens, A., Walters, A., Grabow, J.-U., & Schlemmer, S. 2011, *J. Mol. Spectr.*, 267, 100
Müller, H. S. P., Zingsheim, O., Wehres, N., *et al.* 2017, *J. Phys. Chem. A*, submitted
Neill, J. L., Muckle, M. T., Zaleski, D. P., *et al.* 2012, *ApJ*, 755, 153
Neill, J. L., Crockett, N. R., Bergin, E. A., Pearson, J. C., & Xu, L.-H. 2013, *ApJ*, 777, 85
Ordu, M. H., Müller, H. S. P., Walters, A., *et al.* 2012, *A&A*, 541, A121
Polehampton, E. T., Baluteau, J.-P., Ceccarelli, C., Swinyard, B. M., & Caux, E. 2002, *A&A*, 388, L44
Öberg, K. I. 2016, *Chemical Reviews*, 116, 9631
Ott, J., Weiß, A., Staveley-Smith, L., Henkel, C., & Meier, D. S. 2014, *ApJ*, 785, 55
Parise, B., Ceccarelli, C., Tielens, A. G. G. M., *et al.* 2006, *A&A*, 453, 949
Pickett, H. M., Poynter, R. L., Cohen, E. A., Delitsky, M. L., Pearson, J. C., & Müller, H. S. P. 1998, *J. Quant. Spectr. Rad. Transf.*, 60, 883
Qin, S.-L., Schilke, P., Rolffs, R., *et al.* 2011, *A&A*, 530, L9
Reid, M. J., Menten, K. M., Brunthaler, A., *et al.* 2014, *ApJ*, 783, 130
Sánchez-Monge, Á., Schilke, P., Schmiedeke, A., *et al.* 2017, *A&A*, 604, A6
Schmiedeke, A., Schilke, P., Möller, T., *et al.* 2016, *A&A*, 588, A143
Snyder, L. E., Kuan, Y.-J., & Miao, Y. 1994, Lecture Notes in Physics 439: The Structure and Content of Molecular Clouds, Eds. T. L. Wilson and K. J. Johnston, Springer-Verlag, 439, 187
Taquet, V., Charnley, S. B., & Sipilä, O. 2014, *ApJ*, 791, 1
Weinreb, S., Barrett, A. H., Meeks, M. L., & Henry, J. C. 1963, *Nature*, 200, 829

Astrochemistry VII – Through the Cosmos from Galaxies to Planets
Proceedings IAU Symposium No. 332, 2017
M. Cunningham, T. Millar & Y. Aikawa, eds.

doi:10.1017/S1743921317007037

Methanol formation in TW Hya and future prospects for detecting larger complex molecules in disks with ALMA

Catherine Walsh[1,2], Shreyas Vissapragada[2,3] and Harry McGee[1]

[1]School of Physics and Astronomy
University of Leeds, Leeds LS2 9JT, UK
email: c.walsh1@leeds.ac.uk

[2]Leiden Observatory
Leiden University, P.O. Box 9513, Leiden 2300 RA, Netherlands

[3]Columbia Astrophysics Laboratory
Columbia University, New York, NY 10027, USA

Abstract. Gas-phase methanol was recently detected in a protoplanetary disk for the first time with ALMA. The peak abundance and distribution of methanol observed in TW Hya differed from that predicted by chemical models. Here, the chemistry of methanol gas and ice is calculated using a physical model tailored for TW Hya with the aim to contrast the results with the recent detection in this source. New pathways for the formation of larger complex molecules (e.g., ethylene glycol) are included in an updated chemical model, as well as the fragmentation of methanol ice upon photodesorption. It is found that including fragmentation upon photodesorption improves the agreement between the peak abundance reached in the chemical models with that observed in TW Hya ($\sim 10^{-11}$ with respect to H_2); however, the model predicts that the peak in emission resides a factor of $2-3$ farther out in the disk than the ALMA images. Reasons for the persistent differences in the gas-phase methanol distribution between models and the observations of TW Hya are discussed. These include the location of the ice reservoir which may coincide with the compact mm-dust disk ($\lesssim$ 60 au) and sources of gas-phase methanol which have not yet been considered in models. The possibility of detecting larger molecules with ALMA is also explored. Calculations of the rotational spectra of complex molecules other than methanol using a parametric model constrained by the TW Hya observations suggest that the detection of individual emission lines of complex molecules with ALMA remains challenging. However, the signal-to-noise ratio can be enhanced via stacking of multiple transitions which have similar upper energy levels.

Keywords. astrochemistry, planetary systems: protoplanetary disks, molecular processes, stars: individual (TW Hya)

---

1. Introduction

The detection of complex organic molecules (COMs) in protoplanetary disks has been considered a key science goal for the Atacama Large Millimeter/submillimeter Array (ALMA). In astrochemistry, a COM is loosely defined as a carbon-containing molecule containing six or more atoms (see, e.g., Herbst & van Dishoeck 2009). Many families of COMs (e.g., amines, aldehydes, and carboxylic acids) are thought to be stepping stones towards prebiotic molecules such as amino acids and simple sugars. COMs typically possess a complex rotational spectrum; hence, emission from COMs is weak, especially from small astrophysical objects such as protoplanetary disks ($\sim 1''$). An additional challenge for protoplanetary disks in particular is that most of the disk material is cold ($\lesssim$ 100 K) and COMs are expected to reside on and within the water-dominated ice mantles on dust grains. However, the detection of cold gas-phase water and ammonia in the nearby disk around TW Hya has confirmed that non-thermal desorption of ice species into the gas-phase does occur in disks. This helps to maintain a low, yet detectable, amount in the gas (Hogerheijde *et al.* 2011; Salinas *et al.* 2016).

The superior sensitivity and spatial resolution of ALMA has allowed, for the first time, the possibility to see relatively complex molecules in nearby protoplanetary disks (e.g., Öberg *et al.* 2015). Such detections are vital for probing the link between the chemical complexity observed towards the warm and dense gas in the immediate environment of forming stars (e.g., Herbst & van Dishoeck 2009; Caselli & Ceccarelli 2012) and that seen in comets in the Solar System (e.g., Mumma & Charnley 2011; Le Roy *et al.* 2015). It remains unclear to what extent chemistry during disk formation and evolution shapes the chemical complexity of icy planetesimals. Recent theoretical models of complex chemistry during disk formation and over the disk lifetime have suggested that (i) chemical complexity in ices is enhanced en route into the disk (Drozdovskaya *et al.* 2014, 2016; Yoneda *et al.* 2016), and (ii) gas-phase COMs should be present in sufficiently detectable quantities in nearby protoplanetary disks (e.g., Furuya & Aikawa 2014; Walsh *et al.* 2014; Parfenov *et al.* 2016). Methanol is an abundant interstellar ice (see, e.g., Boogert *et al.* 2015) which has been shown in the laboratory to be an important feedstock for larger and more complex molecules (e.g., Öberg *et al.* 2009; Modica & Palumbo 2010; Chen *et al.* 2013); hence, gas-phase methanol is a key target for dedicated observational studies in nearby protoplanetary disks.

2. Detection of gas-phase methanol in TW Hya

Four rotational transitions of A-type methanol were targeted during ALMA Cycle 2 observations of TW Hya (Walsh *et al.* 2016). This spectral set-up, within which four strong lines could be simultaneously observed, was motivated by simulations of both the chemical structure and the emergent spectrum across the full ALMA frequency range (Walsh *et al.* 2014). This work predicted that methanol emission from the cold, outer disk would dominate, arising from photodesorption of the ice reservoir. Hence, the targeted lines possessed relatively low-lying upper level energies ranging from ≈ 20 K to ≈ 100 K.

Data reduction and imaging was conducted as described in Walsh *et al.* (2016). The individual lines were not detected in the synthesised images; however, a detection was made possible through stacking of the three transitions at 304.208 GHz ($E_{\rm up} = 21.6$ K), 305.473 GHz ($E_{\rm up} = 28.6$ K), and 307.166 GHz ($E_{\rm up} = 38.0$ K). The gas-phase methanol was detected with a signal-to-noise ratio of 5.5 in the stacked channel maps, reaching a sensitivity of 2 mJy beam^{-1} in a 0.15 km s^{-1} channel. A more recent analysis of this dataset using a matched filter to search for emission in the *uv* domain showed that the three lines listed above were robustly detected with signal-to-noise levels of 4.4, 6.2, and 3.4, respectively (Loomis *et al.* 2017).

The stacked emission was then simulated using a physical model of TW Hya constrained by a wealth of observations (Kama *et al.* 2016) and using a parametric model for the gas-phase methanol distribution and abundance. This suggested that the gas-phase methanol arises in ring, peaking at ≈ 30 au, and with an abundance ranging from $3 \times 10^{-12} - 4 \times 10^{-11}$ with respect to H_2. The derived emission morphology and abundance differed from that predicted by previous models. First, the methanol emission in TW Hya appeared more compact than the predictions, and second, the peak abundance was $2 - 3$ orders of magnitude lower than expected (Walsh *et al.* 2014). Figure 1 shows the line profile in orange dashed lines with the best-fit model overlaid in green.

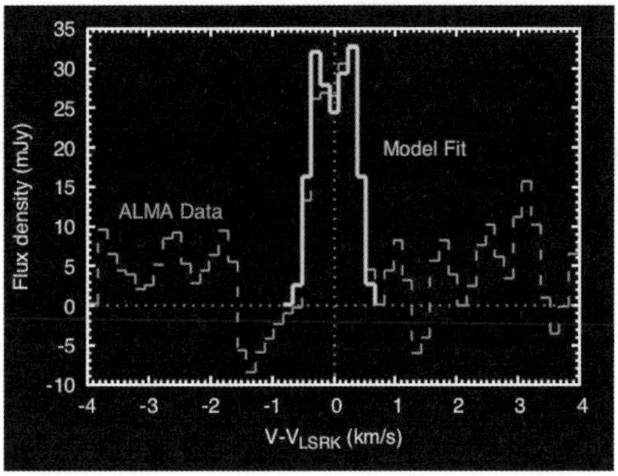

Figure 1. Methanol line profile from TW Hya extracted from the stacked channel maps within the 3σ contour of the 317 GHz continuum emission (orange dashed line) compared with the model fit line profile (green solid line). Figure adapted from Walsh *et al.* (2016).

3. An updated chemical model for methanol gas and ice

The apparent discrepancy between predictions and the observations of methanol in TW Hya may be reconciled by modelling the complex chemistry using a disk structure specific to TW Hya. TW Hya is a colder and more settled disk that that used for the previous predictions (from Nomura *et al.* 2007). In addition, since the models presented in Walsh *et al.* (2014), there have been numerous laboratory investigations into methanol ice chemistry. Studies into methanol ice photodesorption showed that methanol does not desorb intact at low temperatures (Bertin *et al.* 2016; Cruz-Diaz *et al.* 2016). The upper limit for the methanol photodesorption was found to be $\lesssim 10^{-5}$ molecules photon^{-1} for methanol mixed with CO. This alone could help to reconcile the low methanol abundances seen in TW Hya because previous estimates of intact methanol photodesorption suggested a yield of $\sim 10^{-3}$ molecules photon^{-1} (Öberg *et al.* 2009).

To investigate the abundance and distribution of methanol gas and ice an updated chemical model was developed and the chemistry was computed using an observationally constrained physical model of the TW Hya protoplanetary disk (from Kama *et al.* 2016). The chemical model presented in Walsh *et al.* (2015) was updated with the fragmentation pathways for methanol ice photodesorption (Bertin *et al.* 2016). The current model also includes an extended surface chemistry network for methanol and its related compounds from Chuang *et al.* (2016) including methyl formate ($HCOOCH_3$), glycolaldehyde ($HOCH_2CHO$), glyoxal (HC(O)CHO), and ethylene glycol ($(CH_2OH)_2$).

Figure 2 shows the fractional abundances of methanol ice (left) and gas (right) as a function of disk height, z/r, at a radius of 30 au and at four different time steps (0.5, 1.0, 5.0, and 10 Myr) for a model in which photodesorption only is included as a non-thermal desorption mechanism. More details on these calculations are presented in an upcoming publication by Ligterink *et al.* (2017). When fragmentation upon photodesorption is included, the gas-phase abundance drops by $2-3$ orders of magnitude in the molecular layer present at the methanol snow surface ($z/r \approx 1.5$). The snow surface marks the boundary in the disk atmosphere at which a molecule transitions from the ice phase to the gas phase. Further, the location of the methanol snow surface becomes deeper in time: this is because the methanol ice now needs to reform from its fragments rather than simply refreeze from the gas phase once desorbed.

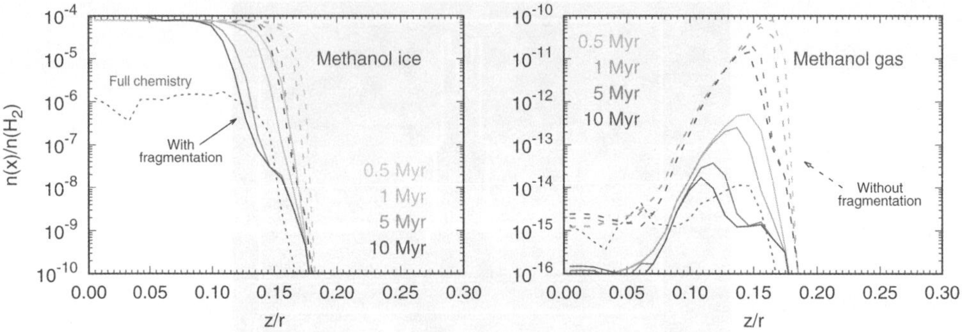

Figure 2. Fractional abundance of methanol ice (left) and gas (right) with respect to H_2 as a function of disk vertical height (z/r) and at a radius of 30 au. The results with (solid) and without (dashed) fragmentation upon photodesorption are shown. The y-axis scales differ due to the low abundance of the gas relative to the ice. Results at four time steps are plotted (0.5, 1.0, 5.0, and 10 Myr). The dotted gray lines show the results at 1 Myr using a more comprehensive chemical model (see text for details).

4. The chemical origin of gas-phase methanol in TW Hya

In Figure 3 the two-dimensional fractional abundances of methanol ice (left) and gas (right) are shown as a function of disk radius, r, and height scaled by the radius, z/r. These have been computed using the updated chemical model as described previously. Note that the colour-bar scales differ due to the difference in peak abundance between the gas and the ice. This model is more complex than that used for the results presented in Figure 2 in that it also includes all sources of non-thermal desorption (photodesorption and reactive desorption) as well as chemistry induced by bulk ice photodissociation (see Walsh *et al.* 2014, for full details). The fractional abundances are shown at a single time step (1 Myr).

As expected, the methanol ice is confined to the midplane throughout the disk, lying below $z/r \approx 0.1 - 0.15$ within 100 au and below $z/r \approx 0.05$ beyond 100 au. This is due to the increasing penetration of stellar and interstellar UV radiation in the outer disk. The gas-phase methanol has a different distribution to that in previous models that predicted a it resides in the outer disk atmosphere ($z/r \gtrsim 0.2$, Semenov & Wiebe 2011; Furuya & Aikawa 2014; Walsh *et al.* 2014). The peak abundance of $\sim 10^{-11}$ with respect to H_2 is reached in the outer disk beyond $\approx$ 100 au with an abundance of $\gtrsim 10^{-12}$ reached beyond $\approx$ 80 au. The bulk of the gas-phase methanol also lies close to the midplane ($z/r \lesssim 0.15$) and mirrors the distribution of methanol ice by becoming more confined to the midplane ($z/r \lesssim 0.05$) beyond $\approx$ 100 au. The peak abundance reached is of the same order of magnitude as that derived from the observations suggesting that the inclusion of fragmentation of methanol ice upon photodesorption does help to reduce the peak gas-phase abundance in the outer disk. The distribution of methanol ice and gas in this model are very similar to that presented in recent work by Parfenov *et al.* (2017) for the case where they assume that the barrier for diffusion (across the grain/ice surface) is 0.4 times that for desorption. Here we use a value of 0.3.

Figure 2 shows how the more complex chemical model results (dotted lines) compare with the more simple models which neglect bulk ice photodissociation and reactive desorption (solid and dashed lines). The ice abundance is lower in the midplane ($\sim 10^{-6}$ with respect to H_2) demonstrating the impact of ice photodissociation on the processing of feedstock ice species such as methanol. The gas-phase distribution demonstrates that (i) the combination of reactive desorption and photodesorption does not help to boost

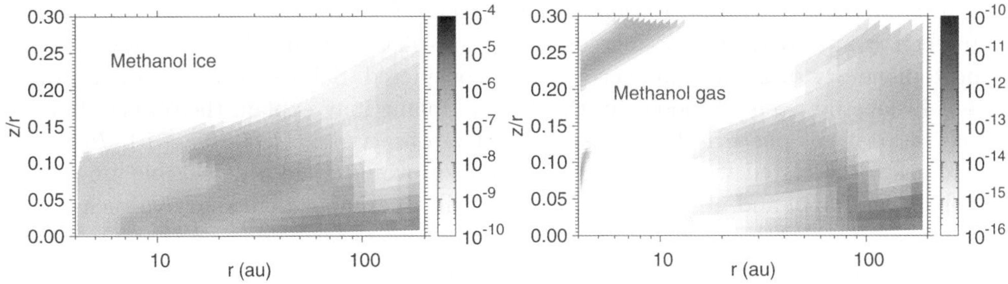

Figure 3. Fractional abundance of methanol ice (left) and gas (right) with respect to H_2 as a function of disk radius and height (scaled by the radius) and at a time of 1 Myr.

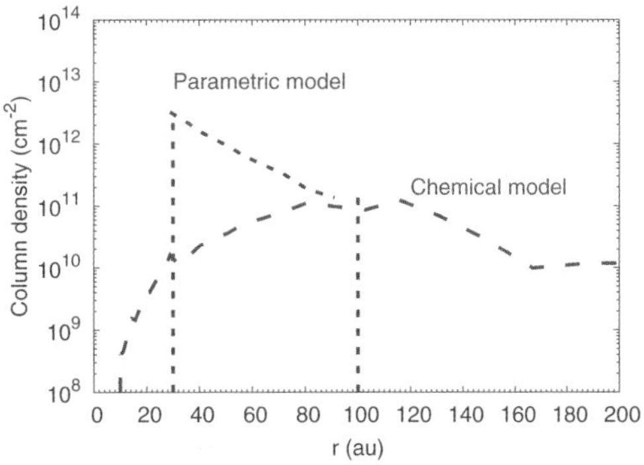

Figure 4. Column density of methanol gas as a function of radius for the parametric model from Walsh *et al.* (2016) and the chemical model (blue and red lines respectively).

the peak gas-phase abundance of methanol in the molecular layer ($z/r \approx 0.1 - 0.15$) and (ii) reactive desorption does help boost the abundance in the midplane ($z/r \lesssim 0.1$) to higher than that from models which include photodesorption (with fragmentation) only. The overall lower peak abundance reached in the more complex model is related to the depletion of methanol ice over time.

Figure 4 shows the column density of gas-phase methanol as a function of radius from the full chemical model compared with the column density derived from the parametric models used in Walsh *et al.* (2016). The peak column density for the chemical model is $\sim 10^{11}$ cm^{-2} and is reached farther out in the disk ($\approx 80 - 120$ au) than that derived from the parametric models. Given that the line emission is optically thin and in local thermodynamic equilibrium (LTE), this suggests that the emission peak for the chemical model also resides at this location which is a factor of ≈ 3 farther from the star than suggested by the observations. Parfenov *et al.* (2017) find similar results.

5. On the distribution of methanol gas in TW Hya

The results for TW Hya using the full and updated chemical model show that the peak abundance of gas-phase methanol matches well that constrained by the observations. Thus, the inclusion of fragmentation upon photodesorption of methanol ice helps to reconcile the relatively high abundances of gas-phase methanol predicted by the previous generation of protoplanetary disk models with complex chemistry ($\sim 10^{-10} - 10^{-9}$ with respect to H_2, e.g., Semenov & Wiebe 2011; Furuya & Aikawa 2014; Walsh *et al.* 2014). However, the modelled radial distribution does not agree with the observations: the peak emission is predicted at a radius that is a factor of $2 - 3$ beyond that seen in the ALMA data (see also Parfenov *et al.* 2017). This suggests that the models, despite their complexity, are missing sources of gas-phase methanol in the inner regions of protoplanetary disks beyond those already considered. Salinas *et al.* (2016) speculate that collisions between icy bodies in the disk midplane may explain the observed abundance ratio of gas-phase NH_3 to gas-phase H_2O observed in TW Hya with *Herschel.* It remains to be tested whether such a mechanism would yield a detectable amount of gas-phase methanol and also if larger molecules like methanol would survive such a process. Öberg *et al.* (2017) analysed ALMA observations of formaldehyde (H_2CO) in TW Hya and found two reservoirs: an inner warm component within ≈ 10 au and an outer colder component beyond ≈ 15 au arising from the disk atmosphere. Given the strong chemical association between methanol and formaldehyde it is possible that these two molecules share a similar emission morphology; however, methanol is less volatile than formaldehyde. Hence, a chemical process that is able to release methanol in this cold region is still required. One such process is co-desorption with a less volatile yet abundant ice such as CO. This has recently been investigated in experiments by Ligterink *et al.* (2017). Gas-phase formation of methanol may also need to be revisited.

An alternative explanation is that the ice reservoir may not be present throughout the full radial extent of the disk midplane. Clues pointing to this are present in recent high spatial resolution ALMA images of dust continuum emission from TW Hya which show that the $\sim$ mm-sized dust grains are confined to $\lesssim 60$ au (Andrews *et al.* 2016; Nomura *et al.* 2016). This dust morphology shows that the large dust grains in TW Hya have settled to the midplane and have drifted inwards due to the head wind and resultant loss in angular momentum induced by the velocity differences in gas and mm-sized dust in the outer disk (Weidenschilling 1977). Complementary observations of line emission from both single-dish observations and spatially resolved observations with ALMA have provided additional support that these large dust grains are host to the bulk of the ice reservoir (see the recent studies by Kama *et al.* 2016, Salinas *et al.* 2016, Bergin *et al.* 2016, and Du *et al.* 2017). Sequestration and concentration of the ice reservoir towards the inner disk midplane ($\lesssim 60$ au) may also help to explain the emission morphology of gas-phase methanol as seen with ALMA. This process would help increase the volume of methanol ice in the inner midplane ($\lesssim 60$ au) and would yield an emission morphology which peaks closer to the central star.

6. Detection of COMs higher up the ladder of complexity

The possibility to detect COMs of a similar or higher complexity as methanol in disks around young low-mass stars like TW Hya can now be better quantified in light of the detection of methanol in this source. The disk-integrated rotational spectra of multiple COMs across the full ALMA frequency range have been computed assuming LTE holds, that the disk is face on to the line of sight (as also done in Walsh *et al.* 2014), and that the disk is at the distance of TW Hya (54 pc). The assumption of LTE for methanol emission was recently tested and verified for TW Hya in Parfenov *et al.* (2016, 2017). The parametric model which reproduces the methanol detection in TW Hya is used as a template (Walsh *et al.* 2016). The methanol is confined to the midplane ($z/r \lesssim 0.1$) between a radial range of $30 - 100$ au: this resulted in a best-fit methanol abundance of 3×10^{-12} with respect to H_2. It is assumed that each COM has this abundance in the computation of the spectra. This is an optimistic estimate because in other astrophysical environments gas-phase methanol is often the most abundant COM. This also assumes that all COMs have a similar production mechanism to methanol in the disk.

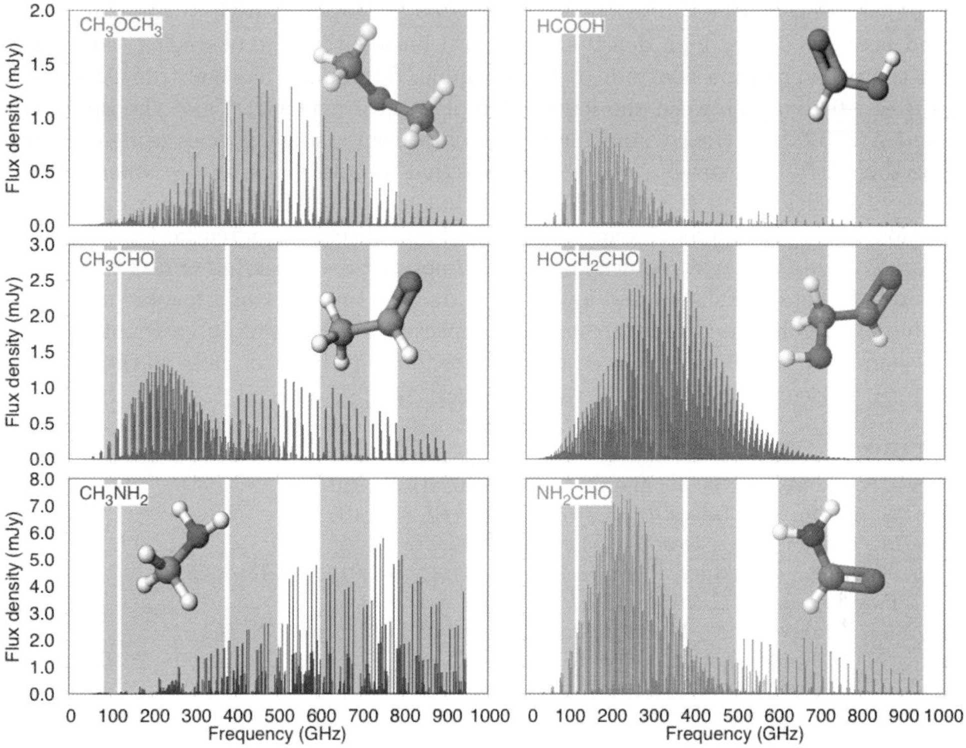

Figure 5. Disk-integrated rotational spectra of six COMs generated using the physical model of TW Hya and confining the gas-phase abundance to the midplane $z/r \lesssim 0.1$ and within $30 \leqslant r \leqslant 100$ au. The assumed abundance is that which best fits the TW Hya methanol data, 3×10^{-12} with respect to H_2. The gray shaded regions are the current ALMA observing bands. Line lists are from the Cologne Database for Molecular Spectrocopy (`http://www.astro.uni koeln.de/cdms/`) or JPL (`https://spoc.jpl.nasa.gov/`).

Figure 5 shows the disk-integrated rotational spectra for six COMs: dimethyl ether (CH_3OCH_3), formic acid (HCOOH), acetaldehyde (CH_3CHO), glycolaldehyde ($HCOCH_2CHO$), methylamine (CH_3NH_2), and formamide (NH_2CHO). The spectra are approximately ordered by peak line flux from top to bottom. The grey shaded regions show the ALMA observing bands available in ALMA Cycle 5 (the current observing cycle). The frequency of the strongest rotational transitions vary according the properties of the molecule (see, e.g., Herbst & van Dishoeck 2009). Formic acid, acetaldehyde, and formamide all have their strongest transitions towards the lower frequency bands (4, 5, and 6), glycolaldehyde peaks in band 7, and dimethyl ether and methylamine have their strongest transitions in band 8 and bands 9 and 10, respectively. Because of the increase in ALMA sensitivity towards the lower frequencies (for a fixed integration time), formamide is a strong contender for detection in protoplanetary disks with a peak line flux density of ≈ 7 mJy. Recall that the rms noise of the ALMA data for TW Hya was 2 mJy beam^{-1} channel^{-1}. Detections of individual transitions of formic acid, acetaldehyde, and glycolaldehyde may also be possible with a sufficiently deep integration. However, dimethyl ether and methylamine may lie beyond the capabilities of the telescope, at least in candidate disks around young low-mass stars.

These line spectra are computed for a disk at the same distance as TW Hya (54 pc). For disks at a distance of ≈ 150 pc which is more typical of nearby star-forming regions, the line emission will be diluted by a factor of ≈ 9. However, each of the species considered here has a very high density of spectral lines within the observing band within which the peak emission is attained. For example, formamide has 660 transitions (with $\log(I_\nu) \gtrsim -6$) lying between the lower edge of band 6 (211 GHz) and the upper edge of band 7 (373 GHz). Hence, stacking of line data from a single species using data from unbiased spectral line surveys may enable a signal-to-noise boost of $\sqrt{n}$ where n is the number of stacked lines. It should be noted that care must be taken to avoid stacking transitions of vastly different excitation energies and checks for potential line blends will be required. On this latter point, the typical line widths of disk-integrated line profiles from protoplanetary disks around low-mass stars is of the order of a few km s^{-1}; hence, blending of neighbouring line transitions is expected to be less of an issue than it is towards sources in which the observed spectra are found to be at or close to the confusion limit (e.g., hot cores, see Herbst & van Dishoeck 2009).

References

Andrews, S. M., Wilner, D. J., Zhu, Z., *et al.* 2016, *ApJL*, 820, L40

Bergin, E. A., Du, F., Cleeves, L. I., *et al.* 2016, *ApJ*, 831, 101

Bertin, M., Romanzin, C., Doronin, M., *et al.* 2016, *ApJL*, 817, L12

Boogert, A. C. A., Gerakines, P. A., & Whittet, D. C. B. 2015, *ARA&A*, 53, 541

Caselli, P. & Ceccarelli. C. 2012, *A&ARv*, 20, 56

Chen, Y.-J., Ciaravella, A., Muñoz Caro, G. M., *et al.* 2013, *ApJ*, 778, 162

Chuang, K.-J., Fedoseev, G., Ioppolo, S., van Dishoeck, E. F., & Linnartz, H. 2016, *MNRAS*, 455, 1702

Cruz-Diaz, G. A., Martín-Doménech, R., Muñoz Caro, G. M., & Chen, Y.-J. 2016, *A&A*, 592, A68

Drozdovskaya, M., Walsh, C., Visser, R., Harsono, D., & van Dishoeck, E. F. 2014, *MNRAS*, 445, 913

Drozdovskaya, M., Walsh, C., van Dishoeck, E. F., *et al.* 2016, *MNRAS*, 462, 977

Du, F., Bergin, E. A., Hogerheijde, M. R., *et al.* 2017, *ApJ*, 842, 98

Furuya, K. & Aikawa, Y. 2014, *ApJ*, 790, 97

Herbst, E. & van Dishoeck, E. F. 2009, *ARA&A*, 47, 427

Hogerheidje, M. R., Bergin, E. A., Brinch, C., *et al.* 2011, *Science*, 334, 338

Kama, M., Bruderer, S., van Dishoeck, E. F., *et al.* 2016, *A&A*, 592, A83

Le Roy, L., Altwegg, K., Balsiger, H., *et al.* 2015, *A&A*, 583, A1

Ligterink, N., Walsh, C., Bhuin, R. G., *et al.* 2017, *A&A*, to be submitted

Loomis, R. A., Öberg, K. I., Andrews, S. M., *et al.* 2017, *ApJ*, submitted

Modica, P. & Palumbo, M. E. 2010, *A&A*, 519, A22

Mumma, M. J. & Charnley, S. B. 2011, *ARA&A*, 49, 471

Nomura, H., Aikawa, Y., Tsujimoto, M., Nakagawa, Y., & Millar, T. J. 2007, *ApJ*, 661, 334

Nomura, H., Tsukagoshi, T., Kawabe, R., *et al.* 2016, *ApJL*, 819, L7

Öberg, K. I., Garrod, R. T., van Dishoeck, E. F., & Linnartz, H. 2009, *A&A*, 504, 891

Öberg, K. I., Guzmán, V. V., Furuya, K., *et al.* 2015, *Nature*, 520, 198

Öberg, K. I., Guzmán, V. V., Merchantz, C. J., *et al.* 2017, *ApJ*, 839, 43

Parfenov, S. Yu., Semenov, D. A., Sobolev, A. M., & Gray, M. D. 2016, *MNRAS*, 460, 2648

Parfenov, S. Yu., Semenov, D. A., Henning, Th., *et al.* 2017, *MNRAS*, 468, 2024

Salinas, V. N., Hogerheijde, M. R., Bergin, E. A., *et al.* 2016, *A&A*, 591, A122

Semenov, D. & Wiebe, D. 2011, *ApJS*, 196, 25

Walsh, C., Millar T. J., Nomura, H., *et al.* 2014, *A&A*, 563, A33

Walsh, C., Nomura, H., & van Dishoeck, E. F., 2015, *A&A*, 582, A88

Walsh, C., Loomis, R. A., Öberg, K. I., *et al.* 2016, *ApJL*, 823, L10

Weidenschilling, S. J. 1977, *MNRAS*, 180, 57

Yoneda, H., Tsukamoto, Y., Furuya, K., & Aikawa, Y. 2016, *ApJ*, 833, 105

Astrochemistry VII – Through the Cosmos from Galaxies to Planets
Proceedings IAU Symposium No. 332, 2017
M. Cunningham, T. Millar & Y. Aikawa, eds.

doi:10.1017/S174392131700775X

Simulations of branched carbon-chain chemistry in star-forming regions

Robin T. Garrod

Depts. of Chemistry & Astronomy, University of Virginia,
Charlottesville, VA 22904, USA
email: rgarrod@virginia.edu

Abstract. The detection of *iso*-propyl cyanide (i-C_3H_7CN) toward the Galactic Center hot-core source Sgr B2(N) (by Belloche *et al.* 2014) marked the first interstellar detection of an aliphatic molecule with a branched carbon-chain structure. Surprisingly, this branched form was found to have an almost equal abundance with its straight-chain homologue, normal-propyl cyanide (i:n = 0.40 ± 0.06). The detection of this first example of an interstellar molecule with a side-chain raises the question as to how prominent such structures may be in interstellar chemistry, and whether the large branched-to-straight chain ratio is maintained for even larger molecules.

Here are presented recently published models that simulate the chemistry occurring in Sgr B2(N) using a chemical network that explicitly includes the straight-chain and branched forms of propyl cyanide (*normal*/*iso*) and butyl cyanide (*normal*/*iso*/*sec*/*tert*), as well as butane (n/i) and pentane (n/i/*neo*). Formation is assumed to occur on dust-grain surfaces, but a full complement of destruction mechanisms is included both on the grains and in the gas phase.

The models suggest that branched structures become increasingly dominant as molecular sizes increase. In the case of butyl cyanide, the *sec* form is at least ∼2 times more abundant than the straight-chain *normal* form, and together the branched forms dominate *normal*-butyl cyanide by a factor of at least 3. The results for the larger alkanes suggest similarly large ratios of branched to straight-chain molecules. A key set of reactions in the surface/ice chemistry of interstellar nitriles is found to be the addition of the CN radical to unsaturated hydrocarbons, especially acetylene and ethene. The models also predict that the dominant, *sec* form of butyl cyanide reaches a peak abundance equal to that of n-propyl cyanide, albeit with a smaller emission radius. This makes s-C_4H_9CN a good candidate for detection. New ALMA observations to search for this molecule are ongoing.

Keywords. astrochemistry, ISM: abundances, ISM: individual (Sagittarius B2N), ISM: molecules, molecular processes

---

1. Introduction

Up to now, close to 200 different molecules have been detected in interstellar or circumstellar regions, with many of these containing multiple carbon atoms. However, until recently, almost all such species fell into two groupings: either cyclic molecules (both aromatic and aliphatic), of which relatively few have been detected, with C_{60} and C_{70} being the largest so far (Cami *et al.* 2010); or straight-chain molecules, for example propylene (C_3H_6; Marcelino *et al.* 2007). The latter grouping includes molecules that are actually linear, such as the cyanopolyynes, of which cyanoacetylene (HC_3N) is the simplest; but the underlying structure of most straight-chain organics consists simply of an open, unbranched chain of carbon atoms.

One of the largest unbranched carbon-chain molecules found in the ISM is *normal* propyl cyanide (n-C_3H_7CN), which was detected toward the Galactic Center hot-core region Sgr B2(N) (Belloche *et al.* 2009). Its branched carbon-chain isomer, *iso*-propyl

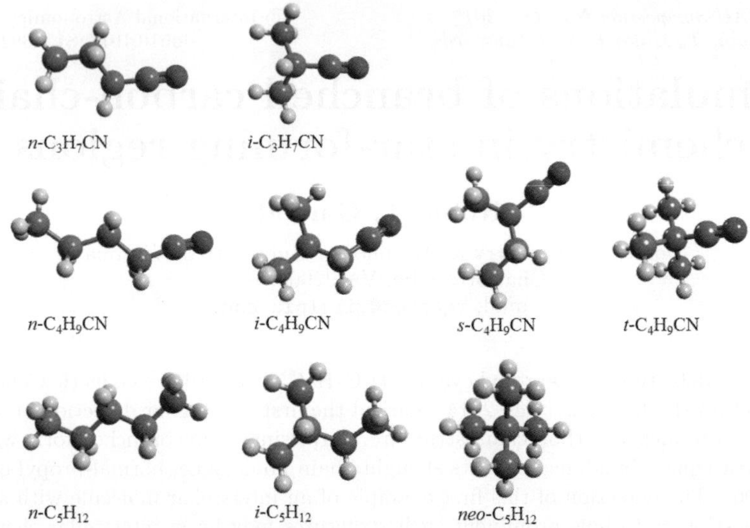

Figure 1. The structures of the various forms of propyl cyanide, butyl cyanide and pentane. Each of these molecules is explicitly considered in the chemical models presented by Garrod *et al.* (2017). *Image credit: Eric Willis*

cyanide (i-C_3H_7CN) was recently detected with the ALMA telescope toward the same source, as part of the *EMoCA* 3-mm line survey of Sgr B2(N) (Belloche *et al.* 2014). This was the first detection of a branched carbon-chain molecule in the ISM. The structures of both forms of propyl cyanide are illustrated in Fig. 1.

The formation of branched structures naturally requires at least four carbon atoms to be present, meaning that a significant degree of chemical complexity must be achieved in the source for branching to arise as an observable feature. Intriguingly, the observed ratio between branched and straight-chain propyl cyanide in Sgr B2(N) is close to unity, at a value $[i{:}n]=0.40 \pm 0.06$.

As well as observing propyl cyanide emision, Belloche *et al.* (2014) used the chemical kinetics model *MAGICKAL* (Garrod 2013) to investigate the production of both forms of propyl cyanide in Sgr B2(N) – this itself constituted the first attempt to model branched carbon-chain chemistry in the ISM. It was assumed in this model that, as with many other complex organics, propyl cyanide could be formed on dust grains during the warm-up of the hot core from ~10 K to the hundreds of Kelvin achieved in hot cores (Garrod & Herbst 2006), to then desorb into the gas phase at yet higher temperatures. In that model, propyl cyanide was produced within or upon the ice mantles that formed on the grains at low temperatures (~10 K), through the addition of radicals such as CN and C_3H_7, which meet via thermal diffusion at somewhat higher temperatures (>20 K). In order to treat explicitly the formation of the different forms of propyl cyanide, the structures of the pre-cursor radicals and the position of the radical site (i.e. an unpaired electron) also were distinguished. For example, C_3H_7 may take two forms; a primary radical form, where the unparied electron is on an end carbon, and a secondary form, where it is on the middle carbon atom. The addition of the CN radical to either form of C_3H_7 would then produce the corresponding straight-chain or branched form of propyl cyanide. Other formation mechanisms were also included, such as the addition of a methyl group to one or other of the C_2H_4CN radicals, or the addition of radicals C_2H_5 and CH_2CN.

The production mechanisms of pre-cursor radicals on grains and their resultant ratios are critical to the $i{:}n$ ratio achieved for propyl cyanide in the Belloche *et al.* (2013) model.

Table 1. Reactions incorporated into the model of Garrod *et al.* (2017) for the formation of propyl cyanide and butyl cyanide.

CH_3	+	CH_2CH_2CN	→	n-C_3H_7CN
CH_2CH_3	+	CH_2CN	→	n-C_3H_7CN
$CH_2CH_2CH_3$	+	CN	→	n-C_3H_7CN
CH_2	+	CH_2CH_2CN	→	$CH_2CH_2CH_2CN$
H	+	$CH_2CH_2CH_2CN$	→	n-C_3H_7CN
CN	+	C_3H_6	→	CH_3CHCH_2CN
H	+	CH_3CHCH_2CN	→	n-C_3H_7CN
H	+	CH_3CH_2CHCN	→	n-C_3H_7CN
CH_3	+	CH_3CHCN	→	i-C_3H_7CN
CH_3CHCH_3	+	CN	→	i-C_3H_7CN
CH_2	+	CH_3CHCN	→	$CH_2CH(CH_3)CN$
H	+	$CH_2CH(CH_3)CN$	→	i-C_3H_7CN
H	+	$CH_3C(CH_3)CN$	→	i-C_3H_7CN
CH_3	+	$CH_2CH_2CH_2CN$	→	n-C_4H_9CN
CH_2CH_3	+	CH_2CH_2CN	→	n-C_4H_9CN
$CH_2CH_2CH_3$	+	CH_2CN	→	n-C_4H_9CN
$CH_2CH_2CH_2CH_3$	+	CN	→	n-C_4H_9CN
CN	+	C_4H_8	→	$CH_3CH_2CHCH_2CN$
H	+	$CH_3CH_2CHCH_2CN$	→	n-C_4H_9CN
CH_3	+	CH_3CHCH_2CN	→	i-C_4H_9CN
CH_3CHCH_3	+	CH_2CN	→	i-C_4H_9CN
$CH_2CH(CH_3)CH_3$	+	CN	→	i-C_4H_9CN
CH_3	+	CH_3CH_2CHCN	→	s-C_4H_9CN
CH_3	+	$CH_2CH(CH_3)CN$	→	s-C_4H_9CN
CH_2CH_3	+	CH_3CHCN	→	s-C_4H_9CN
$CH_3CHCH_2CH_3$	+	CN	→	s-C_4H_9CN
CH_3	+	$CH_3C(CH_3)CN$	→	t-C_4H_9CN
$CH_3C(CH_3)CH_3$	+	CN	→	t-C_4H_9CN

The mechanisms included for pertinent radicals consist of (i) the direct photodissociation of stable molecules by UV photons induced by cosmic-ray collisions with gas-phase H_2; (ii) the addition of a hydrogen atom to a carbon double- or triple-bond; and (iii) the abstraction of a hydrogen atom from a stable molecule by another radical, often OH or NH_2, which are derived from the major ice components H_2O and NH_3 and which have relatively low barriers to hydrogen abstraction from many other species.

The Belloche *et al.* (2013) model indicated that, with radical production determined as above, the strong bias – especially through pathways (ii) and (iii) – to produce secondary radicals, which are energetically favored, meant that the slight dominance of the straight-chain form of propyl cyanide in the observations could not be reproduced. Indeed, the closest that could be reached was a ratio of ~2:1 in favor of i-C_3H_7CN. The only mechanism that was effective in producing the *normal* form in any significant quantity was the reaction $C_2H_5 + CH_2CN \rightarrow n\text{-}C_3H_7CN$, which involves radicals for which only a primary form is available, and which therefore has no analog that would allow production of the branched form.

2. Chemical modeling of butyl cyanide

The above-described model by Belloche *et al.* (2014) included propyl cyanide and butane as the only branched species. In order to expand the treatment of branched interstellar molecules, Garrod *et al.* (2017) constructed a larger chemical network that included the next homolog in the series of nitriles, namely butyl cyanide, as well as the equivalent alkane, pentane. Butyl cyanide (C_4H_9CN) contains five carbon atoms, and may exist in one of four forms (*normal*, *sec*, *iso*, and *tert*), illustrated in Fig. 1. Pentane may take three forms (*normal*, *iso*, and *neo*).

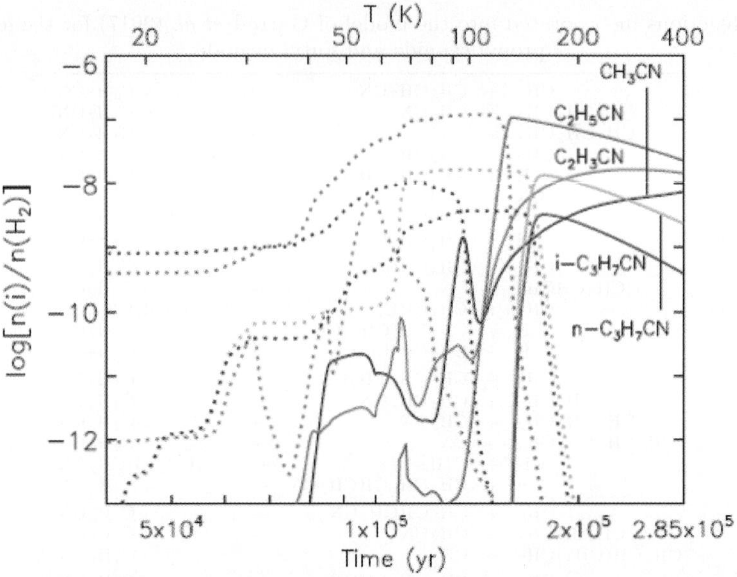

Figure 2. Time- and temperature-dependent chemical abundances for a selection of nitriles during the warm-up of a model hot core, taken from Garrod *et al.* (2017).

As well as including various mechanisms that produce all the possible pre-cursor radicals for the grain-surface/ice formation of butyl cyanide and pentane (as well as gas-phase and grain-surface destruction mechanisms for all new species), another type of reaction was included that was not previously present in the surface chemistry: the addition of the CN radical to species with a carbon double- or triple-bond, producing a radical in which the cyanide group is attached to the terminal carbon in the molecule, leaving a secondary radical that can form a stable, straight-chain molecule by the addition of a hydrogen atom. It was found by Gannon *et al.* (2007) that such processes can occur in the gas phase with no substantial activation energy barrier. The Garrod *et al.* (2017) model assumed that the barriers would be the same for surface/mantle reactions. The full set of reactions included in the model that lead to the various forms of propyl and butyl cyanide are shown in Table 1. All the reactions shown are assumed to have no activation energy barriers.

3. Model results

The inclusion of the above mechanisms for CN to attack carbon multiple bonds allows an appropriate [*i*:*n*] ratio to be achieved. While the C_2H_5 + CH_2CN reaction is still important, the addition of CN to propylene, followed by the addition of a hydrogen atom to the resulting radical, allows yet more *n*-C_3H_7CN to be formed. The results for an "intermediate" warm-up timescale (corresponding to a time period of 2×10^5 yr to go from 10–200 K) are shown in Fig. 2; the plot shows the progression of fractional abundances with respect to H_2 over time, during the warm up stage, with dotted lines indicating mantle abundances and solid lines of the same color indicating the gas-phase abundance of the same species. The branched form of propyl cyanide, *i*-C_3H_7CN, is found to be formed predominantly by the hydrogenation of vinyl cyanide (C_2H_3CN), preferentially forming a secondary radical to which a methyl group (CH_3) can then attach. The *iso* to *normal* ratio for propyl cyanide achieved in the specific model shown, achieved at the time of peak propyl cyanide abundance, is [*i*:*n*] = 0.24.

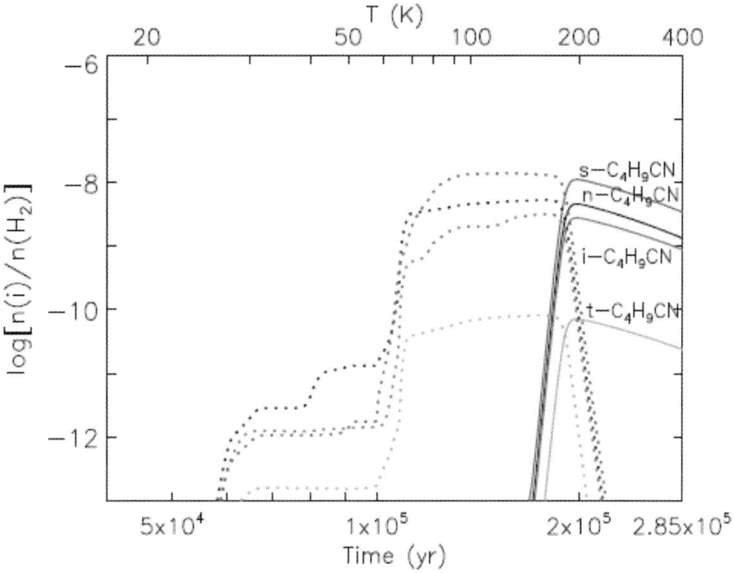

Figure 3. Time- and temperature-dependent chemical abundances for the various forms of butyl cyanide during the warm-up of a model hot core, taken from Garrod *et al.* (2017).

The results for butyl cyanide are shown in Fig. 3. The most abundant forms of butyl cyanide are found to be *sec* and *normal*, in that order; these are produced through a similar mechanism to *iso*-propyl cyanide, i.e. the hydrogenation of vinyl cyanide to either a primary or a secondary radical, followed by the addition of an ethyl group (C_2H_5), forming either the *normal* or *sec* form, respectively. The *sec* form is seen to be $\sim$2 times more abundant than the *normal*. The *iso* form is marginally less abundant than the *normal*, by around a factor of 2, although the range of models by Garrod *et al.* (2017) generally produce a ratio of around unity. The *tert* form, however, is never very abundant, and is therefore unlikely to be detected easily. Summing the abundances of all branched molecules gives a ratio $[(i{+}s{+}t){:}n] = 3$, which is the lowest value achieved in the several model runs tested by Garrod *et al.* (2017), with others producing values as high as 5.8.

It is found that, for the alkanes (see Fig. 4), *normal*-butane strongly dominates the branched *iso* form, whereas *iso*-pentane dominates over *normal*-pentane. *Neo*-pentane, which, like *tert*-butyl cyanide, is doubly branched, is an order of magnitude less abundant that the straight-chain form.

4. Conclusions

Several conclusions may be drawn from the models of propyl and butyl cyanide. The inclusion of reactions in which CN radicals may attack carbon multiple bonds, with no significant activation energy barrier to overcome, allows appropriate quantities of the straight-chain form of propyl cyanide to be produced on dust grains. In general, one might expect that in cases where branched forms of molecules may be produced, they will tend to dominate, due to the propensity to form secondary radicals over primary radicals. The nitriles (and indeed butane) appear to be an exception to this outcome, but only for propyl cyanide; for the next largest homolog, butyl cyanide, the branched forms together dominate, and *sec* in particular is the most abundant form. The *normal* form of butyl cyanide was searched for in the *EMoCA* survey data; no detection was

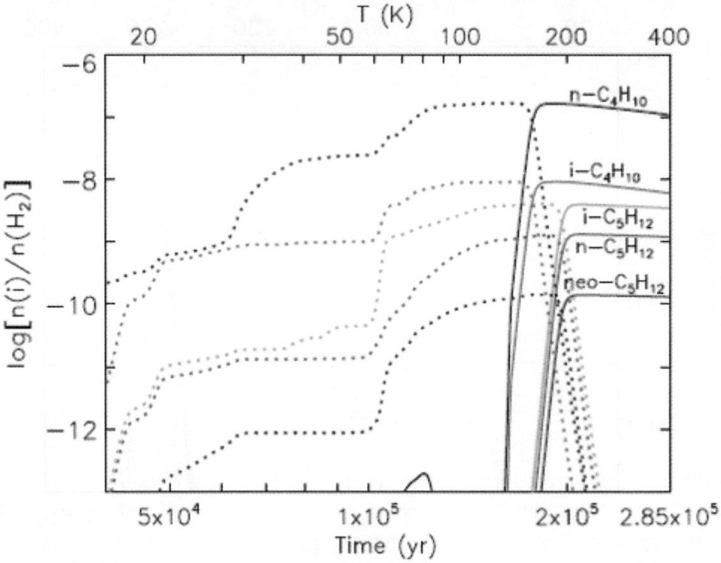

Figure 4. Time- and temperature-dependent chemical abundances for the various forms of butane and pentane during the warm-up of a model hot core, taken from Garrod *et al.* (2017).

forthcoming, but an upper limit ratio with *n*-propyl cyanide of <0.59 was obtained. This value is consisitent with the range produced by the models of $\sim 0.1 - 1$.

Sec butyl cyanide is predicted to be the most abundant form of this molecules, and may be detectable as a result; the *n* and *i* forms are also plausible detection targets, but the abundance of *t*-C_4H_9CN predicted in the models makes it a poor candidate for detection. Observational efforts to detect the former three forms are ongoing.

It should be noted that vinyl cyanide (C_2H_3CN) appears to be important to the formation of larger nitriles, as it may be hydrogenated to produce radicals (both primary and secondary) to which methyl and ethyl groups can attach. In fact, almost all of the grain surface-produced vinyl cyanide in the models is converted to another larger nitrile (mainly ethyl cyanide, C_2H_5CN). The gas-phase vinyl cyanide that peaks at high temperatures is actually formed through the destruction of gas-phase ethyl cyanide, following its desorption from grains.

The models presented here show that branching in organic molecules is likely to be an important feature in interstellar chemistry, especially in star-forming regions, where the degree of observable chemical complexity is usually greatest.

References

Belloche, A., Garrod, R. T., Müller, H. S. P., Menten, K. M., Comito, C. & Schilke, P. 2009 *A&A*, 499, 215

Belloche, A., Garrod, R. T., Müller, H. S. P. & Menten, K. M. 2014 *Science*, 345, 1584

Cami, J., Bernard-Salas, J., Peeters, E., & Malek, S. E. 2010 *Science*, 659, 1180–1182

Garrod, R. T. & Herbst, E. 2006 *A&A*, 457, 927

Garrod, R. T. 2013 *ApJ*, 765, 60

Garrod, R. T., Belloche, A., Müller, H. S. P. & Menten, K. M. 2017 *ApJ*, 601 A48

Marcelino, N., Cernicharo, J., Agúndez, M., Roueff, E., Gerin, M., Martín-Pintado, J., Mauersberger, R. & Thum, C. 2007 *ApJ*, 665, L127–L130

Astrochemistry VII – Through the Cosmos from Galaxies to Planets
Proceedings IAU Symposium No. 332, 2017
M. Cunningham, T. Millar & Y. Aikawa, eds.

doi:10.1017/S1743921317008729

The first detections of the key prebiotic molecule PO in star-forming regions

Víctor M. Rivilla[1], **Francesco Fontani**[1], **Maite Beltrán**[1], **Anton Vasyunin**[2], **Paola Caselli**[2], **Jesús Martín-Pintado**[3] and **Riccardo Cesaroni**[1]

[1]Osservatorio Astrofisico di Arcetri
Largo Enruco Fermi, 50125, Florence, Italy
email: rivilla@arcetri.astro.it

[2]Max Planck Institute for Extraterrestrial Physics,
Giessenbachstrasse 1, 85748, Garching, Germany

[3]Centro de Astrobiología (CSIC-INTA)
Ctra. de Torrejón a Ajalvir km 4, 28850, Torrejón de Ardoz, Spain

Abstract. Phosphorus is a crucial element in prebiotic chemistry, especially the P–O bond, which is key for the formation of the backbone of the deoxyribonucleic acid. So far, PO had only been detected towards the envelope of evolved stars, and never towards star-forming regions. We report the first detection of PO towards two massive star-forming regions, W51 e1/e2 and W3(OH), using data from the IRAM 30m telescope. PN has also been detected towards the two regions. The abundance ratio PO/PN is 1.8 and 3 for W51 and W3(OH), respectively. Our chemical model indicates that the two molecules are chemically related and are formed via gas-phase ion-molecule and neutral-neutral reactions during the cold collapse. The molecules freeze out onto grains at the end of the collapse and desorb during the warm-up phase once the temperature reaches $\sim$35 K. The observed molecular abundances of 10^{-10} are predicted by the model if a relatively high initial abundance of 5×10^{-9} of initial phosphorus is assumed.

Keywords. ISM: molecules, ISM: abundances, stars: formation

1. Introduction

The detection of new interstellar molecules related with prebiotic chemistry in star-forming regions will allow us to make progress on understanding how the building blocks of life could originate in the interstellar medium (ISM). However, there is a key ingredient that still evades detection: phosphorus, P. This element is essential for life (Maciá *et al.* 1997), because it plays a central role in the formation of P-bearing macromolecules such as phospholipids, which are the structural components of all cellular membranes, or adenotryphosphate (ATP), responsible for the transfer of energy in cells (Pasek & Lauretta 2005). Especially important to basic biochemistry is the P–O bond, fundamental for many relevant biological molecules such as phosphate esters, which are essential for the formation of the backbone of the genetic macromolecule deoxyribonucleic acid (DNA).

Phosphorus is thought to be synthesized in massive stars and injected to the ISM through supernova explosions (Koo *et al.* 2013). It has a cosmic abundance of P/H$\sim$ 3×10^{-7} (Grevesse & Sauval 1998). It has been detected towards atmospheres of stars (Caffau *et al.* 2011; Roederer *et al.* 2014; Caffau *et al.* 2016), but it is barely detected in the ISM. The ion P^+ has been detected in several diffuse clouds (Jura & York 1978), and only a few simple P-bearing species (PN, PO, CP, HCP, C_2P, PH_3) have been identified towards the circumstellar envelopes of very evolved objects (Tenenbaum *et al.* 2007; De Beck *et al.* 2013; Agúndez *et al.* 2014). In star-forming regions, only PN has

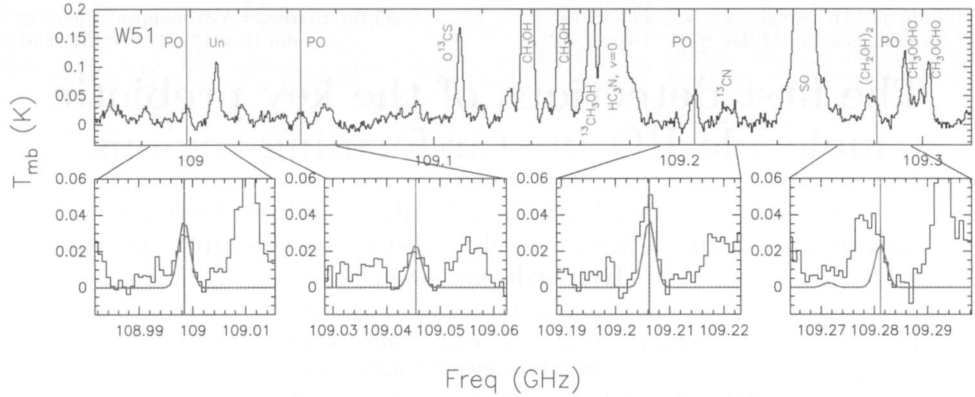

Figure 1. Spectrum observed at 3 mm towards W51. The PO transitions are indicated with blue vertical lines. The lower panels show zoom-in views of the PO transitions. The curve is the LTE fit.

been detected so far (Turner & Bally 1987; Ziurys & Friberg 1987; Fontani *et al.* 2016). Previous searches of PO towards star-forming regions (Sutton *et al.* 1985; Matthews *et al.* 1987) were unsuccesful.

Despite the prebiotic importance of P-bearing molecules, their chemistry is still poorly understood in the ISM. The few theoretical models devoted to P-chemistry disagree in both the chemical formation pathways and the predictions of the abundances of PO and PN. While some works (Millar *et al.* 1987; Adams *et al.* 1990; Charnley & Millar 1994) suggest that PN should be more abundant than PO, other studies involving theoretical modeling and laboratory experiments (e.g. Thorne *et al.* 1984) predict that P should be found mainly in the form of PO. To constrain the chemical models and understand the chemistry of phosphorus in the ISM, astronomical detections of PO and PN in star-forming regions are required.

We present observations searching for PN and PO towards two massive star-forming regions: the W51 e1/e2 and the W3(OH) complexes (with luminosities of 1.5×10^6 and 1.0×10^5 $L_\odot$, respectively).

2. Observations and Results

We used data from different observing runs performed with the IRAM 30m telescope at Pico Veleta (Spain) on 2012 and 2015. The W51 e1/e2 (hereafter W51) region was observed at 1, 2 and 3 mm, while the W3(OH) region was observed at 3 mm. The spectra were exported from the software package CLASS of GILDAS to MADCUBA (Madrid Data Cube Analysis on ImageJ is a software developed in the Center of Astrobiology, Madrid) to visualize and analyze single spectra and datacubes; see (Rivilla *et al.* 2017), which was used for the line identification and analysis.

2.1. *Detection of PN*

We detected the J=2−1 and J=3−2 transitions of PN towards W51 (left and middle panels of Fig 2).

Using an angular diameter of ∼12 arcsec, the LTE simulated spectra reproduce the PN J=2−1 and J=3−2 transitions for an excitation temperature of ∼35 K (Fig. 2), as previously found by Turner & Bally 1987. We derived a PN source-averaged column density of 2.1×10^{13} cm^{-2}. This translates into a molecular abundance with respect to molecular hydrogen of 1.1×10^{-10}.

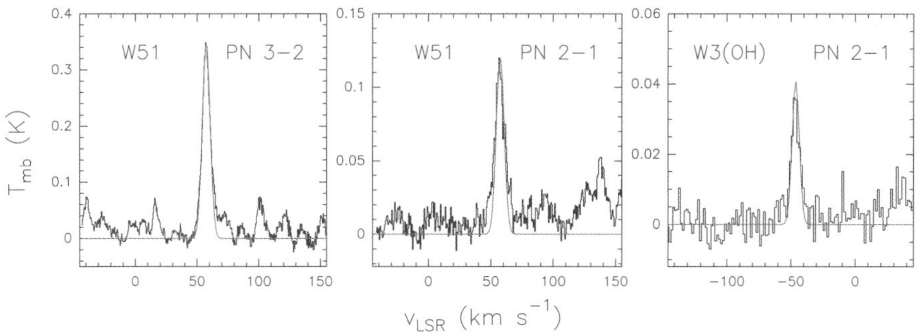

Figure 2. PN transitions observed towards W51 and W3(OH). The red line is the LTE fit.

We also detected the J=2−1 PN transition towards W3(OH) (right panel of Fig. 2). Assuming the same excitation temperature and linear diameter as that of W51 (i.e. 30 arcsec at the distance of W3(OH)), we obtained a source-averaged PN column density of 0.2×10^{13} cm^{-2}, which implies a molecular abundance of 0.4×10^{-10}.

2.2. *Detection of PO*

The PO radical is a diatomic molecule with a dipole moment of μ=1.88 D (Kanata *et al.* 1988). Its millimeter spectrum was measured in the gas phase in the laboratory by Kawaguchi *et al.* (1983). PO has a $^2\Pi_r$ ground electronic state, and therefore each J-transition splits into a doublet because of Λ coupling. The spin of I=1/2 of the phosphorus atom produces a further splitting of each Λ doublet into two hyperfine levels.

We used MADCUBA to identify the PO transitions in the observed spectra. We assumed the same temperature ($T_{\rm ex}$=35 K) and source size as that of PN to produce the LTE spectrum. We show the 3 mm spectra of W51 and the LTE fit in Fig. 1. We clearly identified the PO quadruplet towards W51. The four transitions are well fitted by the LTE model. The lines are detected above 6σ. These PO transitions were already detected in the wind of the oxygen-rich AGB star IK Tauri by De Beck *et al.* (2013).

The LTE fit of the four PO spectral features at 3 mm gives a source-averaged PO column density of 4×10^{13} cm^{-2} and a molecular abundance of 2×10^{-10}. The predicted PO lines at at 1 and 2 mm are consistent with the observed spectra. In summary, we have detected all the 12 detectable PO transitions towards W51 with intensities consistent with LTE emission: two of them (at 3 mm) are unblended, two (at 3 mm) are partially blended, while the others (all the 2 mm and 1 mm transitions) are heavily blended.

For W3(OH), PO was also detected. We have fitted the lines with the LTE model, assuming $T_{\rm ex}$=35 K and the same diameter assumed for PN (30 arcsec), and obtained a column density of 0.6×10^{13} cm^{-2} and a molecular abundance of 1.2×10^{-10}.

3. Discussion: the chemistry of phosphorus

We investigated the formation of the phosphorus-bearing molecules PO and PN in massive star-forming regions using the model based on Vasyunin & Herbst (2013), which simulates the chemical evolution of a parcel of gas and dust with time-dependent physical conditions. First, the formation of a dark dense clump from a translucent cloud during free-fall collapse is simulated (cold starless phase). During this phase, the density increases from 10^3 cm^{-3} to 10^6 cm^{-3} and visual extinction rises from A_v=2 to $A_v \gg$100 mag. Gas and dust temperatures are both decreasing slightly from 20 K to 10 K. On a second phase, the dense dark clump warms up from 10 K to 200 K during 2×10^5 years, thus developing into a hot core (warm-up protostellar phase). We defined as time 0 the moment when the protostar starts to heat up the environment.

We used two sets of initial chemical abundances to explore the influence of the poorly constrained P depletion factor: i) the low metals model (EA1, Wakelam & Herbst 2008), where the atomic P is depleted and has an abundance of 2×10^{-10} with respect to hydrogen, typically used in dark cloud chemical models; and ii) the high metals model, where the phosphorus abundance is increased by a factor of $\sim$25 to 5×10^{-9}.

The results of the modeling are shown in Fig. 3, which shows the temporal evolution of the temperature (Fig. 3a), gas abundance of water (Fig. 3b), gas abundances of PO and PN (Fig. 3c) and the PO/PN ratio (Fig. 3d). The two P-bearing species are chemically related and are formed purely in a sequence of gas-phase ion-molecule and neutral-neutral reactions. PO is efficiently formed during the cold collapse phase in a sequence of reactions:

$$H_3O^+ + P \longrightarrow HPO^+ + H_2 \tag{3.1}$$

$$HPO^+ + e^- \longrightarrow PH + O \tag{3.2}$$

$$HPO^+ + e^- \longrightarrow PO + H \tag{3.3}$$

$$P^+ + H_2 \longrightarrow PH_2^+ \tag{3.4}$$

$$PH_2^+ + e^- \longrightarrow PH + H \tag{3.5}$$

$$O + PH \longrightarrow PO + H \tag{3.6}$$

PN is mainly formed from PO via:

$$N + PO \longrightarrow PN + O \tag{3.7}$$

During the starless phase the model predicts that PO is more abundant than PN (Fig. 3c). Both species freeze out to grains at the end of the cold collapse phase and consequently the gas abundances sharply decrease (Fig. 3c). PO has an abundance $\sim$5 times higher than PN at that time.

At the warm-up phase, once the temperature gradually increases from 10 K to 200 K (Fig. 3a), the species evaporate to the gas phase in the order determined by their desorption energies. PO and PN desorb simultaneously when the temperature reaches $\sim$35 K ($\sim0.7\times10^5$ yr, Fig. 3c), because in our model they have the same evaporation energy of 1900 K .

Some PO is converted to PN via gas-phase reactions, and the abundances of the two species become almost equal until $\sim1.2\times10^5$ yr (Fig. 3c). Once the temperature reaches $\sim$100 K, water ice evaporates and the gas-phase abundance of water reaches a high value of 10^{-4} (see Fig. 3b). Then, the abundance of protonated water H_3O^+ correspondingly increases, and PN reacts with H_3O^+:

$$PN + H_3O^+ \longrightarrow HPN^+ + H_2O \tag{3.8}$$

In turn, HPN^+ has two equally probable channels of dissociative recombination:

$$HPN^+ + e^- \longrightarrow PN + H \tag{3.9}$$

$$HPN^+ + e^- \longrightarrow PH + N \tag{3.10}$$

Since the abundance of atomic oxygen is higher by almost an order of magnitude than the abundance of atomic nitrogen, PH is converted to PO rather than PN. As such, PN is gradually destroyed, while PO is additionally produced, which significantly increases the PO/PN ratio (Fig. 3d).

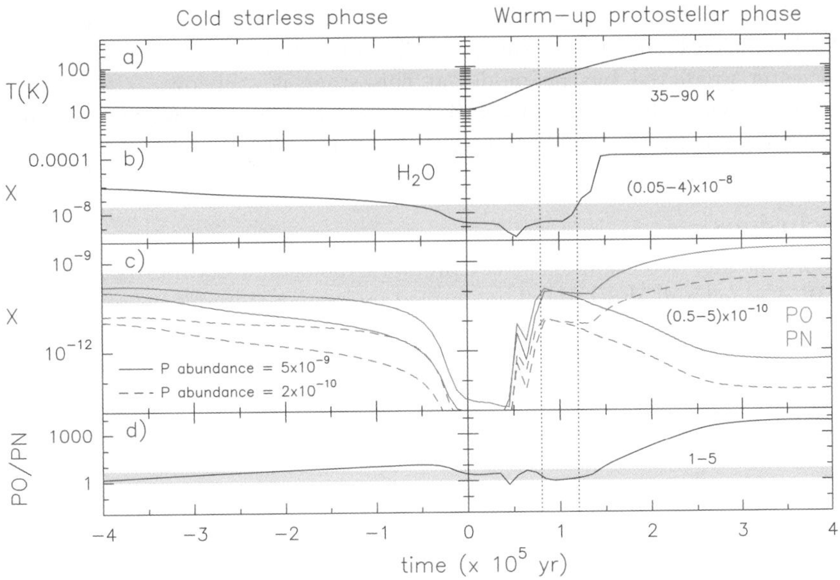

Figure 3. Results of our chemical model. The left panels correspond to the cold collapse phase, and the right panels to the subsequent warm-up phase. We show the evolution during 0.8×10^5 yr, being time 0 the moment when the protostar starts to heat up the environment. The two vertical dotted lines in the right panels indicate the temporal range for which the observational constraints are fulfilled (see text). *a)* Evolution of the temperature assumed by the model. The gray band corresponds to the temperature range between 35 and 90 K which is compatible with the observational constraints (see text). *b)* H_2O abundance. The gray band corresponds to the range found towards massive star-forming regions by Marseille *et al.* (2010). *c)* PO and PN abundances predicted by the model (red and blue lines, respectively). We have considered two different initial abundances of depleted atomic phosphorus: 5×10^{-9} (solid lines, high metals model), and 2×10^{-10} (dashed lines, "low metals" model, usually assumed in dark cloud models). The gray band indicates a conservative range of one order of magnitude around the molecular abundances derived from our observations of PO and PN (see text). *d)* Ratio of the molecular abundances of PO and PN. The gray band shows a conservative range (1−5) around the ratio found by the observations.

To compare the results of the observations with the predictions of our chemical models, we show the values of the abundances and the PO/PN ratio derived from the observations in Figs. 3c and 3d, respectively. The main source of uncertainty of the estimated abundances arises from the uncertainty of the size of the emitting region and the H_2 column density. We have considered a conservative range of one order of magnitude uncertainty for the abundances, i.e. $(0.5–5)\times10^{-10}$ (see Fig. 3c). In the case of the PO/PN ratio, which is independent of the H_2 column density, we have considered a lower uncertainty of a factor of 5, i.e., values in the range 1−5. These observational constraints can only be explained by the "high metals" model (Fig. 3c), and during a period of $\sim5\times10^4$ yr (between 0.8 and 1.2×10^5 yr in Fig. 3). This suggests that the value of the initial depleted atomic abundance of P is a factor of ∼25 higher (i.e. 5×10^{-9}) than that commonly assumed by the standard "low metal" model.

The temporal window that fulfills the observational constraints translates in a temperature range between the desorption temperature of the P-bearing molecules (35 K) and ∼90 K (see Fig. 3a). The former temperature is the same than that derived from PN transitions towards W51, which may indicate that the observations have detected the bulk of PN (and PO) that have recently desorbed from dust grains. The latter temperature is just below the sublimation temperature of the bulk of water (100 K), when PN is efficiently destroyed and partially converted to PO. Then, the gas abundance of water predicted by the model at this stage is still low, $\sim10^{-9}$, which is in good agreement with the values of 5×10^{-10} – 4×10^{-8} found towards several massive star-forming regions by Marseille *et al.* (2010) (see Fig. 3b).

4. Conclusions

We report the first detection of the key prebiotic molecule PO towards two star-forming regions: W51 e1/e2 and W3(OH). The derived molecular abundances of PO are $\sim10^{-10}$ in both sources. We have found an abundance ratio PO/PN of 1.8 and 3 for W51 e1/e2 and W3(OH), respectively, in agreement with the values estimated for evolved stars. Our chemical modeling indicates that the two molecules are chemically related and are formed via gas-phase ion-molecular and neutral-neutral reactions during the cold collapse. The molecules freeze out onto grains at the end of the collapse, and evaporate during the warm-up phase once the temperature reach $\sim$35 K. Similar abundances of PO and PN are expected during a period of $\sim5\times10^4$ yr at the early stages of the warm-up phase, when the temperature is in the range 35–90 K. The observed molecular abundances require a relatively high initial abundance of atomic phosphorus of 5×10^{-9}, 25 times higher than the "low-metal" P-abundance typically used in dark cloud chemical models.

References

Adams, N. G., McIntosh, B. J., & Smith, D., 1990, *A&A*, 232, 443
Agúndez, M., Cernicharo, J., Decin, L., Encrenaz, P., & Teyssier, D., 2014, *ApJ*, 790, L27
Caffau, E., Bonifacio, P., Faraggiana, R., & Steffen, M., 2011, *A&A*, 532, A98
Caffau, E., Andrievsky, S., Korotin, S., Origlia, L., Oliva, E., *et al.* 2016, *A&A*, 585, A16
Charnley, S. B. & Millar, T. J., 1994, *MNRAS*, 270, 570
De Beck, E., Kamiński, T., Patel, N. A., Young, K. H., Gottlieb, *et al.* 2013, *A&A*, 558, A132
Fontani, F., Rivilla, V. M., Caselli, P., Vasyunin, A., & Palau, A., 2016, ArXiv e-prints:1604.02565, accepted in ApJ,
Grevesse, N. & Sauval, A. J., 1998, *SSRv* 85, 161
Jura, M. & York, D. G., 1978, *ApJ*, 219, 861
Kanata, H., Yamamoto, S., & Saito, S., 1988, *JMoSp*, 131, 89
Kawaguchi, K., Saito, S., & Hirota, E., 1983, *J. Chem. Phys.*
Koo, B.-C., Lee, Y.-H., Moon, D.-S., Yoon, S.-C., & Raymond, J. C., 2013, *Science*, 342, 1346
Maciá, E., Hernández, M. V., & Oró, J., 1997, *OrLi*, 27, 459
Marseille, M. G., van der Tak, F. F. S., Herpin, F., Wyrowski, F., Chavarría, L., *et al.* 2010, *A&A*, 521, L32
Matthews, H. E., Feldman, P. A., & Bernath, P. F., 1987, *ApJ*, 312, 358
Millar, T. J., Bennett, A., & Herbst, E., 1987, *MNRAS*, 229, 41P
Pasek, M. A. & Lauretta, D. S., 2005, *AsBio*, 5, 515
Rivilla, V. M., Beltrán, M. T., Cesaroni, R., Fontani, F., Codella, C., & Zhang, Q., 2017, *Astronomy & Astrophysics*, 598, A59
Roederer, I. U., Jacobson, H. R., Thanathibodee, T., Frebel, A., & Toller, E., 2014, *ApJ*, 797, 69
Sutton, E. C., Blake, G. A., Masson, C. R., & Phillips, T. G., 1985, *ApJS*, 58, 341
Tenenbaum, E. D., Woolf, N. J., & Ziurys, L. M., 2007, *ApJ*, 666, L29
Thorne, L. R., Anicich, V. G., Prasad, S. S., & Huntress, Jr. W. T., 1984, *ApJ*, 280, 139
Turner, B. E. & Bally, J., 1987, *ApJ*, 321, L75
Vasyunin, A. I. & Herbst, E., 2013, *ApJ*, 762, 86
Wakelam, V. & Herbst, E., 2008, *ApJ*, 680, 371
Ziurys, L. M. & Friberg, P., 1987, *ApJL*, 314, L49

Astrochemistry VII – Through the Cosmos from Galaxies to Planets
Proceedings IAU Symposium No. 332, 2017
M. Cunningham, T. Millar & Y. Aikawa, eds.

doi:10.1017/S1743921317006937

Chemical modelling of N-bearing COMs in star-forming regions

David Quénard[1], Izaskun Jiménez-Serra[1], Serena Viti[2] and Jonathan Holdship[2]

[1]School of Physics and Astronomy, Queen Mary University of London, Mile End Road, London E1 4NS, UK
email: d.quenard@qmul.ac.uk

[2]Department of Physics and Astronomy, University College London, Gower Street, London, WC1E 6BT, UK

Abstract. The study of complex organic molecules, and more specifically those of prebiotic interest, is important to understand the chemical richness of star-forming regions. The chemistry of nitrogen bearing molecules such as formamide or methyl isocyanate is poorly constrained. We study different chemical pathways to form and destroy these molecules from both the gas phase and grain surface chemistry. From comparison with observations of four different relevant astrophysical regions, we show that both the gas phase and grain surface chemistry are required to explain the observed abundances of these species.

Keywords. astrochemistry, molecular processes, ISM: molecules

1. Introduction

Complex organic molecules (COMs), and more specifically nitrogen-bearing COMs, are of great interest for their potential role in prebiotic chemistry (Saladino *et al.* 2012). Several COMs have been detected in comets (Crovisier *et al.* 2004, Goesmann *et al.* 2015), including species such as formamide (NH_2CHO, Biver *et al.* 2014).

Detailed modelling of the chemistry of N-bearing COMs in star-forming regions such as pre-stellar cores (L1544) and hot corinos (IRAS16293–2422) are needed to fully characterise the main formation and destruction routes of these species, establishing their expected abundances, and compare them to available observations. This study focuses on the study of 3 important N-bearing COMs: isocyanic acid (HNCO and its isomers), formamide (NH_2CHO) and methyl isocyanate (CH_3NCO).

2. Chemical modelling and results

The chemistry of N-bearing COMs in the ISM, and of its precursors, is highly uncertain under the physical conditions found in pre-stellar cores or low-mass proto-stars. We have used the gas-grain chemical code UCLCHEM (Holdship *et al.* 2017) to model the chemistry of N-bearing COMs in these regions. We have updated UCLCHEM with processes such as surface diffusion (Hasegawa *et al.* 1992), chemical reactive desorption (Minissale *et al.* 2016) and reaction-diffusion competition (Garrod & Pauly 2011; Ruaud *et al.* 2015) to study the formation of N-bearing complex organics in those regions.

The chemical network of formamide and methyl isocyanate are highly uncertain. In this study, we have included several possible chemical formation and destruction route of formamide and methyl isocyanate. To test their chemical pathways, we have confronted the result given by UCLCHEM with observations of these species in the proto-typical

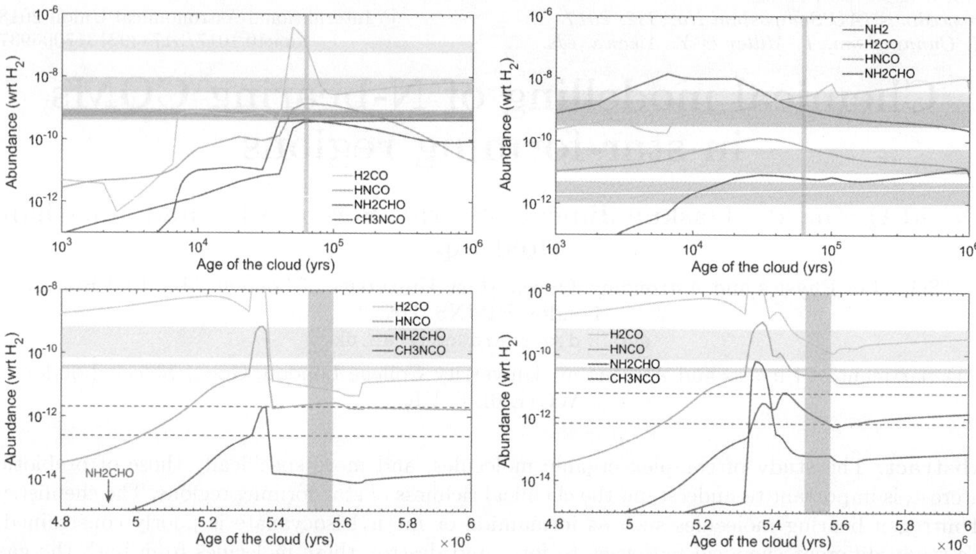

Figure 1. *Top panels:* Modelled abundances from UCLCHEM for the hot corino (left) and the cold envelope (right) of IRAS16293–2422. The grey vertical area shows the best agreement for an age of ~(6–7)$\times10^4$ yrs. *Bottom panels:* Modelled abundances from UCLCHEM for the core centre (left) and the methanol peak (right) of L1544. The grey vertical area shows the best agreement for an age of ~(5.5–5.6)$\times10^6$ yrs. This age corresponds to ~(2–3)$\times10^5$ yrs after the end of the contraction. Coloured areas and dashed lines represent observations errors bars and upper limits, respectively.

low-mass proto-star IRAS16293–2422. We have modelled the hot corino (T_{kin}=250 K, n_H=5$\times10^8$ cm^{-3}) and the cold envelope (T_{kin}=20 K, n_H=2$\times10^6$ cm^{-3}) of the source using a 3-steps chemical modelling: initial diffuse cloud (T_{kin}=10 K, n_H=100 cm^{-3}) followed by a collapse and a warming-up phase.

We have also used recent observations toward the pre-stellar core L1544 to test the chemical network. Jiménez-Serra *et al.* (2016) reported several observational constraints for various N-bearing COMs (including NH_2CHO and CH_3NCO) in both the core centre and the *methanol peak* (Bizzocchi *et al.* 2014) of L1544. The physical conditions of these two regions are respectively: T_{kin}=10 K, n_H=5$\times10^6$ cm^{-3} and T_{kin}=10 K, n_H=4$\times10^5$ cm^{-3} using a 2-steps chemical modelling: initial diffuse cloud followed by a contraction phase of the core (collapse).

The results for these four regions is shown in Fig. 1. We obtain a good agreement between the observed abundances and the predicted values, for all four regions. In our network, formamide is mainly formed via gas-phase reactions at low temperature (~10-20 K) and from both grain surface and gas phase chemistry at higher temperature ($\gtrsim$50 K). Methyl isocyanate is equally formed from both the grain surface and the gas phase, even though it is destroyed efficiently on the grain surface to form N-methylformamide, as suggested by Belloche *et al.* (2017).

References

Belloche, A., Meshcheryakov, A. A., Garrod, R. T., *et al.* 2017, *A&A*, 601, A49

Biver, N., Bockelée-Morvan, D., Debout, V., *et al.* 2014, *A&A*, 566, L5

Bizzocchi, L., Caselli, P., Spezzano, S., & Leonardo, E. 2014, *A&A*, 569, A27

Crovisier, J., Bockelée-Morvan, D., Colom, P., *et al.* 2004, *A&A*, 418, 1141

Garrod, R. T. & Pauly, T. 2011, *ApJ*, 735, 15
Hasegawa, T. I., Herbst, E., & Leung, C. M. 1992, *ApJS*, 82, 167
Holdship, J., Viti, S., Jiménez-Serra, I., Makrymallis, A., & Priestley, F. 2017, *AJ*, 154, 38
Jiménez-Serra, I., Vasyunin, A. I., Caselli, P., *et al.* 2016, *ApJ* (Letters), 830, L6
Minissale, M., Dulieu, F., Cazaux, S., & Hocuk, S. 2016, *A&A*, 585, A24
Ruaud, M., Loison, J. C., Hickson, K. M., *et al.* 2015, *MNRAS*, 447, 4004
Saladino, R., Crestini, C., Pino, S., Costanzo, G., & Di Mauro, E. 2012, *Physics of Life Reviews*, 9, 84

Astrochemistry VII – Through the Cosmos from Galaxies to Planets
Proceedings IAU Symposium No. 332, 2017
M. Cunningham, T. Millar & Y. Aikawa, eds.

doi:10.1017/S1743921317008079

Destruction of $C_2H_4O_2$ isomers in ice-phase by X-rays: Implication on the abundance of acetic acid and methyl formate in the interstellar medium

Marina G. Rachid[1], K. Faquine[1] and S. Pilling[1,2]

[1]Laboratorio de Astroquimica e Astrobiologia, Universidade do Vale do Paraiba, Av. Shishima Hifumi, 2911, Urbanova, CEP: 12244000, So Jos dos Campos, SP, Brazil

[2]Departamento de Fisica, Instituto Tecnologico de Aeronautica, ITA - DCTA Vila das Acacias, So Jos dos Campos, 12228-900 SP, Brazil

Abstract. $C_2H_4O_2$ isomers, methyl formate ($HCOOCH_3$), acetic acid (CH_3COOH) and glycoaldehyde ($HOCH_2CHO$), have been detected in a lot of sources in ISM. However, their abundances are very different, with methyl formate much more abundant than the other two isomers. This fact may be related to the different destruction by ionizing radiation of these molecules. The goal of this work is experimentally study the photodissociation processes of methyl formate and acetic acid ices when exposed to broadband soft X-ray from 6 up to 2000 eV. The experiments were performed coupled to the SGM beamline in the Brazilian Synchrotron Light Source (LNLS/CNPEM) at Campinas, Brazil. The simulated astrophysical ices (12K) were monitored throughout the experiment using infrared vibrational spectroscopy. The analysis of processed ices allowed the determination of the effective destruction cross sections of the parent molecules as well as the effective formation cross section of daughter molecular species. The relative abundance between acetic acid and methyl formate ($N_{CH_3COOH}/N_{HCOOCH_3}$) in different astronomical scenarios and their column density evolution in the presence of X-rays were calculated and our results suggests that such radiation field can be one of the factors that explain the difference in the isomers $C_2H_4O_2$ abundances. We also quantified the daugther species after the establishment of a chemical equilibrium in the samples.

Keywords. Astrochemistry, X-rays, $C_2H_4O_2$, Methyl formate, Acetic acid.

---

1. Introduction

Of all the organic molecules detected so far in the interstellar medium, special attention has been paid to the $C_2H_4O_2$ isomers: methyl formate ($HCOOCH_3$), acetic acid (CH_3COOH) and glycoaldehyde ($HOCH_2CHO$). These species are involved in the synthesis of key biomolecules and they have been detected in a large variety of astronomical sources (Shiao *et al.* 2010 and references therein).

Besides the same molecular formula, the abundance of these especies ae quite different. To account the observed abundances of $C_2H_4O_2$ isomers three process must be considered for each molecule: (I) the formation pathways, (II) the desorption rates and (III) the different destruction of the molecules exposed to a radiation field. These molecules are formed in the surface of dust grains, that can be found in denser regions of clouds and protoplanetary disks. Then, in order to investigate the destruction of these molecules under a radiation field, we have to consider agents that can penetrate large column of gas, because these agents can influence the chemistry inside denser regions of protoplanetary disks and molecular clouds. X- ray and cosmic rays have low absorption cross-section, penetrating deeply inside dense molecular clouds and become an important source of photodissociation and photoionization, even in embedded regions ISM (Pilling & Andrade 2012).

The present work examined the photodissociation of methyl formate and acetic acid frozen samples by broadband X-ray from 6 eV up to 2000 eV. The experiments were performed at 12K in attempt to minimize photodessorption processes induced by X-rays and to guarantee that photolysis products remains in the solid phase. Such low temperatures may be found in highly embedded regions of molecular clouds and some Young Stellar Objects (YSOs). The forming daughter species produced by radiolysis and the parent species were monitored by infrared spectroscopy. We also studied the difference in the isomers survival and the relative abundance in ice grain mantles and compared the experimental result with the available observational data.

2. Experimental methodology

In an attempt to simulate the photodissociation process triggered mainly by soft X-rays (broadband photon beam) on the $C_2H_4O_2$ isomers, methyl formate ($HCOOCH_3$) and acetic acid (CH_3COOH), we use a broadband soft X-ray (6-2000 eV). We utilized the facilities of the Brazilian Synchrotron Light Laboratory (LNLS) located in Campinas, Brazil. The experiments were performed using a high-vacuum portable chamber from the Laboratrio de Astroqumica e Astrobiologia (LASA/ UNIVAP) coupled to the spherical grating monochromator (SGM) beamline, which was operated in the off-focus and white beam mode, producing a wide spectral range of ionizing photons (mainly from 6 eV up to 2000 eV). The beamline details can be found elsewhere (Castro *et al.* 1997, Rodrigues *et al.* 1998).

Briefly, the samples were deposited on a polished and clean ZnSe substrate (transparent to IR) previously cooled to 12K. For methyl formate experiment, the icy film was prepared by depositing 5 mbar of methyl formate gas during roughly 120s. For the acetic acid experiment, the sample was prepared by depositing roughly 3 mbar of acetic acid gas during around 120s. The vapor pressure of methyl formate and acetic acid at room temperature was also measured and the values were around 200 mbar and 15.5 mbar, respectively. In situ chemical analyses of the samples were performed by a portable Fourier transform infrared (FTIR) spectrometer (Agilent Inc., model Cary 630) coupled to the experimental chamber. The spectra were taken from 4000 cm^{-1} to 600 cm^{-1} with resolution of 2 cm^{-1}. The pressure inside the vacuum chamber during the irradiation process was below 10^8 mbar.

As discussed by Pilling and Bergantini (2015), the integrated photon flux in the sample, correspondent to 6-2000eV was estimated as 1.4 10^{14} photons cm^{-2} s^{-1} and the integrated energy flux was approximately 7.6 10^4 erg cm^{-2} s^{-1} . The photon flux profile from SGM beamline is similar to the radiation distribution of YSOs due to the large X-ray component produced by infalling material in protostars. In addition, the radiation distribution of YSOs in the soft X-ray range, as observed for T-Tauri stars (e.g., Imanishi *et al.* 2002), has a maximum around 1 keV, which is similar to the radiation spectra of the employed synchrotron beamline. In these experiment we simulated the effects of induced fast photoelectrons (keV) and slow (eV) secondary electrons, produced by X-rays, in the pure ice samples.

3. Results

The Figures 1 and 2 present the infrared spectra of methyl formate and acetic acid before (uppermost line) and after irradiation, at different irradiation times. The daughter

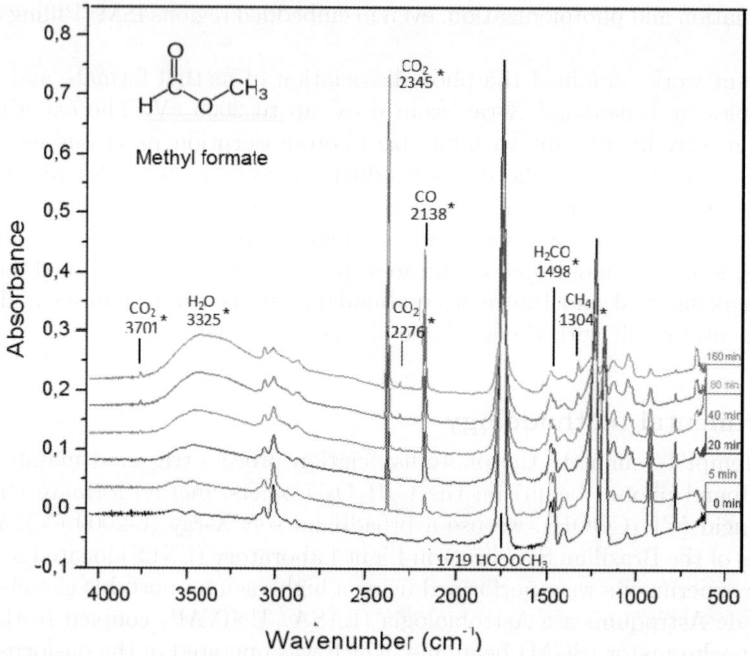

Figure 1. Infrared spectra of frozen methyl formate at 12K during isothermal irradiation with broadband soft X-ray (6 - 2000 eV) at different irradiation times. In all panels the bottom spectrum is the non-irradiated ice, and the uppermost is the spectra obtained at the maximum irradiation time (180 min).

species were not the same for the two isomers, for example, H_2CO was formed only in the methyl formate photolysis while C_2H_6 and C_5H_{10} were formed only in acetic acid photolysis (For the complete list of daughter species and the bands used to identified them see Rachid *et al.* 2017 and references therein) .

The evolution of the abundance of a given molecular species during the irradiation is related to the radiation fluence by:

$$A - A_0 = A_\infty(1 - e^{-\sigma_{d,f} \times F}), \tag{3.1}$$

where F is the fluence, in units of cm^{-2} , A, A_0 , and $A\infty$ are the areas of the infrared band related to a specific vibrational mode of a given specie at a given fluence (A), for the unirradiated sample (A_0), and at the highest fluence (A_∞). In this equation, $\sigma_{d,f}$ represents the effective destruction cross section (σ_d) of parent species or effective formation cross section (σ_f) of the new formed species, both in units of cm^2 (see also Pilling and Bergantini, 2015). We measure the band area varying with the fluence and fit Equation 3.1 to calculate the effective destruction cross section of methyl formate and acetic acid when exposed to soft X-rays. The obtained values where 4.8 $x10^{-18}$ cm^{-2} and 4.2 $x10^{-18}$ cm^{-2} for acetic acid and methyl formate, respectively. The effective formation cross section of the formed species were also calculated by the same method and values range from 1-6 x 10^{-18} cm^2.

4. Discussion

To obtain the physicochemical parameters of the $C_2H_4O_2$ isomers in conditions similar to those where they were detected, we adopted three astrophysical scenarios with different

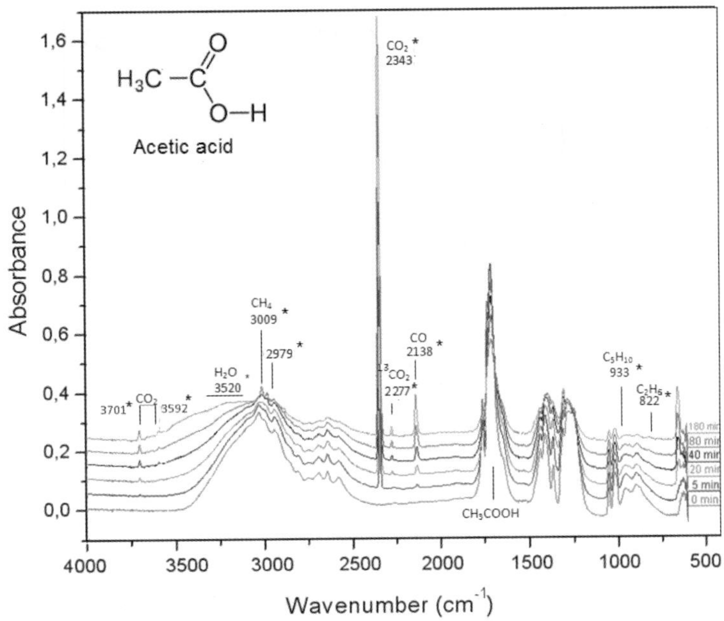

Figure 2. Infrared spectra of frozen methyl formate at 12K during isothermal irradiation with broadband soft X-ray (6 - 2000 eV) at different irradiation times. In all panels the bottom spectrum is the non-irradiated ice, and the uppermost is the spectra obtained at the maximum irradiation time (160 min).

X-ray field: a circumstellar environment (typical T Tauri stars) and two molecular clouds associated with star forming regions (e.g. Sgr B2 and W51). The modeling of soft-Xray flux in each of these scenarios can be found in Rachid *et al.* (2017).

The half-lives of the parent species were determined following the equation:

$$\tau_{1/2} = \frac{ln2}{\sigma_d \times \phi} \tag{4.1}$$

where ϕ indicates the photon flux in units of photons $cm^{-2}s^{-1}$. Due to the small difference in the effective destruction cross section, the half-lives are not so different. An interesting way to see how these small differences can help to explain the relative abundances is studying the changes in the column density of the molecules as a function of time and well as the evolution of the relative column density between the two isomers in time . From the effective dissociation cross section we can calculate the X-ray photodissociation rate (k):

$$k = \sigma_d \times \phi, \tag{4.2}$$

where ϕ is the photon flux in units of cm^{-2} s^{-1} and σ_d is the effective dissociation cross section, in cm^2.

Using the calculated values of k and considering a first-order reaction we can calculate the evolution of the column densities of each molecule when exposed to different X-ray field:

$$N(t) = N_0 e^{-kt} \tag{4.3}$$

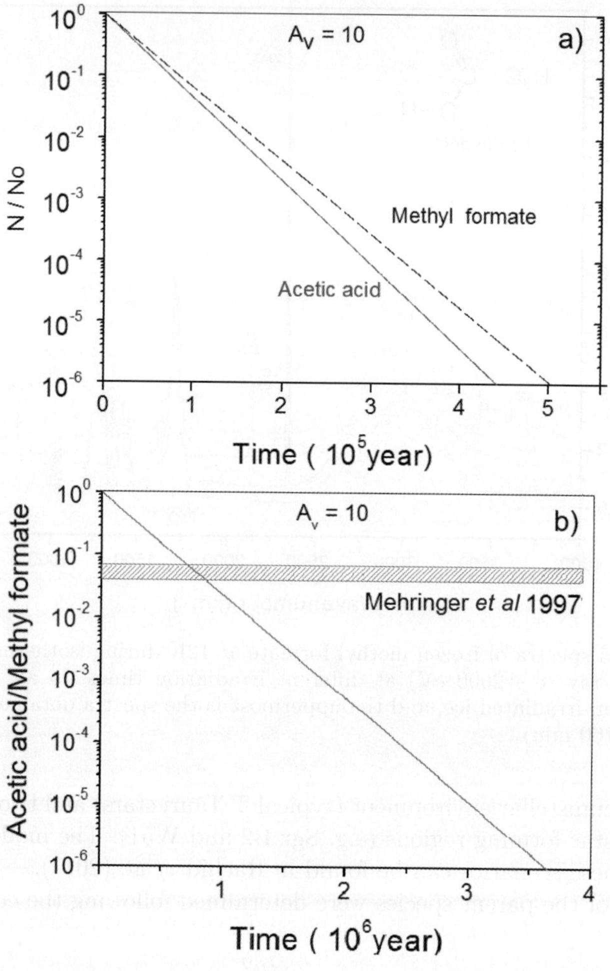

Figure 3. a)Evolution of the column density (N(t)/N$_0$) for methyl formate (dashed black line) and acetic acid (blue line) for a scenario similar to the giant molecular cloud Sgr B2 at A_v = 10; b) Abundances ratio ($N_{CH_3COOH}/N_{HCOOCH_3}$) evolving in time. Horizontal red bar shows the observed values in Sgr B2 (Mehringer *et al.* 1997)

where k is the constant rate, in s^{-1}, N$_0$ is the initial column density and N(t) is the column density in time, both in molecules cm^{-2}.

Using N(t), we calculated the relative abundance ($N_{CH_3COOH}/N_{HCOOCH_3}$) evolving in time for different visual extinction. Figure 3a present the temporal evolution of the column density (N(t)/N$_0$) for acetic acid and methyl formate, at A_v = 10, in a X-ray radiation field similar to those found in Sgr B2, following Equation (4.3). Figure 3b shows the relative abundance between these two species in time. The red dashed bar shows the observed value of the relative abundance in Sgr B2 (Mehringer *et al.*1997).

We also considering the attenuation of the X-ray flux in order to simulate the effects of this radiation in inner parts of the clouds and protoplanetary disks. For simplification, we assume that all photons in the 6-2000 eV range are equally attenuated by the gas/dust. The attenuation of the photon flux as a function of the visual extinction is given by (adapted from Mendoza *et al.* 2013):

$$\phi_x(A_v) = \phi_x(A_v = 0)e^{-2.3\times10^{-2}A_v} \quad (4.4)$$

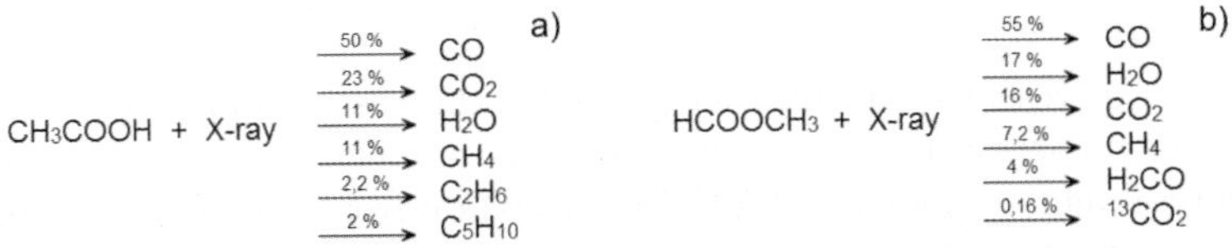

Figure 4. Main products from methyl formate photodissociation by broadband soft X-ray (6-2000eV) and their branching ratios of a) Acetic acid and b) Methyl formate

where A_v is the visual extinction, $\phi_x(\mathrm{Av} = 0 \text{ mag})$ is the non-attenuated flux in units of photons cm^{-2} s^{-1}.

We observed that at moderate visual extinction, the observed relative abundances of Sgr B2 could be reach after an exposition to X-ray radiation of nearly 10^6 years. For high values of visual extinction (highly embedded ices), the observed relative abundances of Sgr B2 could be reach after an exposition to X-ray radiation of approximately 10^8 years. Considering that a typical molecular cloud has an age of approximately 10^7 years, the time to reach the observed relative abundances in such radiation field can be reach for some value of visual extinction between 10 and 300 magnitudes (considering only the effects of X-ray photodissociation). This results indicates that maybe the X-ray photodestruction of methyl formate and acetic acid in the inner part of Sgr B2 can help to explain the difference in abundances of these two isomers.

We performed the same analysis for different values of the X-ray field and visual extinction. We concluded that considering an YSO scenario the differences in the abundances can also be better explained taking into account the X-ray photodestruction of the molecules in highly embedded regions. In the case of W51, we find that the exposed time to X-ray has to be long to reach the observed values of visual extinction, this can be due to an underestimation of the X-ray field in our model. For more details of the analysis and the models see Rachid *et al.* (2017).

In this work we could also quantify the daughter species through the Equilibrium Branching Ratio of the formed species (EBR). The EBR is the relative abundance of the daughter species after the system reach chemical equilibrium (that is, the regions in the difference area graph where the column densities of formed/destructed species don't vary so much). These parameters are important and can be used as input of astrochemical modeling of the gas and ice phase chemistry. The EBR of each formed specie can be calculated by the following expression:

$$EBR(\%) = \frac{n_i}{\sum n_i} \times 100(\%), \tag{4.5}$$

where n_i is the column density of the specie i and $\sum n_i$ is the sum of the column density of the identified species. The column density (N) of a specie can be determined using the area of a band in the infrared spectra:

$$N = \frac{2.3}{A} \int Abs_\nu d\nu, \tag{4.6}$$

where A is the band strength for the used band and Abs_ν is the absorption at a given frequency. The integral in this equation gives the band area. Figure 4a and 4b give the EBR of the identified species for acetic acid and methyl formate, respectively. The main photodissociation product is CO (EBR > 50% of the assigned species), for both isomers, that is also the major product of dissociation in gas phase (Fantuzzi *et al.* 2011).

5. Conclusions

In this work, we present the results of experimental investigation of the effects of broad-band X-rays with energy from 6 to 2000 eV (plus fast electrons, low energy secondary electrons and a minor component of VUV photons) on frozen methyl formate and acetic acid thin films (at 12K) . Our main result and conclusions are as follows:

New molecules produced during photolysis were identified on the basis of their characteristics infrared absorption bands.

The effective dissociation cross section of methyl formate and acetic acid were determined: 4.2 x 10^{-18} cm^2 and 4.8 x 10^{-18} cm^2, respectively. This difference in the half-lives implies in great difference in the survival of molecules in large scales of time.

We calculated the evolution of column density and the of the relative abundances ($N_{CH_3COOH}/N_{HCOOCH_3}$) considering an X-ray field similar to the one existent at the main sources where the $C_2H_4O_2$ isomers were observed (around YSOs and in molecular clouds) to analyze their survival. We conclude that the photodissociation induced by X-ray in highly embedded ices can account for the relative abundance of methyl formate and acetic acid in regions with high visual extinction.

We determined the equilibrium branching ratios (EBR) from methyl formate and acetic acid. The main photodissociation product is CO (EBR > 50% of the assigned species), for both isomers.

6. Acknowledgement

The authors acknowledge the Brazilian agencies FAPESP (Projects JP 2009/18304-0, IC 2014/08643-0, DR 2012/17248-2 and MS 2016/11334-5) and CNPq (Research fellowship 304130/2012-5 and 306145/2015-4) as well as the FVE/ UNIVAP for financial support. We thank Dr. A. Bergantini for help during the experiments, and the staff of UNIVAP and LNLS for technical support.

References

Castro, A. R. B., Fraguas, G. B., Pacheco, J. G., *et al.* 1997, *Commissioning of the SGM beamline, LNLS Tech Memorandum CT 06/97 (Memorando Tecnico), LNLS, Campinas, Brazil.*

Fantuzzi, F., Pilling, S., Santos, A. C. F., Baptista, L., Rocha, A. B., & Boechat Roberty, H. M. 2011, *MNRAS*, 417(4), 2631-2641.

Imanishi, K., Tsujimoto, M., & Koyama, K. 2002, *ApJ*, 572, 300

Mehringer, D. M., Snyder, L. E., Miao, Y., & Lovas, F. J. 1997, *The Astrophysical Journal Letters*, 480(1), L71.

Mendoza, E., Almeida, G. C., Andrade, D. P. P., *et al.* 2013, *MNRAS*, 433, 3440

Modica, P. & Palumbo, M. E. 2010, *Astronomy & Astrophysics*, v. 519, p. A22.

Pilling, S., Andrade, D. P. P., do Nascimento, E. M., *et al.* 2011a, *MNRAS*, 411, 2214

Pilling, S., Baptista, L., Andrade, D. P. P., & Boechat-Roberty, H. M. 2011b, *AsBio*, 11, 883

Pilling, S., Seperuelo Duarte, E., Domaracka, A., *et al.* 2011c, *PCCP*, 13, 15755.

Pilling, S. & Andrade, D. P. P. 2012, *X-Ray Spectroscopy*, ed. S. K. Sharma (Rijeka: InTech), 185.

Pilling, S. & Bergantini, A., 2015, *The Astrophysical Journal*, 811(2), 151.

Rodrigues, A. R. D., Craievich, A. F., & Gonclves da Silva, C. E. T. 1998, *J. Synchrotron Rad.* 5, 1157

Rachid, M. G., Faquine, K., & Pilling, S. 2017, *Planetary Space Science*, in press.

Shiao, Y. S. J., Looney, L. W., Remijan, A. J., Snyder, L. E., & Friedel, D. N., 2010, *The Astrophysical Journal* 716(1), 286

Astrochemistry VII – Through the Cosmos from Galaxies to Planets
Proceedings IAU Symposium No. 332, 2017
M. Cunningham, T. Millar & Y. Aikawa, eds.

doi:10.1017/S1743921317007712

Modelling the emission of nitrogen-bearing molecules towards the solar type protostar IRAS16293-2422

A. Hernández-Gómez[1,2], E. Caux[2], L. Loinard[1] and S. Bottinelli[2]

[1]Instituto de Radioastronomía y Astrofísica, Universidad Nacional Autónoma de México, Morelia 58089, Mexico

[2]IRAP, Université de Toulouse, CNRS, UPS, CNES, Toulouse, France

Abstract. Nitrogen, one of the most abundant elements in the Universe is a fundamental element of molecules which are crucial for life. We present here the modelling of the emission of two of the simplest nitrogen-bearing molecules, CN and NO, with a non-LTE radiative transfer code in IRAS16293-2422, a class 0 low-mass protostar. In this model, we assumed IRAS16293-2422 is formed by 2 compact, hot and dense sources embedded in a three layers of envelope.

Keywords. Protostars, molecular emission, non-LTE modelling.

---

1. Introduction

Nitrogen is one of the most abundant elements in the Universe. Nitrogen-bearing molecules are found in several astrophysical environments such as star-forming regions, protoplanetary disks, pre-stellar cores, comets, etc. In particular, a considerable amount of molecules associated with life contain nitrogen in their structure. For this reason, it is of vital importance to understand the formation and evolution of such molecules in early stages of star formation. The protostar IRAS16293-2422 is a very well known protostar located in the rho Ophiuchus star-forming region at 120 pc and has been identified as a binary system from interferometric observations. Large molecular surveys toward this source have revealed thousands of line transitions from tens of molecular species towards both binary components and the large envelope surrounding them (including many nitrogen-bearing species). Its proximity and its chemical wealth makes IRAS16293-2422 the ideal laboratory to probe not only the chemistry, but also the clustered star formation in detail. Two of the simplest nitrogen-bearing molecules are CN and NO. These molecules are crucial in the chemical network for the formation of molecular nitrogen, and other important molecules such as HCN, HNC, N_2H+, NH_3, etc. We present here the modelling of the emission of CN and NO with a non-LTE radiative transfer code considering that IRAS16293-2422 is formed by 2 compact, hot and dense sources and has three layers composing the envelope.

2. The physical model of IRAS16293-2422

To model the line emission of both CN and NO, we need information about the line profiles of different transitions over a wide range of excitation energies. For that, we have used data from the TIMASSS molecular survey (Caux *et al.* 2011) carried out toward this source (covering a frequency range between 80 – 500 GHz) using the IRAM, APEX and JCMT single-dish radiotelescopes, and the CHESS survey (Ceccarelli *et al.* 2010) (in the 0.5 – 1 THz frequency range) with the HIFI instrument onboard the ESA Herschel Space

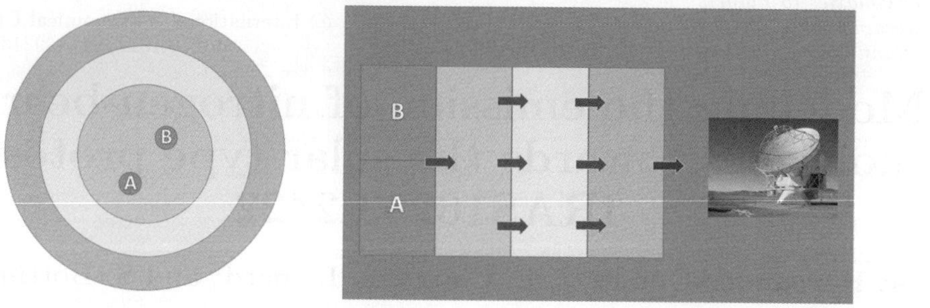

Figure 1. Left. Illustration of the physical structure of IRAS16293 used for the radiative transfer modelling. Both components of the binary system A and B are shown, along with the three components associated with the inner and external parts of the envelope of the protostar. **Right**. Model of the five physical components as slabs used for the CASSIS-RADEX modelling. The inner sources A and B (red) are within the same slab and are not interacting with each other. The output spectra of these two components thus interact with the internal warm layer of the envelope (orange), and the resulting spectrum subsequently interacts with the external cold layers of the envelope (yellow and green). The purple arrows indicates that the emission is calculated along all the slabs, while the black arrow indicates the final computed emission though all the slabs which is then diluted in the corresponding telescope beam.

Observatory. We assumed that the emission originates from five physical components as described in Figure 1.

3. Cyanide (CN) and nitric monoxide (NO) emission

In order to model the CN and NO emission, we have used the non-LTE code RADEX (van der Tak *et al.* 2007) inside CASSIS†, performing a Monte Carlo Markov Chain Method minimisation to compute the best-fit spectra. Each physical component is constrained by 6 physical properties, the column density N, the kinetic temperature T_{kin}, the linewidth FHWM, the source size, the source V_{LSR} and the H_2 density, for a total of 30 free parameters. For CN, we have fixed the size and the H_2 density for all the components and the T_{kin} of the 2 hot cores from data in the literature. The size of the cold layers has been set to 100″ to ensure they fill the telescope beams. For NO, we have first used the same free parameters, but as we found physical properties very close to the CN ones, we finally fixed them for all components to the values obtained for the CN best-fit. Results are summarized in Table 1, Figure 2 and Figure 3.

4. Discussion

Some of the CN physical properties (V_{LSR}, FWHM) are strongly constrained by the very high resolution spectra of the CN (3–2) lines observed with APEX around 340 GHz (see Figure 2). But some others are either degenerated (size, N(CN)), or not well constrained (T_{kin}) because of the lack of spectral resolution for the CN (2–1) and (1–0) lines, that can bias the model. Additional observations at high resolution have been requested for both series of lines. With these data in hands, we will try to model simultaneously all N-bearing species observed either in the cold layers (NH, NH_2) or in the warm envelope (HCN, HNC, HC_3N, HC_5N, CH_3CN, N_2H^+) of IRAS16293, or in both (CN, NO, NH_3, HNCO, see poster from Hernández-Gómez, this conference). If detected we will also use the isotopes (^{13}C and ^{15}N) and deuterated counterparts (for NH, NH_2, NH_3, HCN, HNC,

† http://cassis.irap.omp.eu

Component	N(CN) (cm^{-2})	T_{kin} (K)	FWHM ($km\ s^{-1}$)	V_{LSR} ($km\ s^{-1}$)	Size ($''$)	$n(H_2)$ (cm^{-3})
IRAS16293A	$(2.2 \pm 0.45) \times 10^{14}$	100.0	2.2±0.05	6.0	2.0	1.0×10^8
IRAS16293B	$(9.7 \pm 0.2) \times 10^{14}$	100.0	0.62 ±0.05	3.2	1.0	1.0×10^8
Warm envelope	$(4.5 \pm 0.27) \times 10^{13}$	25.0 ± 1.0	2.13 ±0.06	3.9 ± 0.04	17.3±0.6	1.0×10^7
Cold layer 1	$(2.0 \pm 0.2) \times 10^{14}$	12.5 ± 0.7	0.27 ±0.01	4.0 ± 0.01	100.0	3.0×10^4
Cold layer 2	$(1.2 \pm 0.16) \times 10^{14}$	16.5 ± 0.5	0.62 ±0.01	4.1 ± 0.02	100.0	3.0×10^4

Component	N(NO) (cm^{-2})	T_{kin} (K)	FWHM ($km\ s^{-1}$)	V_{LSR} ($km\ s^{-1}$)	Size ($''$)	$n(H_2)$ (cm^{-3})
IRAS16293A	$(1.6 \pm 0.25) \times 10^{16}$	100.0	2.2	6.0	2.0	1.0×10^8
IRAS16293B	$(2.4 \pm 0.25) \times 10^{16}$	100.0	0.62	3.2	1.0	1.0×10^8
Warm envelope	$(1.0 \pm 0.1) \times 10^{15}$	25.0	2.13	3.9	17.3	1.0×10^7
Cold layer 1	$(8.8 \pm 4.3) \times 10^{13}$	12.5	0.27	4.0	100.0	3.0×10^4
Cold layer 2	$(6.1 \pm 2.4) \times 10^{13}$	16.5	0.62	4.1	100.0	3.0×10^4

Table 1. Assumed physical parameters and best fitted ones using the CASSIS-RADEX MCMC minimisation for all 5 components. CN (top) and NO (bottom).

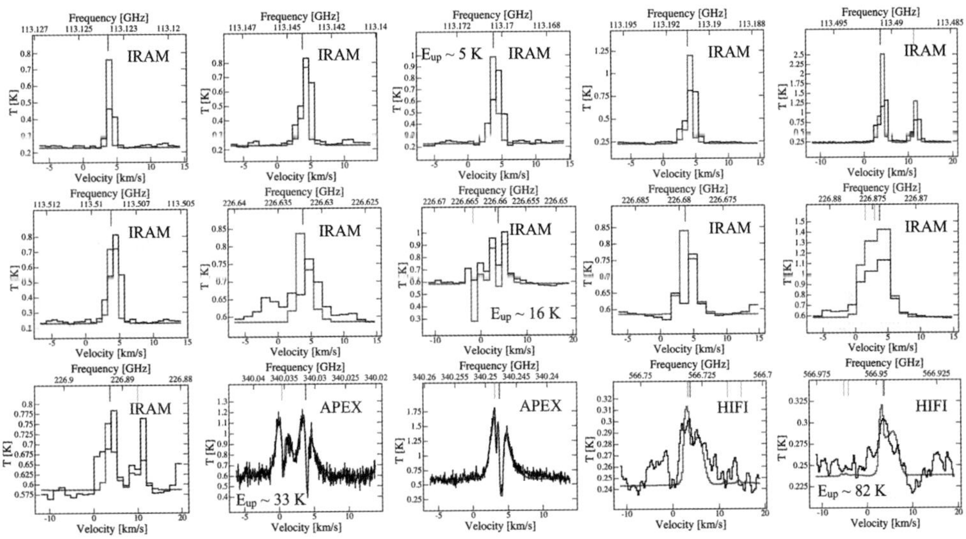

Figure 2. Observed (black) and predicted (red) CN lines modelled with CASSIS-RADEX. CN line frequencies are from CDMS (Müller *et al.* 2001). IRAM : CN (1–0) & (2–1). APEX : CN (3–2). Herschel-HIFI : CN (5–4). We used the collisional coefficients with He scaled by 1.37 as though they are for H_2 from Lique *et al.* (2009).

HC_3N, CH_3CN and HNCO). Interferometric data (SMA, ALMA) will help to constrain the physical properties of the 2 hot corinos, I16293A and I16293B. Finally, we will run the Nautilus chemical code (Ruaud *et al.* 2016) and the LIME 3D radiative transfer code (Brinch & Hogerheijde, 2010) using GASS (Quénard *et al.* 2017) to fully constrain the chemical and physical properties of the 5 identified components.

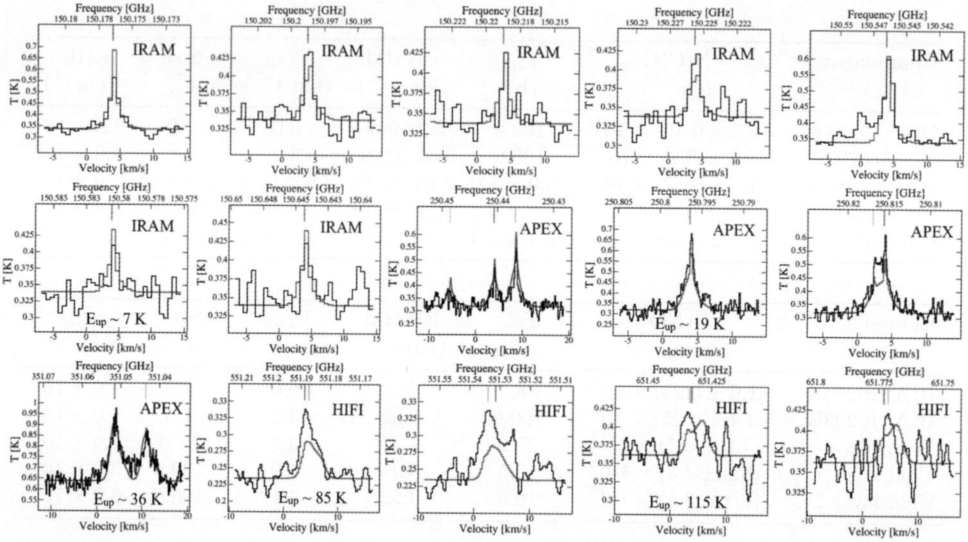

Figure 3. Observed (black) and predicted (red) NO lines modelled with CASSIS-RADEX. NO line frequencies are from CDMS. IRAM : NO (2–1). APEX : NO (3–2) & (4–3). Herschel-HIFI : NO (6–5) & (6–6). We used the collisional coefficients with He scaled by 1.37 as though they are for H_2 from Lique *et al.* (2009).

References

Brinch, C. & Hogerheijde, M. R. 2010, *A&A*, 523, A25

Caux, E., Kahane, C., Castets, A., *et al.* 2011, *A&A*, 532, A23

Ceccarelli, C., Bacmann, A., Boogert, A., *et al.* 2010, *A&A*, 521, L22

Lique, F., van der Tak, F. F. S., Kłos, J., Bulthuis, J., & Alexander, M. H. 2009, *A&A*, 493, 557

Lique *et al.* 2010, *J. Chem. Phys.*, 132, 024303

Müller, H. S. P., Thorwirth, S., Roth, D. A., & Winnewisser, G. 2001, *A&A*, 370, L49

Quénard, D., Bottinelli, S., & Caux, E., 2017, *MNRAS*, 468, 685

Ruaud, M., Wakelam, V., & Hersant, F., 2016, *MNRAS*, 459, 3756

van der Tak, F. F. S., Black, J. H., Schöier, F. L., Jansen, D. J., & van Dishoeck, E. F., 2007, *A&A*, 468, 627

Astrochemistry VII – Through the Cosmos from Galaxies to Planets
Proceedings IAU Symposium No. 332, 2017
M. Cunningham, T. Millar & Y. Aikawa, eds.

doi:10.1017/S1743921317007888

Solid-state production of complex organic molecules: H-atom addition versus UV irradiation

K.-J. Chuang

Sackler Laboratory, Leiden Observatory, Leiden University,
PO Box 9513, 2300 RA Leiden, The Netherlands
email: chuang@strw.leidenuniv.nl

Abstract. Complex organic molecules (COMs) have been observed in comets, hot cores and cold dense regions of the interstellar medium. It is generally accepted that these COMs form on icy dust grain through the recombination reaction of radicals triggered by either energetic UV-photon or non-energetic H-atom addition processing. In this work, we present for the first time laboratory studies that allow for quantitative comparison of hydrogenation and UV-induced reactions as well as their cumulative effect in astronomically relevant $CO:CH_3OH=4:1$ ice analogues. The formation of glycolaldehyde (GA) and ethylene glycol (EG) is confirmed in pure hydrogenation experiments at 14 K, except methyl formate (MF), which is only clearly observed in photolysis. The fractions for MF:GA:EG are 0 : (0.2-0.4) : (0.8-0.6) for pure hydrogenation, and 0.2 : 0.3 : 0.5 for UV involving experiments and can offer a diagnostic tool to derive the chemical origin of these species. The GA/EG ratios in the laboratory (0.3-1.5) compare well with observations toward different objects.

Keywords. astrochemistry, atomic processes, methods: laboratory, ISM: atoms, ISM: molecules, ultraviolet: ISM

1. Introduction

COMs have been identified in low and high-mass proto-stellar envelopes and in dark cloud cores (Herbst & van Dishoeck 2009). Among the observed COMs, glycolaldehyde (GA), the sugar-related molecule, and its related species, i.e. methyl formate (MF) and ethylene glycol (EG), have been detected toward solar mass proto-stars (Jorgensen *et al.* 2012, Jørgensen *et al.* 2016) and in comets (Crovisier *et al.* 2004, Biver *et al.* 2014, Goesmann *et al.* (2005), LeRoy *et al.* 2015). It is widely accepted that the O-bearing COMs form on icy dust grains though the recombination of reactive intermediates, e.g. HCO, CH_2OH and CH_3O that are generated by energetic processing such as UV photolysis (Oberg 2016), electron impact (Maity *et al.* 2015) and ion bombardment (Moore & Hudson 2005), as well as non-energetic processing, e.g. hydrogenation (Fedoseev *et al.* 2015, Chuang *et al.* 2016). The formations of GA, EG and MF are the most common COM products in above processing, indicating that the completely different triggers result in the qualitatively similar results. However, the quantitative comparison of COM abundances is not necessarily the same in different scenarios. Hence, it can be used as a tool to relate the role of specific chemical processes to specific astronomical environments. We present for the first time laboratory studies that aim to quantitatively study COM formation, i.e. GA, EG and MF, through hydrogenations and UV photolysis, as well as their cumulative effect in $CO:CH_3OH$ ice mixtures. The results of MF/EG, MF/GA and GA/EG solid-sate abundance ratios for different experiments can also be compared with those found for the gas-phase abundance ratio in different astronomical environments.

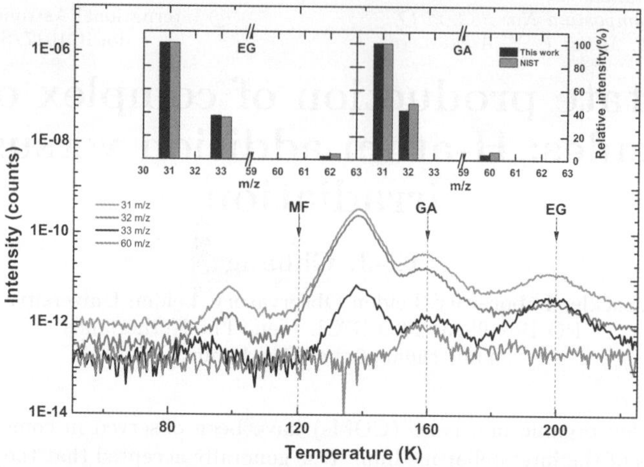

Figure 1. The TPD QMS spectra obtained after co-deposition of CO:CH_3OH = 4:1 ice mixtures with H-atom bombardment at 14 K. Product fragmentation patterns are compared to the NIST database values.

2. EXPERIMENTAL

All experiments are performed under ultra-high vacuum conditions, using SURFRESIDE2. The details of the set-up and procedure have been described in literature (Ioppolo *et al.* 2013, Chuang *et al.* 2017). The icy sample of CO and CH_3OH are introduced simultaneously through two molecular dosing lines, and face a Hydrogen Atom Beam Source and a UV-photon source generated by a Microwave Discharge Hydrogen flowing Lamp (MDHL). Ices are monitored in situ by Fourier Transform Reflection-Absorption InfraRed Spectroscopy (FT-RAIRS) in the range from 700 to 4000 cm^{-1}, with a 1 cm^{-1} resolution. After the co-deposition, the icy sample is studied by a temperature-programmed desorption experiment using a Quadrupole Mass Spectrometer (TPD QMS). Analyzing the unique desorption temperature and ionization fragment pattern for each species allows for identification of the newly formed COM products, i.e. GA, EG and MF.

3. RESULTS AND DISCUSSION

Figure 1 presents a QMS TPD spectra of a typical experiment after the hydrogenation of a CO:CH_3OH ice mixture at 14 K. Two desorption profiles at 100 and 140 K can be assigned to H_2CO and CH_3OH, respectively. Two other peaks at 160 and 200 K are identified as GA and EG, respectively. Furthermore, the comparison of the electron ionization (70 eV) fragment pattern for GA and EG shows that they are largely consistent with the NIST database values. A trace signal of 60 m/z centered around 120 K can be assigned to MF. The similar QMS TPD spectra are observed in pure UV photolysis and cumulative experiments.

In addition to the qualitative identification of the newly formed COMs, the integration of the QMS TPD desorption intensity can also give further quantitative information. The absolute formation abundances of three COMs are presented on the upper panel of Figure 2 revealing that (1) the hydrogenation experiment (+H) produces the lowest yield of COMs when using the relative high H-atom flux, which is intentionally chosen to reach H-atom/UV-photon=1. However, performing an experiment with the same total H-atom fluence as other experiments in a longer timescale (+H(slow)), the COMs yields are comparable with those of the pure UV-photolysis experiment (+hν) except for MF;

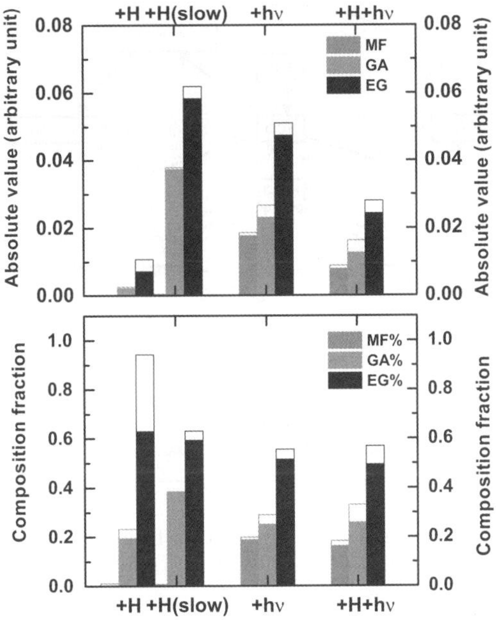

Figure 2. Absolute abundances (upper panel) and composition fractions (lower panel) for MF, GA and EG as obtained in QMS TPD experiments for pure hydrogenation (+H,+H(slow)), pure UV photolysis (+hν) and their cumulative effect (+H+hν), all for 14 K CO:CH_3OH = 4:1 ice mixtures.

(2) the three COMs have the most abundant yields in UV-photolysis experiment; (3) the combination of H-atoms and UV-photons experiment (+H+hν) only reach half the amount of those of the pure UV-photolysis experiment.

In the bottom panel of Figure 2, the total composition fraction of total COMs is given as 0.0 for MF, 0.2-0.4 for GA and 0.8-0.6 for EG, respectively, in the hydrogenation experiments. The longer exposition experiment has a higher GA fraction (lower EG fraction) than those of the higher H-atom flux experiment. The unambiguous detection of MF is only found in the experiments when UV-photons are involved. Although the absolute abundances of COMs in the cumulative experiment only reach 50 per cent of the pure UV photolysis experiment, the composition fractions in both experiment are 0.2 for MF, 0.3 for GA and 0.5 for EG. It shows that the COM yields are rather uniformly reduced in the combination experiment.

In the Figure 3, the extended formation scheme for MF, GA, EG and other complex species from the CO hydrogenation is presented (Chuang *et al.* 2016). The three COM formations at low temperature can be explained by the barrier-less recombination of radicals, i.e. HCO, CH_2OH, CH_3O, that are either generated by H-atom addition/abstraction reactions on the CO-H_2CO-CH_3OH network or by UV photo-dissociation of CH_3OH (H_2CO) (Oberg 2016). For the hydrogenation scenario, ices go through successive H-addition forming fully hydrogen saturated molecule and H-abstraction reactions resulting in partially hydrogen saturated radicals or simple molecule. In these forward/backward reactions, the C-O bond is always preserved. In contrast, the UV-photons applied here ($\leqslant$ 10.2 eV) cannot directly dissociate the CO ice, but the photon energy is above the dissociation energy of C-O single bond of CH_3OH (3.91 eV) resulting in CH_3 and OH radicals. This gives the UV-photolysis scenario an extra mechanism to form new products, i.e. CH_4 and H_3CC-bearing COMs (Oberg *et al.* 2009). In the combination experiment

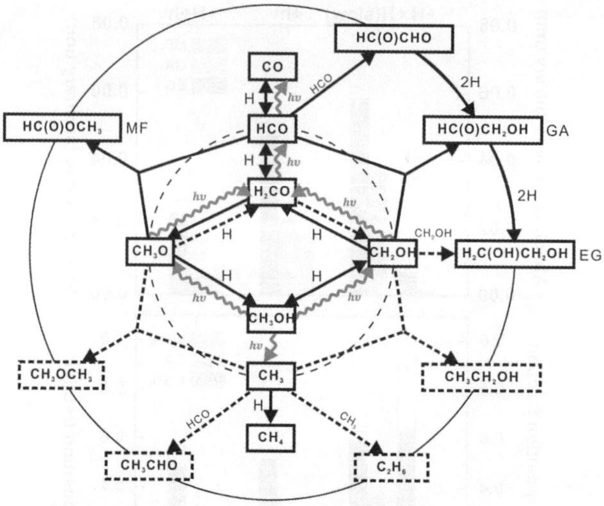

Figure 3. Extended COM formation scheme as obtain from hydrogenation and UV-photolysis experiments in CO-H_2CO-CH_3OH reaction network.

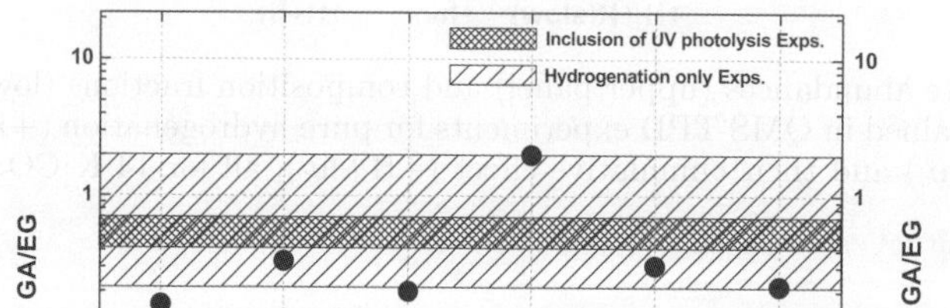

Figure 4. The GA/EG ratio derived from astronomical observations. The data are taken from (1) Crovisier *et al.* (2004) (upper limits), (2) Biver *et al.* (2014) (upper limits), (3) Goesmann *et al.* (2005), (4) Jørgensen *et al.* (2016), (5) Coutens *et al.* (2015).

of H-atoms and UV-photons, the H-atom not only has an effect on enhancing CH_4 formation, but also reduces the three COMs formation by converting the C-bearing radicals into simple H_2CO and CH_3OH.

4. ASTROCHEMICAL IMPLICATIONS AND CONCLUSIONS

Figure 4 compares the laboratory results with the data derived from observations in different astronomical environments. The GA/EG relative ratio in observations is found to be in good agreement with the experimental values of hydrogenation and UV-photolysis processing (0.3-1.5), supporting a solid-state formation pathway for GA and EG. The other ratios, MF/EG and MF/GA, are about 1–3 orders of magnitude greater than those found in hydrogenation and UV-photolysis experiments. This is likely due to the lack of H_2CO in the initial ice (Chuang *et al.* 2016), or suggests that MF is produced via other reactions in the gas phase (Taquet *et al.* 2015). For more details on this work see (Chuang *et al.* 2017).

The take home message of these findings is that:

• In pure hydrogenation experiments, COMs can be formed without UV photons at a low T. Their yields and fractions depend on H-atom/CO:CH_3OH ratios and accretion rates of the involved species.

• The fractions for MF:GA:EG are 0: (0.2-0.4) : (0.8-0.6) for pure hydrogenation, and 0.2 : 0.3 : 0.5 for UV photolysis experiments.

• The GA/EG ratio observed in comets and in solar-mass proto-stars is consistent with the laboratory values (0.3-1.5), suggesting a solid-state formation pathway.

Acknowledgment

This work has been made possible by valuable contributions from G. Fedoseev, D. Qasim, S. Ioppolo, E. F. van Dishoeck, and H. Linnartz.

References

Biver, N., Bockelée-Morvan, D., Debout, V., Crovisier, J., Boissier, J., Lis, D. C., Dello Russo, N., Moreno, R., Colom, P., Paubert, G., Vervack, R., & Weaver, H. A. 2003, *ARAA*, 566, L5

Chuang, K.-J., Fedoseev, G., Ioppolo, S., van Dishoeck, E. F., & Linnartz, H. 2016, *MNRAS*, 467, 2552

Chuang, K.-J., Fedoseev, G., Qasim, D., Ioppolo, S., van Dishoeck, E. F., & Linnartz, H. 2017, *MNRAS*, 540, L49

Coutens, A., Persson, M. V., Jørgensen, J. K., Wampfler, S. F., & Lykke, J. M. 2015, *ARAA*, 576, 5

Crovisier, J., Bockelée-Morvan, D., Biver, N., Colom, P., Despois, D., & Lis, D. C. 2004, *ARAA*, 418, L35

Fedoseev, G., Cuppen, H. M., Ioppolo, S., Lamberts, T., & Linnartz, H. 2015, *MNRAS*, 448, 1288

Goesmann, F., Rosenbauer, H., Bredehöft, J. H., Cabane, M., Ehrenfreund, P., Gautier, T., Giri, C., Krüger, H., Le Roy, L., MacDermott, A. J., McKenna-Lawlor, S., Meierhenrich, U. J., Caro, G. M.M., Raulin, F., Roll, R., Steele, A., Steininger, H., Sternberg, R., Szopa, C., Thiemann, W., & Ulamec, S. 2005, *Science*, 349, 2

Herbst, H. & van Dishoeck, E. F. 2009, *Annu. Rev. Astron. Astrophys*, 47, 427

Ioppolo, S., Fedoseev, G., Lamberts, T., Romanzin, C., & Linnartz, H. 2013, *Rev. Sci. Instrum.*, 84, 073112

Jørgensen, J. K., Favre, C., Bisschop, S. E., Bourke, T. L., van Dishoeck, E. F., & Schmalzl, M. 2012, *ApJ*(Letter), 757, L4

Jørgensen, J. K., van der Wiel, M. H.D., Coutens, A., Lykke, J. M., Müller, H. S.P., van Dishoeck, E. F., Calcutt, H., Bjerkeli, P., Bourke, T. L., Drozdovskaya, M. N., Favre, C., Fayolle, E. C., Garrod, R. T., Jacobsen, S. K., Öberg, K. I., Persson, M. V., & Wampfler, S. F. 2016, *ARAA*, 595, 117

Maity, S., Kaiser, R. I., & Jones, B. M. 2015, *Phys. Chem. Chem. Phys.*, 17, 3081

Moore, M. H., & Hudson, R. L. 2005, *Proceedings of the International Astronomical Union*, 1, 247

Öberg, K. I., Garrod, R. T., van Dishoeck, E. F., & Linnartz, H. 2009, *ARAA*, 504, 891

Öberg, K. I. 2016, *Chemical Reviews*, 116, 9631

Le Roy, L., Altwegg, K., Balsiger, H., Berthelier, J.-J., Bieler, A., Briois, C., Calmonte, U., Combi, M. R., De Keyser, J., Dhooghe, F., Fiethe, B., Fuselier, S. A., Gasc, S., Gombosi, T. I., Hässig, M., Jäckel, A., Rubin, M., & Tzou, C.-Y. 2015, *ARAA*, 583, 1

Taquet, V., López-Sepulcre, A., Ceccarelli, C., Neri, R., Kahane, C., & Charnley, S. B. 2015, *ApJ*, 804, 81

Astrochemistry VII – Through the Cosmos from Galaxies to Planets
Proceedings IAU Symposium No. 332, 2017
M. Cunningham, T. Millar & Y. Aikawa, eds.

doi:10.1017/S1743921317006809

Modelling the emission of HNCO towards IRAS16293-2422 using a new set of collisional coefficients

Hernández-Gómez A.[1,2], **Sahnoun E.**[3], **Caux E.**[2], **Wiesenfeld L.**[4], **Loinard L.**[1], **Bottinelli S.**[2] **and Hammami K.**[3]

[1]Instituto de Radioastronomía y Astrofísica, Universidad Nacional Autónoma de México, Morelia 58089, Mexico

[2]IRAP, Université de Toulouse, CNRS, UPS, CNES, Toulouse, France

[3]Université de Tunis El Manar, Tunisia

[4]Université Grenoble Alpes, CNRS, IPAG, F-38000 Grenoble, France

Abstract. Isocyanic acid (HNCO) is a simple molecule containing the four main atoms essential for life and can be considered as a prebiotic molecule. To model the HNCO emission in the IRAS16293-2422 class 0 low-mass protostar, we used a new set of HNCO collisional coefficients with ortho-H_2 and para-H_2, computed from a set of rotational excitation quenching rates between HNCO and H_2 based on a novel potential energy surface for the rigid molecules interactions. We present here the HNCO Potential Energy Surface used to compute this new set of collisional coefficients and the result of the IRAS16293-2422 HNCO spectrum modelling using them.

Keywords. Protostars, molecular emission, non-LTE modelling.

1. Introduction

Isocyanic acid (HNCO) has been observed in different physical environments in the interstellar medium and in external galaxies. As a simple molecule containing the four main atoms essential for life as we know it, isocyanic acid can be considered as a prebiotic molecule. Modelling the emission of HNCO in star-forming regions is therefore of prime importance. IRAS16293-2422, the Class 0 low-mass protostar with the richest line spectrum is the template source for astrochemistry and has been the subject of many spectroscopic studies. We have in hands the broadest spectral surveys conducted towards this source with single-dish instruments, the TIMASSS survey (Caux *et al.* 2011) (with IRAM, JCMT and APEX in the 80 – 500 GHz range), and the CHESS survey (Ceccarelli *et al.* 2010) (in the 0.5 – 1 THz range with the HIFI instrument onboard the ESA Herschel Observatory). Because in star-forming environments the HNCO emission is not always at LTE, collisional coefficients are needed to use non-LTE transfer codes as RADEX to predict the line emission. The HNCO-He system has been calculated by Green (1996), but the published collisional coefficients fail to reproduce the observed emission in IRAS16293-2422. To model the HNCO lines observed in this source, we used new collisional coefficients of HNCO with ortho-H_2 and para-H_2, computed from a set of rotational excitation quenching rates between HNCO and H_2 based on a novel potential energy surface for the rigid molecules interactions (Sahnoun *et al.* 2017; in prep.).

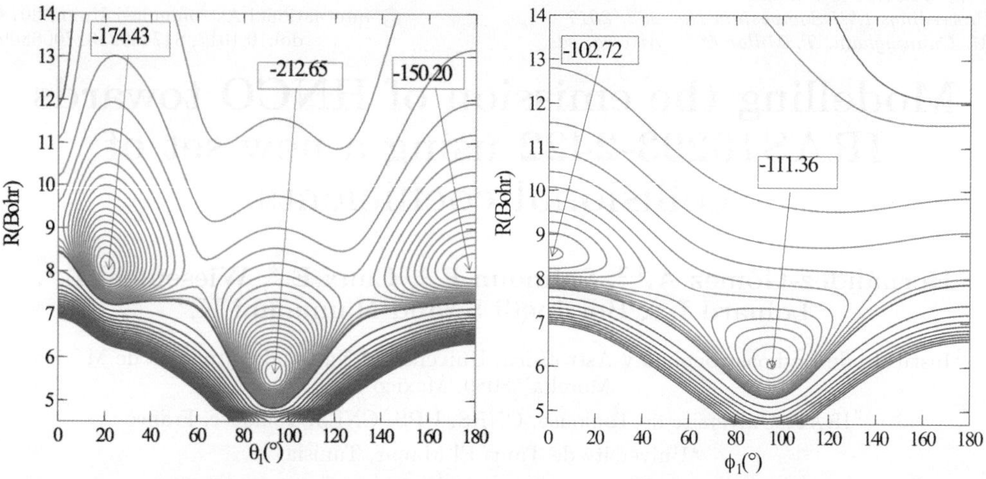

Figure 1. Left. Planar configuration with H_2 along the intermolecular axis, and on the side opposite to H(NCO) (ϕ = 180°). Dependence on the angle θ in the HNCO plane. PES extrema in cm^{-1}. **Right**. Planar configuration with H_2 perpendicular to the intermolecular axis. Dependence on the angle ϕ perpendicular to the HNCO plane. θ = 0°. PES extrema in cm^{-1}.

2. Potential energy surface (PES) and scattering

The interaction of HNCO and H_2 has been computed for both molecules taken as rigid. HNCO is planar, with the angles <HNC>= 128.0° and <CNO>= 170.0°. It is thus an asymmetric rotor. Interaction energies were computed in the supermolecule approach, with the help of the MOLPRO ab initio suite. We made use of the CCSD(T)-F12 method, with the moderately large basis set augcc-pAVDZ/AVTZ. Basis set superposition error has been corrected. The potential is highly anisotropic and fitting procedures required up to 2500 points per distance. A view of the PES is given in Figure 1, with attractive part in red and repulsive part in blue. Subsequent fits in terms of hyperspherical functions lead to 509 terms in the expansion, with angular terms up to $l = 15$. The complex structure in the PES reveals itself by the multiple extrema.

Scattering computations were performed from threshold to $E_{tot} = 500$ cm^{-1} (for para-H_2). For para-H_2 (basis set $J = 0, 2$), computations were performed at the coupled-channel (CC) level up to 20 cm^{-1} and at the coupled states (CS) level above. For ortho-H_2, computations were performed at the CS level. Rotational basis for HNCO is up to $J = 21$, maximum rotational energy is 600 cm^{-1}. Examples of cross-sections are given in Figure 2. The computations cover all rotational HNCO levels up to 144 cm^{-1}, and T $\leqslant$ 205 K, still an incomplete set for a full spectral simulation.

3. IRAS16293-2422 HNCO model

To model the HNCO emission of IRAS16293-2422, we used the CASSIS† code with RADEX (van der Tak *et al.* 2007) for a two-component model, one originating from the warm internal part of the envelope and one originating from the cold external part of the envelope. We performed a MCMC minimisation computing synthetic spectra with 3 free parameters for each component (size, T_{kin}, FWHM). The best fitted physical parameters are shown in Table 1.

† http://cassis.irap.omp.eu

Table 1. Best fitted physical parameters obtained from the CASSIS-RADEX MCMC minimisation.

Component	N(HNCO) (cm^{-2})	T_{kin} (K)	FWHM (km s^{-1})	V_{LSR} (km s^{-1})	Size ($''$)	$n(H_2)$ (cm^{-3})	X
Inner envelope	$(6.3 \pm 0.9) \times 10^{14}$	75.0 ± 3.0	5.5 ± 0.1	3.7 ± 0.1	3.5 ± 0.2	1.0×10^{8}	1.0×10^{-9}
Outer envelope	$(6.2 \pm 0.9) \times 10^{12}$	14.0 ± 1.0	1.3 ± 0.1	4.1 ± 0.1	60.0	4.0×10^{5}	1.4×10^{-10}

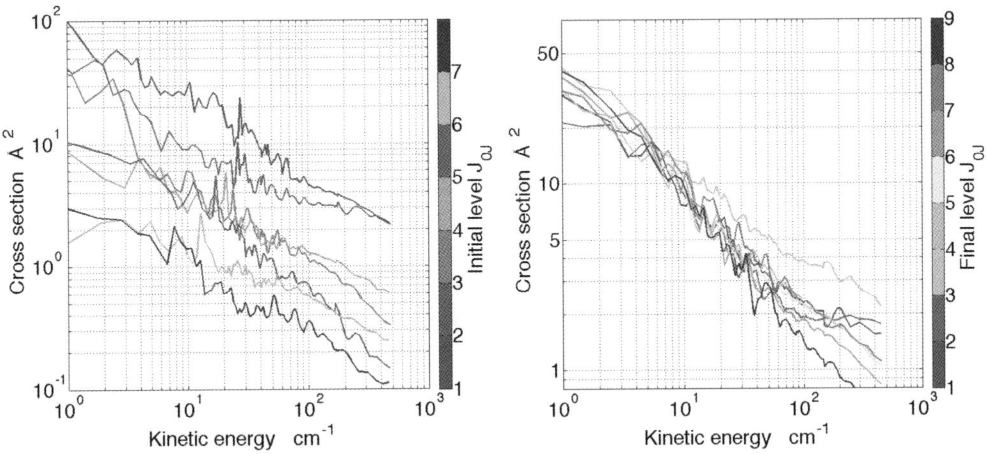

Figure 2. Left. Rotational quenching cross sections HNCO – ortho-H_2 ($J = 1$). Initial state as shown. Final state, ground state 0_{00}. **Right**. Rotational quenching cross sections HNCO – ortho-H_2 ($J = 1$). Initial state 1_{01}. Final state as shown.

We assumed a H_2 ortho/para ratio of 3 for both components, but we checked that smaller values do not change the results. We could not find such a good agreement with the old collisional coefficients with He from Green (1996).

4. Results

In Figure 3 we show the spectra computed with the physical parameters obtained with our model and we compare them directly with the observations from IRAM, APEX, JCMT and HIFI. We are able to reproduce most of the line profiles at all frequencies. In particular, the absorption features observed with HIFI are much better reproduced with our new computed collisional coefficients. In order to understand the distribution of HNCO along the envelope of IRAS16293, we have estimated the radial abundance profile of HNCO using the Nautilus chemical model (Ruaud *et al.* 2016) and the static physical structure computed by Crimier *et al.* (2010). The obtained abundances are shown as a grey rectangle in Figure 4. We note that the abundance corresponding to the obtained sizes in our model for both inner and outer envelopes are consistent with the Nautilus model.

To have a better idea of the structure of IRAS16293-2422 and to refine the results we have obtained, we plan to use the LIME 3D radiative transfer code (Brinch & Hogerheijde) to model the line emission using a three-dimensional structure. To achieve this, we will use GASS (Quénard *et al.* 2017), which will allow us to define a physical 3D structure with several components to be used as input for LIME.

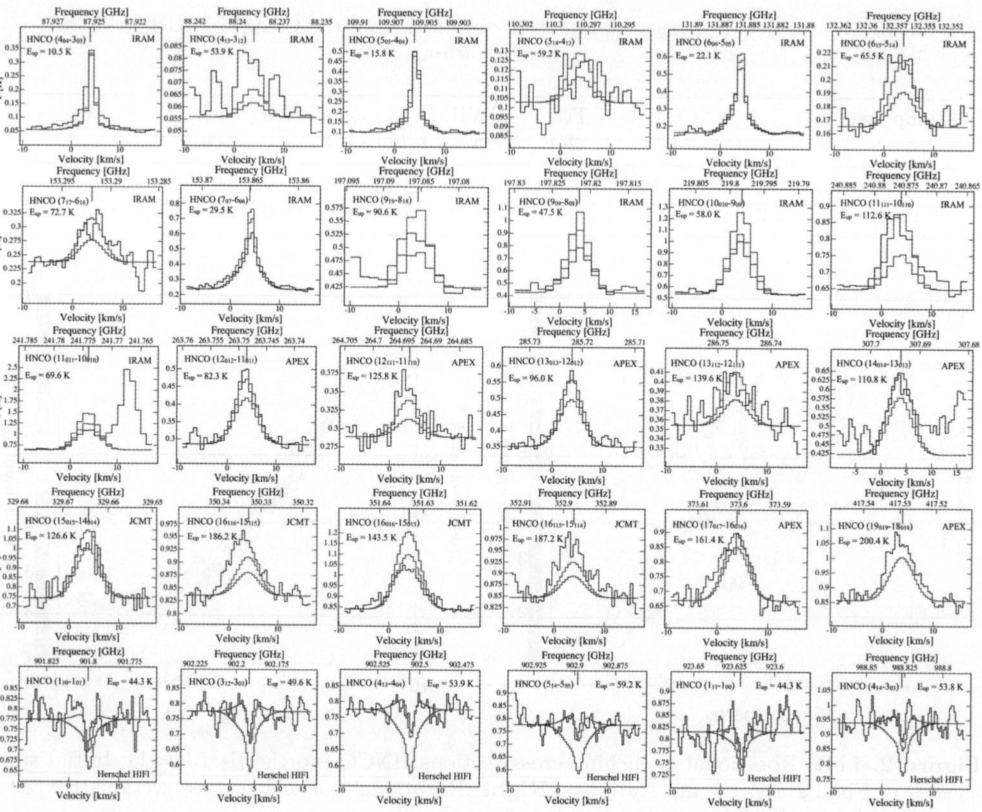

Figure 3. Observed HNCO lines (black) and modelled ones with new collisional coefficients (red) and Green collisional coefficient (blue). HNCO line frequencies are from CDMS (Müller *et al.* 2001).

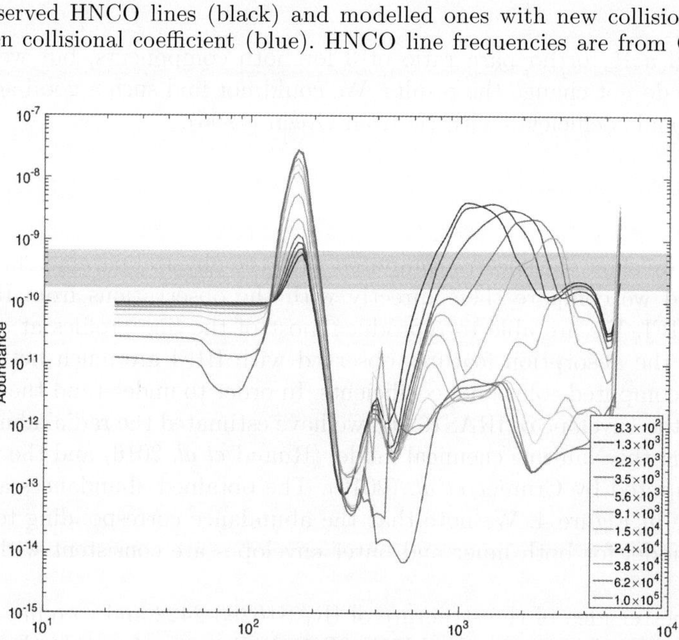

Figure 4. HNCO abundance profiles computed with Nautilus.

References

Brinch, C. & Hogerheijde, M. R. 2010, *A&A*, 523, A25
Caux, E., Kahane, C., Castets, A., *et al.* 2011, *A&A*, 532, A23
Ceccarelli, C., Bacmann, A., Boogert, A., *et al.* 2010, *A&A*, 521, L22
Crimier, N., Ceccarelli, C., Maret, S., *et al.* 2010, *A&A*, 519, A65
Green 1986, NASA Tech. Memo. TM 87791
Müller, H. S. P., Thorwirth, S., Roth, D. A., & Winnewisser, G. 2001, *A&A*, 370, L49
Quénard, D., Bottinelli, S., & Caux, E., 2017, *MNRAS*, 468, 685
Ruaud, M., Wakelam, V., & Hersant, F., 2016, *MNRAS*, 459, 3756
Sahnoun *et al.* 2017, in preparation
van der Tak, F. F. S., Black, J. H., Schöier, F. L., Jansen, D. J., & van Dishoeck, E. F., 2007, *A&A*, 468, 627

Author index